Advances in Conservation Agriculture

Volume 3: Adoption and Spread

It is widely recognised that agriculture is a significant contributor to global warming and climate change. Agriculture needs to reduce its environmental impact and adapt to current climate change whilst still feeding a growing population, i.e. become more 'climate-smart'. Burleigh Dodds Science Publishing is playing its part in achieving this by bringing together key research on making the production of the world's most important crops and livestock products more sustainable.

Based on extensive research, our publications specifically target the challenge of climate-smart agriculture. In this way we are using 'smart publishing' to help achieve climate-smart agriculture.

Burleigh Dodds Science Publishing is an independent and innovative publisher delivering high quality customer-focused agricultural science content in both print and online formats for the academic and research communities. Our aim is to build a foundation of knowledge on which researchers can build to meet the challenge of climate-smart agriculture.

For more information about Burleigh Dodds Science Publishing simply call us on +44 (0) 1223 839365, email info@bdspublishing.com or alternatively please visit our website at www.bdspublishing.com.

Related titles:

climate•SMART•publishing
IN AGRICULTURAL SCIENCE

BURLEIGH DODDS SERIES IN AGRICULTURAL SCIENCE
NUMBER 104

Advances in Conservation Agriculture

Volume 3: Adoption and Spread

Edited by Professor Amir Kassam, University of Reading, UK and Moderator, Global Conservation Agriculture Community of Practice (CA-CoP), FAO, Rome, Italy

burleigh dodds
SCIENCE PUBLISHING

Published by Burleigh Dodds Science Publishing Limited
82 High Street, Sawston, Cambridge CB22 3HJ, UK
www.bdspublishing.com

Burleigh Dodds Science Publishing, 1518 Walnut Street, Suite 900, Philadelphia, PA 19102-3406, USA

First published 2021 by Burleigh Dodds Science Publishing Limited

Library of Congress Control Number: 2019951731

British Library Cataloguing in Publication Data
A catalogue record for this book is available from the British Library

ISBN 978-1-78676-475-1 (Print)
ISBN 978-1-78676-478-2 (PDF)
ISBN 978-1-78676-477-5 (ePub)
ISSN 2059-6936 (print)
ISSN 2059-6944 (online)

DOI 10.19103/AS.2021.0088

Typeset by Deanta Global Publishing Services, Dublin, Ireland

Contents

Series list xv
Dedication xxiii
Foreword xxiv
Preface xxvii
Acknowledgements xxx

1 State of the global adoption and spread of Conservation Agriculture 1
Amir Kassam, University of Reading, UK; Theodor Friedrich, Retired Food and Agriculture Organization of the United Nations (FAO) Expert on Conservation Agriculture, Germany; and Rolf Derpsch, Consultant, Paraguay

1 Introduction 1
2 Global and regional adoption and spread 1
3 Conservation Agriculture adoption at national level 3
4 Global Conservation Agriculture cropland goal for 2050 5
5 Concluding remarks 11
6 Where to look for more information 11
7 References 12

2 Conservation Agriculture in West and Central Canada: an integrated review of adoption 15
Tom Goddard, Consultant, Canada; and Adam Hayes, Soil Management Consultant, Canada

1 Introduction 15
2 Farmers 19
3 Organizations 22
4 Research and extension 25
5 Policy and programmes 29
6 Cumulative effect: issues and adoption 34
7 Current state 38
8 Conclusion 40
9 Acknowledgements 41

10 Where to look for further information 42
11 References 44

3 Conservation Agriculture in the USA 49
Sjoerd W. Duiker, Penn State University, USA

1 Introduction 49
2 The changing United States climate 49
3 Land use in the United States 52
4 History of Conservation Agriculture 53
5 Current state of Conservation Agriculture 62
6 The future of Conservation Agriculture 67
7 Conclusion 70
8 Acknowledgments 71
9 Where to look for further information 71
10 References 72

4 Conservation Agriculture in Central America, the Caribbean and Mexico 79
Jose R. Benites, Independent Consultant on Land and Water, and Conservation Agriculture, Peru; Theodor Friedrich, Retired Food and Agriculture Organization of the United Nations (FAO) Expert on Conservation Agriculture, Germany; Roberto Azofeifa Rodríguez, Head of the Department of Agro-environmental Production of the National Directorate of Agricultural Extension, Costa Rica; Ariel Espinoza Salinas, Catholic Relief Services (CRS), Nicaragua; J. C. Juárez López, FAO Technical Officer, Guatemala; Carlos Andrés Zelaya Elvir, Consultant, Food and Agriculture Organization of the United Nations (FAO) (retired), Honduras; and Jose Escobar, Consultant ASA, CRS, El Salvador

1 Introduction 79
2 Conservation Agriculture in Central America 81
3 Conservation Agriculture in the Caribbean 100
4 Mexico 103
5 Conclusion 108
6 Where to look for further information 109
7 References 109

5 Conservation Agriculture in South America 113
Rafael Fuentes-Llanillo, IDR-Paraná IAPAR-EMATER and FEBRAPDP, Brazil; Marie Luise Carolina Bartz, University of Coimbra, Portugal and FEBRAPDP, Brazil; Tiago Santos Telles, IDR-Paraná IAPAR-EMATER, Brazil; Ademir Calegari, formerly IAPAR and FEBRAPDP, Brazil; Augusto Guilherme Araújo,

formerly IAPAR, Brazil; Amir Kassam, University of Reading, UK; David Roggero, AAPRESID, Argentina; Dimas Soares Junior, IDR-Paraná IAPAR-EMATER and FEBRAPDP, Brazil; Edgard Ramírez, AAPRESID, Argentina; Facundo Capandeguy, AUSID, Uruguay; Herbert Arnold Bartz, Farmer, Doctor Honoris Causa UEL and FEBRAPDP, Brazil (in memoriam); Jaime Hernández-Zamora, ANAPO, Bolivia; Ken Moriya, FEPASIDIAS, Paraguay; Luciano Dabalá, AUSID, Uruguay; María Cecilia Ginés, AAPRESID, Argentina; Martín Maria Cubilla, FEPASIDIAS, Paraguay; Ricardo Ralisch, formerly UEL Professor and FEBRAPDP, Brazil; Richard Trujillo Mendoza, ANAPO, Bolivia; Roberto Peiretti, Global Farmer Network, USA; Rolf Derpsch, formerly GTZ Paraguay Consultant, Germany; Telmo Jorge Carneiro Amado, UFSM, Brazil; and Theodor Friedrich, Retired Food and Agriculture Organization of the United Nations (FAO) Expert on Conservation Agriculture, Germany

1 Introduction 113
2 The no-tillage system and Conservation Agriculture in Brazil 116
3 The no-tillage system in Argentina 127
4 Evolution and current state of the no-tillage system in Paraguay 130
5 No-tillage and Conservation Agriculture in Uruguay 138
6 No-tillage and Conservation Agriculture in Bolivia 139
7 Future prospects for Conservation Agriculture in South America 142
8 Where to look for further information 144
9 References 144

6 Conservation Agriculture in the agri-environmental European context 149
Emilio J. Gonzalez-Sanchez, Universidad de Córdoba, Spain, European Conservation Agriculture Federation (ECAF), Belgium and Asociación Española Agricultura de Conservación Suelos Vivos (AEACSV), Spain; Gottlieb Basch, Universidade de Évora, Instituto de Ciências Agrárias Mediterrânicas (ICAM), Portugal and European Conservation Agriculture Federation (ECAF), Belgium; Julio Roman-Vazquez, European Conservation Agriculture Federation (ECAF), Belgium; Elizabeth Moreno-Blanco, European Conservation Agriculture Federation (ECAF), Belgium; Miguel Angel Repullo-Ruiberriz de Torres, European Conservation Agriculture Federation (ECAF), Belgium and Centro Ifapa 'Alameda del Obispo', Spain; Theodor Friedrich, Retired Food and Agriculture Organization of the United Nations (FAO) Expert

on Conservation Agriculture, Germany; and Amir Kassam, University of Reading, UK

1 Introduction 149
2 Soil threats in Europe 152
3 Benefits of Conservation Agriculture in brief 159
4 Adoption of Conservation Agriculture in Europe: A brief history and current situation 161
5 The role of Conservation Agriculture in the European Policies: The CAP, EU, Climate Action and the Green Deal 173
6 Conclusion 177
7 Where to look for further information 178
8 References 178

7 Adoption and spread of Conservation Agriculture in North Africa 185
Rachid Mrabet, National Institute of Agricultural Research (INRA), Morocco; Haithem Bahri, National Research Institute for Rural Engineering, Water, and Forestry (INRGREF), Tunisia; Omar Zaghouane, Consultant, Algeria; Hatem Cheikh M'Hamed, National Institute of Agricultural Research of Tunisia (INRAT), Tunisia; Sherif Ragab Mohamed El-Areed, Beni-Suef University, Egypt; and Moamen Mohamed Abou El-Enin, Al Azhar University, Egypt

1 Introduction 185
2 Geography, soil, climate and farming environment 187
3 Agriculture and food security 190
4 Conservation Agriculture: History and development 193
5 Major research achievements and advances in knowledge: Re-aligning development, economy and environment 199
6 Adoption barriers, constraints and gaps 209
7 Adoption drivers and opportunities 214
8 Networks, partner institutes and distributional effects 217
9 Conclusion 223
10 Where to look for further information 227
11 Acronyms and abbreviations 228
12 References 230

8 Conservation Agriculture in West and Central Africa 247
Lamourdia Thiombiano, Agriculture and Environmental Research Institute, Burkina Faso; Kalifa Coulibaly, Nazi BONI University, Burkina Faso; Saidi Mkomwa, African Conservation Tillage Network (ACT), Kenya; Kofi Boa, CA Center, Ghana; Robert B. Zougmoré, CCAFS/ICRISAT/WCA, Mali; Oumarou Balarabe,

CIRAD, France; Traoré Salifou, Joseph Ki Zerbo University, Burkina Faso; Sylla Halimatu, Ministry of Agriculture, Burkina Faso

1 Introduction 247
2 Diversity of Conservation Agriculture practices within the region 248
3 Promoting Conservation Agriculture in the region 249
4 Conservation Agriculture adoption and spread in the region 251
5 The contribution of Conservation Agriculture to food production and security 260
6 The contribution of Conservation Agriculture to the environment and combatting climate change 262
7 Conclusions 264
8 Where to look for further information 265
9 References 266

9 Conservation Agriculture in Eastern and Southern Africa 271
Saidi Mkomwa, African Conservation Tillage Network (ACT), Kenya; Sepo Marongwe, African Conservation Tillage Network (ACT), Zimbabwe; Raymond Nazare, Freelance Consultant, Zimbabwe; and Weldone Mutai, African Conservation Tillage Network (ACT), Kenya

1 Introduction 271
2 Background: Agriculture, livelihoods and climate change 274
3 Conservation Agriculture, Agenda 2063 and the Malabo Declaration 275
4 Conservation Agriculture as practised in Eastern and Southern Africa 279
5 Case studies on mainstreaming Conservation Agriculture in Tanzania and Zimbabwe 285
6 Conclusions and future prospects 297
7 Where to look for further information 298
8 Acknowledgements 299
9 References 299

10 From theory to practice: key lessons in the adoption of Conservation Agriculture in South Africa 303
H. J. Smith, ASSET Research, South Africa; G. Trytsman, Independent Researcher, South Africa; A. A. Nel, Independent Agronomist, South Africa; J. A. Strauss, Western Cape Department of Agriculture, South Africa; E. Kruger, Mahlathini Development Foundation, South Africa; R. K. Mampholo, National LandCare Program Coordinator, Department of Agriculture, Land Reform and Rural Development, South Africa; J. N. Van Coller, Conservation Agriculture Farmer - Libanon Farm, South Africa; H. Otto, Conservation Agriculture Farmer - Korannafontein Farm, South Africa; J. G. Steyn,

Conservation Agriculture Farmer - Humanskraal Farm, South Africa; I. D. Dreyer, Conservation Agriculture Farmer - Skulpspruit Farm, South Africa; D. Slabbert, Conservation Agriculture Farmer - Van Rooyenswoning Farm, South Africa; R. Findlay, No-Till Club of KwaZulu-Natal, South Africa; E. Zunckel, Conservation Agriculture Farmer - Zunckels Farm, South Africa; and L. Visser, Agriculture Research Council – Small Grains, South Africa

1 Introduction 303
2 The status of Conservation Agriculture adoption and spread in South Africa 306
3 Challenges in Conservation Agriculture adoption and spread in South Africa 308
4 Appropriate theories supporting Conservation Agriculture innovation 310
5 Farmers as intuitive innovators 314
6 The spontaneous adoption of Conservation Agriculture in a South African context 316
7 Engaging with Conservation Agriculture farmer innovators 318
8 Conclusion and future trends 326
9 References 330

11 Conservation Agriculture in West Asia 337
Isam Bashour, Roula Bachour, Nicolas Haddad and Razan Dbaibo, American University of Beirut, Lebanon; Kassem Jouni, World Food Programme, Lebanon; Faten Adada, FAO Regional Office for the Near East and North Africa, Egypt; Yahya Shakhatreh, Yahya Bani-Kalaf, Abedaraheem Bawaliz, Iyad Musallam, Faddel Ismael, Mahmud Huwaian and Nabeel Bani Hani, National Agricultural Research Center, Jordan; Mina Devkota, International Center for Agricultural Research in the Dry Areas (ICARDA), Morocco; Atef Haddad and Yaser Musa, Aga Khan Foundation, Syria; Rabea Al Hayek, Agricultural Research Station, Syria; Abdulsattar Asmair Alrijabo, University of Mosul and Local Leader of the Conservation Agriculture Programme in Nineveh Province and North of Iraq, Iraq; Irfan Gultekin, Bahri Dagdas International Agricultural Research Institute, Turkey; Mohammad Esmaeil Asadi, Crop, Soil and Conservation Agriculture Systems, Golestan Agricultural and Natural Resources Research and Education Center, Iran; and Amir Kassam, University of Reading, UK

1 Introduction 337
2 Lebanon 338
3 Jordan 346
4 Syria 358

5 Iraq 365
6 Turkey 370
7 Iran 375
8 Conclusion 387
9 Where to look for further information 388
10 References 389

12 Adoption of Conservation Agriculture in Central Asia 395
Aziz Nurbekov, Tashkent State Agrarian University, Uzbekistan; Alisher Mirzabaev, ZEF-University of Bonn, Germany; Muratbek Karabayev, CIMMYT, Kazakhstan; Nurali Asozoda, President Tajik Academy of Agricultural Science, Tajikistan; Nasriddin Khalilov, Samarkand Institute of Veterinary Medicine, Uzbekistan; and Dossymbek Sydyk, South-Western Research Institute of Livestock and Crop Production, Kazakhstan

1 Introduction 395
2 Research and conservation agriculture systems in Central Asia 402
3 Adoption of Conservation Agriculture in Central Asia 412
4 What is next: challenges and opportunities 419
5 Conclusion 420
6 Where to look for further information 421
7 References 422

13 Conservation Agriculture in Eurasia 429
Hafiz Muminjanov, Food and Agriculture Organization of the United Nations (FAO), Italy; Tatiana Semenova, World Food Programme, Kyrgyzstan; Amir Kassam, University of Reading, UK; Theodor Friedrich, Retired Food and Agriculture Organization of the United Nations (FAO) Expert on Conservation Agriculture, Germany; Avetik Nersisyan, International Plant Protection Convention, Italy; Turi Fileccia, formerly Food and Agriculture Organization of the United Nations (FAO), Italy; Gagik Mkrtchyan, Armenian Technology Group Foundation, Armenia; Armen Zakaryan, Food and Agriculture Organization of the United Nations (FAO), Armenia; Imran Jumshudov and Yagub Guliyev, National Agrarian Research Centre, Azerbaijan; Alena Abarava, Lyakhovichi State Agricultural College, Belarus; Allan Pineda Burgos and Demna Martsvaladze, Food and Agriculture Organization of the United Nations (FAO), Georgia; Muratbek Karabayev, formerly International Maize and Wheat Improvement Center (CIMMYT), Kazakhstan; Gulnaz Kaseeva, Public Organization "AgroLead", Kyrgyzstan; Omurbek Mambetov and Cholpon Alibakieva, Food and Agriculture Organization of

the United Nations (FAO), Kyrgyzstan; Boris Boincean, Research Institute of Field Crops, Moldova; Mihail Rurac, State Agricultural University, Moldova; Vadim Bandurin, Association of the Supporters of No-till Planting, Russia; Luidmila Orlova, National Movement for Conservation Agriculture, Russia; Alexandr Toigildin, Ulyanovsk State Agricultural University, Russia; Khalil Safin, Bashkortostan State Agricultural University, Russia; Muhammadi Muminov, Jamshed Sangynov and Shavkat Juraev, Cooperative of Agricultural Extension Specialists "Sarob", Tajikistan; Askarsho Zevarshoev, Public Organization "Eco Consultancy", Tajikistan; Nicola Kosolap, Ukraine Agrarian University, Ukraine; Oksana Davis and Oleksandr Zhuravel, Food and Agriculture Organization of the United Nations (FAO), Ukraine; Michail Draganchuk, YouTube channel "No-tiller", Ukraine; Alim Pulatov, Tashkent Institute of Irrigation and Agricultural Mechanization Engineers, Uzbekistan; Aziz Nurbekov, Tashkent State Agrarian University, Uzbekistan; Josef Kienzle and Sandra Corsi, Food and Agriculture Organization of the United Nations (FAO), Italy; and Emilio J. Gonzalez-Sanchez, Universidad de Córdoba, Spain

1 Introduction 430
2 Conservation Agriculture in Eastern Europe 433
3 Conservation Agriculture in the Caucasus 452
4 Conservation Agriculture in Central Asia 461
5 Conclusion 485
6 Where to look for further information 487
7 References 488

14 Conservation Agriculture in South Asia 491
Yashpal Singh Saharawat, International Fertilizer Development Centre (IFDC), India; Mushtaq Gill, SACAN Services, Pakistan; Mahesh Gathala, CIMMYT, Bangladesh; Tika Bahadur Karki, Nepal Agricultural Research Council, Nepal; D. B. T. Wijeratne, Food and Agriculture Organization of the United Nations (FAO), Sri Lanka; Sayed Samiullah Hakimi, Kabul University, Afghanistan; Neelam Chaudhary, Directorate of Plant Protection and Quarantine, India; Md. Enamul Haque, Murdoch University, Australia; Richard W. Bell, Murdoch University, Australia; C. M. Parihar, Indian Agricultural Research Institute, India; Harisankar Nayak, Indian Agricultural Research Institute, India; Rajbir Singh, ICAR-Agricultural Technology Application Research Institute, India; R. K. Malik, CIMMYT, India; Upendra Singh, International Fertilizer Development Centre, USA; Raj Paroda, Trust

for Advancement of Agricultural Sciences, India; and Amir Kassam, University of Reading, UK

1 Introduction 491
2 History and status of Conservation Agriculture in South Asia 493
3 Crop productivity and income in Conservation Agriculture 510
4 Soil health in Conservation Agriculture 511
5 Water productivity under Conservation Agriculture 513
6 Climate change mitigation and adaptation in Conservation Agriculture 514
7 Nutrient management in Conservation Agriculture 515
8 Weed management in Conservation Agriculture 516
9 Further upscaling of Conservation Agriculture in South Asia 518
10 Where to look for further information 519
11 References 520

15 Conservation Agriculture in Southeast Asia 533
Yuji Niino, FAO Regional Office for Asia and the Pacific, Thailand; Victor B. Ella, University of The Philippines Los Baños, The Philippines; Florent Tivet, Centre de Coopération Internationale en Recherche Agronomique pour le Développement (CIRAD), Agroécologie et Intensification Durable des Cultures Annuelles (AIDA), University of Montpellier, France and General Directorate of Agriculture (GDA), Department of Agricultural Land Resources Management (DALRM), Cambodia; Vira Leng, General Directorate of Agriculture (GDA), Department of Agricultural Land Resources Management (DALRM), Cambodia; Hoa Tran Quoc, Centre de Coopération Internationale en Recherche Agronomique pour le Développement (CIRAD), University of Montpellier, France, and Department of Agricultural Land Management (DALaM), Laos; Pascal Lienhard, Centre de Coopération Internationale en Recherche Agronomique pour le Développement (CIRAD), Agroécologie et Intensification Durable des Cultures Annuelles (AIDA), University of Montpellier, France and Northern Mountainous Agriculture and Forestry Science Institute (NOMAFSI), Vietnam; Pham thi Sen, Northern Mountainous Agriculture and Forestry Science Institute (NOMAFSI), Vietnam; Thatheva Saphangthong, Department of Agricultural Land Management (DALaM), Laos; Vang Seng, General Directorate of Agriculture (GDA), Department of Agricultural Land Resources Management (DALRM), Cambodia; Lyda Hok, Royal University of Agriculture, Cambodia; Manuel Reyes, Kansas State University, USA; Stéphane Boulakia,

Centre de Coopération Internationale en Recherche Agronomique pour le Développement (CIRAD), Agroécologie et Intensification Durable des Cultures Annuelles (AIDA), University of Montpellier, France; and Sim Choon Cheak, Sime Darby Plantation Research Sdn. Bhd., Malaysia

1 Introduction 533
2 Current status of Conservation Agriculture in Southeast Asia 536
3 Regional networking 568
4 Conclusion 570
5 Where to look for further information 571
6 References 571

16 Adoption and spread of Conservation Agriculture in East Asia 577
Hongwen Li and Jin He, National Institute of Conservation Tillage, China Agricultural University and Conservation Tillage Research Centre, Ministry of Agriculture and Rural Affairs, China

1 Introduction 577
2 History of Conservation Agriculture 581
3 Policy and actions for Conservation Agriculture promotion 586
4 Adoption scale 591
5 Advice for Conservation Agriculture promotion 592
6 Conclusion 593
7 Where to look for further information 594
8 References 594

17 Conservation Agriculture in Australian dryland cropping and in New Zealand: the lessons of 70 years 599
Jean-Francois Rochecouste, Grains Research Development Corporation, Australia; John Baker, Cross Slot IP Ltd, New Zealand; and Bill Crabtree, Farmer and Author, Australia

1 Introduction 599
2 Dealing with the negative consequences of no-till systems 602
3 Challenges from stubble retention 605
4 Cropping diversity and cover cropping 607
5 Compaction, controlled traffic farming and precision agriculture 609
6 Evolving concepts in Conservation Agriculture: recycled organics and carbon sequestration 613
7 The New Zealand Conservation Agriculture story 615
8 Acknowledgement 619
9 Where to look for further information 619
10 References 619

Index 625

Series list

Title	Series number
Achieving sustainable cultivation of maize - Vol 1 From improved varieties to local applications Edited by: Dr Dave Watson, CGIAR Maize Research Program Manager, CIMMYT, Mexico	001
Achieving sustainable cultivation of maize - Vol 2 Cultivation techniques, pest and disease control Edited by: Dr Dave Watson, CGIAR Maize Research Program Manager, CIMMYT, Mexico	002
Achieving sustainable cultivation of rice - Vol 1 Breeding for higher yield and quality Edited by: Prof. Takuji Sasaki, Tokyo University of Agriculture, Japan	003
Achieving sustainable cultivation of rice - Vol 2 Cultivation, pest and disease management Edited by: Prof. Takuji Sasaki, Tokyo University of Agriculture, Japan	004
Achieving sustainable cultivation of wheat - Vol 1 Breeding, quality traits, pests and diseases Edited by: Prof. Peter Langridge, The University of Adelaide, Australia	005
Achieving sustainable cultivation of wheat - Vol 2 Cultivation techniques Edited by: Prof. Peter Langridge, The University of Adelaide, Australia	006
Achieving sustainable cultivation of tomatoes Edited by: Dr Autar Mattoo, USDA-ARS, USA; and Prof. Avtar Handa, Purdue University, USA	007
Achieving sustainable production of milk - Vol 1 Milk composition, genetics and breeding Edited by: Dr Nico van Belzen, International Dairy Federation (IDF), Belgium	008
Achieving sustainable production of milk - Vol 2 Safety, quality and sustainability Edited by: Dr Nico van Belzen, International Dairy Federation (IDF), Belgium	009
Achieving sustainable production of milk - Vol 3 Dairy herd management and welfare Edited by: Prof. John Webster, University of Bristol, UK	010
Ensuring safety and quality in the production of beef - Vol 1 Safety Edited by: Prof. Gary Acuff, Texas A&M University, USA; and Prof. James Dickson, Iowa State University, USA	011
Ensuring safety and quality in the production of beef - Vol 2 Quality Edited by: Prof. Michael Dikeman, Kansas State University, USA	012
Achieving sustainable production of poultry meat - Vol 1 Safety, quality and sustainability Edited by: Prof. Steven C. Ricke, University of Arkansas, USA	013
Achieving sustainable production of poultry meat - Vol 2 Breeding and nutrition Edited by: Prof. Todd Applegate, University of Georgia, USA	014

Achieving sustainable production of poultry meat - Vol 3 015
Health and welfare
Edited by: Prof. Todd Applegate, University of Georgia, USA

Achieving sustainable production of eggs - Vol 1 016
Safety and quality
Edited by: Prof. Julie Roberts, University of New England, Australia

Achieving sustainable production of eggs - Vol 2 017
Animal welfare and sustainability
Edited by: Prof. Julie Roberts, University of New England, Australia

Achieving sustainable cultivation of apples 018
Edited by: Dr Kate Evans, Washington State University, USA

Integrated disease management of wheat and barley 019
Edited by: Prof. Richard Oliver, Curtin University, Australia

Achieving sustainable cultivation of cassava - Vol 1 020
Cultivation techniques
Edited by: Dr Clair Hershey, formerly International Center for Tropical Agriculture (CIAT), Colombia

Achieving sustainable cultivation of cassava - Vol 2 021
Genetics, breeding, pests and diseases
Edited by: Dr Clair Hershey, formerly International Center for Tropical Agriculture (CIAT), Colombia

Achieving sustainable production of sheep 022
Edited by: Prof. Johan Greyling, University of the Free State, South Africa

Achieving sustainable production of pig meat - Vol 1 023
Safety, quality and sustainability
Edited by: Prof. Alan Mathew, Purdue University, USA

Achieving sustainable production of pig meat - Vol 2 024
Animal breeding and nutrition
Edited by: Prof. Julian Wiseman, University of Nottingham, UK

Achieving sustainable production of pig meat - Vol 3 025
Animal health and welfare
Edited by: Prof. Julian Wiseman, University of Nottingham, UK

Achieving sustainable cultivation of potatoes - Vol 1 026
Breeding improved varieties
Edited by: Prof. Gefu Wang-Pruski, Dalhousie University, Canada

Achieving sustainable cultivation of oil palm - Vol 1 027
Introduction, breeding and cultivation techniques
Edited by: Prof. Alain Rival, Center for International Cooperation in Agricultural Research for Development (CIRAD), France

Achieving sustainable cultivation of oil palm - Vol 2 028
Diseases, pests, quality and sustainability
Edited by: Prof. Alain Rival, Center for International Cooperation in Agricultural Research for Development (CIRAD), France

Achieving sustainable cultivation of soybeans - Vol 1 029
Breeding and cultivation techniques
Edited by: Prof. Henry T. Nguyen, University of Missouri, USA

Achieving sustainable cultivation of soybeans - Vol 2 030
Diseases, pests, food and non-food uses
Edited by: Prof. Henry T. Nguyen, University of Missouri, USA

Achieving sustainable cultivation of sorghum - Vol 1 031
Genetics, breeding and production techniques
Edited by: Prof. William Rooney, Texas A&M University, USA

Achieving sustainable cultivation of sorghum - Vol 2 032
Sorghum utilization around the world
Edited by: Prof. William Rooney, Texas A&M University, USA

Achieving sustainable cultivation of potatoes - Vol 2 033
Production, storage and crop protection
Edited by: Dr Stuart Wale, Potato Dynamics Ltd, UK

Achieving sustainable cultivation of mangoes 034
Edited by: Prof. Víctor Galán Saúco, Instituto Canario de Investigaciones Agrarias (ICIA), Spain; and Dr Ping Lu, Charles Darwin University, Australia

Achieving sustainable cultivation of grain legumes - Vol 1 035
Advances in breeding and cultivation techniques
Edited by: Dr Shoba Sivasankar et al., formerly International Crops Research Institute for the Semi-Arid Tropics (ICRISAT), India

Achieving sustainable cultivation of grain legumes - Vol 2 036
Improving cultivation of particular grain legumes
Edited by: Dr Shoba Sivasankar et al., formerly International Crops Research Institute for the Semi-Arid Tropics (ICRISAT), India

Achieving sustainable cultivation of sugarcane - Vol 1 037
Cultivation techniques, quality and sustainability
Edited by: Prof. Philippe Rott, University of Florida, USA

Achieving sustainable cultivation of sugarcane - Vol 2 038
Breeding, pests and diseases
Edited by: Prof. Philippe Rott, University of Florida, USA

Achieving sustainable cultivation of coffee 039
Edited by: Dr Philippe Lashermes, Institut de Recherche pour le Développement (IRD), France

Achieving sustainable cultivation of bananas - Vol 1 040
Cultivation techniques
Edited by: Prof. Gert H. J. Kema, Wageningen University and Research, The Netherlands; and Prof. André Drenth, University of Queensland, Australia

Global Tea Science 041
Current status and future needs
Edited by: Dr V. S. Sharma, formerly UPASI Tea Research Institute, India; and Dr M. T. Kumudini Gunasekare, Coordinating Secretariat for Science Technology and Innovation (COSTI), Sri Lanka

Integrated weed management 042
Edited by: Emeritus Prof. Rob Zimdahl, Colorado State University, USA

Achieving sustainable cultivation of cocoa 043
Edited by: Prof. Pathmanathan Umaharan, Cocoa Research Centre – The University of the West Indies, Trinidad and Tobago

Robotics and automation for improving agriculture 044
Edited by: Prof. John Billingsley, University of Southern Queensland, Australia

Water management for sustainable agriculture 045
Edited by: Prof. Theib Oweis, ICARDA, Jordan

Improving organic animal farming 046
Edited by: Dr Mette Vaarst, Aarhus University, Denmark; and Dr Stephen Roderick, Duchy College, UK

Improving organic crop cultivation 047
Edited by: Prof. Ulrich Köpke, University of Bonn, Germany

Managing soil health for sustainable agriculture - Vol 1 048
Fundamentals
Edited by: Dr Don Reicosky, Soil Scientist Emeritus USDA-ARS and University of Minnesota, USA

Managing soil health for sustainable agriculture - Vol 2 049
Monitoring and management
Edited by: Dr Don Reicosky, Soil Scientist Emeritus USDA-ARS and University of Minnesota, USA

Rice insect pests and their management 050
E. A. Heinrichs, Francis E. Nwilene, Michael J. Stout, Buyung A. R. Hadi & Thais Freitas

Improving grassland and pasture management in temperate agriculture 051
Edited by: Prof. Athole Marshall; and Dr Rosemary Collins, IBERS, Aberystwyth University, UK

Precision agriculture for sustainability 052
Edited by: Dr John Stafford, Silsoe Solutions, UK

Achieving sustainable cultivation of temperate zone tree fruit and berries - Vol 1 053
Physiology, genetics and cultivation
Edited by: Prof. Gregory A. Lang, Michigan State University, USA

Achieving sustainable cultivation of temperate zone tree fruit and berries - Vol 2 054
Case studies
Edited by: Prof. Gregory A. Lang, Michigan State University, USA

Agroforestry for sustainable agriculture 055
Edited by: Prof. María Rosa Mosquera-Losada, Universidade de Santiago de Compostela, Spain; and Dr Ravi Prabhu, World Agroforestry Centre (ICRAF), Kenya

Achieving sustainable cultivation of tree nuts 056
Edited by: Prof. Ümit Serdar, Ondokuz Mayis University, Turkey; and Emeritus Prof. Dennis Fulbright, Michigan State University, USA

Assessing the environmental impact of agriculture 057
Edited by: Prof. Bo P. Weidema, Aalborg University, Denmark

Critical issues in plant health: 50 years of research in African agriculture 058
Edited by: Dr Peter Neuenschwander and Dr Manuele Tamò, IITA, Benin

Achieving sustainable cultivation of vegetables 059
Edited by: Emeritus Prof. George Hochmuth, University of Florida, USA

Advances in breeding techniques for cereal crops 060
Edited by: Prof. Frank Ordon, Julius Kuhn Institute (JKI), Germany; and Prof. Wolfgang Friedt, Justus-Liebig University of Giessen, Germany

Advances in Conservation Agriculture - Vol 1 061
Systems and Science
Edited by: Prof. Amir Kassam, University of Reading, UK and Moderator, Global Conservation Agriculture Community of Practice (CA-CoP), FAO, Rome, Italy

Advances in Conservation Agriculture - Vol 2 062
Practice and Benefits
Edited by: Prof. Amir Kassam, University of Reading, UK and Moderator, Global Conservation Agriculture Community of Practice (CA-CoP), FAO, Rome, Italy

Achieving sustainable greenhouse cultivation 063
Edited by: Prof. Leo Marcelis and Dr Ep Heuvelink, Wageningen University, The Netherlands

Achieving carbon-negative bioenergy systems from plant materials 064
Edited by: Dr Chris Saffron, Michigan State University, USA

Achieving sustainable cultivation of tropical fruits 065
Edited by: Prof. Elhadi M. Yahia, Universidad Autónoma de Querétaro, Mexico

Advances in postharvest management of horticultural produce 066
Edited by: Prof. Chris Watkins, Cornell University, USA

Pesticides and agriculture 067
Profit, politics and policy
Dave Watson

Integrated management of diseases and insect pests of tree fruit 068
Edited by: Prof. Xiangming Xu and Dr Michelle Fountain, NIAB-EMR, UK

Integrated management of insect pests: Current and future developments 069
Edited by: Emeritus Prof. Marcos Kogan, Oregon State University, USA; and Emeritus Prof. E. A. Heinrichs, University of Nebraska-Lincoln, USA

Preventing food losses and waste to achieve food security and sustainability 070
Edited by: Prof. Elhadi M. Yahia, Universidad Autónoma de Querétaro, Mexico

Achieving sustainable management of boreal and temperate forests 071
Edited by: Dr John Stanturf, Estonian University of Life Sciences, Estonia

Advances in breeding of dairy cattle 072
Edited by: Prof. Julius van der Werf, University of New England, Australia; and Prof. Jennie Pryce, Agriculture Victoria and La Trobe University, Australia

Improving gut health in poultry 073
Edited by: Prof. Steven C. Ricke, University of Arkansas, USA

Achieving sustainable cultivation of barley 074
Edited by: Prof. Glen Fox, University of California-Davis, USA and The University of Queensland, Australia; and Prof. Chengdao Li, Murdoch University, Australia

Advances in crop modelling for a sustainable agriculture 075
Edited by: Emeritus Prof. Kenneth Boote, University of Florida, USA

Achieving sustainable crop nutrition 076
Edited by: Prof. Zed Rengel, University of Western Australia, Australia

Achieving sustainable urban agriculture 077
Edited by: Prof. Johannes S. C. Wiskerke, Wageningen University, The Netherlands

Climate change and agriculture 078
Edited by Dr Delphine Deryng, NewClimate Institute/Integrative Research Institute on Transformations of Human-Environment Systems (IRI THESys), Humboldt-Universität zu Berlin, Germany

Advances in poultry genetics and genomics 079
Edited by: Prof. Samuel E. Aggrey, University of Georgia, USA; Prof. Huaijun Zhou, University of California-Davis, USA; Dr Michèle Tixier-Boichard, INRAE, France; and Prof. Douglas D. Rhoads, University of Arkansas, USA

Achieving sustainable management of tropical forests 080
Edited by: Prof. Jürgen Blaser, Bern University of Life Sciences, Switzerland; and Patrick D. Hardcastle, Forestry Development Specialist, UK

Improving the nutritional and nutraceutical properties of wheat and other cereals 081
Edited by: Prof. Trust Beta, University of Manitoba, Canada

Achieving sustainable cultivation of ornamental plants 082
Edited by: Emeritus Prof. Michael Reid, University of California-Davis, USA

Improving rumen function 083
Edited by: Dr C. S. McSweeney, CSIRO, Australia; and Prof. R. I. Mackie, University of Illinois, USA

Biostimulants for sustainable crop production 084
Edited by: Youssef Rouphael, Patrick du Jardin, Patrick Brown, Stefania De Pascale and Giuseppe Colla

Improving data management and decision support systems in agriculture 085
Edited by: Dr Leisa Armstrong, Edith Cowan University, Australia

Achieving sustainable cultivation of bananas - Volume 2 086
Germplasm and genetic improvement
Edited by: Prof. Gert H. J. Kema, Wageningen University, The Netherlands; and Prof. Andrè Drenth, The University of Queensland, Australia

Reconciling agricultural production with biodiversity conservation 087
Edited by: Prof. Paolo Bàrberi and Dr Anna-Camilla Moonen, Institute of Life Sciences - Scuola Superiore Sant'Anna, Pisa, Italy

Advances in postharvest management of cereals and grains 088
Edited by: Prof. Dirk E. Maier, Iowa State University, USA

Biopesticides for sustainable agriculture 089
Edited by: Prof. Nick Birch, formerly The James Hutton Institute, UK; and Prof. Travis Glare, Lincoln University, New Zealand

Understanding and improving crop root function 090
Edited by: Emeritus Prof. Peter J. Gregory, University of Reading, UK

Understanding the behaviour and improving the welfare of chickens 091
Edited by: Prof. Christine Nicol, Royal Veterinary College - University of London, UK

Advances in measuring soil health 092
Edited by: Prof. Wilfred Otten, Cranfield University, UK

The sustainable intensification of smallholder farming systems 093
Edited by: Dr Dominik Klauser and Dr Michael Robinson, Syngenta Foundation for Sustainable Agriculture, Switzerland

Advances in horticultural soilless culture 094
Edited by: Prof. Nazim S. Gruda, University of Bonn, Germany

Reducing greenhouse gas emissions from livestock production 095
Edited by: Dr Richard Baines, Royal Agricultural University, UK

Understanding the behaviour and improving the welfare of pigs 096
Edited by: Emerita Prof. Sandra Edwards, Newcastle University, UK

Genome editing for precision crop breeding 097
Edited by: Dr Matthew R. Willmann, Cornell University, USA

Understanding the behaviour and improving the welfare of dairy cattle 098
Edited by: Dr Marcia Endres, University of Minnesota, USA

Defining sustainable agriculture 099
Dave Watson

Plant genetic resources 100
A review of current research and future needs
Edited by: Dr M. Ehsan Dulloo, Bioversity International, Italy

Developing animal feed products 101
Edited by: Dr Navaratnam Partheeban, formerly Royal Agricultural University, UK

Improving dairy herd health 102
Edited by: Prof. Émile Bouchard, University of Montreal, Canada

Understanding gut microbiomes as targets for improving pig gut health 103
Edited by: Prof. Mick Bailey and Emeritus Prof. Chris Stokes, University of Bristol, UK

Advances in Conservation Agriculture – Vol 3 104
Adoption and Spread
Edited by: Professor Amir Kassam, University of Reading, UK and Moderator, Global Conservation Agriculture Community of Practice (CA-CoP), FAO, Rome, Italy

Advances in Precision Livestock Farming 105
Edited by: Prof. Daniel Berckmans, Katholieke University of Leuven, Belgium

Achieving durable disease resistance in cereals 106
Edited by: Prof. Richard Oliver, formerly Curtin University, Australia

Seaweed and microalgae as alternative sources of protein 107
Edited by: Prof. Xin Gen Lei, Cornell University, USA

Microbial bioprotectants for plant disease management 108
Edited by: Dr Jürgen Köhl, Wageningen University & Research, The Netherlands; and Dr Willem J. Ravensberg, Koppert Biological Systems, The Netherlands

Improving soil health 109
Edited by: Prof. William Horwath, University of California-Davis, USA

Improving integrated pest management (IPM) in horticulture 110
Edited by: Prof. Rosemary Collier, Warwick University, UK

Climate-smart production of coffee 111
Achieving sustainability and ecosystem services
Edited by: Prof. Reinhold Muschler, CATIE, Costa Rica

Developing smart agri-food supply chains 112
Using technology to improve safety and quality
Edited by: Prof. Louise Manning, Royal Agricultural University, UK

Advances in integrated weed management 113
Edited by: Prof. Per Kudsk, Aarhus University, Denmark

Understanding and improving the functional and nutritional properties of milk 114
Edited by: Prof. Thom Huppertz, Wageningen University, The Netherlands; and Prof. Todor Vasiljevic, Victoria University, Australia

Energy-smart farming: efficiency, renewable energy and sustainability 115
Edited by: Emeritus Prof. Ralph Sims, Massey University, New Zealand

Understanding and optimising the nutraceutical properties of fruit and vegetables 116
Edited by: Prof. Victor Preedy, King's College – University of London, UK; and Dr Vinood Patel, University of Westminster, UK

Advances in plant phenotyping for more sustainable crop production 117
Edited by: Prof. Achim Walter, ETH Zurich, Switzerland

Optimising pig herd health and production 118
Edited by: Prof. Dominiek Maes, Ghent University, Belgium; and Prof. Joaquim Segalés, Universitat Autònoma de Barcelona, Spain

Optimising poultry flock health 119
Edited by: Prof. Sjaak de Wit, University of Utrecht, The Netherlands

Advances in seed science and technology for more sustainable production 120
Edited by: Dr Julia Buitink, INRAE, France; and Prof. Olivier Leprince, L'Institute Agrocampus Ouest, France

Understanding and fostering soil carbon sequestration 121
Edited by: Dr Cornelia Rumpel, CNRS, France

Advances in sensor technology for sustainable crop production 122
Edited by: Dr Craig Lobsey, University of Southern Queensland, Australia; and Dr Asim Biswas, University of Guelph, Canada

Achieving sustainable cultivation of bananas - Vol 3 123
Diseases and pests
Edited by: Prof. André Drenth, The University of Queensland, Australia; and Prof. Gert H. J. Kema, Wageningen University and Research, The Netherlands

Developing drought-resistant cereals 124
Edited by: Prof. Roberto Tuberosa, University of Bologna, Italy

Achieving sustainable turfgrass management 125
Edited by: Prof. Michael Fidanza, Pennsylvania State University, USA

Promoting pollination and pollinators in farming 126
Edited by: Prof. Peter Kevan and Dr Susan Willis Chan, University of Guelph, Canada

Improving poultry meat quality 127
Edited by: Prof. Massimiliano Petracci, University of Bologna, Italy; and Dr Mario Estévez, University of Extremadura, Spain

Advances in monitoring of native and invasive insect pests of crops 128
Edited by: Dr Michelle Fountain, NIAB-EMR, UK; and Dr Tom Pope, Harper Adams University, UK

Advances in understanding insect pests affecting wheat and other cereals 129
Edited by: Prof. Sanford Eigenbrode and Dr Arash Rashed, University of Idaho, USA

Understanding and improving crop photosynthesis 130
Edited by: Dr Robert Sharwood, Western Sydney University, Australia

Dedication

This book is dedicated to the global Conservation Agriculture movement and all the stakeholders engaged in this movement who have successfully generated and sustained an international momentum to transform the unsustainable and out-of-date conventional tillage agriculture into the alternate paradigm of Conservation Agriculture, representing an ecosystem approach to regenerative sustainable agriculture. The book is particularly dedicated to all the Conservation Agriculture pioneer and champion farmers and their families, researchers and extension agents as well as all the champions in the national and international public, private and civil sectors who have dedicated themselves to mobilizing policy and institutional support for making Conservation Agriculture a global mainstream reality for sustainable development in the 21st century.

Foreword

These volumes are a timely celebration of the most progressive change in farming practices that has been set in motion during the past 60 years and which is gathering momentum around the world at an extraordinary pace.

For thousands of years, soil inversion, whether by hoe or plough, has been almost universally applied by farmers and seen as essential for the successful growing of annual crops. Tilling the soil buries weeds, loosens the soil to let roots penetrate easily and allows rainfall to sink in and become available to the crop. By burying crop residues, it may also interrupt the life cycles of crop diseases and pests.

However, as the frequency and depth of tillage has increased, the negative effects have become more obvious. Every time soil is dug or ploughed, its structure is broken up and it becomes increasingly at risk to water and wind erosion. The speed with which soil organic matter content falls is accelerated, causing the surface to become susceptible to crusting, thereby reducing rainfall infiltration and increasing run-off while also restricting moisture retention capacity in the plant rooting zone below the surface. These processes tend to make crops and soils more vulnerable to drought and can ultimately lead to farmland becoming so degraded that it is abandoned.

As we understand more about the processes of climate change, we are becoming increasingly aware of the extent to which frequent tillage also contributes to global warming. As the organic matter content of soil falls this reduces the capacity of farmed land to serve as a carbon sink. Moreover, soil inversion, whether manual or mechanical, is very heavy in its energy requirements, with tractor use forploughing accounting for a large share of the fossil fuel consumption in food production.

We are also belatedly learning that frequent soil inversion, especially when associated with heavy applications of pesticides, reduces soil biotic activity and undermines soil health with a corresponding fall in productivity.

The 'dust bowl' in the United States in the 1930's awoke farmers and scientists to the damaging effects of excessive tillage but it was not until the 1960s that American farmers began to adopt various no-till systems to reduce wind and water erosion. These set a precedent for the progressive emergence in the following decades of Conservation Agriculture (CA) which consists of a combination of continuous no tillage with year-round biomass soil cover and crop associations and/or rotations involving cover crops – often legumes.

The USA continues to be the country with the largest area under CA (43.5 million ha in 2015/16 or 35.1% of its arable farming area). Its farmers' associations are very active in pioneering new technologies, especially those related to the better use of cover crops in CA systems to increase Nitrogen availability and to cut herbicide dependence

One could claim that CA was ahead of its time in that it contains the main elements of what is now termed 'sustainable agriculture'. Indeed, there is much to be learnt from the CA story of the past 60 years which is relevant to inducing the global shift that must urgently be made from the currently unsustainable food production and consumption systems to ones that are truly sustainable.

Five of such lessons from the CA experience are:

First. CA shows that fundamental changes in farming technologies - in this case, getting rid of the plough - can spread very rapidly throughout the world, in both developed and developing countries in which it is being taken up by large and small-scale farmers alike. The area of arable land under CA has grown from about 2 million ha in the USA in the early 1970s to 180 million ha (12.5% of global cropland) in 78 countries in 2015-16.

Secondly, the growth in CA uptake and the adaptation of CA methods to different ecological and societal conditions has been driven largely by practitioners, especially innovative farmers and machinery and hand-tool manufacturers. Approaches to CA are constantly evolving and new developments are being openly shared between all those involved. The more that farmers find that CA can boost their incomes, the faster will be the pace of change.

Thirdly, formal research has been important in developing new approaches to CA but most studies have been focussed on identifying the impact of the shift to CA on crop performance; the physical and biological conditions of soils; global warming, and farm incomes. Such validation studies have help boost the case for policy support for CA.

Fourthly, in most countries, however, CA has so far spread between farmers without explicitly supportive government policies. It seems certain that the rate of diffusion can be accelerated by targeted incentives (for instance, subsidies on appropriate machinery and equipment, payments to farmers for soil carbon accumulation and enhancement of water resources) and improvements in agricultural extension.

Lastly, although international agencies such as FAO have only invested quite small resources in CA, they have played a valuable catalytic role, mainly by promoting exchanges of experiences between practitioners, countries and regions and nurturing the emergence of regional promotional institutions. FAO has placed CA at the core of its vision for sustainable food and agriculture which calls for "a world in which food is nutritious, safe and accessible for everyone, where natural resources are managed sustainably, and where rural dwellers have decent livelihoods and contribute actively to economic development". FAO has also sponsored the foundation of a CA Community of Practice (COP) which shares new developments between its many members.

Although my own country, China, carried out trials on zero tillage from the 1980s, it was only after the turn of the century that the promotion of CA was adopted as a national priority. As a result, there has been a very rapid growth in the area under CA from just a few ha in 2002 to 9 million ha in 2015. The results have been good in terms of yield increases, especially for maize, and of reductions in erosion and river sedimentation. The government, therefore, intends to take additional measures to boost farmer uptake of CA including the recent creation of the China Institute for Conservation Tillage, technical support for the manufacture of improved machinery, subsidies for the provision of eco-systems services by farmers, and improvements in agricultural extension. It also intends to enhance the benefits of CA by linking its promotion to other aspects of its sustainable development agenda including Integrated Pest Management (IPM).

This edifying book is a compendium of much of what we have learnt across the world about CA. Its authors have all been deeply and enthusiastically involved in one way or another in the remarkable evolution and spread of Conservation Agriculture in recent decades. It will serve as an immensely valuable source of reference - and inspiration - for all those who are committed to putting the world's food systems on a truly sustainable footing.

It is with great pleasure, therefore, that I commend this book to you,

by Qu Dongyu,
Director General, UN Food and Agriculture
Organization (FAO), Rome

Preface

Conservation Agriculture (CA) and its community of practice have made tremendous strides in the last two decades, particularly since the modern version of CA as we know it today was defined by FAO at its first Regional Workshop on CA in Harare, Zimbabwe in 1998. The term CA was globalized through the 1st World Congress on Conservation Agriculture organized by ECAF in partnership with FAO in 2001 in Madrid, Spain. In 1999, the global spread of CA stood at about 45 M ha of cropland with some 10 countries in which CA was being practiced and promoted. In 2008/09, the global spread of CA was about 106 M ha of cropland across some 36 countries, corresponding to an annual rate of expansion of about 6.1 M ha. In 2015/16, the global spread of CA was some 180[1] M ha of cropland, split about equally between the Global South and the Global North, involving 78 countries. This corresponds to an annual rate of increase of some 10.5 M ha, a spectacular rate of transformation, led mainly by farmers and their associations with support from national and international champions and enthusiasts, many on a voluntary basis, and some national and international institutions. Although the spread of CA in the 1990s and 2000s was led by countries in North and South America and Australia, the spread of CA since 2008/09 across Europe (including Russia and Ukraine), Asia and Africa has been accelerating, making the transformation from conventional tillage agriculture to CA a truly global phenomenon.

The Green Revolution of the 1960s and 1970s was based on production intensification of wheat and rice, and later maize, relying on the unsustainable intensive tillage-based agriculture and expensive production inputs which brought short-term benefits to some types of farmers only in a handful of countries. The Green Revolution agriculture lost its effectiveness and appeal in the 1990s due to the unacceptably high negative economic, environmental and social impact as well as due to the loss of control by farmers of their own affairs related to production management and capital investment. The alternate CA revolution on the other hand has involved all types of smallholder and large-scale farmers, men and women, and rich and poor farming households, in all major land-based agro-ecologies in all continents. It has brought to the rural communities and society at large a wide range of productivity, economic, environmental and social benefits which cannot be harnessed with the conventional Green Revolution agriculture.

The expansion in area of CA across the world and the increase in number of smallholders adopting CA has benefitted from the growing support of the

[1] In 2018/19, the total CA cropland area increased to 205 M ha, spread across 102 countries.

international research and development community including FAO, IFAD, World Bank, EU, CIRAD, and CGIAR as well as from many national and local level research and extension systems including NGOs, farmer associations and private sector service providers. Overall, the spread of CA since the very beginning has largely been led by farmers. Initially this was in response to the need to minimize soil erosion and land degradation, but soon this became a strategy to build and maintain soil health and productivity, reduce the high cost of production and diminishing returns, harness ecosystem services for society and nature, address climate change, and support pro-poor sustainable agricultural development strategies.

The CA Community of Practice is made up of many stakeholders and champions including farmers and their families, extension workers, development experts, researchers and academics, heads of institutions, policy analysts and decision-makers, as well as national and regional CA associations. Collectively, they have provided the pioneers and champions and support that keeps generating and sustaining the momentum to what has now become a global phenomenon, transforming conventional tillage-based agriculture into CA as the basis for sustainable agriculture and land use intensification. At the same time, they have generated enormous amount of new knowledge, formal and experiential, as well as scientific and empirical evidence, regarding CA science and systems, CA practice and benefits and CA adoption and spread. All this knowledge and evidence constitutes the new understanding about regenerative and sustainable agriculture and represents the desire by the CA Community of Practice to move away for the degrading paradigm of tillage-based Green Revolution agriculture to the alternative paradigm of CA.

In light of the above, I was easily persuaded by the publisher Burleigh Dodds Science Publishing to take up the challenge of editing this book 'Advances in Conservation Agriculture' to bring together the latest state-of-the-art global knowledge and development-oriented information about CA science and systems, practice and benefits and adoption and spread. This book is purposely not designed to be a theoretical debate about what scientists and academics with no practical experience of real farming or of CA think about CA and how CA is performing internationally. Global scientific and empirical evidence speaks volumes about the productivity, economic, environmental and social attractiveness of CA to farmers in all continents. Globally, the rate of adoption of CA is accelerating but much remains to be explained in terms of the superior performance of CA, and also much remains to be innovated in the coming decades to maximize the wide range of benefits offered by CA to farmers, society and the natural world.

The past and current research on CA reviewed in this book is aimed at being of value to all CA stakeholders, including students, especially in the practical context of addressing global challenges related to sustainable development with effective solutions. This is particularly true in the adoption of strategies dealing with: sustainable production intensification, climate smart agriculture, regenerative agriculture, agroecology, and restoration of degraded lands including biodiversity and land-mediated ecosystem services. Thus, this book is a contribution to making sustainable agriculture development real globally. This would not have been possible without the extraordinary help of colleagues and field experts from the global CA Community of Practice who have been involved over many years in championing and bringing about the ongoing and accelerating global CA revolution. More than 120 authors based in more than 30 countries have made this book possible, building their respective chapters on the vast amount of global evidence and knowledge that is now available regarding the superior global performance of the alternate ecologically sustainable paradigm of CA.

In the longer run, the expanding knowledge and education system globally must become fully engaged in generating the future human and institutional capacity and appropriate new mindset to underpin and sustain the process of mainstreaming the CA paradigm as a core component of the much-needed sustainable global food and agriculture system. Mainstreaming of any alternate paradigm in any field requires that all the relevant public, private and civil sector institutions and policies must align themselves behind it and ensure the necessary strategic and practical support to sustain its evolution. It is hoped that the information contained in this book will contribute to more rapid mainstreaming of CA globally by inspiring and generating increasing number of stakeholders and champions, particularly youngsters with practical training of CA, to become engaged in the promotion of CA as a foundation for sustainable agricultural development.

Acknowledgements

This book builds on the vast amount of innovative and pioneering development-oriented work done over the past several decades in the public, private and civil sectors by those individuals and organizations who make up the global Conservation Agriculture (CA) Community of Practice. These CA stakeholders include farmers, their families and organizations, service providers along the value chain, scientific and empirical researchers, educators and trainers at all levels in schools, colleges, universities and extension services, technical assistance and development organizations and their experts at the national, regional and international levels, manufacturers of inputs, policy analysts and policy-makers, politicians and all the innovators and champions. I express my very special and sincere gratitude to all the stakeholders of this CA Community who are making sustainable agriculture real in all continents through the successful adoption and spread of the new paradigm of CA and the institutional capacity to sustain the transformation.

I would like to recognize the enormous assistance and cooperation extended to me in the preparation of this book from a large numbers of colleagues and organizations. First of all I would like to extend my sincere thanks to the Publisher Burleigh Dodds Scientific Publishing for inviting me to edit this book, and specifically to Francis Dodds who was actively engaged throughout in overseeing the project, with the able assistance from Amanda Renwick at Burleigh Dodds and from Rennie Alphonsa (Volumes 2 and 3) and Arun Rajakumar (Volume 3) at Deanta who oversaw production of the book.

Most importantly, I would like to express my deep gratitude to every lead author and all the co-authors for their valuable contributions and for spending their precious time and resources to bring together the relevant and latest global knowledge and information on CA science and systems, practice and benefits, and adoption and spread.

I acknowledge with many thanks the external reviewers of the different chapters of the book for their help to ensure that the chapters comprise the best quality material. They include: Theodor Friedrich, Rolf Derpsch, Sjoerd Duiker, Jeff Mitchell, Ivo Mello, Rafael Fuentes, Paulo Sérgio Pavinato, Denizart Bolonhezi, Telmo Amado, Enamul Haque, Frank Place, D. Pedro Freitas, Robelio Marchão, Julio César Salton, Luis Carlos Balbino, Brian Sims, Jacky Desbiolles, Willem Hoogmoed, Andrés Sylvestre Begnis, Sebastián Senesi, Gottlieb Basch, Hendrick Smith, Don Reicosky, Timothy Reeves, Yashpal Singh Saharawat, Vicenzo Tabaglio, Gualbert Gbehounou, Kevin Gallagher, Johnnie Van den Berg, Mário Carvalho, Jean-Pierre Sarthou, Emilio Gonzalez, Daryl Domitruk, David Coates, Tom Goddard, Julian Dumanski, Michele Pisante, Francis

Shaxson, Ademir de Oliveira Ferreira, Zete Stumpf, Len Kryzanowski, Keith Reid. Wade Thomason, Randall Reeder, Pat Wall, Brian Sims, August Araujo, Telmo Amado, Sandra Corsi, Stephan Loss, Harun Cicek, Pascal Rushemuka, Willy Mulimbi B., Rachid Mrabet, Raymond Nazare, Edward Chuma, Sieglinde Snapp, Alex Morgounov, Turi Fileccia, Richard Bell, Olivier Husson, Delgermaa Chuluunbaatar, Jeff Tullberg, Peter Hobbs, C. M. Parihar, A. R. Sharma and Raymond Nazare.

Finally, I express my very special thanks to Dr. Qu Dongyu, the Director General of the Food and Agriculture Organization of the United Nations, for his inspiring and illuminating Foreword. Dr. Dongyu's contribution is all the more important because the promotion CA in China is a national priority that has led to a very rapid growth in the area under CA and in the institutional capacity to sustain it.

Cover image acknowledgements

- Top left: Theodor Friedrich, Retired Food and Agriculture Organization of the United Nations (FAO) Expert on Conservation Agriculture, Germany, Italy.
- Top middle: Jamil Abdalla Fayad, Agricultural Research and Rural Extension Company of Santa Catarina (EPAGRI), Brazil.
- Top right: Foundations for Farming, Zimbabwe.
- Bottom left: International Centre of Insect Physiology and Ecology (ICIPE), Kenya.
- Bottom middle: Abdulsattar Asmair Alrijabo, Mosul University, Iraq.
- Bottom right: Ngo Tien Dung, Center for Initiatives on Community Empowerment and Rural Development (ICERD), Vietnam.

Chapter 1

State of the global adoption and spread of Conservation Agriculture

Amir Kassam, University of Reading, UK; Theodor Friedrich, Retired Food and Agriculture Organization of the United Nations (FAO) Expert on Conservation Agriculture, Germany; and Rolf Derpsch, Consultant, Paraguay

1 Introduction
2 Global and regional adoption and spread
3 Conservation Agriculture adoption at national level
4 Global Conservation Agriculture cropland goal for 2050
5 Concluding remarks
6 Where to look for more information
7 References

1 Introduction

Chapter 2 in Volume 1 provided a detailed description of the development of Conservation Agriculture (CA) globally and included national CA cropland areas for 2008/09, 2013/14 and 2015/16 (Kassam et al., 2020). The total cropland area of CA for these three update years was 106.5 M ha, 156.7 M ha and 180.4 M ha, respectively. This brief chapter provides an update of the CA cropland area for 2018/19 based on the national CA area information provided in the national and regional chapters of this third volume and the values published at the CA-Global website (https://www.ca-global.net).

This volume is about the adoption and spread of CA at the national and regional level covering all the area in the world where CA has been promoted to replace tillage-based agriculture. In this context, this chapter provides a brief overview of the global, regional and national status of the adoption and spread of CA including an update of the CA adoption estimates.

2 Global and regional adoption and spread

The historical chart of CA uptake at the global level is shown in Fig. 1. The transformation of conventional tillage-based agriculture began in the 1930s

http://dx.doi.org/10.19103/AS.2021.0088.01

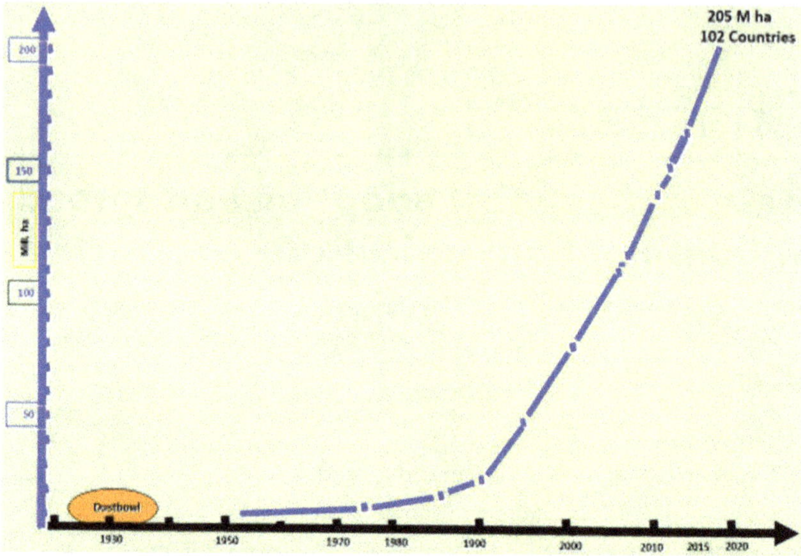

Figure 1 Historical chart of CA uptake at the global level (Source: Kassam et al., 2021).

after the 'Dust Bowl' that shook the farming communities in the mid-west of the United States and the scientific community to rethink what was not going right with farming, particularly with regards to soil conservation.

Minimization of soil disturbance with stubble mulching was a major breakthrough in the understanding of how the objective of crop production intensification could be combined with the objective of soil and water conservation at the practical level by farmers.

Initially, the intentions were to eliminate the erosion problem of tillage, for which the term conservation tillage became popular determining the minimal necessary soil cover with crop residues to reduce the erosion to acceptable levels. It took few more years before the concept of tilling the soil was questioned *per se*, not only for the erosion problems it created but also for other types of soil degradation processes it accelerated. Only in the late 1960s, it became obvious that no-till seeding through stubbles and crop biomass cover was the way to avoid or eventually reverse soil degradation and erosion. Yet, like the problems with climate change, the general global public is not aware that the ongoing soil degradation caused by tillage is putting the survival of humankind on this planet at risk.

The global total CA cropland area in 2018/19 is about 205.4 M ha, corresponding to about 14.7% of the total global cropland, with the spread being more or less equally split between the Global South (50.5%) and the Global North (49.5%). Overall, the increase in the global CA cropland area since 2008/09 has continued at an annual rate of about 10 M ha, from 106.5 M ha in 2008/09 to 205.4 M ha in 2018/19. Prior to that, the annual rate was about

5 M ha during the period 1990–2008/09. In 1990, the CA area of cropland was 11 M ha, and in 2000, the CA area was 67 M ha.

The information on the global spread of CA cropland area by regions in 2008/09, 2013/14, 2015/16 and 2018/19 is shown in Table 1 (Kassam et al., 2021). The global CA cropland area increased by 98.9 M ha or 92.9% from 106.5 M ha in 2008/09 to 205.4 M ha in 2018/19. The change in the CA cropland area in the different continents since 2008/09 has been 67.5% (from 49.6 M ha to 83.0 M ha) in South America, 64.8% (from 40.0 M ha to 65.9 M ha) in North America, 91.5% (from 12.2 M ha to 23.3 M ha) in Australia and New Zealand, 68.00% (from 0.1 M ha to 6.9 M ha) in Russia and Ukraine, 259.0% (from 1.6 M ha to 5.6 M ha) in Europe, 566.5% (from 2.6 M ha to 17.5 M ha) in Asia and 547.8% (from 0.5 M ha to 3.1 M ha) in Africa.

The global CA cropland increased by some 48.6 M ha or 31.0% since 2013/14 from 157 M ha to 205.4 M ha in 2018/19. The change in the CA cropland area in the different continents since 2013/14 has been 25.0% (from 66.4 M ha to 83.0 M ha) in South America, 22.2% (from 54.0 M ha to 65.9 M ha) in North America, 30.4% (from 17.8 M ha to 23.3 M ha) in Australia and New Zealand, 32.7% (from 5.2 M ha to 6.9 M ha) in Russia and Ukraine, 169.8% (from 2.1 M ha to 5.6 M ha) in Europe, 70.4% (from 10.3 M ha to 17.5 M ha) in Asia and 216.4% (from 1.0 M ha to 3.1 M ha) in Africa.

The global CA cropland increased by some 25 M ha or 13.9% since 2015/16 from 180 M ha to 205.4 M ha in 2018/19. The change in the CA cropland area in the different continents since 2015/16 has been 18.7% (from 69.9 M ha to 83.0 M ha) in South America, 4.4% (from 63.2 M ha to 65.9 M ha) in North America, 2.8% (from 22.7 M ha to 23.3 M ha) in Australia and New Zealand, 21.1% (from 5.7 M ha to 6.9 M ha) in Russia and Ukraine, 57.4% (from 3.6 M ha to 5.6 M ha) in Europe, 25.8% (from 13.9 M ha to 17.5 M ha) in Asia and 108.3% (from 1.5 M ha to 3.1 M ha) in Africa.

In 2008/09, some 4.77 M ha or 4.48% of the global CA area was in Europe (including Russia and Ukraine), Asia and Africa. By 2018/19, the area increased to 33.17% or 16.18% of total, on a larger global total CA area. During the past decade, larger percentage increases occurred in these regions, while the CA area continued to increase in the Americas and in Australia and New Zealand. In South America, the total CA area is approaching 70% of the total regional cropland area, and in Australia and New Zealand, the CA area is approaching 75% of the total cropland area.

3 Conservation Agriculture adoption at national level

Historical development of no-till systems and the modern version of CA that has been promoted over the past three decades are documented in the study by Kassam et al. (2019, 2020) and updated at the national level in the CA-Global

Table 1 Global spread of CA cropland area ('000 ha) in different regions for 2008/09, 2014/15 and 2018/19 and corresponding percent change (Kassam et al., 2021)

Region	CA cropland area 2008/09	CA cropland area 2013/14	CA cropland area 2015/16	CA cropland area 2018/19	Percent change in CA area since 2015/16	Percent change in CA area since 2013/14	Percent change in CA area since 2008/09	Percent CA cropland area in the region 2018/19
South and Central America	49 564.10	66 377.00	69 895.00	82 996.18	18.7	25.0	67.5	68.7
North America	40 003.80	53 967.00	63 181.00	65 937.22	4.4	22.2	64.8	33.6
Australia and New Zealand	12 162.00	17 857.00	22 665.00	23 293.00	2.8	30.4	91.5	74.0
Russia and Ukraine	100.00	5200.00	5700.00	6900.00	21.1	32.7	6800.0	4.5
Europe	1560.10	2075.97	3558.20	5601.53	57.4	169.8	259.0	5.2
Asia	2630.00	10 288.65	13 930.20	17 529.02	25.8	70.4	566.5	3.6
Africa	485.23	993.44	1509.24	3143.09	108.3	216.4	547.8	1.1
Total	**106 505.23**	**156 759.06**	**180 438.64**	**205 400.04**	**13.8**	**31.0**	**92.9**	**14.7**

website. Extent of adoption of CA area by country for 2008/09, 2013/14 and 2018/19 is shown in Table 2.

In 2008/2009, the global CA area was spread over 36 countries. In 2013/14, the number of countries with CA area had increased to 53 and in 2015/16 to 77. In 2018/19, the number of countries with CA area had increased to 102. The corresponding increases in the number of countries from 2008/09 to 2018/19 were from 8 to 13 in South and Central America, from 11 to 31 in Europe, from 2 to 26 in Asia and from 9 to 25 in Africa. This shows that much greater interest is being shown for CA systems in recent years by farmers globally in every region. Also, the extent of support from public sector institutions and governments, although modest, has been increasing steadily. New countries taking up CA adoption are mainly in Central America, Europe, Africa, West Asia, South Asia and Southeast Asia. This trend is expected to continue, benefitting increasing number of smallholder farmers.

Globally, the 10 lead CA countries are the United States, Brazil, Argentina, Australia, Canada, China, Russia, India, Paraguay and Kazakhstan. In South and Central America, the lead five countries are Brazil, Argentina, Paraguay, Bolivia and Uruguay; in Europe, the lead countries are Spain, France, Romania, United Kingdom and Italy; in Africa, South Africa, Zambia, Mozambique, Ghana and Malawi; and in Asia, the lead countries are China, India, Kazakhstan, Pakistan and Iran.

4 Global Conservation Agriculture cropland goal for 2050

The focus of the eighth World Congress on Conservation Agriculture in June 2021 was on the 'Future of Farming: Sustainable and Profitable Farming with Conservation Agriculture'. The Congress was convinced that CA must be the mainstay of the shift that the world has to make urgently towards sustainable farming and food systems and that CA must become synonymous with sustainable farming for the future.

The Congress set a national goal for itself to increase the global CA cropland area to 50% of the total cropland by 2050, in particular to respond to the global challenge to mitigate the advancing climate change and land degradation. This represents an area of 700 M ha (WCCA, 2021). Achieving this goal would require increasing the annual rate of expansion in CA area by some 50%, to 15 M ha. It would also require a massive boost to the momentum of the global CA Community's activities with a concentration on the following six themes:

1 Catalysing the formation of additional farmer-run CA groups in countries and regions in which they do not yet exist and enabling all groups to accelerate CA adoption and enhancement, maintaining high-quality standards.

Table 2 Extent ('000 ha) of adoption of CA worldwide by country in 2008/09, 2013/14, 2015/16 and 2018/19

No	Country	CA area 2008/09	CA area 2013/14	CA area 2015/16	CA area 2018/19
1	United States	26 500.00	35 613.00	43 204.00	44 049.00
2	Brazil	25 502.00	31 811.00	32 000.00	43 000.00
3	Argentina	19 719.00	29 181.00	31 028.00	32 907.00
4	Canada	13 481.00	18 313.00	19 936.00	21 739.00
5	Australia	12 000.00	17 695.00	22 299.00	22 927.00
6	Paraguay	2400.00	3000.00	3000.00	3158.00
7	Kazakhstan	1300.00	2000.00	2500.00	3000.00
8	China	1330.00	6670.00	9000.00	9000.00
9	Bolivia	706.00	706.00[a]	2000.00	1858.03
10	Uruguay	655.10	1072.00	1260.00	1278.00
11	Spain	650.00	792.00	900.00	1000.00
12	South Africa	368.00	368.00[a]	439.00	1607.08
13	Germany	354.00	200.00	146.00	352.89
14	Venezuela	300.00	300.00[a]	300.00[b]	300.00+
15	France	200.00	200.00[a]	300.00	720.00
16	Finland	200.00	200.00	200.00	120.00
17	Chile	180.00	180.00[a]	180.00[b]	180.00+
18	New Zealand	162.00	162.00[a]	366.00	366.00+
19	Colombia	102.00	127.00	127.00[b]	127.00+
20	Ukraine	100.00	700.00	700.00[b]	900.00
21	Italy	80.00	380.00	283.92	432.00

22	Zambia	40.00	200.00	316.00	552.67
23	Kenya	33.10	33.10[a]	33.10[b]	33.10+
24	UK	24.00	150.00	362.00	562.00
25	Portugal	25.00	32.00	32.00[b]	47.05
26	Mexico	22.80	41.00	41.00[b]	149.22
27	Zimbabwe	15.00	90.00	100.00	100.00[c]
28	Slovakia	10.00	35.00	35.00[b]	365.00
29	Sudan	10.00	10.00[a]	10.00[b]	10.00+
30	Mozambique	9.00	152.00	289.00	289.00[c]
31	Switzerland	9.00	17.00	17.00[b]	11.02
32	Hungary	8.00	5.00	5.00[b]	24.29
33	Tunisia	6.00	8.00	12.00	14.00
34	Morocco	4.00	4.00	10.50	12.83
35	Lesotho	0.13	2.00	2.00	2.00+
36	Ireland	0.10	0.20	0.20	3.66
37	Russia	-	4500.00	5000.00	6000.00
38	India	-	1500.00	1500.00[b]	3500.00
39	Malawi	-	65.00	211.00	211.00[c]
40	Turkey	-	45.00	45.00	100.00
41	Moldova	-	40.00	60.00	180.00
42	Ghana	-	30.00	30.00[b]	235.00
43	Syria	-	30.00	30.00[b]	30.00+
44	Tanzania	-	25.00	32.60	32.60[c]
45	Greece	-	24.00	24.00[b]	110.50

(Continued)

Table 2 (*Continued*)

No	Country	CA area 2008/09	CA area 2013/14	CA area 2015/16	CA area 2018/19
46	Korea, DPR	-	23.00	23.00[b]	23.00+
47	Iraq	-	15.00	15.00[b]	12.00
48	Madagascar	-	6.00	9.00	9.00[c]
49	Uzbekistan	-	2.45	10.00	120.00
50	Azerbaijan	-	1.30	1.30[b]	37.50
51	Lebanon	-	1.20	1.20[b]	1.20
52	Kyrgyzstan	-	0.70	50.00	60.00
53	Netherlands	-	0.50	7.35	17.50
54	Namibia	-	0.34	0.34[b]	0.80
55	Belgium	-	0.27	0.27	0.27+
56	Pakistan	-	-	600.00	1320.00
57	Romania	-	-	583.82	583.82[c]
58	Poland	-	-	403.18	403.18[c]
59	Iran	-	-	150.00	300.00
60	Estonia	-	-	42.14	42.14[c]
61	Czech Republic	-	-	40.82	40.82[c]
62	Austria	-	-	28.33	28.33[c]
63	Lithuania	-	-	19.28	19.28[c]
64	Croatia	-	-	18.54	18.54[c]
65	Bulgaria	-	-	16.50	16.50[c]
66	Sweden	-	-	15.82	26.00

67	Latvia	–	–	11.34	11.34[c]
68	Uganda	–	–	7.80	7.80[c]
69	Algeria	–	–	5.60	7.00
70	Denmark	–	–	2.50	38.50
71	Slovenia	–	–	2.48	26.00
72	Bangladesh	–	–	1.50	1.50
73	Swaziland	–	–	1.30	0.80
74	Tajikistan	–	–	1.20	5.00
75	Vietnam	–	–	1.00	1.00[c]
76	Cambodia	–	–	0.50	0.50[c]
77	Laos	–	–	0.50	0.50[c]
78	Luxemburg	–	–	0.44	0.44[c]
79	Cyprus	–	–	0.27	0.27[c]
80	Peru	–	–	–	2.89
81	Nepal	–	–	–	1.00
82	Armenia	–	–	–	0.30
83	Georgia	–	–	–	0.19
84	Belarus				400.00
85	Jordan	–	–	–	0.20
86	El Salvador	–	–	–	4.05
87	Guatemala	–	–	–	10.00
88	Honduras	–	–	–	171.00
89	Nicaragua	–	–	–	0.21
90	Sri Lanka	–	–	–	0.05

(Continued)

Table 2 (*Continued*)

No	Country	CA area 2008/09	CA area 2013/14	CA area 2015/16	CA area 2018/19
91	Burkina Faso	-	-	-	1.00
92	Cameroon	-	-	-	2.00
93	Myanmar	-	-	-	0.02
94	Rwanda	-	-	-	0.25
95	Burundi	-	-	-	0.20
96	Malaysia	-	-	-	7.50
97	Philippines	-	-	-	6.75
98	Timor Leste	-	-	-	1.00
99	Guinea	-	-	-	0.40
100	Ethiopia	-	-	-	7.50
101	DR Congo	-	-	-	2.06
102	Niger	-	-	-	5.00
	Total	**106 505.23**	**156 738.96**	**180 438.64**	**205 400.04**
	Percent difference		**47.17 since 2008/09**	**69.42 since 2008/09** **15.12 since 2013/14**	**92.85 since 2008/09** **31.05 since 2013/14** **13.83 since 2015/16**

[a] 2013/14 values taken from 2008/09.
[b] 2025/16 values taken from 2013/14.
[c] 2018/19 values taken from 2015/16.Source: Estimates obtained from the CA-Global website (https://www.ca-global.net/ca-stat).

2 Greatly speeding up the invention and mainstreaming of a growing array of truly sustainable CA-based technologies, including through engaging with other movements committed to sustainable farming.
3 Embedding the CA Community in the main global efforts to shift to sustainable food and governance systems and replicating the arrangements at local levels.
4 Assuring that CA farmers are justly rewarded for their generation of public goods and environmental services.
5 Mobilizing recognition, institutional support and additional funding from governments and international development institutions to support good-quality CA programme expansion.
6 Building global public awareness of the steps being taken by our CA Community to make food production and consumption sustainable.

5 Concluding remarks

The global burden of chronic crises includes food insecurity, climate change, loss of biodiversity, environmental degradation and unsustainable diets and human ill health. CA systems have a role to play in contributing to addressing all these crises. Increasingly, CA must be seen to be a central part of sustainable food systems and sustainable environmental management.

CA global community must continue its effort to not only improve the quality and performance of CA systems but also undertake strategic research that would allow CA systems to operate biologically or organically, utilizing minimum input of synthetic agrochemicals or avoiding them where possible. Already there are promising signs that such biological CA systems are possible, thus making it possible for more farmers to adopt CA-based organic farming.

Equally important is the need to support smallholder farmers transform their conventional systems into CA systems with sustainable mechanization support. Already many more smallholders globally are practicing CA than large-scale farmers. However, the needs of smallholder farmers must be given greater attention than in the past. Equally important is the need to make farming an area of opportunity for women and youth, and transformation of conventional agriculture to CA has much to offer towards this goal, particularly when integrating precision GPS-based practices and robotics in making farming more efficient and profitable as shown in volumes 1 and 2.

6 Where to look for more information

A general source of information on CA is the Food and Agriculture website (www.fao.org/conservation-agriculture) and so are the websites of the European Conservation Agriculture Federation (https://ecaf.org/), the Africa Conservation

Tillage Network (http://www.act-africa.org/) and the Conservation Agriculture Alliance for Asia and Pacific (http://www.caa-ap.org/).

More information regarding the development of CA systems globally can be found in books and journals and on websites of national and regional CA organizations.

Books include Goddard et al. (2006), Baker et al. (2007), FAO (2011, 2016), Jat et al. (2014), Farooq and Siddique (2014), Chan and Fantle-Lepczyk (2015) and Dang et al. (2020). Nationally oriented information on CA development is available for several countries including Australia (Crabtree, 2010), Canada (Lindwall and Sonntag, 2010), Brazil (Junior et al., 2012; de Freitas and Landers, 2014), Argentina (Peiretti and Dumanski, 2014) and United States (Lessiter, 2018).

Lessiter (2011) provides a narrative on the 40 legends of the past in no-till farming in North America. The International Soil and Water Conservation Research published a special issue on pioneers in soil conservation and CA, which provides good information on the adoption of CA in several countries (Dumanski et al., 2014).

Proceedings of World Congresses on CA are a good source of historical and current information on CA research and adoption. Similarly, proceedings of Africa Congresses on CA provide good information on research and development work in Africa (Kassam et al., 2017; Mkomwa and Kassam, 2021). Websites of national and regional CA associations are a good source of CA information on adoption and spread as well as on research.

7 References

Baker, C. J., Saxton, K. E., Ritchie, W. R., Chamen, W. C. T., Reicosky, D. C., Ribeiro, M. F. S., Justice, S. E. and Hobbs, P. R. (Eds.) 2007. *No-Tillage Seeding in Conservation Agriculture* (2nd edn.). CABI Publishing, Wallingford, UK.

Chan, C. and Fantle-Lepczyk, J. 2015. *Conservation Agriculture in Subsistence Farming: Case Studies from South-Asia and Beyond*. CABI Publishing, Wallingford. 278p.

Crabtree, W. L. 2010. *Search for Sustainability with No-Till Bill in Dryland Agriculture*. Crabtree Agricultural Consulting, Western Australia, 204p. Available at: www.no-till .com.au.

Dang, Y., Dalal, R. and Menzies, N. 2020. *No-Till Systems Farming Systems for Sustainable Agriculture: Challenges and Opportunities*. Springer, Cham, Switzerland.

de Freitas, P. L. and Landers, J. N. 2014. The transformation of agriculture in Brazil through development and adoption of zero tillage conservation agriculture. *International Soil and Water Conservation Research* 2(1), 35–46.

Dumanski, J., Reicosky, D. C. and Peiretti, R. A. 2014. Pioneers in soil conservation and conservation agriculture. *International Soil and Water Conservation Research* 2(1), 1–4.

FAO 2011. Save and grow: a policymakers guide to the sustainable intensification of smallholder crop production. Available at: www.fao.org/ag/save-and-grow/. Food and Agriculture Organization, Rome.

FAO 2016. *Save and Grow in Practice: Maize, Rice, Wheat - A Guide to Sustainable Production*. Food and Agriculture Organization, Rome.

Farooq, M. and Siddique, K. H. M. (Eds.) 2014. *Conservation Agriculture*. Springer International, Switzerland.

Goddard, T., Zoebisch, M. A., Gan, Y. T., Ellis, W., Watson, A. and Sombatpanit, S. (Eds.) 2006. *No-Till Farming Systems*. Special Publication No. 3. World Association of Soil and Water Association, Bangkok, Thailand.

Jat, R. A., Sahrawat, K. L. and Kassam, A. H. (Eds.) 2014. *Conservation Agriculture: Global Prospects and Challenges*. CABI Publishing, Wallingford, UK.

Junior, R. C., de Araujo, A. G. and Llanillo, R. F. 2012. *No-Till Agriculture in Southern Brazil: Factors That Facilitated the Evolution of the System and the Development of the Mechanization of Conservation Farming*. FAO and IAPAR, Rome, Italy.

Kassam, A. H., Mkomwa, S. and Friedrich, T. (Eds.) 2017. *Conservation Agriculture for Africa*. CABI Publishing, Wallingford, UK.

Kassam, A., Friedrich, T. and Derpsch, R. 2019. Global spread of conservation agriculture. *International Journal of Environmental Studies* 76(1), 29–51.

Kassam, A., Derpsch, R. and Friedrich, T. 2020. Ch. 02. Development of conservation agriculture systems globally. In: *Advances in Conservation Agriculture*: *Systems and Science* (Vol. 1) Kassam, A. (Ed.). Burleigh Dodds Science Publishing, Cambridge, UK.

Kassam, A., Friedrich, T. and Derpsch, R. 2021. Successful experiences and learnings from Conservation Agriculture Worldwide. *Keynote Address, Sub-Theme*. 8th World Congress on Conservation Agriculture, 21–23 June 2021, Bern, Switzerland (vol. 1).

Lessiter, F. 2011. 40 legends of the past. *40th Anniversary Issue of No-Till Farmer*, November 2011.

Lessiter, F. 2018. From maverick to mainstream: a history of no-till farming. *No-Till Farmer*. 416p.

Lindwall, C. W. and Sonntag, B. (Eds.) 2010. *Landscape Transformed: The History of Conservation Tillage and Direct Seeding*. Knowledge Impact in Society, Saskatoon, University of Saskatchewan, Canada.

Mkomwa, S. and Kassam, A. (Eds.) 2021. *Conservation Agriculture in Africa: Climate Smart Agricultural Development*. CAB International, Wallingford, UK.

Peiretti, R. and Dumanski, J. 2014. The transformation of agriculture in Argentina through soil conservation. *International Soil and Water Conservation Research* 2(1), 14–20.

WCCA 2021. Declaration. The 8th World Congress on Conservation Agriculture 21–23 June. Bern, Switzerland.

Chapter 2

Conservation Agriculture in West and Central Canada: an integrated review of adoption

Tom Goddard, Consultant, Canada; and Adam Hayes, Soil Management Consultant, Canada

1 Introduction
2 Farmers
3 Organizations
4 Research and extension
5 Policy and programmes
6 Cumulative effect: issues and adoption
7 Current state
8 Conclusion
9 Acknowledgements
10 Where to look for further information
11 References

1 Introduction

Canada is the second-largest country in the world with a band of agricultural land spanning across North America along the southern border of the country. Nearly 193 000 farms manage 38 million hectares of cropland (about 2.4% of total global cropland). The central portions mostly exhibit a humid continental climate with a slight semi-arid climate, while the west and east coastal regions exhibit a marine coastal climate. The largest block of cropland is the northern extent of the Great Plains, which constitutes 87% of the cropland in Canada. The heterogeneity of the agroclimatic areas and farm types results in diverse farming systems across the country.

As early as 1605, wheat, flax and vegetables were grown and livestock was pastured in the eastern provinces. Forests were cleared in southern Quebec and Ontario largely for mixed farming of livestock, pasture, cereals as well as fruits and vegetables through the 1700s and 1800s. The prairie lands were primarily a source of furs for Europe and eventually were used for grazing by cattle herds

http://dx.doi.org/10.19103/AS.2021.0088.02

moving north from the United States. Cultivated agriculture started in the late 1800s with the development of a trans-national railway (completed in 1885), which brought settlers to the prairie region. Settlers in the initial period were found to follow the practice of summer fallow to provide more reliable crop yields in years of drought. This observation, along with further experimentation, prompted one of the early federal research station managers to promote the use of summer fallow and 'dust mulch' (Janzen, 2001). The alternate years of cultivation and summer fallow promoted the mineralization of organic matter and the uptake of nutrients by the commercial crops. Janzen's (2001) review of early literature shows that one-quarter to one-third of topsoil organic matter was lost in the first few decades after the cultivation of the prairie grasslands.

Along with intensive cultivation for seedbed preparation, weed control, nutrient release and dust mulch came soil degradation, which exacerbated losses from wind, water and tillage erosion depending on the location in Canada and the landscape morphology. This resulted in topsoil being degraded, moved downslope or removed from fields altogether. Gregorich and Anderson (1985) examined prairie toposequences in native and varying ages of cultivation and found nitrogen losses in the order of 80% and carbon losses of greater than 600 kg ha^{-1} yr^{-1}. The erosion and degradation were a matter of concern for governments and some farmers. Prompted by the rapid increase in the erosion of the degraded soils and resultant desertion of farm land in the 'Dirty Thirties', the Canadian government formed the Prairie Farm Rehabilitation Administration (PFRA) through an Act of Parliament in 1935. The main management practices to mitigate erosion were strip-cropping and shelterbelts. Cultivation and cropping continued much the same as before.

At the same time, another national institution that promoted summer fallow on the prairies through its policy was the Canadian Wheat Board (CWB). Farmers on the prairies were required to sell wheat and barley to the CWB, which then became a single-selling desk to attract best prices on international markets. Farmers were assigned an annual sales quota based upon the amount of cultivated area they raised crops on. If some of that land was in summer fallow, they could quickly sell all of the crops from the seeded acres, a key driver in years of high yields. Hence, keeping land in fallow ensured that farmers sold high-yielding crops without delays. Although the name implied national coverage, the jurisdiction focus was on the prairie region. The CWB was disbanded, and controlling interest was sold to the private sector in 2015 in response to the political and farmer pressure wanting expanded choices to market grains.

In the dry areas of the prairies, summer fallow was practised on alternate years, barren of crops half of the time (one crop per year), thus keeping the mining or mineralization of nutrients as well as the structural degradation of the soil significant. In more moist regions of the prairies, the land was often summer fallowed every third year. After decades, soils became infertile and poorly structured. Erosion processes removed topsoil completely or redistributed it within fields, increasing field variability. By 1980, it was estimated that non-eroded cultivated soils had lost up to 30% of the original organic matter and eroded soils lost up to 53% (Anderson, 2008). Although summer fallow was predominant in the prairie region, it was not unheard of elsewhere. The prairie region was known for erosive wind and rainy weather, while Ontario had more erosive rainstorms and tillage erosion from plowing and row crop culture.

In Ontario, agriculture up until the 1940s consisted largely of mixed farming operations of both livestock and crops. Farm fields were small, and the crop mix was largely pasture and small grains. Crop rotations were long, and fields were plowed or tilled infrequently. Soil organic matter levels were relatively high, and thus soil erosion was not a widespread issue. Even though this was generally the case, the soil surveyors of those days began to have concerns about the soil. The two southwestern counties experienced more intensive cropping than the rest of the province. The Essex Soil Report, surveyed in 1939, expressed concern about the geographically dominant fine-textured soils, stating that 'in an area where there is a high concentration of row crops (tobacco, tomatoes, corn, etc.) the organic matter content of the surface soil has been lowered (Richards et al., 1949). This in turn has destroyed the soft, mellow crumb and granular structure that prevailed under natural conditions'. Soon after the Second World War, there was an increased emphasis on soil conservation.

Van Vliet (1981) presented reasons why soil erosion was on the increase in central (Ontario) Canada:

- increase in the provincial corn acreage by 45% from the late 1960s to the mid-70s;
- the farm consolidation programme in the 1970s that removed fence rows to produce longer fields;
- larger equipment leading to soil compaction and reduced infiltration; and
- the intensive farming of leased land or land adjacent to urban centres without regard for soil loss.

Annual fall plowing to establish corn (i.e. continuous corn) began to deplete organic matter levels and leave soils exposed over winter and early spring.

By the 1970s, soil erosion became a significant issue. Water erosion created rills and gullies in fields; in the worst-case scenario, it washed soil across roads into other fields. Every spring, the rills and gullies had to be filled and levelled through cultivation before the crop could be planted. Wind erosion was also a significant problem, especially in the lighter textured areas of the province. In some springs, wind erosion was so severe that the soil drifted across roads - requiring it to be cleared off by snow removal equipment. Traffic had to be slowed to a crawl as visibility was reduced to near zero at times. The damage to crops was often devastating, especially to horticultural crops. Young plants were severely sandblasted or cut off at the ground.

In addition to wind and water erosion, farmers and extension staff had noticed the evidence of acute soil loss from knoll and crest slope positions on fields with variable topography. Research in the 1980s confirmed that annual primary and secondary tillage passes had scraped topsoil (A horizons) and subsoil (B horizons), exposing (most often - high lime and high pH) parent materials (C horizons) on fields with rolling, hummocky and undulating slope configurations. In most cases, the surface soils were re-deposited in level and depressional areas downslope from the points of removal, as evidenced by buried topsoil in these field locations. However, water erosion did work in concert with tillage erosion, causing the soil to be lost from the field and, where surface waters were close, into adjacent watercourses causing pollution of the watercourses and ultimately the lower Great Lakes.

Driver and Wall (1982) reported study results indicating that on row crop land the total annual erosion costs attributable to yield and nutrient and pesticide losses are approximately $68 million, with more than 80% of the costs occurring in the southern and western regions of Ontario. Water erosion costs represent about 98% of the damages, with wind erosion accounting for the remainder. An additional $6 million was added for drainage maintenance costs.

As a strong response to the significant impact of erosion on the long-term productivity of soils and on water quality, the society and the agricultural community adopted and strengthened its approaches for conserving soil and nutrients to improve productivity and reduce its impact off-site. The need to change agricultural practices to protect soils and the environment became a social concern of the public at large. Farmers became aware of the long-term and complex impacts of degraded fields and that their reputation as stewards of the land was diminishing in the eyes of the public.

This brief glance back into history to the beginning of cultivation brings us to the latter part of the twentieth century when no-till (NT) and diverse cropping appeared. The chapter will review the roles played by farmers, organizations, researchers and policy in the adoption of Conservation Agriculture (CA) in Canada. It will end with a brief look at the current and future state of Conservation Agriculture.

We acknowledge the value of conversations we have had over the decades with farmers, academics, extension specialists and others involved in the growth of CA across Canada. Although the scientific literature does not cover all aspects or answer all questions regarding CA, we have augmented our knowledge on CA with personal experiences working as soil management specialists for provincial governments during the rapid growth of CA in Canada.

2 Farmers

Although the early movement of NT appeared in the 1980s, there is a record of a pioneer much earlier than that. C. S. Noble was a farmer in the southwestern prairie region farming from about 1904 onwards who at one point had the largest farm in the British Empire (nearly 15 000 hectares) but lost much of it to financers. He did recognize at that time how the soil degraded quickly under intensive cultivation and said, 'take your soil seriously', and being also a businessman and entrepreneur, he set about to develop a low-angle, wide-blade cultivator ('Noble Blade') that would cut off weeds but not turn the soil over, thus leaving the plant residue on the surface (Lindwall, 2010; Lafond et al., 2014; Smith, 2018). This saw some success in the dry parts of the prairie region but did not effectively control weeds in the moister regions.

Farmers are generally innovative but what can be a point of discussion is which comes first? Are farmers by nature innovative in order to excel at farming or does farming create innovative farmers? Farmers manage to achieve profitability through opportunities that increase production, that are of low cost, that save labour overall or at peak times, and that have fewer risks and lower stress. All the information they collect needs to be fitted to their particular circumstances of climate, natural resources and built infrastructure. It is up to them to do the fitting. Farmers need to look at the whole farming system, the integrated whole over space and time. The information they receive from institutions is often on specific or narrow topics - farmers are responsible for the integrative function. Innovative farmers kept the integrated view when looking at early proofs of NT and residue management practices.

Their integrative views made them very concerned about the land degradation caused by first-, second- and third-order effects of cultivation - the loss of organic matter, the loss of soil cover, increased erosion and landscape soil variability, weed issues and ultimately farm revenue and environmental risks. The 'Dirty Thirties' were still within the reach of memories of farmers or their grandparents , and they were sensitive either to the slowness in recovery or the possibility of a reoccurrence of catastrophic events. They remained alert to management practices or technologies that would mitigate the risks.

The early pioneers became aware of what one another was doing through word of mouth, farm organizations, extension generalists and the farm press and related media (the pre-Internet era). In Canada, there were a few staff at research stations, universities and extension services that saw possibilities with NT and tried to support the pioneer farmers within their existing capabilities for exploration. There was not much fanfare in the early days. Farmers and specialists quietly worked away exchanging ideas and experiences, recording successes and failures. In doing so, they expanded their unofficial networks in both numbers and geographies, which affirmed that innovation was widespread, and NT had a wide range of variations and applications.

In Ontario, early conservation tillage focused on spring plowing, modified moldboard plows, disc/coulter chisel plow (and various modifications) or a light disking on the sandier soils and some ridge-till on the clay soils. A few farmers experimented with NT in the latter part of the 1960s, but, generally, adoption was limited (Wall et al., 1985). Lobb (2011) reports an estimate of about 4000 acres to 6000 acres of zero tillage in 1970.

Lobb (2011) notes that serious erosion, financial hardship and better herbicides brought renewed interest in NT in the early 1980s. Progress towards NT made use of university residue removal experiments where some added that in-row tillage was found to be helpful. The necessity for crop rotation was confirmed by farmers. Farmers also learned that soil type was not usually a limiting factor for NT where drainage and soil condition was good. The focus shifted from tillage and tillage comparisons to tillage 'systems development'.

The early adopters of NT faced many challenges. Tillage buried crop residues but with NT, there could be significant amounts of residue to plant through. The residue made it difficult for many conventional planters and drills to achieve good seed-to-soil contact and uniform depth of planting. This resulted in uneven emergence and reduced stands. The residue also kept the soils with higher clay content wetter and cooler in the spring delaying planting. Initially, NT equipment and/or modifications were not available or were in limited supply.

Early adopters met this challenge by modifying their equipment, such as adding coulters to the front of drills. With these modifications and the development of NT drills by several manufacturers, the acreage of NT cereals and soybeans began to increase across Ontario. NT winter wheat acreage increased rapidly as farmers saw the benefits of reduced labour requirements and timely planting after soybeans (Wall et al., 1985).

Planting success increased as farmers adopted a system of three coulters ahead of planter seed and fertilizer openers developed by farmer Ray Rawson of Michigan, United States,. A variety of coulter types, widths and configurations were used based on what worked best for the farm conditions. All of this improved the success of NT corn. Over time, as the soil improved, the crop could be planted with fewer coulters and less aggressive press wheels. Even with this, NT corn acreage did not get much above 10% of total corn acres. This was due to a perceived and sometimes real yield reduction for NT corn.

Many growers across the province and on all soil types were able to make NT work for most or all of their crops, and where NT corn was a challenge, growers used a rotational tillage system: 2 years planting soybeans and a cereal NT followed by minimum tillage or in some cases conventional tillage in the cereal stubble before corn. Others used some form of vertical tillage or in-row strip-tillage before corn.

Another challenge was a shift in weed spectrum from annual weeds to winter annuals and perennials. The significant reduction in the cost of glyphosate herbicide, used to kill weeds ahead of planting, and the development of new herbicides helped contribute to NT success.

NT adoption increased through the late 1980s and through the 1990s due to improved equipment and planter modifications, government incentives, research and extension support, reduced labour, fuel and often equipment costs. Adoption slowed in the 2000s as financial incentives and extension support were reduced.

Lobb (2011) quotes Jack Rigby, one of the early adopters, 'In the early days, some likeminded people from across Southern Ontario were interested in preserving and improving their soil by changing the way they farmed. These were people who were prepared to throw away the rule book, examine every aspect of raising a crop and question the traditional wisdom, not afraid to make mistakes, willing to share their successes and their failures and in the end write a new rule book.'

Early farmers had little help with research and technical information or support in adopting CA, which changed quickly as more became involved. It is informative to be aware of positive and negative drivers that influenced early adopters.

2.1 Positive drivers

- Observation and experience of success with NT and new cropping management (e.g. weed control) provided positive reinforcement to farmers.
- Affirmation of doing the right thing by peers, researchers and specialists.
- Support from local groups of innovative farmers as well as the farm organizations with a mandate for education and promotion of sustainable practices. The chance to connect across greater distances with like-minded farmers.
- Recognition by scientists, specialists and governments (local, province, etc.) for their leading innovation at the farm level and at the research level.
- Enhanced partnerships and communication between farmers, subject matter specialists and researchers.
- Financial incentives to support the adoption, development and research into soil conservation practices.
- Farm equipment manufacturers and pesticide companies working to adopt or adapt new concepts to support conservation cropping systems.

2.2 Negative drivers

- Cognitive biases: There are many at play and farmers or rural society as a whole have well-entrenched biases. Confirmation bias is 'I told you it wouldn't work' as soon as a failure is seen - no need to investigate further. Confirmation and status quo biases were evident where farmers observed smooth clean-tilled fields as good management and signs of patchy residue, uneven stands or shorter crops as proof that CA was not working, regardless that yields at the end of the year were higher - yields are not as visible. Anchor bias exists after decades of tillage experience and is the tendency to reject any other option even when evidence indicates otherwise. Awada et al. (2014) found these biases were prominent in the prairie region.
- Risk aversion: Farmers have an appetite for risk but also a risk aversion. The latter can be significant when a farmer is faced with something new whose risks cannot be easily calculated. As well some farmers have a lower 'risk capacity', that is, the ability to take on higher likelihood or magnitudes of risk, often related to finances. One way to mitigate this was to conduct a strip trial or trialling NT on only one field at first. Eventually, farmers could attend field days to obtain knowledge applicable to their farm circumstances.

3 Organizations

As farmers experienced early successes, they were eager to share that information with others they thought would be interested in it or with those that they perceived as innovative farmers. That sparked the growth of unofficial networks across Canada and into the United States. It should be noted that since much of the settlement and agriculture in Canada is near the Canada–USA border, the farmers often have similar local farming conditions in north-south geography rather than east-west across provinces. Thus, we have a history of trading agriculture ideas in both directions.

Canada has a provincial government system responsible for economic development, natural resources and agriculture that tend to favour provincial delineation of conservation farmer associations, especially with the larger provinces. Farmers realized there was potential for the growth of an organization; as a result, they often created some form of a governance structure - board, secretary, etc. Eventually, they became official or affiliated with a parent organization in order to handle external funding. A small amount of funding came from memberships, while more significant sources originated from government grants and contributions from industry.

Their roles were primarily to provide education and share knowledge between farmers. They supported field-scale research and demonstrations, produced extension documents and farmer field days and field tours (Lindwall and Sonntag, 2010). One innovative approach was equipment comparison at field days whereby manufacturers were invited to bring their latest planting equipment to a field location where they could be operated side-by-side, and farmers could walk the field and evaluate its performance first-hand. Another innovation in the early days of the internet was the development of a NT farmer registry (password-protected for registered users) to enable others to talk directly to farmers with the same field/climate conditions and equipment that applied to their circumstances.

Organizations also provided roles of linking with and encouraging researchers, academics and specialists. Farmers needed laboratory and research plot information to be applicable and relevant to field scales. Organizations could link the few interested researchers to farmers with experience and resolved their concerns or questions to guide research directions. Likewise, they could connect farmers to researchers through extension publications or facilitating tours to research stations. Education institutions (colleges and universities) had a point of contact and synthesis of farmer opinions and management changes, all of which were used to enrich curricula.

The enthusiasm and collaboration along with the challenging questions and need for objective assessments precipitated some developments at the provincial extension and research, educational institution level. One was the development of farmer field schools or diagnostic schools where participants (e.g. farmers, industry, students) could go to a field location to see a wide range of agronomic treatments and discuss them with the specialists. The presentation of all agronomic aspects encouraged systems thinking, an essential component in CA innovation. Another development was the resurgence of interest in field trialling. CA farmers wanted to do a better job of testing and evaluating on their own farms. The emergence of precision agriculture and complementary technologies in the 1990s carried that momentum to where we are today - a rich armory of tools to conduct field-scale research and demonstration and collaborate in data analysis.

In the late 1970s, the ManDak Zero Tillage Farmers Association was formed in Manitoba in association with North Dakota, United States, and generated the full range of extension material, including a comprehensive binder called the Zero Tillage Production Manual. In 1978, the Alberta Conservation Tillage Society was formed from the collaboration initiative of a leading farmer, federal scientist and provincial extension agent. Alberta also had a couple of other ad hoc organizations created for the administration of targeted partnership funding from the provincial government and industry sources, which accelerated adoption through extension and grants. In 1987, the Saskatchewan Soil Conservation Association (SSCA) was formed with government and industry support, collaborating with university and federal research scientists.

Several organizations contributed to the advancement of CA in Ontario. Two stand out, the Ontario Soil and Crop Improvement Association (OSCIA) and the Innovative Farmers Association of Ontario (IFAO).

In the early 1980s, OSCIA and their 52 local county associations began to evaluate conservation tillage. They worked together with the Ontario Ministry of Agriculture and Food (OMAF) to put in tillage trials and host several significant events. The Save our Soil Conference in 1980 was the first such conference organized by OSCIA, OMAF and the Ontario Agricultural College of the University of Guelph. It was followed by the Soil Today Food Tomorrow conference in 1983. 'Till and Save' conferences were put on by OMAF in 1982, 1983 and 1984 with guest speakers from Ontario, Canada and the United States to cover topics on NT, ridge-till, minimum till, crop rotation, cover crops, windbreaks, economics, weed, insect and disease control, fertility, erosion control and more. The organization continues to be active with its mission 'To facilitate responsible economic management of soil, water, air and crops through the development and communication of innovative practices'. In 1987, the OSCIA was asked to administer the Land Stewardship Programme, a multi-million dollar programme, to encourage farmers to adopt conservation tillage with cost-share incentives. This was the first large-scale programme with a significant portion of the funding delivered by a farm organization. The programme division of the organization has gone on to deliver over 25 soil and environment-related programmes for the federal and provincial governments as well as other organizations.

At the same time, there were several individual farmers or groups across Ontario experimenting with conservation tillage. They began to influence others in their areas. Their thirst for knowledge on how to make NT and ridge-till work led to the formation of the Innovative Farmers Association of Ontario (IFAO) in 1986. Lobb (2011) reported that in the winter of 1986, an invitation-only workshop was organized by one of these farmers and OMAF. Invitations went to 'Innovative Conservation Tillage Farmers' where 22 farmers shared their experience on NT and ridge-till, and 24 researchers and extension staff provided support to the discussion. The workshops continued each year and eventually grew into an annual conference format and attendance peaked around 500 people in 1993 and 1994. The conference continues till today. The interactive sessions between farmers and speakers are often the most productive and have been a hallmark of IFAO. From early on until a number of years ago, summer tours were held visiting farms, research facilities, and agriculture industry facilities in Ontario, Quebec and several U.S. states. Many industry sponsors helped support the conferences and the organization. The organization continues to educate and promote sustainable agriculture through enhancing the soil, and it also works closely with researchers to help improve the quality of NT farming.

> The real value of the Innovative Farmers has been their willingness to set the agenda for innovation, research and priorities for soil management. By collaborating and sharing, we accelerated the learning process and reduced the risk of failure. This fresh, positive approach attracted support from all quarters in agriculture.
>
> *Don Lobb, farmer*

4 Research and extension

Canada has a history of good collaboration of farmers with research institutions (federal, provincial, universities). There have been joint memberships on advisory and development committees and cooperative reports and recommendations directed to policy and programme developers for support. There are seldom directives or official structures between the various institutions, collaboration relied upon the leadership of individuals.

Historically, each province had one agriculture university, at least in the early adoption period of CA. Presently, there are a few other universities that have some agriculture components. The universities had no official mandate to provide extension services as in some other countries. The staff of the prairie region universities was involved in NT planting, fertility, weed, pathology and rotation effect experiments. Network communications were maintained through contacts amongst the schools as well as federal and provincial research institutions. Each prairie province had annual soils and agronomy technical conference where information was shared and networks reinforced across all institutions. The Canadian Society of Soil Science (CSSS) also has annual meetings which fostered cross-institutional information and collaboration. Provincial soils and crops annual meetings include Alberta Soil Science Workshop, Saskatchewan Soils and Crops Workshop, Manitoba Soil Science Society meetings (links in Section 10, Where to look for further information).

Agriculture and Agri-Food Canada (AAFC) has a research branch with scientists distributed across Canada at various research facilities. They too developed research programmes for CA components and collaborated amongst their institutions as well as others (Acton and Gregorich, 1995). Provincial governments were more variable in their policies towards conducting or leading research, however, all were linked to researchers in other institutions through formal cooperative experiments or informal networks.

The provinces have agriculture colleges with 2-year programmes emphasizing practical and applied knowledge along with field skills. Although not significant centres of research, they collaborated with research institutions and were a very important conduit of information to students and leaders in curricula development.

Some provinces had their own research facilities co-located with federal sites or located in different areas. The three prairie provinces also collaborated on machinery testing facilities. It had been felt for some time that agriculture machinery needed comprehensive field testing to validate performance under farm conditions and to suggest improvements to the manufacturers. The Prairie Agriculture Machinery Institute had a facility in each province and divided the specialties amongst themselves - one would specialize in testing harvesters, another in seeders, etc. (https://pami.ca/who-we-are/). Their testing reports catalysed equipment improvements for CA by providing evaluation frameworks for farmers, comparative and standardized test results and, objective recommendations for improvements to the manufacturers.

Extension services have been the domain of provincial departments of agriculture, usually with a network of field/district and regional offices with specialists. Extension agents worked will all of the above players and farmer organizations to develop appropriate extension materials for CA across the media platforms accessed by farmers. Extension services also organized farmer tours to other provinces and developed farmer field schools and diagnostic field schools. CA information needs to underline the importance of developing these systems-approach extension platforms.

Cressman (1981) documented the concern about the need for erosion control on agricultural lands to protect soil productivity resulted in the Farm Planning service of the Ontario Agricultural College in the 1950s and early 1960s. Technical guidance was available to any farmer interested in applying conservation practices to his field operations. However, interest in the programme waned when chemical fertilizers emerged as the main source of plant nutrients. It was generally believed that it cost substantially less to replace with fertilizer the nutrients eroded from the field than it did to implement erosion control practices.

The University of Guelph, Ontario, conducted some zero-till (no crop residue moved off the row) research at the beginning in 1969 and shifted to exploring the NT residue removed from the row concept. This produced a distinct yield improvement, and the work also noted a crop rotation benefit for NT. Long-term tillage and rotation plots were established at the University of Guelph's Elora research station in the early 1980s and, in 1995, at the Ridgetown research station. Weeds and problem weed control research in reduced and NT systems conducted by the University of Guelph (U of G) was a significant contributor to the success of these systems. Plant disease/tillage relationships were also studied. C.S. Baldwin conducted research in the 1970s and 1980s into the effects of wind erosion, the use of windbreaks and their yield benefits. Research continues to show that profitability for NT is the same or better than other forms of tillage even if the yield is slightly lower. Research continues with an emphasis on the benefits of cover crops in conservation agriculture as well as soil biological and physical studies.

The University of Guelph established a centre for soil and water conservation in 1988, which brought a sharp focus to the tillage-soil health relationship. In

1990, the Land Stewardship chair was established to support soil and water improvement activity.

Agriculture and Agri-Food Canada research at Delhi, Ottawa and Harrow stations in Ontario supported CA advancement as well. Long-term tillage and rotation plots were established at the three stations as well as weed control and cover crop trials. Changes over time in carbon and other soil properties were measured. Comparisons of energy use were also conducted for different tillage and planting operations. The research was used by provincial extension staff to support the adoption of conservation practices.

Research funding from cooperative federal-provincial programmes, such as the Soil and Water Environmental Enhancement Programme, helped fund conservation agriculture research at all the research and extension institutions as well as and private-sector researchers. Provinces also had their own agriculture funding sources (proposal-based projects), which could be leveraged with federal funding and private sector support from CA farmer organizations or the private sector.

Canada has been fortunate to have willing collaboration and leadership of staff across the various institutions of research, education and extension across governments and the private sector. Perhaps the fact that any one institution did not have many staff or adequate resources was the catalyst to encourage collaboration, share resources and think and reason together. The results of the collective synergies were far greater than the additive results without collaboration.

4.1 An example of research diffusion

Passive (diffusion) or active (extension) transference of knowledge can occur through individual actions or through a planned campaign. The latter can be expensive, while the former is often cheaper but can have far-reaching and effective impact. One example of an individual's efforts is that of Ben Dyck, a research and development engineer at a national agriculture research station in Swift Current, Saskatchewan during the early days of NT and CA development in Canada.

Ben used to develop plot seeding equipment for plant breeders and conduct agronomic research in CA while situated in the dry-prairie region of western Canada. In the late 1970s, he worked with his predecessor Hank Anderson and new research engineer Wayne Lindwall to develop a planter with minimal soil disturbance. They ended up with a prototype that worked reasonably, a double disk arrangement with a 12-inch disk mounted back of centre against a 15-inch disk. They took it to the provincial farm exhibition (Regina, Saskatchewan) in the early 1980s, where it was exposed to the curious and innovative farmers. They then built a 10 foot (3 m) wide planter from the ground up that had a hydraulic trip opener and separate mid-row fertilizer bander. Another five were built and distributed to research stations and a university so that NT agronomic plot studies could be carried out effectively. By the mid to late 1980s, the thoughts, designs and concepts had been disseminated to leading farmers and local equipment manufacturers, while plot scale equipment had been delivered to agronomic researchers at several institutions (McLaughlin and Dyck, 1986). Key contacts and influencers were connected with the research, industry and farm communities across several provinces. Those, in turn, were connected to farm organizations, industry associations and government research agenda decision-makers.

In the late 1980s and early 1990s, agricultural scientists in Canada had an international development agreement with Brazil for which Ben did some design work (planting directly into maize stover) and sent them a prototype on a 3-point hitch tool bar. In 1995, the year after retirement, Ben travelled to Australia (Western Australia and New South Wales) to give a series of presentations to farmers. Foreign researchers and farmers came to the prairies throughout the 1980s and 1990s to visit Ben and researchers at other research stations, universities and farmers who were 'thought leaders'. By 1980, Wayne Lindwall had moved to the Lethbridge research station where he carried out independent and collaborative work with many other institutions. This leadership of a few was leveraged to many through extension.

4.2 An example of collaborative applied research: Tillage 2000

Tillage 2000 (T2000) was a long-term (5 years), on-farm field-scale research and demonstration project, which began in 1985 and was fully implemented during 1986 (Aspinall and Kachanoski, 1993). It was provincial in scope, allowing for the testing and demonstration of innovative conservation practices across most regions of Ontario. Cooperators were selected based on field site suitability (soil type, topography, access, location), and the interest and capabilities of the cooperator. The main objective was to develop and evaluate conservation farming systems, which maximized economic productivity and minimized soil degradation for specific soil types. The project was conducted on 40 farms across Ontario. T2000 was unique to soil conservation in Ontario by virtue of its methodology and process. The project included both a research and demonstration component within an economic farm unit framework. The process was both investigative and developmental over several years. The programme was designed to introduce concepts of conservation tillage systems to a larger number of producers and to provide a way to distribute known information and experience through field-scale demonstration.

T2000 was a cooperative effort by the Ontario Ministry of Agriculture and Food (OMAF), the Department of Land Resource Science, the University of Guelph, and the Ontario Soil and Crop Improvement Association. In addition, several conservation authorities (watershed-based governance) had joint agreements with OMAF to deliver the T2000 programme as part of an agricultural soil and water conservation programme. In southwestern Ontario, T2000 was part of the federal-provincial Soil and Water Environmental Enhancement Programme. Support also came from agribusiness who provided expertise and donated products to the project. The project was managed by the OMAF soil conservation advisors. These staff were largely new hires at the time. There was little technical information available about conservation tillage systems, so this project provided them with valuable hands-on experience working with farmer cooperators as they learned together. The University of Guelph involvement also meant that the learnings were being shared with others in the department providing current conservation tillage information for lectures. It also brought field-scale research to the farm involving the cooperator so they could see research first hand and they were able to express their challenges and research needs to the researcher.

The T2000 sites provided a great opportunity for local farmers to see a conservation tillage system working on a farm near them. The farmer cooperators supported a farmer-to-farmer extension effort. The project provided the cooperators with field-scale research data, providing results to support them to speak and promote practices to producers at farm tours, meetings and extension events. Many became mentors to those wishing to adopt conservation agriculture.

NT did particularly well considering that the farmers and their advisors were novices and the soils in these plots were in a state of biological and physical transition. A strong contributing factor to the success of this project was the commitment of the farmers to make it work. They truly applied a systems approach for making it work. The data in the T2000 project indicated that it should be possible to implement a conservation tillage system with no loss in yield productivity or economic return, in all but the heavier textured soils. This is especially true for sandy textured soils where increases in yield are likely under conservation tillage.

5 Policy and programmes

The early plowing of the prairie grasslands, subsequent intensive tillage and the successive drought period of the 1920s and 1930s brought large erosion events, poor crops and failing farm economics. After one drought period in 1931–32, more than one-third of the population of the province of Saskatchewan received government 'relief' payments (Marchildon, 2009). Prairie provinces and the federal government were both concerned and developed policy responses to the disaster with dedicated organizations. Political differences and strained relationships between provinces and the federal government over the political power of prairie farmers created discord in the formation of institutions (Marchildon, 2009). Alberta created Special Areas Boards that administered and rehabilitated eroded and abandoned lands while the federal Prairie Farm Rehabilitation Administration (PFRA), although prairie mandated, focused their presence and power in Saskatchewan (Marchildon et al., 2008). The Alberta Special Areas Boards emphasized a range of conservation practices but emphasized efforts on returning cropland to perennial forages (more drought resistance) and cattle farming. The PFRA had a wide spectrum of research, technology and extension efforts with land management practices focusing on crop-summer fallow strip cropping and shelterbelt tree plantings. PFRA would pay certain farms to change practices to become 'demonstration farms' (Marchildon et al., 2008). PFRA continued operations until the 1990s although their focus became more centred on water infrastructures. such as farm storage and regional irrigation systems.

Soil degradation issues and their consequences (erosion, salinization, loss of fertility) persisted into the 1970s. In the 1980s, the Senate of Canada held consultation meetings across the country and produced a seminal document, 'Soil at Risk: Canada's eroding future' in 1984 (Senate, 1984). This prompted both a bottom-up push from farm organizations involved in the consultations as well as top-down attention and nudging from the Senate to the national and all provincial governments. The Senate had a champion senator from Saskatchewan (Senator Sparrow). The report and champion garnered media attention and public interest and support for policy change. This was the period of the emergence of pioneer farmers and NT technology in Canada when farmers felt empowered, governments wanted to support them or at least felt compelled to and progress forward.

The 'Soil at Risk' report made 20 recommendations, including the development of a comprehensive soil and water conservation policy at the national and provincial levels of government. They further made a number of recommendations on research, including the establishment of regional institutes for applied research. Educational institutions were encouraged to increase training in conservation and that a national soil conservation week be created for public awareness and education. Senator H. Sparrow, the chair of the Senate committee, was a tireless advocate for the report recommendations from coast to coast. This high-level report prompted policy developments at the national and provincial levels for more than a decade, which aligned well with the emergent groundswell of farmer interest and adoption of CA principles. Drivers for policy change came from international markets, trade agreements, the government's desire to reduce policy costs and changes in perspectives.

Ontario Chapter, Soil Conservation Society of America (OCSCSA, 1983) Studies by the Pollution from Land Use Activities Reference Group (PLUARG) of the International Joint Commission (IJC) have estimated that 60% to 70% of the sediments delivered to the Great Lakes by streams are from agricultural sources and 30% to 40% are from urban sources. The principal sources of soil erosion by water are sheet and rill erosion (70–100%) and stream-bank erosion (0–30%). The IJC urged Canada and the United States to adopt measures to address this issue.

Historically, Canadian agriculture policy was focused on price support often categorized under the title of stabilization policies, which protected farm gate commodity revenues or costs. In a review of policies up to 1995, Barichello (1995) noted the historic stability policies focus was in response to Canada's significant climate variability as well as global markets, thin margins and few insurance options. Under the lesser title of 'Other Policy Instruments', Barichello (1995) listed institutional policies of research/development/extension and sustainable agriculture. He did estimate that the expenditures in all these categories were about 2% of market sales for Canada. Later, Gellner and Rattray (2001) noted the continued emphasis on direct payment policies to farmers (listing nine programmes) but perceived policy frameworks are shifting from farm gate to the larger integrated food system and rural economy. At the same time, there was a significant downward trend in support of agriculture and the desire of government for the farm and commodity organizations to undertake more decision-making and programme delivery roles.

Against the historical inertia cloud of price support and stabilization policy, the 'Soils at Risk' report was disruptive but aligned with international trade agreements wanting 'green' policy, public interest in environmental issues and the drive for farmers to find economic environmentally sustainable production practices. There was a range of soil conservation reports that appeared in the mid-1980s accompanied with some national incentive funding initially through existing programmes and eventually a National Soil Conservation Programme created in 1989 (Ward et al., 2010). This programme, often with matching funds from provinces, funded farmer organizations across Canada as well as some applied research efforts by education and research institutions. The condition or option of leveraging funding from other governments and institutions nudged innovation and new directions at a range of institutions as well as collaborations of staff between them. Collaborative extension was a key component to move CA adoption.

An example of collaborative extension

In the early 1980s, the Ontario Ministry of Agriculture and Food responded by introducing the Ontario Soil Conservation and Environmental Protection programme (OSCEPAP) to provide financial incentives for farmers to install erosion control structures and improve manure storage. The early uptake of the programme was slow. To promote and help deliver the programme and to provide soil conservation assistance, 12 soil conservation advisors were hired by the province. They were partnered with conservation authorities (community-based watershed management agencies; Conservation Ontario, 2020) in what was called the Joint Agricultural Soil and Water Conservation Programme. The advisors worked with conservation authority staff, OMAF agricultural engineers, farm organizations, individual farmers, input suppliers, machinery dealers, university and government researchers and others to promote the use of erosion control structures, windbreaks and better manure storages. In a little over a year, the OSCEPAP programme was fully subscribed, thanks to this extension effort. The programme provided financial assistance for erosion control structures and technical advice for that and improved tillage practices, but it did not provide financial incentives for the adoption of reduced tillage.

In 1988, the province introduced the Land Stewardship Programme to address this and hired 14 more conservation advisors. This increased the number of conservation advisors on farms sharing information on CA, helping farmers access conservation equipment, setting up tillage comparison plots, and travelling to other jurisdictions to bring back new information. They also helped organize NT and conservation workshops, conferences and summer tours and wrote numerous articles, newsletters and publications. In 1991, the decision was made to have many of the conservation advisors become soil and crop specialists so all the commodity specialists would be promoting soil conservation along with their crop production advice. These staff working with enthusiastic early adopters along with OSCEPAP and the Land Stewardship Programmes led to a significant increase in the adoption of CA in Ontario. Many conservation authorities have continued with agriculture-based programmes, which provide technical assistance and funding to farmers to help protect and enhance their soils.

Policy and public movements converged and aligned from national to local scales through intersecting networks and implementation mechanisms. Perhaps similar to the internet, there were different routes that could be used or optimized to meet the dissemination of knowledge needs. If one path did not perform well, adaptative management and implementation could move to other routes. There were various reports of the costs of land degradation and the need for conservation from national and provincial governments and research and educational institutions. Federal-provincial funding became targeted to the need and available to agencies on the ground, which ranged from research, education and farm organizations. This was key in supporting and leveraging provincial organizations, such as the Alberta Conservation Tillage Society (ACTS), Reduced Tillage Linkages (RTL), Saskatchewan Soil

Conservation Association (SSCA), the Eastern Conservation Association and the Manitoba-North Dakota NT Association (ManDak).

A full range of extension efforts was deployed by these organizations. The conventional media of radio, television, print were all used along with event-extension at agriculture fairs, large and small and research plot tours. Specific field days were produced to show the latest technology with planters operating side-by-side in field conditions. Research stations focused on peer-reviewed publications and extension was limited to periodic newsletters or annual field days. Extension organizations gathered and synthesized information from various sources (e.g. Baig and Gamache, 2011) and connected farms and farmers through numerous field days and databases. One early internet database from an Alberta organization, RTL, allowed password access to a database of farmers sharing information on their equipment types, cropping system, soil and climate condition, etc. so that other farmers seeking practical knowledge and experience could contact the appropriate peers directly. The farmer organizations were effective in getting information to farms in part because their board members were farmers and they were held accountable by them and the larger population of CA innovators.

OMAF continued to support CA programmes providing significant technical support in the early 1990s to the development and implementation of the Environmental Farm Plan (EFP), which was led by the Ontario Farm Environmental Coalition (Rudy, 2018). The EFP Workbook was a comprehensive compendium that addressed key topics ranging from handling and storage of pesticides to nutrient management to milking centre wash water to woodlands to wildlife. Completion of the EFP became a requirement for farmers to access government environmental incentive programmes. One challenge with the programme was that the farmer chose the issues they wished to address, not necessarily the highest risk issues.

At the same time, OMAF partnered with the Ontario Federation of Agriculture to create cross-sector committees for the production of a series of best management practices books. These books presented sensible, affordable options for the protection of soil and water resources on the farm while harmonizing productivity, business objectives and the environment. The series was recently expanded to include a number of information sheets on diagnosing soil degradation issues and factsheets on soil health topics.

In the 2010s, the focus shifted from CA to soil health in what is now called the 'Ontario Ministry of Agriculture, Food and Rural Affairs'. A new programme was developed, the Great Lakes Agricultural Stewardship Initiative (GLASI), to support the adoption of soil health practices on the farm and help reduce phosphorus from entering Lake Erie and southeastern Lake Huron. Crop advisors worked with farmers to assess soil health on a field and created a simple soil health plan called the Farmland Health Check-up. In face of reduced government extension staff, this broadens the reach of extension efforts and encourages the discussion of good soil management between industry and farmers. Completion of the check-up gave farmers access to financial incentives.

The increased interest in soil health by OMAFRA, farmers and their organizations resulted in OMAFRA forming an Agricultural Soil Health and Conservation Working Group in 2015. The working group consisted of representatives from many of the key farm organizations, Agriculture and Agri-Food Canada, Conservation Ontario, academics, and the Ontario Certified Crop Advisor Association. OMAFRA (2018) New Horizons: Ontario's Agricultural Soil Health and Conservation Strategy is a long-term framework to guide collaborative soil health research, investments and activities until 2030. The strategy's vision is 'Healthy agricultural soils contribute to a vibrant agricultural sector, productive economy, sustainable environment and thriving society.' The strategy's goals, objectives and actions are divided into four themed areas to address different aspects of the issues: soil management, soil data and mapping, soil evaluation and monitoring, and soil knowledge and innovation.

At a recent (2020) national meeting to explore ag-environmental policy needs, a question was posed to the group assembled, 'what characteristics of past and present policy are the greatest benefit?' The popular answers were extension and place-based implementation characteristics. CA national (or provincial) policies were adapted or customized for geographic conditions (agro-climatic as well as socio-economic conditions).

6 Cumulative effect: issues and adoption

Canada has a national agriculture census every 5 years that all farms participate in. Prior to the 1990 census NT or minimum-till, land management practices were not recorded. There had been a few sub-provincial level surveys that sampled some farms before and after that for NT practices or other special interests (e.g. type of planter or fertilizer management) but none with the geographic scope and control of the Stats Canada agriculture survey (Statistics Canada, 2020).

The 1991 survey (of 1990 practices) indicated a national total of 1.95 million hectares of NT planting with 95% from western Canada (69% from Saskatchewan) and 5% from Ontario. Of course, this reflects the uneven distribution of agricultural land across Canada, not the interest or degree of adoption. Most of the agricultural land is located in western Canada (85%), the northern portion of the Great Plains region of North America. This is reflected in graphed data where the western provinces of Saskatchewan and Alberta influence the national total over time (Fig. 1).

If the national total and the two large western provinces are removed from the data, one is able to get a better picture of the adoption rate in the other

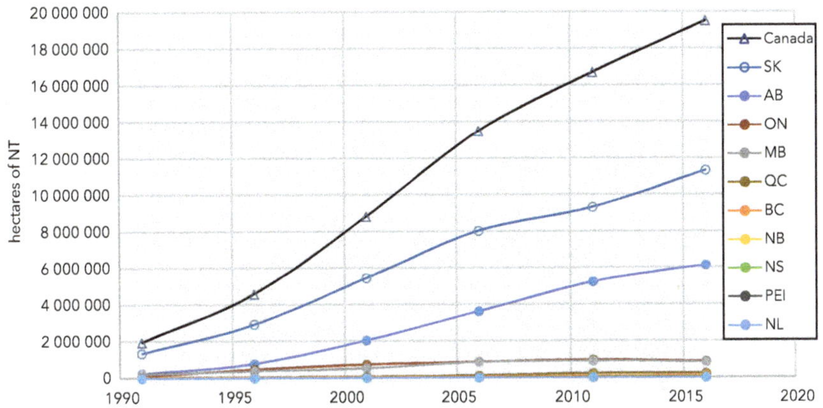

Figure 1 National and provincial change in no-till (NT) cropping practice from 1990 to 2015 (census data is collected in the following year).

provinces (Fig. 2). Ontario had a rate of increase greater than all the provinces in the early years and then experienced a decline since 2010. The adjacent province had a lag in the increase of NT but a similar levelling off after 2010. Quebec has many livestock farms (dairy, hogs) and manure, which requires incorporation (regulations) making consistent NT difficult in many fields.

The hectares of NT practice can be normalized across provinces if the NT hectares are expressed as a percentage of total seeded hectares (Fig. 3) where then Saskatchewan, Alberta and Ontario all have similarly large increases in NT for the first decade at which point the prairie provinces accelerate adoption, while Ontario starts to level off. Currently, both the largest prairie provinces have twice the adoption level as Ontario and the other provinces.

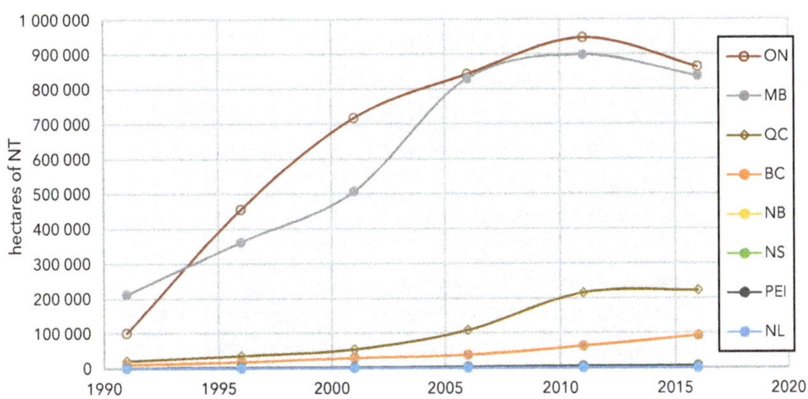

Figure 2 Provincial change in no-till (NT) cropping practice from 1990 to 2015 in provinces with lesser amounts of agricultural land (census data is collected in the following year).

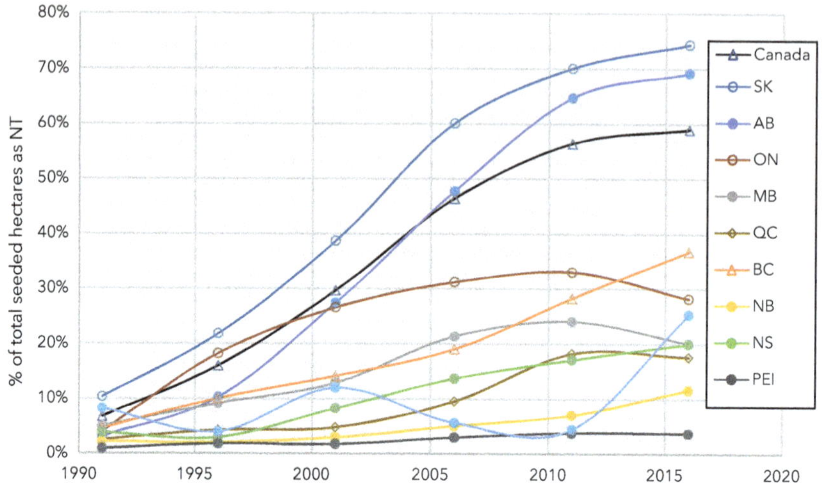

Figure 3 National and provincial change in no-till (NT) cropping practice from 1990 to 2015 as a percentage of total land prepared for seeding (census data is collected in the following year).

Adoption of CA started with NT but became so much more as farmers experienced issues with biomass, pest and fertility management. Many were of a positive nature like increased yields, elimination of erosion, improved hydrology and nutrient cycling. The issues varied according to agroclimatic conditions within and across provinces. Other publications have gone into depths of crop performance, agronomic issues and derivative interests and specific conditions. A summary of some of the common concerns and solutions reveals that as well as specific issues, the realization that CA is a system and should also be considered as such for long-term agronomic and economic conclusions (Table 1). Plot research is often designed to isolate factor effects so putting all the pieces together falls to the farmers. They make observations and when unexpected or unexplained responses occur in the field, they channel that back to researchers, often through informal channels of collaboration at field days, extension meetings, phone calls (now using more social media channels). Results of research plots get published in the scientific literature where others summarize and synthesize them into larger scales of systems or geography (e.g. Baig and Gamache, 2011).

Caution was needed in the early work on NT and cropping systems plot research as scientists did not have the proper equipment, they did not document methodology completely or plot sizes were constraining for the desired treatments. Lack of documentation of methodology persists to the present time and the cost of field research is discouraging for scientists studying system levels of agriculture. A more recent article in the NT Farmer magazine titled '29

Table 1 A brief summary of technical, economic and social issues and solutions for CA in Canada

Problems	Solutions	References
Excessive erosion	No till. Residue management.	McConkey (2010); Baig and Gamache (2011)
Cultivated summerfallow	Chemical weed control. Low disturbance seeding.	McConkey (2010); Awada et al. (2016)
Equipment designs	Dedicated research engineers. Machinery institute testing and reports. Farmer innovation. Air seeder, narrow knife development.	Lindwall (2010); McConkey (2010)
Equipment penetration under a range of conditions	Diverse crop rotations	McConkey (2010)
Equipment and input cost barriers. Field demonstration costs.	Government grants or subsidized equipment and materials. Market developments.	Ward et al. (2010); Awada et al. (2016)
Social resistance of farmers	National reports of concern (by a senator). Public awareness. Neighbour interactions. Extension.	Awada et al. (2016)
Relevant research results on cropping systems and agronomy	National research stations changed all plot research from conventional tillage to no till management	McConkey (2010); Baig and Gamache (2011)
Agronomic issues of diverse crops and rotations	Collaboration among research stations to quantify more aspects of the system. Farmer organizations, extension. New cropping systems.	McConkey (2010); Awada et al. (2016); Baig and Gamache (2011); Lafond et al. (2014); Prairie Soils and Crops (2012); Prairie Soils and Crops (2009).
Excessive crop residue from cereals under NT	More diverse crop sequences. Varietal selection.	McConkey (2010)
Economics of CA systems	Provincial and national experts working with farmers. Complimentary plot analyses.	Baig and Gamache (2011)
Diseases, weeds	No effect or +/− impacts of NT. Rotation and residue management are important.	Baig and Gamache (2011); Lafond et al. (2014); Prairie Soils and Crops (2012); Prairie Soils and Crops (2009)
Relevant information extension to farmers, agriculture specialists and policy makers	Farmer led CA organizations - funding from governments, research grants, industry. Collaboration among organizations. Field days, popular media, conferences, newsletters.	McClinton and Polegi (2010); Baig and Gamache (2011); Ugochukwu and Phillips (2018)

Reasons Why Many Growers Are Harvesting Higher NT Yields in Their Fields Than Some University Scientists Find in Research Plots' summarized the exasperation farmers have in reading and interpreting researcher results (Lessiter, 2015). Nevertheless, farmers need systems information to make informed business decisions and filling that gap was left to the farmer organizations and some provincial specialists to ensure 'the last mile' was developed for information needed by farmers. Perhaps the expanse of Canadian geography between research plots or stations has fostered extrapolation methods and relative comfort with uncertainty. The use of models has also become popular across Canada to provide relevant information at scale.

The national agri-environmental reporting system (Clearwater et al. 2016) incorporates the national land management changes to show the agri-environmental well-being of the country to OECD and other international organizations. The Alberta greenhouse gas offset system benefited from the national science and modelling programme to have the first soil carbon offset system in the world (Goddard et al., 2008). The science development efforts continue, adding new research results to the previous compendiums that refine carbon storage rates for different conditions and over time (Liang et al., 2020). The few sites where long-term research was allowed to continue have helped to quantify the storage curve which starts at a carbon accumulation rate of 1.3% yr^{-1} then declines to 0.23% yr^{-1} for periods longer than 20 years as a new steady-state for annual cropping is approached. Modelling the land use change over the prairie region has shown the combined effect of carbon changes, crop residue and diverse cropping have all added up to a negative radiative forcing on climate. The land use change of CA adoption from an albedo influence context has had an equivalent atmospheric drawdown of about 8 MT of CO_2 yr^{-1} since 1981 (Liu et al., 2020).

7 Current state

The PFRA was instrumental in encouraging CA developments early on, but the national government, in efforts to redirect focus and reduce budgets, tried unsuccessfully to evolve it into a National Soil and Water Conservation Service and both institutions have disappeared. The federal government now has a 5-year, federal-provincial cost-sharing programme with multilateral established goals and priority areas. Provinces can develop regionally appropriate implementations of the priority areas that suit local conditions or constraints of natural resources, institutions and farm socio-economic characteristics. As Gellner and Rattray (2001) mentioned nearly two decades ago, the expectations are for local-level organizations to take on more programme delivery roles.

Provinces were the main agents for extension services to farms and the agri-food industry. Those institutions have been reduced or eliminated in all provinces. This has put the onus on farm organizations and consultants to take over the extension and technology transfer roles. Consultants/agronomists attached to input suppliers are often viewed with suspicion as biased in the business alliances with particular input brands, and they are specialized in cash crop production or regulatory compliance requirements of farms (e.g. nutrient management plans). They are also poorly equipped for research and have no incentives for public good research. Considering the huge value of extension in the past, it will be a challenge to shape the future frameworks and implementation to meet the needs of innovative farms.

Ontario shows evidence (above) that they were able to evolve with changing government organizations and funding to develop other collaborative organizations between government, industry and farmers to leverage efforts and deliver extension programming and incentive grants, perhaps not dominantly focused on CA but bringing it along while it deals with tangential issues of nutrient pollution, water resources and biodiversity.

The grassroots organizations do not have the institutional capacity to support extension services or to leverage research programmes. Thus, farms now need to do more of these roles themselves, increasing their operating costs. The internet has been a huge benefit for farm operators to access a wide variety of information. On the downside, it can be difficult at times to perceive the biases or credibility in the information and/or sources. The rapid development of technologies along with the growth of consultants and farmer organizations brings the opportunity for field trialling and citizen science (Goddard et al., 2020). Alberta has local, farmer-directed, applied research associations that move on to these opportunities. Central Canada provinces have Watershed Authorities that can do the same (Manitoba, Ontario, Quebec). Some crop commodity organizations are collaborating across commodities and across geopolitical boundaries to derive efficiency of scales to tackle remaining CA production issues and take advantage of market opportunities for sustainable production.

Technology has blossomed on all fronts - crop breeding with a range of specialized traits, new fertilizer formulations combined with slow-release or inhibitor properties, pesticide mixtures for concerns of weed resistance. Equipment has more options for seed row openers and precise placement of more than one fertilizer product streams, with similar custom mix options and on/off controllers for pesticide applicators, along with large price tags. Precision farming technology creates algorithmic prescriptions of sub-field zones. The panoply of tech advancements and tools presents a dazzling array of opportunities for CA farmers to explore - if only they had more time or money.

The diversity of farming systems has enlarged as well. CA has encouraged diverse crop rotations. Most recently the search for cover crop mixes is at the forefront of interest. Farming can range from implementing all of the technology possible to little or none. Organic farming is seeing steady increases and its variant, regenerative agriculture, is a recent newcomer to the lexicon of farming systems (Mitchell et al., 2019). Since many of them follow the principles of CA then they present opportunities for farmers to do trialling and assess the benefit/cost and economics of them.

In Ontario, there has been some slippage backwards on the tillage front for the last number of years. High corn yields and the resulting high levels of residue have made some farmers feel they require tillage. High land prices have also contributed to the feeling that they must produce on every hectare. They may be using as little as light vertical tillage all the way up to moldboard plowing. There are also fast heavy discs and other forms of 'vertical tillage' that often masquerades as conservation tillage. These can do a good job if used properly. This has resulted in the return of significant erosion on some fields. Strip tillage, tillage of about one-third of the row spacing, has increased meaning generally less tillage on those fields. Cover crops usage is on the rise helping to protect soils and improve their health.

The current federal-provincial agriculture framework the Canadian Agricultural Partnership (CAP) continues to provide financial incentives for soil and water improvement. The proportion dedicated to issues is determined by provinces. Ontario has a significant soils component which includes funding for soil health technology transfer materials such as videos and written material, events, a research/demonstration project to study the effect of cover crops and organic amendments on soil health and to support the continued monitoring of land management impacts on water quality in Lakes Erie and Huron by conservation authorities and researchers. This funding will help support OMAFRA as they work with agricultural organizations, agribusiness, universities and the federal government to implement Ontario's Agricultural Soil Health and Conservation Strategy.

Long-term tillage trials continue to provide useful data and several long-term cover crop trials have been established to study the benefits to soil health. Research also continues to learn more about soil life, the effectiveness and economics of soil health best management practices and related impacts on water quality.

8 Conclusion

Early settlement in Canada was not kind to soils. The land became intensely cultivated to prepare the seedbed for the planting technology of the times, control pests and mineralize nutrients for plant growth. The resulting degradation and increasing erodibility became recognized and concern of farmers, governments and society at large.

Innovative farmers knew the status quo was not sustainable and worked hard at discovering alternatives. NT became the new focus of attention, trials and experimentation. Along the path farmers, researchers and extension agents realized cropping systems also played significant roles and a systems approach was the future.

Policy frameworks emerged from the concerns of erosion and degradation of land from the 1930s up to and including the 1970s. The early policy was top-down national and provincial level legislation dedicating resources to address issues. NT interest from farmers then became the driver for bottom-up policy for research priorities and extension investments by governments. What evolved was continuous intersections of farmer organizations, researchers, industry and governments to design, implement and reassess programmatic CA policies.

The provinces in Canada had a healthy mixture of research institutions - educational (universities and colleges), federal and provincial research centres, farm organizations and extension specialists and private consultants. Funding and staffing capacity was stretched over large geographies which limited the power of any one institution. The inherent diversity of the range of institutions was sparked into collaborative action by leaders from all domains resulting in synergies to move the CA dial quickly and overcome the technical issues of implementing CA into local conditions.

Farmers continue to seek knowledge and solve CA problems. They now have more technology in their hands which shortens distances in space and time to find solutions. The spirit of collaboration persists but is strained as budget and staff are cut or redeployed in extension and research institutions. CA interests need to develop new alliances with larger trade, environmental and social interests. The direction of CA will remain positive, but there will be more pathways of opportunities.

9 Acknowledgements

We would like to acknowledge all the people who catalysed the adoption of CA over the decades. As evidenced by the collaboration across institutions and farmers there are literally thousands of leaders who all played key roles. Here are a few that we recollect.

Some of the farmer pioneers that started the ball rolling and in many cases whose progeny continue their efforts: Gordon Hilton, Dick Middleton (Alberta - Alberta Conservation Tillage Association), Jim Halford (Saskatchewan - SSCA), Jim McCutcheon (Manitoba - ManDak No-Till Association), Don Lobb (Innovative Farmer's Association of Ontario). Farmer organizations: Peter Gamache (Alberta), Blair McClinton (Saskatchewan), (Manitoba), Harold Rudy (Ontario).

The scientists that helped with experiments and systems development: From Agriculture and Agri-food Canada: Wayne Lindwall (engineering), George Clayton (agronomy), Bob Blackshaw (weeds), Neil Harker (weeds, agronomy), Bob Zentner (economics), Wally Findlay (field trialling), Ron Beyaert (agronomy), Denis Angers (agronomy, soils), Guy Lafond (agronomy, cropping systems).

From universities: Marvin Nyborg, and Cesar Izaurralde (University of Alberta), Darwin Anderson, Glen Hass, and Jeff Schoeneau (University of Saskatchewan), Don Flaten (University of Manitoba), Gary Kachanoski, Tony Vyn, and Bill Deen (University of Guelph).

Many leaders did not have their names in the papers or spoke at programmes, but they were absolutely essential to make things happen: all the people that were not on the executive but volunteered tirelessly on organizations, produced events and created extension materials. All the farmers that we observed intensely focused at field days and engaged in animated discussions with their peers - these were the champions to spread the knowledge and skills across rural Canada.

The extension agents and specialists with provincial departments of agriculture and the numerous CA or other agriculture organizations were the glue that connected and strengthened the intertwined networks. Executive managers with governments either recognized their staff leaders might be on to something and at least removed barriers and allowed them to pursue a new work trajectory or, they shared the vision and need and pushed for programming changes.

The private equipment and agronomic input companies were collaborators as well by helping support research, demonstrations and extension. Some of the large machinery companies were unhelpful as they saw CA as a threat to their 'heavy metal' industry. The smaller machinery manufactures often had farmers on their executive and illustrated the rapid cycles of development, characteristic of innovation. Fertilizer companies supported new technologies of side-banding for increased efficiency (John Harapiak, Westco Fertilizers). Agricultural chemical companies worked with farmers, researchers and others to find solutions to shifting weed populations and one company lowered the cost of a key herbicide which was significant in moving NT forward.

We have missed many more but this gives an idea of the number of people from many institutions that worked together to make things happen. The transformation could not have happened without the leadership and collaboration across organizations and industry.

10 Where to look for further information

Several farmer-directed groups exist across Canada organized on a provincial geographic basis and one national organization:

- Alberta Reduced Tillage Linkages ceased operations in 2009 but have kept their website alive with archived 'Resource Centre' information. https://reducedtillage.ca/.
- Saskatchewan Soil Conservation Association (SSCA), whose mission is: "to promote conservation agriculture systems that improve the land and environment for future generations." https://ssca.ca/.
- Manitoba North Dakota Zero Tillage Farmers Association (ManDak) has changed its name to Northern Prairies Ag Innovation Alliance effective January 2016, with an office at Bismarck, North Dakota, USA. NPAIA (@0Till) Twitter.
- Ontario Soil and Crop Improvement Association is a not-for-profit farm organization whose mission is to facilitate responsible economic management of soil, water, air and crops through the development and communication of innovative farming practices. https://www.ontariosoilcrop.org.
- Innovative Farmers Association of Ontario is a not-for-profit farm organization that works to educate and promote sustainable agriculture through enhancing the soil. https://www.ifao.com.
- Quebec farmer group covering part of the province is active with extension projects ("My Soil"). https://www.monsol.ca/.
- New Brunswick Soil and Crop Improvement Association is a not-for-profit farm organization whose mission is to stimulate innovation, research and leadership for the advancement of the agricultural industry in New Brunswick. https://www.nbscia.ca/home/.
- Eastern Canada Soil and Water Conservation Centre, New Brunswick. https://ccnb.ca/entrepreneurship-et-innovation/research-and-industrial-services/centres-and-specialized-services/ecswcc.aspx.
- Soil Conservation Council of Canada is the face and voice of soil conservation and health at a national level in Canada. https://soilcc.ca/.

The prairie provinces have annual science meetings with the participation of researchers and extension specialists from all organizations. Conservation agriculture is a reoccurring topic of interest:

- Alberta Soil Science Workshop (https://soilsworkshop.ab.ca/).
- Saskatchewan Soils and Crops Workshop (https://agbio.usask.ca/soilsncrops/).
- Manitoba Soil Science Society (https://www.mbsoils.ca/).

The websites of all provincial agriculture ministries contain varying amounts of information on conservation agriculture, usually current contexts rather than historical archives. Reductions in provincial government support for applied research have resulted in lessened amounts of current research activities in many provinces. In some cases, activities have migrated to universities, colleges or the national agriculture research service (Agriculture and Agri-Food Canada).

- Archive of Agri-Environmental Programs in Ontario: https://atrium.lib.uoguelph.ca/xmlui/handle/10214/13065.
- Quebec. A provincial government research agency has a number of soil health projects underway: https://www.irda.qc.ca/en/research-projets/?s=1320.

The Prairie Agriculture Machinery Institute (PAMI) was originally a collaboration of three facilities across the prairie provinces to test and improve agriculture machinery (government and industry supported). Loss of government support necessitated consolidation and program prioritization, to one institute now, http://pami.ca/about-us/history/. Their digital Resource Library contains all their historical equipment reports as well as other useful connections for farmers.

11 References

Acton, D. F. and Gregorich, L. J. (Eds.) 1995. The health of our soils–towards sustainable agriculture in Canada. Centre for Land and Biological Resources Research, Research Branch, Agriculture and Agri-Food Canada, Ottawa, 138 pp. Available at: https://atrium.lib.uoguelph.ca/xmlui/bitstream/handle/10214/15123/FDMR_Health_of_Our_Soils.pdf?sequence=1&isAllowed=y.

Anderson, D. 2008. Organic matter in prairie soils. *Prairie Soils and Crops* 1, 21–26. Available at: http://www.prairiesoilsandcrops.ca.

Aspinall, J. D. and Kachanoski, R. G. 1993. Tillage 2000: 1985–1990 final report. Ontario Ministry of Agriculture and Food, Soil and Water Environmental Enhancement Program, Ontario Soil and Crop Improvement Association, University of Guelph.

Awada, L., Gray, R. S. and Nagy, C. 2016. The benefits and costs of zero tillage RD&E on the Canadian prairies. *Canadian Journal of Agricultural Economics/Revue Canadienne d'Agroeconomie* 64(3), 417–438. https://doi.org/10.1111/cjag.12080.

Awada, L., Lindwall, C. W. and Sonntag, B. 2014. The development and adoption of conservation tillage systems on the Canadian Prairies. *International Soil and Water Conservation Research* 2(1), 47–65. https://doi.org/10.1016/S2095-6339(15)30013-7.

Baig, M. N. and Gamache, P. 2011. The economic, agronomic and environmental impact of no-till on the Canadian Prairies. Reduced Tillage Linkages, Canada. Available at: http://www.reducedtillage.ca/ImpactnotillrtlaugFinalAug30optomized.pdf.

Barichello, R. R. 1995. Overview of Canadian agricultural policy systems. Proceedings of the First Canada/U.S. Agricultural and Food Policy Systems Information Workshop. Friesen Printers, Winnipeg, Canada, pp. 37–59.

Clearwater, R. L., Martin, T. and Hoppe, T. (Eds.) 2016. Environmental sustainability of Canadian agriculture: agri-environmental indicator report series–report #4. Agriculture and Agri-Food Canada, Ottawa, ON. Available at: http://publications.gc.ca/site/eng/9.812827/publication.html.

Conservation Ontario 2020. About Conservation Ontario. Available at: www.conservationontario.ca.

Cressman, D. R. 1981. Perceptions and attitudes of resource managers towards erosion and sedimentation problems in Ontario. Erosion and Sedimentation in Ontario: A Time for Action, 1981. Ontario Chapter Soil Conservation Society of America. pp. 84–96. Available at: https://sis.agr.gc.ca/cansis/publications/surveys/on/onm-3/index.html.

Driver, G. and Wall, G. 1982. Cropland soil erosion: estimated cost to agriculture in Ontario. Ontario Ministry of Agriculture and Food. Ontario Institute of Pedology.

Gellner, J. and Rattray, B. 2001. Policy developments in Canadian agriculture since 1986, in: trade Liberalization under NAFTA: report card on agriculture. Proceedings of the 6th Agric. Food. Policy Systems Info Workshop. Friesen Printers, Winnipeg, Canada, pp. 105–117.

Goddard, T., Basch, G., Derpsh, R., Hongwen, L., Jin, H., Karabayev, M., Kassam, A., Moriya, K., Peiretti, R. and Smith, H. 2020. Chapter 12. Institutional and policy support for conservation agriculture uptake. In: Dodds, B. (Translator) Sci Publ. Vol. 1, pp. 513-564.

Goddard, T., Haugen-Kozyra, K. and Ridge, A. 2008. Alberta: North America's first compliance offset carbon market. Proceedings of the Conservation Agriculture Carbon Offset Consultation. UN -FAO and CTIC, West Lafayette, Indiana, October 28–30, 2008. 9p. Available at: http://www.fao.org/ag/ca/Carbon%20Offset%20Consultation/CARBONMEETING/3FULLPAPERSBYCONSULTATIONSPEAKERS/PAPERGODDARD.pdf.

Gregorich, E. G. and Anderson, D. W. 1985. Effects of cultivation and erosion on soils of four toposequences in the Canadian prairies. *Geoderma* 36(3–4), 343–354. https://doi.org/10.1016/0016-7061(85)90012-6.

Janzen, H. H. 2001. Soil science on the Canadian prairies–peering into the future from a century ago. *Canadian Journal of Soil Science* 81(4), 489–503.

Lafond, G. P., Clayton, G. W. and Fowler, D. B. 2014. Conservation agriculture on the Canadian prairies. In: *Conservation Agriculture: Global Prospects and Challenges* Jat, R. A., Sahrawat, K. L. and Kassam, A. H. (Eds). CAB International, Wallingford, UK, pp. 89–107.

Lessiter, F. 2015. 29 reasons why many growers are harvesting higher no-till yields in their fields than some university scientists find in research plots. *No-Till Farmer Magazine Blog*, USA. Available at: https://www.no-tillfarmer.com/articles/4038-reasons-why-many-growers-are-harvesting-higher-no-till-yields-in-their-fields-than-some-university-scientists-find-in-research-plots.

Liang, B. C., VandenBygaart, A. J., MacDonald, J. D., Cerkowniak, D., McConkey, B. G., Desjardins, R. L. and Angers, D. A. 2020. Revisiting no-till's impact on soil organic carbon storage in Canada. *Soil and Tillage Research* 198. https://doi.org/10.1016/j.still.2019.104529, 104529.

Lindwall, C. W. 2010. Chapter 4. Breakthroughs in engineering and equipment development. In: *Landscapes Transformed: The History of Conservation Tillage and Direct Seeding* Lindwall, C. W. and Sonntag, B. (Eds). University of Saskatchewan, Canada.

Lindwall, C. W. and Sonntag, B. 2010. *Landscapes Transformed: The History of Conservation Tillage and Direct Seeding*. University of Saskatchewan, Canada, 37–43.

Liu, J., Worth, D. E., Desjardins, R. L., Haak, D., McConkey, B. and Cerkowniak, D. 2020. Influence of two management practices in the Canadian Prairies on radiative forcing. *Science of the Total Environment* 765, 142701. https://doi.org/10.1016/j.scitotenv.2020.142701.

Lobb, D. 2011. Innovative farmers Association of Ontario: 25 years of innovation (1986-2011). Available at: www.ifao.com.

Marchildon, G. P. 2009. The prairie farm rehabilitation administration: climate crisis and federal-provincial relations during the great depression. *Canadian Historical Review* 90(2), 275–301.

Marchildon, G. P., Kulshreshtha, S., Wheaton, E. and Sauchyn, D. 2008. Drought and institutional adaptation in the Great Plains of Alberta and Saskatchewan, 1914–1939. *Natural Hazards* 45(3), 391–411. https://doi.org/10.1007/s11069-007-9175-5.

McClinton, B. and Polegi, J. 2010. Chapter 6. Saskatchewan soil conservation association. In: *Landscapes Transformed: The History of Conservation Tillage and Direct Seeding* Lindwall, C. W. and Sonntag, B. (Eds). University of Saskatchewan, Canada, pp. 51–66.

McConkey, B. 2010. Chapter 3. Foundations for growth–the role of researchers and AAFC research stations. In: *Landscapes Transformed: The History of Conservation Tillage and Direct Seeding* Lindwall, C. W. and Sonntag, B. (Eds). University of Saskatchewan, Canada, pp. 25–36.

McLaughlin, N. B. and Dyck, F. B. 1986. Directory of research equipment developed at the Swift Current Research Station. Technical Bulletin, 1986-199E, Agriculture Canada. Available at: https://ia800507.us.archive.org/30/items/directoryofresea19869mcla/directoryofresea19869mcla.pdf.

Mitchell, J. P., Reicosky, D. C., Kueneman, E. A., Fisher, J. and Beck, D. 2019. Conservation agriculture systems. *CAB Reviews: Perspectives in Agriculture, Veterinary Science, Nutrition and Natural Resources* 14(1), 1–25. https://doi.org/10.1079/PAVSNNR201914001.

OCSCSA (Ontario Chapter of the Soil Conservation Society of America) 1983. *Soil Erosion: Its Agriculture and Environmental Implications for Southern Ontario*. Ontario Chapter of the Soil Conservation Society of America, Don Mills.

Ontario Ministry of Agriculture, Food and Rural Affairs 2018. New horizons: Ontario's agricultural soil health and conservation strategy. Available at: www.omafra.gov.on.ca.

Prairie Soils and Crops 2009. Volume 2–weeds, herbicides and management (8 papers). Saskatchewan Soil Conservation Association, Canada. Available at: https://prairiesoilsandcrops.ca/volume2.php.

Prairie Soils and Crops 2012. Volume 5–long-term crop rotations in the Canadian Prairies (18 papers). Saskatchewan Soil Conservation Association, Canada. Available at: https://prairiesoilsandcrops.ca/volume5.php.

Richards, N. R., Caldwell, A. G. and Norwich, F. F. 1949. Soil survey of Essex County, report no. 11 of the Ontario Soil Survey. Ontario Ministry of Agriculture and Food, pp. 63.

Rudy, H. B. 2018. *The Soil Fixers: Land Stewards Committed to the Cause*. Friesen Press, Victoria, Canada.

Senate 1984. Soil at risk: Canada's eroding future. Standing Committee on Agriculture, Fisheries and forestry, The Senate of Canada, Ottawa. Available at: http://parl.canadiana.ca/view/oop.com_SOC_3202_1_3/1?r=0&s=1.

Smith, S. R. 2018. Deep in the furrows = a four-generation Noble venture. *Alberta Views*, June 1, 2018. Available at: https://albertaviews.ca/deep-in-the-furrows/.

Statistics Canada 2020. *Tillage Practices Used to Prepare Land for Seeding*. https://doi.org/10.25318/3210040801-eng.

Ugochukwu, A. I. and Phillips, P. W. B. 2018. Technology adoption by agricultural producers: a review of the literature. In: *From Agriscience to Agribusiness: Theories, Policies and Practices in Technology Transfer and Commercialization* Kalaitzandonakes, N., Carayannis, E. G., Grigoroudis, E. and Rozakis, S. (Eds). Springer International Publishing, Cham, pp. 361–377. https://doi.org/10.1007/978-3-319-67958-7_17.

Van Vliet, L. J. P. 1981. Agricultural land use and erosion in Ontario. Erosion and Sedimentation in Ontario: A Time for Action, Ontario Chapter Soil Conservation Society of America, Don Mills, Canada, pp. 97–104. Available at: https://sis.agr.gc.ca/cansis/publications/surveys/on/onm-3/index.html.

Wall, G. J., Vaughan, E. E. and Driver, G. 1985. *Cropping, Tillage and Land Use Management Practices in Southwestern Ontario - 1984*. Ontario Institute of Pedology, Guelph, Canada. https://longpointbiosphere.com/download/Environment/Cropping-tillage-land-management-practices-in-southwestern-ontario-Wall-and-Vaughan-1985.pdf.

Ward, B., Smith, D., Shaw, G., Haak, D. and Fredette, J. 2010. Policy and program response to land management issues. In: *Landscapes Transformed: The History of Conservation Tillage* Lindwall, W. and Sonntag, B. (Eds). University of Saskatchewan, Canada, pp. 15–23.

Chapter 3

Conservation Agriculture in the USA

Sjoerd W. Duiker, Penn State University, USA

1 Introduction
2 The changing United States climate
3 Land use in the United States
4 History of Conservation Agriculture
5 Current state of Conservation Agriculture
6 The future of Conservation Agriculture
7 Conclusion
8 Acknowledgments
9 Where to look for further information
10 References

1 Introduction

Climate and soils determine where crop production is concentrated in the United States. The climate is changing and Conservation Agriculture (CA) will be critical to guarantee sustainable soil and water management in the future. In this chapter we will review evidence of a foundation of CA-based agriculture in pre-Columbian agriculture, how this changed when European colonists introduced plow-based monoculture systems that led to the Dust Bowl, the largest environmental and social disaster in U.S. history. This disaster led to the formation of an institutional support infrastructure to develop and implement more sustainable soil management and led to the development of modern CA systems. We will show statistics of the adoption of CA in the previous decades and discuss threats and opportunities for CA and directions for future research.

2 The changing United States climate

Climate determines where agriculture can be practiced and also impacts the adoption of CA. CA is attractive in areas where water deficits are common because infiltration is typically improved and evaporation losses reduced. If water is excessive, on the other hand, soils may dry out more slowly under

http://dx.doi.org/10.19103/AS.2021.0088.03

heavy mulch cover, while soil temperatures stay lower. This makes CA more challenging under cooler climates with short growing seasons (Defelice et al., 2006). The climatic zones of the United States are marine, dry and moist (Fig. 1). The western part of the United States is predominantly arid or semi-arid with extensive mountain regions, plateaus and valleys. The west coast has a Mediterranean climate with dry hot summers and cool moist winters that transitions to a fully humid temperate marine climate in the northwest. The semi-arid Great Plains are in the foothills east of the Rocky Mountains, while the climate becomes humid east of the N-S line from North Dakota-Texas with hot or warm summers. The climate is progressively colder as one moves from south to north.

The climate in the United States is changing, and the scientific consensus is that it will continue to change as a result of global greenhouse gas emissions. These changes will affect agriculture profoundly - making greater adoption of CA more urgent than ever to conserve soil and water. The following discussion of climate change in the United States comes from the Fourth National Climate Assessment published by the U.S. Global Change Research Program (Wuebbles et al., 2017).

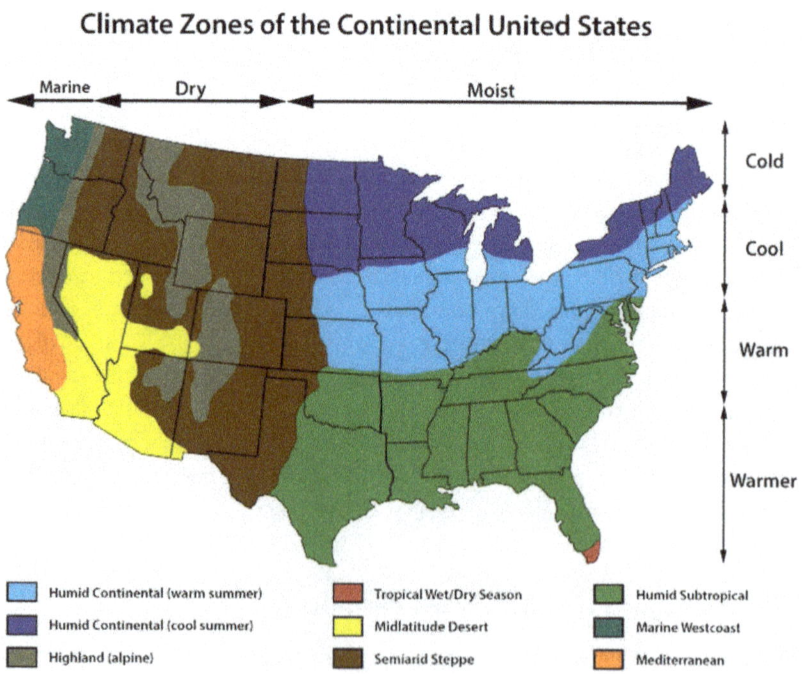

Figure 1 The climatic regions of the United States according to the Köppen-Geiger Climate Classification. Image credit goes to Permanence Institute (http://www.imagepermanenceinstitute.org). Used with permission.

The annual average temperature in the United States increased by 1.0°C between 1901 and 2016 and is expected to increase another 1.4°C between 2021 and 2050 (relative to the average of 1976–2005). Heat waves have become more frequent since the 1960s, while cold waves and extremely cold temperatures have become less frequent. Average annual precipitation between 1901 and 2015 decreased in much of the Southwest and Southeastern United States but increased in the northern Great Plains, Midwest and Northeast. These trends are expected to continue. Decadal or multi-decadal droughts are expected in the U.S. central Plains and Southwest in the second half of this century (Cook et al., 2015).

Drought can cause reduced yields or crop failure or increase demand for irrigation. Unsustainable irrigation and conflicts for water resources are already common in the United States. The Ogallala Aquifer underlying part of the Great Plains is an example of what may lie ahead for agriculture in the southwestern United States. The aquifer is used for 30% of irrigation in the United States and supports one-sixth of world grain production (Frankel, 2018). Much of the water in this aquifer is fossil water. Since the Second World War, irrigated agriculture has boomed in this area, resulting in withdrawal rates much greater than recharge rates (Braxton, 2009). Groundwater levels have dropped more than 50 m in some areas, especially in the southern part of the region (Frankel, 2018). A drier southwest would mean even greater demands on irrigation and a further decrease in groundwater levels. Eventually, irrigated agriculture is expected to disappear from much of this region forcing a return to dryland farming (Braxton, 2009). Reduced spring snowmelt in the Sierra Nevada due to increasing temperatures will affect irrigation agriculture in California. Snowmelt from the Sierra Nevada supplies an estimated 75% of agricultural water for California, and its reduction will cause stress on irrigation water supplies (Huning and AghaKouchak, 2018). The future of agriculture will hinge in large part on improving water use efficiency in the western United States to which CA can contribute.

Increased water scarcity in the western United States contrasts with increased precipitation in other parts of the United States. A warming climate and higher precipitation have caused an increase in maize and soybean acreage in North Dakota, an area that was marginal for those crops in the past (EPA, 2016). Spring precipitation increased in parts of the northwest, northern Great Plains and Midwest, while it decreased in most of the south. Excess spring precipitation can cause major crop losses due to delays in planting. The record year for prevented planting was 2019 when more than 8.1 million hectares of maize and soybeans in the Midwest (12% of all corn/soybean acres in the United States) could not be planted due to flooding and wet soils in spring (Farm Bureau, 2019). Excess precipitation also increases the threat of soil compaction by heavy farm machinery and grazing cattle. It will be very important in these areas to improve soil structure and internal drainage. CA is therefore important in areas of excess moisture too, which was confirmed by USDA statistics. Percent prevent planting in South Dakota, the heaviest-hit state for excess spring precipitation in 2019, was only 20% of no-till, but 39% of mulch till, 38% of reduced till and 27% of conventional till (Dobberstein, 2020). Fall precipitation is increasing in the eastern United States and the northern United States but decreasing in the northwestern United States. Excess moisture in the fall can cause delays in the harvesting of maize and soybeans, and soil damage by harvesting equipment. Again, greater resiliency to soil compaction due to improved soil structure suggests that CA is important for the future of crop production in these areas. Finally, heavy downpours are becoming more frequent as the warmer atmosphere is able to contain more water. This poses increased risks for soil erosion and runoff, another reason why CA will be important for the future of U.S. agriculture.

3 Land use in the United States

To understand the distribution of CA it is important to have a concept of the geographical distribution of crop production in the United States. The total land area of the United States is 9.2 million km^2, with approximately half (480 million ha) used for agriculture and the other half (435 million ha) for non-agricultural purposes (Table 1). Cropland, including that laying idle and that used for pasture, is almost 160 million ha (17% of total area), and grazed pastures and range (primarily in the arid west) cover 265 million ha (29%). The balance is grazed forest land and farmsteads and farm roads. Most non-agricultural land is in forests and recreation and wildlife areas. Urban land constitutes 28 million ha. A large portion of the United States is not suited for crop production due to an arid climate, steep topography or marginal, shallow soils (Fig. 2).

Important concentrations of cropland are the Great Plains stretching from Montana and North Dakota in the north to Texas in the south; the Corn Belt (Midwest) ranging from Ohio to Nebraska; the Palouse in eastern Washington, western Idaho and northeast Oregon; the Mississippi Delta; the coastal plains in the east; and valleys on the West Coast and the sandy soils of the Florida peninsula. Large expanses of U.S. cropland are high organic matter Mollisols derived from prairie vegetation and alluvial soils along rivers and streams. Mollisols are common in the Great Plains, the Corn Belt and the Palouse and are among the most fertile and deep soils in the world. Alluvial soils are found along streams and in coastal plains.

A great diversity of crops is grown in the United States. The 23 crops grown on more than 100 000 ha, altogether planted on 126 million ha (79% of total cropland) are shown in Table 2. The 'Big Five' are maize, soybeans, wheat, hay and cotton, grown on 115 million ha, or 72% of total cropland, while the other

Table 1 Land use in the United States in 2012

	Area (ha)	Area (%)
Agricultural		
Cropland used for crops	137.7	15.0
Idle cropland	15.8	1.7
Cropland used only for pasture	5.3	0.6
Grassland pasture and range	265.3	29.0
Forest land grazed	52.7	5.7
Farmsteads, farm roads	3.2	0.4
Total agricultural	480.3	52.5
Non-agricultural		
Forest-use land not grazed	203.3	22.2
Transportation	10.9	1.2
Recreation and wildlife areas	102.9	11.2
National defense areas	10.9	1.2
Urban land	28.4	3.1
Miscellaneous other land	79.4	8.7
Total non-agricultural	435.4	47.5
Total land area	915.3	100.0

Source: Bigelow and Borchers (2017).

18 crops planted on more than 100 000 ha occupied 11 million ha (7% of total). There are many other crops grown on smaller acreages on the remaining 21% of cropland, many of them specialty crops such as fruits and vegetables. The dominance of the 'Big Five' causes a lack of diversity in-crop rotations because they are grown in different regions, except for corn and soybeans that are often grown in rotation.

4 History of Conservation Agriculture

It is commonly thought that native Americans practiced shifting cultivation. It is assumed they would burn an area of natural vegetation, plant a polyculture of crops (such as the famous 'three sisters' – maize, beans and squash) until the soil was depleted or weeds took over and then moved on to a new piece of land that would be treated the same way. A long fallow period would be needed to replenish soil fertility before an area could be cultivated again. Crop yields were thought to be low. It was assumed that large expanses of fallow land allowed this system to continue for centuries, although the shortening of fallow periods is believed to have led to the collapse of some native American societies such as the Mayans when population pressure increased (Diamond, 2005). This

Table 2 Major crops planted in the United States (2019)

Crop	Area ('000 ha)	Comments
Maize (all purposes)	37 139	7% silage, 90% GMO (herbicide resistant)
Soybean	32 416	94% GMO (herbicide resistant)
Hay (harvested)	21 373	
Wheat	18 472	
Cotton	5557	98% GMO (herbicide and pest resistant)
Sorghum (all purposes)	2076	
Barley	1157	
Rice	1116	
Oats	1032	
Canola	817	
Rye	759	
Sunflower	559	
Peanut	552	
Dry edible beans	529	
Sugarbeet	456	
Dry edible pea	416	
Potato	392	
Sugarcane	371	
Chickpea	227	
Lentil	217	
Proso millet	175	
Flaxseed	144	
Tobacco	118	

Source: NASS (2019).

narrative is so engrained in the American psyche that it is believed to be fact. However, new research suggests that native Americans used CA to practice agriculture that was permanent, intensive and environmentally sustainable. This was possible because (1) fertile soils were plentiful (Mollisols, Alfisols and Inceptisols); (2) maize was their staple crop, a C-4 crop with high yield potential but moderate nitrogen needs; and (3) soils were managed without plowing (Mt. Pleasant, 2015). Mt. Pleasant convincingly shows that maize yields of 25–50 bushels/A (1500–3000 kg/ha) were possible, sustaining the population in Cahokia and Moundville, the largest Mississippian sites in North America in the 17th–19th centuries. The use of CA by these native Americans led to the slow release of nitrogen from decomposing organic matter and low nitrogen losses through leaching and erosion, making permanent agriculture possible.

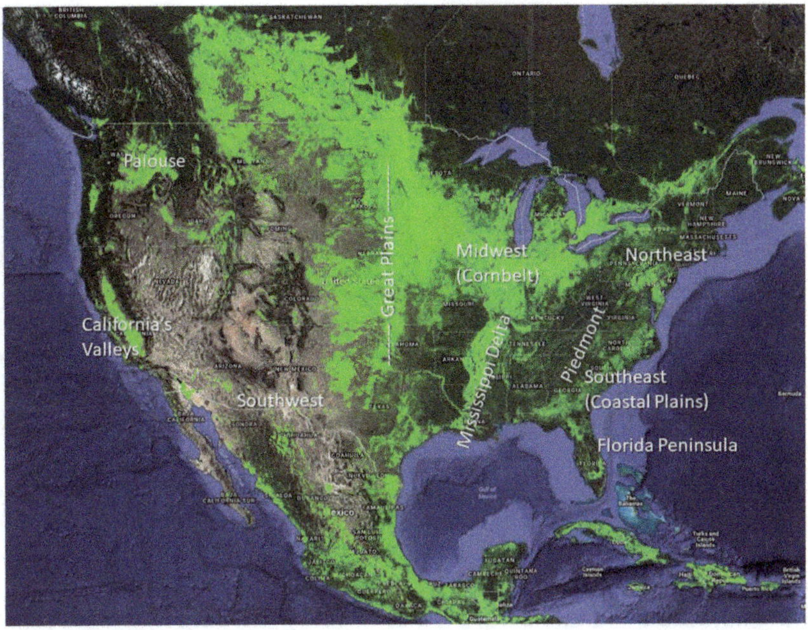

Figure 2 Map showing crop production in the United States by the bright green coloring and important crop production areas mentioned in the text. Credit goes to U.S. Geological Survey. https://www.usgs.gov/media/images/map-croplands-united-states. Used with permission.

The face of agriculture changed with European colonization and the introduction of plow-based monocultures starting in the 1600s (Haystead and Fite, 1955). Clearing the original forest and subsequent tillage-based farming in Maryland and Virginia led to a quick release of nutrients for 2–3 years of tobacco, followed by a few years of corn and wheat. After exhausting the old, farmers moved on to new land. This system resulted in tremendous environmental degradation. It was estimated that the Potomac carried annually 770 million kg of dissolved nutrients and 213 million kg of sediment to the Chesapeake Bay and the James River carried 3–4 million cubic meters of soil from north of Richmond and the Roanoke 3–4 million tons of soil from the Piedmont region (Craven, 1926). Thus pioneers moved further south and west. Beginning in the early 1800s, forests in the south were cleared for cotton production. A few years of continuous cotton quickly exhausted the soil and resulted in nutrient loss and water erosion due to exposure of the plowed soil to high rainfall intensity in this region. It is estimated that southern Piedmont cropland lost on average 19 cm of topsoil (Trimble, 1974). The devastation of soil erosion can still be witnessed today in revegetated gullies and extensive pine plantations occupying the ravaged cotton soils.

Thus farmers moved further west to what seemed like endless expanses of fertile prairie soils, where they grew corn, oats, wheat and hay. But the tough prairie sod was hard to eradicate with wooden or cast-iron plows that stuck to the clayey soils. This caused John Deere to invent the steel moldboard plow in 1837 that did not stick to the soil. John Deere's steel plow became very popular, with annual sales of 2000 in 1849 which increased to 60 000 in 1873 (Drache, n.d.). Besides John Deere, there were 420 plow manufacturers in the United States in 1859 (Drache, n.d.). The tillage revolution enabled the conquest of the West but at a price. Approximately 90 000 kg ha^{-1} of soil organic carbon was lost in 50 years of continuous corn production under moldboard plowing in the Morrow plots in Champaign, Illinois, and approximately 70 000 kg ha^{-1} in the Sanborn Field in Columbia, Missouri (Huggins et al., 1998). With the organic carbon 3000 kg ha^{-1} of N, 88 kg ha^{-1} of P and 432 kg ha^{-1} of S would be released, assuming a C:N, C:P and C:S ratio of organic matter of 0.039, 0.0011 and 0.0054, respectively, for soils with high SOC content (Tipping et al., 2016; these are conservative estimates because nutrient-rich soil organic matter has corresponding ratios of 0.12, 0.016 and 0.016, respectively). Due to the high organic matter content of these prairie soils, decades of crop production were possible without fertilizer or manure application, but the largely invisible processes of sheet erosion and nutrient depletion took its toll. Sheet erosion on this gently rolling land continues today at unsustainable levels (Johnson, 1987).

Pioneers moved further west in the late 1800s and early 1900s. However, crop production is more precarious in the dryer climate of the Great Plains. Years of good rainfall may be followed by years of drought. The power of the plow led to fencerow-to-fencerow wheat fields that produced bumper crops in years of favorable precipitation. But crop failure followed during dry years, resulting in wind erosion of the pulverized soil devoid of vegetative cover. Between 1925 and 1931 drought hit the northern Plains, and large numbers of farmers went bankrupt (Alston, 1983). Drought hit the southern Plains between 1930 and 1936, resulting in the infamous 'Black Blizzards'. Years of crop failure led to the largest human displacement in the United States, with 2.5 million people moving to states like California where they were recruited as cheap seasonal farm laborers living in decrepit conditions, famously chronicled in the Nobel-prize-winning novel *Grapes of Wrath* (Steinbeck, 1939). Since then 10% of the Palouse (a region of highly erodible undulating loess) has lost all topsoil, and 60% of the land lost 1/4–3/4 of its topsoil (USDA, 1978). Finally, there was no more virgin land to put under the plow when the shores of the Pacific were reached, signalling the end of the conquest of the land in the United States.

It was not until a national disaster struck in the form of the Dust Bowl in the 1930s that the American people and its government awoke to the severity of the problem of plow-based farming. In his treatise 'Soil Erosion, a National Menace', Hugh Bennett, the Father of Soil Conservation, estimated that annually more than 3 billion tons of soil washed out of fields, causing annual losses of at least $400 000 000 to farmers, not counting costs to roads, railways, reservoirs, streams and ditches (Bennett, 1934). Mr. Bennett estimated that 125 million acres (51 million ha) of land had lost the better part of its topsoil due to water erosion and that wind erosion was taking its toll on semi-arid lands. He mentions a dust storm earlier that year originating in the Great Plains that blotted out the sun on May 11, 1934, over a vast area of the northeastern United States. This storm is believed to have swept away 300 million tons of topsoil from Great Plains farmland. The cause: breaking of the grass cover, soil loosening and loss of organic matter and soil structure by continuous cultivation.

Bennett's campaign led to the government's realization of the urgency to protect soil on private as well as public land. In 1930 resources were appropriated to establish the first soil erosion research and demonstration stations. In 1933 the Soil Erosion Service was set up in the Interior Department with Bennett as its Director. This agency became the Soil Conservation Service (now Natural Resources Conservation Service) of the U.S. Department of Agriculture. President F. D. Roosevelt was receptive to Bennett's message and supportive of these conservation efforts. In 1937 he sent a letter to the state governors encouraging them to adopt the State Soil Conservation Districts Law as a part of the agricultural program in their state (Roosevelt, 1937). This led to the formation of Soil Conservation Districts that coordinated assistance from the public, private, local, state and federal sources to develop locally-driven solutions to natural resource concerns. Today, there are almost 3000 Conservation Districts, about one per county in the United States. They collaborate with federally funded USDA-NRCS offices assisting farmers with conservation planning and implementation, with research and education support from the USDA-Agricultural Research Service and the land-grant universities. Every state has at least one land-grant university with an extension service. The extension service is composed of specialists and extension educators, the latter being predominantly stationed in rural areas. The development and promotion of CA in the United States have been deeply impacted by the institutional support provided by these organizations.

Soil conservation assistance is given on a voluntary basis in the United States. Subsidies or technical assistance may be provided, but these agencies do not have the power to enforce conservation. The voluntary nature of soil conservation promotion is by intent. However, although soil conservation is voluntary, the 1985 Farm Bill institutionalized 'cross compliance' requiring farmers with HEL (Highly Erodible Land) to produce according to an approved conservation plan or stand to lose eligibility for program benefits for any commodity they produce on any land they own or operate. Similarly, farmers are not allowed to produce on land classified as wetlands converted after December 23, 1985, or lose all program benefits (Luzar, 1988). Program benefits include Farm Home Administration loans, crop insurance, a variety of price and income supports, Commodity Credit Corporation storage payments and farm storage loans. Cross compliance became a powerful motivation for farmers in states with high proportions of HEL cropland to take soil conservation more seriously. In Pennsylvania, for example, 60% of cropland is classified as HEL, and the 1985 Farm Bill triggered a fresh push in the Cooperative Extension Service to teach farmers about the benefits and techniques of conservation tillage (Duiker, 2019). Besides federal efforts in soil conservation, there may be local or state regulations that require farmers to have and implement a conservation plan.

Promotion of stewardship ethic and establishment of a government support structure for soil conservation did not automatically mean endorsement of CA. However, the public awareness of the importance of soil conservation as a result of the Dust Bowl and the existence of an institutional framework to investigate soil erosion, develop solutions and work collaboratively to apply them sets the United States apart from other countries and is the reason the United States became the cradle of the modern CA movement.

Over the years, the agricultural community has had a hate-love relationship with the plow (Lal et al., 2007). The moldboard plow long remained a symbol of agricultural conquest, being featured prominently on many agricultural buildings in the United States, and even in the flag of the U.S. Department of Agriculture. But it is also recognized that the plow caused devastating soil erosion by wind and water, sped up the decomposition of organic matter, destroyed soil structure, released large amounts of carbon to the atmosphere, led to the loss of a large pool of plant nutrients stored in soil organic matter and promoted loss of soil biological diversity. But the belief that plowing is an integral and necessary part of farming has proven difficult to leave behind. In 1943 Edward Faulkner published his controversial 'Plowman's Folly' (Faulkner, 1943) in which he postulated that 'no one has ever advanced a scientific reason for plowing'. Faulkner proposed a form of shallow disk tillage that left large amounts of residue at the soil surface. At the same time, Walter Thomas Jack (1946) published 'The Furrow and Us' in which he defends the use of the moldboard plow to stimulate decomposition of crop residues and manure to supply nitrogen to crops. The controversy between plow-based and no-tillage farming continues until today and slowed down the rate of adoption of no-tillage in the United States.

In the first half of the 20th-century scientific discoveries and technological advances revolutionized U.S. agriculture, and four technologies particularly (positively or negatively) affected the development and adoption of CA: (1) mechanization; (2) nitrogen fertilizer production and use; (3) development of chemical weed control; (4) herbicide-resistant crops. In 1876 Nikolaus A. Otto built a successful four-stroke engine. In 1889 Gottlieb Daimler introduced a four-stroke engine with two cylinders arranged in a V. This led to the development of the first American tractor with a two-cylinder gasoline engine in 1903 by

Charles W. Hart and Charles H. Parr. During World War I there were more than 200 companies producing tractors, but due to much-needed quality control, this decreased to 9 in 1933 (Goering, 2008). The Farmall tractor, introduced in 1924, was the first tractor that could be used for in-crop cultivation. Tractors reduced the need for labor and draft animals in agriculture. Since as much as 25% of the land was used to produce feed for draft animals (Goering, 2008), tractors reduced the need to grow perennial forages and feed crops such as oats, and hence crop diversity declined.

Another major development was the development of nitrogen manufacturing capacity during the Second World War (Russell and Williams, 1977). The Haber–Bosch process to fix atmospheric nitrogen to make synthetic ammonia was developed in the early 1900s. During the Second World War several plants were built to supply ammonium nitrate to make ammunition, including one at the Tennessee Valley Authority (TVA) in Muscle Shoals, Alabama. The TVA was established in 1933, with the national responsibility to increase fertilizer manufacture and use. After the Second World War part of the ammonium nitrate production capacity was converted to produce ammonium nitrate fertilizer. The TVA began a large research and education program on nitrogen fertilizer. Nitrogen fertilizer use on U.S. cropland, therefore, increased from about 5 kg ha^{-1} in 1950 to 50 kg ha^{-1} in 1980 (Van Grinsven et al., 2015). For many years American farmers had relied on mineralization of organic matter to supply nitrogen to crops, stimulated by plow-based agriculture. Soil organic matter content had stabilized at a new, much lower level, and hence its nitrogen supplying capacity decreased. This would have necessitated greater use of leguminous cover crops and perennial forages in crop rotations, and the use of composts and manures to supply nitrogen to crops had it not been for the development of manufactured nitrogen fertilizer. The use of readily available nitrogen fertilizers allows U.S. farmers to produce high yields, but this comes at an economic and environmental cost. The energy to produce nitrogen fertilizer is responsible for more than one-third of the greenhouse gas emission of crop production and nitrate leaching from excessive N-fertilizer use threatens surface water quality (Van Grinsven et al., 2015). Widely available nitrogen fertilizers, therefore, have led to a reduction in the use of cover crops, legume forages and animal manure.

Third, the discovery and commercialization of herbicides were important to the development of CA in modern agriculture. Weed control is probably the major reason farmers use tillage (Duiker, 2015). Primary tillage is the first tillage pass to kill standing vegetation, loosen and fracture soil and incorporate soil amendments. The moldboard plow is a particularly effective tool to eradicate weeds by burying them completely. As already discussed, the steel moldboard plow enabled the eradication of very tough grasses and forbs with extensive root systems and worked well in heavy clay soil common to the prairies. There are other primary tillage tools such as chisel and disk plows that are less effective for weed control but may be used if weed pressure is low, or if, for some reason like the presence of tree roots or rocks close to the soil surface, moldboard plows are difficult to operate. After primary tillage, secondary tillage is needed to kill any remaining weeds, crush clods and prepare a suitable seedbed. In-crop weeds are subsequently controlled by tertiary tillage such as rotary hoeing and between-row cultivation. Tillage has traditionally been seen as the 'Big Hammer' against weeds. Therefore, it was generally considered impossible to grow crops without tillage, until the discovery of herbicides. Therefore, the early no-till movement heavily relied on the availability of herbicides for weed control (Triplett and Dick, 2008). The first widely used modern herbicide in the United States was 2,4-D (2,4-D dichlorophenoxyacetic acid) in the 1940s. This systemic growth hormone-disrupting herbicide is sprayed over the top in cereal crops such as wheat or barley, killing most broadleaf weeds. It continues to be a widely used herbicide. Another very influential herbicide was atrazine (1-chloro-3-ethylamino-5-isopropylamino-2,4,6-triazine) discovered in 1952. This photosynthesis-inhibiting herbicide is sprayed on the soil before or shortly after planting corn and has residual activity providing 1-2 months of weed control while not affecting corn. Two other important herbicides that affected the adoption of no-tillage were the non-selective herbicides paraquat (1,1′-dimethyl-4,4′-bipyridiumium chloride) discovered in 1958 and glyphosate (N-(phosphonomethyl)glycine) discovered in 1971. These herbicides are non-selective so have lethal action on almost any plant. They do have major differences, however, paraquat being a contact herbicide that quickly kills primarily leaves it comes in contact with, while glyphosate is a systemic enzyme inhibiting herbicide that is taken up through the leaves and is translocated throughout the plant. These 'burndown' herbicides allowed quick termination of existing vegetation and made primary tillage for weed control unnecessary. Since 1940, many different herbicides have been developed. Kraehmer et al. (2014) list commercially available herbicides with 19 different modes of action. In the past 30 years, however, no new modes of action have been brought on the market due to public resistance to pesticides and the high cost of commercialization and the limited ability to recover costs. Kraehmer et al. (2014) list 15 available modes of action (many of which were discovered in the last 30 years) that have not been commercialized. It seems, therefore, that there are still available options to expand chemical weed control, although environmental effects and public safety and opinion also have to be given due attention.

A fourth factor that has impacted the adoption of no-tillage has been the development and commercialization of herbicide-resistant crops. GMO crops (Genetically Modified Organism crops, such as herbicide-resistant and insect pest-resistant crops) have become very common in the United

States primarily in soybeans, corn (maize) and cotton (Fig. 3). GMO adoption started in the 1990s. The GMO traits that have been widely used are insect resistance through the incorporation of genes from the soil bacterium *Bacillus thuringiensis* (in cotton and corn), and resistance against certain broad-spectrum herbicides (most significantly glyphosate) in soybean, corn, cotton, canola and sugarbeets (CAST, 2012). Adoption of GMO crops has been rapid. Today, 94% of soybean, 98% of cotton and 90% of corn acreages grown in the United States are herbicide-tolerant, whereas 83% of corn and 92% of U.S. cotton acres are insect-resistant. A survey of farmers in Iowa, Illinois, Indiana, Mississippi, North Carolina and Nebraska showed that 25% of the farmers using conventional tillage transitioned to no-till and 31% to reduced till after adoption of herbicide-resistant crops (Givens et al., 2009). Of the farmers using reduced tillage, 25% transitioned to no-till after the adoption of herbicide-resistant crops (Table 3). The change toward conservation tillage was greatest among cotton and soybean growers, and smaller among corn growers, possibly because conservation tillage was already common among corn growers before the introduction of glyphosate-resistant crops. The transition from conventional tillage to no-till and reduced till with the adoption of glyphosate-resistant crops was most common among smaller farmers and was attributed to cost and labor savings. Therefore, glyphosate-resistant crops have stimulated the adoption of

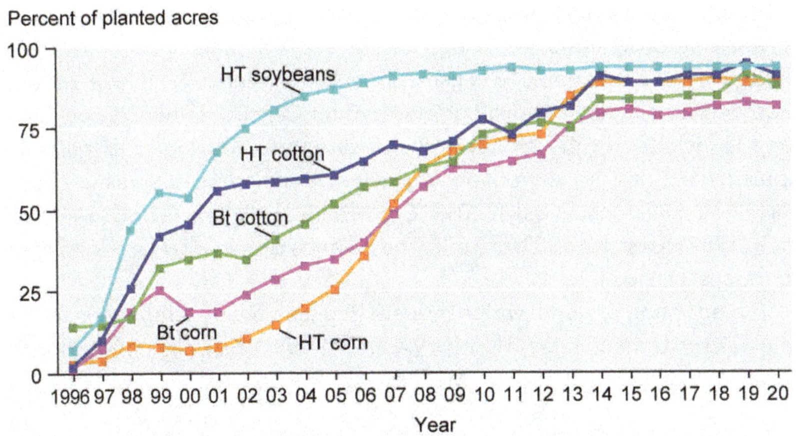

Note: HT indicates herbicide-tolerant varieties; Bt indicates insect-resistant varieties (containing genes from the soil bacterium Bacillus thuringiensis). Data for HT/Bt corn and cotton are not mututally exclusive, as HT and Bt categories include those varieties with overlapping (stacked) HT and Bt traits.
Source: USDA, Economic Research Service using data from the 2002 ERS report, *Adoption of Bioengineered Crops* (AER-810) for the years 1996-99 and National Agricultural Statistics Service, (annual) June Agricultural Survey for the years 2000-20.

Figure 3 Adoption of genetically engineered crops in the United States, 1996–2019. https://www.ers.usda.gov/data-products/adoption-of-genetically-engineered-crops-in-the-us/recent-trends-in-ge-adoption.aspx. Credit USDA-ERS. Used with permission.

Table 3 Tillage systems before and after implementation of glyphosate-resistant cropping systems

Tillage system before GR crop	Tillage system after GR crop			
	No-till	Reduced till	Conventional-till	Total
	%			
Conventional-till	25	31	44	37
Reduced till	25	74	2	38
No-till	92	6	3	25
Total	41	41	18	100

Source: From Givens et al. (2009).

no-till and reduced tillage, specifically in cotton and soybeans, and particularly among smaller farmers.

5 Current state of Conservation Agriculture

5.1 No-till adoption

Tillage intensity in the USA was monitored nationwide using transect surveys with the assistance of the U.S. Department of Agriculture's Natural Resource Conservation Service (USDA-NRCS) until 2004, with ad-hoc transect surveys since then. Recent tillage data were obtained from USDA's Agricultural Census that takes stock of the state of agriculture every 5 years. Tillage data were included in the 2012 and 2017 census, although only intensive tillage, reduced tillage and no-tillage were distinguished in the Census. In the Census, mulch tillage is included in the reduced tillage category. No-tillage has gained in popularity and has grown from approximately 5% of planted acres in 1989 to 37% (42.30 million hectares) in 2017. Conventional tillage has decreased from 50% in 1989 to less than 30% in 2017. The balance is shared by full-width mulch and reduced tillage (Fig. 4).

The adoption of no-till varies by state (Fig. 5a). No-till adoption is highest in the eastern United States (TN, MD, VA, KY, PA, WV, DE), while no-till adoption is greater than 50% as well in MT, NE, AL, SD and NC. States with no-till percentages between 30% and 50% are KS, OH, SC, MO, CO, IN, OR, CT, NJ, OK, IA, ND and WI. No-till adoption is less than 30% in the remaining states, with the lowest (<10%) adoption in Arizona, California and Minnesota. Reduced tillage is common (>40%) in Midwestern states (ND, MN, WI, IA, IL), New York, Washington, Louisiana and Alaska and is fairly common (>30%) in many states in the central part of the country (Fig. 5b). Conventional tillage use is highest (>60%) in states in the extremities of the United States: RI, CA, HI, NV, FL, ME and AZ (Fig. 5c). Conventional tillage represents less than 20% of planted acres in many states now and is less than 10% of planted acres in MT, TN, VA and MD.

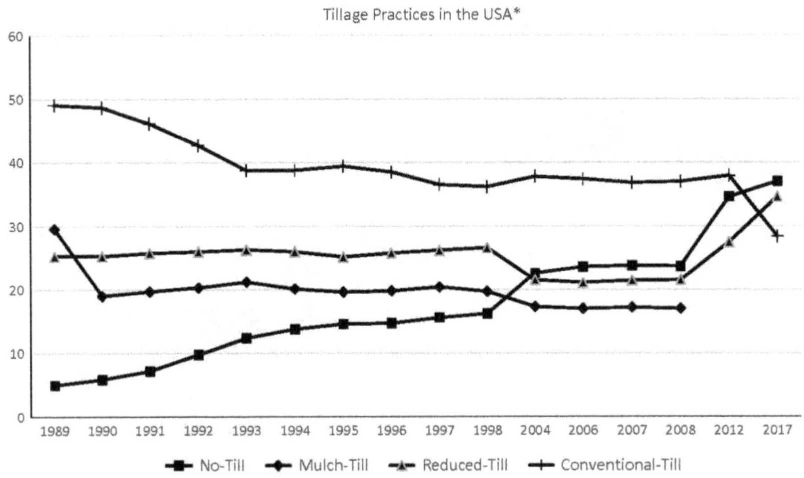

Figure 4 Tillage adoption as a percentage of planted acres in the USA. 2012 and 2017 data are from the Agricultural Census, while earlier data were derived from transect surveys. Tillage definitions varied somewhat with only three classes (no till, reduced till and conventional till) recognized in the Census. Reduced till includes mulch till in 2012 and 2017. From Conservation Technology Information Center https://www.ctic.org/CRM and 2017 Census of Agriculture https://www.nass.usda.gov/AgCensus

In general, the states with the highest % adoption of no-till had the lowest % conventional tillage.

No-till (or strip-till, a modification of no-tillage) adoption varies by crop: no-till is most widely used for soybeans and wheat but less for corn and cotton. In the period 2012–2017 no-till was used for 40% of soybean, 45% of wheat, but only 27% of corn and 18% of cotton (Claassen et al., 2018). Variability in no-till use between crops has led to low adoption of continuous no-tillage – only 20% of corn, soybean, wheat and cotton acreage was planted with no-till for four consecutive years (Claassen et al., 2018). In the Midwest, it is common to no-till soybeans, but to use tillage prior to corn planting. Crop differences and where the crops are grown impact no-till adoption. For example, both soybean and corn are grown on deep soils in areas with relatively favorable precipitation so that soil and water conservation is not a primary concern, but soybean is less sensitive than corn to reduced soil temperatures and variable plant populations that are more common under no-till conditions, while wheat is grown primarily in semi-arid regions where water conservation is crucial. Although there are many different reasons for the adoption of the tillage system, low adoption in states such as CA and FL can be ascribed to the high acreage of specialty crop production in these states. No-till technology for these crops such as herbicides for weed control and planting machinery are not widely available. Low adoption of no-till in northern states such as Minnesota is probably due

(a)

(b)

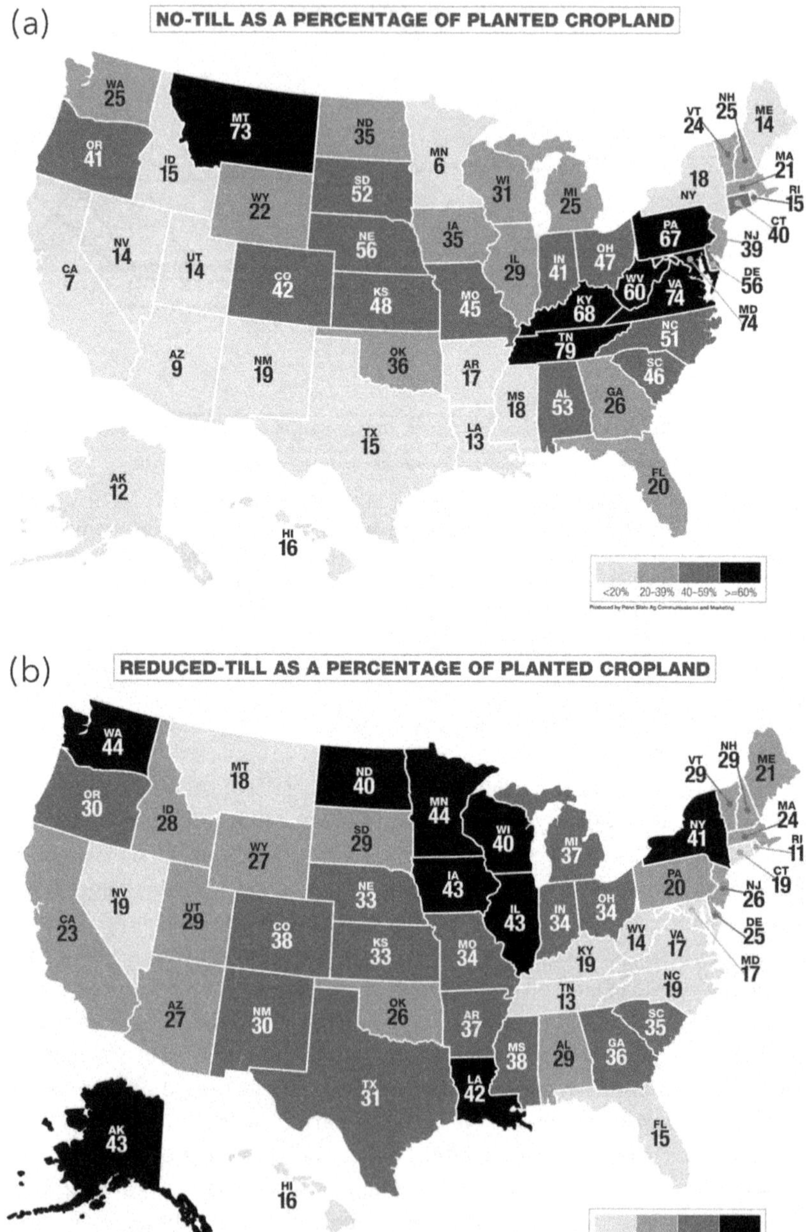

Figure 5 No-till (a), reduced till (b) and conventional-till (c) adoption in the states of the USA (data from 2017 Agricultural Census).

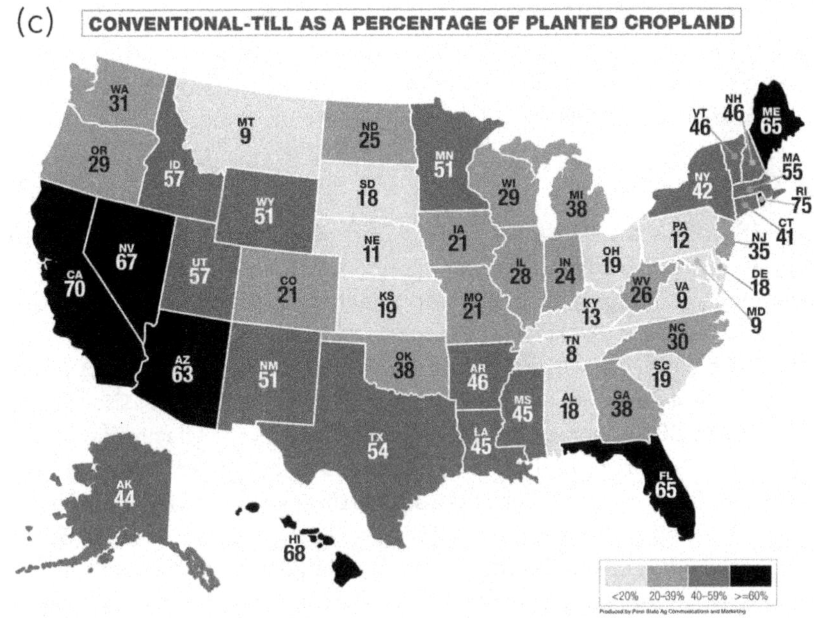

Figure 5 (*Continued*).

to the cold climate and widespread poorly drained soils (Defelice et al., 2006). The soil under no-till remains cool in the spring due to the insulating effect of crop residue, while the residue also conserves moisture, so soils dry out slowly. However, locally adopted CA systems can be developed to accommodate these problems if there is a vision among researchers and practitioners that CA is critically important. For example, planting green is a new cover crop management technique for corn or soybeans that helps to dry out excessively wet soil (Reed et al., 2019). This is just one adaptation to create CA systems that work in more poorly drained soils. One would expect high NT adoption in all semi-arid areas, but, nonetheless, its adoption is low in states such as AZ, TX, UT and ID. It is likely that social and perhaps economic factors explain low CA adoption in these states.

5.2 Cover crop adoption

The organic cover is the second pillar of CA, cover crops being important to harvest additional radiative energy, store it in carbohydrates and add it to the soil. Cover crops are typically planted after the harvest of annual crops and occupy the land during a period when the soil would otherwise be devoid of living vegetation. Cover crop adoption is still in its infancy in the United States. According to the 2017 Agricultural Census, cover crops were used on only 5%

of the planted cropland in the United States, although an increase of 50% since 2012. However, there were large differences in cover crop use between states. Cover crop use is much higher in the eastern states, with the highest adoption in the northeastern states (Fig. 6).

5.3 Crop diversity

Crop diversity is the third pillar of CA. Crop diversity can be in-crop (mixed cropping), which is relatively uncommon in the United States (except for in cover crops or forages), or in the crop rotation. Crop diversity is affected by many factors, be they climate, soils, pests, diseases, consumer preferences, producer preferences, food system structure, economies of scale, input costs, scientific advances or government policies. Aguilar et al. (2015) showed that from 1978 to 2012, average crop diversity (defined as the effective number of crop species grown in U.S. Counties) decreased in the United States. However, there were differences between regions. Crop diversity decreased in the Eastern Uplands (the region from northern Alabama to southwestern Pennsylvania), the Heartland (Midwest) and the southeastern United States, primarily because of the increasing popularity of corn and soybeans. On the other hand, cropping diversity increased in the Mississippi Delta, due to decreasing cotton production in this area. Despite the heavy reliance on only five major crops, there are many

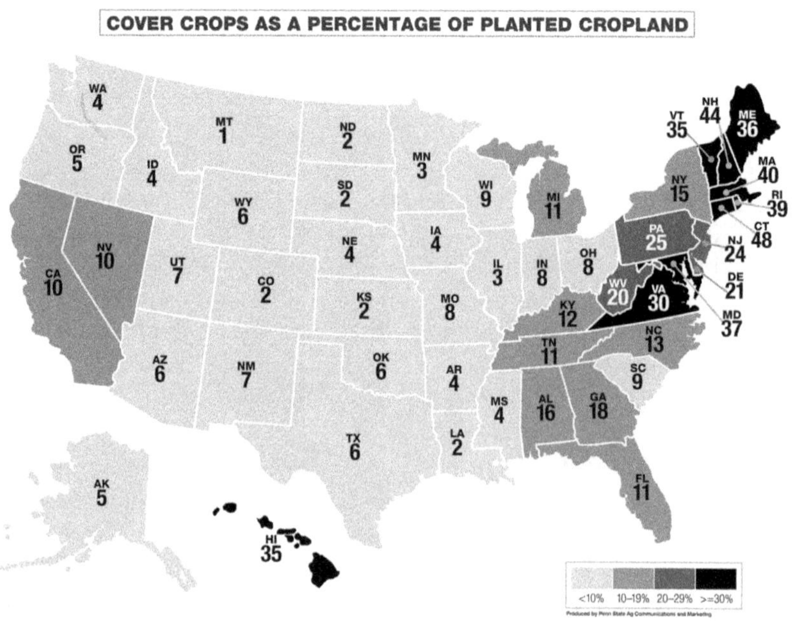

Figure 6 Cover crop as a percentage of planted cropland by state (data from 2017 Agricultural Census).

other crops being grown in the United States as was shown in Table 2. However, most crops are only grown on small acreages.

6 The future of Conservation Agriculture

6.1 Weed control

While herbicides and herbicide-resistant crops have been helpful in stimulating the adoption of no-tillage in the United States, heavy reliance on herbicides for weed control is not sustainable. There is much public concern about the effects of herbicides and other pesticides on the environment and on human health. There is an increasing number of weed species that have developed resistance to herbicides. In 2014, there were 220 weed species resistant to one or more herbicides (Heap, 2014). Twenty-four weed species were resistant to glyphosate, many of them becoming troublesome weeds in Roundup Ready cropping systems. In 2020, the number had increased to 262 weed species (152 dicots and 110 monocots) with resistance to herbicides. Weeds with resistance against 23 of the 26 known herbicide sites of action and to 167 different herbicides have been found (International Herbicide-Resistant Weed Database, weedscience.org/home.aspx, accessed November 23, 2020). Weeds resistant to herbicides are becoming a threat to the future of continuous no-tillage in the United States. According to the Weed Science Society of America, conventional tillage is the dominant tillage practice in glyphosate-resistant cotton because of the presence of glyphosate-resistant weeds (CAST, 2012). Because weed scientists are running out of chemical options, the Weed Science Society of America has called on USDA-NRCS to allow greater flexibility in its programs to allow farmers to use occasional (moldboard) tillage for weed control (Shaw et al., 2018).

The problem is that it takes years to build soil health with continuous no-tillage and that this can quickly be nullified by intensive inversion tillage. But sole reliance on one herbicide for weed control is, in the long term, a losing proposition, and even relying on different herbicides will only delay the eventual development of herbicide resistance in weed populations. But there are many successful farmers who have used no-tillage continuously for over 30 years without problems with herbicide resistance or out-of-control weed infestations. The author manages a 42-year-old continuous no-tillage field with excellent weed control. So what is the secret to successful weed control in CA? Diversity in management strategies is key, such as crop rotation sequence; irrigation schedules and methods; cover crops to suppress weeds; variable planting time and crop seeding rates; planting configuration; choice of cultivar; nutrient management optimization; heavy and long-lasting mulch cover; intercrop/relay systems; grazing integration in cropping systems; tank mixing herbicides with different modes of action; rotation of herbicide mode of action; and using pre-emergence herbicides besides glyphosate (CAST, 2012). Rotations of four different crops that include cool- and warm-season crops can be a key component of integrated weed management in the Great Plains (Anderson et al., 2007). By using a dualistic approach of prevention and control farmers have reduced herbicide inputs and costs in CA by 50% (Anderson, 2007). It is clear that agronomists and farmers need to focus not on one technology such as tillage or herbicides as the emphasis of their weed control program, but on a multitude of 'little hammers' to control weeds (Liebman and Gallandt, 1997). This means farmers and service providers need to be continuously learning about new ways to control weeds, taking advantage of all support services available.

6.2 Soil health

The emphasis of the early CA movement in the United States was on no-tillage to save fuel, time and soil (Triplett and Dick, 2008). Comparison studies have shown that no-till reduces soil erosion 2.5–1000 times, enough to bring soil erosion into line with natural soil regeneration rates (Montgomery, 2007). Soil erosion reduction, infiltration increase and evaporation reduction are primarily due to greater crop residue cover and soil health improvement in CA (Dabney et al., 2004; Truman et al., 2003; Truman et al., 2005; Jones et al., 1994; Mitchell et al., 2012; Unger and Baumhardt, 1999). Runoff reduction and soil erosion prevention with CA continue to be among the prime reasons why CA adoption is so important. Carbon sequestration was another attribute of no-tillage suggested as a way to mitigate greenhouse gas emissions (West and Post, 2002; Six et al., 2000; Reicosky, 1997; Lal and Kimble, 1997). But universal carbon sequestration with no-till is now being questioned (VandenBygaart, 2016). Organic matter accumulates close to the soil surface in no-till compared with mixing it in the plow layer with tillage. Tillage impacts on root growth below tillage depth and soil density may have caused overestimation of carbon sequestration benefits due to shallow sampling depth (Baker et al., 2007; Olson and Al-Kaisi, 2015). Consensus is emerging that carbon sequestration due to conversion from conventional tillage to no-till is occurring under certain climate and soil conditions, but that results are subject to large uncertainties (Ogle et al., 2019). Instead of carbon sequestration, the effect of CA on other soil health parameters needs to be emphasized (Karlen et al., 1994; Moebius-Clune et al., 2008). Among critical soil health parameters are soil porosity and nutrient cycling (Franzluebbers, 2002). Although total pore volume may not be different, continuity and depth of macropores for root growth and water percolation and greater quantities of mesopores that hold plant available water is critical. Deeper moisture penetration in no-till in drier climates due to improved infiltration and reduced evaporation and higher numbers of deep-burrowing earthworm species in no-till in humid climates cause deeper root growth, which allows crops to withstand drought better and improve uptake of nutrients from deeper soil layers (Kemper et al., 2011). A better understanding of the effect of the three pillars of CA on soil health and crop performance, including root architecture needs to be a priority.

6.3 Nutrient management

Improving nutrient use efficiency is a key priority for future crop production, with nitrogen being of greatest importance. Nitrogen fertilizer represents more than 50% of fossil energy use in corn production, and greenhouse gas emissions from crop production are first and foremost because of nitrous oxide emissions and CO_2 emissions from natural gas to make nitrogen fertilizer (Camargo et al., 2013). Nitrogen is a key factor in the eutrophication of estuaries in the Chesapeake Bay and the Gulf of Mexico leading to extensive dead zones (Erisman et al., 2013). No-tillage reduces soil erosion and increases infiltration and that affects nutrient losses. No-tillage may have a limited impact on nitrate loss (Fraser et al., 2013; Van Es et al., 2019) or increase nitrate loss due to greater deep drainage volume (Daryanto et al., 2017). Therefore, better nutrient management practices using the 4 R's (right fertilizer source, right rate to match crop needs, right time for crop demand and right place for maximum uptake) are crucial (Van Es et al., 2019). However, a transition from reliance on highly water-soluble fertilizers to greater internal nutrient cycling has to be part of future CA development. There is evidence that, if soil organic matter content is high, soil N-supplying capacity far exceeds that of high-yielding corn (Osterholz et al., 2017). Increasing organic matter content with CA and ensuring greater crop uptake of mineralized nitrogen is, therefore, a promising avenue to reduce N-fertilizer needs. Residual effects of legumes in crop rotations can meet all or a large part of N-demands of high-yielding corn in CA, even for two consecutive seasons (Fox and Piekielek, 1988; Levin et al., 1987; Yost et al., 2014). Leguminous cover crops can function as green manures that supply all nitrogen needs of following crops and have a crop rotation effect boosting yields further (Clark et al., 1997; Ku et al., 2018; Parr et al., 2011). Therefore, greater efforts are needed to find promising legumes that fit in CA cropping systems, which also includes adjustments to cash crop rotations and farmer expectations. Cover crops also take up soluble nutrients from the subsoil and cycle those nutrients to the topsoil, where the nutrients are released upon degradation of the cover crop (Fraser et al., 2013; Kladivko et al., 2014; Malone et al., 2014; Staver and Brinsfield, 1998; Wang and Weil, 2018). The high availability of phosphorus in surface soil has contributed to soluble phosphorus losses in runoff and drain tile that is contributing to the eutrophication of Lake Erie (Baker et al., 2017). A better understanding of nutrient cycling and nutrient transformation in CA cropping systems is needed to reduce fertilizer nutrient use and minimize environmental impacts.

6.4 Water-use efficiency

There is a great need to improve water-use efficiency in rainfed and irrigated agriculture in the United States as farmers adapt to a changing climate. CA has an important role to play in 'weatherizing' cropping systems by increasing infiltration, improving soil water holding capacity, reducing evaporation and increasing crop transpiration. CA is helping to transform cropping systems in dryland areas. But the benefits of no-tillage need to be combined with increasing crop diversity in crop rotations. To rely only on no-till to improve water-use efficiency (for example by replacing tillage with chemical fallow in wheat-fallow systems) is shortsighted thinking because of other aspects of crop management such as weed control, nutrient management and soil health. Peterson and Westfall (2005) showed that no-till practices enabled cropping diversification of the traditional wheat-fallow system increasing water-use efficiency 30% in the western central Plains, boosting annualized grain and crop residue yields 75–100% and improving farmer's profit 25–45%, while also increasing carbon return to the soil. Similarly, water-use efficiency in North Dakota increased with increasing crop rotation diversity: corn–soybean–spring wheat > corn–soybean–oat/pea hay > corn–soybean–oat/pea hay companion seeded with alfalfa > corn–soybean>continuous corn (Pikul Jr. et al., 2012). Forage systems have high precipitation use efficiency (on a price-received basis) in the southern Great Plains (Nielsen et al., 2005). Livestock needs to be part of the equation to utilize (cover) crops that cannot be grown for grain. Tanaka et al. (2005) showed how integrated crop-livestock systems including deep-rooted pigeon pea [*Cajanus cajan* (L.) Millsp.] after winter wheat improved cropping system water-use efficiency in the southern Great Plains. They showed that the water-use efficiency of winter wheat was improved by 18–56% by including a broadleaf crop in the rotation in the central Great Plains. Flax yields increased three-fold if rotated with safflower (*Carthamus tinctorius* L.) instead of grown in monoculture in the northern Great Plains. Grain amaranth (*Amaranthus* spp.), an alternative crop for the northern Great Plains, was found to respond to water stress by increasing its rooting depth (Johnson and Henderson, 2002).

7 Conclusion

In a way, U.S. agriculture has come full circle from a past of indigenous farming practices based on CA to a period of intense exploitation and environmental degradation based on plow-based, monoculture crop production to a future that goes back to the roots of ecologically-based farming. Our modern society is different from those existing before European colonization, being highly urbanized and science-based, so modern CA systems will be different from those before Columbus. The United States has the benefit of a solid institutional support system of research, financial, technical and educational support that helped birth the CA movement and continues to support discovery, adaptation and implementation of conservation practices. However, this future is under threat due to continued lack of diversity in crop rotations and a mere 5% of cover crop use on planted acres resulting in extensive periods where the soil is devoid of living above- and below-ground vegetation. Herbicide-resistant crops led to easy weed control in no-till production, but nature is throwing farmers and agricultural scientists a curve-ball of 'monster weeds' resistant to several modes of action threatening a reversion to plow-based farming such as that experienced in cotton production and advocated by some weed scientists.

At the same time, GMO crops have led to reduced diversity in crop selections because of the cost of commercialization of GMO crops. Lack of herbicide options in many crops is also inhibiting widespread adoption of CA in specialty crops that resulted in low no-till adoption in states such as California and Florida. There is a lack of affordable planting and harvesting equipment that can plant and harvest multiple crop types or the availability of custom service operators offering these services to farmers. But in the end crop diversity will not increase with the dominance of the 'Big Five' – corn, soybeans, wheat, hay and cotton. There are hopeful signs in research and development that employ ecological principles of continuous no-tillage, crop diversity and continuous organic cover to guarantee a bright future for CA. But as has been the case since the beginning of the CA movement, there are also other forces threatening a return to plow-based, non-diverse crop production.

8 Acknowledgments

Special thanks to Keith R. Diehl for creating Figs 5 and 6.

9 Where to look for further information

An introduction to soil health improvement with CA is:

Duiker, S. W., Myers J. C. and Blazure, L. C. (2015), *Soil Health in Field and Forage Crop Production,* Harrisburg, PA: USDA-NRCS.

A great resource, though a bit dated, is:

Reeder, R. (Ed) (2000), *Conservation Tillage Systems and Management,* MWPS-45 2nd Edition, Ames, IA: Midwest Plan Service.

9.1 Seminal articles/books

Although dated, I consider this book to be among the best introduction to principles of CA:

- Phillips, R. E. and Phillips, S. H. (1984), *No-Tillage Agriculture: Principles and Practices,* New York: Van Nostrand Reinhold Company Inc.

This study gives a great overview of CA in the United States:

- Claassen, R., Bowman, M., McFadden, J. Smith, D. and Wallander, S. (2018), *Tillage Intensity and Conservation Cropping in the United States*, Washington, DC: Economic Information Bulletin Number 197, USDA-ERS.
- The article by Defelice et al. (2006) is a great summary of yields with conventional and no-tillage in research trials with corn and soybeans in different areas of the United States and Canada.

Key societies, professional organizations or other websites:

- No-Till Farmer is the go-to place for practical information and trends in CA among practitioners. This organization publishes a journal and comprehensive Conservation Tillage Guides. Its focus is on the Midwest.
- No-Till on the Plains is another great organization for CA information in the Great Plains Region. Both No-Till Farmer and No-Till on the Plains organize annual conferences, mostly for practitioners.
- *Soil and Tillage Research* and the *Journal of Soil and Water Conservation* are both good sources of scientific information about CA.

10 References

Aguilar, J., Gramig, G. G., Hendrickson, J. R., Archer, D. W., Forcella, F. and Liebig, M. A. (2015). Crop species diversity changes in the United States: 1978–2012, *PLoS ONE* 10(8), e0136580. DOI: 10.1371/journal.pone.0136580.

Alston, L. J. (1983). Farm foreclosures in the United States during the interwar period, *J. Eco. History* 43(4), 885-903.

Anderson, R. L. (2007). Managing weeds with a dualistic approach of prevention and control, *Agron. Sustain. Dev.* 27(1), 13–18.

Anderson, R. L., Stymiest, C. E., Swan, B. A. and Rickertsen, J. R. (2007). Weed community response to crop rotations in western South Dakota, *Weed Technol.* 21(1), 131–135.

Baker, J. M., Ochsner, T. E., Venterea, R. T. and Griffis, T. J. (2007). Tillage and soil carbon sequestration: what do we really know?, *Agric. Ecosyst. Environ.* 118(1-4), 1-5.

Baker, D. B., Johnson, L. T., Confesor, R. B. and Crumrine, J. P. (2017). Vertical stratification of soil phosphorus as a concern for dissolved phosphorus runoff in the Lake Erie basin, *J. Environ. Qual.* 46(6), 1287–1295.

Bennett, H. H. (1934). Soil erosion: a national menace, *Sci. Mon.* 39, 385–404.

Bigelow, D. P. and Borchers, A. (2017). *Major Uses of Land in the United States*, vol. 2012, Washington, DC: Economic Research Service, United States Department of Agriculture, EIB-178.

Braxton, J. (2009). Saving the Ogallala aquifer, *Sci. Am.* Special Editions 19, 1s, 32–39 DOI: 10.1038/scientificamericanearth0309-32.

Camargo, G. G. T., Ryan, M. R. and Richard, T. L. (2013). Energy use and greenhouse gas emissions from crop production using the farm energy analysis tool, *BioScience* 63(4), 263–273.

CAST (2012). *Herbicide-Resistant Weeds Threaten Soil Conservation Gains: Finding a Balance for Soil and Farm Sustainability*, Issue Paper 49, Ames, IA: The Council for Agricultural Science and Technology (CAST).

Claassen, R., Bowman, M., McFadden, J., Smith, D. and Wallander, S. (2018). *Tillage Intensity and Conservation Cropping in the United States*, Washington, DC: USDA-ERS, Economic Information Bulletin Number 197.

Clark, A. J., Decker, A. M., Meisinger, J. J. and McIntosh, M. S. (1997). Kill date of vetch, rye, and a vetch-rye mixture: II. Soil moisture and corn yield, *Agron. J.* 89(3), 434–441.

Cook, B. I., Ault, T. R. and Smerdon, J. E. (2015). Unprecedented 21st century drought risk in the American Southwest and Central Plains, *Sci. Adv.* 1(1), e1400082. DOI: 10.1126/sciadv.1400082.

Craven, A. O. (1926). *Soil Exhaustion as a Factor in the Agricultural History of Virginia and Maryland*, vol XIII, 1st ed, Urbana, IL: University of Illinois Studies in the Social Sciences, University of Illinois.

Dabney, S. M., Wilson, G. V., McGregor, K. C. and Foster, G. R. (2004). History, residue, and tillage effects on erosion of loessial soil, *Trans. ASAE* 47, 767–775.

Daryanto, S., Wang, L. and Jacinthe, P. A. (2017). Impacts of no-tillage management on nitrate loss from corn, soybean and wheat cultivation: a meta-analysis, *Sci. Rep.* 7(1), 12117. DOI: 10.1038/s41598-017-12383-7.

Defelice, M. S., Carter, P. R. and Mitchell, S. B. (2006). Influence of tillage on corn and soybean yield in the United States and Canada, *Crop Manag.* 5(1) DOI: 10.1094/CM-2006-0626-01-RS.

Diamond, J. (2005). *Collapse: How Societies Choose to Fail or Succeed*, New York: Penguin Books.

Dobberstein, J. (2020). No-Till Lessons from 'Mudagageddon'. In: *No-Till Farmer Conservation Tillage Guide*, No-Till Farmer: Brookfield, WI, vol 49, 2nd ed, pp. 20–28.

Drache, H. M. (n.d.). The impact of John Deere's plow, Available at: https://www.lib.niu.edu/2001/iht810102.html.

Duiker, S. (2015). Soil physical properties and management. In: White, C., Barbercheck, M. and Curran, W. (Eds), *Penn State Organic Production Guide*, University Park, PA: Pennsylvania State University, 17–42.

Duiker, S. W. (2019). Soil management. In: Williamson, J. and Lingenfelter, D. D. (Eds), *The Agronomy Guide 2019-2020*, University Park, PA: Penn State Extension, pp, 1–18.

EPA (2016). *What Climate Change Means for North Dakota*, EPA 430-F-16-036, Washington, DC: United States Environmental Protection Agency.

Erisman, J. W., Galloway, J. N., Seitzinger, S., Bleeker, A., Dise, N. B., Petrescu, A. M. R., Leach, A. M. and De Vries, W. (2013). Consequences of human modification of the global nitrogen cycle, *Philos. Trans. R. Soc. London. B* 368(1621), 20130116. DOI: 10.1098/rstb.2013.0116.

Farm Bureau (2019). Prevent plantings set record in 2019 at 20 million acres, Available at: https://www.fb.org/market-intel/prevent-plantings-set-record-in-2019-at-20-million-acres.

Faulkner, E. H. (1943). *Plowman's Folly*, Norman, OK: University of Oklahoma Press.

Fox, R. H. and Piekielek, W. P. (1988). Fertilizer N equivalence of alfalfa, birdsfoot trefoil, and red clover for succeeding corn crops, *J. Prod. Agric.* 1(4), 313–317.

Frankel, J. (2018). Crisis on the high plains: the loss of America's largest aquifer: the Ogallala, *Univ. Denver Water Law Rev.*, Available at: http://duwaterlawreview.com/crisis-on-the-high-plains-the-loss-of-americas-largest-aquifer-the-ogallala/.

Franzluebbers, A. J. (2002). Soil organic matter stratification ratio as an indicator of soil quality, *Soil Till. Res.* 66(2), 95–106.

Fraser, P. M., Curtin, D., Harrison-Kirk, T., Meenken, E. D., Beare, M. H., Tabley, F., Gillespie, R. N. and Francis, G. S. (2013). Winter nitrate leaching under different tillage and winter cover crop management practices, *Soil Sci. Soc. Am. J.* 77(4), 1391–1401.

Givens, W. A., Shaw, D. R., Kruger, G. R., Johnson, W. G., Weiler, S. C., Young, B. G., Wilson, R. G., Owen, M. D. K. and Jordan, D. (2009). Survey of tillage trends following the adoption of glyphosate-resistant crops, *Weed Technol.* 23(1), 150–155.

Goering, C. E. (2008). A century of tractor development: 1907–2007, *Trans. ASABE* 51(2), 379–383.

Haystead, L. and Fite, G. C. (1955). *The Agricultural Regions of the United States*, Horman: University of Oklahoma Press.

Heap, I. (2014). Global perspective of herbicide-resistant weeds, *Pest Manag. Sci.* 70(9), 1306–1315.

Huggins, D. R., Buyanovsky, G. A., Wagner, G. H., Brown, J. R., Darmody, R. G., Peck, T. R., Lesoing, G. W., Vanotti, M. B. and Bundy, L. G. (1998). Soil organic C in the tallgrass prairie-derived region of the corn belt: effects of long-term crop management, *Soil Till. Res.* 47(3–4), 219–234.

Huning, L. S. and AghaKouchak, A. (2018). Mountain snowpack response to different levels of warming, *Proc. Natl Acad. Sci. U. S. A.* 115(43), 10932–10937.

Jack, W. T. (1946). *The Furrow and Us: Essays on Soil and Sentiment*, Philadelphia, PA: Dorrance & Company.

Johnson, B. L. and Henderson, T. L. (2002). Water use patterns of grain amaranth in the northern Great Plains, *Agron. J.* 94(6), 1437–1443.

Johnson, L. C. (1987). Soil loss tolerance: fact or myth?, *J. Soil Water Cons.* 42, 155–160.

Jones, O. R., Hauser, V. L. and Popham, T. W. (1994). No-tillage effects on infiltration, runoff, and water conservation on dryland, *Trans. ASAE* 37, 473–479.

Karlen, D. L., Wollenhaupt, N. C., Erbach, D. C., Berry, E. C., Swan, J. B., Eash, N. S. and Jordah, J. L. (1994). Crop residue effects on soil quality following 10-years of no-till corn, *Soil Till. Res.* 31(2–3), 149–167.

Kemper, W. D., Schneider, N. N. and Sinclair, T. R. (2011). No-till can increase earthworm populations and rooting depths, *J. Soil Water Cons.* 66(1), 13A–17A.

Kladivko, E. J., Kaspar, T. C., Jaynes, D. B., Malone, R. W., Singer, J., Morin, X. K. and Searchinger, T. (2014). Cover crops in the upper midwestern United States: potential adoption and reduction of nitrate leaching in the Mississippi River Basin, *J. Soil Water Cons.* 69(4), 279–291.

Kraehmer, H., Laber, B., Rosinger, C. and Schulz, A. (2014). Herbicides as weed control agents: state of the art: I. Weed control research and safener technology: the path to modern agriculture, *Plant Physiol.* 166(3), 1119–1131.

Ku, H. H., Jeong, C. and Colyer, P. (2018). Modeling long-term effects of hairy vetch cultivation on cotton production in Northwest Louisiana, *Sci. Total Environ.* 624, 744–752.

Lal, R. and Kimble, J. M. (1997). Conservation tillage for carbon sequestration, *Nutr. Cycl. Agroecosyst.* 49(1/3), 243–253.

Lal, R., Reicosky, D. C. and Hanson, J. D. (2007). Evolution of the plow over 10,000 years and the rationale for no-till farming, *Soil Till. Res.* 93(1), 1–12.

Levin, A., Beegle, D. B. and Fox, R. H. (1987). Effect of tillage on residual nitrogen availability from alfalfa to succeeding corn crops 1, *Agron. J.* 79(1), 34–38.

Liebman, M. and Gallandt, E. R. (1997). Many little hammers: ecological management of crop-weed interactions. In: Jackson, L. E. (Ed), *Ecology in Agriculture*, San Diego, CA: Academic Press, 291–343.

Luzar, J. E. (1988). Natural resource management in agriculture: an institutional analysis of the 1985 Farm Bill, *J. Econ. Issues* 22(2), 563–570.

Malone, R. W., Jaynes, D. B., Kaspar, T. C., Thorp, K. R., Kladivko, E., Ma, L., James, D. E. and Singer, J. Morin, X. K. and Searchinger, T. (2014). Cover crops in the Upper Midwestern United States: simulated effect on nitrate leaching with artificial drainage, *J. Soil Water Cons* 69, 292–305.

Mitchell, J. P., Singh, P. N., Wallender, W. W., Munk, D. S., Wroble, J. F., Horwath, W. R., Hogan, P., Roy, R. and Hanson, B. R. (2012). No-tillage and high-residue practices reduce soil water evaporation, *Calif. Agric.* 66(2), 55–61.

Moebius-Clune, B. N., Van Es, H. M., Idowu, O. J., Schindelbeck, R. R., Moebius-Clune, D. J., Wolfe, D. W., Abawi, G. S., Thies, J. E., Gugino, B. K. and Lucey, R. (2008). Long term effects of harvesting maize stover and tillage on soil quality, *Soil Sci. Soc. Am. J.* 72(4), 960–969.

Montgomery, D. R. (2007). Soil erosion and agricultural sustainability, *Proc. Natl Acad. Sci. U. S. A.* 104(33), 13268–13272.

Mt. Pleasant, J. (2015). A new paradigm for pre-Columbian agriculture in North America, *Early Am. Stud.* 13(2), 374–412.

National Agricultural Statistics Service (NASS) (2019). *Acreage*, ISSN: 1949-1522, Available at: https://www.nass.usda.gov/Publications/Todays_Reports/reports/acrg0619.pdf, Washington, DC: National Agricultural Statistics Service, United States Department of Agriculture.

Nielsen, D. C., Unger, P. W. and Miller, P. R. (2005). Efficient water use in dryland cropping systems in the Great Plains, *Agron. J.* 97(2), 364–372.

Ogle, S. M., Alsaker, C., Baldock, J., Bernoux, M., Breidt, F. J., McConkey, B., Regina, K. and Vazquez-Amabile, G. G. (2019). Climate and soil characteristics determine where no-till management can store carbon in soils and mitigate greenhouse gas emissions, *Sci. Rep.* 9(1), Article number 11665. DOI: 10.1038/s41598-019-47861-7.

Olson, K. R. and Al-Kaisi, M. M. (2015). The importance of soil sampling depth for accurate accounting of soil organic carbon sequestration, storage, retention and loss, *CATENA* 125, 33–37.

Osterholz, W. R., Rinot, O., Liebman, M. and Castellano, M. J. (2017). Can mineralization of soil organic nitrogen meet maize nitrogen demand?, *Plant Soil* 415(1–2), 73–84.

Parr, M., Grossman, J. M., Reberg-Horton, S. C., Brinton, C. and Crozier, C. (2011). Nitrogen delivery from legume cover crops in no-till organic corn production, *Agron. J.* 103(6), 1578–1590.

Peterson, G. A. and Westfall, D. G. (2005). Managing precipitation use in sustainable dryland agroecosystems, *Ann. Appl. Biol.* 144, 127–138.
Pikul Jr., J. L., Osborne, S. L. and Riedell, W. E. (2012). Corn yield and nitrogen- and water-use under no-tillage rotations, *Commun. Soil Sci. Plant Anal.* 43(21), 2722–2734, DOI: 10.1080/00103624.2012.719980.
Reed, H. K., Karsten, H. D., Curran, W. S., Tooker, J. F. and Duiker, S. W. (2019). Planting green effects on corn and soybean production, *Agron. J.* 111(5), 2314–2325.
Reicosky, D. C. (1997). Tillage-induced CO_2 emission from soil, *Nutr. Cycl. Agroecosyst.* 49(1/3), 273–285.
Roosevelt, F. D. (1937). Letter to all state governors on a uniform soil conservation law, *Am. Presidency Proj.*, Available at: https://www.presidency.ucsb.edu/node/209378.
Russell, D. A. and Williams, G. G. (1977). History of chemical fertilizer development, *Soil Sci. Soc. Am. J.* 41(2), 260–265.
Shaw, D. R., Barrett, M., Schroeder, J., Asmus, A. B., Ervin, D., Jussaume, R. A. and Coble, H. (2018). Critical steps in combating herbicide resistance: our view, *Weed Sci.* 66(5), 559–561.
Six, J., Elliott, E. T. and Paustian, K. (2000). Soil macroaggregate turnover and microaggregate formation: a mechanism for C sequestration under no-tillage agriculture, *Soil Biol. Biochem.* 32(14), 2099–2103.
Staver, K. W. and Brinsfield, R. B. (1998). Using cereal grain winter cover crops to reduce groundwater nitrate, *J. Soil Water Cons* 53, 230–240.
Steinbeck, J. (1939). *The Grapes of Wrath*, New York: The Viking Press.
Tanaka, D. L., Anderson, R. L. and Rao, S. C. (2005). Crop sequencing to improve use of precipitation and synergize crop growth, *Agron. J.* 97(2), 385–390.
Tipping, E., Somerville, C. J. and Luster, J. (2016). The C:N:P:S stoichiometry of soil organic matter, *Biogeochemistry* 130(1–2), 117–131.
Trimble, S. W. (1974). *Man-Induced Soil Erosion on the Southern Piedmont, 1700–1970*, Ankeny, IA: Soil Conservation Society of America.
Triplett Jr., G. B. and Dick, W. A. (2008). No-tillage crop production: a revolution in agriculture!, *Agron. J.* 100(S3), S-153–S-165.
Truman, C. C., Reeves, D. W., Shaw, J. N., Motta, A. C., Burmester, C. H., Raper, R. L. and Schwab, E. B. (2003). Tillage impacts on soil property, runoff, and soil loss variations from a Rhodic Paleudult under simulated rainfall, *J. Soil Water Cons.* 58, 258–267.
Truman, C. C., Shaw, J. N. and Reeves, D. W. (2005). Tillage effects on rainfall partitioning and sediment yield from an ultisol in central Alabama, *J. Soil Water Cons.* 60, 89–98.
Unger, P. W. and Baumhardt, R. L. (1999). Factors related to dryland grain sorghum yield increases: 1939–1997, *Agron. J.* 91(5), 870–875.
USDA (1978). *Palouse Cooperative River Basin Study*, Washington, DC: United States Department of Agriculture, Economics, Statistics, and Cooperatives Service, Forest Service, Soil Conservation Service.
VandenBygaart, A. J. (2016). The myth that no-till can mitigate global climate change, *Agric. Ecosyst. Environ.* 216, 98–99.
Van Es, H., Ristow, A., Nunes, M. R., Schindelbeck, R., Sela, S. and Davis, M. (2019). Nitrate leaching reduced with dynamic-adaptive nitrogen management under contrasting soils and tillage, *Soil Sci. Soc. Am. J.* 84, 220–231.
Van Grinsven, H. J. M., Bouwman, L., Cassman, K. G., Van Es, H. M., McCrackin, M. L. and Beusen, A. H. W. (2015). Losses of ammonia and nitrate from agriculture and their

effect on nitrogen recovery in the European Union and the United States between 1900 and 2050, *J. Environ. Qual.* 44(2), 356–367.

Wang, F. and Weil, R. R. (2018). The form and vertical distribution of soil nitrogen as affected by forage radish cover crop and residual side-dressed N fertilizer, *Soil Sci.* 183(1), 22–33.

West, T. O. and Post, W. M. (2002). Soil organic carbon sequestration rates by tillage and crop rotation: a global data analysis, *Soil Sci. Soc. Am. J.* 66(6), 1930–1946.

Wuebbles, D. J., Fahey, D. W., Hibbard, K. A., DeAngelo, B., Doherty, S., Hayhoe, K., Horton, R., Kossin, J. P., Taylor, P. C., Waple, A. M. and Weaver, C. P. (2017). Executive summary. In: Wuebbles, D. J., Fahey, D. W., Hibbard, K. A., Dokken, D. J., Stewart, B. C. and Maycock, T. K. (Eds), *Climate Science Special Report: Fourth National Climate Assessment*, vol I, Washington, DC: U.S. Global Change Research Program, 12–34. DOI: 10.7930/J0DJ5CTG.

Yost, M. A., Morris, T. F., Russelle, M. P. and Coulter, J. A. (2014). Second-year corn after alfalfa often requires on fertilizer nitrogen, *Agron. J.* 106(2), 659–669.

Chapter 4

Conservation Agriculture in Central America, the Caribbean and Mexico

Jose R. Benites, Independent Consultant on Land and Water, and Conservation Agriculture, Peru; Theodor Friedrich, Retired Food and Agriculture Organization of the United Nations (FAO) Expert on Conservation Agriculture, Germany; Roberto Azofeifa Rodríguez, Head of the Department of Agro-environmental Production of the National Directorate of Agricultural Extension, Costa Rica; Ariel Espinoza Salinas, Catholic Relief Services (CRS), Nicaragua; J. C. Juárez López, FAO Technical Officer, Guatemala; Carlos Andrés Zelaya Elvir, Consultant, Food and Agriculture Organization of the United Nations (FAO) (retired), Honduras; and Jose Escobar, Consultant ASA, CRS, El Salvador

1 Introduction
2 Conservation Agriculture in Central America
3 Conservation Agriculture in the Caribbean
4 Mexico
5 Conclusion
6 Where to look for further information
7 References

1 Introduction

In Central America, the Caribbean and Mexico,[1] there is severe pressure on the soil due to the practices of subsistence agriculture and cash crop agriculture for export, both related to an economy highly dependent on limited economic resources and forced intensive land use. Years of tillage have led to a decrease in yields and soil degradation in many places.

Soil erosion is the most significant ecological restriction to sustainable agricultural production in hilly areas of the region. Unsustainable practices such as conventional inversion tillage, monocropping and heavy use of chemical inputs are leading to land and environmental degradation, decreasing soil fertility, rapidly declining production levels and land abandonment and desertification. The leading cause of these negative consequences is the practice of plough-based or hoe-based agriculture, coupled with poor soil health management and limited crop diversity.

1 Referred to as 'the region' in this chapter.

http://dx.doi.org/10.19103/AS.2021.0088.16

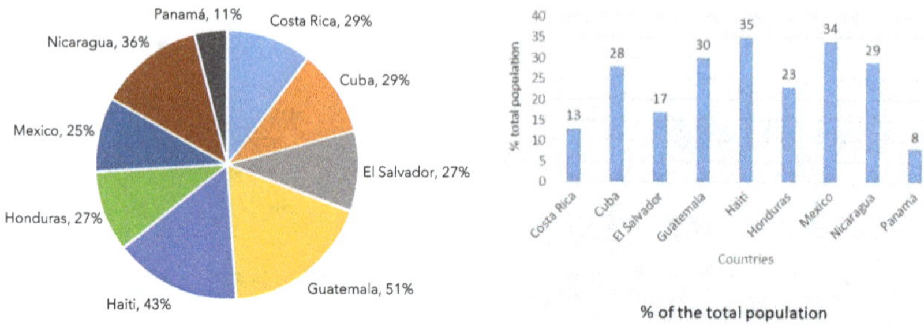

Figure 1 Percentages of the territory and the total population of Costa Rica, Cuba, El Salvador, Guatemala, Haiti, Honduras, Mexico, Nicaragua and Panama affected by severe land degradation (Bai et al., 2008).

Figure 1 shows severe land degradation[2] affects large proportions of the land areas of the countries: Guatemala (51%); Haiti (43%); Nicaragua (36%); Cuba and Costa Rica (29%); El Salvador and Honduras (27%); Mexico (25%); and Panama (11%). Land degradation also detrimentally affects large proportions of the population: more than 30% of people in Haiti, Mexico and Guatemala; between 25% and 29% of the population in Nicaragua, Cuba and Honduras, and between 8% and 17% of the population in El Salvador, Costa Rica and Panama.

Over the past 50 years, the traditional approach to controlling soil erosion was based on the promotion of physical structures but, sadly, the impacts have been minimal. The physical works achievements of many development programmes, often considered a success during their life-time, have frequently disappeared a few years after the staff have departed. It is now apparent that either the procedures used to promote land use development or the new technologies employed were not acceptable to the farmers for whom they were introduced. So, progress has not been sustainable (Shaxson, 1999).

Within the past 30 years, some isolated examples have emerged in the region of more successful procedures and technologies that have led to acceptable and sustainable agricultural development. Field projects in the region now provide proportionately more emphasis on helping farmers to improve land care and soil health and fewer efforts to combat erosion directly through physical structures. This more careful approach to land use, such as through the practice of Conservation Agriculture (CA), provides a more effective response to an old problem as it recognises the desire of farmers explicitly to increase their yields and income as they stabilise or reverse the process of resource depletion and degradation (Barber, 1999).

2 Land degradation comprised six classes: water erosion, wind erosion, soil fertility decline, salinization, waterlogging and lowering of the water table.

The CA principles apply to the diversity of agricultural production systems resulting from a range of topography, soil types and climates. These systems vary not only in the crops cultivated or the animals raised but also in the plot or farm sizes, types and intensities of management, the arrangement in the landscape, orientation for export or internal markets, agroecological conditions and their location. CA also provides governments opportunities to harmonise specific national objectives - notably better management of natural resources and the development of sustainable agriculture and livelihoods - with the primary purpose of benefiting rural families.

However, information regarding the positive effects of CA systems in the region is mainly found in dispersed technical reports of programmes or technical cooperation projects, such as those of the United Nations Food of Agriculture Organization (FAO), Rome, Italy; the Inter-American Institute for Cooperation on Agriculture (IICA), San Jose, Costa Rica; the Tropical Agricultural Research and Higher Education Center (CATIE), Turrialba, Costa Rica; the International Maize and Wheat Improvement Center (CIMMYT), Mexico; and the Programme for Sustainable Agriculture in Central America Hillsides (PASOLAC), San Salvador, El Salvador, often supported by bilateral donors. A few research data and technological notes are published in grey literature and non-referenced publications, training manuals, newsletters, pamphlets and others. The main problem is that there is a shortage of statistics of areas under the different systems, in particular, areas under CA.

Farmer organisations are the main instrument to generate and spread improved knowledge on sustainable land use. Eventually, this has led to mobilising public, private and civil sector support all of which have been vital in promoting the adoption of CA, especially in Mexico, for example through the associations of producers of maize and associated crops in different regions of the country with the support of Mexico's Secretariat of Agriculture and Rural Development (SAGARPA).

The purpose of the chapter is to provide an appreciation of the actual level of CA adoption in the countries of the region.

2 Conservation Agriculture in Central America

2.1 Costa Rica

Costa Rica is a country with a tropical climate and fragmented relief. As the country is a narrow strip of land of 51 100 km^2 with lowlands on its coastlines and a chain of three mountain ranges towards the centre, its exposure to the influence of the Caribbean Sea from the east and the Pacific Ocean from the west, make it a particularly rainy country, and therefore vulnerable to soil degradation by water erosion. For agricultural production to feed a growing population, one of the most significant challenges is the protection of soils against degradation by water erosion and loss of soil fertility and organic matter.

Costa Rica's efforts to produce food and, at the same time, protect its soils go back to the early 1940s. The Inter-American Technical Service for Agricultural Cooperation (STICA) promoted soil conservation based on physical structures between 1942 and 1954 aimed at reducing runoff and soil erosion. The main measures included terraces, irrigation ditches, the establishment of living vegetation barriers, contour bunding and planting and land levelling. These approaches are costly and challenging for producers to establish and maintain.

Since then, the efforts that the country has made through the Agricultural Extension Programme, supported by various projects with international cooperation, have been increasing and diverse in approaches and strategies to improve productivity. However, despite the efforts made, which represent considerable investments from both national public funds and international cooperation resources, the country continues to face the great challenge of protecting its soils against degradation by water erosion, compaction and loss of organic matter.

In the 1990s, the Government of Costa Rica, with the technical cooperation of the FAO and the financial contribution of the Government of the Netherlands, carried out several projects in the fields of CA systems, agroforestry, comprehensive rural development and capacity building for the sustainable development of producer organisations. Where severe soil compaction was present, it was necessary to undertake the deep ripping of the soil before initiating the transformation of the production system to CA (Fig. 2).

The principles of CA are at the base of soil fertility management because they consider the complex interactions between soil, water and vegetation. The promotion of new low-cost and easily accessible technical options offered highly relevant experiences. In addition to the results achieved in various fields of production, skilled extensionists served as the basis for the proposal and approval of land management and incentives under the approach of 'producing conserving and conserving producing'. In this CA-based approach, soil and water conservation is the key to the sustainability of production systems.

In the first decade of the twenty-first century, the development of the agricultural sector underwent changes aimed at strengthening sustainable and organic production. Through the application of good farming practices and the certification of production processes under various standards, a variety of productive activities involving coffee, bananas, pineapples and other crops, oriented mainly towards international markets, made progress in terms of competitiveness. The introduction of policies that recognised environmental

Figure 2 Ripping to alleviate soil compaction before the establishment of CA, 2020 (R. Azofeifa).

benefits in production systems give incentives for investment in environmentally beneficial and sustainable production projects (including organic production).[3]

As well as the government declaration that Costa Rica will become a carbon-neutral country in 2021, various initiatives across different sectors were announced during the 2006–2010 administration, aimed at reducing and compensating for greenhouse gas emissions, as an innovative way of improving competitiveness.

The main progress in terms of climate change mitigation has been achieved through integrated silvopastoral systems,[4] as well as integrating the CA principles into perennial crops such as coffee. In annual crops, a few large-scale farmers are applying CA-related no-till farming practices, but without any specific government support programmes.

2.2 El Salvador

Experience with CA in El Salvador dates back to 1970, when the Ministry of Agriculture launched a technology improvement programme for the maize-sorghum system in the Guaymango municipality, from 1973 to 1983. The plan was based on halting stubble and crop biomass burning, uniform distribution of biomass on the ground surface, proper use of fertilisers and herbicides, use of new varieties of maize, construction of ditches along the contour lines to

3 Costa Rica has an Agricultural Extension Programme made up of more than 250 people who provide technical support in more than 90 offices of the Ministry of Agriculture and Livestock throughout the country.

4 Silvopastoral system comprises the practice of integrating trees, forage crops and the grazing of domesticated animals in a mutually beneficial way. The system follows CA principles.

capture and manage runoff, living and dead vegetation barriers, and planting on contours.

The maize-sorghum system produces no less than 10 t ha^{-1} of stubble biomass in one agricultural calendar year. Although grazing consumed 1-2 t ha^{-1} $month^{-1}$, there was a good volume of stubble for soil protection at the beginning of the following agricultural year. The most important practices for soil conservation management was the non-burning of the stubble biomass and its retention on the soil surface without tillage

One of the positive effects has been a reduction in the incidence of pests and diseases. One of the insect species, the white beetle larva (*Phyllophaga* sp.), a soil pest that had previously been severe in maize crops, is still in the fields but it is no longer a problem.

The Guaymango experience demonstrated that soil and water conservation practices impact producers' socioeconomic systems by improving yields. Table 1 shows that maize yield went from 0.97 to 3.25 t ha^{-1} and sorghum from 0.70 to 2.1 t ha^{-1}, over 15 years.

Between 1994 and 1999, National Center for Agricultural and Forestry Technology 'Enrique Álvarez Córdova' (CENTA) implemented the Project 'Sustainable agriculture in hillside areas', with technical support from FAO and with funds from the Government of the Kingdom of the Netherlands (CENTA-FAO, 2002). The technical proposal of the project followed the principles of CA, which emphasised soil cover with harvest stubble, pasture management, cover crops such as green manures and crop diversification to promote changes in the use of the land. The project placed greater emphasis on the chemical degradation of the soil. For this reason, it was necessary to highlight the integrated management of soil fertility, including chemical, physical and biological degradation (Barber, 1996). The result of these practices was the increased biomass production and soil mulch cover, which improved the conditions of the structure, porosity and aeration of the soil to facilitate root development of plants and achieve increases in infiltration, storage and retention of water for plants (Vieira, 2000).

According to Argueta (1996), the average degree of soil mulch cover of the primary grain areas was 72%, sufficient to keep the risk of erosion at low

Table 1 Evolution of maize and sorghum yields (t ha^{-1}). CA in the municipality of Guaymango, El Salvador (Osorio, n.d.)

Year	Maize	Sorghum
1974	0.97	0.70
1978	2.34	1.50
1983	3.25	2.08
1989	3.25	2.10

levels in the hillside conditions of El Salvador. Figure 3 presents the relationship between the degree of soil cover as a function of the biomass amounts of the maize-sorghum system.

In 5 years of the FAO-Laderas project (2002a), the average soil cover achieved was 55%, which was a lower value than the one estimated by Argueta (1996) in his research as optimal (72%). The consumption of stubble biomass by cattle and burning was the leading cause for the reduction of optimal soil cover, indicating that more extended periods are needed for changes to take place for better management of the stubble biomass and to solve the burning problems and primary consumption by livestock. The amount of immobilised carbon was between 2000 kg ha^{-1} and 2500 kg ha^{-1}. It is estimated that around 60% of the nutrients absorbed by maize, sorghum or bean plants remain in the stubble biomass. According to the calculated macronutrient contents in these stubbles, the estimate of the accumulated amounts of nitrogen, phosphorus, and potassium per hectare are shown in Fig. 4. For its calculation, the data of the average contents of macronutrients in the stubble biomass of these crops were taken, as an average of dozens of samples from different areas of El Salvador, including six micro-watersheds evaluated where the study was carried out.

With 7.0 t ha^{-1} of maize and sorghum stubble biomass left as soil cover during the dry months of the year, it is estimated that there would be 3150 kg ha^{-1} of carbon in the biomass (Fig. 5).

The FAO-Laderas Project as the first to place CA in a comprehensive sustainable development proposal for small hillside producer (FAO 2002b). This project allows them to protect their soils, achieve better crop yields, better

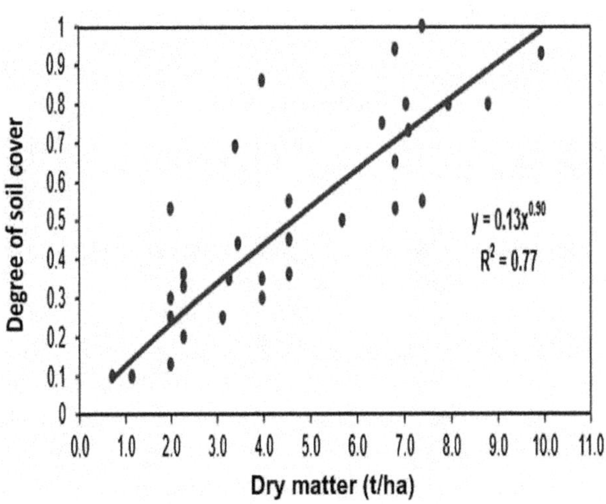

Figure 3 Relationship between the weight of dry biomass of maize and sorghum stubble and degree of ground cover. Reproduced from Argueta (1996).

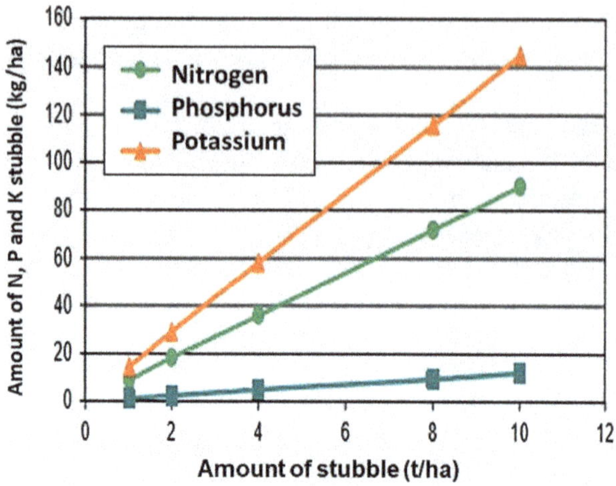

Figure 4 Estimated values of macronutrient contents (NPK) in different quantities of maize and sorghum stubble.

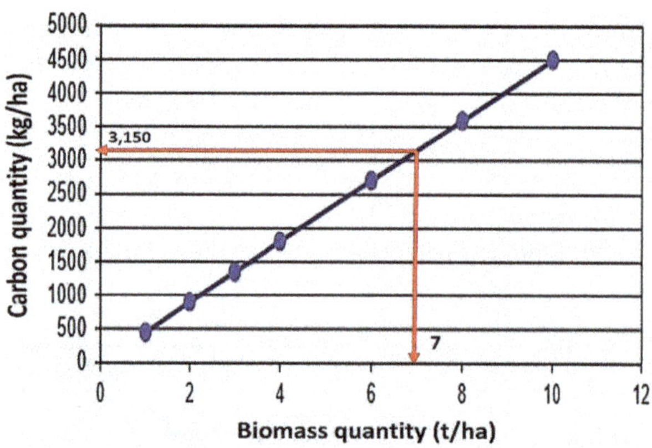

Figure 5 Estimated organic carbon amount immobilised in the stubble biomass as a function of the original amount.

soil quality with higher infiltration rates and storage of water in the soil profile, thus reducing the vulnerability of crops to drought.

With the previous considerations, a first country estimate of the CA areas and crops developed is shown in Table 2. An area under CA is considered one that does not disturb the soil, holds stubble on the surface and practices associations or rotations with cereals and legumes (beans, Canavalia or another green manure).

The CRS Soil and Water for Agriculture (ASA) programme have generated evidence on the impact of CA concerning water retention after 4 years of implementation with staple grain systems compared to plots traditionally

Table 2 Estimated values of areas of CA cropping systems areas, El Salvador

Cropping systems	Area (ha)	Sources of information
Maize-beans, Maize-sorghum	300	• PAES (2007)
Maize-beans, Maize-sorghum	1500	• Osorio (n.d.)
Maize-beans, Maize-sorghum	600	• Casares (2020)
Maize-beans, Maize-sorghum	300	• Rodriguez (2020)
Maize-beans, Maize-sorghum	100	• Chicas (2020)
Maize-beans, Maize-sorghum	250	• Alberto (2020)
Improved pastures	1000	• Alberto (2020); Chicas (2020)
Total	4050	

managed by producers. The traditional management of the producers consists of managing the soils with very little coverage of harvest stubble biomass, mowing their plots and making inefficient use of agricultural inputs. Figure 6 shows the average results of soil moisture measurements during the 2018 drought in maize and beans crops. In summary, the CA-ASA plots capture more water; there is 4.2% more water available in maize crops and 2.1% in beans. The difference of 4.2% in soil moisture, in the case of maize cultivation, is equivalent to 400 m^3 of water per hectare in the CA-ASA plot. The results also showed that the physical characteristics of the soil, such as structure and porosity, improved.

The average results of the sampling of worms in the corn crop in the year 2017 was 92 worms in the ASA plots that include good management practices of harvest stubble and use of green manures planted in association with corn; against 51 worms in a traditional plot that manages very little stubble biomass. The same trend was shown in 2018 in favor of the ASA plots (Fig. 7).

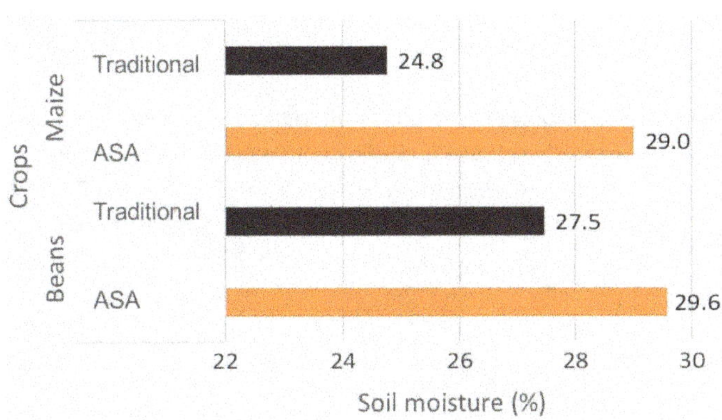

Figure 6 Average soil moisture in maize and bean crops under traditional soil inversion-based agriculture and CA during the 2018 drought period. El Salvador.

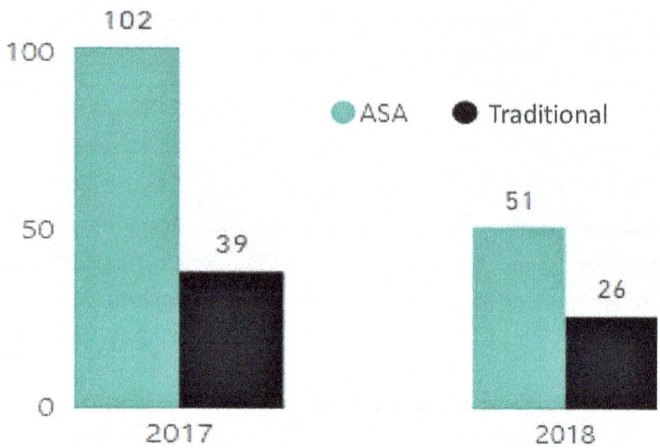

Figure 7 Earthworm population found in the ASA and traditional plots in the maize crop, during the years 2017 and 2018.

The samplings were carried out in an area of 30 cm^3, during the years 2017 and 2018. The results achieved in the ASA field are the product of managing an average of 17 t ha^{-1} of green manure biomass and crop stubble biomass on the soil surface.

The increase in the populations of worms in the soil plays a significant role in the improvement of its physical-chemical characteristics, by helping to mix the soil with organic matter, accelerate the humus formation process, facilitate water infiltration and aeration and participate in the decomposition dynamics of organic matter, thereby improving soil fertility (Fig. 7).

Finally, a significant challenge in El Salvador for CA adoption is land tenure. At least 40% of the producers who plant the staple grains rent the land and, in most cases, do so through negotiations they carry out with ranchers, who agree to rent the land under the condition that the renters plant maize and sorghum crops. The harvest stubbles and biomass of these crops are grazed by the ranchers' cattle, which causes soil compaction and accelerates the processes of soil degradation

The appropriate approach to implementing successful CA projects for restoring soil and landscape health and productivity must therefore have long-term planning horizons of 10–15 years, combined with appropriate public policies which can facilitate expansion and generate economic resources for national programmes. Crop rotations also require adaptation to the traditional relay production system involving staple crops of maize and sorghum and maize and beans.

2.3 Guatemala

The European Union and the FAO decided in November 2010 to carry out a combined technical mission for CA and mechanisation in Guatemala. The objective was to demonstrate the suitability of CA in the current situation of the hillside terrain for the cultivation of basic grains and vegetables and to establish alternatives of improved management through the application of complementary CA-based Good Agricultural Practices (FAO and EU, 2010).

The beneficiary population of the programme was more than ten thousand families who depended on agricultural production of maize, beans, coffee, cocoa, achiote, cardamom, peanuts and chilli as a source of food and some income. In some cases, other commercial foods, such as vegetables and dairy products, were also produced.

Agricultural work is often not profitable, and most people only work on the land to subsist. Guatemala is a country rich in natural resources. However, their deterioration is alarming. The damaging agricultural practices applied by producers in the use of land, water and forests are undoubtedly one of the leading causes of the degradation of these natural resources. For this reason, many rural people must leave their homes and work for private agricultural companies and on oil exploitation to earn an income to purchase food.

In recent years, a topic under discussion and included in the government's plan is climate change. When the situation was analysed, it was found that the cutting and burning of the forests is the leading cause of greenhouse gas emissions (Fig. 8). Without being very aware of the effects of this phenomenon, small-scale farmers continue to carry out the harmful practice of stubble biomass burning before establishing their crops, which produce less and less and require more external inputs.

Figure 8 Traditional maize growing areas in Guatemala before planting (J. Juárez).

The project noted that the target areas had unprotected soils due to uncontrolled grazing, inappropriate use of stubble, burning of crop residues and other biological waste, inappropriate use of pesticides and low investment capacity of subsistence agriculture.

Also, the low productivity of the staple grain crops observed was due to poor agronomic management, particularly in terms of plant spacing, degeneration of the native varieties, poor plant nutrition and post-harvest losses.

Significant progress has been made in increasing short-term productivity by making improvements in these areas through the adoption and promotion of CA principles. The principles of CA applied were: do not disturb or compact the soil by trampling by animals or people, protect the soil with biomass cover and diversified crop rotation. The implementation of the CA principles demonstrated an improvement in the management and conservation of soil, water and inputs, generating a diversification of production and income distribution, and also a higher investment capacity for farmers (FAO, 2016) (see Fig. 9a–d).

To date, 10000 hectares of CA-based systems comprising maize and beans have been established in five municipalities in the northern part of the

Figure 9 (a) Maize growing area with the CA principles adopted by the project farmer (J. Juárez); (b) Farmers seeding maize into heavy stubble using 'chuzos' or dibble sticks (J. Juárez); (c) Rural women being instructed on the use of hand planter for crop planting in CA (J. Juárez); (d) Two women farmers preparing a sprayer for herbicide application that allows CA no-till seeding (J. Juárez).

country using the three principles of CA. Permanent soil cover consists of a cover crop (*Mucuna* sp.) and crop residues. Crop rotation involves no-till beans, maize, cassava and sweet potatoes, planted using the 'chuzo' or dibble stick. On a further 20 ha, a diversity of species, mainly coffee, plantains, trees, fruit trees and pineapple, are managed in CA systems. Also, there are 50 ha of CA agroforestry systems combining the planting of forest tree species with the no-till planting of maize in rotation with beans and cassava. These achievements demonstrate that the government is implementing policies to encourage better land management so that through the municipalities, the decision making of farmers is facilitated in changing the unsustainable practices of conventional agriculture to CA, utilising participatory planning at the community, municipal and regional levels.

Finally, from this FAO project experience which involved the implementation of CA, the country's smallholder food systems have improved, and there is a decrease in the degradation of ecosystems that increasingly puts the food security of the Guatemalan population at risk.

2.4 Honduras

The foundations for CA in Honduras began to be laid in the 1970s and 1980s when World Neighbours promoted improved agricultural practices in the steep lands of the country, introduced minimum soil disturbance or no-tillage in the production of vegetables and reinitiated the method of direct planting. In the 1990s, a farmer network of teaching and learning centres for sustainable agriculture - CEAS - was formed and led by José Elías Sánchez. This was complemented by the work of Roland Bunch (Bunch, 1995), and inspired the foundations for the development of the Quesungual Slash and Mulch Agroforestry System (QSMAS) which, starting from local agroforestry in Lempira, gradually introduced the three principles of CA: no-tilling; elimination of slash and burn with a permanent soil cover with mulch from crop and pruning biomass and green cover in fallow periods; diversification of plant species, including maize, pulses, tubers, vegetables, fruit plants and woody tree and shrub species. CA practices are also being used since 2000, with perennial systems involving fruit tree crops, cocoa, coffee, pastures and industrial crops (Table 3).

Fields, with slopes ranging from 12% to 35% (Fig. 10a and b) and infertile soils, now reach an average production of 2.9 t ha^{-1} for maize (CRS, 2019). In the early 1990s, in the same areas, maize yields averaged 0.7 t ha^{-1}; the highest harvests would not surpass 1.2 t ha^{-1}, and the land had to be left fallow for up to 7 years before using it again for a 2-year production cycle. Farmers are reporting yield increases with QSMAS of at least 60% for maize, coupled with year-on-year stability. In addition to maize and sorghum, one hectare of the QSMAS provides a year's supply of firewood and poles to farmers, which adds value.

Table 3 The estimated area under CA systems in Honduras in 2020

Cropping system	Hectares
Annual crops (irrigated)	3000
Annual crops (dryland)	167 801
Vegetables	200
Fruit trees (bananas, citrus fruits, avocado, vid, etc.)	5000
Cocoa, coffee and oil palm plantation	262 895
Pastures	20 000
Agroforestry (includes Quesungual)	79 000
Forestry with cover crop	5191
Total	543 087

Source: Based on the Honduras forest and land use map (Duarte et al., 2014), Coffee Statistics (IHCAFE, 2018).

Additionally, controlled cattle grazing can be done, usually for no more than 2 months and allowing cattle to consume up to 35% of the stubble biomass (personal interview in Reitoca by Carlos Zelaya), achieving equilibrium between mulching and feed, and still containing runoff and soil erosion.

CA, together with integrated soil fertility management (ISFM), generates a net income per hectare of maize of up to US$843 when planted with a legume crop. This income is US$288 higher than the check plot with standard practices, an equivalent of a 1-month minimum wage (CRS, 2018).

With increased productivity, farm stabilisation and permanent cropping, QSMAS favours more land to go into long-lasting forest cover, generating forest landscape recovery, watershed management and climate change adaptation and mitigation.

Figure 10 (a) Celso Sierra, farmer of Reitoca, Francisco Morazán shows his agroforestry farm: maize, sorghum, beans (Phaseolus) and jocote (Spondias purpurea). November 2011 (C. Zelaya); (b) Valle San Jerónimo, Guarita, Lempira. Harvesting maize that withstood 45 days of drought in the growing season (July to August), December (C. Zelaya).

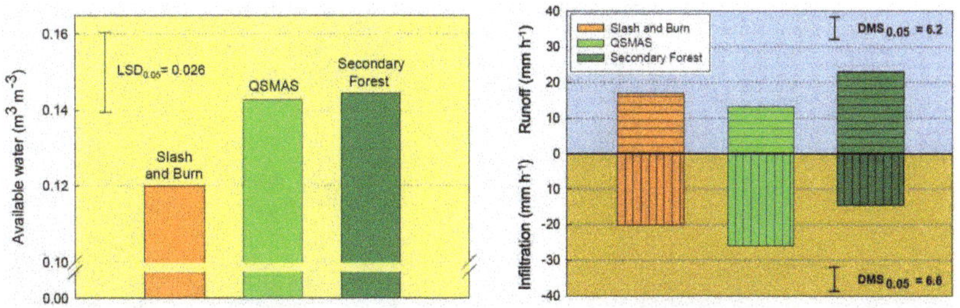

Figure 11 Quesungual (QSMAS) improves dry season and drought adaptation of crops through higher soil water availability, reduced runoff and increased infiltration (Castro et al., 2008).

For farmers, what makes the system more attractive is the possibility of higher yields and reduced vulnerability to a drought that comes with moisture retention (Fig. 11). The agroforestry system retains 15% more water in the soil (8% soil moisture in a traditional field and 23% in a Quesungual area) during the driest month (April) than in the slash and burn system. This difference is equivalent to 20 mm of precipitation, which means that crops can cope for 20 or more days without rain (CRS, 2019).

The increase in soil moisture resulted in its higher content of organic matter that comes from the interaction of soil and mulch biomass. This content was monitored in three different places, for 4 years, showing an increase from −0.7% (slash and burn) to 0.1% in QSMAS plots (Fig. 12). Also, soil erosion diminishes, estimating that the loss of nutrients from erosion is ten times less in the QSMAS than in the slash and burn practice. QSMAS increases the diversity and abundance of soil macrofauna, and, through the effect of these organisms on organic matter breakdown, nutrient cycling, and soil structure, improved soil quality and productivity (Pauli et al., 2005). All this has led to the restoration of large areas of landscapes (Figs 13– 15). Actions impeding the spread of CA and QSMAS are:

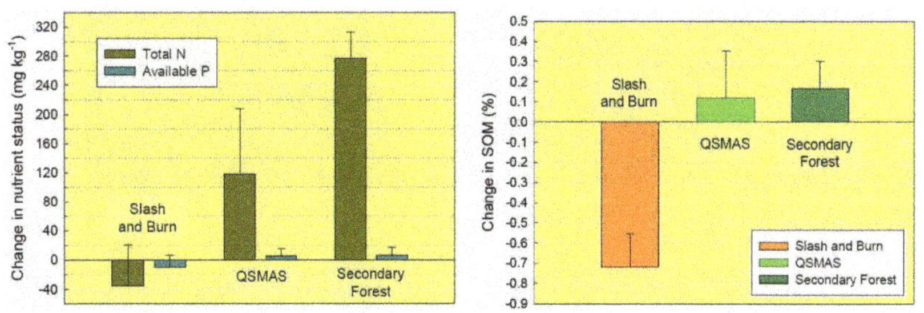

Figure 12 QSMAS improves soil nutrient status and soil organic matter content (Castro et al., 2008).

Figure 13 Landscape restoration is one of the many benefits of CA in Quesungual (Valladolid, Lempira, dry season of 2019) (C. Zelaya).

Figure 14 Landscape restoration in Reitoca (2011) (C. Zelaya).

a Traditionally trained extensionists continue promoting monocropping with conventional tillage. On the other hand, farmers know the benefits of CA, and QSMAS is slowly expanding.

Figure 15 Alonzo Benítez, El Zarzal, San Juan, Intibucá. Integrated agroforestry in three hectares of his coffee farm, February 2019 (C. Zelaya).

b However, a significant challenge for more intense adoption is the lack of incentives to stop the use of slash and burn practice in rented land, even though owners demand a change in the use of this practice.

c More efforts are needed to better integrate livestock production into the CA systems, particularly silvopastoral range management. The adoption of CA in coffee, with a more integral agroforestry approach, is delayed by the driving force of rapid gains. Though sun-exposed coffee plantations reach maturity faster and thus earlier higher yields, the plants have a shorter lifespan.

Lastly, more public research and private investments are needed, as much in the agronomic aspects as in the development and supply of appropriate implements and markets.

2.5 Nicaragua

During the years 2000 to 2004, public and private institutions, projects and NGOs made various efforts to reduce soil degradation processes without achieving significant effects, mainly due to the promotion of practices based on physical structures to control water runoff (Obando and Montalván, 1994).

As a result, in 2005, the Government requested technical assistance from FAO (2005). The project began procuring adequate equipment to do CA such as knife-roller, tractor-mounted no-till seeder, animal traction no-till planters, manual jab-planters known as 'matracas', and animal traction boom sprayers. This equipment, which came from Brazil, was essential to carry out the fieldwork in the demonstration areas located in the districts of Quezalguaque, Posoltega and Telica. Other inputs, such as seeds of cover crops and fertilisers, were distributed to the farmers. Preliminary sub-soiling of the areas that suffered from compaction problems received subsidies. Other operations were carried out by the producers according to their routine schedules.

The 11.6 ha initially planted was with the agreement of small collaborating producers, to reduce their income risk during the adaptation of the CA system on their plots under the conditions of northwestern Nicaragua.

In 2006 the initial plots increased by 61% to 19.1 ha at the expense and risk of the producers. As the plots were planted and the extensive training process developed, Mr Vicente Alegre, a collaborating farmer, decided to buy a Brazilian no-till planter (Fig. 16) entirely independent, and planted 96.9 ha in the surrounding area nearby and provided no-till sowing services to other interested parties (Table 4).

The additional spread in 2006 in the project's area reached up to 190 ha. Thus, the total area under CA reached 209 ha, which was possible to achieve in partnership with NGOs (ALFASOL and MEDA), Project (MST/GEF/OP-15) and INTA's Emergency and Food Security Project (PESA/FAO). Through these approaches, it is expected that Nicaraguan lands can continue to be used and managed sustainably and profitably by farmers.

As shown in Fig. 17a–d the process of training of producers and technicians and dissemination based on CA field plots allowed creating technical bases in knowledge and application of a CA system. Farmer training included field days,

Figure 16 Planting maize in the field of the producer Vicente Alegre using a no-till direct seeder in a CA system.

Table 4 The area under CA promoted by the FAO Project in the northwestern region of Nicaragua

	CA (ha)	
Farmers	2005	2006
Initial plots	11.6	19.1
Additional plots	96.9	190.0
Total	108.5	209.1

assembly and calibration demonstrations of no-till direct seed drills, workshops, equipment and tool exhibitions and fairs, and the establishment of a machinery pool that was used in a farmer training programme to support smallholders who chose to adopt CA.

The availability of water in the soil is one of the most crucial production factors for small-scale agriculture in Nicaragua. The principles of CA were introduced by CRS in the maize-bean systems in Yalagüina, Madriz, in northern Nicaragua, in the harsh Dry Corridor of Central America. As shown in Fig. 18, the gravimetric soil moisture content increased by 2.42% on average between the CA-ASA intervention and conventional farmer management systems, corresponding to 38% greater gravimetric soil moisture. Only 2 years after

Figure 17 (a) Field days to show permanent soil cover protection with harvest residues; (b) Assembly and calibration demonstrations of direct seed drills; (c) Machinery pool to support farmers who choose to adopt Conservation Agriculture and (d) Equipment and tool exhibition fairs.

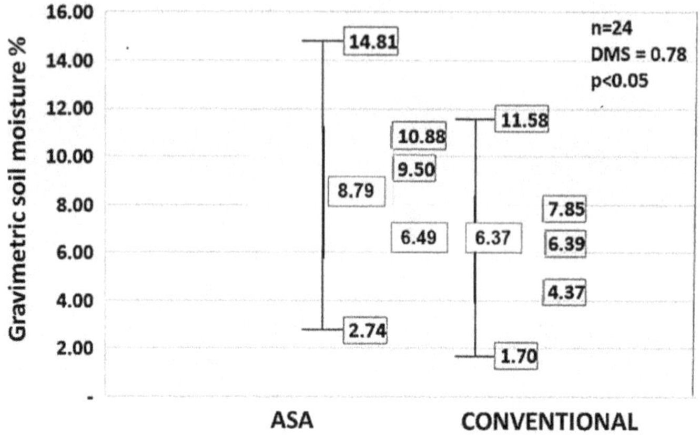

Figure 18 Direct seeding with animal traction in Guantanamo, October 2006 (T. Friedrich).

starting to implement these practices, the results indicate a better water retention capacity in the plots, even in dry season conditions. Native species of green manure/cover crops were included, which were sown at the end of the rainy season and fertilised at the end of the dry season, leaving a dense biomass cover for the next maize crop. Measurements of gravimetric soil moisture (% water by weight of soil) in 24 Yalagüina farms revealed significant improvements in soil moisture content during the dry season, as a result of the implementation of CA practices (Fig. 18).

Figure 18 shows Gravimetric soil water content (%) of 24 ASA plots during the dry season after 2 years of CA (no-till, permanent mulch cover and green manure cover crop) in Yalagüina, Madriz, Nicaragua (Martin, 2017).

A 2% increase in soil moisture content equals more than 90 000 L of additional water per hectare of land. This increase in water holding capacity, after a short period, demonstrates the enormous potential of CA practices to increase resilience to water scarcity in the Dry Corridor and other areas facing increased climatic variability.

Aside from efforts to date adoption and areas planted with CA are still very small as can be seen in Table 5. However, synergies created between the FAO project and the different institutions involved have led to several initiatives focused on the future expansion of CA.[5]

Despite the availability of tools, and the intense training of producers, the main barrier that has prevented further expansion of the area under CA in Nicaragua is a weak institutional framework that did not allow the government to

5 The Ministry of Environment and Natural Resources (MARENA); the MST/GEF/OP-15 project; the Nicaraguan Institute of Agricultural Technology (INTA); the National Forest Institute (INAFOR); the Non-Governmental Organization ALFASOL of the Spanish cooperation AECI; the PESA/FAO Project; the Canadian Mennonite Cooperation MEDA Project and the CRS ASA project.

Table 5 The estimated area under CA and the number of adopting farmers managing the CA system in Nicaragua in 2020

Crop	Area (ha)	Number of farmers
Maize	1375	1198
Beans	405	357
Sorghum	415	380
Rice	50	1
Sugar cane	300	2
Sesame	150	70
Total	2695	2008

approve policies to support the adoption of CA. We hope that the Nicaraguan's government will again encourage farmers the actions that help to continue with the massive adoption of CA.

2.6 Panama

The annual cost of land degradation in Panama is about US$309 million, equal to 1.5% of the country's gross domestic product; therefore, action against land degradation has a return estimated at US$5 for every dollar invested in restoring degraded land (Muñoz et al., 2018). Agricultural production in Panama is comprised mainly of maize, beans, fruits (including bananas, pineapples and oranges) and vegetables (onions and potatoes). Agroforestry systems with coffee (the highest quality coffee beans on the world market) are being promoted.

In Panama, most maize farmers manage mixed crop-livestock systems, and over 85% of landowners use maize residue to feed their livestock, particularly during the dry season.

Panama has received technical cooperation from FAO projects on issues of soil conservation, land and water management since the 1980s. These field projects emphasised helping farmers to improve land care and soil health, but there was less emphasis on combating erosion directly. Farmers adopted crop residue retention on the surface but still depended on minimum tillage.

Survey information on land preparation practices of maize farmers in 1994 allowed Pereira de Herrera and Sain (1999) to characterise current tillage methods and to improve the understanding of the process through which farmers adopt zero and minimum soil disturbance practices. The three most common reasons for using no-till were (1) reduced production costs, (2) improved soil conservation and (3) better weed control (Table 6). The reduction in production costs per hectare with zero tillage compared to conventional tillage varies, depending mainly on the number of passes made with the disc

Table 6 Farmers' reasons for using no-till in maize, in three regions[a] of Azuero, Panama, 1994 (Pereira de Herrera and Sain, 1999)

	Percentage of farmers		
Reason	Region I	Region II	Region III
Reduces costs	46	37	38
Conserves soil	26	32	38
Better weed control	10	5	12
Other	18	26	12

[a] Region I was formed by the districts of Las Tablas, Pedasí, Pocrí and Guararé in Los Santos Province and included 28 localities. Region II consisted of the districts of Los Santos and Macaracas in Los Santos Province and included 12 localities. Region III, formed by the districts of Chitré and Parita in Herrera Province, included ten localities.

harrow in the conventional system and on the type of desiccant herbicide used in the no-till system.

Currently, the most challenging obstacles to increasing CA adoption in Panama are weed infestation and the use of crop biomass or residues for animal feed. However, solutions to these problems can be found through a combination of participatory research on potential solutions increasing awareness of the importance of soil degradation and the value of residue for soil cover, soil health and crop productivity. Labour costs in Panama have risen markedly over recent years, from US$6–8/day to US$22–25/day. As a result, many farmers in Las Tablas, capital of the Province of Los Santos, have bought zero-tillage machinery. Farmers are also beginning to understand the value of residue for soil cover and the importance of crop rotation.

Experiences on CA and considerations for its implementation show the potential benefits that can be harnessed by farmers with support at the district, provincial and national levels and highlight the need for longer-term research on the economics of CA systems. As in other countries, CA needs to be promoted in universities and colleges, and, with help from international agencies, the Agricultural Research Institute of Panama (IDIAP) should develop a centre of excellence on CA and sustainable agricultural systems.

3 Conservation Agriculture in the Caribbean

In 2001 CA was introduced in Cuba as a new farming concept in the context of the AGROMECH congress (Friedrich, 1998–2013). In 2006 Cuba requested support from FAO for CA in the Guantanamo province due to the severe problems of soil degradation along with continuous droughts (Friedrich, 2013; Hernández Lara et al., 2008).

In 2013 CA was again promoted by the FAO and a group of ecological activists who integrated the concept into their Mother Earth festivals and activities. CA was presented in conferences, TV, radio and other media (Friedrich, 2014, 2017). FAO started a new CA project for grain crops, and the University of Villa Clara and Bayamo, as well as a GEF project on soil and water management, bought some no-till seed drills. In 2016 the FAO, together with the Ministry of Agriculture, organised an international workshop for CA adoption in Cuba, which led in 2017 to the formation of an inter-institutional task force for the introduction of Conservation Agriculture. However, CA adoption in Cuba did not continue as planned (Friedrich et al., 2017).

The first CA pilot experiences were under drip irrigation in Guantanamo, including sorghum as soil cover, maize, vegetables and perennial crops such as plantain with mucuna cover crop (Figs 19 and 20). After 2013 the Association of Agricultural and Forestry Technicians (ACTAF) promoted CA, resulting in some spontaneous adoption. One case was on a mixed crop/livestock farm dedicated to annual and perennial crops and pig raising, where crops were seeded into rip lines made with a converted animal traction plough without tillage. Another farmer adopted CA on a recently established fruit farm in Pinar del Rio province with mango and guava, all with different soil covers (Fig. 21). For the smaller trees, soil cover was mulched and concentrated around the trees to suppress undergrowth and protect the soil. Sheep, used for grazing under trees, managing soil cover, provided additional income. By eating fruits from the ground, which had been fallen due to insect attacks, they also helped with pest control, avoiding propagation of insects in the attacked fruit. Production and farm income increased, and the initial plan to install irrigation on the farm was abandoned as it appeared unnecessary under CA.

Besides ACTAF, Universities in Santa Clara and Bayamo promoted CA, and CA was included in FAO and GEF projects. Farmers' organisations such as ACPA, the small livestock producers association, and environmental groups such as Cubanos en la Red, started to promote CA. However, there is still no organisation entirely dedicated to CA. Dependence on food imports and recurring droughts are strong incentives to adopt CA. However, there is still, with a few exceptions, no little significant adoption in Cuba, mainly because of the unavailability of equipment.

In other Caribbean states, a project on Agricultural Mechanisation Policies and Strategies for the Caribbean Community (CARICOM) Countries introduced CA in 1999. However, to date, there is no adoption of CA in CARICOM states, with the possible exception of some Mennonite communities in Belize (Friedrich, 1998–2013).

In Haiti, an NGO project introduced no-till in the 1980s, but it did not lead to any adoption, despite severe soil degradation and drought problems. CA was reintroduced in the north-eastern and central highlands by FAO in 2013 for the cassava-peanut, maize-bean and rice cropping systems (Friedrich, 1998–2013),

Figure 19 Direct seeding with tractor planter in Guantanamo, October 2006 (T. Friedrich).

Figure 20 Direct seeding with tractor planter in Guantanamo, Oct. 2006 (T. Friedrich).

resulting in CA projects from 2015 to 2019 which gave good results, but with no uptake.

A general problem for CA uptake in the Caribbean is the lack of dedicated farmer organisations with strong champions leading the movement as well as a lack of political and institutional support. While farmer groups are vital for promoting CA, lasting government support is essential, which is particularly the case in Cuba.

Figure 21 Farmer Onay Martínez on his CA fruit farm in Pinar del Rio province, September 2016 (T. Friedrich).

4 Mexico

Mexico is facing a severe process of soil erosion. Eighty percent of the country's degraded land surface is affected by different degrees of soil erosion, of which 33% is critically eroded and 27% on a moderate scale (Bai et al., 2008). The leading causes of land degradation and erosion are deforestation, intensive monoculture, inadequate tillage methods, over-grazing and over-cultivation.

Numerous efforts aimed at soil conservation were carried out from 1984 to 1995 to control erosion through the formation of physical conservation works structures with heavy machinery. Still, they were not accompanied by improved agronomic practices to enhance soil health and productivity.

Research on no-till with residue cover in Mexico began in the 1950s through the national agricultural research organisation, Instituto Nacional de Investigaciones Agrícolas (INIA, today INIFAP). Still, it was soon abandoned due to poor results. In 1973, CIMMYT began research on no-till maize, encouraging researchers from INIA and another agricultural research organisation, the Fideicomisos Instituidos en Relación con la Agricultura (Rural Development Trust Funds, FIRA), to start their own CA programmes. In 2011, FIRA provided no-till training for farmers, organised conferences and, most importantly, purchased the necessary equipment (Speratti et al., 2015).

CIMMYT partnered with the Mexican Secretariat of Agriculture (SAGARPA) to carry out a project called the Sustainable Modernisation of Traditional Agriculture, better known as MasAgro (Speratti et al., 2015).

The Ministry of Environment, Natural Resources and Fisheries (SEMARNAP) began fieldwork in 1996 with the support of the FAO to improve agricultural production systems through updated technologies for the use, management and conservation of soils and waters in three micro-basins in the States of San Luis Potosí, Querétaro and Veracruz (Bassi, 1999). This project included a study trip of six participating technicians, from the General Direction of Restoration and Conservation of Soils of the States of Santa Catarina, Brazil to familiarise them with the principles of CA locally called 'Plantio Direto'.

The results of this project influenced the Government of Mexico to update the institutional, legal framework that regulates soil conservation, the development of strategies and policies for the use, management and protection of soils and water (Bassi, 1999). In addition to this, Dr Ramón Claverán, Director of the National Research Center for Sustainable Production (CENAPROS), asked the FAO to be the organiser of the VI Meeting of the Latin American Conservation Tillage Network (RELACO). At this meeting, it was decided to change the name of the Latin American Network for Conservation Tillage to the Latin American Network for Conservation Agriculture, keeping the same acronym (Benites, 1997).

In 2010, the International Maize and Wheat Improvement Centre (CIMMYT) initially launched the MasAgro programme (the Sustainable Modernisation of Traditional Agriculture), to support the use of sustainable technologies, CA amongst them, in maize and wheat production (Camacho-Villa et al., 2016). The Mexican Ministry of Agriculture, Livestock, Rural Development, Fisheries, and Food (SAGARPA) funded this joint effort.

In the first years, MasAgro had to overcome three significant limitations for the adoption of CA: weed control, availability of equipment and tools, and competition for crop residues. No-till weed control, especially for small farmers with limited access to herbicides, is the first critical agronomic barrier for CA adoption (Speratti et al., 2015). CIMMYT (2012) recommended the use of herbicides together with no-till and surface residue retention, Cover crops promoted by the FAO project are a viable option to reduce weed infestation and herbicide cost (Teasdale et al., 2007; Bassi, 1999).

A second limiting barrier to the adoption of CA is a lack of, or insufficient access to, equipment for no-till planting, fertilising and spraying pesticides due to both economic and local technological constraints. Suitable equipment for CA may not be available locally or is relatively more expensive. MasAgro imported prototypes of equipment and machinery from Argentina, Bangladesh, Brazil, China, India, and Nepal, ranging from large tractor-drawn planters to animal-drawn seeders and hand planters, and developed and adapted these to local conditions. CIMMYT and the FAO supported local machine shops to encourage domestic production of equipment and machinery to increase accessibility for farmers.

The third barrier to CA adoption has been the competing use of crop residues for soil cover or animal feed (Fig. 22). However, solutions can be found through a combination of participatory research on potential solutions and by changing perceptions of the importance of soil degradation and the value of residue for soil cover.

The adoption of CA practices in Mexico under the three barriers has been slow, despite the technical support available in the country (Lumpkin and Sayre, 2009). Once the constraints to CA adoption were addressed, MasAgro was able to expand CA in many parts of Mexico based on participatory innovation system thinking, combining research and development with multi-stakeholder approaches (Camacho-Villa et al., 2016). CA is an essential element of the technology menu that the programme offers to Mexican farmers involved with irrigated and rain-fed maize and wheat production. A comprehensive recent survey by MasAgro and CIMMYT (as reported by B. Govaerts of CIMMYT) found that in Mexico there was 149 218 ha under CA systems (Table 7).

Public complaints to politicians and researchers once again proved to be a very effective way for farmer groups to get their voices heard and let them become part of MasAgro (CIMMYT, 2018). At a MasAgro meeting in 2014, with small farmers from Poza Rica in Veracruz, and large farmers from the Bajío, it was found that they all planted mainly maize or sorghum in the summer in rotation or associated with beans (Fig. 23), with irrigation and zero tillage. Some farmers

Figure 22 Competition for the use of crop residues for soil cover or livestock (A. G. Red).

Table 7 Records and surface (ha) where International Development Partners (IDP) collaborators indicated to apply some activity related to Conservation Agriculture (Bram Govaerts, 2020)

	With coordinates		Without coordinates		Data obtained through geoprocessing		Total	
State	No. of records	Surface (ha)	No. of records	Surface (ha)	No. of records	Surface (ha)	No. of records	Surface (ha)
Aguascalientes	1	1	825	3750			826	3751
Baja California	186	2317					186	2317
Campeche	236	526	14	19			250	546
Chiapas	6490	13 643	98	146			6588	13 789
Chihuahua	49	4820					49	4820
Distrito Federal	14	11					14	11
Durango	833	2792	1	3			834	2794
Guanajuato	1259	7277	874	18 372	7537	29 002	9670	54 651
Guerrero	261	335	264	263			525	598
Hidalgo	154	245	4	7			158	252
México	183	465	1	1			184	465
Michoacán de Ocampo	206	625	2	4			208	629
Morelos	396	721	519	1033			915	1754
Oaxaca	8375	8859	121	129			8496	8988
Puebla	114	178	4	6			118	184
Querétaro	117	279	23	53			140	332
Quintana Roo	26	58					26	58
San Luis Potosí	873	6628	1401	2699			2274	9327
Sinaloa	1082	12 533	260	2347			1342	14 880

Sonora	75	2652	4	64			79	2716
Tabasco	1073	1357	12	15			1085	1372
Tamaulipas	18	54					18	54
Tlaxcala	238	521	1	1			239	522
Veracruz	3866	8182	21	45			3887	8227
Yucatán	144	4586	3	3			147	4589
Zacatecas	1356	11 131	75	462			1431	11 593
Grand Total	27 625	90 797	4527	29 420	7537	29 002	39 689	1 49 218

Figure 23 CA adopted in hilly areas of Chiapas, Mexico, by commercial farmers growing maize-beans in association, 2013 (CIMMYT and MasAgro).

use zero tillage only in the summer and planted wheat or barley in winter with tillage, and therefore this is not a CA system.

5 Conclusion

In the region, CA annual cropping systems do not disturb the soil, leave stubble biomass on the surface and include diversified crop rotations or associations including cover crops and a varied mix of legumes, grasses and other species. In the hillside areas in the region, there are traditional systems that follow the principles of CA. These include agroforestry with annual crops, cocoa and coffee plantations, pasture restoration with mixtures of gramineous and legumes plants and controlled grazing to maintain proper soil cover and supply of fodder. There is also increasing adoption of fruit trees with cover crops and establishment of managed forests with various undergrowth strata to increase soil cover and to prevent erosion problems during the establishment phase.

Estimated values in hectares of area under CA as described earlier were obtained from collaborators of El Salvador (1950 ha), Guatemala (10 000 ha), Honduras (543 087 ha), Mexico (149 218 ha) and Nicaragua (2695 ha). The total area under different forms of CA reported in the region is 696 950 ha.

However, besides knowledge, technologies, and supplies, adoption and spread of CA needs a favourable (policy) environment, motivation, and participation of farmers and their communities.

The presence of leaders or champions and farmer organisations is essential for knowledge sharing and capacity building. The combination of all these parameters with CA at the catchment level results in sustainable rural development based on the integrated management of natural resources, efficient production, improved livelihoods and employment, and market access for competitive value chains.

Despite a very long history of good-quality CA projects in Central America, the Caribbean and the collaboration of high-level subject matter experts and champions such as Francis Shaxson, Richard Barber, Marcos Vieira, Guillermo Augusto de Araujo, Valdemar Hercilio de Freitas, Alexandra Bot, Lauro Bassi, Rafael Fuentes, Brian Sims, Luis Alvarez Welchez, Jan van Wambeke, Patrick Wall, Ken Sayre and the authors of this chapter, the lack of national and state policies of the region to support the uptake of CA has resulted in the abandonment of many of these good initiatives as a result of a lack of continuity.

6 Where to look for further information

Among the guidelines for future research on the subject dealt with in this chapter, it is suggested to standardize the detailed information in the methodology of any study related to CA to facilitate comparison between studies in Central America, Mexico, and the Caribbean countries. As in the research work, a standardization of the method for collecting and elaborating surface statistics is necessary. The methodology must include information.

Emphasis on research on the management of crop residues and land cover area, crop rotation, and the area of other herbaceous crops such as legumes. It is essential to investigate the role of agricultural energy and mechanization. Agricultural mechanization is not an isolated activity but is part of a complex set of interactions between many participants. For its correct application requires a higher level of knowledge, better agricultural management systems, and training.

The organizations involved with respect to this topic are the regional universities, farmer´s organizations, the Food and Agriculture Organization of the United Nations (FAO), Rome, Italy; the Inter-American Institute for Cooperation on Agriculture (IICA), San Jose, Costa Rica; the Tropical Agricultural Research and Higher Education Center (CATIE), Turrialba, Costa Rica; the International Maize and Wheat Improvement Center (CIMMYT), Mexico; and the Programme for Sustainable Agriculture in Central America Hillsides (PASOLAC), San Salvador, El Salvador, often supported by bilateral donors.

7 References

Alberto, A. 2020. Consulta personal. Gerente Tecnico de la Mancomunidad la Montañosa. Chalatenango, El Salvador.

Argueta, M. T. 1996. Análisis de la producción y utilización de rastrojos y su relación con riesgos de erosión en el Departamento de Morazán. UCA, San Salvador, El Salvador. 60 pp.

Bai, Z. G., Dent, D. L., Olsson, L. and Schaepman, M. E. 2008. Global assessment of land degradation and improvement. 1. Identification by remote sensing. Report 2008/01. ISRIC–World Soil Information, Wageningen.

Barber, R. 1996. Informe Final. Experto en Manejo de Tierras. Proyecto Agricultura Sostenible en Zonas de Ladera. El Salvador. CENTA-FAO.

Barber, R. 1999. Land and crop management in the hilly terrains of Central America: lessons learned and farmer-to-farmer transfer of technologies. Food and Agriculture Organization of the United Nations. FAO Soils Bulletin 76. Food and Agriculture Organization, Rome.

Bassi, L. 1999. Relación final del proyecto Apoyo al Fomento de la Conservación de Suelos en Mexico, TCP/MEX/4554. Food and Agriculture Organization, Rome.

Benites, J. 1997. VI Conference of the Latin American Network on Conservation Tillage (RELACO), held on 17–20 November in Morelia. Mexico. Unpublished Back To Office Report. Food and Agriculture Organization, Rome.

Bunch, R. 1995. Two ears of maize; A guide to people-centered agricultural improvement. World Neighbors, Oklahoma City.

Camacho-Villa, T. C., Almekinders, C. J. M., Hellin, J., Martinez-Cruz, T. E., Rendon-Medel, R., Guevara-Hernández, F., Beuchelt, T. D. and Govaerts, B. 2016. The evolution of the MasAgro hubs: responsiveness and serendipity as drivers of agricultural innovation in a dynamic and heterogeneous context. *The Journal of Agricultural Education and Extension* 22(5), 455–470.

Casares, F. 2020. Consulta personal. Gerente Tecnico del Programa Raíces Ahuachapan. CRS. El Salvador.

Castro, A., Rivera, M., Ferrerira, O., Pavón, J., García, E., Amésquita, E., Ayarza, M., Barrios, E., Rondón, M., Pauli, N., Baltodano, M. E., Mendoza, B., Wélchez, L. A., Johnson, N., Cook, S. and Rao, I. 2008. Adaptation to climate change: is the Quesungual system an option for smallholders in dry hillsides agroecosystem?". Poster presented in the meeting of the CGIAR Water Challenge Program. Water and Food Challenge Program Theme 1, Rainfed Topic Workshop, Tamale, Ghana, 22–25 September 2008.

CENTA-FAO 2002. Análisis de sostenibilidad. Proyecto CENTA-FAO-HOLANDA.

Chicas, J. 2020. Consulta personal. Getente Tecnico de la Asociación de cuencas del Golfo de Fonseca. Morazan. El Salvador.

CIMMYT 2012. MasAgro. Reporte de Actividades 2011–2012. http://masagro.mx/index.php/ es/2012-06-21-17-47-58/documentos.

CIMMYT 2018. MasAgro web site spanish [WWW Document]. Línea Acción 1 MasAgro Prod. Desarro. Sustentable con el Prod. http://masagro.mx/es/2012-06-21-1747-58/ documentos.

CRS 2018. Semi-annual report 2018, unpublished, for internal use. CRS.

CRS 2019. La cosecha de Aurelio, crónica de cambio significativo, folleto. CRS.

Duarte, E., Orellana Díaz, O., Maradiaga, I., Casco, F., Danhy, F., Jiménez, A., Emanuelli, P. and Milla, F. 2014. Mapa Forestal y de Cobertura de la Tierra de Honduras: Análisis de Cifras Nacionales. 10.13140/RG.2.1.1553.3046 (trabajo realizado para el Instituto de Conservación Forestal, Äreas Protegidas y Vida Silvestre (ICF).

FAO 2002a. Síntesis de los resultados del proyecto. Proyecto CENTA-FAO-HOLANDA. Agricultura Sostenible en Zonas de Ladera–II Fase. El Salvador.

FAO 2002b. Agricultura de Conservación. Estudios de casos en América Latina y África. FAO Naciones Unidas, Roma. 85 pp.

FAO 2005. TCP/NIC/3001. Strengthening local capacities for the sustainable management of natural resources in prioritised micro-basins in northwestern Nicaragua. FAO, Rome.

FAO 2016. Ahorrar para crecer en la práctica. Guía para la producción sostenible de cereales. FAO Naciones Unidas, Roma. 124 pp.

Food and Agriculture Organization & EU 2010. Informe Final de la misión de campo. Programa Extraordinario de Apoyo a la Seguridad Alimentaria y Nutricional. Publicación interna.

Friedrich, T. 2013. Mission reports CARICOM (1998–1999); Haiti (1999, 2013); Cuba (2001–2008); mission reports FAO, unpublished.

Friedrich, T. 2014. Food security: current challenges/La seguridad alimentaria: retos actuals. *Cuban Journal of Agricultural Science*, 48(4), 319–322.

Friedrich, T. 2017. Cuban agriculture, challenge, model, or illusion?; resource: engineering and technology for a sustainable world March/April 2017. *American Society of Agricultural and Biological Engineers (ASABE)*, 24(2), 7–9.

Friedrich, T., Stori, M., Alonso, I., Calero Martín, B. J., Reeves, T., López, P., Goddard, T. and Peiretti, R. (Eds.) 2017. PRODUCE + WITH – INPUTS, Memories of the international expert consultation on conservation agriculture for sustainable development; MINAG and FAO, La Habana, Cuba. 193 pp.

Govaerts, B. 2020. Statistical information on conservation agriculture adoption in Mexico derived from MasAgro and CIMMYT data.

Hernández Lara, O., Cintra Arenciabia, M., Alfonso, C. A., Sánchez Arce, I., Rodríguez Aguilar, Y., Oliva Collazo, R., López Martínez, N., Linares Jímenez, T., Ceballos Prevost, D., San Lois, D. and Velásquez Leiva, C. 2008. Manual de Agricultura de Conservación, Guia de Trabajo; Instituto de Suelos, MINAG y FAO; La Habana, Cuba.

Instituto Hondureño del Café (IHCAFE) 2018. Produccion cosecha 2016–2017 por departamento y municipio (unpublished).

Lumpkin, T. and Sayre, K. 2009. Enhancing resource productivity and efficiency through conservation agriculture. 4th World Congress on Conservation Agriculture: Innovations for Improving Efficiency, Equity, and Environment. 3 pp.

Martin, M. 2017. Conservation agriculture adds water to the ground in Nicaragua. ASA, CRS. https://asa.crs.org/2017/12/la-agricultura-de-conservacion-agrega-agua-al-suelo-en-nicaragua/#:~:text=La%20agricultura%20de%20conservaci%C3%B3n%20agrega%20agua%20al%20suelo%20en%20Nicaragua,-diciembre%2011%2C%202017&text=La%20disponibilidad%20de%20agua%20en,en%20el%20norte%20de%20Nicaragua.

Muñoz, P., Ali, M., Radloff, P. and Yao, Z. 2018. Global mechanism of the UNCCD2018. Country profile of panama. Investing in land degradation neutrality: making the case. An Overview of Indicators and Assessments, Bonn, Germany.

Obando, M. and Montalvan, D. 1994. Technical and economic analysis of a soil conservation project in Nicaragua. Economic and Institutional Analyses of Soil Conservation Projects in Central America and the Caribbean World Bank Environment Paper edited by Ernst Lutz, Stefano Pagiola, Carlos Reiche Number 8.

Osorio, M. J. n.d. Adopción y difusión de la labranza de conservación en Metalio-Guaymango, El Salvador. Centro Nacional de tecnología Agropecuaria y Forestal (CENTA), Ministerio de Agricultura y Ganadería. PowerPoint presentation. See last slide in https://slideplayer.es/slide/5623820/.

Osorio, M. J. (sf) Adopción y difusión de la labranza de conservación en Metalio-Guaymango, El salvador. Centro Nacional de tecnología Agropecuaria y Forestal (CENTA), Ministerio de Agricultura y Ganadería. El Salvador.

PAES 2007. Informe final de evaluación del programa PAES-Trifinio. Programa Ambiental de El Salvador (préstamo bid 886/oc-es) San Salvador, El Salvador, febrero de 2007.

Pauli, N., Barrios, E., Oberthür, T., Conacher, A., Ayarza, M. and Welchez, L. A. 2005. The effects of the Quesungual agroforestry system of Western Honduras on soil macrofauna. *Sociedade and Natureza* 1(1), 466–477.

Pereira de Herrera, A. and Sain, G. 1999. Adoption of maize conservation tillage in Azuero, Panama. CIMMYT Economics Working Paper No. 99-01. Mexico, DF: CIMMYT.

Rodriguez, E. 2020. Consulta personal. Jfe de la Agencia de Extension de CENTA, Ministerio de Agricultura y Ganaderia. Tacuba, Ahuachapan. El Salvador

Shaxson, F. 1999. New concepts and approaches to land management in the tropics with emphasis on steep lands. Food and Agriculture Organization of the United Nations. FAO Soils Bulletin 75.

Speratti, A., Turmel, M. S., Calegari, A., Araujo-Junior, C. F., Violic, A., Wall, P. and Govaerts, B. 2015. Latin, A., Farooq, M. and Siddique, K. H. M. (Eds) *Conservation Agriculture in Latin America*. Springer International Publishing, Switzerland.

Teasdale, J., Brandsaeter, L., Calegari, A. and Skora Neto, F. 2007. Cover crops and weed management. In: Upadhyaya, M. K. and Blackshaw, R. E. (Eds) *Non-Chemical Weed Management: Principles, Concepts, and Technology*. CABI, Reading, 49–64.

Vieira, M. 2000. Manejo integrado de tierras un enfoque para producir y conservar. Proyecto Agricultura Sostenible en Zonas de Ladera. El Salvador: CENTA–FAO.

Chapter 5

Conservation Agriculture in South America

Rafael Fuentes-Llanillo, IDR-Paraná IAPAR-EMATER and FEBRAPDP, Brazil; Marie Luise Carolina Bartz, University of Coimbra, Portugal and FEBRAPDP, Brazil; Tiago Santos Telles, IDR-Paraná IAPAR-EMATER, Brazil; Ademir Calegari, formerly IAPAR and FEBRAPDP, Brazil; Augusto Guilherme Araújo, formerly IAPAR, Brazil; Amir Kassam, University of Reading, UK; David Roggero, AAPRESID, Argentina; Dimas Soares Junior, IDR-Paraná IAPAR-EMATER and FEBRAPDP, Brazil; Edgard Ramírez, AAPRESID, Argentina; Facundo Capandeguy, AUSID, Uruguay; Herbert Arnold Bartz, Farmer, Doctor Honoris Causa UEL and FEBRAPDP, Brazil (in memoriam); Jaime Hernández-Zamora, ANAPO, Bolivia; Ken Moriya, FEPASIDIAS, Paraguay; Luciano Dabalá, AUSID, Uruguay; María Cecilia Ginés, AAPRESID, Argentina; Martín Maria Cubilla, FEPASIDIAS, Paraguay; Ricardo Ralisch, formerly UEL Professor and FEBRAPDP, Brazil; Richard Trujillo Mendoza, ANAPO, Bolivia; Roberto Peiretti, Global Farmer Network, USA; Rolf Derpsch, formerly GTZ Paraguay Consultant, Germany; Telmo Jorge Carneiro Amado, UFSM, Brazil; and Theodor Friedrich, Retired Food and Agriculture Organization of the United Nations (FAO) Expert on Conservation Agriculture, Germany

1 Introduction
2 The no-tillage system and Conservation Agriculture in Brazil
3 The no-tillage system in Argentina
4 Evolution and current state of the no-tillage system in Paraguay
5 No-tillage and Conservation Agriculture in Uruguay
6 No-tillage and Conservation Agriculture in Bolivia
7 Future prospects for Conservation Agriculture in South America
8 Where to look for further information
9 References

1 Introduction

Conservation Agriculture (CA)[1] in South America started in Brazil in 1972 with the first no-tillage (NT) commercial seeding of 200 ha of soybean by the pioneer

1 Definitions for the terms used in this chapter: NT - No-Tillage: The practice of planting/sowing annual crops with minimum soil disturbance(concentrated around the sowing line) on the straw of former crops or over live plants

http://dx.doi.org/10.19103/AS.2021.0088.04

farmer Herbert Bartz (in Rolândia, Paraná State, Brazil). The development, adoption and success of NT, initially led by farmers, revolutionized Brazilian agriculture and spread throughout South America. The expansion from NT practices to include permanent soil cover and crop diversity led to the concept of the no-tillage system (NTS) as a new model of agriculture in Brazil and other South American countries. This evolution led the Food and Agriculture Organization (FAO) to define the concept of CA as any type of farming system that follows three principles:

- minimum soil disturbance;
- soil permanently covered with straw and/or live plants; and
- crop rotation/association and diversity of species.

The main driver of expansion of CA and NTS in South America was the need to stop soil erosion caused by intensive tillage of annual crops. The advantages brought by erosion control, reduction of tillage operations and consequent savings in fuel, labor and time were communicated from farmer to farmer (Fuentes-Llanillo et al., 2020).

The main actors in the adoption and spread of CA/NTS in South America were undoubtedly farmers, supported by advancements made by researchers and manufacturers of equipment and agrochemicals, and by the provision of extension services. The willingness of innovative farmers to share their lessons and experience with other farmers has contributed enormously to the acceptance of NT agriculture systems (Dumanski, Reicosky and Peiretti, 2014). Based on the need to support each other, producers from different regions of the world spontaneously began to work together. The foundation and development of national institutions by farmers had a great impact in facilitating agriculture in its paradigm shift to NT. This development initially took place informally and mainly at a local level but then consolidated to form national, regional and international associations (Peiretti, 2012; Fuentes-Llanillo et al., 2020). The first example was in Brazil, in 1979, when farmers created the Earthworm Club (Clube da Minhoca) in Ponta Grossa town in Paraná State, bringing together farmers, scientists, extension officers, and other professionals to discuss practical problems and solutions affecting NT. This created a ripple effect in Brazilian Southern States with the establishment of many Friends of the Soil Clubs (Clubes Amigos da Terra - CTAs). After the development of local organizations in other Latin American countries, national organizations developed such as:

without tillage operations, even eventually. NTS - No-Tillage System: Farming system of annual crops developed respecting three principles: (i) Continuous no or minimum mechanical soil disturbance, (ii) permanent soil cover with crop biomass and/or live plants; and (iii) crop diversification involving rotations and/or associations, including cover crops. CA - Conservation Agriculture: Any type of farming system that follows the three principles such as no-tillage system of annual crops (NTS), integrated crop-livestock-forest systems (NT-ICLF), no-tillage system in perennial crops (NTP), no-tillage system in vegetables (NTV), among others.

- the Argentinian No-Till Farmers' Association - AAPRESID in 1989;
- the Uruguayan Association of No-Till Farmers - AUSID in 1991; and
- the Brazilian No-Till Farmers' Federation - FEBRAPDP in 1992.

1992 saw the creation of a continental organization, the American Confederation of Farmers' Organizations Toward a Sustainable Agriculture - CAAPAS. In 1993, CAAPAS incorporated the Paraguayan No-Till Farmers' Federation - FEPASIDIAS, the Soil Conservation Society of Chile - SOCOSCHI (which became the Center for Development of Zero Tillage - CEDECELA), the Mexican Association of Conservation Tillage - AMLC and the Bolivian Oil Seeds and Wheat Farmers Association - ANAPO (Peiretti, 2012; Fuentes-Llanillo et al., 2020).

As a result of the economic, social and environmental benefits of the system, and through the leadership of farmers, a synergistic environment was created. With support from public and private research centers, the chemical and machinery industries and, later, with extension and credit support, NT and CA/NTS expanded rapidly to millions of hectares in the late 1990s and 2000s. This is the example that many other countries and regions in the world are trying to reproduce with varying degrees of success. The figures on adoption are shown in Table 1.

According to Table 1, the surface area under CA in South America in 2018/19 was around 83 million hectares, representing about 40% of global CA over the last decade. This has allowed these countries to become responsible for 15% of world grain crop production of soybean, maize and wheat (30% of the world market in total) and 12% of world meat production of cattle, pork

Table 1 Surface under Conservation Agriculture in South America in 2008-2009, 2013-2014, 2015-2016 and 2018-2019.

	Area under CA (million hectares)			
Countries	2008-2009	2013-2014	2015-2016	2018-2019
Brazil	25.50	31.81	32.00	43.00
Argentina	19.72	29.18	31.03	33.00
Paraguay	2.40	3.00	3.00	3.10
Bolivia	0.71	0.71	2.00	1.86
Uruguay	0.66	1.07	1.26	1.28
Venezuela	0.30	0.30	0.30	0.30
Chile	0.18	0.18	0.18	0.18
Colombia	0.10	0.13	0.13	0.13
South America	49.57	66.38	69.90	82.85
World	106.51	156.76	180.44	-

Source: Data from 2008-2009, 2013-2014 and 2015-2016 are of the Kassam et al. (2019) study; and those of 2018-2019 are original data from the authors.

and poultry (28% of the world market) while accounting for only 3.5% of the world population. This chapter describes the origins and current status of CA as well as NT and the NTS in Brazil, Argentina, Paraguay, Uruguay and Bolivia. These represent the top five CA countries with about 99% of the area under CA cultivation in South America.

2 The no-tillage system and Conservation Agriculture in Brazil

2.1 History and present status

The rapid expansion of agriculture in South Brazil at the end of the 1960s was based on conventional soil management, causing soil degradation due to soil erosion losses (Telles et al., 2013), representing up to 10 tons/hectare for each ton of grain produced (Amado and Eltz, 2003). Farmers, technicians and researchers started to look for new ways of establishing crops with reduced soil disturbance. Two potential solutions were suggested: (i) the use of subsoil plows, mainly in the west of Paraná, and (ii) the implementation of NT, based on minimal soil disturbance (IAPAR, 1981). Researchers from the Federal University of Rio Grande do Sul undertook some initial experiments with NT at the end of the 1960s in Santo Ângelo city, in the state of Rio Grande do Sul. However, the US-manufactured 'Buffallo' seeding machine was destroyed in a fire, and the experiment was abandoned.

The pioneer of NT in Brazil was farmer Herbert Bartz. In 1971, facing serious problems with soil erosion, he concluded that he needed to find an alternative approach if he were to continue farming. Bartz became aware of a technique in which furrows were made in the soil for depositing seeds and fertilizers but without turning over the soil. Called 'No-Tillage', the technique and was being tested at the Meridional Institute of Agricultural Research (Instituto de Pesquisas e Experimentação Agropecuária Meridional - IPEAME) - in Londrina in the State of Paraná. IPEAME was the precursor of the Agricultural Research Institute of Paraná State (Instituto Agronômico do Paraná - IAPAR[2]). The first experiments with NT were started in IPEAME in 1971 (Derpsch, 1984). Bartz contacted Rolf Derpsch, from the German Agency for International Cooperation (Deutsche Gesellschaft für Technische Zusammenarbeit - GTZ), who was then based at IPEAME, who first showed him some articles about NT. The experimental station had a German planter that could be modified to plant without disturbing the soil. Derpsch teamed up with Bartz in 1971 to try NT on his farm. Bartz adapted the machines he had available to undertake what he considered minimum tillage but was initially unsuccessful (Santin, 2018).

2 Nowadays Rural Development Institute of Paraná State (Instituto de Desenvolvimento Rural do Paraná IDR-Paraná - IAPAR-EMATER).

In 1972, GTZ and a major chemical manufacturer, Imperial Chemical Industries (ICI), undertook the first demonstrations of NT in the north of Paraná using a soya-wheat rotation, made possible by the launch of the contact herbicide 'Paraquat' (Casão Junior et al., 2012). ICI transferred its NT research team from Australia to Brazil in Brazil. ICI kept an experimental station on Bartz's farm to test Paraquat in NT trials (M. Bartz, personal communication, May 2021). The starting point for NT in Latin America can be traced to 1972 when Herbert Bartz planted the first soybean crop under NT on a commercial scale in Rolândia City, Northern Paraná State, Brazil (Santin, 2018). The team working on NT rapidly established strong relationships with others in the field: Derpsch, Bartz, a few researchers (mainly from IAPAR) and equipment manufacturers. In 1972, Bartz also visited Harry Young in Kentucky, the United States. Young was one of the pioneering NT farmers in North America and had been practicing NT for ten years, under the supervision of Shirley Phillips from the University of Kentucky. IPEAME also established trials of cover crops in 1971/1972.

By 1973, experiments were conducted in Ponta Grossa in Paraná with different soil-preparation systems, including NT. The results were published in 1974 and constitute the first research record on conservation soil management in Brazil (Casão Junior et al., 2012). Also in 1973, the IAC (Instituto Agronômico de Campinas/Campinas Agronomic Institute) undertook experiments on NT in Matão City, São Paulo State, looking at controlling erosion and spacing between terraces.

In 1975, after establishing a partnership with ICI and a number of farmers cooperatives (Cooperativa Arapoti Limitada - CAPAL, Batavo and Castrolanda), IAPAR started a wide-ranging research project on NT in Paraná (in Londrina and Carambeí) with the participation of 25 researchers. Changes were monitored in the soil, crops, weeds and the development of pests and diseases; the economic impacts were also measured. Large experimental plots of conventional and NT crops were studied for over two decades, with and without rotations, for soya, corn, cotton, beans and wheat.

By 1976, the IAPAR researchers had found that use of NT alone would not be sufficient for effective soil and water conservation, since the soil continued to be exposed to erosion from rainfall and other weather conditions. Based on their research and the practical experience of some farmers, the researchers set out the pillars of the no-tillage system (NTS) (IAPAR, 1976, 1978, 1981):

- soil movement only in the sowing or planting line (no tillage);
- the rotation of crops with different species; and
- the retention of crop residues on the soil surface due to the protection they provided to the soil.

This provided the foundation for further research.

In 1977, IAPAR established another important partnership with GTZ, which enabled the involvement of several foreign researchers with different specializations. One of the main contributions of the partnership with GTZ was the use of cover crops to facilitate crop rotation in NTS. The institute studied more than 150 different summer and winter cover crop species. As a result of this work, in 1981, IAPAR published the first national reference on NT, titled 'No-Tillage in the State of Paraná' (IAPAR, 1981). Another important initiative was the agreement in 1977 between ICI and CNPT/EMBRAPA for research and development of NT machinery (Casão Junior et al., 2012).

Many other pioneers followed Bartz's pioneering work in 1972. They included the Uemura family and the Japanese community of Mauá da Serra City, Paraná State, in 1975. In 1976, they were followed by Franke Dijkstra and Manoel Henrique (Nonô) Pereira in the Natural Grasslands Region in Southern Paraná (Campos Gerais)(Borges, 1993; Santin, 2018). The movement was characterized by pioneering farmers supported by researchers. From the 1970s to the early 1990s,there was a synergic movement involving farmers, public and private research groups, equipment manufacturers and agrochemical companies to develop NT. Pioneering research came from IAPAR and EMBRAPA – National Wheat Center (Empresa Brasileira de Pesquisa Agropecuária – Centro Nacional de Pesquisa de Trigo/Brazilian Agricultural Research Corporation – National Wheat Center). This research was supplemented by state and private research institutions, public and private universities and other groups in EMBRAPA. Chemical companies developed glyphosate and other desiccation herbicides for use in NT. Brazilian equipment manufacturers developed appropriate seeders, sprayers and knife-rollers. This collective effort was critical in making NT feasible.

Another crucial advancement in this period was the development of NT in the Brazilian Savannah (Cerrado Biome), where the participation of various pioneer farmers, the consultant John Landers, EMBRAPA's Cerrados Center, French research by Lucien Seguy and Serge Bouzinac, and the contribution of many others was of paramount importance. No-tillage practices and processes developed for subtropical conditions were applied in the Central-West Brazilian region, a tropical region to which farmers migrated because of cheap land and government incentives in the late 1970s. Expansion of land for pasture and crops was enabled initially by treating soils with lime and fertilizers, after which NT was slowly but consistently adopted in place of conventional tillage.

Following two successful short courses organized by John Landers (with agribusiness support), the Farmer Association for the Cerrado Biome (Associação de Plantio Direto no Cerrado – APDC) was established in 1992 to promote NT in the Brazilian Savannah, with headquarters in Goiânia (State of Goiás). It later expanded its network to 47 'Clube Amigos da Terra' and other farmer organizations (Freitas and Landers, 2014). As a result, in only a decade, the area under NT in Brazil increased from about 200 ha in 1972/1973 to 400 000 ha in 1983/1984 (Derpsch, 1984; Freitas and Landers, 2014).

By the early 1990s, only 13% of the farmers in the State of Paraná had adopted the NTS. One initiative to improve adoption was by ITAIPU Binacional, in the extreme west of the state of Paraná. Concerned with minimizing run-off and sedimentation from conventional agriculture into their reservoir, ITAIPU worked with 280 farmers to improve the NTS with IAPAR support. This initiative depended on partnerships with farmers willing to test new technologies to address the problems they faced on their farms. The technologies introduced in these areas included: crop rotation, growing cover crops, reducing input use, using of no-till planters and producing seeds for cover crops. Over five years, a multidisciplinary team from IAPAR worked directly with the farmers, ITAIPU technicians, cooperatives, municipalities and EMATER, among other regional agents. The team organized numerous courses, field days, meetings and working machine demonstrations, and also carried out production and distribution of cover crop seeds among other activities. The main results of this research were summarized in the book 'No-Tillage System with Quality', published by IAPAR and ITAIPU (Casão Junior et al., 2006).

In 1993, there were still only 300 000 ha of NT in Rio Grande do Sul. To increase adoption, EMBRAPA organized a series of three-day courses on topics such as soil liming, phosphate fertilization, the use of machines for small farms and the use of micronutrients and inoculants. This project was named METAS and included EMATER (the rural extension service) as well as seed companies and agrochemical and machinery manufacturers. Four years after the beginning of the project, the area under NT in the state of Rio Grande do Sul had reached 850 000 ha (Casão Junior et al., 2012).

Brazil is among the leading countries adopting NT, particularly in the use of NT in the tropics as well as adoption of NT by smallholder farms (Ralisch et al., 2003; Wall and Ekboir, 2002; Bolliger et al., 2006). The latter is of particular significance since the adoption of true (permanent rather than sporadic) NTS by smallholder farmers worldwide has been poor, remaining relatively marginal outside Brazil, Paraguay (where NT spread from Southern Brazil) and small areas in Central America, where similar systems were already traditional (Buckles et al, 1998; Bolliger et al., 2006). Ribeiro and Milleo (2002) identified that reduced labor requirements and costs from discontinuing plowing and mechanical weeding are major incentives in adoption by smallholder farmers. These advantages enabled Brazil, where the area estimated under zero-till-type land management was less than 200 ha in 1972/1973, to reach 1 million hectares in 1991, and to increase dramatically to nearly 22 million hectares in 2003/04 and 25.5 million hectares in 2005/06 (Fig. 1).

Figure 1 shows the three periods of development:

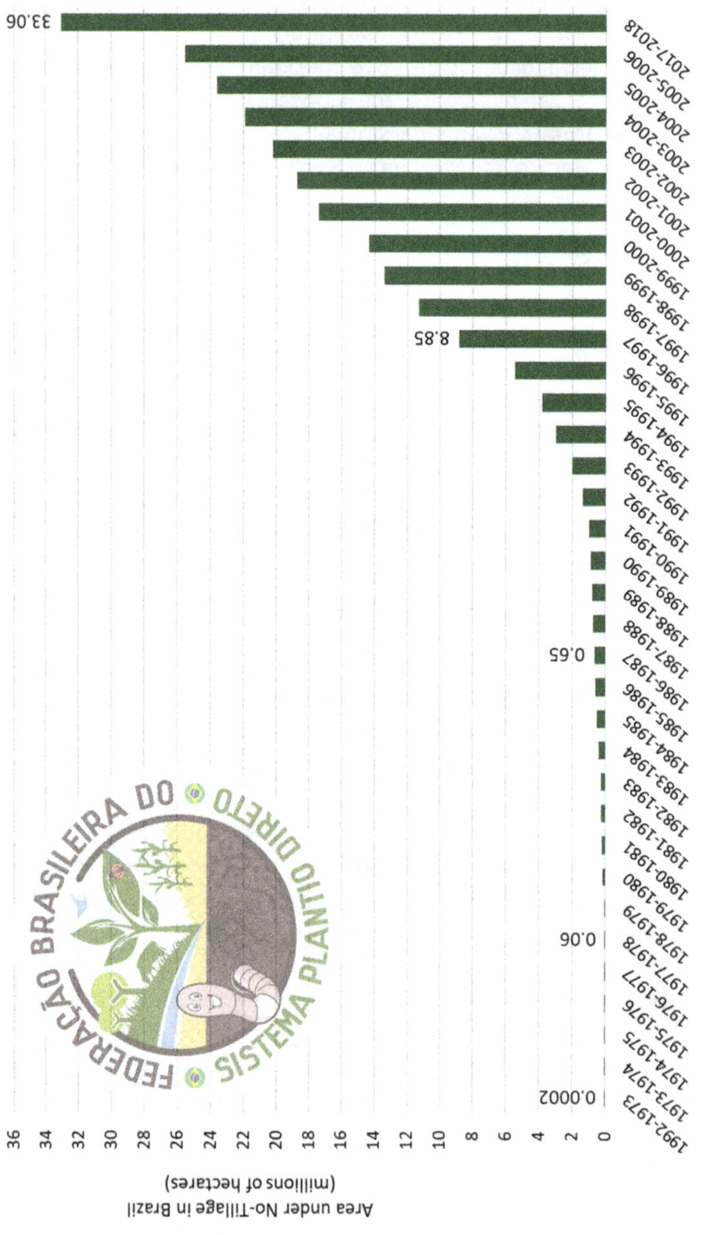

Figure 1 Evolution of the agricultural area of annual crops under no-tillage in Brazil. Source: the 1972–1973 to 2005–2006 crop years data were collected from FEBRAPDP (2021), and the 2016–2017 crop year data were collected from Fuentes-Llanillo et al. (2021).

- the initial period of **adoption by pioneer farmers and development of research** (1972–1980);
- the second period of **synergistic research and technological development,** combining efforts by the private sector and public organizations (1981–1991); and
- the third period of **massive adoption** (1992–2017).

In Brazil, the official data on area and number of farms using NT were first collected in the Agricultural Census of 2006 by the Brazilian Institute of Geography and Statistics - IBGE (Fuentes-Llanillo et al., 2013). This enabled a more detailed assessment of the magnitude of adoption of NT and NTS. Later, comparison has been made possible by the Agricultural Census of 2017 (Fuentes-Llanillo et al., 2021). Table 2 shows the evolution of the adoption of NT in terms of the area and the number of farms between 2006 and 2017, both in the major regions and in Brazil as a whole, comparing the percentage of NT adoption to the total area of annual crop production. A comparison of the 25.5 million-hectare estimates in 2006 to 17.9 million hectares of NT in the census survey probably shows the inclusion of intermediate reduced tillage systems by Fuentes-Llanillo et al. (2013) who found around 6.9 million hectares of minimum-tillage (MT) and other mixed-tillage systems (MTS) resulting in 24.8 million hectares. Annual crops include sugar cane (~10 million ha) and cassava (~1 million ha).

Between 2006 and 2017, an increase of 84.9% in the area under NT was recorded in Brazil, rising from 17.9 million hectares to 33.0 million hectares. The percentage of NT adoption has also risen, from 48.8% to 62.8 %, making it the prevalent soil management system in annual crop (AC) cultivation. This increase in the NT area has been observed in all five major regions of the country (South, Southeast, Central-west, North, and Northeast). The number of farms under NT also rose from 507 000 to more than 553 000, that is, an increase of 9.2% (Table 2). In all regions except in the Northeast, the same national trend was observed. In relating the change in NT area with the change in the number of NT farms, the conclusion is that predominantly larger farms implemented NT during the period. This pattern was observed in all major regions, most clearly in the Northeast and least evidently in the Southeast (Fuentes-Llanillo et al., 2021).

This fourth period can be characterized as one of **NT as a prevalent soil management system** (2005–2017). In the southern region, where the NTS was first implemented in Brazil, a remarkable expansion in area under NT (39.3%) was recorded with an increase from 8.6 million hectares to 11.9 million hectares. The NT system had spread mainly into areas with low agricultural land use. The average farm area under NT in the region rose from 24 to 32 ha in the period, which is the lowest in the country. The adoption of NT increased from 68.2% to 80.3% (Fuentes-Llanillo et al., 2021).

Table 2 Number of farms and area under no-tillage (NT), area under annual crops and proportion of no-tillage, according to major regions and the whole of Brazil in the 2005–2006 and 2016–2017 crop years.

	2005–2006				2016–2017				Variation	
Regions	Farms NT n	Area NT ha	Area Annual Crops ha	NT/AC %	Farms NT n	Area NT a	Area Annual Crops ha	NT/AC %	Farms NT %	Area NT %
North	20 355	220 661	727 281	30.3	28 964	1 170 982	2 021 663	57.9	42.30	430.7
Northeast	81 930	1 170 724	6,425,683	18.2	61 163	3 326 724	7 731 877	43.0	-25.35	184.2
Central west	16 184	6 523 624	9 971 176	65.4	25 823	13 726 366	17 510 845	78.4	70.09	110.4
Southeast	32 753	1 406 496	6 945 671	20.2	63 479	2 916 463	10 566 286	27.6	93.81	107.4
South	355 445	8 550 269	12 531 423	68.2	370 953	11 912 433	14 836 706	80.3	4.36	39.3
Brazil	506 667	17 871 773	36 601 234	48.8	553 382	33 052 969	52 667 377	62.8	9.22	84.9

Source: Elaborated based on data from Fuentes-Llanillo et al. (2013), and Fuentes-Llanillo et al. (2021).

The greatest expansion of the NT area in Brazil between 2006 and 2017 occurred in the Central-west region, from 6.5 million hectares to 13.7 million hectares, an increase of 110.4%. The region has the largest NT area and average NT farm area in the country, surpassing the southern region in that period. It confirms the expansion and consolidation of NT in the Cerrado Biome as the only system able to adapt that environment to crop cultivation. Adoption of NT increased from 65.4% to 78.4% after a decade (Fuentes-Llanillo et al., 2021).

In the southeastern region, the area under NT expanded from 1.4 to 2.9 million hectares, an increase of 107.4%, between 2006 and 2017. This was sizable growth but much below its potential, because of the slow speed of conversion of the sugarcane cultivation to CA. The adoption of NT witnessed a small increase, from 20.2% to 27.6% (Fuentes-Llanillo et al., 2021).

The area under NT in the northeastern region rose from 1.2 to 3.3 million hectares, corresponding to a 184.2% increase. A marked increase in total NT area was recorded in the States of Bahia (BA), Maranhão (MA), and Piauí (PI), which belong to area of the Cerrado Biome known as MATOPIBA, an acronym that includes the three States (MA, PI, BA) together with Tocantins State (TO) located in the northern region. The adoption of NT increased from 18.2% to 43.0%, although there was a decrease of 25% in the number of farms. This shows that expansion occurred through land concentration (Fuentes-Llanillo et al., 2021). The MATOPIBA region is considered the new agriculture frontier of Brazil.

In the northern region, the total area under NT increased from 0.2 to 1.2 million hectares, that is, an increase of 431%. This increase is important, since part of the area belongs to the Amazon Forest Biome. Here, more sustainable crop and soil management is particularly important given the natural fragility of the soils in the area. Adoption of NT increased from 30.3% to 57.9% (Fuentes-Llanillo et al., 2021). Crop cultivation in the Amazon is a sensitive topic. The annual crop area in the Amazon Forest Biome in Northern Brazil increased by 0.22% between the 2005–2006 and the 2016–2017 crop years. In the 2005-2006 crop year, the annual crop area was around 2.0 million hectares; for the 2016-2017 crop year it was more than 2.5 million hectares. The growth in NT observed in the Amazon Forest Biome in Northern Brazil is minute when compared to the total area of agriculture and livestock. Between 2006 and 2017, the NT area represented 0.4% and 1.8% of the total agriculture area, respectively (IBGE, 2021a,b) (Table 2).

Conservation Agriculture of annual crops is limited to the NTS alone. No-till integrated crop-livestock-forest (NT-ICLF) systems, involving direct sowing and permanent soil cover, are growing in Brazil. They include crop-livestock, agroforestry, silvopastoral and agrosilvopastoral systems. Conservation Agriculture also includes fruit tree crop and vegetable cultivation using direct sowing and permanent soil cover. Forest plantations with minimum soil disturbance are important for protection of water resources in watersheds. According to EMBRAPA (2020) and Polidoro et al. (2021), the area under ICLF has risen from 2.60 to 15 million hectares between 2006 and 2018, showing huge expansion of these CA systems in the country, as shown in Figure 2.

Since ICLF systems include part of the NT area surveyed in the census, some overlap in total CA nationally must be taken into account in order to make a reliable estimate of CA adoption. If one takes account of the standard ICLF system of one season of cropping of each three years combined with 22 to 26 months of pasture, the additional area of CA coming from ICLF was 1.7 million hectares in 2006 and 10 million hectares in 2018 that should be considered in the total national CA account in 2006 and 2017. The 17.9 million hectares of NT added to 1.7 million hectares of ICLF in Brazil totaled at least 19.6 million hectares in 2006. In the same way, the 33 million hectares of NTS added to 10 million hectares of NT-ICLF corresponds to an **area under CA in Brazil of least 43 million hectares in 2018-2019**. Unfortunately, there is still little information about the magnitude of other CA systems, such as the no-tillage system in perennial crops (NTSP) and the no-tillage system in vegetables (NTSV).

2.2 Challenges and future trends

Brazil is a huge country with a great diversity of regions and biomes, climatic and soil conditions. However, it is possible to give a general idea of CA

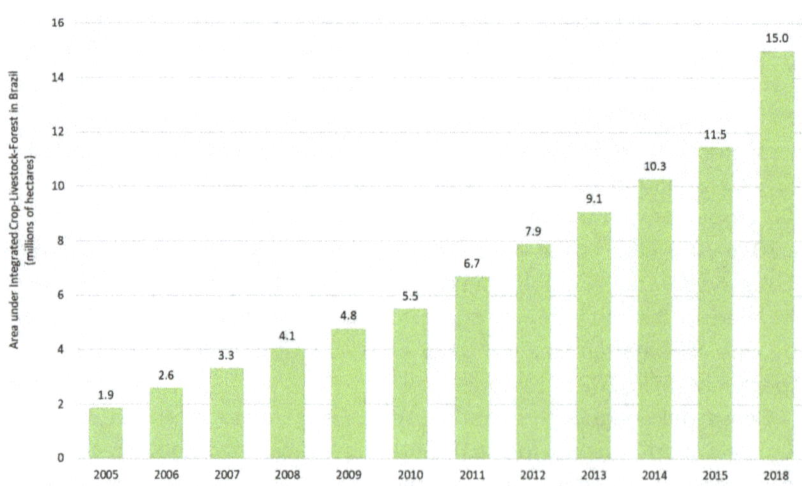

Figure 2 Evolution of area of Conservation Agriculture under integrated crop-livestock forest systems (ICLF) from 2005 to 2018 Brazil. Source: Data from ICLF Development Network (2020).

challenges and trends. The major annual cropping systems are based mainly on soybean. The most common rotations are soybean-maize, soybean-wheat, soybean-cotton, soybean-pasture, soybean-millet and soybean-fallow. Sole crop systems include sugarcane, rice and cassava.

After more than 48 years of introducing NT practices commercially in Brazil, there is still a need to improve observance of the three CA principles:

- continuous minimal soil disturbance;
- permanent soil cover; and
- crop rotations and related practices, including cover crops.

The partial or limited adoption of these three principles has led to several problems. These include: soil compaction, erosion, increase in pests and diseases (especially nematodes) and weed pressure, leading to a higher pesticide use. The rapid uptake of genetically modified herbicide-resistant crops from 2000 led to the overuse of glyphosate and the development of resistance, which is making this practice unsustainable. In addition, the global demand for soybean has also encouraged monocropping in the summer season, resulting in poor rotations without proper use of the second and third principles of CA.

Diverse leguminous cover crop species, such as Leguminosae/Fabaceae, as well crucifers and grasses, are a particularly important part of the CA/NT systems in Brazil. Grasses are also an important component of the CA/ICLF system, which has been validated in different Brazilian agroecological zones in protecting soil, enhancing soil attributes and increasing farm net income in a sustainable way (Moraes et al., 2014; Silva et al., 2014). Increased use of cover crops and crop rotation has improved NT success with higher amounts of biomass and improved soil quality (biological, physical and chemical), resulting in increased crop productivity. Beyond the use of individual cover crop species, properly mixing cover crops (a cocktail/mix of 2 or more species) for a specific use, with appropriate diagnosis, is leading to better results in weed control, decrease of nematode populations and root diseases as well as improved nutrient cycling and biodiversity (Calegari et al., 2020).

The main challenges for CA include the following:

1. Because of the dominance of the soybean crop in summer, there is a lack of crop rotation and diversification and poor permanent soil cover, leading to soil erosion and water contamination.
2. Increase in water and soil losses due the removal of terraces and soil compaction from excessive reliance on NT only and an increase in the size of machinery used.

3. Poor management of transgenic herbicide-resistant cultivars, resulting in resistant weeds.
4. High incidence of nematodes and fungal diseases, leading to high pesticide use – resulting in resistance and loss of biodiversity.

The opportunities include:

1. Improvement and scaling up of NTS for sugarcane, cassava and irrigated rice.
2. Encouraging the beneficial use of biological technologies and cover crop mixes for soil bio-activation together with adoption of biological control of pests and diseases, combined with biological herbicides or non-chemical weed control.
3. Development and adoption of ICLF systems to exploit synergies from the inclusion of animals in cropping and forest systems, permitting CA/NTS expansion into degraded pastures without deforestation.
4. Innovations in machinery design to improve no-till seeding, especially to reduce soil disturbance and exposure, to minimize hairpinning and accumulation of residues in the slot with high levels of residue cover, and to increase seeding depth uniformity and soil-seed contact (Araujo et al., 2020), as well as to control soil compaction with heavy machinery through controlled traffic systems.
5. Use of the NTSV. This is an innovative system developed by EPAGRI (Santa Catarina State Research and Extension Services) from 1998 (Masson et al., 2019). In 2019 there were 1200 farms adopting NTSV, with up to 3000 ha in Santa Catarina State, representing 10% of the area under vegetable cultivation (EPAGRI, 2021). There are technical guides and manuals for diverse crops like tomato, onion, garlic and cucumber among others. There is a potential for expansion in other regions of Brazil and Latin America.

One way to measure and assure the quality of NTS in Brazil is the use of the NTS Participatory Quality Index (PQI or IQP in Portuguese). This has been developed by FEBRAPDP as a tool to qualify and monitor the farm system. The aim of the PQI is to assess NTS in practice and to help farmers make better operational decisions. The PQI is an integrated index, consisting of a set of indicators designed to predict potential impacts on soil quality (Nunes et al., 2020; Roloff et al., 2012).

Overcoming technological and other barriers depends on a mix of public policies, economic incentives for crop diversity and changes in farmers' attitudes. It is necessary to develop biological weed and pest control to replace herbicides and chemical control of pests and diseases. The expansion of integrated crop-livestock-CA systems to replace degraded pastures in the Brazilian savannah can meet food production needs without tropical deforestation.

3 The no-tillage system in Argentina

Historically, Argentina has been strongly identified with grain crop production and with traditional farming methods based on soil tillage. The conventional or 'tillage-based' paradigm in agriculture has resulted in long-term agro-ecosystem degradation due to damaging effects such as soil erosion, loss of organic matter, soil nutrient depletion and desertification. It became necessary to develop new technologies that both preserve natural resources and satisfy the increasing demand for agricultural products.

The NT system is a productive system based on the absence of soil disturbance and the maintenance of soil mulch cover and use of rotations. It also includes good agricultural practices (GAPs) related to soil management, balanced crop nutrition, integrated pest management, responsible use of phytosanitary products, and integrated cattle management. NTS is also referred to as CA and it is the predominant type of crop management in Argentina, where almost all annual crops are under permanent NT cultivation.

The first experiences of NT in Argentina were not always successful. Farmers faced problems with seeding, weeds and declines in yield. The costs were high and the machinery inadequate, involving significant risks. Until the mid-1970s, tillage continued to predominate. However, some farmers and technicians persisted in the development of NT-based systems. The exchange of information and establishment of links with farmers in other countries led local producers to start trials and test new machinery. In 1989, these innovative farmers created the Argentinian No-Tillage Farmers Association (AAPRESID - its acronym in Spanish). From the beginning of the 1990s, this new approach started to spread among Argentinian farmers concerned about soil erosion and improved water capture and storage for rainfed agriculture.

AAPRESID has played a fundamental role in promoting and disseminating NTS, functioning as a bridge between theory and practice by building a network for exchanging information and experiences. Building relationships with scientists from the National Institute of Agricultural Technology (INTA) was also important since it improved management of trials. Collaboration with the agro-industrial sector was also key in developing adapted seeders, sprayers and harvesters suitable for NT farming. The evolution of the NTS in Argentina was based on a virtuous cycle of collaboration led by farmers working with scientists, universities and companies. In these 30 years of constant learning, a new paradigm of agricultural production has been established, with benefits reaching all the Argentinian society.

Some of the benefits of this new method of production are (Albertengo et al., 2014):

- less soil erosion (reduced by up to 96%);
- less fuel use (reduced by up to 66%);
- better maintenance or improvement in soil organic matter;
- higher efficiency of water use;
- increased soil fertility;
- lower production costs;
- higher production stability; and
- higher potential long-term yield.

The following sections look at the evolution and the present status of NTS cultivation in Argentina.

3.1 Evolution of no-tillage adoption in Argentina

Currently, there are around 200 million hectares globally that are cultivated using CA methods. Argentina has around 35 million hectares under NT. In the last ten years, NT in Argentina has grown to cover over 90% of cultivable land for grain crops (Fig. 3).

The implementation of NT has proven to be highly effective in controlling soil erosion and increasing organic matter content. However, it should be noted that isolated use of no-disturbance technologies does not necessarily translate into soil quality improvements. Agricultural production in Argentina still faces problems related to reduction in soil fertility, organic matter and carbon content, as well as development of resistant weed biotypes. Losses in P, Mg, Zn, Br and K levels have been observed in some agricultural regions (Wyngaard and Reussi Calvo, 2019). In addition, the area of soybean monoculture has remained at around 20% (S4 + AAPRESID). This shows that the implementation of sustainable practices such as crop rotation has not developed alongside NT adoption. The main role of AAPRESID is to improve understanding and implementation of the range of CA practices.

3.2 Toward a climate-smart agriculture: Always-Green Agriculture

Alongside existing problems related to the simplification of the system, new challenges have arisen, such as climate change. Based on research by AAPRESID farmers with the support of scientists, the association has set a new challenge: Always-Green Agriculture (ASV), designed to leave fields green for as long as possible.

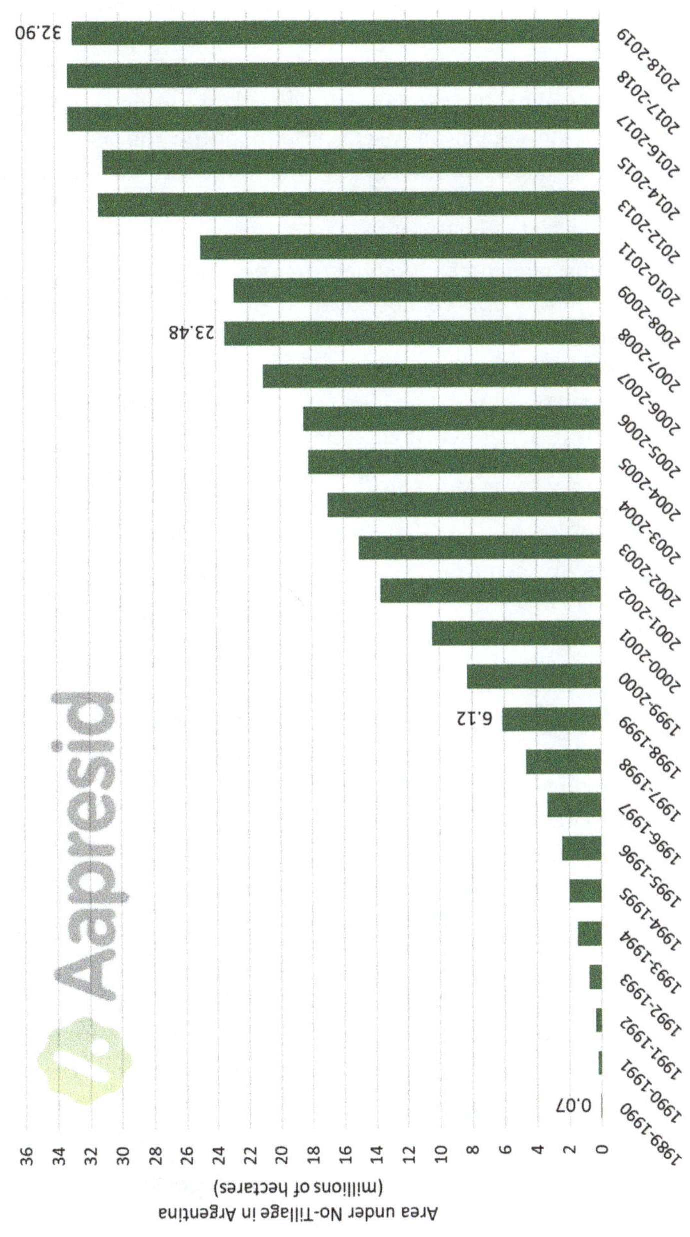

Figure 3 Evolution of area under no-tillage from 1989 to 2019 in Argentina.Source: Bolsa de Cereales de Buenos Aires/AAPRESID, 2019.

ASV is a production model that aims to enhance soil biological activity through the presence of crops growing in the soil for as long as possible. This means having living and green fields whenever agroecological conditions allow it.

According to research underpinning ASV, the rhizosphere and soil microbiomes govern key natural agroecosystem processes. Physical and chemical aspects of soil health, cycling of nutrients and carbon sequestration, as well as crop growth, health and yield all depend on the activity of soil microorganisms. This microbial activity requires interaction with crop roots. Maintaining 'always green' fields with crops growing for as long as possible helps to maintain this microbiological machinery.

There are currently three regional groups, involving farmers and scientists, working on ASV approaches covering almost 50000 hectares. Each has different objectives, including new grain crops to take advantage of water surpluses; or planting new crops to promote ecosystem services such as improved water retention, nutrient cycling, carbon sequestration, weed control etc. The first site (Justiniano Posse) is exploring the availability of surplus water to produce more wheat using greener rotations. The second (Pergamino) and third (Bragado Chivilcoy) sites are seeking to improve on standard rotations by combining grain and service crops specifically designed to promote different ecosystem services, to increase both yields and sustainability. Research at these sites has shown that leaving the fields green for as long as possible can provide the following benefits (AAPRESID, 2020):

a. Higher water-use efficiency and reduction of flood risk by up to 40%;
b. Reduction in the number of herbicide applications by up to 50%;
c. Reduction in environmental impact of herbicides by up to 60%;
d. Reduction in the use of synthetic fertilizers by up to 20%; and
e. Improvement in soil carbon sequestration by up to 60%.

ASV is proving to be not only technically feasible but also economically profitable. We are evolving toward an agriculture that helps to improve (rather than damage) the environment: improving water management by reducing the risk of runoff and floods; decreasing dependence on synthetic inputs (fertilizers and herbicides); and maximizing carbon sequestration and mitigating GHG emissions, therefore reducing net emissions from agricultural production.

4 Evolution and current state of the no-tillage system in Paraguay

Agricultural production in Paraguay during the 1970s and 1980s was characterized by intensive tillage. High levels of soil disturbance occurred with the use of equipment such as motorized disc plows, animal traction moldboard plows and heavy harrows followed by light harrows for weed control. The conventional tillage system promoted soil losses through rill, sheet and gully erosion. This led to depletion of natural fertility, loss of surface horizons, decreased organic matter content and nutrients in the soil. After five or six years of tillage, soil lumps formed that needed two or three harrowing operations and the action of rain to break and allow germination of wheat and soybean seeds. It was even common to drag wooden logs behind the harrows to bury the hard lumps, to facilitate the action of the seeders (Aoyama, 1984). Soil management and conservation techniques were based on lifting of contours and construction of large terraces ('murunduns') where intensive tillage was practiced. However, terraces suffered continued erosion damage from rain. Farmers were forced to look for more efficient methods to control erosion such as NT where crops are sown directly into the residues of a previous crop without tillage.

4.1 The beginning of no-tillage in Paraguay

In a classic text 'AGENDA AND AGRICULTURAL MENTOR: A Farmer's Guide', Bertoni (1927) recommended 'to slash without burn through continuous crop re-establishment without tillage, lowering costs, to make possible intensive and rational agriculture in any country wanting to progress. Slash without burn is the only way to perpetual preservation and improvement of soil fertility with the help of legumes species'.

However, NT in Paraguay did not start until the mid-1970s. A milestone in 1976 was a visit made by a group of farmers from Colonias Unidas in the Itapúa Department in Paraguay to Rolândia in Paraná State in Brazil to see Brazilian NTS pioneer Herbert Bartz. Inspired by him, this group and others started to investigate mechanized NT seeders through visits to relatives using this equipment in Brazil. This created a cycle of transforming conventional seeders into NT seeders. The first seeders from Ibirubá in Rio Grande do Sul State in Brazil arrived in Paraguay in 1977 via the Colonias Unidas Cooperative. The process of development of the NT system had begun. Lack of herbicides and difficulties in adapting seeders were the main initial limitations.

Research into the NTS/CA in Paraguay started in the 1980s. In 1980, the Pirapó Agricultural Experimental Station in Itapúa Department started to include NTS/CA in its research program on cultivation methods. From 1982, the National Agricultural Technology Center in Paraguay (CETAPAR) started an intensive training campaign on NT for Japanese farmers in Paraguay (Romero, 1995). From 1983, Brazilian Fankhauser disc seeders imported from Brazil were used for NTS in Colonia Yguazú in Alto Paraná Department. In that year, the president of the local cooperative, Akinobu Fukami, applied NT on a commercial scale on his farm, becoming the first farmer to practice the system successfully. Farmer cooperatives played a key role in the dissemination of NTS/CA by providing support and/or financing in acquisition of seeders. The early examples included the Colonias Unidas de Itapúa Cooperatives, including the Yguazu Cooperative. Contact between local cooperatives and those in Brazil from 1983 provided access to technical information. Cooperatives such as Pirapó, La Paz, COPRONAR de Naranjal, Raúl Peña, Pindo de San Cristóbal, Friesland, Volendam, Sommerfeld, Bergthal and Río Verde were crucial to the dissemination of NTS/CA.

In 1984, AGRIEX, a Paraguayan agribusiness company, carried out extensive field tests with the first imported seeder from the United States. The initial results were mixed. While the use of seeders involved less labor, time and cost, the yields were lower. The initial tests focused on reduction of soil losses as an alternative to the construction of contours to control water erosion. The first CA field demonstrations were carried out 1986 using white lupine in a pre-maize rotation. In 1986, Eugenio Altenhofen began permanent NTS on a commercial scale on his farm in Lapachal de Colonia Obligado in Itapúa Department. In 1988, Dr Patrick Wall and D. Mohan Kohli of CIMMYT Paraguay organized the first course on NTS/CA in Pirapó Colony. They invited the pioneer Brazilian farmer Manoel Henrique (Nonô) Pereira, who had 12 years of experience in NTS, and the technician Josué Pavei, both from Ponta Grossa in Paraná State in Brazil where NTS had already been adopted in over 400 000 ha (Derpsch and Moriya, 2011).

In 1989, the Regional Agricultural Research Centre (CRIA) and the Agricultural Mechanization Centre (CEMA), both linked to the Paraguay Ministry of Agriculture and Livestock (MAG), began testing NT and rotation systems with the cooperation of CIMMYT (Romero, 1995). Erni Schlindwein began to use NT on his farm in Naranjal in Alto Paraná Department in 1990. Victor Dickel began to use NT on his farm in Natalio in Itapúa Department in 1991, and now has 30 years of continuous use of NT, crop rotations and cover crops. From 1992, the agronomist Nelly Venialgo at the Centre for Agricultural Mechanization (CEMA) of MAG developed methods for measuring soil/sediments lost by water erosion at the field level in different conventional soil preparation systems (disk plow plus light scraping) versus NT. Her results showed the positive effects of NT and the role of cover crops in reducing soil losses.

In 1986/1987, Georg Birbaumer and Miguel Angel Florentín (MAG-GTZ) liaised with Ademir Calegari (IAPAR Pato Branco, Paraná, Brazil), including visits to the IAPAR Experimental Station in Pato Branco in Paraná State in Brazil and the Choré Experimental Station in Paraguay, to investigate the use of green manures by MAG-GTZ through Rolf Derpsch. This validated the use of cover crops as mucuna, rye, vicia (hairy and common), forage pea, forage turnip, crotalarias, lolium and black oats, among others. This was important in the development of NTS/CA for soybean, maize and common bean. Experiments in agroforestry combining these crops with tree species like *Ilex paraguariensis* and *Paulownia tomentosa* were conducted in Edelira in the Itapúa Department by Magín Mesa in the 1990s.

The development of farmers' organizations focused on NT was crucial to its expansion. This included the creation of organizations such as the Colonias Unidas Farmers' Association for Sustainable Agriculture (APASCU) in 1991 with Rudi Dressler as its first president, the Naranjal NT Association (ASIDINAR) in 1991 (with Erni Schlindwein as the first president) and the Animal Traction NT Association (ASDTA) in 1992. These provided opportunities for on-farm training and communication of knowledge between farmers. Other organizations such as the Paz NT Association (ASIDIPAZ), Pindó NT Association, Pirapó NT Association, Yguazú NT Association and Carmen del Paraná NT Association (RECOSID) were created and joined the Paraguayan Federation of No-Tillage for Sustainable Agriculture (FEPASIDIAS) in 1993.

These efforts were complemented by conferences, symposia and congresses such as the First Farmers' Meeting on NTS in 1993 organized by the Colonias Unidas Cooperative at Colonia Obligado in Itapúa Department. A second farmers' meeting on NT in 1994 included contributions from the MAG-GTZ Soil Conservation Project. After the Second Latin American Meeting of No-Tillage in Small Farms in 1996, an on-farm soil-recovery system support project was launched in Edelira in Itapúa Department. Each participating farm developed a 20 000 square meter NTS/CA area in the Department of Paraguarí. This made it possible to validate the role of NTS in the recovery of highly degraded soils over a three-year period. With technical assistance, these practices were disseminated to family farms over the following years.

The Paraguayan Government only became involved in 1992, when the Ministry of Agriculture and Livestock (MAG) began the promotion of the NT system in the Department of Alto Paraná in Colonia Santa Fe Pikyry. In 1993, the MAG initiated the MAG-GTZ Soil Conservation project which validated the use of NT/CA in basalt clay soils through rotation of crops incorporating green manures. Research was conducted at the experimental station of CETAPAR-JICA in Yguazú in Alto Paraná Department involving MAG, JICA and GTZ. Work on sandy soils was carried out at MAG's Choré Experimental Station.

The MAG-GTZ Soil Conservation Project facilitated dissemination of knowledge through tours within the country and farm visits to Paraná and Santa Catarina States in Brazil. Since 1993 awareness/training activities were carried out jointly with the Paraguayan Chamber of Cereal and Oil Exporters (CAPECO), CETAPAR and MAG through the Soil Conservation Project. Training of public and private sector technicians was prioritized from that period. From 1996 to 2001, MAG worked with CETAPAR JICA and CAPECO to provide training in NTS for Annual Crops for more than 150 technicians. From 1999, MAG set up the National Soil Management Conservation and Recovery Program, which, in the following 15 years, has spread NTS to family farms growing grain crops.

Such initiatives have resulted in high levels of adoption of NTS/CA, encompassing more than 90% of the annual crop area by 2000. This success has attracted visitors from South Africa, Cuba, Brazil, Chile, Argentina, Germany, Australia, New Zealand, the United States and other countries. In 2003, Dr Norman Borlaug, winner of the 1970 Nobel Peace Prize, visited NT family farms in the country. A presentation on NT was also made in 2003 to the Paraguayan President, Juan Carlos Wasmosy. The event was organized by CAPECO and included presentations by a number of NT pioneers from Brazil (Manoel Henrique Pereira), and Argentina (Victor Trucco) as well as Rolf Derspch.

According to Derpsch et al. (2000), this rapid expansion of NTS/CA was the result of:

1 efficient and economical control of soil erosion provided by direct planting systems;
2 availability of knowledge on how to practice NT in the region;
3 the widespread use of cover crops;
4 consistent and positive messages by the public and private sectors (avoiding confusion with other forms of 'conservation tillage' such as minimum tillage and reduced tillage);
5 effective communication from farmer to farmer;
6 strong economic performance of NT compared to conventional tillage; and
7 a global market that required highly competitive farmers able to manage production costs.

One key factor in rapid growth has been the technical support provided by GTZ, MAG and CAPECO. The key figures have been Rolf Derpsch, Ken Moriya and Luis Enrique Cubilla who developed strategic alliances, policies and training programs. The dissemination routes include newspapers, radio and television.

The first International Workshop was held at CAPECO in October 2002: 'Standardisation of soil analysis and calibration experiments proposal for fertilizer and limestone recommendations for NTS in Paraguay', organized with the Federal University of Santa Maria (UFSM) in Brazil, together with National Soil Management, Conservation and Recovery Program of the Ministry of Agriculture (MAG). In January 2003, CAPECO and UFSM signed a technical and scientific cooperation agreement to conduct the first NTS experiments to develop fertilizer recommendations for Paraguay's major grain crops. The agreement has generated the first manual of fertilizer recommendations for soybeans, corn, wheat and sunflower under NTS cultivation in Paraguay (Cubilla et al., 2012). This provided recommendations for N, P and K application in wheat, soybeans, maize and sunflower grown under NTS/CA for the eastern region.

4.2 Adoption of no-tillage in Paraguay

In 1992, the area under NT in Paraguay was 20 000 hectares (Fig. 4). Steady progress increased the area under NT, reaching more than 90% of soybean cultivation in 2000, 1 million hectares of NT in 2001, 1.6 million hectares in 2004, 2.1 million hectares in 2007, 2.6 million hectares in 2011, 2.9 million hectares in 2013–2014, around 3.0 million hectares in 2015–2017 and 3.1 million hectares in 2018–2019. In 2013, the percentage of land with uninterrupted NT declined from 90% to 70.5% (2.227 million ha) of the total annual crop area of 3.158 million hectares, due to temporary interruptions in NT from the use of harrows mainly for weed control in wheat cultivation in the winter (Pekholtz, 2018). A key problem has been the emergence of glyphosate-resistant weeds such as capi í pororó (*Digitaria insularis*) and mbu (*Coniza* spp). This problem was been exacerbated by insufficient soil cover due to the lack of crop diversification/rotation with, for example, cover crops and other winter income crops. Lack of crop diversification has also exacerbated the incidence of root diseases and nematodes.

After intensive training organized by FEPASIDIAS and supported by different companies, the percentage area of uninterrupted NT increased from 2014, and by 2017 reached 2.645 million hectares, representing 78.1% of the total annual crop area of 3.389 million hectares (Pekholtz, 2018). In 2016 and 2018, two national NT meetings were held with support from CAAPAS. The success of such activities meant that by 2019/2020, some 3.1 million hectares were estimated to have been planted under NT (90% of the total crop area).

Adoption of NT in traditional family farming has been limited, reaching about 30 000 ha in 2010. The uptake has been supported by inputs and equipment provided by the National Soil Management and Conservation Program Projects organized by the Ministry of Agriculture and Livestock via the Sustainable Management of Natural Resources Project (PMRN 2000–2015), supported by the German KfW (Kreditanstalt fur Wiederaufbau). Other initiatives include the Program for the Modernization of The Public Management of Agricultural Support (PAGRO 2011-2016). The uptake has been dependent on provision of technical assistance on an ongoing basis.

Three milestones give an idea of the expansion of the use of NT and NTS/CA in Paraguay. By the mid-1990s, NT had become the predominant system in Paraguay, representing over 50% of the total annual cropping area. By 2001, Paraguay was considered by CAAPAS as the world leader in adoption of NT, with around 90% of the total cropped area, while in 2019, the NT area reached more than 3.0 million hectares with more than 90% of the total cropland, compared with 2.4 million hectares in 2009 (Calegari et al., 2020).

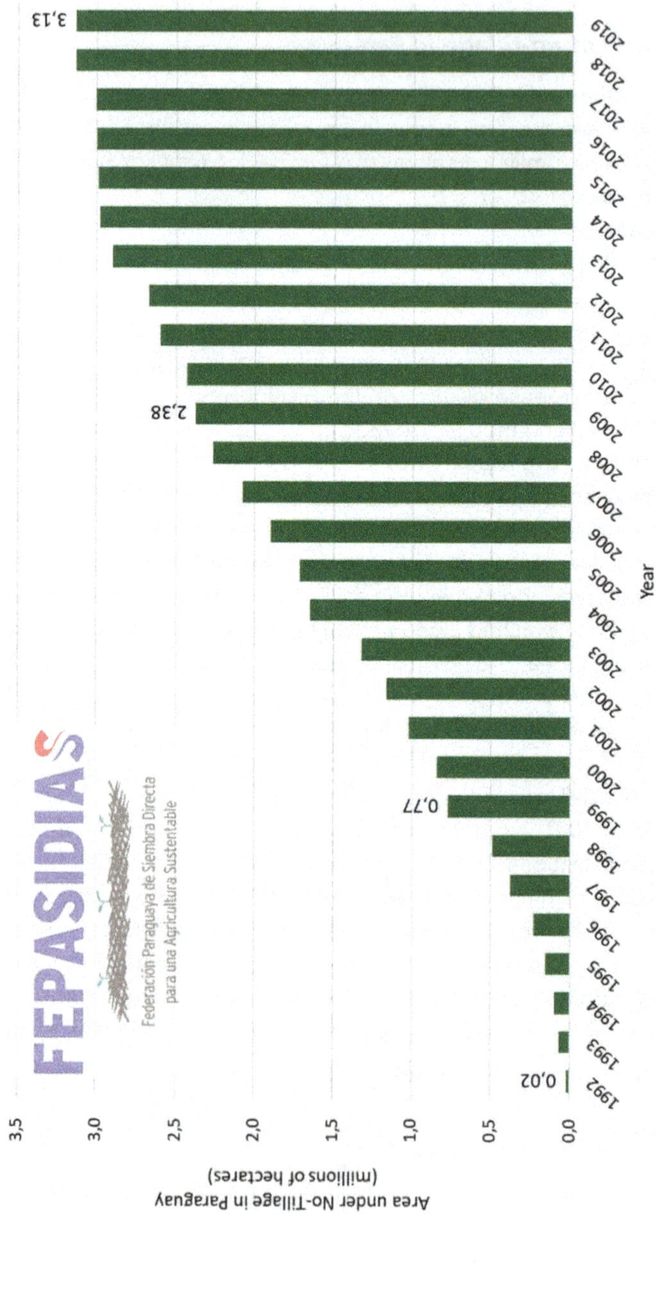

Figure 4 Evolution of the area under no-tillage from 1992 to 2019 in Paraguay.

4.3 Impacts of the adoption of no-tillage

The obvious initial impact has been reduction of soil erosion which was widespread in conventional tillage agriculture. NT has also allowed an improvement in water quality. An example is a study carried out in Itapúa in the Capiibary water basin in which most farms are under the NT system. The Ministry of Environment and Sustainable Development (MADES) and The Federal Institute for Geosciences and Natural Resources of Germany (BGR) investigated the presence of 598 active agrochemical ingredients. No residues of fertilizers, herbicides or other pesticides were found above critical levels in both surface and groundwater. When applied correctly, the environmental impact of herbicides used in the direct planting system is minimal. A key aspect of NTS/CA has been crop rotation and the use of different species of green manures. This has increased organic matter in soil and thus improved carbon sequestration (Calegari et al., 2020; Florentin et al., 2010).

The shift from conventional tillage to NTS has also meant an increase in yields per hectare. This is the case with soybean cultivation, where yields have increased in tandem with the increase in the NT area. This was the case in the 1990s. From 2000 to 2010, soybean cultivation expanded into less productive areas, but as the NT area increased, average yields remained stable. From 2010 to 2020, yields have increased again as both the overall area of soybean planting and area with NT have increased. Higher profitability has resulted both from the higher yields and the reduced need for labor as well as lower fuel and input costs from moving to NT (Calegari et al., 2020). No-tillage, when combined with the use of crop rotations and different species of cover crops/green manures, has also proved beneficial for smallholder farmers, when combined with animal traction equipment and small tractors. These practices have allowed the recovery of degraded soils, reduced costs and increased income in a way that is environmentally sustainable.

4.4 Challenges to the evolution of the no-tillage system

While NTS in Paraguay is an established farming system, it still faces challenges. In the case of traditional family farming, there needs to be greater access to NTS/CA equipment suitable for smallholders, as well access to basic inputs (e.g., cover crop seeds) and technical advice. The amount of soil cover In CA grain farming systems is also still insufficient. There is a need to test/validate/consolidate mixtures of cover crop species suitable to local needs. Increasing the extent and quality of coverage (green manures) is needed so that the soil can achieve its maximum productive potential through accumulated organic matter (Calegari et al., 2020; Florentin et al., 2010).

There is a need to develop NT/CA practices for poorly-draining marginal soils, as well as develop/validate NT for semi-arid and/or high-sodium areas. These conditions require application of all three pillars of NTS/CA for sustainable intensification to be feasible: continuous minimum soil disturbance, permanent soil cover and the use of cover crops and crop rotation. With all the NT experience accumulated over the years in Paraguay, good training is still needed for farmers as well as field technicians, especially those working in rural extension services, facilitating implementation of innovations originating from research or successful farmer experience.

5 No-tillage and Conservation Agriculture in Uruguay

The area used for agriculture in Uruguay is approximately 16 million hectares, with natural and planted pastures occupying 80% of this area (García-Préchac et al., 2004). The Uruguayan economy is dominated by livestock production. The country's main agricultural production system is crop-pasture rotation (CPR) (Díaz-Zorita et al., 2002). According to the Office of Agricultural Statistics (DIEA), part of the Uruguayan Ministry of Livestock, Agriculture and Fisheries, one issue in the implementation of CA in Uruguay is that 65% of crops are sown on leased land with contracts renewed each year. This makes it difficult to plan medium-term crop rotations and other investment strategies in soil and water conservation (Derpsch et al., 2010). This means that farmers keep pastures under intensive grazing for several years until they begin to show signs of degradation. Crops are then grown for several years. The main crops are wheat or barley in winter and sunflower in summer, with pastures using different combinations of grasses and legumes (García-Préchac et al., 2004).

No-tillage (NT) has been used in the CPR system since the 1990s. In the early 1990s, concern about soil erosion during the arable cultivation stage of CPR led farmers to consider CA. This group of Uruguayan farmers founded the 'Asociación Uruguaya Pro Siembra Directa (AUSID)', an NT support organization. They started to communicate with similar farmers' organizations in Brazil, Argentina, Chile and Paraguay to share their own experiences and research. The AUSID is still is a crucial institution in the adoption of NT throughout Uruguay, for example in carrying out research in cooperation with other organizations and in extension work with farmers (Calegari et al., 2020).

The combination of CPR and NT/CA, without soil disturbance during crop cultivation, has been successful in improving soil quality. The strong root systems of dense and fibrous grasses promote improvements in soil aggregation, while soil cover improves moisture retention and soil organic carbon. NT operations also mean reduction in fuel use and labor costs (Calegari et al., 2020). The crop-livestock systems of Uruguay can be considered complete CA systems.

According to Kassam et al. (2019), the area devoted to CA in Uruguay in 2009-2010 was 655 100 ha; in 2013-2014 it was 1.072 million hectares, and in 2015-2016 it was 1.260 million hectares, that is, a 92% increase between 2009-2010 and 2015-2016.

In a survey conducted during 1999-2000, it was found that NT was used in 35% of farms (Scarlato et al., 2001). However, with farms mainly oriented to crop production, NT adoption increased to 52% of farms (Ernst, 2001). In the mid-2000s The 'Federación Uruguaya de Grupos CREA (FUCREA)' reported that about 60% of Uruguayan farmers had adopted NT due to concerns about soil erosion and degradation (García-Préchac et al., 2004).

With the greater internationalization of agribusiness in Uruguay, the area devoted to grain crop production increased from 298 000 ha in 2001-2002 to 1.2 million hectares in 2018-2019, replacing areas previously using CPR systems. In 2018-2019, the area cropped with soybean was one million hectares. Since the 2000s, NT associated with soybean production has been gradually replacing conventional planting in Uruguay. During 2006-2007, approximately 82% of cropland, that is 672 000 hectares, were under NT, compared with 2000-2001 when it was only about 119 000 ha, that is, an increase of 464% (Derpsch et al., 2010). Grain production in Uruguay thus has a high rate of NT adoption, surpassing 90% of the total cultivated area in 2019 (Calegari et al., 2020).

However, according to Arbeletche (2020), the expansion of NT in soybean cropping has occurred without adequate diversification of species in crop rotation, which sometimes exposes the soil to erosion and reduces soil health. These poorly diversified grain production systems have thus become highly dependent on agrochemicals (Guibert et al., 2011). About 80% of the area used with grain production systems in Uruguay has been affected by soil erosion and degradation (Netzeband and Arbeletche, 2016). The adoption of only the no-tillage practices, without all the CA practices, does not ensure soil health and quality, and consequently, restricts yields and sustainability of farming systems. According to Calegari et al. (2020), there is a need for government intervention to encourage cover crop use and provide subsidies for good quality CA to promote the environmental, economic and social sustainability of farming systems.

6 No-tillage and Conservation Agriculture in Bolivia

6.1 History and present situation

No-tillage seeding with a planting stick was the traditional agricultural technique used by native people in Latin America, before European colonization introduced plowing. Conventional agriculture in Bolivia uses plows or disc harrows, roman plows or even wooden ards, employing animal traction to clear and prepare land. No-tillage farming was first introduced in the eastern Andean valleys of Santa Cruz in 1980 via foreign aid projects, but the practice did not catch on initially. In 1990, NT agriculture was again promoted in the lowlands of Santa Cruz by the International Wheat and Maize Improvement Center (CIMMYT) in collaboration with the International Center for Tropical Agriculture (CIAT) and the Association for Oil Crops and Wheat (ANAPO – Asociación de Productores de Oleaginosas y Trigo). Long-term experiments were set up on the ANAPO experimental farm at Cuatro Cañadas, northeast of Santa Cruz. The cropping system was developed to incorporate the three principles of CA. In the early 2000s, ANAPO published direct seeding manuals for farmers. With the help of farmers and machinery companies from Brazil and Argentina, the technology for CA became available. Contacts between ANAPO and AAPRESID, the CA organization in Argentina, further strengthened understanding and implementation of NTS (Campero and Wall, 1999).

The adoption of CA has only occurred in the eastern plains of Bolivia in the department of Santa Cruz, limited to the membership of ANAPO. The main cropping system for CA in the region is soybean-based. The rotational crops are primarily maize, sorghum for grain and forage sorghum. *Brachiaria* is used as a cover crop or intercrop with maize for use as forage. The other crops grown in the area with potential to enrich rotations are wheat, sunflower, rice, sugarcane, beans, ghia and sesame (Calegari et al., 2020).

ANAPO remains the only organization promoting CA in Bolivia. The ANAPO research farm organizes field days and training courses. About 70% of the ANAPO membership has adopted CA. However, so far, neither the government nor other organizations have really taken up CA.

Some local NGOs are becoming interested in CA, mainly through recent initiatives from the Bolivian FAO office. Scientific support for CA is still very low. The CIAT center in Santa Cruz has done some research on direct seeding and no-tillage, including aerobic rice systems, but CA is not a main focus of research.

In practice, farmers' organizations like ANAPO, or for family farming, CIOEC (La Coordinación de Integración de Organizaciones Económicas Campesinas – The Coordination of Integration of Peasant Economic Organizations) are organized around specific commodities or issues such as family farming. Farmers' organizations in Bolivia mostly act as lobby groups. An exception is organic farming which has been promoted mainly by foreign NGOs and has been adopted by some family farmer organizations.

Bolivia is suffering more and more from natural disasters like droughts and floods related to climate change (IBRD/ The World Bank, 2017). The experience of ANAPO suggests that CA provides a potential solution. There have been some initiatives to mainstream CA in Bolivian agriculture. The introduction of CA on a pilot scale has shown promising results, for example, in cotton cultivation in the lower eastern plains, or the cultivation of grains such as quinoa in the Andean highlands.

The NT system is used in the department of Santa Cruz on about 87% of the total annual cropping area (Table 3), most of which can be considered as CA. In other parts of Bolivia, NT or 'direct seeding' might be used as a technique for some crops alternated with conventional tillage (ANAPO, 2019). Farm sizes in the ANAPO membership range from 50 ha to 5000 ha. Very few farmers rotate crops with pasture areas for livestock, and most of the pasture management in Bolivia is extensive, with low productivity. ANAPO tries to maintain, following the Brazilian example, a tree cover of 20% - mostly with windbreaks between plots (Calegari et al., 2020).

Santa Cruz has three different agro ecological zones characterized by average precipitation:

- a dry zone with less than 800 mm rainfall;
- an intermediate zone with precipitation between 800 mm and 1200 mm; and
- a humid zone with more than 1200 mm rainfall.

In normal years, 70% of precipitation falls in the summer months between November and May, and 30% in the winter months (May to November), with some significant changes during the past few years and a general trend toward less precipitation. The climate in the department normally allows two crops per year. The dry zone only uses 30% of the available area for dry-season winter crops, the intermediate zone about 50%, and in the humid area, double cropping is applied on 100% of the area. Most of the area is rain-fed. Irrigation systems for field crops are rare, and in the dry winter period, not very effective due to high evaporation losses (ANAPO, 2019). Bolivia has one of the lowest crop yields in the region, mostly due to severe soil degradation. This has been worsened by increased drought as a result of climate change and deforestation.

The main impact of CA on production is in saving water, which is most visible in drought years. With good quality CA and particularly good ground

Table 3 Tillage systems and main crops grown in the department of Santa Cruz, Bolivia.

	No-Tillage		Conventional Tillage		
Crop	ha	%	ha	%	Total
Maize	187 200	80	46 800	20	234 000
Sorghum	372 020	89	45 980	11	418 000
Sunflower	98 100	90	10 900	10	109 000
Wheat	83 460	78	23 540	22	107 000
Soya	1 117 248	88	152 352	12	1 269 600
Total	1 858 028	87	279 572	13	2 137 600

Source: ANAPO, 2019.

cover in the 2018/19 summer season, the soybean yields reached 3 t ha^{-1} despite severe drought conditions, while adjacent conventionally tilled fields or NT fields with no ground cover yielded less than 2 t ha^{-1} and eventually failed. Changing to CA did not decrease yields, but in the long term, tended to increase and become more stable (ANAPO/CIMMYT/CIAT, 2001) (Table 4). A key factor in the success of CA was improved moisture management over conventionally tilled fields with better water infiltration in soils (González, 2004). The main impact has been more stable and higher yields, under the drought conditions which have prevailed since 2015. However, since CA is not officially promoted, there has been no research to document these effects in a systematic way.

6.2 Challenges and opportunities

Conventional agriculture is practiced mainly in large farms in the eastern plains of Bolivia. Brazil and Argentina have provided technology. Farms use knife rollers and cover crops. No-tillage farming is also practiced, often with soybean monocultures. Roundup Ready soybean has been used in Bolivia since 2005 and this technique makes it difficult to promote good CA with good ground cover. There have been steps to introduce small-scale no-tillage seed drills and hand-jab planters in the Andean highlands. While jab planters are produced locally and adapted for small grains like quinoa, seed drills have been imported from Brazil. The drills have been adapted by adding wings, opening furrows for direct seeding under extreme sandy and dry soil conditions. With an increase in soil organic matter levels and soil cover, wings should no longer be needed and seeding depth can be reduced. Knife rollers are also locally produced on a pilot scale in the highlands. Small-scale mechanized equipment such as single axle tractors are unknown in Bolivia.

7 Future prospects for Conservation Agriculture in South America

Large-scale CA adoption in South America has allowed a permanent increase in agricultural production to meet the growing world demand for food, with well-known economic, environmental and social benefits, including carbon capture and mitigation of the effects of climate change. Nevertheless, it is possible to do much more and much better. Because of the dominance of soybean in the summer months within simplified NTS cropping systems, there is limited crop rotation, insufficient diversification and poor permanent soil cover. This can lead to soil erosion, problems with weeds, nematodes and fungal diseases, requiring high chemical use to control insect pests and diseases. The use of NTS combined with cover crops and crop rotation has increased, together with

Table 4 Soybean, maize, sorghum, wheat and sunflower yields (t ha^{-1}) after 18 years and 36 cropping cycles in three different cropping systems (SUC 1, SUC 2 and ROT) with conventional tillage (CT) and no-tillage (NT) for the 2017–2018 season

	Soybean			Maize	Sorghum	Wheat		Sunflower	
Soil tillage management	SUC 1	SUC 2	ROT	ROT	ROT	SUC 1	ROT	SUC 2	ROT
CT	2.85	2.62	2.93	3.70	2.45	0.97	1.52	0.43	0.50
NT	2.97	3.00	3.19	4.51	2.93	2.22	2.95	0.75	0.82
Difference in favor of NT	0.12	0.38	0.26	0.81	0.48	1.25	1.43	0.32	0.32

SUC 1: succession soybean/wheat; SUC 2: succession soybean/sunflower; ROT: rotation soybean/wheat/maize/sunflower/soybean/sorghum. Source: ANAPO, 2019.

the use of biological control of pests and diseases, with improved effects on soil and plant health. However, further scaling-up needs to occur.

Because of technological advances, there is a great opportunity to expand NTS in cropping systems for sugarcane, cassava and irrigated rice in Brazil, representing more than 12 million hectares. In addition, there is potential for expansion of NTSP and NTSV systems in Brazil and in South and Latin America as a whole.

Conservation Agriculture-based technologies are available, but nation-wide adoption depends on a mix of public policies, economic incentives for crop diversity and changes in farmers' ways of thinking and attitudes. More effort is needed in these areas, in the countries and in the regions where CA has not yet become popular.

8 Where to look for further information

All South American CA Farmers' Organizations can be accessed on their websites:

- AAPRESID http://www.aapresid.org.ar.
- AUSID http://www.ausid.com.uy/.
- FEBRAPDP https://febrapdp.org.br/.
- FEPASIDIAS http://www.fepasidias.org.py/.
- ANAPO https://www.anapobolivia.org/.

The Food and Agriculture Organization of the United Nations (FAO) provides on its website a complete section on the matter and specific papers on CA in South America at: www.fao.org/conservation-agriculture/en/.

The Editor of this book Dr Amir Kassam moderates the Global CA-CoP CONSERVATION AGRICULTURE COMMUNITY OF PRACTICE that can be accessed at: amirkassam786@googlemail.com.

All authors of this chapter are available to provide more information.

9 References

AAPRESID 2020. *Sistema Chacras*. http://eltiempodepergamino.com/12210-agro-aapresid-impulsa-los-sistemas-de-produccion-agricolas-sostenibles?fbclid=IwAR0oX4R_01KUM-clWPbGtcppOsT-OZTZGkcvT_RFPEmol_aQQPlfRXFEgWU.

Albertengo, J., Belloso, C., Giraudo, M. B., Peiretti, R., Permingeat, H. and Wall, L. 2014. Chapter 15. Conservation agriculture in Argentina. In: Jat, R. A., Sahrawat, K. L. and Kassam, A. H. (Eds) *Conservation Agriculture: Global Prospects and Challenges*. p. 372 http://doi.org/10.1079/9781780642598.0352.

Amado, T. J. C. and Eltz, F. L. F. 2003. Plantio direto na palha-rumo à sustentabilidade agrícola nos trópicos. *Ciência e ambiente, Santa Maria* 27(3), 49–66.

ANAPO 2019. *Informe interno sobre avance de la siembra directa en Bolivia.* ANAPO, Santa Cruz, Bolivia. 35 p.

ANAPO/CIAT/CIMMYT 2001. *Guía de Siembra Directa. Recomendaciones hechas por Agricultores para Agricultores*, Santa Cruz, Bolivia. 69 p.

Aoyama, C. 1984. *Guía de Plantio Directo - Datos de Discusión para el Simposio entre los Expertos de la JICA en el sur de Paraguay.* Estacion Experimental Agricola de Pirapó, Paraguay, 24 p.

Araüjo, A., Sims, B., Desbiolles, J., Bolonhezi, D., Haque, E., He, J., Abdalla Fayad, J., Kienzle, J., Wildner, L., Li, H., Zanella, M., El-Gharras, O., Bell, R., Peiretti, R., Mkomwa, S., Duiker, S. and Friedrich, T. 2020. The status of mechanization in conservation agriculture systems. In: Kassam, A. (Ed), *Advances in Conservation Agriculture Volume 1: Systems and Science.* Burleigh Dodds Science Publishing, Cambridge, UK, pp. 427–488. (ISBN: 978 1 78676 264 1; www.bdspublishing.com; http://dx.doi.org/10.19103/AS.2019.0048.11).

Arbeletche, P. 2020. El agronegocio en Uruguay: su evolución y estrategias cambiantes en el Siglo XXI. *Revista RIVAR* 7(19), 109–129. https://doi.org/10.35588/rivar.v7i19.4355.

Bertoni, M. S. 1927. Agenda y mentor agrícola. Guía del agricultor y colono. Con el calendario de todos los Trabajos Rurales. *4ª Edición Ampliada, de la 3a encargada por el Congreso Nacional Paraguayo Reedición Especial Julio*, 2011, 512 p.

Bolliger, A., Magid, J., Amado, J. C. T., Skóra Neto, F., Ribeiro, MdFdS., Calegari, A., Ralisch, R. and de Neergaard, A. 2006. Taking Stock of the Brazilian "Zero-Till Revolution": A Review of Landmark Research and Farmers' Practice. *Advances in Agronomy* 91, 47–110 https://doi.org/10.1016/S0065-2113(06)91002-5.

Borges, G. O. 1993. Resumo histórico do plantio direto no Brasil. In: EMBRAPA, Centro Nacional de Pesquisa de Trigo. EMBRAPA, CNPT, Plantio Direto no Brasil. Passo Fundo, Brasil: Aldeia Norte/FUNDACEP/FECOTRIGO/FUNDAÇÃO ABC. pp. 13-18.

Buckles, D., Triomphe, B. and Sain, G. 1998. *Cover Crops in Hillside Agriculture: Farmer Innovation with Mucuna*, IRDC/CIMMYT, Canada, 230 p.

Calegari, A., Araujo, A. G., Tiecher, T., Bartz, M. L. C., Fuentes-Lanillo, R., Santos, D. R., Capandeguy, F., Zamora, J. H., Jump, J. R. B., Moriya, K., Dabalá, L., Cubilla, L. E., Cubilla, M. M., Carballal, M., Trujillo, R., Peiretti, R., Derpsch, R., Miguel, S. and Friedrich, T. 2020. No-till farming systems for sustainable agriculture in South America. In: Dang, Y., Dalal, R. and Menzies, N. (Eds). *No-Till Farming Systems for Sustainable Agriculture.* Springer, Cham. pp. 533–565. https://doi.org/10.1007/978-3-030-46409-7_30.

Campero, M. and Wall, P. C. 1999. Efectos del rastrojo en la superficie del suelo sobre el balance hídrico y el rendimiento. Memorias. *III Reunión Nacional de Trigo y Cereales Menores.* Cochabamba, Bolivia, 115–121.

Casão Junior, R., Siqueira, R., Mehta, Y. R. and Passini, J. J. 2006. Foz do Iguaçu, ITAIPU binacional. *Sistema Plantio Direto com qualidade.* IAPAR, Londrina. 212 p.

Casão Junior, R., Araújo, A. G. and Fuentes-Llanillo, R. 2012. *No-Till Agriculture in Southern Brazil: Factors That Facilitated the Evoluti on of the System and the Development of the Mechanization of Conservation Farming.* Food and Agriculture Organization, Roma. 78 p.

Cubilla, A. M. M., Wendling, A., Eltz, F. L. F., Amado, T. J. C. and Mielniczuc, J. 2012. *Recomendaciones de fertilización para soja, trigo, maíz y girasol bajo el sistema de siembra directa en el Paraguay.* CAPECO Asunción, Paraguay. 88 pp.

Derpsch, R. 1984. Histórico, requisitos, importância e outras considerações sobre Plantio Direto no Brasil. In: *Pablo Vital Torrado and Raphael R. Aloisi*. Campinas, Cargill Foundation., 1-12.
Derpsch, R., Florentín, M. A. and Moriya, M. 2000. *Importancia de la Siembra Directa para alcanzar la sustentabilidad de la producción agrícola*. Proyecto Conservación de Suelos MAG - GTZ, San Lorenzo, 40 p.
Derpsch, R., Friedrich, T., Kassam, A. and Hongwen, L. 2010. Current Status of Adoption of No-Till Farming in the World and Some of Its Main Benefits. *International Journal of Agricultural and Biological Engineering* 3, 1-25. https://doi.org/10.3965/j.issn.1934-6344.2010.01.001-025.
Derpsch, R. and Moriya, K. 2011. *Historia y evolución de la Siembra Directa en Paraguay (Documento)*, MAG Paraguay, Asunción, 17 p.
Díaz-Zorita, M., Duarte, G. A. and Grove, J. H. 2002. A Review of No-Till Systems and Soil Management for Sustainable Crop Production in the Subhumid and Semiarid Pampas of Argentina. *Soil and Tillage Research* 65(1), 1-18. https://doi.org/10.1016/S0167-1987(01)00274-4.
Dumanski, J., Reicosky, D. C. and Peiretti, R. A. 2014. Preface. Global pioneers in soil conservation: common elements and lessons learned. *International Soil and Water Conservation Research* 2(1), 1-4. Available at: http://www.ecaf.org/. doi:10.1016/S2095-6339(15)30018-6.
EPAGRI–Empresa de pesquisa agropecuária e extensão rural de Santa Catarina. 2021. *Sistema de Plantio Direto de Hortaliças (SPDH) reduz o uso de agrotóxicos*. https://www.epagri.sc.gov.br/ (accessed 19 May 2021).
Ernst, O. 2001. El tiempo de barbecho como variable de manejo (Fallow time as a management factor). In: *9a Jornada Nacional de Siembra Directa. AUSID, EEMAC, Facultad de Agronomía*, Univ. de la Rep. O. del Uruguay, Paysandú, Uruguay.
FEBRAPDP, n. d. https://febrapdp.org.br/.
Florentin, M. A., Peñalva, M., Calegari, A. and Derpsch, R. 2010. *Green manure/Cover Crops and Crop Rotation in Conservation Agriculture on Small Farmers* (vol.12-2010). Food and Agriculture Organization, ROME. ISBN 978-92-5-106839-7 ISSN 1020-4555 . 97p.
Freitas, P. L. and Landers, J. N. 2014. The transformation of agriculture in Brazil through development and adoption of zero tillage conservation agriculture. *International Soil and Water Conservation Research* 2(1), 35-46. https://doi.org/10.1016/S2095-6339(15)30012-5.
Fuentes-Llanillo, R., Telles, T. S., Soares Júnior, D. and Pellini, T. 2013. Tillage systems on annual crops in Brazil: figures from the 2006 Agricultural Census. *Semina: Ciências Agrárias, Londrina, Brazil* 34(6) (Suppl. 1), 3691-3698. https://doi.org/1010.5433/1679-0359.
Fuentes-Llanillo, R., Telles, T. S., Soares Junior, D., Kaweesa, S. and Mayer, A. B. 2020. Chapter 12: Social benefits of conservation agriculture systems. Kassam, A. (Ed.), *Advances in Conservation Agriculture Volume 2: Practice and Benefits*, Burleigh Dodds Science Publishing, Cambridge, UK (ISBN: 978 1 78676 264 1; http://dx.doi.org/10.19103/AS.2019.0049.12.
Fuentes-Llanillo, R., Telles, T. S., Soares Júnior, D., Melo, T. R., Friedrich, T. and Kassam, A. 2021. Expansion of no-tillage practice in conservation agriculture in Brazil. *Soil and Tillage Research* 208, 18. https://doi.org/10.1016/j.still.2020.104877 104877.

García-Préchac, F., Ernst, O., Siri-Prieto, G. and Terra, J. A. 2004. Integrating no-till into crop-pasture rotations in Uruguay. *Soil and Tillage Research* 77(1), 1-13. https://doi.org/10.1016/j.still.2003.12.002.

González, Y. Q. 2004. *Evaluación de la infiltración y escurrimiento del agua determinado con microsimulador de lluvia en un suelo franco-arenoso, bajo el efecto de sistemas de labranzas, UAGRM*. Facultad de Ciencias agrícolas. Tesis de grado. 58 p.

Guibert, M., Grosso, S., Arbeletche, P. and Bellini, M. E. 2011. De Argentina a Uruguay: espacios y actores en una nueva lógica de producción agrícola. *Pampa* 7, 13-38.

IAPAR–Instituto Agronômico do Paraná 1976. *Manual agropecuário para o Paraná. V. 1.* Fundação Instituto Agronômico do Paraná, Londrina. 388 p.

IAPAR–Instituto Agronômico do Paraná 1978. *Manual agropecuário para o Paraná. V. 2.* Fundação Instituto Agronômico do Paraná, Londrina. 742 p.

IAPAR–Instituto Agronômico do Paraná 1981. *Plantio Direto no Estado do Paraná* Fundação Instituto Agronômico do Paraná, Londrina. 244 p.

IBGE 2021a. Instituto Brasileiro de Geografia e Estatítstica. Censo agropecuário 2006. https://sidra.ibge.gov.br. (accessed May 17th, 2021).

IBGE 2021b. Instituto Brasileiro de Geografia e Estatítstica. Censo agropecuário 2017. https://sidra.ibge.gov.br. (accessed May 17th, 2021).

ICLFI Development Network 2020. ICLF in numbers. https://www.redeilpf.org.br/ilpf-em-numeros/ilpf-em-numeros-ingles.pdf. (accessed Dec 10th, 2020).

International Bank for Reconstruction and Development and The World Bank 2017. Resilience or resignation: facing droughts and floods in rural, poor Bolivia. 55 p. https://reliefweb.int/sites/reliefweb.int/files/resources/122085-WP-P150751-Bolivia-Droughts-and-Floods-in-Bolivia-English-Final-Dec-2017%20%281%29.pdf (accessed 31/8/2021).

Kassam, A., Friedrich, T. and Derpsch, R. 2019. Global spread of conservation agriculture. *International Journal of Environmental Studies* 76(1), 29-51. https://doi.org/10.1080/00207233.2018.1494927.

Masson, I., Arl, V. and Wuerges, E. W. 2019. Trajetória, concepção metodológica e desafios estratégicos junto ao sistema de plantio direto de hortaliças (SPDH). In: Fayad, J. A., Arl, V., Comin, J. J., Mafra, A. L. and Marchesi, D. R. (Eds). *Sistema de Plantio Direto de Hortaliças*. EPAGRI, Florianópolis.

Moraes, A., Carvalho, P. C., Anghinoni, I., Lustosa, S. B. C., Andrade Costa, S. E. V. G. and Kunrath, T. R. 2014. Integrated crop-livestock systems in the Brazilian subtropics. *European Journal of Agronomy* 57, 4-9. https://doi.org/10.1016/j.eja.2013.10.004.

Netzeband, N. and Arbeletche, P. 2016. Expansión del cultivo de soja en la región de Young-Uruguay. Un análisis político-ecológico del cambio agrario estructural y sus impactos socioeconómicos. *Economía y Sociedad* 20, 49-66.

Nunes, A. L. P., Cortez, G. L. S., Melo, T. R., Figueiredo, A., Wandscheer, C. A. R., Bortoluzzi, J., Brown, G. G., Bartz, M. L. C., Ralisch, R. and Guimarães, M. F. 2020. Farm systems, soil chemical properties, and clay dispersion in watershed áreas. *Pesquisa Agropecuária Brasileira* 55. https://doi.org/10.1590/S1678-3921.pab2020.v55.01279.

Peiretti, R. A. 2012. La Siembra Directa y las Instituciones de Productores. Sus roles frente al desafío de aumentar la producción agrícola global en forma sustentable. 20th AAPRESID Congress, Rosario, Argentina, 08-10 August, 2012.

Pekholtz, F. 2018. Estimación por Análisis Geoespacial de cobertura de Siembra Directa en 2013 y 2017, Región Oriental. In: *República de Paraguay*. EFEPEK, Asunción, Paraguay. 12 p.

Polidoro, J. C., Freitas, P. L., Hernani, L. C., Anjos, L. H. Cd, Rodrigues, RdA. R., Cesário, F. V., Andrade, A. G. and Ribeiro, J. L. 2021. Potential impact of plans and policies based on the principles of Conservation Agriculture on the control of soil erosion in Brazil. *Land Degradation and Development* 32(12), 3457–3468. https://doi.org/10.1002/ldr.3876.

Ralisch, R., Abi-Saab, O. J. G. and Cainzos, R. L. P. 2003. Bright Spots Research Project drivers affecting no-till system in Brazilian small farms, Report. CIAT, San Salvador.

Ribeiro, M. F. S. and Milleo, R. D. S. 2002. *Referências em Plantio Direto para Agricultura Familiar do Centro-Sul do Paraná*. Instituto Agronômico do Paraná (IAPAR), Londrina, PR, Brasil.

Roloff, G., Lutz, R. A. T. and Mello, I. 2012. Metodologia Participativa para avaliação do Sistema de Plantio Direto implantado na Bacia Hidrográfica Paraná 3. FEDERAÇÃO BRASILEIRA DE PLANTIO DIRETO & ITAIPU BINACIONAL, Foz do Iguaçu, Brasil, 36 p.

Romero, A. 1995. La investigación en Siembra Directa. Trabajo presentado en las Jornadas de Capacitación de Capacitadores en Siembra Directa, Bella Vista y Tuparendá. Proyecto Conservación de Suelos MAG – GTZ con la colaboración de CAPECO. Julio 1995.

Santin, W. 2018. *O Brasil possível: a biografia de Herbert Bartz*. 1st Edition. Author Edition, Londrina, Brazil. 240 pp. ISBN 978-85-922249-1-2.

Scarlato, G., Buxedas, M., Franco, J. and Pernas, A., 2001. Siembra directa en la agricultura del litoral oeste uruguayo (Crop production with no-till in Southwestern Uruguay). In: Adopción y demandas de investigación y difusión en siembra directa (Ed.), *Instituto Nacional de Investigación Agropecuaria*. No. 6. Serie FPTA-INIA-Uruguay, pp. 8–102.

Silva, F. Dd, Amado, T. J. C., Ferreira, A. O., Assmann, J. M., Anghinoni, I. and Faccio Carvalho, P. CdF. 2014. Soil carbon indices as affected by 10 years of integrated crop-livestock production with different pasture grazing intensities in southern Brazil. *Agriculture, Ecosystems and Environment* 190, 60–69.

Telles, T. S., Dechen, S. C. F. and Guimarães, MdF. 2013. Institutional landmarks in Brazilian research on soil erosion: a historical overview. *Revista Brasileira de Ciência do Solo* 37(6), 1431–1440. https://doi.org/10.1590/S0100-06832013000600001.

Wall, P. C. and Ekboir, J. 2002. Conservation Agriculture for small farmers: challenges and possibilities, ASA, CSCA, SSSA Meeting, Indianapolis, USA, November 10–14.

Wyngaard, N. and Reussi Calvo, N. 2019. Avances en la interpretación de análisis de suelo. 14p. Conference paper. https://www.researchgate.net/profile/Nahuel-Reussi-Calvo/publication/335240601_Avances_en_la_interpretacion_de_analisis_de_suelo/links/5d5aa394458515210250f78f/Avances-en-la-interpretacion-de-analisis-de-suelo.pdf (accessed August 31, 2021).

Chapter 6

Conservation Agriculture in the agri-environmental European context

Emilio J. Gonzalez-Sanchez, Universidad de Córdoba, Spain, European Conservation Agriculture Federation (ECAF), Belgium and Asociación Española Agricultura de Conservación Suelos Vivos (AEACSV), Spain; Gottlieb Basch, Universidade de Évora, Instituto de Ciências Agrárias Mediterrânicas (ICAM), Portugal and European Conservation Agriculture Federation (ECAF), Belgium; Julio Roman-Vazquez, European Conservation Agriculture Federation (ECAF), Belgium; Elizabeth Moreno-Blanco, European Conservation Agriculture Federation (ECAF), Belgium; Miguel Angel Repullo-Ruiberriz de Torres, European Conservation Agriculture Federation (ECAF), Belgium and Centro Ifapa 'Alameda del Obispo', Spain; Theodor Friedrich, Retired Food and Agriculture Organization of the United Nations (FAO) Expert on Conservation Agriculture, Germany; and Amir Kassam, University of Reading, UK

1 Introduction
2 Soil threats in Europe
3 Benefits of Conservation Agriculture in brief
4 Adoption of Conservation Agriculture in Europe: A brief history and current situation
5 The role of Conservation Agriculture in the European Policies: The CAP, EU, Climate Action and the Green Deal
6 Conclusion
7 Where to look for further information
8 References

1 Introduction

For millennia, and until relatively recently in human history, the cultivation of agricultural soils has been based on some form of mechanical soil disturbance. Especially, since WWII, the intensification of agricultural production in Europe and in the industrialized regions has led to intensive mechanical soil disturbance by tillage to prepare a seed bed and to control weeds. This has led to progressive and generalized degradation of soil quality translated into soil erosion, soil organic matter (SOM) decline, loss in soil biodiversity and

http://dx.doi.org/10.19103/AS.2021.0088.05

soil compaction, as well as an increase in greenhouse gas (GHG) emissions. Additionally, cropping systems that promote soil aggregate breakdown, exposing it to the climatic elements, favour water and wind soil erosion leading to considerable losses of topsoil (a non-renewable resource in the human time scale), while at the same time increasing the dependency on external inputs such as mineral fertilizers and crop protection chemicals. In his well-researched book *Dirt: Erosion of Civilizations*, David Montgomery has shown that tillage has been a major root cause of land degradation and soil erosion over centuries throughout the world. The resulting loss of soil productivity and functions has affected human migration patterns, caused human conflicts and wars, and led to the collapse of civilizations (Montgomery, 2007).

Both at global and European levels, agriculture will play a key role in the coming decades and will have to deal with strategic concerns. Among the most serious environmental threats related to European agriculture can be found the degradation of soils, biodiversity decline, reduction in water quality and availability and poor mitigation and adaptation of agricultural production systems to climate change (EEA, 2020). These environmental pressures are interlinked as has been shown by global experimental and empirical evidence and in the scientific literature (Kassam et al., 2012).

Farming and the environment must be in harmony in order to prevent the degradation of natural resources. The pursuit of sustainability is a common objective all over the world, including in Europe. In the well-known report of the World Commission on Environment and Development: Our Common Future, also known as the Brundtland Report (United Nations, 1987), it was shown that the road taken by society was destroying the environment, at the same time leaving more people vulnerable and in poverty. Indeed, the Brundtland Report was the first attempt to eliminate the conflict between development and sustainability.

To make sustainable agriculture real in Europe, a new model is needed. Conservation Agriculture (CA) is an alternate paradigm of agricultural production and land use that is applicable to all land-based agricultural production systems in rainfed and irrigated farming, including annual, perennial and mixed systems, orchards, agroforestry and plantation systems, crop-livestock systems and pasture and rangeland systems. In Europe, the European Conservation Agriculture Federation (ECAF) endorses the three principles of CA as proposed by the FAO (http://www.fao.org/conservation-agriculture). CA is described as an ecosystem approach to regenerative sustainable agriculture and land management based on the practical application of context-specific and locally adapted three interlinked principles, namely (Kassam et al., 2019, 2020) (Figs 1 and 2):

i **Continuous no or minimum mechanical soil disturbance:** This principle is implemented by the practice of no-till seeding or broadcasting of crop seeds and direct placement of planting

Figure 1 Conservation Agriculture in annual crops. Winter wheat in no-tillage. Source: Spanish Conservation Agriculture Association – Living Soils (AEACSV, 2021).

Figure 2 Conservation Agriculture in perennial crops. Groundcovers between rows of orange trees. Source: AEACSV (2021).

material into untilled soil, causing the minimum soil disturbance from any cultural operation, harvest operation or farm traffic and no-till weeding. Sowing seed or planting crops directly into untilled soil reduces erosion; reduces the loss of SOM and disruptive mechanical cutting and smearing of pressure faces; promotes soil biodiversity and microbiological processes; protects soil structure and connected pores; favours movement of gases and water through the soil; and promotes overall soil health and functions including improved retention of soil moisture, plant nutrients and soil carbon; reduces labour and energy requirements and reduces GHG emissions; and contributes to integrated weed, insect pest, pathogen and nutrient management as well as overall resilience and sustainability.

ii **Permanent maintenance of a vegetative mulch cover on the soil surface:** This principle is implemented by retaining crop biomass, root stocks and stubbles and biomass from cover crops. Use of crop residues (including stubbles) and cover crops reduces soil erosion; protects the soil surface; conserves water and nutrients; supplies organic matter and carbon to the soil system; promotes soil biodiversity and microbiological activity to enhance and maintain soil health and functions including the structure and aggregate stability (resulting from glomalin production by mycorrhiza), improved capture and retention of water, plant nutrients and soil carbon; provides a habitat for animals; and contributes to integrated weed, insect pest, pathogen and nutrient management as well as overall resilience and sustainability.

iii **Diversification of species in cropping systems:** This principle is implemented by adopting economically, environmentally and socially adapted crops in rotations, and/or sequences and/or associations involving annual and perennial crops, including a balanced mix of legume and non-legume crops and cover crops. The use of diversified cropping systems contributes to diversity in rooting morphology and root composition; enhances soil biodiversity and microbiological activity; builds up SOM; enhances crop nutrition and crop protection through the suppression of pathogens, diseases, insect pests and weeds; and contributes to integrated weed, insect pest, pathogen and nutrient management as well as overall resilience and sustainability. Crops can include annuals, short-term perennials, trees, shrubs, nitrogen-fixing legumes and pastures, as appropriate. By including livestock and trees in the cropping system, nutrient recycling and environmental services are enhanced.

2 Soil threats in Europe

About 95% of the food globally is directly or indirectly produced from soils. Indeed, soils are essential to life, and the relevance has been recently emphasized by the European Parliament in a motion for a resolution on soil protection (2021/2548(RSP)) (EP, 2021), which highlights some key facts about soils in Europe.

- Soil erosion affects 25% of agricultural land in the EU and increased by about 20% between 2000 and 2010.
- More than half of agricultural land in the EU has average erosion levels higher than what can be naturally replaced, underlining the need for the use of sustainable management techniques for soils.
- Approximately 60–70% of EU soils are unhealthy due to the current management practices, with a further, yet uncertain, percentage of soils being unhealthy due to poorly quantified pollution issues.
- Carbon stocks in top soils are declining, which entails clear climate change implications due to the release of CO_2 into the atmosphere.
- With regard to compaction, 32–36% of European subsoils are highly susceptible to it.
- Agricultural intensification and overuse of pesticides are causing soil contamination by pesticide residues, including due to some pesticides' high soil persistence and toxicity to non-target species, and have lasting effects on soil health. Besides, diffuse pollution by agrochemicals poses a threat to the soil.
- Large parts of southern Europe are likely to become desertified by 2050 as a result of climate change and inappropriate agricultural and agronomic practices if strong action is not taken; and this threat is not being coherently, efficiently and effectively addressed in the EU.
- It is estimated that soil erosion causes a loss of agricultural production of €1.25 billion in the EU annually. The costs of inaction on soil degradation exceed the Union-estimated €50 billion per year. Investing in avoiding land degradation and in restoring degraded land makes sound economic sense as the benefits generally far exceed the cost, whereas the costs of restoration are estimated to be ten times higher than prevention costs.

Therefore, there is a need for urgent action incorporating into EU policies CA principles to tackle those challenges with a view to ensuring a high level of protection to soils and the rest of natural resources. Major soil-related problems in Europe are elaborated in the following sections.

2.1 Soil erosion

Soil erosion has long been neglected as a serious threat to soils in Europe. In the 1990s, 105 Mha, or 16% of Europe's total land area (excluding Russia) were estimated to be affected by water erosion. More recently, the modelled surface affected by erosion through water in the EU-27 was estimated at 130 Mha, of which around 20% suffer from soil loss rates of >10 t ha^{1} y^{1} (European Environment Agency, 2017). Further, in 2015, in a public release on 2 September of their erosion study, the European Commission Joint Research Centre (EC-JRC, 2015) announced that water runoff is estimated to erodes 970 million tonnes of soil every year in the EU, corresponding to 1 cm loss from an area twice the size of Belgium (around 30 000 km^{2}). Whatever area is affected by water erosion, we consider erosion rates far beyond 1 t ha^{1} y^{1} as irreversible within a time span of 50–100 years (Huber et al., 2008). According to the report of EC-JRC (2012), the Mediterranean region is particularly prone to water erosion, but arable land in northern Europe is also susceptible to erosion, especially loamy soils after ploughing (Bielders et al., 2003). Data about soil erosion in Europe are difficult to understand or to interpret when average soil loss rates of 2.46 t ha^{1} yr^{1} on erosion-prone lands (agricultural, forests and semi-natural areas) are mentioned in the same context with 'only' 4 million ha of croplands which currently have unsustainable soil loss rates of more than 5 t ha^{1} yr^{1} (Panagos et al., 2015). Besides setting the threshold for unsustainable soil loss rates to a level of more than 5 t ha^{1} yr^{1} this interpretation neglects the fact that it is mainly agricultural, arable and permanent cropland that contributes to the phenomenon of soil erosion and not forests or semi-natural areas. Tolerable versus actual soil erosion rates in Europe are comprehensively documented by Verheijen et al. (2009). They state: 'A recent assessment carried out by the JRC estimates that water erodes 970 million tonnes of soil every year in the EU. This would mean a 1 m depth loss of soil from an area corresponding to the size of the city of Berlin, or a 1 cm loss from an area twice the size of Belgium. The fact that it takes 100 years to form 1 cm of new soil under natural temperate grasslands gives an idea of the magnitude of the problem of soil loss in the EU'.

Wind erosion is estimated to affect some 42 Mha of Europe's total land area (EC-JRC, 2012), of which 1 Mha is considered to be severely affected (European Environment Agency, 2003) (Fig. 3). High wind erosion rates of around 9.5 t ha^{1} y^{1} were found to occur from arable land (Goossens et al., 2001). Strong winds and sandy soils from freshly ploughed adjacent fields caused in 2011 a mass pile-up of around 80 cars and several deaths on a highway in Germany due to extremely reduced visibility caused by the sandstorm (Spiegel International, 2011).

2.2 Soil organic matter decline

Organic matter is considered a key element of soils essential for the main soil functions related to crop performance and environmental services. Without adequate levels of SOM soils cannot function optimally (a minimum threshold

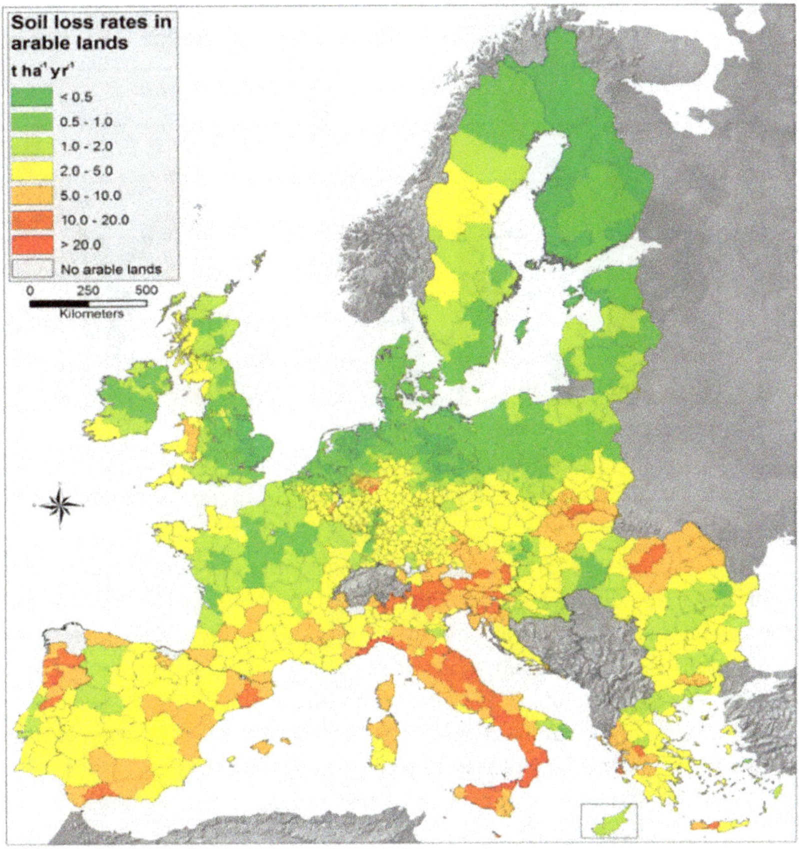

Figure 3 Soil loss from arable land in EU-28. (Source: European Commission Joint Research Centre). Source: EC-JRC (2015).

of 3.4% SOM or 2% SOC, is widely accepted as a base reference value for soil quality). However, only levels of less than 2% SOM are considered low, and less than 1% SOM as very low. Low SOM levels can be widely found in southern European countries, but also in some regions of France, UK, Germany, Belgium and Norway (Jones et al., 2012). Apart from climate conditions, it is also well known that land use and land use change can significantly deplete carbon stocks in soils, especially when forests, grasslands or native ecosystems are converted to cropland. Yet, astonishingly, these carbon losses in the form of CO_2 emissions from soils do not account for the contribution of agriculture to the overall emission of greenhouse gases (GHG). A generalized SOM loss on cropland all over Europe, especially when intensively cultivated, is well documented (Fig. 4) (Balesdent et al., 2000; Arrouays et al., 2001; Gobin, 2005; Bellamy et al., 2005; Verheijen et al., 2005; Gobin et al., 2011).

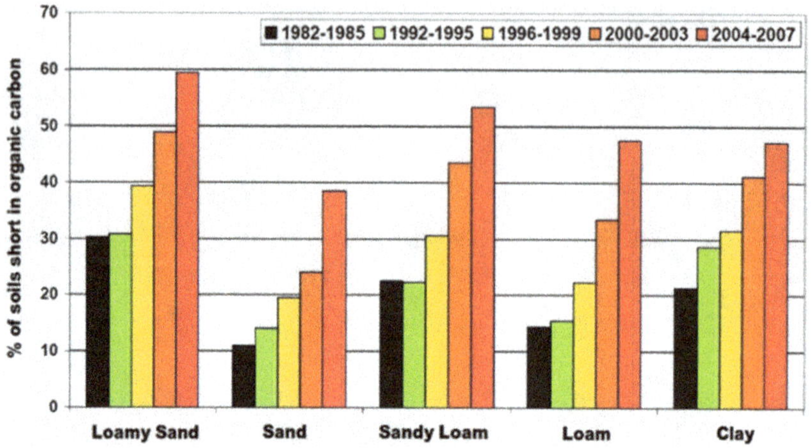

Figure 4 Evolution of percentage of arable soils short in soil organic carbon, based on more than 60 000 samples taken every 3 years to a depth of 23 cm. Source: Gobin (2005).

The results of the SmartSoil FP-7 project also corroborate the overall negative SOC balance of soils in Europe (−100 kg SOC ha^{1} y^{1} on average), with most of the areas showing neutral to slightly negative SOC balance, but some regions with high negative balances, mostly where carbon stocks are still rather high (Merante et al., 2014). This report also highlights the association of conventional soil management, such as the intensive use of tillage practices, not only to losses of SOC but also to soil erosion and overall soil degradation processes. Another interesting result of this project is the demonstration of the combined effects of SOC levels and their balance on crop yields (Fig. 5). It is not by chance that soils under pasture have a higher SOC content than those under cropping (Blair et al., 2006). The absence of soil disturbance along with practices that increase the return of plant biomass as soil mulch must be considered the key protection pathway, which contributes to increased soil biology and soil aggregate stability, thus physically protecting SOC from oxidation and microbial decomposition.

2.3 Loss in soil biodiversity

Soil biodiversity is characterized by the diversity, amount and activity of different soil-living species belonging to the groups of macro, meso and micro fauna and to the microflora composed of the bacterial and fungal communities and other microorganisms. The status of soil biodiversity can be estimated through a) quantification of soil microorganisms and b) diversity of soil macro fauna. The first is normally done through the assessment of the carbon content in the microorganism pool (microbial biomass carbon) (Serna-Chavez et al.,

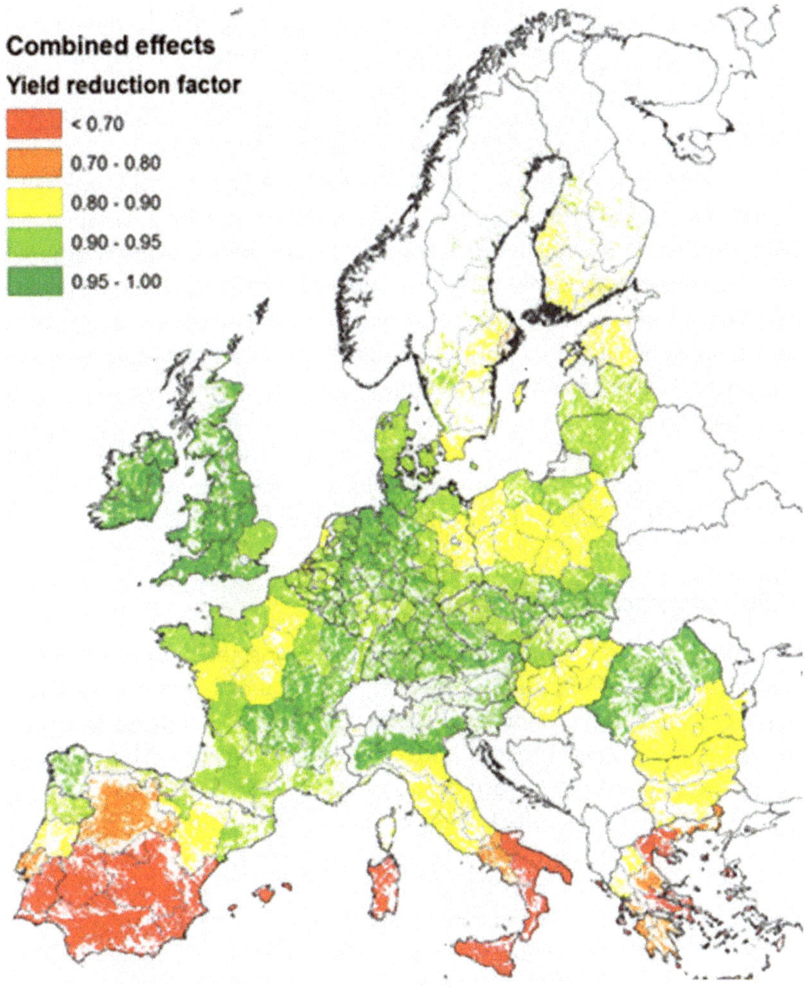

Figure 5 Yield impact of SOC losses. Source: Blair et al. (2006).

2013). The macro fauna diversity can be determined by identifying the number of different macro fauna groups co-existing in a given soil (earthworms, ants, termites, spiders, millipedes, centipedes, isopods, fly larvae, cockroaches and mantids, moth and butterfly larvae, grasshoppers and crickets, gastropods, beetles) (Orgiazzi et al., 2016).

All soil threats and degradation processes that impact soil quality or health and its functioning also impact soil biodiversity. On the other hand, it is the soil biota itself that plays a fundamental role in the soil's capacity to function normally and deliver important ecosystem services and goods, such as nutrient cycling, building up and maintaining soil structure (e.g. glomalin production), water and

GHG fluxes and retention in soil, etc. (Lavelle and Spain, 2001; Mulder et al., 2011; Kassam, 2020). Over the past two decades numerous studies have been conducted to better understand how soil organisms react to human activities and to improve the comprehension of their biogeography (Gardi et al., 2008, 2009; Keith et al., 2012; Tsiafouli et al., 2015; Orgiazzi et al., 2016).

Hence, land use change, land use and agricultural intensity, with all its associated impacts on other soil threats, are considered to be the main driving force for the decline in soil biodiversity at the ecosystem, species and genetic levels (European Environment Agency, 2004). The Global Soil Biodiversity Atlas developed by JRC (Orgiazzi et al., 2016) provides a comprehensive understanding of the diversity of soil organisms, their contribution to ecosystem functions and services and their geographical and spatial distribution. It also points out interventions needed to preserve soil biodiversity, which should be considered as one of the major natural resources. In addition, soil biodiversity connected through the food webs also affects general biodiversity above the soil. This is true for soil biodiversity worldwide (FAO, 2019).

2.4 Soil compaction

Soils, although very rarely, may be or become compacted naturally as a result of the way they were formed in place. In agronomic terms, soils are considered compacted when the total porosity (especially, the air-filled porosity) restricts aeration, and its pores become so small, thus limiting root penetration, infiltration and drainage. On agricultural land, soil compaction can be and is often caused by either livestock, when grazing under wet conditions, but much more frequently by heavy machinery, including tractors, harvesting equipment and tillage implements directly, or consequently by destroying the soil's structural stability, thus making it more susceptible to compaction.

With regard to the situation of soil compaction on European cropland, there are no reliable data. Whereas some estimate around 36% of European subsoils being highly or very highly susceptible to compaction, others report around 32% of soils being highly susceptible and 18% being genuinely, but moderately, affected by compaction (Van-Camp et al., 2004). Again, others estimate that around 33 Mha (33%) of European farmland is suffering from soil compaction. Considering the actual axle loads of major agricultural machinery, exceeding in some cases the permissible axle loads of trucks for road transport, it can be safely assumed, that subsoils beyond the reach of mechanical subsoiling tools are generally compacted with a severely reduced water infiltration capacity (Friedrich, 2019).

Both increasing mechanization in agriculture causing more and deeper soil disturbance and disruption, and bigger and heavier machinery on increasing farm sizes to save time for operations contribute to increasing more and more

Figure 6 Gully generated in one season due to erosion and runoff. Source: AEACSV (2021).

the physical stress over soils often causing subsoil compaction below the normal tillage depth that is difficult to remediate. On the other hand, the awareness about the problem of compaction is also increasingly leading to technical solutions such as low or variable pressure or twin tyres to alleviate the weight load of machinery. However, those solutions protect only the surface soil from compaction and not the subsoil and therefore would only protect from soil compaction for smaller and medium-size machinery. Solutions adopted elsewhere, namely controlled traffic farming (CTF), seem adequate for larger farms and fields existing in some parts of Europe; this includes those cases, where smaller farms make use of machinery contractor services, which usually operate with machinery at the upper limit of available sizes (Fig. 6).

3 Benefits of Conservation Agriculture in brief

Besides effectively addressing the aforementioned soil threats, benefits from CA include water conservation, climate action, contributions to an efficient farm economy and protection of natural and public infrastructure.

3.1 Conservation Agriculture and water conservation

CA has a direct impact on water quality. By avoiding mechanical soil disturbance, the structural stability of the soil is improved. This increases water infiltration and improves the water-holding capacity of the soil. Together with the physical protection that the soil mulch cover provides, water runoff and soil erosion are significantly reduced and so is soil evaporation loss, together providing greater water reserves for the crop (ECAF, 2020).

3.2 Conservation Agriculture and climate action

Agriculture is a sector that has a direct influence on processes related to climate change, being, in 2017, the second-highest sector in the EU in terms of GHG emissions into the atmosphere. CA has the ability to transform agricultural soils from being carbon emitters into carbon sinks because of no-tillage and the return to the soil of diverse crop biomass from above-ground parts of plants and from diverse roots systems and root exudates. Likewise, due to the reduction of agricultural operations, fossil energy use decreases, thus emitting less CO_2 to the atmosphere. Lower GHG emissions in CA also arise from reduced and more efficient use of inputs, particularly of nitrogen fertilizers (ECAF, 2020).

3.3 Conservation Agriculture and farm economy

Regarding the economic terms, CA brings significant benefits to farm economies, especially to smallholders. The production levels obtained using CA techniques are similar or higher to those resulting from conventional agriculture. The main value of CA is in cost and time savings. Over time, as soil health improves, CA significantly reduces the number of farming operations, reducing energy needs by 50–70%, fertilizer requirements by 30–50%, water requirements by 30–40% and tractor operations by 50–70%, allowing farmers greater flexibility in taking decisions. This leads to higher farm income. Also, because of no-tillage, farm power requirements are lower, thus reducing the size of capital investment in tractors and tillage equipment, and considerably increasing their working life and reducing their maintenance (ECAF, 2020).

CA with its integrated management of resources and inputs and lower farm power needs and capital investments offers the best alternative system for adaptation to meet new production needs in terms of sustainability, profitability and food security.

3.4 Conservation Agriculture and the protection of natural and public infrastructure

Sediments are by far the most important contaminants in surface water, affecting aquatic ecosystems by reducing sunlight penetration to aquatic plants and fouling the habitats of fish and other organisms. Transported in waterways, deposited sediments cause damage to roads and sewers, basement siltation, drainage disruption, undermining of foundations and pavements, creating gullies in roads, siltation of harbours, water channels and water reservoirs, causing loss of water storage capacity and damage to electricity generating turbines. Farmers in Europe lose around €1.25 billion per year in agricultural productivity due to soil erosion. Studies estimate that soil erosion damage on the originating farm is nearly 40% of the total cost of the erosion, with 60% being offsite (ECAF, 2020).

Through the application of CA, an effective retention of sediments and potential pollutants in the field can be achieved, thus avoiding the pollution of water bodies and the costs of their eventual necessary removal from the water.

In CA, off-site transport of both nutrients and pesticides is significantly reduced, and the adsorption and degradation of organic and inorganic compounds are improved and accelerated. All these processes and mechanisms help to protect natural and public infrastructure, while allowing for the necessarily sustainable production intensification to which the EU has committed in order to contribute to global food security (ECAF, 2020).

4 Adoption of Conservation Agriculture in Europe: A brief history and current situation

The history of CA varies in Europe from country to country. The process has been mainly farmer driven, by few pioneer farmers throughout Europe, in some cases with the support of researchers. An important milestone for CA in Europe, resulting from the developments in the different European countries, was the foundation of the ECAF in 1999, which, together with the United Nations Food and Agriculture Organization (FAO), organized the First World Congress on Conservation Agriculture in 2001 in Madrid (Spain), initiating a series of congresses that are still running (2003 Brazil, 2005 Kenya, 2009 India, 2011 Australia, 2014 Canada, 2017 Argentina and 2021 Switzerland) and promoting CA at European policy levels. The financial instrument LIFE from the European Commission has been a support for ECAF in the development of CA projects over the past 20 years. In 2017, the CA-based Agricarbon project was awarded as the Best LIFE project by the European Commission (www.agricarbon.eu).

In northern European countries, the combined negative effects of excessive tillage, particularly in wet soils, the decline in the rural population and increased costs of machinery, led many researchers to consider a reduction of tillage and to start experiments in Germany (Bäumer, 1970), the Netherlands (Van Ouwerkerk and Perdok, 1994) and the UK (Christian, 1994). The first real attempts to introduce some kind of conservation or zero(no)-tillage were triggered by economic objectives, that is to reduce production costs with tillage, namely with machinery, fuel and labour (Basch et al., 2015). These early attempts were assisted by the availability of recently discovered plant growth regulators and herbicides (Phillips and Phillips, 1984). After an initial take-off of no-till, mainly in the UK, where the area of adoption reached almost 300 000 ha in the early 1980s, increasing problems with weed control and volunteer cereals and the straw burn ban practically stopped the use of no-till (Christian, 1994). Before the 2000s only a few European countries registered noteworthy areas under CA. Not by chance those were amongst the founding members of the European Conservation Agriculture Federation, namely, Spain, the UK, France, Germany, Italy and Portugal. Other countries such as Finland (2008/09), Romania and Poland (2015/16) joined the list of CA-practising countries with already promising areas under CA.

In countries of western Europe, with the exception of the Netherlands, it was mainly scientists who believed and insisted in no-till-based crop establishment, and visionaries like Tebrügge (2003) engaged themselves to showcase that no-till was a viable option in annual cropping systems, even under the conditions in Central Europe. The publications of the workshops' proceedings of a Concerted Action EU-project on the 'Experience with the Applicability of No-tillage Crop Production in the West-European Countries' provide a credible testimony of these efforts (Tebrügge and Böhrnsen, 1994, 1995, 1996, 1997, 1998). Whereas some countries show a recent steady increase of their area under CA, others show stagnation or declining figures. Surprisingly, in Germany, where very positive research findings would support a strong adoption, the area between 2008/09 and 2015/16 decreased by almost 60%. In Spain, where we find the largest area under CA, the first studies on CA in annual crops date back to 1976 and were performed in the 'Haza del Monte' farm in Seville. These studies aimed at anticipating the sowing date of a second crop, which was NT soybeans sown into the residues of a previous winter crop (Fernández-Quintanilla, 1997). In February 1995, a group of farmers, technicians and scientists, many of them participants of the above-mentioned experiments, founded the Spanish Association for Conservation Agriculture and Living Soils (AEACSV). Due to the development of European funded projects, such as LIFE 99ENV/E/308 and LIFE 96ENV/E/338, and the support of manufacturers of NT machinery and the industry of plant protection products, a number of activities that required technical-scientific knowledge were conducted with a high degree of regularity, that is still on-going currently. The considerable areas under CA in Romania and Poland reported for the first time in 2015/16 are likely to be the result of the restructuring of land tenure and CAP farm support, creating large-scale enabling conditions for the implementation of CA.

Figures 7 and 8 show the current adoption of CA in Europe and the share of soil management systems per country.

Even if progress has been made in the adoption of zero tillage, conventional tillage was the most widespread tillage practice in EU-27 in 2016, applied on

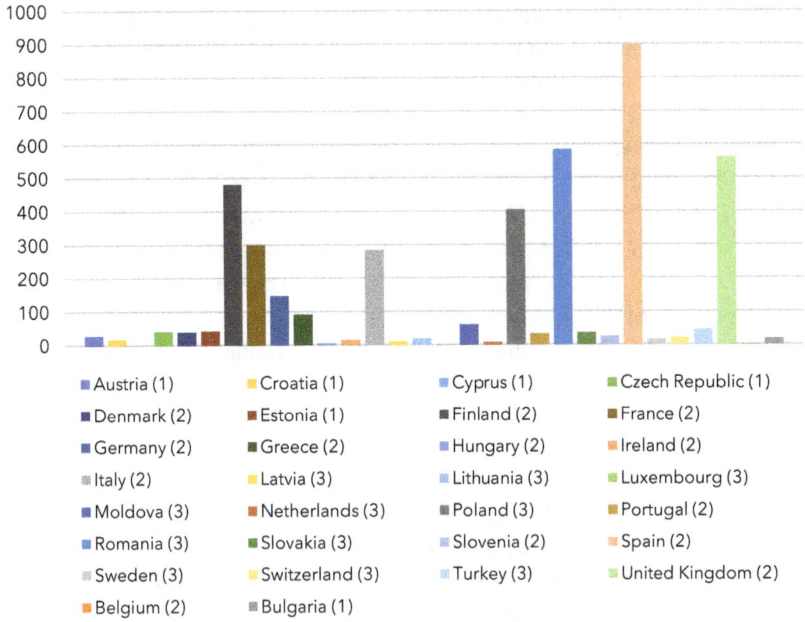

Figure 7 Adoption of Conservation Agriculture in Europe (×1000 ha). Sources: (1) Eurostat (2018). (2) Data provided by ECAF members at its General Assembly, 2020. (3) FAOstat (2018).

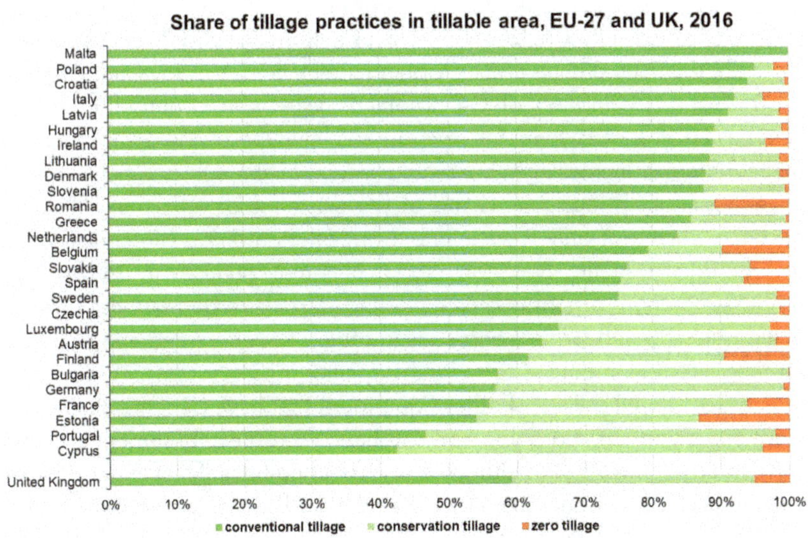

Figure 8 Share of tillage practices in agricultural land, EU-27 and UK, 2016. Source: Eurostat (2020).

two-thirds of the total arable area, implying intensive ploughing and harrowing with its negative impacts on the environment (Eurostat, 2020). The share of zero tillage practices ranged only from 0% to 28%. The highest shares of ≥20% were observed in the following regions: province Luxembourg in Belgium, Bolzano in Italy and Helsinki-Uusimaa in Finland (Map 1).

In 2008/09, the CA cropland area in Europe was 1.56 M ha (across 11 countries). By 2015/16, the CA area increased to 3.56 M ha (across 29 countries), corresponding to 3.42% of the arable land. Thus, over the period 2008/09 to 2015/16, the CA area increased by 127%. This corresponds to an average annual increase of 285 714 ha (Table 1).

An interim update for the year 2018/19 shows that CA cropland area in Europe has increased to at least 5.17 Mha, corresponding to 4.97% of the arable land, and a further increase of almost 45% since 2015/16, or 230% since 2008/09 (Kassam, personal communication). Thus, since 2015/16, the CA cropland area in Europe has increased annually by 538 300 ha, indicating a much higher level of CA uptake between 2015/16 and 2018/19 than during the period between 2008/09 and 2015/16.

In Europe, CA is being practised in all agroclimatic regions from the arctic environment in the north to dryland Mediterranean environment in the South, from temperate maritime environment in the west to continental temperate environment. They cover humid, sub-humid and semi-arid and dryland moisture agroclimatic areas and soils varying from 90% sand to soils with 90% clay.

The countries in Europe that have more than 250 000 ha of CA cropland are Spain, France, Italy, UK, Germany, Slovakia, Romania and Poland. Countries with more than 100 000 ha of CA cropland are Finland, Moldova and Greece.

The rate of adoption of CA cropland systems in countries in Europe is increasing in recent years for several reasons. One of the main reasons is that farmers are increasingly becoming aware of the sub-optimal nature of conventional tillage agriculture and of the existence of CA as an alternative approach which not only is sustainable but offers a range of economic and environmental benefits. Secondly, about half the countries in Europe have non-governmental organizations including farmer organizations that are working directly with farmers and assisting them to understand and adopt CA and innovate new methods to raise the quality of CA systems. Thirdly, there are excellent no-till drills and other equipment that have become available in recent years that are affordable and can handle crop establishment and strip tillage in any type of soil in either wet or dry conditions. Machine companies have intensified their demonstration activities and service provisioning efforts in recent years so that risks of failure are low and manageable within a process-based transformation from tillage agriculture to CA.

Future prospects for increased CA adoption across Europe appear promising in that the European Commission has become aware that tillage

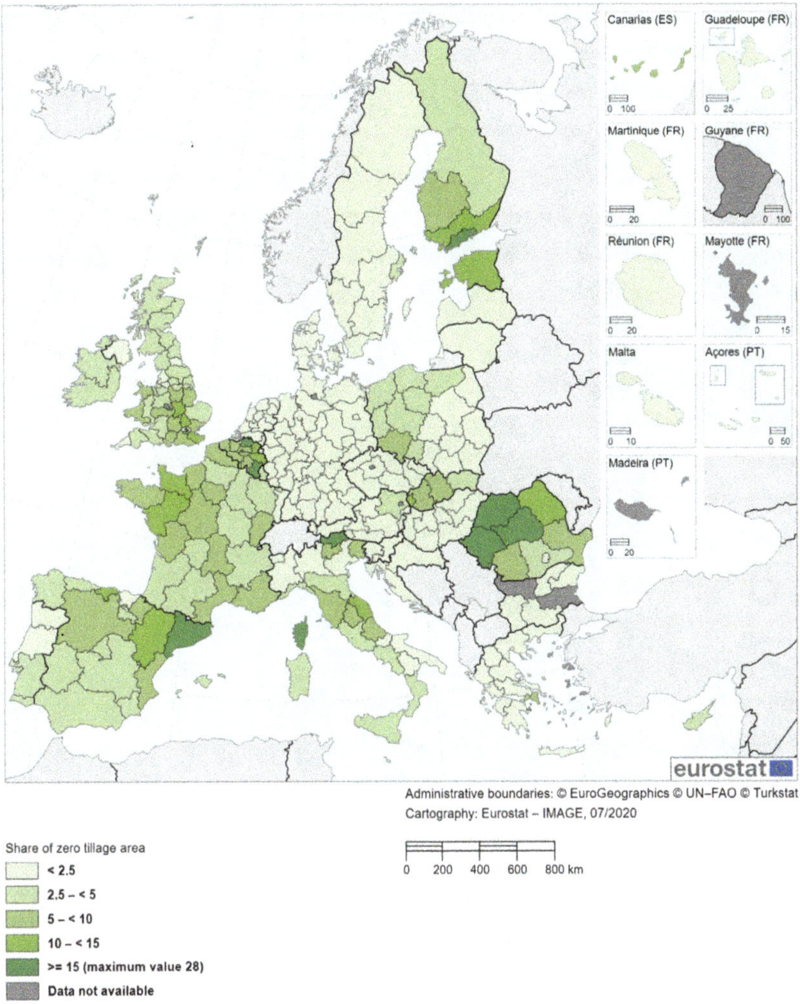

Map 1 Share of zero tillage in the tillable area by NUTS 2 regions, EU-27 and UK, 2016. Source: Eurostat (2020).

agriculture cannot any longer be considered to be sustainable, and that farmers are increasingly adopting CA as an alternative. Indeed, in the last reform of the CAP, which was aimed at greening, one of the core CA practices, that is diversified rotation, was included in the list of green practices under Pillar 1. Under Pilar 2, more site-specific CA efforts have been supported locally and this has worked particularly well in Spain, Italy and France where CA practices are being supported financially by local governments. In the UK, for example, there is government support available for first-time buyers of no-till drills. In these countries, a national association that makes up the membership of ECAF,

Table 1 Extent of CA adoption ('000 ha) in annual cropland in Europe in 2008/09, 2013/14, 2015/16 and 2018/19

Country	CA area 2008/09	CA area 2013/14	CA area 2015/16	CA area 2018/19
Spain	650.00	792.00	900.00	1000.00
Italy	80.00	380.00	283.92	432.00
Finland	200.00	200.00	200.00	120.00
France	200.00	200.00*	300.00	720.00
Germany	354.00	200.00	146.00	352.89
United Kingdom	25.00	150.00	362.00	562.00
Slovakia	10.00	35.00	35.00#	365.00
Portugal	28.00	32.00	32.00#	17.81
Switzerland	9.00	17.00	17.00#	11.02
Hungary	8.00	5.00	5.00	24.29
Ireland	0.10	0.20	0.20	3.66
Moldova	-	40.00	60.00	180.00
Greece	-	24.00	24.00#	110.50
The Netherlands	-	0.50	7.35	17.50
Belgium	-	0.27	0.27	0.27+
Romania	-	-	583.82	583.82+
Poland	-	-	403.18	403.18+
Estonia	-	-	42.14	42.14
Czech Republic	-	-	40.82	40.82+
Austria	-	-	28.33	28.33+
Lithuania	-	-	19.28	19.28+
Croatia	-	-	18.54	18.54+
Bulgaria	-	-	16.50	16.50+
Sweden	-	-	15.82	26.00
Latvia	-	-	11.34	11.34+
Denmark	-	-	2.50	38.50
Slovenia	-	-	2.48	26.00
Luxembourg	-	-	0.44	0.44+
Cyprus	-	-	0.27	0.27+
Total	**1564.10**	**2035.97**	**3557.20**	**5172.10**
% difference		**30.1 since 2008/09**	**127.4 since 2008/09 74.7 since 2013/14**	**230.68 since 2008/09 154.04 since 2013/14 45.40 since 2015/16**

Source: Kassam et al. (2019) plus interim estimates for 2018/19, Kassam, personal communication.* Figure carried forward from 2008/09; # figure carried forward from 2013/14; + figure carried forward from 2015/16.

as well as other farmer organizations, is playing an effective role in both CA research and development, and increasingly in CA education and training.

Most countries in Europe also have orchards and vineyards under CA management systems. The area covered by these perennial CA systems is substantial and increasing because of the productivity, economic and environmental benefits. Countries that are leading in this transformation are Spain, with about 1.3 Mha of CA orchards and vineyards, and Greece, with some 0.5 Mha (Table 1). Other countries with expanding areas of perennial CA systems include Italy, France and Portugal.

Adoption of new farming systems or agricultural practices is normally the response to driving forces or pressures as to improve the current state or to avoid or alleviate impact (negative) with regard to changes occurring whether of natural or other origins (technological, knowledge, etc.). In other parts of the world, the need for cost reduction was also an important driver to maintain agricultural production feasibly. Western Australia with almost 95% of CA on annual cropland (Rochecouste and Crabtree, 2014) is certainly a clear example of how only the shift to CA-based farming practices allowed annual cropping systems to survive. In most European territories, such extreme and adverse natural conditions and events are rarely experienced, thus not triggering the need for change. In general, agroecological conditions, especially rainfall patterns and variability and soil status, can be considered rather favourable when compared to other regions of the planet. However, the recognition and gradual perception that European soils are threatened and that climate change is more and more affecting the European continent too may slowly increase the pressure to change. With regard to the economic situation becoming a driver or a motivation or even creating a pressure to adopt more cost-efficient farming practices, it appears there is still a long way to go considering that European agriculture continues to be subsidized without consideration of the environmental impacts.

Often, in Europe, the diversity of soil and climatic conditions and farming systems is blamed for the difficulties in adapting the seeders and the resistance to adoption of no-tillage. These systems are frequently found not to be suitable for certain crops, the timing of sowing (winter- or spring-sown cereals), soil types, variable seasonal conditions, etc. Such unfavourable conditions include spring-sown cereals, badly drained clay soils, crops not suited to no-till in the rotation due to soil removal at harvest (i.e. potatoes), etc. (Soane et al., 2012). Unstable, weakly structured soils, whether sandy or sandy loams, especially when low in SOC (Munkholm et al., 2003; Ehlers and Claupein, 1994) or even loams and clays (Van Ouwerkerk and Perdok, 1994) have been excluded from soils considered to be suitable for no-till. However, CA can contribute to the rehabilitation of those types of soils, building a better structure and improving SOC due to permanent vegetal soil covers and a higher return of crop biomass.

The Southeastern European region was one of the cradles of agriculture and arable farming. Thus, history may play a role in the deep cultural entrenchment of ploughing and preparing a clean seedbed for sowing. Conventional tillage in modern European agriculture is still widely based on the mouldboard plough, which replaced the ancient Roman plough in the mid-17th century. The total inversion of the upper soil layer controlled effectively perennial grass weeds and the plough was, therefore, the preferred tillage implement. Still today yearly ploughing championships are held in several European countries.

Another frequently referred challenge to the well-functioning of CA is the issue of crop residues. While some regions with low productivity levels (e.g. semi-arid rainfed condition in the Mediterranean region) do not even produce enough residues to provide a good soil cover for erosion protection and soil moisture conservation, other regions, with quantities of residues of 10 tonnes of straw per hectare or more, often experience difficulties in warming and drying the soil sufficiently in spring, thus delaying crop emergence or initial development. While in former times the alternative uses of straw were for animal bedding, less for feed and only locally for other uses such as the mushroom industry; today plants, for producing bioenergy, are an important competitor for straw being used as a carbon source for soil carbon sequestration. An additional challenge of excess residues is their adequate management. Leaving high stubble, where possible, and uniform distribution of the chopped residues are key to facilitate the performance of the no-till seeding equipment to establish the following crop. Row cleaners in wide-row crops may also facilitate seed-soil contact, the warming up of the soil and, overall, emergence.

Although highly mechanized the European market for agricultural machinery and equipment lacks solutions for CA-specific implements, especially those capable to handle higher amounts of residues in the wetter regions of Europe. Most of the pure direct drilling seeders available on the European market are brands from overseas. European manufactures have been reluctant to focus on this type of equipment as it could reduce the sales of other tillage implements which most of them were also producing. Only a strong demand for no-till seeders could probably overcome this situation, which in fact is now happening in some cases. Surprisingly, some European enterprises focussed almost exclusively on the manufacturing of NT seeders. Already in the 1990s a Finish brand was quite successful in selling its equipment boosting the uptake of CA in Finland quite early to over 7% of the annual cropland under CA. More recently, a manufacturer in the UK appeared on the market with an inclined disc-coulter-type NT seeder, which seems to help boost CA adoption in the UK. In Spain, NT seeders are included in the renewal plan for machinery, and in 2019 and 2020 have accounted for more than 60% of the subsidies. Figure 9 shows the evolution of the official register of the Spanish machinery, where the NT seeder trend is positive.

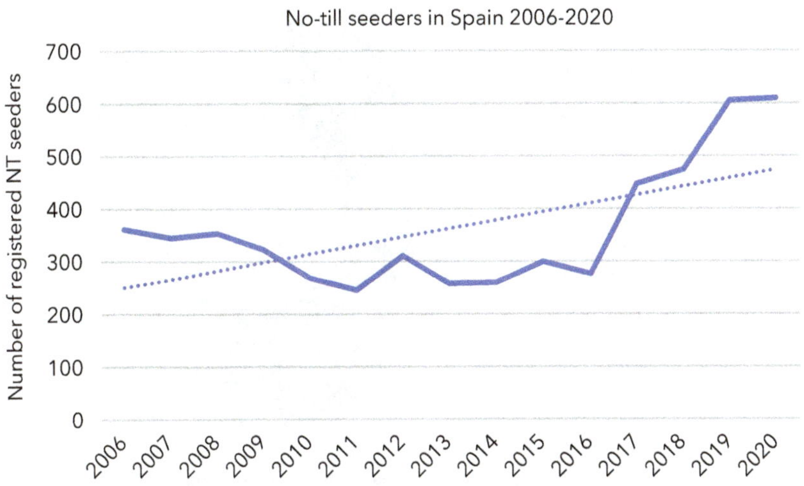

Figure 9 Evolution of the no-till seeders official register of the Ministry of Agriculture of Spain. Source: Mapa (2021).

For long, the small average farm size in many European countries also limited the investment of individual farmers in the substantially more expensive NT seeding technology. Today, increasing farm sizes and the broader availability of service providers are allowing us to overcome this challenge (Fig. 10). A similar situation exists when it comes to other inputs that could improve the performance of CA-established crops. Varieties/genotypes of the main crops better adapted to the different NT conditions (lower temperatures, N dynamics, higher initial bulk density, etc.) as well as site- and rotation-specific cover crops could further help in creating more options and interest in CA adoption.

In Europe, the evaluation and acceptance of new farming practices and technologies still strongly depends on their impact on the yield performance of the crops and less on the overall performance of the system or the net profit created by production. Therefore, yields are compared on a unit area productivity basis, but rarely on a cost basis (costs per unit of produce). Also, the fear of potential yield losses, even if production costs are reduced considerably, weighs heavily in the decision-making process of whether to try to shift or not towards the 'unknown'. Whereas several authors report some yield penalties in the first years after adoption, due to soil structural conditions, but with clear improvement after 3 years (Ball et al., 1989; Christian and Ball, 1994; Six et al., 2004; Anken et al., 2006), others conclude that crop yields can be similar right after the shift to CA (Basch and Teixeira, 2002). Under drier climatic conditions yields of CA-based systems frequently are higher than conventional yields (Soane et al., 2012). Yield reductions under CA have often been reported to be originated by problems related to weed control,

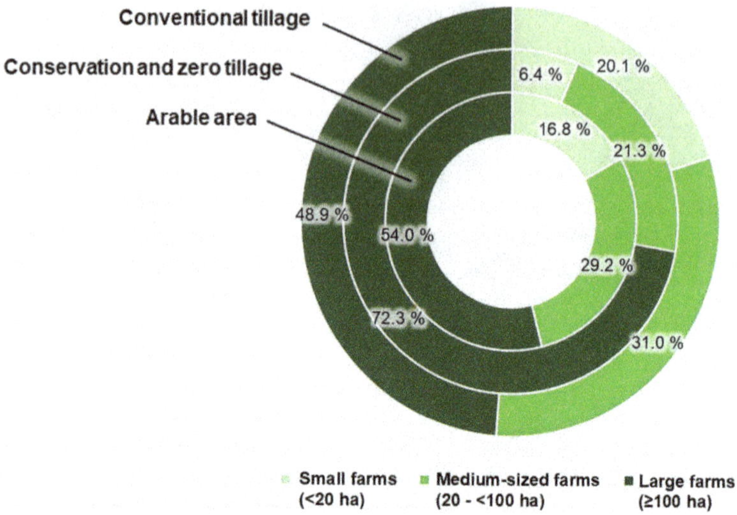

Figure 10 Share of soil management practices by the size of holding in the EU-27 (2016). Size refers to the cropland area of the holding. Source: Eurostat (2020).

compaction or residue management. In general, yield penalties can be avoided or minimized in the transition phase when good advice is available, or when soil structural deficiencies can be overcome and restored without the help of tillage. Over time, however, initial yield reduction disappears, even under less favourable conditions (Soane et al., 2012), or improves and surpasses conventional yields with lower inputs as soil structure, N availability and other soil physical, chemical and biological properties improve (Carvalho and Lourenço, 2014).

One of the major challenges reported with pest or disease problems under CA is the higher incidence of slugs. Although manageable with molluscicides, increased costs and their detrimental effects on beneficial fauna that could help control this very problem recommend restricting this solution to very extreme cases. Recent studies, however, found that insect and slug pests were not more abundant in reduced-tillage systems than in high-disturbance tillage systems and that foliar pests were more abundant in systems with more intense tillage practices (Rowen et al., 2020). Slugs were found to be even more abundant in conventional plots than in conservation plots, possibly due to the lower presence of natural enemies such as ground beetles (Scaccini et al., 2020). Regardless of the intensity of tillage adopted, which can result in higher or lower antagonistic pressure on different pests and diseases, multiple control mechanisms should be employed to avoid yield loss of economic consequence (Leake, 2003). Crop residues remaining at the soil surface may bear the risk of carry-over of pathogens from one season to the other (Mikkola et al., 2005). On the other hand, the concentration of organic matter at the soil surface was suggested to suppress some soil-borne fungal crop diseases (Ehlers and Claupein, 1994) and that residues at the soil surface favour the presence of beneficial polyphagous predators (Jordan et al., 1997). Minimum soil disturbance, together with well-chosen crop rotations, has been reported to help develop disease-suppressive soils (Peters et al., 2003).

A very specific and important challenge for the acceptance and adoption of CA in Europe is the widespread opinion that minimum soil disturbance is synonymous with increased herbicide use. This misinformation has led to a kind of distancing of the broad public, administration, decision-makers and politicians and even farmers to consider CA as a promising farming approach and a solution to a number of challenges now faced by agriculture in Europe. Especially critical in several central and northern European countries is the issue of glyphosate, which became a red flag for all environmentalist, green and alternative pretensions, representing all the evil modern agroindustry has brought about, while ignoring completely what is needed to take care of the soil, to be resource-efficient (water, nutrients, energy etc.), to be the steward for entire landscapes and ecosystems and to mitigate climate change and help adapt to it.

There is now enough scientific and empirical evidence confirming that applying the principles of CA does not increase the need for more chemical weed control compared to conventional farming. At most, a shift from post-emergence to pre-emergence or pre-seeding application may occur, but over time even a reduction of herbicide inputs (number of applications and rates) is being observed (Bräutigam and Tebrügge, 1997; Barros et al., 2007, 2008; Calado et al., 2010). While overall weed numbers normally decline under CA, certain weeds, mainly perennials and some grass weed species may become more abundant (Basch et al., 2015). Methods for their effective management in CA systems have been comprehensively described by Basch et al. (2020). In general, CA cropping systems facilitate the integration of several integrated weed control strategies, based on crop rotation, residue cover, cover cops combined with not soil engaging mechanical control, better timing of application, thus leading over time to less herbicide use.

Figure 11 shows the results of a recent survey of 1700 European farmers, which confirm that not more glyphosate is used in no-tillage compared to conventional tillage. This goes against the general understanding that CA is more dependent on herbicides than tillage-based farming. In fact, with CA, the converse is true because the practice of judicious use of herbicides is established. Soil tillage has been used for centuries to control weeds that compete with crops for water and nutrients. An integrated weed management approach is established in CA systems. Since farmers avoid soil tillage in CA, one of the keys to success in CA is to be able to effectively reduce weed pressure. In CA, farmers need to approach weed management holistically, particularly at the pre-sowing and pre-emergence stages, as both are key moments for

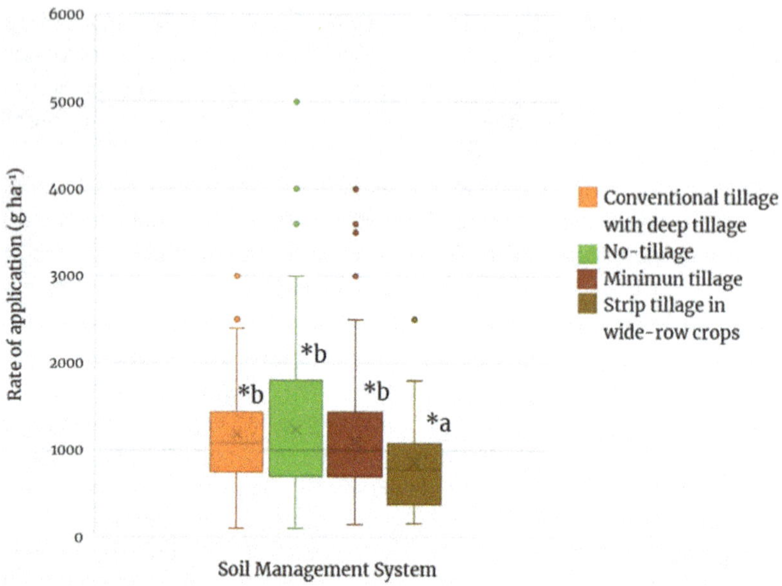

Figure 11 Glyphosate application rate for preemergence uses in different soil management practices in Europe. *Different letters indicate differences between soil management systems ($P < 0.05$, Duncan). Source: ECAF (2020).

the optimal establishment of crops. However, an effective tillage-free post-emergence weed control measure is also required at the latter stages of crop development before the crop canopy closes (ECAF, 2020).

All three principles of CA help to reduce weed infestation density and pressure. In practical terms, a field in a no-tillage system that integrates several crops in the multiannual rotation, always leaving at least 30% of the soil covered throughout the season, is in a better condition than an exposed ploughed field with a poor crop mix in the rotation to confront weed infestations. Abundant crop residues reduce weed infestation, as sunlight is not directly reaching weed seeds and soil temperature is lower than in an exposed ploughed field. Therefore, weed germination is generally not favoured and particularly perennial grass weeds can be better suppressed by heavy soil cover than by tillage. Additionally, no soil disturbance with no-tillage negatively affects the weed seed bank. Once weed seeds residing at or near the soil surface are controlled, weed seeds buried at lower depths are not able to germinate. Conversely, in tillage-based agriculture, soil mechanical disturbance moves, and 'hides' weed seeds in the soil profile, creating a situation of a never-ending weed problem. This happens especially with inversion tillage operations, but also with light tillage operations. Finally, crop diversification with particular crop species can make an allelopathic contribution to management of specific weeds and also allows different chemical strategies to control weeds, helping to prevent herbicide resistance in weeds. Unlike in tillage agriculture, monoculture is not a valid practice in CA (ECAF, 2020).

5 The role of Conservation Agriculture in the European Policies: The CAP, EU, Climate Action and the Green Deal

Whereas soil conservation farming practices have been broadly adopted in North and South America and Australia, Europe still struggles today to embrace effective soil conserving and improving cropping systems in order to achieve the goals established as the key objectives of the past Common Agricultural Policy (CAP): viable food production; sustainable management of natural resources and climate action; economic competitiveness/farm income, productivity/production, biodiversity and the fight against natural constraints.

During a period of decades in the last century, the CAP promoted production intensification through production-oriented income support to the agricultural sector favouring high productivity levels, obtained with massive external inputs, instead of promoting competitiveness and sustainability. The consequences were the eventual overproduction of some commodities, increasing the damage of the environment and natural resources, as well as low input use efficiencies and thus low cost-efficiency and competitiveness of farming. Only since the early 1990s did some concerted efforts begin to link agricultural and environmental policy, and at the beginning of this century developments started towards a transition from production-linked subsidies to a farm payment system that is separate from production. At the same time, the preservation of the environment and natural resources, above all, the soil, water, air and biodiversity, the need for climate action through carbon sequestration in soils, GHG emission reductions and the replacement of non-renewable energy sources became the focus of agricultural and rural development policy.

Europe took very long to recognize that its agricultural soils also suffer from severe degradation due to intensive land use practices. Yet, these initial environmental concerns derived more from off-site impacts caused by farming activities rather than concerns about the soil itself as a key and irreplaceable agricultural resource. Soil started to be looked upon as a threatened resource with the initiation of the so-called EU Soil Forum in 2002, to discuss the Soil Thematic Strategy ratified by the European Commission in September 2006. This strategy was thought to 'create a common legal framework to ensure that EU soils stay healthy for future generations and remain capable of supporting the ecosystems on which our economic activities and our well-being depend' (Van-Camp et al., 2004). Despite not resulting in the envisaged 'Soil Framework Directive ' as wished, this strategy led towards manifold efforts in both agricultural and research policies putting a strong focus on soils and their sustainable management and helped create awareness about the major European soil threats.

In Europe, albeit performing a shift from production-oriented income support to an 'area-based' support through so-called direct payments, there continues to be an enormous welfare transfer from consumers and taxpayers to the farming sector, justified by the high standards (environmental and animal welfare) European farmers have to comply with, but without real performance or result-based linkage of the subsidies paid. As a matter of fact, these standards apparently have been increased over the programming periods, from Statutory Management Requirements (SMRs), Good Agricultural and Environmental Condition (GAECs), Cross-Compliance to the Greening Measures that will transform into the new 'Enhanced Conditionality' in the next CAP. Analysing, however, these 'mandatory' standards for farmers receiving the direct payments and knowing about the high discretionary power of EU member states (MS) to interpret and implement these standards, the effects of these measures to satisfy the EU requirements for climate action and the preservation of water, soil, biodiversity and landscapes remain questionable. Additionally, the European Parliament resolution declaring a climate and environmental emergency in Europe and globally, the European Green Deal (EU, 2019), the forthcoming Farm to Fork strategy for sustainable food, the achievement of the Sustainable Development Goals (SDGs) and the 4 per Thousand Initiative (https://www.4p1000.org/), all pose additional challenges for European agriculture, asked to maximize yields while being required to operate in an environmentally sustainable manner. All these challenges are ultimately linked to, and mediated by, well-functioning and sustainably managed soils, along with water and biodiversity. These concerns are also reflected in the mission-oriented policy for Horizon Europe with the European Commission assigning one of its five specific missions to 'Soil Health and Food' (European Commission, 2020), namely in what are regarded as Clean Oceans and Rivers (soil erosion) and Climate Change (role and management of soils and their contribution to carbon sequestration).

For more than five decades, European agriculture has been subject to the strong influence of the CAP framework. Until the early 2000s income support to the farming sector was mostly production-oriented favouring high productivity levels, obtained with massive external inputs, instead of promoting competitiveness and sustainability. Since 2003 the so-called decoupling process led to the shift of income support to an 'area-based' support system through direct payments based on the historical income support of farms. Still, the enormous welfare transfer to the farming sector, justified by the high standards (environmental and animal welfare) European farmers have to comply with continues. On top of the direct payments, which supposedly oblige all farmers to respect minimum standards, also when it comes to soil conservation, some of the EU MS promoted and still promote the practice of CA with additional payments from the Rural Development Funds. In some MS or regions within MS, these additional supports resulted indeed in increased CA adoption rates. However, there are examples (e.g. Portugal) where considerable initial adoption rates were triggered by attractive temporary support schemes, but which decreased substantially after the continuous support ceased. This clearly indicates that temporary financial support alone, with the objective to face potential risks during the transition phase, may not sustain an enduring shift towards CA (Goddard et al., 2020).

Despite the fact that awareness about soil threats and the need for its conservation have risen clearly at the European policy and administration level, and numerous studies and research projects on soil quality and sustainable soil management have been financed, there appears to be little feedback and effect on the definition of clearly soil-oriented Ag Policy. However, unless the public in general and often even the farming community do not recognize clearly the extent of damage, loss of ecosystem services and costs of tillage-based agriculture, whether conventional or organic, it will be difficult to convince politicians and decision-makers to embrace CA as a preferred option to achieve SDGs, climate goals, ecosystem services, food security, competitiveness and many more objectives. Perhaps, the newly created EU Mission Board for Soil health and food with its slogan 'Caring for soil is caring for life' and the objective to ensure that 75% of soils are healthy by 2030 can make a difference through its attempts of citizen engagement and finally to propose actions, hopefully not only for the upcoming research agenda but also for a sound soil-oriented CAP.

The latest endeavour to improve the European agri-environment is presented in the Green Deal. It consists of several lines of action, aimed at acting in all sectors of the economy, investing in environmentally friendly technologies. The policy areas in the Green Deal are:

- Biodiversity: Measures to protect our fragile ecosystem.
- From Farm to Fork: Ways to ensure more sustainable food systems.
- Sustainable agriculture: Sustainability in EU agriculture and rural areas thanks to the CAP.
- Clean energy.
- Sustainable industry: Ways to ensure more sustainable, more environmentally respectful production cycles.
- Building and renovating: The need for a cleaner construction sector.
- Sustainable mobility: Promoting more sustainable means of transport.
- Eliminating pollution: Measures to cut pollution rapidly and efficiently.
- Climate action: Making the EU climate neutral by 2050.

Some of the elements of the Farm to Fork strategy are:

- The Commission will take action to reduce the use and risk of chemical and more hazardous pesticides by 50%.
- The Commission will act to reduce nutrient losses by at least 50%, while ensuring no deterioration in soil fertility and reduce fertilizer use by at least 20%.
- The Commission will help the EU's organic farming sector to grow, with the goal of 25% of total farmland being used for organic farming by 2030.

The European Green Deal is also the EU lifeline out of the COVID-19 pandemic. One-third of the 1.8 trillion Euro investments from the Next Generation EU Recovery Plan and the EU's 7-year budget will finance the European Green Deal (EU, 2021).

The adoption of CA principles in locally adapted production systems for the growth of annual and perennial crops, pastures and forages, together with good-quality seeds and optimally integrated nutrient, water and pest management, would realize the goals outlined in the Green Deal by:

- providing similar or even higher yields through improvements in soil structure, organic matter and overall soil fertility;
- lowering production costs through reduced inputs of energy, labour and machinery in the short and long term, and fertilizers, water and pesticides in the medium and long term, thus raising related productivity and efficiency;
- mitigating CO_2 emissions through reduced fuel consumption and sequestration of atmospheric carbon into SOM, and reducing N_2O and CH_4 emissions through reduced use of mineral nitrogen and improved soil drainage;
- reducing runoff and erosion through better soil aggregate stability and improved water infiltration, and protective cover of the soil by crops and/or crop residues;
- diminishing off-site damage of infra-structures and pollution of water bodies through less runoff with a much-reduced sediment load;
- maintaining in-field and off-site biodiversity through the absence of destructive soil disturbance, protective soil shelter and less off-site transport of contaminants;
- maintaining the diversity of rural landscape through enhanced crop and species diversity and cover crops; and
- maintaining less favoured rural areas under production through the adoption of economically and environmentally viable production methods.

The characteristics of locally adapted CA production systems together with the rational and responsible use of external inputs will optimize crop yields, farm income, competitiveness and (bio)diversity and minimize any negative ecological impacts associated with intensive farming. The optimized use of herbicides to facilitate weed control and soil cover management is an option to reduce production costs and to avoid the aforementioned negative effects associated with soil tillage, including the stimulation of further weed emergence and spread. There is no evidence of an increase in the use of herbicides under CA systems when compared to conventional tillage farming. Instead, there is a shift in application timing and towards the use of contact herbicides. The latter are less persistent in the environment than the more frequently used residual herbicides in conventional farming. In countries such as Canada (Blackshaw and Harker, 2010) and Australia (Crabtree, 2010), which have agroecological conditions similar to Europe, herbicide use per tonne of output is lower in CA systems with integrated weed management than in conventional tillage farming.

When CA systems are adopted over large areas, it is possible to harness much needed environmental services such as clean water, erosion control, carbon sequestration, reduced GHG emissions, reduced risks of floods and drought, biodiversity protection, etc., that have heretofore not been fully possible with conventional intensive tillage-based agricultural land uses in Europe (Kassam et al., 2010, 2020). Thus, CA principles and systems would provide a solid bridge between the two Pillars of CAP and make eco-schemes, agri-environmental schemes, environmentally and financially meaningful on an EU-wide basis. Adoption of CA will also provide a foundation for developing environmental service schemes such as carbon sequestration and trading, clean water provision, soil erosion control and biodiversity enhancement, in which incentives and payments can be linked to specific production systems and services. Such schemes exist elsewhere such as in Canada where a cap and trade scheme, started in 2007, enables regulated industries to purchase carbon offsets from the agriculture sector based on a CA (no-till) production protocol adopted by farmers.

6 Conclusion

More and more farmers are realizing that they need a better alternative to tillage agriculture in Europe. Those farmers who have discovered CA as an alternative and are practising it successfully are inspired by it. Every CA farmer has an exciting story to tell about the discovery, learning by doing, on-farm innovations and most importantly, becoming connected again to the land as an effective steward for the benefit of his/her family, society and nature. More importantly, CA farmers realize the significance of their achievement, which has been won largely without much policy or institutional support, and with little support from a large part of the agriculture industry. CA farmers are part of a global movement which they have created by networking at all levels and driving further developments in CA and land use through sharing of experiences, knowledge and innovations.

Some CA farmers in Europe have been practising this alternate system for some 40 years now and many from over 20 years. They will all say that CA for them has meant that they can now farm with the knowledge and confidence that they are no longer degrading and destroying their land resource base, rather the opposite. A judicious use of inputs allows farmers to implement high-quality and productive CA, whilst addressing challenges in soil, water, biodiversity and climate. Indeed, these farmers have become even more caring stewards of their farmlands to contribute to national and international food security, to harness a range of sustainable ecosystem services for society and to provide a decent way of life for themselves and for future generations.

If CA can achieve all this, what more do we want? If we provide CA farmers with the necessary tools, they will deliver what the CAP, and other key policies, has set as priority goals for European agriculture.

7 Where to look for further information

The best single source is the European Conservation Agriculture Federation website: www.ecaf.org.

8 References

Anken, T., Stamp, P., Richner, W., Walther, U., Weisskopf, P. and Rek, J. 2006. Nitrate leaching and soil structural properties under conventionally cultivated and no-till crops. *Proceedings of the 17th Conference of ISTRO*, ISTRO, Kiel, Germany, pp. 1535–1540.

Arrouays, D., Deslais, W. and Badeau, V. 2001. The carbon content of topsoil and its geographical distribution in France. *Soil Use and Management* 17(1), 7–11.

Balesdent, J., Chenu, C. and Balabane, M. 2000. Relationship of soil organic matter dynamics to physical protection and tillage. *Soil and Tillage Research* 53(3–4), 215–230.

Ball, B. C., Lang, R. W., O'Sullivan, M. F. and Franklin, M. F. 1989. Cultivation and nitrogen requirements for continuous winter barley on a gleysol and a cambisol. *Soil and Tillage Research* 13(4), 333-352.

Barros, J. F. C., Basch, G. and Carvalho, M. 2007. Effect of reduced doses of a post-emergence herbicide to control grass and broad-leaved weeds in no-till wheat under Mediterranean conditions. *Crop Protection* 26(10), 1538–1545.

Barros, J. F. C., Basch, G. and Carvalho, M. 2008. Effect of reduced doses of a post-emergence graminicide to control *Avena sterilis* L. and *Lolium rigidum* G. in no-till wheat under Mediterranean environment. *Crop Protection* 27(6), 1031–1037.

Basch, G., Friedrich, T., Kassam, A. and Gonzalez-Sanchez, E. J. 2015. Conservation agriculture in Europe. In: Farooq, M. and Siddique, K. H. M. (Eds.) *Conservation Agriculture*. Springer International Publishing, Switzerland, pp. 357–389.

Basch, G. and Teixeira, F. (Eds.) 2002. *Mobilização de Conservação do Solo*. APOSOLO, Évora, Portugal (in Portuguese).

Basch, G., Texeira, F. and Duiker, S. 2020. Weed management practices and benefits in CA systems. In: Kassam, A. (Ed.) *Advances in Conservation Agriculture, Volume 2: Practice and Benefits*. Burleigh Dodds Science Publishing, Cambridge, UK, pp. 105–141.

Bäumer, K. 1970. First experiences with direct drilling in Germany. *Netherland Journal of Agricultural Science. Papers on Zero-Tillage* 18(4), 283-292.

Bellamy, P. H., Loveland, P. J., Bradley, R. I., Murray Lark, R. M. and Kirk, G. J. 2005. Carbon losses from all soils across England and Wales 1978-2003. *Nature* 437(7056), 245–248.

Bielders, C. L., Ramelot, C. and Persoons, E. 2003. Farmer perception of runoff and erosion and extent of flooding in the silt-loam belt of the Belgian Walloon Region. *Environmental Science and Policy* 6(1), 85–93.

Blackshaw, R. E. and Harker, K. N. 2010. Evolution of effective and economic weed control. In: Lindwall, W. and Sonntag, B. (Eds.) *Landscapes Transformed: the History of Conservation Tillage and Direct Seeding*. Knowledge Impact in Society, Saskatoon, Saskatchewan, pp. 141–148.

Blair, N., Faulkner, R. D., Till, A. R. and Crocker, G. J. 2006. Long-term management impacts on soil C, N and physical fertility: part III: Tamworth. *Soil and Tillage Research* 91(1-2), 48–56.

Bräutigam, V. and Tebrügge, F. 1997. Influence of long-termed no-tillage on soil borne plant pathogens and on weeds. In: Tebrügge, F. and Böhrnsen, A. (Eds.) *Experience with the Applicability of No-Tillage Crop Production in the West-European Countries. Proceedings of the EC-Workshop-III*. Wissenschaftlicher Fachverlag, Giessen, Germany, pp. 17–29.

Calado, J. M. G., Basch, G. and Carvalho, M. 2010. Weed management in no-till winter wheat (*Triticum aestivum* L.). *Crop Protection* 29(1), 1–6.

Carvalho, M. and Lourenço, E. 2014. Conservation agriculture–a Portuguese case study. *Journal of Agronomy and Crop Science* 200(5), 317-324.

Christian, D. G. 1994. Experience with direct drilling cereals and reduced cultivation in England. In: Tebrügge, F. and Böhrnsen, A. (Eds.) *Experience with the Applicability of No-Tillage Crop Production in the West-European Countries. Proceedings of the EC-Workshop-I*. Wissenschaftlicher Fachverlag, Langgöns, Germany, pp. 25–31.

Christian, D. G. and Ball, B. C. 1994. Reduced cultivations and direct drilling for cereals in Great Britain. In: Carter, M. R. (Ed.) *Conservation Tillage in Temperate Agroecosystems*. Lewis Publishers, Boca Raton, pp. 117–140.

Crabtree, B. 2010. Weed control is superior with no-till. In: Crabtree, B. (Ed.) *Search for Sustainability with No-Till Bill in Dryland Agriculture*. Crabtree Agricultural Consulting Publishing, Beckenham, WA, pp. 27–36.

ECAF 2020. *Making Sustainable Agriculture Real in Europe with Conservation Agriculture: Judicious Use of Glyphosate in Integrated Weed Management*. European Conservation Agriculture Federation, Brussels, p. 150.

EC-JRC. 2012. *The State of Soils in Europe*. A contribution of the JRC to the European Environment Agency's Environment State and Outlook Report–SOER. JRC. JRC Reference Report 2010.

EC-JRC. 2015. Soil area the size of Berlin lost each year due to water erosion in the EU. Available at: https://www.eurekalert.org/pub_releases/2015-09/ecjr-sat090215.php (accessed on 15 April 2020).

EEA 2003. Assessment and reporting on soil erosion. *EEA Technical Report 94*. European Environment Agency, Copenhagen.

EEA 2004. The state of biological diversity in the European Union. Malahide Information Paper. *Malahide/Inf2*. European Environment Agency, Copenhagen.

EEA 2017. Soil erosion, Indicator Assessment. *Prod-ID: IND-194-en*. Available at: https://www.eea.europa.eu/data-and-maps/indicators/soilerosion-by-water-1/assessment (accessed on 17 April 2020).

EEA 2020. Towards a 'sustainable agriculture'. Available at: https://www.eea.europa.eu/themes/agriculture (accessed on 16 July 2021).

Ehlers, W. and Claupein, W. 1994. Approaches toward conservation tillage in Germany. In: Carter, M. R. (Ed.) *Conservation Tillage in Temperate Agroecosystems*. Lewis Publishers, Boca Raton, FL, pp. 141–165.

EP 2021. Motion for a resolution further to questions for oral answer *B9-0010/2021 and B9-0011/2021* pursuant to Rule 136(5) of the rules of procedure on soil protection. Available at: https://www.europarl.europa.eu/doceo/document/B-9-2021-0221_EN.html (accessed on 16 July 2021).

EU 2019. A European green deal. Available at: https://ec.europa.eu/info/strategy/priorities-2019-2024/european-green-deal_en (accessed on 15 July 2021).

EU 2021. A European green deal. Available at: https://ec.europa.eu/info/strategy/priorities-2019-2024/european-green-deal_en#documents (accessed on 16 July 2021).

European Commission 2020. Mission area: soil health and food. Available at: https://ec.europa.eu/info/horizon-europe-next-research-and-innovation-frameworkpro-gramme/mission-area-soil-health-and-food_en (accessed on 16 April 2020).

Eurostat 2018. Agricultural practices. Available at: https://ec.europa.eu/eurostat/databrowser/view/ef_mp_prac/default/table?lang=en (accessed on 15 April 2021).

Eurostat 2020. Agri-environmental indicator – tillage practices. Available at: https://ec.europa.eu/eurostat/statistics-explained/index.php?title=Agri-environmental_indicator_-_tillage_practices (accessed on 16 July 2021).

FAO 2019. *The State of the World's Biodiversity*. Food and Agriculture Organization, Rome.

FAOstat 2018. *Land Use*. Available at: http://www.fao.org/faostat/en/#data/RL (accessed on 14 April 2021).

Fernández-Quintanilla, C. 1997. Historia y evolución de los sistemas de laboreo. El laboreo de conservación. In: García Torres, L. and González Fernández, P. (Eds.) *Agricultura de Conservación: Fundamentos Agronómicos*. Medioambientales y Económicos. Asociación Española Laboreo de Conservación/Suelos Vivos, Córdoba, Spain, pp. 1–12.

Friedrich, T. 2019. Chapter 4. Advances in conservation agriculture. In: Kassam, A. (Ed.). *The Role of No or Minimum Mechanical Soil Disturbance in Conservation Agriculture Systems* (vol. 1), Burleigh Dodds. UK, pp. 155–178.

Gardi, C., Menta, C., Montanarella, L. and Cenci, R. 2008. Main threats to soil biodiversity: the case of agricultural activities impacts on soil microarthropods. In: Toth, G., Montanarella, L. and Rusco, E. (Eds), *Threats to Soil in Europe*. Office for the Official Publications of the European Communities, Luxembourg, pp. 100–110.

Gardi, C. et al. 2009. Soil biodiversity monitoring in Europe: ongoing activities and challenges. *Euro. J. Soil Science* 60: 807–819.

Gobin, A. 2005. *Dalend organisch stofgehalte in de landbouwbodem: Bedreiging voor landbouw en biodiversiteit. Rapport in opdracht van MIRA-T.* (in German).

Gobin, A., Campling, P., Lavelle, P. and Berman, S. 2011. Soil organic matter management across the EU – best practices, constraints and trade-offs. *Annex II Case Studies. Final Report for the European Commission's DG Environment*, September 2011.

Goddard, T., Basch, G., Derpsch, R., Hongwen, L., Jin, H., Kassam, A., Moriya, K., Peiretti, R. and Smith, H. 2020. Institutional and policy support for conservation agriculture uptake. In: Kassam, A. (Ed.) *Advances in Conservation Agriculture, Volume 1: Systems and Science*. Burleigh Dodds Science Publishing, Cambridge, UK, pp. 513–564.

Goossens, D., Gross, J. and Spaan, W. 2001. Aeolian dust dynamics in agricultural land areas in Lower Saxony, Germany. *Earth Surface Processes and Landforms* 26(7), 701–720.

Huber, S., Prokop, G., Arrouays, D., Banko, G., Bispo, A., Jones, R. J. A., Kibblewhite, M. G., Lexer, W., Möller, A., Rickson, R. J., Shishkov, T., Stephens, M., Toth, G., Van den Akker, J. J. H., Varallyay, G., Verheijen, F. G. A. and Jones, A. R. (Eds.) 2008. Environmental assessment of soil for monitoring: volume I indicators and criteria. EUR, p. 23490 EN/1. Office for Official Publication of the European Communities, Luxembourg. 339 pp.

Jones, A., Panagos, P., Barcelo, S., Bouraoui, F., Bosco, C., Dewitte, O. and Montanarella, L. 2012. The state of soil in Europe. A contribution of the JRC to the European Environment Agency's Environment State and Outlook Report-soer 2010. In: The European Environment–State and Outlook 2015, SOER 2010.

Jordan, V. W. L., Hutcheon, J. A. and Kendall, D. A. 1997. Influence of cultivation practices on arable crop pests, diseases and weeds and their control requirements. In: Tebrügge, F. and Böhrnsen, A. (Eds.) *Experience with the Applicability of No-Tillage Crop Production in the West-European Countries. Proceedings of the EC-Workshop-I*, Wissenschaftlicher Fachverlag, Langgöns, Germany, pp. 43–50.

Kassam, A. 2020. *Advances in Conservation Agriculture Volume 2: Practice and Benefits*. Burleigh Dodds Science Publishing, Cambridge, UK.

Kassam, A., Friedrich, T. and Derpsch, R. 2010. Conservation agriculture in the 21st century: a paradigm of sustainable agriculture. *Proceedings of the European Congress on Conservation Agriculture: Towards Agro-Environmental Climate and Energetic Sustainability*. Madrid, pp. 19–68.

Kassam, A., Friedrich, T. and Derpsch, R. 2019. Global spread of conservation agriculture. *International Journal of Environmental Studies* 76(1), 29–51.

Kassam, A., Friedrich, T., Derpsch, R., Lahmar, R., Mrabet, R., Basch, G., González-Sánchez, E. J. and Serraj, R. 2012. Conservation agriculture in the dry mediterranean climate. *Field Crops Research* 132, 7–17.

Kassam, A., González-Sánchez, E. J., Goddard, T., Hongwen, L., Mello, I., Mkomwa, S., Shaxson, F. and Friedrich, T. 2020. Harnessing ecosystem services with conservation agriculture. In: Kassam, A. (Ed.) *Advances in Conservation Agriculture (vol. 2): Practice and Benefits*. Burleigh Dodds Science Publishing, Cambridge, UK, pp. 391–418.

Keith, A. M., Boots, B., Hazard, C., Niechoj, R., Arroyo, J., Bending, G. D., Bolger, T., Breen, J., Clipson, N., Doohan, F. M., Griffin, C. T. and Schmidt, O. 2012. Cross-taxa congruence, indicators and environmental gradients in soils under agricultural and extensive land management. *European Journal of Soil Biology* 49, 55–62.

Lavelle, P. and Spain, A. V. 2001. *Soil Ecology*. Kluwer Publishers Scientific, Amsterdam.

Leake, A. R. 2003. Integrated pest management for conservation agriculture. In: García-Torres, L., Benites, J., Martínez-Vilela, A. and Holgado-Cabrera, A. (Eds.) *Conservation Agriculture: Environment, Farmers Experiences, Innovations, Socio-Economy, Policy*. Springer, Dordrecht, The Netherlands, pp. 271–279.

MAPA 2021. Registro de maquinaria agrícola. Available at: https://www.mapa.gob.es/es/agricultura/estadisticas/estadisticas-registros-maquinaria-agricola.aspx (accessed on 15 July 2021).

Merante, P., Dibari, C., Ferrise, R., Bindi, M., Lesschen, J. P., Kuikman, P., Sanchez, B. and Iglesias, A. 2014. Report on critical low soil organic matter contents, which jeopardise good functioning of farming systems. SmartSOIL (Grand Agreement No 289694): Deliverable, 2.4. Available at: https://projects.au.dk/smartsoil/ (accessed on 16 July 2021).

Mikkola, H. J., Alakukku, L., Känkänen, H., Jalli, H. and Lindroos, M. 2005. Direct drilling in Finland, a review. *Proceedings of the 4th* International *Scientific and Practical Conference*, Ecology *and* Agricultural Machinery, St Petersburg (vol. 2), pp. 141–151.

Montgomery, D. R. 2007. *Dirt, the Erosion of Civilizations*. University of California Press, Berkeley, Los Angeles and London, 285 pp.

Mulder, C., Boit, A., Bonkowski, M., de Ruiter, P. C., Mancinelli, G., van der Heijden, M. G. A., van Wijnen, H. J., Vonk, J. A. and Rutgers, M. 2011. A belowground perspective on Dutch agroecosystems: how soil organisms interact to support ecosystem services. *Advances in Ecological Research* 44, 277–357.

Munkholm, L. J., Schjønning, P., Rasmussen, K. J. and Tanderup, K. 2003. Spatial and temporal effects of direct drilling on soil structure in the seedling environment. *Soil and Tillage Research* 71(2), 163–173.

Orgiazzi, A., Bardgett, R. D., Barrios, E., Behan-Pelletier, V., Briones, M. J. I., Chotte, J.-L., De Deyn, G. B., Eggleton, P., Fierer, N., Fraser, T., Hedlund, K., Jeffery, S., Johnson, N. C., Jones, A., Kandeler, E., Kaneko, N., Lavelle, P., Lemanceau, P., Miko, L., Montanarella, L., Moreira, F. M. S., Ramirez, K. S., Scheu, S., Singh, B. K., Six, J., van der Putten, W. H. and Wall, D. H. (Eds.) 2016. Global soil biodiversity atlas. Available at: https://esdac.jrc.ec.europa.eu/content/global-soil-biodiversity-atlas (accessed on 4 May 2020).

Panagos, P., Borrelli, P., Poesen, J., Ballabio, C., Lugato, E., Meusburger, K., Montanarella, L. and Alewell, C. 2015. The new assessment of soil loss by water erosion in Europe. *Environmental Science and Policy* 54, 438–447.

Peters, R. D., Sturz, A. V., Carter, M. R. and Sanderson, J. B. 2003. Developing disease-suppressive soils through crop rotation and tillage management practices. *Soil and Tillage Research* 72(2), 181–192.

Phillips, E. R. and Phillips, S. H. 1984. *No-Tillage Agriculture, Principles and Practices*. Van Nostrand Reinhold Co, New York.

Rochecouste, J. and Crabtree, B. 2014. Conservation agriculture in Australian dryland cropping. In: Jat, R. A., Ram, A. J., Sahrawat, K. L. and Kassam, A. (Eds.) *Conservation Agriculture: Global Prospects and Challenges*. CABI Publishing, Cambridge, pp. 108–126.

Rowen, E. K., Regan, K. H., Barbercheck, M. E. and Tooker, J. F. 2020. Is tillage beneficial or detrimental for insect and slug management? A meta-analysis. *Agriculture, Ecosystems and Environment* 294, 105–113.

Scaccini, D., Panini, M., Chiesa, O., Nicoli-Aldini, R., Tabaglio, V. and Mazzoni, E. 2020. Slug monitoring and impacts on the ground beetle community in the frame of sustainable pest control in conventional and conservation agroecosystems. *Insects* 11(6), 380.

Serna-Chavez, H. M., Fierer, N. and Van Bodegom, P. M. 2013. Global drivers and patterns of microbial abundance in soil. *Global Ecology and Biogeography* 22(10), 1162-1172.

Six, J., Ogle, S. M., Breidt, F. J., Conant, R. T., Mosier, A. R. and Paustian, K. 2004. The potential to mitigate global warming with no-tillage management is only realized when practised in the long term. *Global Change Biology* 10(2), 155–160.

Soane, B. D., Ball, B. C., Arvidsson, J., Basch, G., Moreno, F. and Roger-Estrade, J. 2012. No-till in northern western and south-western Europe: a review of problems and opportunities for crop production and the environment. *Soil and Tillage Research* 118, 66–87.

Spiegel International 2011. Sandstorm causes deadly autobahn crash. Available at: https://www.spiegel.de/international/germany/mass-pileup-in-northern-germany-sandstorm-causes-deadly-autobahncrash-a-755945.html (accessed on 16 July 2021).

Tebrügge, F. 2003. No-tillage visions: protection of soil, water and climate and influence on management and farmers' income. In: Garcia-Torres, L. et al. (Eds), Conservation Agriculture: Environment, *Farmers' Experience, Innovation, Socio*-economy and *Policy*. Springer Science, Dordrecht, The Netherlands.

Tebrügge, F. and Böhrnsen, A. (Eds.) 1994. Experience with the applicability of no-tillage crop production in the West-European countries. *Proceedings of the EC-Workshop-I, Giessen, Germany*, Wissenschaftlicher Fachverlag, Langgöns, Germany.

Tebrügge, F. and Böhrnsen, A. (Eds.) 1995. Experience with the applicability of no-tillage crop production in the West-European countries. *Proceedings of the EC-Workshop-II, Silsoe, UK*, Wissenschaftlicher Fachverlag, Langgöns, Germany.

Tebrügge, F. and Böhrnsen, A. (Eds.) 1996. Experience with the applicability of no-tillage crop production in the West-European countries. *Proceedings of the EC-Workshop-III, Évora, Portugal*, Wissenschaftlicher Fachverlag, Langgöns, Germany.

Tebrügge, F. and Böhrnsen, A. (Eds.) 1997. Experience with the applicability of no-tillage crop production in the West-European countries. *Proceedings of the EC-Workshop-IV, Boigneville, France*, Wissenschaftlicher Fachverlag, Langgöns, Germany.

Tebrügge, F. and Böhrnsen, A. (Eds.) 1998. Experience with the applicability of no-tillage crop production in the West European countries. Final report, review papers, summary and conclusions of the Concerted Action AIR 3 CT 93-1464. Fachverlag Köhler, Giessen, Germany.

Tsiafouli, M. A., Thébault, E., Sgardelis, S. P., de Ruiter, P. C., van der Putten, W. H., Birkhofer, K., Hemerik, L., de Vries, F. T., Bardgett, R. D., Brady, M. V., Bjornlund, L., Jørgensen, H. B., Christensen, S., Hertefeldt, T. D., Hotes, S., Gera Hol, W. H., Frouz, J., Liiri, M., Mortimer, S. R., Setälä, H., Tzanopoulos, J., Uteseny, K., Pižl, V., Stary, J., Wolters, V. and Hedlund, K. 2015. Intensive agriculture reduces soil biodiversity across Europe. *Global Change Biology* 21(2), 973–985.

United Nations 1987. Our common future. Brundtland Report. Oxford University Press, p. 204.

Van-Camp, L., Bujarrabal, B., Gentile, A. R., Jones, R. J. A., Montanarella, L., Olazabal, C. and Selvaradjou, S. K. 2004. Reports of the technical working groups established under the thematic strategy for soil protection. EUR 21,319 EN/1. Office for Official Publications of the European Communities. Luxembourg 872 pp.

Van Ouwerkerk, C. and Perdok, U. D. 1994. Experiences with minimum and no-tillage practices in the Netherlands. I: 1962–1971. In: Tebrügge, F. and Böhrnsen, A. (Eds.) *Experience with the Applicability of No-Tillage Crop Production in the West-European*

Countries. Proceedings of the EC-Workshop-I, Wissenschaftlicher Fachverlag, Langgöns, Germany, pp. 59–67.
Verheijen, F. G. A., Jones, R. J. A., Rickson, R. J. and Smith, C. J. 2009. Tolerable versus actual soil erosion rates in Europe. *Earth-Science Reviews* 94(1–4), 23–38, 23–38.
Verheijen, F. G. A., Verheijen, F. G. A., Bellamy, P. H., Kibblewhite, M. G. and Gaunt, J. L. 2005. Organic carbon ranges in arable soils of England and Wales. *Soil Use and Management* 21(1), 2–9.

Chapter 7

Adoption and spread of Conservation Agriculture in North Africa

Rachid Mrabet, National Institute of Agricultural Research (INRA), Morocco; Haithem Bahri, National Research Institute for Rural Engineering, Water, and Forestry (INRGREF), Tunisia; Omar Zaghouane, Consultant, Algeria; Hatem Cheikh M'Hamed, National Institute of Agricultural Research of Tunisia (INRAT), Tunisia; Sherif Ragab Mohamed El-Areed, Beni-Suef University, Egypt; and Moamen Mohamed Abou El-Enin, Al Azhar University, Egypt

1 Introduction
2 Geography, soil, climate and farming environment
3 Agriculture and food security
4 Conservation Agriculture: History and development
5 Major research achievements and advances in knowledge: Re-aligning development, economy and environment
6 Adoption barriers, constraints and gaps
7 Adoption drivers and opportunities
8 Networks, partner institutes and distributional effects
9 Conclusion
10 Where to look for further information
11 Acronyms and abbreviations
12 References

1 Introduction

In North Africa, the highly diverse agro-ecological systems are a complex mixture of pastoral, agro-pastoral, rainfed and irrigated farming systems (Nefzaoui et al., 2012). In all these landscapes, desertification and degradation processes are steadily creeping ahead. In addition, weather-related degradation events are growing in intensity and frequency and are weakening economic growth (Sivakumar et al., 2013). Agriculture and food security have been the hardest hit by climate change and the impacts of economic and social crises (Selvaraju, 2013; Khouri et al. (2017). Consequently, social and economic factors have contributed to unsustainable land use and management practices.

http://dx.doi.org/10.19103/AS.2021.0088.06

Soil erosion and dust storms induced by changes in land use and low soil cover are among the main causes of land degradation and reduced productivity (Montanarella et al., 2016). In agricultural lands, soil tillage is also a major cause of soil erosion and land degradation (Kassam et al., 2012). In fact, soil loss and induced impacts are well known and documented, as are its mechanics (Borrelli et al., 2017; Lagacherie et al., 2017).

The water productivity of rainfed crops has generally plummeted due to technological constraints, low and variable rainfall in some areas, reduced soil fertility and quality and improper management of crops with respect to changes in rainfall patterns (Mrabet, 2017). Consequently, the production of food to meet demand comes at considerable environmental and social costs. In addition, intensification and industrialization of agriculture continue to increase off-farm environmental degradation (e.g. sedimentation, eutrophication of water ways or flooding) and especially the siltation of reservoirs necessary for energy production, irrigation and human consumption (Sartori et al., 2019).

There is a growing global recognition of the urgent need to reverse increasing land degradation and environmental footprints, or at least to prevent their relentless progression across millions of hectares of agricultural land (Lagacherie et al., 2017). In the light of land degradation, the failure to meet nutritional requirements of the growing population is another gargantuan task. Consequently, to revitalise soils and cropping (food) systems, North African governments should shift to more sustainable food systems. Efficient and resilient production systems are required to produce adequate, high-quality food and feed while containing costs, tolerating drought and seasonal variability, as well as minimizing environmental impacts.

Powerful reminders of the detrimental impacts of tillage-based agriculture are deterioration of land attributes, increasing food insecurity, and more reliance on external inputs and food imports (FAO et al., 2017). To revert these trends, a herculean task ecologically, options that are climate sensitive with harmful greenhouse gas emission have been developed and adopted worldwide (Roe et al., 2019).

According to the IPCC (2018), to keep global warming below 2°C or 1.5°C, there is an urgent and immediate need to shift to land-based mitigation measures among which is Conservation Agriculture (CA). CA is a set of sustainable land management systems that aim to increase agricultural production while balancing production with nature preservation, regulating trade-offs, and minimizing side effects. According to the FAO (2018), CA is a farming system that promotes continuous no-tillage (NT), or minimum soil disturbance, maintenance of a permanent soil mulch cover, and diversification of plant species (crop rotations, sequences and associations).

CA systems were also reported to permit multiple benefits and deliver the three principles of Climate-Smart Agriculture (CSA), which are (1) increased adaptive capacity and drought buffering to climate effects and reduced ex-ante vulnerability, (2) climate change mitigation through soil carbon sequestration and reduced greenhouse gas (GHG) emissions and (3) improved food security through reduced yield gaps and productivity enhancement and stability (Lipper et al., 2014; FAO, 2013).

Scenario analysis by Prestele et al. (2018) suggests a future potential of CA in the range of 533–1130 Mha (38%–81% of global arable land). However, the actual adoption of such systems is still limited and widely debated. Global-scale adoption of CA is about 180 million, which represents 12% of world arable land (Kassam et al., 2020). Prestele et al. (2018) showed that CA adoption processes could be sped up by alleviating institutional, social and economic barriers.

In North Africa, CA adoption is the lowest in Africa. Traditional tillage-based practices are still the dominant production systems in North Africa, but there are ongoing discussions at several levels about the setbacks, efficacy, success, and scalability of CA systems. Hence, it is important to understand why the adoption of CA systems by the farmers in North Africa continues to be considerably weak. A comprehensive and accelerated effort is needed to narrow the gap in CA uptake and encourage voluntary adoption by farmers.

Accumulating challenges and difficulties of adoption of CA by smallholder farmers in North Africa arise from two main issues: (1) CA is dynamic, meaning that it should respond to simultaneous changes in environment, social and/or economic contexts, (2) CA is also holistic based on a system-wide approach to solving farm management shifts and problems while considering the integrity of the food system. In addition, agriculture functions are changing over time and getting more complex with increasing socio-economic and environmental stresses and social and institutional shocks.

This chapter examines key issues related to the resilience-enhancing properties of CA and factors limiting its adoption in North Africa. It is intended to shed new light on CA research progress and achievements and analyse barriers, drivers and actions needed to encourage farmers to implement CA. It also provides insights on changing potential options to workable solutions for CA uptake and widespread adoption. The chapter draws upon a mix of published and unpublished literature across North Africa, but it may not be exhaustive.

2 Geography, soil, climate and farming environment

North Africa is a region covering the northern portion of the continent and includes six countries (Algeria, Egypt, Libya, Morocco, Sudan and Tunisia). Egypt and Sudan also form the River Nile basin. The region has a land area of 7.9 million sq. km, about 6% of the world total land area, and 3% of the world's population (Table 1). Broadly, countries in the region have different sustainable

Table 1 Population, Employment and Food Security (FAOSTAT)

Country	Population (Million) 2017	Employment in agriculture (% of total)	GFSI* and ranking	GFSI adjustment for Natural Resources and Resilience risks	Gini Index**
Algeria	42.2	9.4	59.8 [70]	52.1	0.749
Egypt	98.4	25.0	64.5 [55]	57.9	0.756
Libya	6.7	8.0	-	-	0.659
Morocco	36.0	38.3	62.8 [59]	54.6	0.766
Sudan	41.8	43.2	45.7 [99]	40.2	0.687
Tunisia	11.6	17.0	60.1 [69]	52.5	0.705

*: https://foodsecurityindex.eiu.com (rank between brackets over 113 countries).
**: https://worldpopulationreview.com/country-rankings/gini-coefficient-by-country.

development goals (SDGs), priorities and pathways in achieving food security, as dictated by their respective national conditions and capabilities (Tables 1 and 2).

Due to their location in the confluence of several bio-geographical regions, the North African countries are home to a wide variety of ecosystems and terrestrial and aquatic life (Zdruli, 2014). Hence, the region is characterised by very different socio-economic conditions as well as behavioural patterns in farming (Ryan, 2011). Furthermore, climate variability, drought events, desertification and erosion constitute threats to ecosystems (Thomas, 2008; Lagacherie et al., 2017).

The Sahara and other deserts occupy more than 75% of land area. Soils are diverse, poor in quality, highly vulnerable to erosion and have been mined of nutrients for millennia (Ryan et al., 2006; Zdruli et al., 2011; Jones et al., 2013). The annual costs of soil degradation were estimated at about 1% of Gross Domestic Product (GDP) in Algeria and Egypt, and about 0.5% in Morocco and Tunisia (Réquier-Desjardins and Bied-Charreton, 2006).

The total agricultural land is 36.8 million hectares, about 77% of which is rainfed. The expansion of irrigation is limited with the exception of the Nile Valley of Egypt and Sudan (Table 1). Nonetheless, land availability is also very limited and soil resources are becoming scarce with increasing population (Ibrahim and Lal, 2013). In terms of employment agriculture's share is variable among countries and it is the highest in Sudan and Morocco (Table 1).

In all North African countries, farmers are experiencing weather patterns in the last few decades outside of climatic norms that have severely impacted food production and income. According to Nefzaoui et al. (2012) precipitation will undergo a reduction of 20% up to the end of the century and an increase in temperature of 0.2°C per decade.

Soil erosion is causing a drop in productivity and increased poverty (Elnemr, 2019). In the case of Tunisia, desertification and land degradation

Table 2 Land and agricultural area, irrigation and fertiliser use by country in North Africa (FAOSTAT, AQUASAT)

Country	Land area (sq km)	Area harvested, crops (million hectares)	Agricultural area per capita (ha)	Land area equipped for irrigation (1, 000 ha)	Ratios of groundwater withdrawals as percentage of total withdrawals (%)	Rainfed land as percentage of total agricultural area	Fertiliser use* (kg/ha of arable land)
Algeria	2,381,740	5.1	0.12	1,360	34	98.6	22.3
Egypt	995,450	5.5	0.06	3,714	10	0.1	649.2
Libya	1,759,540	0.8	0.12	400	100	97.0	11.7
Morocco	712,550	7.9	0.22	1,530	22	95.2	71.1
Sudan	1,886,068	14.0	0.33	1,855	4	-	7.2
Tunisia	163,610	3.5	0.30	476	63	96.0	59.3

*: https://data.worldbank.org/indicator/AG.CON.FERT.ZS?view=chart.

affect about 75% of the national territory (DGACTA, 2017). Each year around 23 000 ha are degraded, including 13 000 ha, which are almost irreversibly lost, corresponding to 0.4% of arable land (de Lattre-Gasquet et al., 2017).

In the case of Algeria, in the north part of the territory, Benslimane et al. (2008) showed that 75% of the risk zones are already desertified or undergoing desertification. The total area threatened by water erosion is estimated at around 10 million ha (Benguerai, 2011) which have induced siltation of reservoirs and reduced soil production capacity.

Morocco's agricultural GDP (and indeed total GDP) is closely correlated with precipitation and hence, imports of commodities (Schilling et al., 2012). For Egypt, Elnemr (2019) stated that there is a wide gap between people's needs from agricultural lands and the availability of resources. However, he concluded that Egypt's transition to sustainable agriculture is possible if policy, governance, science and farmer communities are strongly linked and institutional strength developed. For Sudan, the agriculture sector accounts for more than a third of GDP. In Algeria, agriculture ranks third in terms of contribution to GDP. It employs almost 23% of the total active population and occupies nearly 41 million hectares, or just over 17% of the 238 million hectares of the total area (Chabane, 2011).

The North African region has a Mediterranean climate with highly variable and generally low rainfall that is further exacerbated by unfavourable climate change (Sivakumar et al., 2013; Mrabet, 2011c; Zouabi and Peridy, 2015; Mahgoub, 2014). Predictive models indicate that temperatures will increase between 2°C and 3°C and that rainfall will decrease by up to 30% by 2050 (Driouch et al., 2013; Sivakumar et al., 2013; Dubrovský et al., 2014; Lionello et al., 2014; Schilling et al., 2012). The authors also expect a significant increase in drought risk in the forthcoming decades. Admittedly, these transformative trends in climate are massive and may last for an extended period unless global actions are able to reverse them (Bazza et al., 2018). Sudan has a tropical climate with summer rainfall. The mean temperature is projected to increase by up to 3°C by 2050, and precipitation, to increase by 4% per decade. Most importantly, the agriculture sector will suffer from increasing rainfall variability associated with an increased frequency of both droughts and floods (USAID, 2016).

3 Agriculture and food security

Agriculture and livestock are key sectors in the North African economy. In addition, ensuring food security and nutrition is a global and strategic goal for all countries (AfDB, 2016) (Table 1). Moderate to severe food insecurity has fluctuated during the last five years, varying from 22.9% in 2015 to 35.2% in 2017 (the last record in 2018 was 29.5%). Due to drought, population growth, urbanization, import dependence and income inequity, North Africa is witnessing an increase in the prevalence of undernourishment during the last few decades (increasing from 6.2% in 2005 to 7.1% in 2018) (FAO et al., 2017). In 2018, North Africa had 16.5 and 72.1 million people registered as undernourished and severe and moderately food insecure, respectively. In addition, the average food share in the household budget in north Africa varies between 35.8% in Tunisia and 43.8% in Algeria as reported by AfDB (2016).

There is spatial heterogeneity in the share of agricultural land, with the lowest being 4% in Egypt (Table 2). The agriculture sector is characterised by a variety of different farm structures and agro-management practices combined with pronounced differences in environmental conditions. These require a substantial variation in agricultural inputs (e.g. energy, labour, nutrients, pesticides, water for irrigation) and produce a range of diverse outputs (e.g. crop yields and quality, feed and fibre, income, employment). Gini index, which represents income inequality or wealth distribution of a country, varies between 0.659 and 0.766 for North Africa (Table 1). These high values show social and income inequality and are associated with high unequal distribution of land and wealth in north African countries (Cuffaro and D'Agostino, 2017; FAO, 2019).

Fallowing is an ancient practice in North Africa by which farmers aimed to recover soil fertility, improve production and integrate crop-livestock components while reducing costs. In particular, in the northern part of Algeria, fallow area represents 3.34 million hectares, when the total agricultural area is only 8.4 million hectares, a rate of 40% for the period 2004–2014. However, with tillage, weed infestation and excessive grazing, soils under fallows are losing their fertility and hydrological functions. The replacement of fallow in favourable areas by food legumes and forages including cover crops within CA cropping systems should contribute to improving the food national requirements and to substantially improving soil fertility and health (DSASI-MADR, 2004–2014).

From the Neolithic time, cereals were produced in North Africa. Together, wheat, corn, barley and rice cover, for almost all countries, more than 90% of the entire cereal production, with rice having a significant share (>3%) only in Egypt. Divers vegetables (e.g. tomatoes and potatoes) are cultivated while permanent crops mainly consist of fruit, olives, grapes and dates. The highest crop production comes from Egypt with 23 million tons of cereals and 30 million tons of fruits and vegetables (FAOSTAT, 2021).

Despite productivity increases in recent decades, there are still large differences in the region, for instance, wheat yields range from approximately 1 to almost 7 t ha^{-1} (FAOSTAT, 2021). These differences are also reflected in large, estimated yield gaps for wheat, corn and barley (Mueller et al., 2012).

The region is a net importer of animal feed and staple food, importing varying from 36% in Egypt to 92% in Libya of cereal grain requirements (AfDB, 2016). For example, Sudan only produces about 24% of its total demand of about 2.2 million tons of wheat and the deficit is imported at the average cost of US$ 450 million annually. In Algeria, cereal imports reached 7.4 million tons in 2011 and 6.9 million tons in 2012 (Touchan et al., 2016).

The share of imported agricultural products in the Tunisian food consumption (in calories) is about 40% and has increased since the early 1960s (de Lattre-Gasquet et al., 2017). For the period 2002–2012, on average, the dependency on imported cereals was almost 70% with significant fluctuations depending on local rainfall and productivity (Rastoin and Benabderrazik, 2014; Khaldi and Saaidia, 2017).

Variable weather, low productivity, high and volatile prices, and uncertain availability of food commodities affect the livelihoods of the region's poor. Hence, the cause of food insecurity in the region cannot be attributed to any one factor, but an interaction of a wide range of economical, physical and human variables (AfDB, 2016; Table 1).

Water scarcity is the common denominator for problems affecting farmers in North African countries. The situation is worsening as demand increases beyond sustainable limits, with climate change projected to further stress already scarce resources (United Nations Economic and Social Commission for West Asia, 2019). In addition, agricultural systems face accumulating economic, ecological and societal challenges, raising concerns about their resilience to shocks and stresses. Warmer and drier climate conditions, with more prolonged, widespread, frequent and intense extreme events, together with increasing soil erosion and land degradation pose a threat to the food production system in the region (Bazza et al., 2018).

Most farming systems in North Africa are based on multigenerational family farms. Agricultural landscapes and their use have changed little over millennia in the region, however, the rate of change increased substantially in the second half of the 20th century. Rising populations forced farmers to increase the amount of land utilised for arable farming and the supply of arable land per person has been declining steadily. This has led to the cultivation of marginal areas. In addition, North African agriculture is becoming more resource intensive, consuming fossil fuels, water, soil and other external inputs (e.g. fertilisers) at unsustainable rates (Table 2). Ironically, due to increasing reliance on off-farm inputs to recover from declining soil fertility and disease-sensitive crops, farmers' profit margins have shrunk dramatically.

Livestock plays a key role in food security and income generation in the region. Animal production is characteristically interrelated with cropping systems through weedy fallows, residue and stubble grazing and use of woodlands and rangelands (Magnan, 2015; Moujahed et al., 2015). Crop residues are a key dry season feed resource for ruminants in mixed crop-livestock systems. Communal grazing rights, which often include the right to graze crop residues or cover crops after their harvest, create conflicts that make it difficult for the uptake of CA practices (Magnan, 2015; Moujahed et al., 2015; Guesmi et al., 2019).

In North Africa, traditional agriculture and smallholder farming is the norm. Rainfed and irrigated crops are sown after seedbed preparation with motorised cultivation using a disk plough or offset disks, except in mountain areas where animal-powered subsistence farming persist. If traditional agriculture remains the norm in North Africa, the region's ability to produce food, feed and fibre will be restricted due to scarcity of resources and impacts of climate change and population growth.

To achieve SDGs, increase farmers' incomes and well-being should be a priority for North African governments and agricultural policy makers. According to Tables 1 and 2, Global Food Security Index (GFSI) and the agricultural area per capita are low. The GFSI is a holistic measure of a country's food system, taking into account not only its ability to supply enough calories to the population, but also how the food system is affected by factors ranging from political stability to climate risks.

4 Conservation Agriculture: History and development

Worldwide the CA development is a complex interplay of science, economics, policy, farming and environment. However, in North Africa, such interplay is still relentless. CA is mainly based on NT systems of annual cropping systems, but CA-based agroforestry is also under progress. CA based on NT systems is practiced in major crops (i.e. cereals, food legumes, forages, oil seed crops). In the region, national research and development initiatives on CA were implemented at different times and with varying visions.

Literature on the adoption and adaptation of CA are limited across the world and even more so in North Africa. Despite a large development effort across the region, there is no comprehensive assessment of CA adoption in North Africa. According to Kassam et al. (2020), the pace of adoption of CA is still slow and increasing slowly with time in some countries. In 2008/2009, the CA area was estimated at 4000 ha, 6000 ha, 100 ha and 10 000 ha in Morocco, Tunisia, Algeria and Sudan, respectively. In 2015/16, the CA area was 10 500 ha, 12 000 ha, 5600 ha and 10 000 ha for the same countries (Kassam et al., 2020). The CA adoption is also fragmented throughout the region (Table 3).

4.1 Algeria

In Algeria the adoption of CA or NT systems started in 2002 in the high plains of Sétif province with cereals and legumes crops (Bouzerzour et al., 2006) and was then promoted to other regions (Constantine, Oum El Bouaghi, Guelma ou Annaba). According to Belaid (2015), in five years NT systems were adopted by large farmers (typically more than 100 ha) using 20 direct seed drills and

Table 3 Conservation Agriculture Adoption Trend in North Africa (hectares)

Country	2000	2006	2010	2015	2018
Algeria	-	-	5600	7000[a]	5600
Morocco	50	1200	4000	10500	12826[b]
Sudan	-	-	10000	10000	10000
Tunisia	57	4000	12000	14000	14000

a: Belaid (2015).
b: Estimated to be around 25000 ha in 2019/2020.

an area of 7000 ha was under CA systems. Much of this interest was a result of the availability of NT planters, and the technical and financial support provided by extension services. Since then, several government subsidies programmes have been implemented to increase adoption. However, CA adoption process faced constant socioeconomic and technical constraints (Rouabhi et al., 2016). It was estimated that 71% of farmers dropped NT when subsidies were terminated at the end of the programme (i.e. Conservation Agriculture in North Africa (CANA) project (CANA, 2015); Rouabhi et al., 2019). Paradoxically, the remaining CA adopters increased their land devoted to direct seeding systems from 6 ha per farm to 32 ha per farm.

With technical backstopping and coordination of support from International Center for Agricultural Research in Dry Areas (ICARDA), and funding from the Australian Centre for International Agricultural Research (ACIAR), CA has increased substantially in Algeria to reach 5600 ha involving 60 farmers across 12 provinces. The key to this increase was the engagement of private machinery manufacturers to develop low-cost, locally made, NT seed drills (CANA, 2015) (e.g. the direct seed drill called 'Boudour' from the CMA-SOLA in Sidi Bel Abbes, and the Retoufi seed drill from Sétif). The design of the Boudour drill was inspired by Australian direct seed drills and Retouf from low-cost Syrian drills (El Gharras et al., 2017).

4.2 Egypt

The first NT system in Egypt probably originates from ancient times when seeds were dropped into holes made in the soil with sharp sticks. Today excessive tillage and the removal of crop residues are causing accelerated soil erosion and declining soil fertility. Recognizing that productivity is declining due to mismanagement of soils and crops, several researchers have identified reduction of tillage intensity as means of recovering such declines while increasing crop yields (Harb et al., 2015, 2016; Abou El-Enin and Abo-Remalia, 2015 and recently El-Areed, 2019). However, CA is not yet adopted by any farmers. Egypt needs more research on other crops and demonstration fields across all Egypt.

Egypt, a country almost exclusively reliant on irrigated agriculture, should be undertaking CA research and related aspects of their soil-water-crop systems. According to Loss et al. (2015) based on experience in Iraq and Syria, CA can be promoted to enhance productivity and efficiency of irrigated agriculture. Gomez-Macpherson et al. (2009) and Mrabet et al. (2010) found that under CA systems in North Africa, supplementary irrigation can improve and stabilise crop productivity and reduces drought impacts in dry years and areas. In fact, there are several favourable experiences across the world with CA in irrigated systems (i.e. irrigated rice-wheat systems in the Indo-Gangetic Plains and cotton-wheat growing areas of Central Asia). Beni-Suef University has launched two CA research programmes funded by (i) Academy of Scientific Research and Technology (ASTR) in May 2017 and (ii) PRIMA under horizon 2020 called CerealMed which started in May 2020 aiming at releasing new genotypes of wheat, chickpea and lentil suitable for CA systems.

4.3 Libya

Libya's irrigated agriculture is mainly dependent on groundwater resources (Table 2) and it is hindered by a set of constraints including wind and water erosion, salinization and improper land management (Zurqani et al., 2019). Overall, 55% of Libyan soils are degraded, 30% due to salinization, 10% due to water erosion, and 15% due to wind erosion (Abagandura et al., 2017).

Most agricultural activities are limited to a northern strip along the Mediterranean coast, some mountainous regions, and several dispersed oases in the desert region. In fact, only 5% of the land is cultivated, and it is mostly exhausted, rendering sustainable agriculture a great challenge. As a consequence, Libya imports up to 90% of cereal and 80% of food for its consumption requirements (Abagandura and Park, 2016; Zurqani et al., 2019). The later authors proposed land management options to recuperate soil quality and prevent soil erosion, among which integrated farming systems, crop rotations, crop residue retention, use of cover crops and precision agriculture. All these are components for sustainable agriculture, however, CA systems were not yet tested or implemented.

4.4 Morocco

In Morocco, research on sustainable land use in agriculture dates to 1982 with the creation of dryland research centre through the joint USAID/INRA Morocco project. Among research priorities was the study of NT and cropping systems for water conservation in rainfed areas. The first adoption studies were initiated in 1998 with support from the Food and Agriculture Organization of the United Nations (FAO), the Centre de Coopération Internationale en Recherche

Agronomique pour le Développement (CIRAD) and government's extension services and development agencies. In 2006, the Arab Authority for Agricultural Investment and Development (AAAID), in collaboration with INRA, launched a large-scale CA diffusion programme starting with 900 ha and ending with 1200 ha within two years of implementation (El Gharras et al., 2017).

By 2011, the area sown with direct sowing did not exceed 5000 ha for the whole of Morocco and was almost exclusively large farms who purchased specially imported seed drills. The National Institute of Agricultural Research (INRA) also designed and produced low-cost and versatile direct seed drills for use by smallholder farmers (El Gharras and Idrissi, 2006). These direct seed drills were successfully used by farmers with advice from INRA experts but were not cost-effective due to the limited number produced (El Gharras et al., 2017).

With the mobilization of resources from international financing agencies (i.e. World Bank, Global Environment Facility (GEF) and AFD) and support from Morocco's Agency for Agricultural Development (ADA), CA was promoted as an adaptation strategy to reduce climate change impacts in cereal-based production areas. The projects Integrating Climate Change in the Implementation of The Plan Maroc Vert (PICCPMV) and Adaptation to Climate Change of Agriculture in the Maghreb (ACCAGRIMAG) promoted CA in major agricultural regions. These projects were led by INRA to apply CA principles and spread local direct seed drills in farmer's communities and adoption reached 5000 ha in 2015.

According to El Gharras et al. (2017), CA adoption based on direct seeding technologies has been adopted over roughly 9000 ha in Morocco. However, the latest estimates of CA adoption are 12 826 ha, with 8033 ha being managed by 800 smallholder farmers (with less than 10 ha) and 4793 ha run by 120 large farmers (with more than 10 ha) (Mrabet, 2018).

In 2019/2020, the El Moutmir initiative from 'Office Cherifien des Phosphates' (OCP Group) and Mohammed VI Polytechnic University launched with the technical support from INRA, a large-scale CA systems promotion in most agricultural regions of Morocco targeting 2000 farmers with a crop area of 10 000 ha. This programme mobilised 35 direct seed drills across 600 innovation platforms in 18 provinces. These platforms were designed to be a social learning interface, connecting local Non-Government Organizations (NGO) and Civil Society Organizations (CSO)s, as well as extension services, researchers and experts from research centres as development agents. The latest CA area in Morocco is estimated to be around 25 000 ha (AMAC, 2021). For 2020/2021, the expected CA area by this initiative is 20 000 ha (OCP, 2020).

Since 2014 the Morocco Green Plan is supporting the adoption of CA systems to cope with drought, climate change and variability. In fact, NT drills are subsidised as one of the incentives to farmers to adopt CA systems. In addition, the 22nd Session of the Conference of Parties (COP22) to the United Nations

Framework Convention on Climate Change (UNFCCC) organised in Marrakech helped to further highlight the paramount importance of changing traditional resource degrading agricultural practices to land-based climate solutions. The recently launched Green Generation Policy for the agricultural sector (2020–2030) is further emphasizing mainstreaming ecological intensification including spreading CA and climate-smart agriculture.

CA is already identified as one of the Nationally Appropriate Mitigation Actions (NAMAs) to be developed and it should be included as the main component of the Least Emissions Development Strategy and in the National Adaptation Plan to climate change. For instance, in the Central Morocco region, the objective is to extend CA from the current 500 ha to 60000 ha between now and 2025 (The World Bank, 2014a). The model by Bonzanigo et al. (2016) is targeting 40% of farmers in the Central Morocco region to adopt CA systems.

4.5 Sudan

Since 2000, the Arab Authority for Agricultural Investment and Development (AAAID) with the support from government institutions in Sudan, has been active in the promotion and extension of direct seeding technologies and reduced tillage practices (Rasheed et al., 2006; AAAID, 2018). A project funded by International Fund for Agricultural Development (IFAD) and with technical support from ICARDA, was designed to extend CA technologies to new areas and cropping systems in Sudan. The area under CA has increased to about 10000 ha since 2009 with the adoption by large mechanised farms (Kassam et al., 2020).

Degrande and Benoudji (2017) recommended spreading CA in Sudan through the 'lead farmer- trainees' approach, provided that the training is practical and supplemented with close monitoring by expert trainers. The Building Resilience and Adaptation to Climate Extremes and Disasters (BRICS) project has used CA to increase population resilience to weather extremes while diversifying products and incomes. CA has been reinforced by soil and water conservation techniques, integrated soil fertility management, and the use of improved varieties of millet, sorghum and cowpea and agroforestry. The authors reported an important acceptance of farmers to the project as well as improved land productivity and off-farm externalities or ecosystem services. The major constraints to adoption are limited farmer education levels, lack of financial resources, shortage of labour for farm and crop management, lack of knowledge, seed availability and crop residue maintenance and mulching.

4.6 Tunisia

According to Bahri et al. (2018), CA is mainly based on minimizing or eliminating tillage and seedbed preparation without necessarily retaining crop residue and promoting diversification of crops. According to Cheikh M'hamed et al. (2018), CA based on direct seeding was initiated in Tunisia during the period from 1970 to 1980 through an acquisition of no-till seed drills from the United States. In 1999 an integrated agricultural and rural development project sourced CA technologies from the Brazilian experience. From 2001 to 2011, the 'Projet d'appui au développement de l'agriculture de conservation' (PADAC) project funded by French Facility for Global Environment and supported by a range of universities, public, private sector and international organizations (National Institute of Agricultural Research in Tunisia (INRAT), INGC, ESEK, CIRAD, Cotugrain etc.) aimed to contribute to key development challenges (i.e. food security, environmental sustainability, economic development) through dissemination of CA systems (Raunet et al., 2004). This project targeted the farmers who have large farms exceeding 100 ha in the north of the country.

Since 2006, interest was focused on the small farm with a first project on CA for smallholders funded by the Arab Authority for Agricultural Investment and Development (AAAID), and implemented by CTC and ESAK. From 2012 to 2016, CANA project 'Rapid adoption of Conservation Agriculture in North Africa for smallholder' was implemented in Fernana region (sub-humid region with an annual rainfall of 750 mm which is a sloping area with high vulnerability to water erosion). The project was funded by ACIAR and was managed by the INRAT, the INGC and the ICARDA. In the framework of this project, a new local prototype of NT seed drills (low-cost no-till drills for smallholders) was designed and manufactured.

In order to resolve the major problem of CA adoption in the country which is the conflict between permanent soil cover with residues and stubble grazing, especially during summer period, the CLCA project (Integrated Crop-Livestock under Conservation Agriculture for Sustainable Intensification of Cereal-based Systems in North Africa and Central Asia) funded by International Fund for Agricultural Development (IFAD) and managed by INRAT-INGC-ICARDA was implemented (2013–2016). As a result of the project, a stubble grazing model 30:30 was developed (Guesmi et al., 2019) based on the stocking rate of 30 animals ha^{-1}, during 30 days of stubble grazing period. This model allows to keep adequate crop residues (mulch) in the soil surface (more than 0.4 t ha^{-1} of residue in the soil surface or 40% of the initial biomass of residues on soil surface) and in the same time to preserve body conditions of animals (Moujahed et al., 2015; Guesmi et al., 2019). In 2018, the second phase of CLCA project 'Use of Conservation Agriculture in Crop Livestock Systems (CLCA) in the drylands for enhanced water efficiency, soil fertility and productivity in NEN and LAC countries' had been launched and aimed to up-scaling CLCA technologies in the semi-arid regions of the North part of the country (Rekik et al., 2021).

In Tunisia, the adoption of CA progressed rapidly during the first years of using imported direct drills from Brazil with support from Grain Tunisian

Company (Cotugrain) (AFD, 2006). The CA area was only 27 ha and 55 ha in 1999 and 2000, respectively, reaching 2893 ha in 2005, 6000 ha in 2007, 8000 ha in 2008 and then 12 000 ha in 2010 (Richard, 2005; Angar et al., 2010; FERT, 2016). However, in 2016, the areas recorded a significant decrease (Bahri et al., 2018; FERT, 2016). This decrease was mainly explained by the unavailability of low-cost direct seed drills and the lack of CA comprehensive adoption studies. From the last estimates by Cheikh M'hamed et al. (2018), the CA area in Tunisia is around 14 000 ha and operated by more than 200 farmers and using around 107 NT seed drills (which represents 0.93% of the total cereal production area). Most of the CA is in the semi-arid climate zone. The farming is predominantly mixed cereal-livestock, and the majority of farmers are smallholders (70% having less than 10 ha).

Since 1999, several projects were then launched in Tunisia to promote CA and remedy constraints to CA adoption by farmers and their cooperatives. Among these projects, the most important are given Table 2. These CA projects have seen claims and endorsements about the performance and potential value of CA in the country (M'hedhbi et al., 2003; Ben-Hammouda et al., 2007, 2010; Jemai et al., 2012; Bahri et al., 2018, 2019; Chibani et al., 2018).

5 Major research achievements and advances in knowledge: Re-aligning development, economy and environment

CA has been tested in North Africa on research stations and within farmers' fields in medium to long-term field trials with a blend of financial support from government and international organizations. A wide range of agronomic, biophysical, environmental and socio-economic benefits have been reported from shifting to CA systems as shown in Table 4.

All the three principles of CA (minimal/no soil disturbance, soil cover and crop diversity and rotation) need to be applied by farmers for optimal benefits and minimal side-effects (Pittelkow et al., 2015). Unfortunately, often this is not possible in North Africa since fences rarely exist and maintenance of crop residues is still not respected by livestock producers. Farmers generally seed bare no-till fields which may negatively impact crop productivity and soil functions. Brown et al. (2017) reported that due to limitation in resources and difficulty in assuring enabling drivers to farmers, CA adoption in Africa is rather partial. Farmers, with support from researchers and extension specialists, need to develop a different mindset towards crop and soil management under CA systems, and adapt CA principles to local conditions without compromising any of them (Carmona et al., 2015; Souissi et al., 2018b).

From the research carried out during the last four decades, consortia of farmers, researchers and extension specialists should assess and understand all edaphic and biotic constraints (climatic factors, soil type and quality, disease,

Table 4 Agronomic, biophysical, environmental, and socio-economic benefits from Conservation Agriculture in North Africa

Agronomic and biophysical benefits	References
• Water conservation and use efficiency	Bouzza (1990); Mrabet (1997); Mrabet (2002); Aboudrare et al. (2006); Chennafi et al. (2011); Jemai et al. (2013); Fellahi et al. (2013); Mekhlouf et al. (2013); Cheikh M'hamed et al. (2014); Amara et al. (2015); Cheikh M'hamed et al. (2016); Bahri et al. (2019); El-Areed (2019).
• Increased/stabilised crop yields	Bouzza (1990); Mrabet (1997); Mrabet (2000); Mrabet (2002); M'Hedhbi et al. (2004); Abdellaoui et al. (2006); Aboudrare et al. (2006); Abdalla et al. (2007); Ben-Hammouda et al. (2007); Ramdani et al. (2010); Mrabet (2011a,c); Zaghouane and Makhlouf (2011); Benkherbache et al. (2012); Sassi et al. (2012); Fellahi et al. (2013); Cheikh M'hamed et al. (2014); Harb et al., (2015, 2016); Abou El-Enin and Abo-Remalia (2015); Zaghouane et al. (2015); Labad and Hartani (2016); Souissi et al. (2018a); Cheikh M'hamed et al. (2016); Bahri et al. (2019); Chekali et al. (2019); El-Areed (2019).
• Carbon sequestration and soil quality	Mrabet et al. (2001a); Bessam and Mrabet (2003); Ben-Hammouda et al. (2004); Mrabet et al., (2004); Ibno-Namr and Mrabet (2004); Mrabet (2011c); Jemai et al. (2012); Mekhlouf et al. (2013); Abou El-Enin and Abo-Remalia (2015); Labad and Hartani (2016); Souissi et al. (2018a); Chibani et al. (2018); Bahri et al. (2019).
• Fertiliser use optimisation	Mrabet et al. (2001b); Harb et al. (2015, 2016); Souissi et al. (2018a).
• Increased biological activity	Errouissi et al., (2011); Ben Moussa-Machraoui (2004); Hamim et al. (2012); Labad and Hartani (2016); Labad et al. (2018a)); El-Areed (2019); Taibi et al. (2020).
• Weed management	Aibar (2006); Abdellaoui et al. (2006); Rahali et al. (2011); Djennadi et al. (2014); Hajjaj et al. (2016a,b); Tanji et al. (2017a,b); Hajjaj et al., (2018); Laouar et al. (2018); Labad and Hartani (2016); Labad et al. (2018a,b; 2020).
Environmental/off-farm benefits	
• Erosion and sedimentation control	Moussadek et al. (2011a); Labad and Hartani (2016); Souissi et al. (2018a); Chibani et al. (2018); Bahri et al. (2019).
• Reduced greenhouse gas emissions	Mrabet (2011b); Moussadek et al. (2011b); Sanz-Cobena et al. (2017).
• Energy efficiency & costs	Bourarach (1989); El Gharras et al. (2004); Mrabet (2008a); Mrabet et al. (2012).
Socio-economic benefits	
• Farm income and profitability	Ben-Salem et al. (2006); Boughlala et al. (2011); Zaghouane et al. (2006; 2015); Labbaci et al., (2015); Labad and Hartani (2016).

(Continued)

Table 4 (*Continued*)

Agronomic and biophysical benefits	References
• Decreased labour and operation time	Mrabet (2008b); Boughlala et al. (2011); Labbaci et al., (2015); Zaghouane et al. (2015).
• Farmer organizations and community development	Zaghouane et al. (2015); Labbaci et al. (2015); FERT (2016, 2018).
• Awareness, training and technical development	Vadon et al. (2006); Labbaci et al., (2015); Zaghouane et al. (2015).
• Public (co-)benefits	Vadon et al. (2006); Vadon and Marionneau (2012); Zaghouane et al. (2015); FERT (2016, 2018).
• Technology development (NT drill design and production)	Bahri and Bansal (1992); Bahri et al. (1993); Bourarach et al. (1998); El Gharras and Idrissi (2006); Zaghouane et al. (2015); El Gharras et al. (2017).

pest and weeds) and social challenges (land tenure, livestock integration, etc.) before implementing CA systems in new fields.

5.1 *Crop yields and diversification*

In North Africa, lack of rainfall limits the opportunity for crop production. But CA can be effective in mitigating yield loss in dry environments prone to variable weather patterns (Cheikh M'hamed et al., 2014; Mrabet, 2011a; Sun et al., 2020). In other terms, from research and on-farm data, CA can be used by smallholder farmers to increase yields while also making crops more resilient to changing climatic conditions (Mrabet, 2011b; Bahri et al., 2019; Cheikh M'hamed et al., 2016; El-Areed, 2019). CA can reduce drought effects through better water storage and availability during crop growing season, which was found in early 1990 by Bouzza (1990) in wheat-based systems. Another important operational benefit from CA is the flexibility for the implementation of field crop management that allows timely planting and input application, despite unfavourable field conditions that do prevent such operations in conventional agriculture (e.g. wet soil at planting time) (Mrabet, 2011a; Nefzaoui et al., 2012).

Yields of cereals, food legumes, forage crops (vetch and oats) can be improved or lowered depending on the context and years, but generally tend to improve after a few years of CA practice. CA was particularly effective in reducing drought severity in cereal production (Mrabet et al., 2012). According to the French Association for International Cooperation for Agricultural Development in Developing and Emerging Countries (FERT, 2018), yield increases of 8% to 20% were observed by CA farmers in the Maghreb compared to conventional practices.

A meta-analysis conducted by Souissi et al. (2018b) revealed that CA systems in Tunisia mainly based on suppression of tillage and seedbed preparation decreased crop yields in semi-arid regions, while in wetter climatic regions, the effects on crop yields were similar to conventional tillage systems. Souissi et al. (2018b) concluded that all three CA principles (including retention of crop residues and crop rotation) are mandatory to improve production. Fellahi et al. (2013) recommended that 60% residue cover is needed to improve durum wheat yields while conserving water in soil compared to conventional tillage systems and that NT systems without residue retention are detrimental to wheat growth and development. This was also reported by Mrabet et al. (2003). Amara et al. (2015) found that conventional tillage (based on mouldboard plough & chiselling) favoured root development and soil porosity which helped wheat to produce higher than under CA. However, such impacts were short terms and CA did not have mulch cover at seeding. Weed control and management seem to be another problem associated with CA systems.

For medium and long-term agro-ecosystem functioning and improvement, reinforcing the principle of crop rotation represents one of the pillars of the CA system. Crop diversity is a component of CA and it was reported that several crops can be successfully grown in sequence without tillage and seedbed preparation in various contexts of North Africa. Among such crops, oilseed crops (sunflower and sesame), food legumes (chickpea, bean, lentil, faba bean and soybean), forage crops (vetch) and most cereals (corn, wheat, barley, sorghum and oat) were included (Aboudrare et al., 2006; M'hedhbi et al., 2004; Ben-Hammouda et al., 2007; Mrabet, 2011a,c; Harb et al., 2015; El-Areed, 2019). Forage crops (vetch, triticale, etc.) and crop mixtures (triticale-vetch, oats-vetch) were included for improved integration of livestock within cereal systems and were identified as highly suitable options for the marginal wheat systems (Cheikh M'hamed et al., 2016).

Under CA systems, cover crops and/or crop residues have many advantages, including reduced erosion and soil evaporation, carbon storage, biological diversity, nutrient cycling and weed suppression. However, in North Africa crop residues are limited due to overgrazing and biomass removal, and cover crops are difficult to promote and apply under field conditions. The typical dry weather from spring to fall, and competition for water during the rest of the year, strongly limit the use of cover crops by farmers practising CA. Cover crops were tested and successfully applied in North of Tunisia under CA (Bahri et al., 2018). It is necessary to investigate further different cover crop varieties, sowing rates and sowing times to find the most suitable combinations for CA systems. Using judicious cereal-based rotations including cover crops, farmers could both produce grains and forage for livestock (FERT, 2018).

5.2 Input, costs and profits

In North Africa, farmers are faced with rising prices of inputs (seeds, fertilisers, pesticides and machinery). Hence to be economically attractive, CA systems should provide a net economic benefit in terms of reduced costs or/and higher returns through increased crop and livestock yields.

CA benefits include savings in inputs, time, fuel and machinery wear; more timely sowing and less labour (Chouen et al., 2004; Ben-Salem et al., 2006; Angar et al., 2010; Rouabhi et al., 2019; Boughlala et al., 2011; Magnan et al., 2011; Fenni, 2013; Bahri et al., 2018; Labad and Hartani, 2016). CA systems are about eight times more efficient in fuel consumption than conventional tillage and eliminating tillage and relying on herbicides to manage weeds can be more energy efficient (El Gharras et al., 2004). With direct seeding, CA farmers are encouraged to reduce certified seed rates and optimise fertiliser use as soil quality improves.

CA systems lead to reduced energy consumption through the elimination of tillage operations, but also through reduced indirect energy consumption from more efficient use of all inputs (e.g. seeds, fertilisers, pesticides and machinery). In fact, traditional tillage-based systems require extra labour and energy costs and produce lower yields than CA systems, resulting in lower net economic benefits for farmers.

5.3 Soil functionality and ecological benefits

Soil functionality and productivity are interlinked, and both should be improved to feed the growing population while sustaining the adaptive capacity of farming systems vis-à-vis climate change. CA approaches, particularly NT management systems, have higher adaptive capacity over conventional tillage systems in dry areas (Sun et al., 2020). CA was found to improve the productivity of various crops and deliver key ecosystem services (e.g. increase soil aggregation, reduce erosion rates, and increase soil organic matter) across a range of soil types, cropping systems, and climates in the Mediterranean basin, and particularly in North Africa (Mrabet et al., 2004; Kassam et al., 2012; Chibani et al., 2018; Degrande and Benoudji, 2017). These findings globally confirm the trends in literature on the improvement of soil health and functions under CA.

Several authors have shown growing evidence of the many biological and economic benefits of CA (Mrabet, 2008a,b,c; Zaghouane et al., 2006, 2015; Ben-Hammouda et al., 2007; Jemai et al., 2012; Mrabet et al., 2012; Boulal et al., 2014; Harb et al., 2015; Mouelhi et al., 2016; Chibani et al., 2018). Interestingly, many studies and experimentations proved the benefits of NT strategy on water use efficiency, soil moisture conservation and regenerating soil quality and health (Aboudrare et al., 2006; Abdalla et al., 2007; Ben-Hammouda et al.,

2007; Ben Moussa-Machraoui et al., 2010; Mrabet, 2008a,b,c; Chennafi et al., 2011; Jemai et al., 2013; Mekhlouf et al., 2013; Oulbachir et al., 2014; Cheikh M'hamed et al., 2014, 2016; Labad and Hartani, 2016; Degrande and Benoudji, 2017; Souissi et al., 2018a; Bahri et al., 2019) and others were more in favour to reduced tillage systems (Kribaa et al., 2001; Fellahi et al., 2013; Belagrouz et al., 2016) or conventional tillage (Souissi et al., 2018b; Amara et al., 2015).

Additionally, soil organic carbon (SOC) concentrations have been widely reported to increase in the upper soil layer (Jemai et al., 2012, Angar et al., 2010 and Chibani et al., 2018; Ben-Hammouda et al., 2004; Lal, 2002; Oulbachir et al., 2014; Mrabet, 2018; Mrabet, 2011a,b; Souissi et al., 2018a; Labad and Hartani, 2016). Bessam and Mrabet (2003); Ibno-Namr and Mrabet (2004) and later Moussadek et al. (2014) found that CA systems improved both SOC levels and the quality of soil organic matter. Using the Soil & Water Assessment Tool (SWAT) model, CA was found to be the best adaptation strategy against climate change for Moroccan systems (Brouziyne et al., 2018). Mrabet et al. (2001a) and Abou El-Enin and Abo-Remalia (2015) found the retention of crop residues on the soil surface and the absence of tillage operations boosted SOC accumulation in the top layer (0–5 cm) of CA systems. Field studies that were undertaken in Morocco and Tunisia by Mrabet et al. (2001a) and Jemai et al. (2012) showed 14% and 28% increases in the SOC from the surface soil layer after 11 and 7 years of switching from conventional tillage to conservation agriculture practices, respectively. Carbon-rich soils created through the adoption of CA favoured co-evolution of enhanced ecological functions (i.e., soil biological activities, nutrient availability, water circulation, storage and retention, erosion and sediment control, etc.). However, Bahri et al. (2017) did not observe an increase in soil carbon content under NT after 17 years of adoption due to excessive removal of crop residues, either for direct livestock feeding or collected and sold off the farm for cash income.

Nutrient availability is a major constraint to the productivity of crops in North Africa since soils are naturally infertile. In CA, nutrient management can be implemented through changes in fertiliser rate, placement, type, or timing, and consequently, avoid losses of applied fertilisers that cause environmental pollution. Development of direct seeding drills tried to take into consideration appropriate rates and banding of fertilisers to satisfy crop requirements in N, P and K. Mrabet et al. (2001b) found that with time less amounts of fertilisers were needed with CA due to the enrichment of the soil through decomposition of crop residues and return of nutrients to the soil in dry areas. For balanced plant nutrition, optimum fertilisation is more critical with CA, and soil analysis is necessary before applying fertiliser.

Harb et al. (2015) found that CA systems helped to reduce by 50% NPK fertiliser requirements of wheat compared to conventional tillage systems in a wheat-soybean rotation. Improved SOC and nitrogen availability were also reported by Oulbachir et al. (2014) for CA compared to conventional and reduced tillage systems for a loamy-sand soil (Tiaret, Algeria). In contrast, Souissi et al. (2018b) reported reduced nitrogen use efficiency and nitrogen deficit in wheat under CA compared to conventional systems due to immobilization of surface-applied fertiliser nitrogen in untilled soils, and hence lower nitrogen availability during the early stages of crop growth. To overcome this shortfall, they suggested increased N rates are required under CA systems compared to traditional tillage-based systems. Split N application may increase efficiency, and precise banding to separate fertiliser from crop residues that can offset N immobilization. Much research is still needed to better understand the changes in nutrient cycling and efficiencies under CA cropping systems. The focus should be on developing nutrient management strategies that meet the CA principles while fitting the farmer's circumstances. In limited rainfed North Africa, these strategies should enhance crop productivity and biomass production and quality.

In North Africa, soil microbiology and biochemistry under CA systems is a fertile area of scientific research. Changes in soil biodiversity under contrasting tillage and cropping systems were not intensively studied. Oulbachir et al. (2014) found that conventional tillage favoured the development of microbial diversity of a loamy-sandy soil compared to CA and reduced tillage in the first years. They recommended to quantify soil microbial over time due to the high complexity of biochemical processes in soils. However, recently, the research by Taibi et al. (2020) highlighted the importance of NT systems in restoring the microbiome in soils disturbed by tillage in semi-arid regions of Algeria. In semiarid Morocco, Hamim et al. (2012) tried to quantify the importance of symbiotic arbuscular mycorrhizal fungi (AMF) under CA and found that CA soils were enriched by AMF compared to conventional tillage. El-Areed (2019) found improved soil biological activity and aggregation under CA in irrigated corn fields compared to traditional systems. Errouissi et al. (2011) and Ben Moussa-Machraoui (2004) studied soil invertebrates under a wheat cropping system and compared CA to conventional systems. Mulch from crop residues helped to improve particulate organic matter content which is a signal of increased and diversified microbial activity (Bessam and Mrabet, 2003). However, more profound research is needed to understand the soil-microorganism-plant system and relationships when shifting to and implementing CA under dryland conditions of North Africa.

5.4 Weed infestation and management

Weed management is a major challenge in implementing CA in dryland farming of North Africa (Aibar, 2006; Abdellaoui et al., 2006; Dridi et al., 2010; Mrabet, 2011a; Djennadi et al., 2014; Bahri et al., 2018; Rouabhi et al., 2018).

In CA, weed control is often laborious and costly in the first few years. Farmers will need good knowledge of herbicides, weeds and application technology to manage crops successfully.

Increased dependence and overreliance on herbicides can be a limitation for small farmers to shift to CA systems and for the public at large as concerns for health and the environment are emerging (Friedrich, 2005). In addition, herbicide and weed mismanagement induce resistance and shift in weeds.

Several research studies assessed weed infestation, dynamics and biology, and searched for best practices for managing weeds without suppressing crops yields and harming the environment and human health (Tanji et al., 2017a,b; Hajjaj et al., 2016a,b; Laouar et al., 2018; Labad and Hartani, 2016; Labad et al., 2018b). From such studies, the focus should be on managing weed populations and flora under CA with integrated practices rather than eliminating weeds solely with herbicides.

In the early adoption of CA, monocotyledons dominate compared to conventional and reduced tillage systems (Rahali et al., 2011) and weed seeds accumulate in surface horizon in CA. Seed accumulation in CA soils may cause high weed germination and trigger large infestations.

Change in soil temperature and moisture regimes due to in situ crop residue retention (mulching) affect weed seed germination and seedling emergence, and hence, could either reduce, delay or favour weed growth and infestation. Overall, CA has been found to influence weed flora by favouring development over time of grass weeds and deep-rooted perennials if not appropriately controlled (Aibar, 2006; Hajjaj et al., 2018). According to the work by Hajjaj et al. (2018), the main reason for decrease in weed infestation under CA over time is due to a reduction in weed seed banks in the soil.

In Algeria, Rouabhi et al. (2018) related the lack of CA adoption to difficulties of weed management, particularly the proliferation of *Bromus* spp. At Setif experimental research field, Labad et al. (2020) found that pre-planting applications of glyphosate reduced weed infestations in cereals but accumulation in soil may be harmful to the environment and consumers as both soil and seeds are contaminated. Labad et al. (2018a) did not report any significant impact of glyphosate on soil respiration and biological activity in NT soils. Labad et al. (2018b) found that it is possible to reduce glyphosate without significantly reducing wheat yields and that glyphosate-based direct seeding systems are feasible in the drylands of Algeria.

For the prospect of effective weed control, Hajjaj et al. (2016a,b) conducted herbicide field trials in no-till systems for wheat and food legumes. Sulfonylurea herbicides such as amidosulfuron and iodosulfuron and sulfonamide herbicide (Pyroxsulame) were found more effective in controlling broad-leaved weeds and some annual grasses (i.e. bromus) in wheat. In food legumes, Bentazone herbicide was effective on broad-leaved weeds. Hajjaj et al. (2016b) combined this herbicide with post-emergence one (i.e. Pendimethalin) for better efficacy and higher yields.

Integrated weed management (IWM) under CA includes early seeding, combination or rotation of herbicides, diversified cropping systems, plant geometry, nutrient management, and improved crop varieties that are competitive with weeds (Mrabet, 2008b; 2011c; Rahali et al., 2011; Tanji et al., 2017a,b). In these regards, developing crop genotypes with strong allelopathic potential against associated weeds is an important research issue. In addition, using IWM successful weed management can be achieved over time with low dependence on herbicide to improve the environmental footprint of CA systems (Sims et al., 2018). However, in order to successfully and sustainably manage weeds in CA, farmers should be advised and supported by extension services and service providers as well as encouraged to meet the long-lasting challenge from weeds through incentives and subsidies.

5.5 Livestock integration

Mixed crop-livestock farming systems are highly prevalent in North Africa, offering a range of benefits and challenges for landholders, farmers and breeders (Hajer et al., 2020). Crops and livestock are ethnically, functionally and operationally linked enterprises. However, they are extremely vulnerable to climate variability and associated with the reduced availability of feed production (Nefzaoui et al., 2012; Tibbo and van de Steeg, 2013). In these systems, weeds and crop residues (i.e., stubble and straw) are important forage sources for livestock.

Historically, fallows and rangelands were subjected to increasingly heavy continuous grazing by livestock leaving little ground cover for soil conservation. However, a conflict of interest exists between mulch for covering soil surface and stubble grazing, especially during the summer period. Thereby, trade-offs between the use of stubbles for livestock feeding or to cover the soil have to be resolved, particularly in drylands where fodder potential is low (Hajer et al., 2020). In this context, a stubble grazing model (30:30 model) was developed in Tunisia in the framework of CLCA project in order to give farmers adopting CA some solutions for reasonable stubble grazing during the summer period (Moujahed et al., 2015; Guesmi et al., 2019). The 30:30 model developed based on the stocking rate of 30 animals ha^{-1}, during 30 days of stubble grazing period. This model allows to keep adequate crop residues (mulch) in the soil surface (more than 0.4 t ha^{-1} of residue in the soil surface or 40% of the initial biomass of residues on the soil surface) and at the same time to preserve body conditions of animals (Moujahed et al., 2015; Guesmi et al., 2019).

In a study by Tanji et al. (2017a,b), for better management of weeds and successful integration of livestock in CA systems, weedy fallow was proposed to be replaced by a barley plus field pea forage mixture in wheat rotations.

In his review book chapter, Mrabet (2011a) reported that it is necessary to combine diversified crop rotation with controlled and improved grazing management (partial removal of crop residues, flexible seasonal controlled grazing, introduction of forage legume and dual-purpose crops, temporary displacement of animals to pastures etc.) for strong integration of livestock and CA. This can be effective at preserving or even enhancing soil function and health. Bahri et al. (2018) reported that for sustainable adoption of CA in Tunisia, livestock should be considered as an integrated component. Environmental considerations including the physical and mechanical properties of the soil changes with grazing options under CA and should be further investigated (Hajer et al., 2020).

5.6 Insect-pest and disease dynamics and management

Changes in seeding depth and quality and alteration of soil surface due to crop residues and lack of tillage disturbance have effects on pest and disease dynamics and pressure on crops. Tillage disturbs disease-infected residues (e.g. yellow spot and Fusarium crown rot), and soil-borne pathogens (e.g. Pythium spp., *Rhizoctonia solani* and Pseudomonas spp.), reducing their prevalence (Simpfendorfer et al., 2019). Research by Chekali et al. (2019) has demonstrated increased risks from pathogens favoured by CA compared to conventional tillage systems. Their results also showed that for the wheat, barley and oats, CA increased the incidence of *Fusarium culmorum* infection by 58–60% in dry seasons, and by 32% in a wetter season. However, the incidence of *Fusarium pseudograminearum* infection, the primary cause of crown rot in cereals, increased by 80% in dry seasons under CA. Also, CA systems increased the percentage of whiteheads in durum wheat (73%), but did not affect disease severity, which was estimated by the discolouration on stem bases and grain yield. The experiment did not report significant differences in crop yield and recommended further medium- to long-term studies including the impact of improved tolerant cultivars. In fact, stubble-borne pathogens can survive multiple years on crop residues. *Fusarium pseudograminearum* survives as mycelium for up to three years in infected cereal residues and *Bipolaris sorokiniana*, the cause of common root rot in cereals, survives on wheat residues at the soil surface for at least two years.

More studies are needed on disease profiles and incidence within CA systems for major crops and cropping systems (including leaf diseases, nematodes and viruses) to develop strategies to reduce disease incidence. Among such innovation tools are improved genetic tolerance, site-specific crop management, precision agriculture, pesticide management and disease risk prediction.

5.7 Off-farm benefits

Among the most important off-farm benefits concern climate change mitigation and natural resources protection and conservation including biodiversity preservation. CA systems were found to reduce greenhouse gas emissions from soils with the potential to reduce the effects of climate change on crop yields (Moussadek et al., 2011b; Sanz-Cobena et al., 2017; González-Sánchez et al., 2019). Thus, CA is a mitigation adaptation strategy to cope with climate change.

CA was reported to increase organic matter contents in soils which have important environmental externalities such as improved soil structure and reduced soil pollution. Organic matter build-up in CA soils also contributes to the enhanced supply of nutrients (i.e. nitrogen, phosphorus, potassium and sulphur) which over time reduce dependence on synthetic fertilisers (Mrabet et al., 2001b). Such changes lead to reduced fuel use, large-scale environmental restoration and less soil emissions of nitrous oxide, a potent greenhouse gas. In addition, various studies presented evidence that CA alleviates soil erosion and reduces runoff and sediment loss while improving water entry in soils and hence contributes to improve watershed health (Moussadek et al., 2011a; Mekhlouf et al., 2013; Labad and Hartani, 2016; Souissi et al., 2018a; Chibani et al., 2018; Bahri et al., 2019). Dimanche (1997) found that an NT system reduced the runoff volume by 30–50% and sediment loss by 50–70% compared with disk ploughing on a sandy clay loam soil at Ras Jerri (Meknes, Morocco). In comparison, with chisel ploughing, no-till reduced runoff volume by 24–53% and sediment loss by 43–65%.

In Tunisia, despite low residue cover, CA systems reduced average annual soil losses by 30–40% compared to conventional tillage (2–4 t/ha/year vs 3–7 t/ha/year). Water infiltration rates were 65 mm/h versus 45 mm/h for no-till and conventional tillage, respectively (Raunet et al., 2004). Mrabet et al. (2012) also reported increased water capture and reduced soil water evaporation under CA which ensure a higher level of drought proofing during dry periods. The extra water stored in soil under CA is an important insurance to reduce yield gaps due to extreme, delayed and variable rainfall events (Benkherbache et al., 2012).

6 Adoption barriers, constraints and gaps

CA research has been conducted in North Africa over the last 40 years but the knowledge developed was not significantly translated into farmer adoption. Over the same period in Australia, where the climate and soils are almost identical to North Africa, the adoption of CA has been rapid and widespread, with up to 90% of farmers using minimum tillage in many districts (Llewellyn and Ouzman, 2019).

Drivers and underlying constraints or obstacles to CA adoption are presented in Table 5. A complex web of causality may underlie CA adoption or abandonment attributable to diverging views among stakeholders, including farmers.

Several authors have argued that smallholder farmers face several constraints and obstacles when adopting CA which include limitations to resources and knowledge (Lalani et al., 2016, 2017; Mrabet et al., 2017). CA practices require additional efforts by farmers and knowledge of how to use and maintain new equipment and inputs (e.g. use of direct seed drills, herbicide applications, crop residue and stubble management, etc.). Some farmers experienced increased weed populations and soil compaction problems under CA, which increased the need for additional management, particularly during the first years of adoption. In other terms, a lack of skills and advices from extension services on technical or maintenance CA requirements may lead to management difficulties, a decline of yields, and hence a negative farmer's perception of the effectiveness of CA practices (Cheikh M'hamed et al., 2018; Souissi et al., 2018a). The scope for the adoption of CA in North Africa is limited by low crop biomass production and its competitive use by livestock (Mrabet et al., 2012; Magnan, 2015; Moujahed et al., 2015; Guesmi et al., 2019), but current evidence suggests that even small amounts of residue retention can significantly decrease soil erosion losses (Moussadek et al., 2011a; Bahri et al., 2019).

A variety of approaches were conceived and used in North Africa to promote CA at local levels including agricultural research for development (AR4D), farmer-led actions, farming field schools (FFS), integrated platforms and systems-oriented approaches. CA projects were instigated to fill the gap between farmers, CA scientists and policy-making authorities. However, these approaches may have obscured some trade-offs and negative side effects that emerged at local scales. Hence, more comprehensive and spatially explicit approaches merit investigation to mitigate barriers and guarantee wide CA uptake. The development of social learning mechanisms is of great importance to share knowledge, capacities and skills and raise the social reputation of CA farmers.

Bonzanigo et al. (2016) found that in Central Morocco, CA adoption rates and reactions depended tremendously between different typology of farmers, and within specific pedoclimatic and socio-economic conditions. Hence, there is no 'one size fits all' in CA diffusion, and appropriately designed CA adoption strategy should consider both climate change impacts and farm category and resources (Hunt et al., 2020).

The World Bank study in Morocco reported that the three major constraints for the large-scale implementation of CA are: (1) difficulties with crop rotation and weed control; (2) the integration between cropping and livestock and

Table 5 Summary of barriers and drivers influencing the adoption of CA by farmers

Barriers/ constraints	• Insufficient endorsement and encouragement by governments at national and local levels of CA systems. • Scarce and limited international cooperation for CA system adoption and upscaling. • Characteristics of farms - geophysical and socio-political indicators (farm size, crops & livestock, land tenure, income, type of ownership, soil fertility, slope, altitude, etc.). • Cultural - Farmers' demographics, knowledge, and attitudes (gender, poverty, land ownership, family size and structure, social class, farmer's profile). • Social norms: Social influence and conformity, neighbour's acceptance of CA, trust decisions, generational renewal at the farm. • Lack of information and awareness. • Availability, access and costs of machinery (especially NT drills and sprayers) and inputs (seeds, fertilisers and pesticides). • Changes in weed, disease and pest profiles, infestation and management, and emerging diseases and pests. • Livestock integration and scarcity of pastures placing pressure on crop residues. • Short-term economic gains and profitability - Farmer's attitude to risk taking and uncertainty. • Site-specific conditions, field-scale effectiveness, and difficulty for upscaling.
Drivers/ enablers	• Polycentric policy approach. • Insertion of CA principles in multi-level governance and policy: e.g. environmental and climate change programmes, ecological and energy transition, youth and gender initiatives, political incentives and supports. • Links and compliance with CA world science community, international organisations and financial agencies e.g. South-South Cooperation mainly Brazilian CA experience. • Creation of centres of excellence on CA research, development and extension. • CA long-term capacity development initiatives and high standard educational platforms. • Farmer to farmer learning - sharing experiences, leadership, and education, including synergetic and supportive process of communication. • Farmer's and other stakeholder's willingness & compliance with CA principles. • Local development of machinery manufacturing (drill, tractor, sprayer, etc.) including public-private partnerships. • Developing institutional capacity: Strengthening and enforcement of extension and technical advisory services. • Creation of employment and generation of wealth in rural communities. • Financial investments, insurance, incentives, and subsidies.

(*Continued*)

Table 5 *(Continued)*

Facilitators/ motivators	• Reinforcement and fine-tuning of research institutions and teams (innovative applied research approaches). • Gender mainstreaming in CA projects and initiatives. • Adoption of CA can be facilitated by the development of artificial intelligence, eResearch and high-tech. • Payments for environment benefits and services associated with CA spread and adoption as environmentally friendly practices. • Households engagement in supplementary or complementary off-farm activities to generate incomes to cushion the effects of CA risks and poverty. • Social learning and training and capacity building. • CA curricula development at different levels of the education system. • Cost and benefit-sharing among farmers using CA systems. • Social reputation, commitment, sustainable attitudes, positive behavioural change of CA adopters. • Ease of access to technology, technical assistance, and knowledge dissemination. • Access to permanent and skilled labour (family or hired) for machinery operation. • Market integration of CA productions. • Encouragement of service and input providers (outsourcing) and crowdsourcing. • Consumer demands for sustainable products and certification of CA systems. • Business-science-community cooperation. • Land stewardship and ethics.

competitive uses of crop residues; and (3) a lack of suitable NT seed drills (The World Bank, 2014b).

According to a survey conducted among 28 Algerian farmers in 2016, Rouabhi et al. (2019) highlighted the main motivations and problems encountered by farmers when they adopt CA practices. The major constraints were availability and cost of seed drills, livestock competition for crop residues, weed infestation and glyphosate-based weed management, and short-term decline in crop growth and yields partly due to poor crop establishment. Globally, the most important predictors of poor adoption of CA are limitations in financial resources, as well as attitudes and perceptions of small farmers vis-à-vis CA principles as reported by Lalani et al. (2016).

As reported by Ben-Salem et al. (2006) for Siliana and Zaghouane regions, key constraints to CA adoption are mainly related to low farming experience, high off-farm income, and livestock units per hectare. A more recent study by Fouzai et al. (2018) found that in erosion-prone regions of Tunisia (i.e. Siliana), the adoption of CA is positively correlated with the level of education of farmers, availability of extension services and technical support, and landownership rather than lease.

In their review paper, Cheikh M'hamed et al. (2018), summarised major constraints of CA adoption by Tunisian farmers as (1) conflict between livestock (grazing) and maintained residue in the soil surface as mulch, (2) high cost of NT seed drills, (3) weed management problems, especially in legume crops, (4) soil compaction, (5) limited crop rotation (cereal monoculture), and (6) limited species for cover crops, especially over the dry summer period.

According to a study by FERT (2016), the adoption by Tunisia's pioneer CA farmers was mainly motivated by reducing costs and higher yields rather than by environmental considerations (e.g., soil erosion or soil quality improvement). However, more recent CA adopters are motivated by both soil quality and conservation benefits as well as higher returns and increased productivity with CA systems. Most farmers are constrained by weed infestations, the high cost or lack of suitable NT seed drills and poor knowledge and skills relating to CA. Since most farmers have integrated livestock in their farming enterprises, overgrazing is also a significant barrier to CA uptake (Moujahed et al., 2015). From another study by Dhehibi et al. (2018), most farmers were aware of the benefits of NT, yet a minority had adopted CA practices, likely due to the countervailing strength of perceived risks. When faced with these barriers, a proportion of CA adopters shift to reduced tillage or move completely back to traditional tillage systems. However, there are avenues for reviving the CA practice as reported from the last report of FERT (2018). Disadoption of CA systems is also related to farmer's not perceiving potential ecosystem services permitted by the systems. In addition, short CA projects do not allow farmers time to become aware of such externalities and other private benefits from CA technologies.

In North Africa, CA research and early efforts to promote adoption were mainly based on using imported complex NT drills from USA or Brazil. However, such seed planters were not affordable by farmers, not adapted to soils that are hard and dry at seeding time and were difficult to set up and maintain. Hence, several initiatives were carried by research and technical institutions to produce relatively simple and cheap seed drills to overcome major barriers of CA adoption (El Gharras et al., 2017).

In Morocco, the first attempt to design no-till drills dates to the research work by Bahri and Bansal (1992), Bahri et al. (1993) and Bourarach et al. (1998). El Gharras and Idrissi (2006) report on the design and production of the first commercial NT seed drill manufactured in Morocco by a joint public-private partnership with the industrial manufacturer 'Les Ateliers Marocains' (ATMAR). However, the quality control in manufacturing was not satisfactory and more work was needed to reduce costs and versatility of the direct seed drills to guarantee better appeal to farmers. Other types of direct seed drills, mainly of simple low-cost types designed and made in Morocco and Syria were made available through international projects (FERT, CANA, PICCPMV, ACCARIMAG) and expertise sharing from ICARDA, CIRAD and ACIAR.

Mrabet (2011a) reported that poor land tenure security and limited access to credit are additional disincentives for farmers to invest in CA systems, and hence, they should be addressed by public authorities to aid adoption.

7 Adoption drivers and opportunities

Breeding improved varieties and increased reliance on external inputs (fertilisers and pesticides) were the basis for the 'Green Revolution' and were routine ways of reducing yield gaps in tillage-based systems for over half a century or more. Given this mindset, barriers to the adoption of CA are likely to be significant and changing farmer attitudes will be difficult. However, alternative drivers and solutions may influence farmers, government, institutions, international agencies and NGOs to ensure the efficacy and scalability of CA systems in North Africa. These are discussed herein (Table 5).

Adoption of CA is a multi-faceted, knowledge and learning process, and hence, requires an effective farm advisory network to mainstream research achievements and communicate socio-economic benefits to farmers and other CA systems stakeholders and beneficiaries. As shown by the study conducted by FERT (2016, 2018), there are increasing numbers of innovative (elite) farmers and associations who are implementing CA systems with positive mindsets and results. These are demonstrating effective approaches to CA systems, and providing valuable advice to other interested, neighbouring farmers. This farmer leadership should be encouraged to reduce and overcome barriers for CA adoption.

CA is not merely a package of prescribed methods applicable to everyone, and farmers need to be convinced of the benefits to test and develop their own CA system suited to their individual farming system, environment, and markets, while considering their personal preferences. Most importantly, it is a change in the mindset of farmers, community leaders, researchers, extension agents, natural resources managers, manufacturers, retailers and service providers, managers, and policy makers which is shaped by tradition, and in some cases, prejudices. For all these stakeholders, CA brings changes in behaviour and lifestyle as well as new insights and paradigms.

Gender integration, a cross-cutting issue, is a prerequisite for addressing land and water challenges and ensuring equity, efficiency and sustainability in CA policies and projects. A gender-sensitive approach should be proposed by governing agencies to build awareness, confidence, leadership, communication and networking among women and youth.

Existing evidence for CA climate-smart properties mainly relies on agronomic, environmental and economic analysis. High-quality analysis of whole-farm economics, social constraints and behaviour to management change are also needed, largely associated with farmer perceptions in short- to long-term horizons. Factors related to whole farm management have a significant influence on farmer's decisions and behaviour vis-à-vis adoption of CA. In addition, to better use benefits from CA, synergetic and supportive process of farmer-to-farmer communication should be developed and accompanied to comprehensively communicate on risk, complexity and uncertainty and to alleviate gender, language and illiteracy barriers. This type of dissemination channel will merge expert farmers and opinion leaders within the community which should facilitate CA adoption.

As shown in Table 5, several NGOs and CSOs were created in the region to develop CA and implement related projects. However, more durable social integration and farmer's organisations should be further strengthened and reinforced to ensure wider and sustainable adoption and diffusion of CA systems. CSOs and NGOs should be consolidated and integrated into all initiatives and programmes that concern climate change, environmental protection, biodiversity preservation, circular economy, ecological and energy transition, youth and gender and technology transfer. In addition, national NGO networks should be linked to international confederations that promote CA systems and CSA at large. This should help to intensify knowledge sharing, generate funds, and enforce the position of NGOs in the adoption of CA.

North African farmers operate in a highly variable semi-arid climate that is expected to become even more variable with global change. Thus, in assessing CA systems, performance should be of multiple objectives, and hence, based on a set of criteria (Findlater et al., 2019). In other words, CA performance should take into consideration short-term technical (e.g. crop yields) and economic outputs (e.g. farmers' incomes, cost reduction, contribution to food security), and long-term social and environmental benefits (e.g. poverty alleviation, job creation, mindset changes, carbon sequestration, biodiversity conservation, regulation of water quality or of polluting fluxes, soil fertility). In addition, carbon sequestration, a long duration process in CA soils, has important co-benefits such as improved soil structure, reduced erosion, enhanced soil moisture retention and increased available water and nutrient storing capacity.

When applying and implementing CA, business as usual is no longer an option, and in-depth transformations to farming systems are needed. CA systems must fit into emerging farming environments, as well as multi-government policy, and both farmers and decision-makers should be motivated to transform current unsustainable farming systems into more durable forms. A common language is needed among all stakeholders to mainstream CA principles and upscale benefits. In other words, high institutional strength (in terms of durability and enforcement of research/extension and academic institutions, government agencies, donors, commercial organisations and agribusiness, NGOs, etc.) is needed to foster adoption pathways and channel social, societal and economic concerns of farmers through the decision process. CA feasibility is intrinsically linked to motivation and incentive mechanisms mainly through subsidies, micro-finance, carbon trading, access to markets, payments for ecosystem services and co-benefits, and compensatory costs (Peiretti and Dumanski, 2014).

For effective adoption of CA systems to address land degradation challenges, it is a priority to repurpose policy and implement reforms to benefit productivity and resilience for environmental sustainability simultaneously. For CA systems to permit restoring lands while enhancing ecosystem services (including productivity) and copying with the induced magnitude of change, public policies, private sectors and development organisations should implement and support multiple instruments: risk distortion, financial incentives, subsidies and social spurs.

For CA and CSA to be effectively and widely implemented in rural communities, the accompanying policy should promote rural prosperity and secure remunerative employment mainly among youth. In this sense, CA systems should be connected to markets, value chains, entrepreneurship, off-farm opportunities in connection with urban neighbours and supporting services to further create jobs in rural communities.

The following governments in North Africa have established a subsidy programme in order to encourage farmers to adopt CA but impact and success are still to be analysed when mass and voluntary adoption is conceived:

1 The Moroccan government has established a subsidy programme for farmers to acquire no-till seed drills and launched a national programme for CA promotion in 2016.
2 Starting from 2017, the Tunisian government increased the subsidy on agricultural machinery up to 60%, which may encourage and help farmers to acquire no-till seed drills.
3 The Algerian government established a subsidy programme to purchase local agricultural machinery including no-till drills (Cicek et al., 2017).

According to Bahri et al. (2019), all the cereal areas in Tunisia are suitable for the implementation of CA, which is a high-priority area estimated at 260 000 ha. In these fragile lands, CA can be adopted to improve soil fertility and to combat soil erosion, while providing input savings and yield benefits. In Morocco, Moussadek and Mrabet (2017) estimated that up to 60% of cereal cropping lands can be suitable to CA and could be shifted into CA without many technical constraints. However, CA adoption is a long and complex process, and requires substantial efforts in examining mechanisms driving adoption which require designing feasibility frameworks, road map and development programmes (Glover et al., 2016).

North African farmers are embedded in multiple layers of institutional environment which affect their commitment, capacity and decision to adopt new technologies and to change their paradigms. Securing resources for survival and growth are of first importance to farmers. Hence, the willingness of 'innovative' farmers and stakeholders to adopt and promote CA systems is a prerequisite for large-scale diffusion in a sustainable way, ensuring the creation of a social movement for CA acceptance at all levels of the society.

In order to build science-policy interfaces, raise awareness, communicate, advocate, leverage and promote CA in North Africa, it is necessary to create centres of excellence in CA. Other functions of these centres include providing advice to national institutions, international cooperation, technical assistance, policy analysis, formulation, and monitoring, and project management. These centres should consolidate dialogue and enrich public responsiveness (beliefs and values) to CA development. These can be a core reference on CA vis-à-vis worldwide centres (i.e. CTIC in USA). The African Conservation Tillage Network (ACT) has created such a centre in different African sub-regions (ACT, 2020). Financial assistance to the centres should be primarily from government and national public and private enterprises, but also from international organisations, and NGOs. Even though the climates and farming systems in South America are completely different to North Africa, South–South cooperation with countries like Brazil might be prioritised to share successes and drivers of change (Lal, 2019).

8 Networks, partner institutes and distributional effects

North African agriculture much undergo significant improvements in productivity and environment stewardship in order to meet the complex and interconnected challenges of food security, climate change and population aspirations. Such improvements are possible through systemic CA adoption and diffusion involving all stakeholders - farmers, land managers, researchers, NGOs, businesses, decision-makers, communication media and consumers. However, this ambitious objective for limiting climate change while ensuring food security and environmental sustainability through CA and climate-smart agriculture (CSA) trigger deep changes and generate differential effects on the food and fibre production systems in the region (Mbow et al., 2019).

Extensive cooperative networks and a high level of international collaboration exist in North Africa but merit further development to implement strategic actions that spread CA among farmers. Table 6 presents CA institutional networks developed in North Africa while Table 7 lists major projects associated with CA research and adoption studies and analysis.

To achieve large-scale restoration and climate mitigation, COP 22 held in Marrakech in 2016 saw the launch of the Adaptation of African Agriculture to climate change (AAA) by the Government of Morocco. Among the major pillars of this initiative is to promote the diffusion of CA at large scales in Africa through South-South Cooperation between developing countries (Badraoui et al., 2018;

Lal, 2019). One year earlier, France revelled the 4 per 1000 initiative which advocates for an annual growth rate of 0.4% in the soil carbon stocks to revert soil degradation process and secure food for world population (Chabbi et al., 2017; Minasny et al., 2017). The initiative consists of federating all voluntary stakeholders of the public and private sectors (national governments, local and regional governments, companies, trade organisations, NGOs, research facilities, etc.) under the framework of the Lima-Paris Action Plan (LPAP). The aim of the initiative is to demonstrate that agriculture, and in particular, agricultural soils can play a crucial role where food security and climate change are concerned through the implementation and promotion of CA, agroecology and agroforestry (Lal, 2019).

The ICARDA is in the front line of developing and promoting CA systems to improve food security and livelihoods in North Africa (Lal, 2018; Mrabet et al., 2012). Two mega-projects were led by ICARDA in the region among which (http://www.icarda.org/) (Table 7):

- Conservation Agriculture in North Africa (CANA) project: This project aimed at adapting CA for rapid adoption by smallholder farmers in North Africa. The geographic spread of the project included three North African platforms (Sétif-Algeria), (Chaouia-Ouardigha-Morocco) and (Fernana-Tunisia), with financial support and scientific cooperation from the ACIAR and ICARDA. This project was focus on the rapid adoption of CA by smallholder farmers (CANA, 2015).
- CLCA project: Integrated crop-livestock conservation agriculture for sustainable intensification of cereal-based systems in North Africa and Central Asia https://mel.cgiar.org/projects/clca2. This project concerns Algeria and Tunisia (2015–2022 in two phases). (CLCA, 2016; Rekik et al. 2021).

During the diffusion process of CA in North Africa and with the support of international organisations and NGOs, APAD, ATU-PAM and AMAC were created respectively in Tunisia (2006), Algeria (2005) and Morocco (2016). Bouzerzour and Mahnane (2006) provide an explanation of the creation of ATU-PAM in Algeria.

Human capacity building and development on CA systems was the main area of assistance of the projects. It included guided field visits, on-farm training for farmers as well as researchers, extension agents, and service providers. It is necessary to implement long-term CA capacity development initiatives through high-standard educational platforms. The other project objectives are to develop and undertake applied and multidisciplinary research in CA systems and to equip decision-makers with the necessary knowledge and tools to promote CA. Degree programmes were also offered through universities and should be further improved for students with national and international special schools and universities and research institutions.

Table 6 CA Institutional Networks in North Africa

Country	National research, higher education and extension organisation	NGOs and CSOs	International organisation and knowledge partners	Financial and development agencies	Private sector
Algeria	INRAA, ITGC, Universities (Setif and Oum El Bouaghi), ENSA	FERT, ATU-PAM, EAC	FAO, ICARDA, CIMMYT, ACSAD	FERT, IFAD, ACIAR, PRIMA	
Egypt	ARC, Al-Azhar University - Cairo, Beni-Suef University		ICARDA	PRIMA, ASRT	
Morocco	INRA, ENA, IAV Hassan II, ONCA, ADA, Mohammed VI Polytechnic University	FERT, Coopérative Khemisset Chaouia, AMAC, AGENDA, OCP Foundation	FAO, ICARDA, ACIAR, IRD, CSIC, IFS, CIRAD	FFEM, AFD, AAAID, GIZ, IBRD World Bank/GEF, USAID, CICYT, European Commission, PRIMA, OCP Group	ATMAR
Tunisia	INRAT, INRGREF, ESAM, INGC, ESAK, IRESA, INAT, ESIER, OEP, CTC, CRDA	APAD, Tunisian Farmers' Union, FERT, SMSA-Ettaouen, SMSA - Meliane, ATAE, Coopérative 'El Manel'	ICARDA, CIRAD, IRD, IFPRI, CIMMYT	FFEM, AFD, IFAD, European Union, AAAID, ACIAR, FERT, CGIAR, European Commission, USAID, PRIMA	Société de Production Métallique du Sud (SPMSud), COTUGRAIN
Sudan	ONDR		ICRAF, Feinstein International Center (Tufts University)	AAAID, DFID	Al Muzdana Zero Tillage Agricultural Company

Table 7 Major CA Projects and Initiatives in North Africa

Country	Initiatives/Projects
Algeria	• CLCA project https://mel.cgiar.org/projects/clca2 • CANA – Conservation agriculture in North Africa (2012–2015) • CAMA: Conservation Agriculture in Mediterranean Area. https://intranet.iamz.ciheam.org/Practicas/www.camamed.eu/index (PRIMA project)
Egypt	• CerealMed: Enhancing diversity in Mediterranean cereal farming systems. https://primaobservatory.unisi.it/en/projects/enhancing-diversity-in-mediterranean-cereal-farming-systems (PRIMA project)
Morocco	• AAA initiative (Foundation) – https://www.aaainitiative.org/organization • CANA – Conservation agriculture in North Africa (2012–2015) • FERT – Agriculture de Conservation au Maghreb • KASSA Project, Knowledge Assessment and Sharing on Sustainable Agriculture, http://Kassa.cirad.fr • 2004–2006. • CA2AFRICA – Conservation agriculture in Africa: Analyzing and foreseeing its impact – comprehending its adoption. (2009–2011) https://cordis.europa.eu/project/id/245347/reporting • Azehar Programme (Comisión Interministerial de Ciencia y Tecnología, Plan Nacional I + D + I) – Conservation agriculture in irrigated agricultural systems (2006–2008) (University of Cordoba, CSIC). • PICCPMV – Integrating Climate Change in the implementation of The Plan Maroc Vert » (2011–2015): https://projects.worldbank.org/en/projects-operations/project-detail/P117081?lang=fr (World Bank, 2014a,b) • ACCAGRIMAG – Adapatation to climate change of agriculture in the Maghreb (2013–2016) https://www.ffem.fr/en/carte-des-projets/adaptation-climate-change-agriculture-maghreb-region-accagrimag • CRP-Dryland Systems for North Africa and West Asia (2012–2015) : Saiss, region – rainfed systems https://cas.cgiar.org/sites/default/files/pdf/Dryland-Systems-CRP-CCEE-Final-Report-VOL-1_October-2015_0.pdf & http://drylandsystems.cgiar.org/node/432 • Integrated Natural Resource Management – INRM project funded by INRA Morocco. • Arab Authority of Agricultural Investment and Development (AAAID). • OCP & Mohammed VI Polytechnic University, INRA – Initiative El Moutmir: Climate change integration in Agricultural sector (2019–on-going) • CAMA – Conservation Agriculture in Mediterranean Area. https://intranet.iamz.ciheam.org/Practicas/www.camamed.eu/index (PRIMA project)

Tunisia	• PDRAI – Integrated Rural and Agricultural Development Project (Siliana, Kef and Mateur) (1999–2014) (FIDA-AFD) • PADAC – Projet d'appui au développement de l'agriculture de conservation (2001–2004; 2008–2011) (FFEM-AFD) • FAO – Conservation Project (2005–2007; 2013–2014) with ESAK and CTC • Arab Authority of Agricultural Investment and Development (AAAID) (2007–2009) – small farmers in Siliana (semi-arid regions) and Bizerte (sub-humid region) • CRP-Dryland Systems (2013–2015) – Béni-Kedache-Sidi-Bouzidi transect, Tunisia – agro-pastoral systems: https://cas.cgiar.org/sites/default/files/pdf/Dryland-Systems-CRP-CCEE-Final-Report-VOL-1_October-2015_0.pdf • FERT – RCM Network (2001–2008) then Agriculture de Conservation au Maghreb (2015–2017) • CANA – Conservation agriculture in North Africa (2012–2015). (ACIAR-ICARDA) • CLCA project-phase I – Integrated Crop-Livestock under Conservation Agriculture (CLCA) for Sustainable Intensification of Cereal-based Systems in North Africa and Central Asia (2013–2016) (IFAD/ICARDA) • https://apps.icarda.org/wsInternet/wsInternet.asmx/DownloadFileToLocal?filePath=ICARDA-CLCA.pdf&fileName=ICARDA-CLCA.pdf • PAPS – Eau – Programme d'appui aux politiques publiques de gestion des ressources en eau pour le développement rural et agricole (Agriculture de Conservation en pluvial : 2015–2018). (EU) • CLCA project-phase II – Use of Conservation Agriculture in Crop-Livestock Systems (CLCA) in the Drylands for Enhanced Water Use and Soil Fertility in NEN and LAC Countries (2018–ongoing) (IFAD/ICARDA/CIMMYT) • https://mel.cgiar.org/projects/clca2. (IFAD/ICARDA) • CAMA – Conservation Agriculture in Mediterranean Area. https://intranet.iamz.ciheam.org/Practicas/www.camamed.eu/index (PRIMA project)
Sudan	• Arab Authority of Agricultural Investment and Development (AAAID) • Building Resilience and Adaptation to Climate Extremes and Disasters (BRICS) project

The transition to CA or CSA requires considerable investments and implication of funding agencies either through blended finance or loans to governments or institutions (FAO, World Bank (GEF); French Agency for Development (AFD); AAAID; GIZ; USAID; IFAD, European Commission etc.) and NGOs (i.e., Association Française de coopération internationale pour le développement Agricole des pays en développement et émergents (FERT)).

FERT, with support from local civil society organisations (CSOs), launched the project 'Agriculture de Conservation au Maghreb' which helped to address issues related to CA adoption barriers in the Maghreb (Morocco, Algeria and Tunisia) (Vadon et al., 2006). This project lasted from 2015 to 2017 and aimed to develop innovative practices related to CA, through farmer discussion groups (FERT, 2016). Several local CSOs were created to strengthen institutional aspects of CA systems and leverage their promotion through farmer leadership (Coopérative Khemisset Chaoua (Morocco); Groupe Mateur & Coopérative 'El Manel' (Tunisia) and Exploitation Agricole Collective (EAC) (Algeria).

According to Cusser et al. (2020), long-term research and adoption platforms (over a decade) are needed to get consistent sustainability attributes of CA systems (i.e. slow-to-change attributes of soil quality and environment). In other terms, the main challenge with projects listed in Table 7 is their exit-strategy when farmers do not rely on resources from projects (Brown et al., 2017). Their limited duration, socio-economic barriers and constrained financial opportunities for research and extension restrict continuous appropriation of CA systems by farmers, and hence, their long-term benefits and success of lasting adoption. Decadal implementation of CA research and information platforms should avoid spurious and misleading trends while revealing consistent dynamics in benefits and outcomes. However, such interfaces require closer collaboration between scientists, stakeholders and farmers, stronger private-public partnerships, effective policies, and enhanced public awareness and engagement. On the other hand, replicating successful CA projects requires institutional drivers and enabling conditions that promote social learning, strong leadership, transformational coalition and durable mobilisation of resources and funds.

Transition to CA is a knowledge-intensive process (Caron et al., 2014). Hence, to be able to continue adoption and up/out scale of CA system, the science-policy gap should be filled, and even narrowed, and new arrangements and approaches should be investigated while strengthening the science-policy interface.

North African governments and institutions have started advocating and promoting sustainable land management systems or climate-smart agriculture. According to Nationally Determined Contributions (NDC) of North African countries, CA is presented as a mitigation option for reducing GHG emissions and contributes to a low-carbon economy and as a sustainable land management option for natural resource management. In the case of Sudan, the NDC document reports agroforestry as an adaptation strategy for revitalizing degraded and vulnerable agricultural areas.

9 Conclusion

Climate change is a reality in North Africa, and it is undermining the food systems, socioeconomic and ecological basis of life in the region, especially if the global average temperature rise exceeds 1.5–2°C. These effects are compounded by environmental degradation and the exhaustion of natural resources caused by a sole focus on production and short-term profits. Smallholder farmer households are especially susceptible to these shocks.

North African agriculture is seen as an important sector for economic growth and poverty reduction, and therefore, sustainable productivity growth should play a key role in the development process and for attaining SDGs. Hence, governments are under increasing pressure to develop sustainable strategies. Concrete land-based climate solutions and measures such as CA are needed urgently to alleviate future crises of food insecurity or scarcity of water resources.

CA systems are ecologically sustainable and economically enticing and underpin strategies of climate change adaptation and mitigation. Governments and development agencies are encouraging the adoption of CSA technologies, such as CA. Hence, the wide-scale and rapid adoption of CA in the region is required to revert land degradation process while setting a climate-resilient agricultural development model.

In North Africa, CA research started with support from the USAID and Ministry of Agriculture of Morocco in 1982. By the end of 1990s, CA research had expanded across the region, highlighting the benefits of CA strategies for the delivery of key production and ecosystem services. A review of the literature establishes that CA enforces efficient and sustainable land management in the region through carbon sequestration, crop improvement and climate change and land degradation mitigation. Since 1980, much of the pioneering assessment of CA was started in Morocco, and then later in Algeria, Tunisia and Sudan, and more recently in Egypt. CA systems have enormous potential to improve soil security and sustainability of cropping systems in diversified cultivated lands. Beyond the successful results obtained on experimental farms and farmer plots, the widespread adoption of CA within north African agro-systems remains a challenge.

A number of interconnected barriers to wider adoption of CA have been identified in the farming context of North Africa. Under a rapidly changing climate, drivers of change from policymakers to the farming community are essential to overcome the complexity associated with CA adoption. A polycentric policy approach should be implemented in order to permit increased cooperation, communication, interaction among local, national and international actors and instigate learning-by-doing to improve initiatives and policies at different levels of decision-making and overtime.

In addition, to our best knowledge, there was no study on local-scale co-benefits and trade-offs from CA adoption and diffusion in North Africa. This is required for watershed to large-scale CA dissemination.

Despite ample evidence of the benefits of CA systems, and increased effort by international institutions to promote their use, adoption is low across North Africa. Research programmes can only have an impact if they are accompanied by long-term investments and appropriate policies are put in place. An agricultural research consortium cannot provide the answers to all the societal concerns. Here, it is of paramount importance to create strategic partnerships and platforms for shared learning and knowledge co-production and communication.

In spite of the relevance to farmers, researchers, extension specialists and policymakers that CA systems are gaining over the years, several aspects are misunderstood, and gaps remain underinvestigated. Among these aspects, CA concepts and principles are counterintuitive and contradict the traditional tillage-based farming experience, which has worked for generations and contributed to cultural values and rural traditions. The term and definition as well as the principle and practices of CA should be made more inclusive and precisely defined for the general public and decision-makers and practitioners. In addition, CA widely overlaps with sustainable intensification, climate-smart agriculture, and agro-ecology. Another barrier is insufficient experiential knowledge about CA.

CA initiatives should be context specific and consider technological innovation as well as governance, policy, and socio-institutional changes. The transition from conventional to CA systems requires an understanding of important determinants of social-ecological nexus to pursue changes over long periods. Plurality of north African farms and hence diversity of preferences and views from farmers on CA and soil stewardship are important determinants to consider.

For a CA adoption mechanism to be successful and comprehensive, productivity and economic benefits should relate to enhancing the resilience of natural resources and ecosystem services, and meeting the food and nutrition needs of a growing population. In other words, for CA adoption to ensure long-term sustainability and ecological resilience of agro-ecosystems, agricultural production should be guided by policies and regenerative management protocols. Decision-makers, political influencers and economic actors must be involved as determinants of CA adoption scenarios. Increasing incentives for the private sector mainly machinery manufacturers, seed companies and funding agencies, on its own or through private-public partnerships, to invest in all segments of CA can be one important way to promote rapidly CA systems.

In addition to scientific and technical knowledge developed during the last four decades in North Africa, a comprehensive study of the agricultural situation and policy environment of each country should be carried before large-scale up and out-scaling of CA. This study will be used to design CA adoption pathways that better address emerging drivers and risks, synergies, and opportunities. It is also important to carry out detailed investigations to plan investments and development projects across the diversified farming systems and value chains. It is important to rely on the CA suitability analysis already completed in Morocco and Tunisia (Moussadek and Mrabet, 2017; Bahri et al., 2018). It is also necessary to take into consideration all strengths and weaknesses associated with CA systems (Table 5).

CA adoption/diffusion projects use multiple sources of finance which require a better governance framework at the national level. This framework should value institutional capacities, enhance the efficiency of employed resources and avoid duplicated uses of funds. In addition, it should prepare for post-project legacy plans.

It is also important to certify commodities from CA farms which may give a better market access and satisfy consumer demand for such sustainable and better-quality products. The CA certification experience of Argentina can be of interest to North Africa (Albertengo et al., 2014; Peiretti and Dumanski, 2014; Albertengo, 2020).

CA is recognised as reducing GHG emissions and should be used as a climate mitigation option in agricultural policies for all North African countries (NDC, NAMA, etc.). It should be proposed for 'green financing' from GEF and other channels. In addition, CA systems in combination with improved cultivars and fertilisers help to reduce crop vulnerability to drought and heat stress via better water conservation and use efficiency. CA is proposed by Mbow et al. (2019) as among food security-related adaptation options to address climate risks. Hence, CA systems can also be promoted through projects financed by the Adaptation Fund (2018). This type of project was mainstreamed by ACIAR in north Africa and termed CASI as Conservation Agriculture based Sustainable Intensification (Dixon et al., 2019). CA systems were found to reduced land degradation (i.e. erosion and soil losses) and regenerate soil functions and services, and hence, can be prioritised and financed through Land Degradation Neutrality (LDN) Fund proposed by United Nations Convention to Combat Desertification (Lal et al., 2012; Stavi and Lal, 2015; Sanz et al., 2017).

It is also important to break the long-lasting stagnation of crop productivity and farm output in semiarid rainfed areas of North Africa. In conventional tillage systems, farmers are managing crops in unsustainable ways vis-à-vis environmental and economic challenges. Under the better agroecological potentials of CA systems, more effective matching of the phenology of crops to the environment (rainfall pattern, water deficit, heat, drought) and to disease or pest outbreaks is important for avoiding or minimizing detrimental effects of unfavourable and stressful environmental conditions. Judicious or optimal time of seeding and use of resistant varieties to biotic stresses are important for managing susceptible crop stages. This requires the development of improved cultivars through improved genetics and innovation in plant breeding (molecular marker screening) and phenotyping. In other words, multi- or trans-disciplinary research should be encouraged that connects crop physiology, agronomy and genetics. Ramdani et al. (2010) have coupled supplementary irrigation and cultivars to further improve wheat yields and narrow the productivity gap in semiarid Morocco under CA. El-Areed (2019) also found increased corn productivity and performances as well as water economy using CA systems under sandy soil for irrigated corn and yellow corn hybrids were selected for tolerance to water deficits.

In the New Digital Age or Era of Big Data, CA systems should be aligned with advanced technologies in artificial intelligence, digitalisation, high-tech communication and knowledge management tools, dynamic modelling, and precision agriculture for a deeper understanding of impacts to speed up uptake and improve science-based decisions by all users and practitioners and increase the scale of adoption across value chains.

In conclusion, the following is a summary of the key requirements to promote CA adoption in North Africa:

- Carrying out in-depth social, economic and risk analysis at farmer, community, and country scales (including livelihood resilience to shocks and stresses).
- Ensuring an integral ex-ante assessment (monitoring) of success and adoption (multi-criteria, multi-stakeholder evaluation including co-benefits and trade-offs) for tailoring CA to a sustainable intensification pathway, and hence, ensuring long-lasting and large-scale commitments by CA users.
- Government's initiatives to create a conducive environment for the adoption and development of CA systems including tackling major barriers while creating inclusive and equitable learning, financial and institutional mechanisms in addition to policy support.
- Development of instruments in co-addressing societal challenges and financial incentives in CA diffusion.
- Guaranteeing sharing costs and risks in addressing CA uptake and up-scaling.

- Encouraging and supporting CA-based farmer organisations (associations and cooperatives).
- Use of innovative tools for knowledge management and information sharing will be needed to further stimulate the demand for the technology from wider range of clients (e.g. small and large farms, public and private, NGOs and other environmental activists).
- Transforming CA research and science-based benefits to skills, curricula, awareness and positive changes aligned with the anticipated needs of users.
- Develop efficient and quality research and innovation systems and provide system thinking forum to further embed CA in development contexts and constraining climate and global changes.
- Creation and enforcement of centres of excellence, technology consortiums or competence centres.
- Develop strong and durable human capital and scientific capacity.
- Harnessing and influencing partners and resource mobilisation to achieve the best resource allocation to farmers and their organisations to improve production capacity and environmental stewardship over time and ensure focused effectiveness and tangible impacts of funds.
- Mounting innovative public-private partnership for converging visions and priorities.
- For confirming impacts, North African farmers should be mobilised in CA research, promotion and development pathways and foresight processes.
- Inclusion of CA systems in multi-government policies and programmes.

10 Where to look for further information

- Kassam, A. H. (Ed). 2020. *'Advances in Conservation Agriculture Volume 1: Systems and Science'*, Burleigh Dodds Science Publishing, Cambridge, UK, 602p.
- Kassam, A. H. (Ed). 2020. *'Advances in Conservation Agriculture Volume 2: Practice and Benefits'*, Burleigh Dodds Science Publishing, Cambridge, UK, 498p.
- Pratley, J. E. and Kirkegaard, J. A. (Eds). 2019. *Australian Agriculture in 2020: From Conservation to Automation*. Published on behalf of Agronomy Australia by Graham Centre for Agricultural Innovation, Charles Sturt University, Wagga Wagga, NSW, 444p.
- Kassam, A. H., Mkomwa, S. and Friedrich, T. (Eds). 2017. *'Conservation Agriculture for Africa: Building Resilient Farming Systems in Changing Climate'*, CABI Publishing, Wallingford, UK, 318p.
- Farooq, M. and Siddique, K. M. H. (Eds). 2015. *'Conservation Agriculture'*, Springer, 665p.

- Jat, R. A., Sahrawat, K. L. and Kassam, A. H. (Eds). 2014. *'Conservation Agriculture'. Global Prospects and Challenges'*, CABI Publishing, Wallingford, UK, 393p.
- Sivakumar, M. V. K., Lal, R., Selvaraju, R. and Hamdan, I. (Eds). 2013. *'Climate Change and Food Security in West Asia and North Africa'*, Springler, 423p.
- CA articles from Cornell University at: http://conservationagriculture.mannlib.cornell.edu.
- Global CA-CoP Conservation Agriculture Community of Practice Alerts.
- FAO CA website - http://www.fao.org/conservation-agriculture/en/.
- African Conservation Tillage Network (ACT): http://www.act-africa.org/.
- "4 per 1000" Initiative: https://www.4p1000.org.
- Agriculture de Conservation website - https://agriculture-de-conservation.com/.
- Conservation Technology Information Center: https://www.ctic.org.
- CIMMYT Publications Repository / Sustainable Intensification: https://repository.cimmyt.org/xmlui/handle/10883/2?locale-attribute=en.

11 Acronyms and abbreviations

AAAID Arab Authority of Agricultural Investment and Development
ACIAR Australian Centre for International Agricultural Research
ACSAD Arab Center for the Studies of Arid Zones and Dry Lands
ACT African Conservation Tillage Network: www-act-africa.org
ADA Agence pour le Développement Agricole
AFD Agence Française pour le Développement
AGENDA AGriculture, ENvironnement et Développement pour l'Avenir
AMAC Association Marocaine d'Agriculture de Conservation
APAD Association Pour une Agriculture Durable/Association for Sustainable Agriculture
ARC Agricultural Research Center (Egypt)
ATAE Association Tunisienne d'Agriculture Environnementale
ASRT Academy of Scientific Research & Technology, Egypt
ATMAR Les Ateliers Marocains
ATU-PAM Association Trait d'Union Pour une Agriculture Moderne
BRICS Building Resilience and Adaptation to Climate Extremes and Disasters Project
CA Conservation Agriculture
CANA Conservation Agriculture in North Africa
CASI Conservation Agriculture based Sustainable Intensification
CICYT Comisión Interministerial de Ciencia y Tecnología
CGIAR Consultative Group for International Agricultural Research

CIHEAM International Centre for Advanced Mediterranean Agronomic Studies
CIMMYT International Maize and Wheat Improvement Center
CIRAD Centre de Coopération Internationale en Recherche Agronomique pour le Développement
CLCA Conservation Agriculture in Crop-Livestock Systems
COTUGRAIN Compagnie Tunisienne du Grain
COP Conference of the Parties
CRP CGIAR Research Programme
CSA Climate-Smart Agriculture
CSIC Consejo Superior de Investigaciones Científicas
CSO Civil Society Organization
CTC Technical Center of Cereals
DFID UK - Department for International Development
EAC Exploitation Agricole Collective
ESAK High School of Agriculture of Kef
FAO Food and Agriculture Organization of the United Nation
FFEM French Facility for Global Environment: https://www.ffem.fr/en
FERT Association française de coopération internationale pour le développement Agricole des pays en développement et émergents
ENA National School of Agriculture – Meknès
ENSA High National School of Agronomy (ENSA) - Algeria
ESAM Ecole Supérieure d'Agriculture de Mograne
ESIER Ecole Supérieure des Ingénieurs de l'Equipement Rural
GIZ German Agency for International Cooperation
GEF Global Environmental Facility
IAV Hassan II Agronomy and Veterinary Sciences Institute
IBRD International Bank for Reconstruction and Development
ICARDA International Center for Agricultural Research in the Dry Areas
ICRAF World Agroforestry
IFPRI International Food Policy Research Institute
IFAD International Fund for Agricultural Development
IFS International Foundation for Science
INAT Institut National Agronomique de Tunis
INGC Institut National des Grandes Cultures
INGREF National Research Institute of Rural Engineering, Water and Forest
INRA National Institute of Agricultural Research - Morocco
INRAA National Institute of Agricultural Research - Algeria
INRAT National Institute of Agricultural Research - Tunisia
IPCC Intergovernmental Panel on Climate Change
IRD Institut de Recherche pour le Développement

IRESA Institution of Agricultural Research and Higher Education
ITGC Institut Technique des Grandes Cultures (Technical Institute of Cereals)
NAMA Nationally Appropriate Mitigation Action
NDC Nationally Determined Contributions
OEP Office d'élevage et de pâturage
ONCA Office National du Conseil Agricole
OCP Group Office Cherifien des Phosphates
ONDR National Office for Rural Development
PRIMA Partnership for Research and Innovation in Mediterranean Area
RCM Mediterranean Cereal Newtork
SDGs Sustainable Development Goals
SMSA Société Mutuelle de Service Agricole
SOC Soil Organic Carbon
USAID US Agency for International Development
WANA West Asia and North Africa

12 References

AAAID 2018. Annual Report 2018. Available at: https://www.aaaid.org/wp-content/uploads/2019/11/annual_report_2018_E_compressed.pdf.

Abagandura, G. O. and Park, D. 2016. Libyan agriculture: a review of past efforts, current challenges and future prospects, *Journal of Natural Sciences Research* 6(18), 57–67.

Abagandura, G. O., Park, D., White, D. and Bridges, W. C. 2017. Modelling soil degradation in Libya, *Journal of Natural Sciences Research* 7(24), 30–40.

Abdalla, M. A., Mohamed, A. E. and Makki, E. K. 2007. The response of two-sorghum cultivars to conventional and conservation tillage systems in central Sudan, *AMA, Agricultural Mechanization in Asia, Africa and Latin America* 38, 67–71.

Abdellaoui, Z., Fettih, S. and Zaghouane, O. 2006. Etude comparative de l'effet du semis direct et du labour conventionnel sur le comportement d'une culture de blé dur en conditions pluviales, *Options Méditerranéennes, Série A* 69, 183–187.

Aboudrare, A., Debaeke, P., Bouaziz, A. and Chekli, H. 2006. Effects of soil tillage and fallow management on soil water storage and sunflower production in a semi-arid Mediterranean climate, *Agricultural Water Management* 83(3), 183–196.

Abou El-Enin, M. M. and Abo-Remalia, S. H. 2015. Effect of conservation agriculture on broad bean and maize productivity in Egypt, *Int. J. Innov. Res. Creative Technol.* 1, 447–452.

ACT 2020. Conservation agriculture centres of excellence. Available at: http://www.act-africa.org/library.php?com=5.

Adaptation Fund 2018. Adaptation fund mid-term strategy 2018–2022. Available at: https://www.adaptation-fund.org/wp-content/uploads/2018/03/Medium-Term-Strategy-2018-2022-final-03.01-1.pdf.

AFD 2006. *Le semis direct sur couverture végétale permanente (SCV)*, Paris, France: AFD, 68p.

AfDB 2016. *Agricultural Production, Food Security and Higher Value in North Africa*. Working Paper, North Africa Policy Series, Abidjan, Côte d'Ivoire: AfDB, 26p.

Aibar, J. 2006. La lutte contre les mauvaises herbes pour les céréales en semis direct: principaux problèmes, *Options Méditerranéennes, Série A* 69, 19-26.

Albertengo, J. 2020. Certification scheme for Conservation Agriculture systems, in Kassam, A. H. (Ed), *Advances in Conservation Agriculture Volume 1: Systems and Science*, Cambridge, UK: Burleigh Dodds Science Publishing, pp. 497-512.

Albertengo, J., Belloso, C., Giraudo, M. B., Peiretti, R., Permingeat, H. and Wall, L. 2014. Conservation agriculture in Argentina, in Jat, R. A., Sahrawat, K. L. and Kassam, A. H. (Eds), *Conservation Agriculture: Global Prospects and Challenges*, Boston, MA: CABI Publishing, pp. 352-374.

AMAC 2021. Rapport sur la dissémination du semis direct au Maroc, in *Atelier pour la présentation de la feuille de route systèmes de semis direct sur 1.5 Mha (2020-2030)*, INRA-Rabat, 28 Janvier, 2021.

Amara, M., Feddal, M. A. and Hamani, A. 2015. *Analyse du comportement du sol sous l'action de trois techniques de mise en place d'un blé dur (Triticum durum). Effet sur le développement des racines et conséquences sur le rendement* (vol. 12), Nature Publishing & Technologie, pp. 130-141.

Angar, H., Ben Haj Salah, H. and Ben Hammouda, M. 2010. Développement du semis direct en Tunisie : approches et stratégies mises en œuvre, in 4èmes Rencontres Méditerranéennes du semis direct, Setif, Algeria.

Badraoui, M., Balaghi, R. and Mouchfi, N. 2018. *Rapport du comité scientifique de l'initiative sur l'adaptation de l'agriculture africaine au changement climatique (Initiative AAA) : Etat des lieux et propositions de solutions*. INRA.

Bahri, A. and Bansal, R. K. 1992. Evaluation of different combinations of opners and press wheels for no-till seeding, *Hommes Terre & Eaux* 22(86), 55-66.

Bahri, A., Von Bargen, K. and Bansal, R. K. 1993. Furrow openers and presswheels evaluation for no-tillwheat sowing, *Al Awamia* 83, 197-207.

Bahri, H., Annabi, M., Saoueb, A., Cheikh M'hamed, H., Souissi, A., Chibani, R. and Bahri, B. A. 2017. Can conservation agriculture sequester soil carbon in the long termin northern Tunisia? In Euro-Med Conference: Environment Integrability Recent Advances in Environmental Science from the Euro-Mediterranean and Surrounding Regions, Sousse, Tunisia, pp. 69-71.

Bahri, H., Annabi, M., Cheikh M'hamed, H., Chibani, R., Chtourou, M., Riahi, N. and Ben Becher, L. 2018. Evaluation de l'impact agro-environnemental et de la durabilité de l'adoption du semis direct au Nord de la Tunisie, *Annales de l'INRAT* 91, 98-111.

Bahri, H., Annabi, M., Cheikh M'hamed, H. and Frija, A. 2019. Assessing the long-term impact of conservation agriculture on wheat-based systems in Tunisia using APSIM simulations under a climate change context, *Science of the Total Environment* 692, 1223-1233.

Bazza, M., Key, M. and Knuston, C. 2018. Drought characteristics and management in North Africa and the Near East, *FAO Water Reports* 45. Available at: http://www.fao.org/3/CA0034EN/ca0034en.pdf.

Belagrouz, A., Chennafi, H., Hakimi, M., Soualili, N., Razem, R., Boutalbi, W. and Ferras, K. 2016. Conductivité hydraulique et densité apparente sous labours de conservation du sol, *Revue Agriculture Numéro Spécial* 1, 100-105.

Belaid, D. 2015. Algérie: le semis direct, une révolution technique en agriculture. Available at: http://agriculturealgerie.blogspot.com/2015/07/algerie-le-semis-direct-une-revolution.html?m=1.

Benguerai, A. 2011. *Evolution du phénomene de désertification dans le Sud oranais (Algérie)*, PhD thesis, Abou Bekr Belkaïd de Tlemcen University, 138 pp.
Ben-Hammouda, M., Guesmi, M., Nasr, K. and Kammassi, M. 2004. Evolution de la matière organique en semis direct et conventionnel, in Deuxièmes rencontres méditerranéennes sur le semis direct, Tabarka, Tunisia, pp. 104–107.
Ben-Hammouda, M., M'Hedhbi, K., Kammassi, M. and Gouili, H. 2007. Direct drilling: an agro-environmental approach to prevent land degradation and sustain production, in International Workshop on Conservation Agriculture for Sustainable Land Management to Improve the Livelihood of People in Dry Areas, Damascus, Syria, pp. 37–48.
Benkherbache, N., Benniou, R., Merat, M., Mekhalfia, T. and Makhlouf, M. 2012. From minimum tillage to no tillage, reaction of Waha, Variety of Durum Wheat in Algerian semi-arid region, *Online Journal of Science and Technology* 2(3), 17–22.
Ben Moussa-Machraoui, S. 2004. Semis direct et méso-faune du sol, in *Deuxièmes rencontres méditerranéennes sur le semis direct*, Tabarka, Tunisia, pp. 82–86.
Ben Moussa-Machraoui, S., Errouissi, F., Ben Hammouda, M. and Nouira, S. 2010. Comparative effects of conventional and no-tillage management on some soil properties under Mediterranean semi-arid conditions in northwestern Tunisia, *Soil and Tillage Research* 106(2), 247–253.
Ben-Salem, H., Zaibet, L. and Ben-Hammouda, M. 2006. Perspectives de l'adoption du semis direct en Tunisie. Une approche économique, *Options Méditerranéennes* 9, 69–75.
Benslimane, M., Hamimed, A., El Zerey, W. E., Khaldi, A. and Mederbal, K. 2008. Analyse et suivi du phénomène de la désertification en Algérie du nord, *VertigO* 8(3), 3. Available at: http://id.erudit.org/iderudit/039594ar.
Bessam, F. and Mrabet, R. 2003. Long-term changes in soil organic matter under conventional and no-tillage systems in semiarid Morocco, *Soil Use and Management* 19(2), 139–143.
Bonzanigo, L., Giupponi, C. and Moussadek, R. 2016. Conditions for the adoption of conservation agriculture in Central Morocco: an approach based on Bayesian network modelling, *Italian Journal of Agronomy* 10(1s), 665, doi:10.4081/ija.2016.665.
Borrelli, P., Robinson, D. A., Fleischer, L. R., Lugato, E., Ballabio, C., Alewell, C., Meusburger, K., Modugno, S., Schütt, B., Ferro, V., Bagarello, V., Oost, K., Van Montanarella, L. and Panagos, P. 2017. An assessment of the global impact of 21st century land use change on soil erosion, *Nature Communications* 8(1), 1–13.
Boughlala, M., El Gharras, O. and Dahan, R. 2011. *Economic Comparison between Conventional and No-Tillage Farming Systems in Morocco*, Hommes: Terre et Eaux, pp. 149–150, 35–39.
Boulal, H., El Mourid, M., Ketata, H. and Nefzaoui, A. 2014. Conservation agriculture in North Africa: past and the future, in Jat, R. A., Sahrawat, K. L. and Kassam, A. H. (Eds), *Conservation Agriculture: Global Prospects and Challenges*, Boston, MA: CABI Publishing, pp. 293–310.
Bourarach, E. H. 1989. *Mécanisation du travail du sol en céréaliculture pluviale: performances techniques et aspects économiques dans une région semi-aride au Maroc*, PhD thesis, IAV Hassan II, Rabat, 123p.
Bourarach, E. H., Bouzza, A. and Nousfi, A. 1998. Développement d'un système d'enterrage de semoir direct pour le travail en sol sec, *Hommes Terre Eaux* 28, 5–10.

Bouzerzour, H. and Mahnane, S. 2006. Une association pour une agriculture de conservation sur les hautes plaines orientales semi-arides d'Algérie, *Options Méditerranéennes, Série A* 69, 107–111.

Bouzerzour, H., Mahnane, S. and Makhlouf, M. 2006. Une association pour une agriculture de conservation sur les hautes plaines orientales semi-arides d'Algérie, *Options Méditerranéennes* 69, 107–111.

Bouzza, A. 1990. *Water Conservation in Wheat Rotations under Several Management and Tillage Systems in Semi-Arid Areas*. PhD dissertation, Lincoln, NE: University of Nebraska.

Brouziyne, Y., Abouabdillah, A., Hirich, A., Bouabid, R., Zaaboul, R. and Benaabidate, L. 2018. Modeling sustainable adaptation strategies toward a climate-smart agriculture in a Mediterranean watershed under projected climate change scenarios, *Agricultural Systems* 162, 154–163.

Brown, B., Nuberg, I. and Llewellyn, R. 2017. Stepwise frameworks for understanding the utilisation of conservation agriculture in Africa, *Agricultural Systems* 153, 11–22.

CANA (Conservation Agriculture in North Africa) 2015. Rapport final du projet CANA, 2015. Available at: http://www.cana-project.org/.

Carmona, I., Griffith, D. M., Soriano, M. A., Murillo, J. M., Madejon, E. and Gomez-Macpherson, H. 2015. What do farmers mean when they say they practice conservation agriculture? A comprehensive case study from southern Spain, *Agriculture, Ecosystems and Environment* 213, 164–177.

Caron, P., Biénabe, E. and Hainzelin, E. 2014. Making transition towards ecological intensification of agriculture a reality: the gaps in and the role of scientific knowledge, *Current Opinion in Environmental Sustainability* 8, 44–52.

Chabane, M. 2011. 'L'agriculture de conservation : voie de sécurité alimentaire dans les pays du Maghreb? *Options Méditerranéennes : Série A* 96, 189-208.

Chabbi, A., Lehmann, J., Ciais, P., Loescher, H. W., Cotrufo, M. F., Don, A., SanClements, M., Schipper, L., Six, J., Smith, P. and Rumpel, C. 2017. Aligning agriculture and climate policy, *Nature Climate Change* 7(5), 307–309.

Cheikh M'hamed, H., Angar, H. and Annabi, M. 2014. Conservation agriculture as alternative to reduce impact of climate change for smallholder in North Africa: Tunisian case, in 6ème Congrès Mondial Sur l'Agriculture de Conservation, Winnipeg, Canada, 22-25 Juin 2014.

Cheikh M'hamed, H., Annabi, M., Ben Youssef, S. and Bahri, H. 2016. L'agriculture de conservation est un système de production permettant d'améliorer l'efficience de l'utilisation de l'eau et de la fertilité du sol, *Annales de l'INRAT* 89, 68–71.

Cheikh M'hamed, H., Bahri, H. and Annabi, M. 2018. Conservation agriculture in Tunisia: historical, current status and future perspectives for rapid adoption by smallholder farmers, in Second African Congress on Conservation Agriculture, Johannesburg, South Africa, 9-12 October 2018, pp. 57–60.

Chekali, S., Gargouri, S., Ben Hamouda, M., Cheikh M'hamed, H. and Nasraoui, B. 2019. Incidence of Fusarium foot and root rot of cereals under conservation agriculture in north west Tunisia, *Phytopathologia Mediterranea* 58(1), 95–102.

Chennafi, H., Hannachi, A., Touahria, O., Fellahi, Z. E. A., Makhlouf, M. and Bouzerzour, H. 2011. Tillage and residue management effect on durum wheat [*Triticum turgidum* (L.) Thell. ssp. turgidum conv. durum (Desf.) MacKey] growth and yield under semi-arid climate, *Advances in Environmental Biology* 5(10), 3231-3240.

Chibani, R., Bahri, H., Annabi, M. and Cheikh M'hamed, H. 2018. 'L'agriculture de conservation comme alternative pour améliorer la résistance des sols à l'érosion hydrique dans le Nord de la Tunisie, *Annales de l'INRAT* 91, 86-97.

Chouen, S., Quillet, J. C. and Rojat, D. 2004. Semis direct et techniques conventionnelles en Tunisie: comparison des couts de production sur des exploitations types et éléments d'analyse économique, in AFD/Cirad/CTC/ESAK/ICARDA, *Deuxièmes rencontres méditerranéennes sur le semis direct*, Tabarka, Tunisia, pp. 116-120.

Cicek, H., Rekik, M., Dheibi, B. and Frija, A. 2017. Some insights on conservation agriculture research in North Africa: current situation and opportunities, *Newsupdate*. ACT Network. Available at: http://www.act-africa.org/news.php?com=68&com2=6&item=384#.YH8FAmf7RPY.

CLCA (Integrated Crop-Livestock under Conservation Agriculture) 2016. *Rapport final du projet CLCA, 2016*. Available at: https://apps.icarda.org/wsInternet/wsInternet.asmx/DownloadFileToLocal?filePath=ICARDACLCA.pdf&fileName=ICARDA-CLCA.pdf.

Cuffaro, N. and D'Agostino, G. 2017. Land inequality and growth: meta-analysis and relevance for contemporary development in africa, Working Paper no 222, Rome: Università di Roma Tre.

Cusser, S., Bahlai, C., Swinton, S. M., Robertson, G. P. and Haddad, N. M. 2020. Long-term research avoids spurious and misleading trends in sustainability attributes of notill, *Global Change Biology* 21(6), 3715-3725, doi:10.1111/gcb.15080.

Degrande, A. and Benoudji, C. 2017. L'agriculture de Conservation Repensée: contextualiser l'innovation pour la Résilience au Tchad et au Soudan, in *BRACED Innovation for Resilience Case Studies*. Available at: https://www.researchgate.net/publication/323799498_Rethinking_Conservation_Agriculture_Contextualisation_of_Innovation_for_Resilience_in_Chad_and_Sudan

de Lattre-Gasquet, M., Moreau, C., Elloumi, M. and Ben Becher, L. 2017. Vers un scénario « Des usages agro-écologiques des terres pour une alimentation diversifiée et de qualité et un système alimentaire territorialisé » en Tunisie en 2050, *OCL* 24(3), D306, doi:10.1051/ocl/2017025.

DGACTA 2017. *Direction Générale de l'Aménagement et de la Conservation des Terres Agricoles: Elaboration de la Stratégie de Conservation des Eaux et des Sols de la Tunisie*, DGACTA, 202p.

Dhehibi, B., Zucca, C., Frija, A., Shinan, N. and Kassam, S. N. 2018. Biophysical and Econometric analysis of adoption of soil and water conservation techniques in the semiarid region of Sidi Bouzid (Central Tunisia), *New Medit* 2, 15-28.

Dimanche, P. H. 1997. *Impacts des différents itinéraires techniques de travail du sol sur la dégradation des sols argileux dans la région de Meknès, Maroc*. PhD thesis, Belgique: Louvain-La-Neuve, 268p.

Dixon, J., Huttner, E., Reeves, T., Nyagumbo, I., Timsina, J., El Mourid, M., Loss, S. and Tan, D. K. Y. 2019. Producing food while protecting the environment: inter-disciplinary research methods for international research on conservation agriculture based sustainable intensification (CASI), in Dixon, J. M. and Coffey, S. G. (Eds), *Agricultural Science* Special Issue: ACIAR at Work: Interdisciplinary Research into Smallholder Farming Systems. Combined Volumes 30(2) and 31(1), 64-81.

Djennadi, F. A., Houassin, D. A., Khammassi, M. B., Laouar, A. A. and Souissi, T. 2014. Weed incidence, control and grain yield losses in conservation cropping systems of North Africa, in 6th World Congress on Conservation Agriculture, Winnipeg, Manitoba.

Dridi, N., Mekki, M., Cheikh M'hamed, H. and Ben-Hammouda, M. 2010. Impact de Deux Modes de Semis (Conventionnel vs. Direct) sur la Flore Adventice des Cultures, in 4ème Rencontres Méditerranéennes du Semis Direct, Sétif, Algeria, pp. 39-47.

Driouch, F., Rached, S. and El Hairech, T. 2013. Climate variability and change in North African countries, in Sivakumar, M.V.K., Lal, R., Selvaraju, R. and Hamdan, I. (Eds). *Climate Change and Food Security in West Asia and North Africa*, New York: Springer, pp. 161-172.

DSASI-MADR 2004-2014. *Statistiques agricoles. Série B. Direction des statistiques agricoles et systèmes d'information*. Alger, Algeria: Ministère de l'agriculture et du développement rural.

Dubrovský, M., Hayes, M., Duce, P., Trnka, M., Svoboda, M. and Zara, P. 2014. Multi-GCM projections of future drought and climate variability indicators for the Mediterranean region, *Regional Environmental Change* 14(5), 1907-1919.

El-Areed, S. R. H. 2019. Improvement of yellow corn productivity under water deficit stress using conservation agriculture system, *Egyptian Journal of Plant Breeding* 23(1), 11-27.

El Gharras, O. and Idrissi, M. 2006. Contraintes technologiques au développement du semis direct au Maroc, *Options Méditerranéennes: Série A* 69, 121-124.

El Gharras, O., Ait Lhaj, A. and Idrissi, M. 2004. Développement d'un semoir non-labour industriel, in 2èmes rencontres méditerranéennes sur le semis direct, Tabarka, Tunisia.

El Gharras, O., El Mourid, M. and Boulal, H. 2017. Conservation agriculture in North Africa: experiences, achievements and challenges, in Kassam, A. H., Mkomwa, S. and Friedrich, T. (Eds), *Conservation Agriculture for Africa: Building Resilient Farming Systems in a Changing Climate*, Wallingford, UK: CAB International, pp. 127-138.

Elnemr, M. 2019. Applicability of sustainable agriculture in Egypt, in Negm, A. M. and Abu-hashim, M. (Eds), *Sustainability of Agricultural Environment in Egypt: Part I: Soil-Water-Food Nexus, The Handbook of Environmental Chemistry*, New York: Springer, Chamber of Commerce, Vol. 76, pp. 3-31.

Errouissi, F., Ben Moussa-Machraoui, S., Ben-Hammouda, M. and Nouira, S. 2011. Soil invertebrates in durum wheat(*Triticum durum* L.) cropping system under Mediterranean semi-arid conditions: a comparison between conventional- and no-tillage management, *Soil and Tillage Research* 112(2), 122-132.

FAO 2013. *Climate-smart Agriculture Sourcebook*, Technical Report, Rome: Food and Agriculture Organization of the United Nations.

FAO 2018. *The State of Agricultural Commodity Markets 2018*, Rome: Agricultural Trade, Climate Change and Food Security.

FAO 2019. *World Food and Agriculture: Statistical Pocketbook 2019*, Rome: FAO.

FAO, IFAD and UNICEF, WFP, WHO 2017. *The State of Food Security and Nutrition in the World 2017: Safeguarding against Economic Slowdowns and Downturns*, Technical Report, Rome: Food and Agriculture Organization of the United Nations. Available at: http://www.fao.org/3/ca5162en/ca5162en.pdf.

FAOSTAT 2021. Available at: http://www.fao.org/faostat/en/#home.

Fellahi, Z., Hannachi, A., Chennafi, H., Makhlouf, M. and Bouzerzour, H. 2013. Effets des résidus et du travail du sol sur la production de la biomasse et le rendement du blé dur (*Triticum durum* Desf., variété MBB) en lien avec l'utilisation de l'eau dans les conditions semi-arides des Hautes Plaines Sétifiennes, *Revue de l'Agriculture* 06, 03-11.

Fenni, M. 2013. Socio-economic and environmental benefits of direct seeding of wheat in Sétif high plains (North East of Algeria), in WEI International Academic Conference, Istanbul, Turkey. Westchester, PA: West East Institute (WEI), p. 47.

FERT 2016. *Le semis direct en Tunisie: Situation actuelle et perspectives*, Fert.-APAD-TAE-INGC, 66p.

FERT 2018. L'agriculture de conservation au Maghreb: les agriculteurs font évoluer leurs pratiques. Available at: https://www.agricord.org/sites/default/files/2018_med_capi_ac_maghreb_fr.pdf.

Findlater, K. M., Kandlikar, M. and Satterfield, T. 2019. Misunderstanding conservation agriculture: challenges in promoting, monitoring and evaluating sustainable farming, *Environmental Science and Policy* 100, 47–54.

Fouzai, A., Smaoui, M., Frija, A. and Dhehibi, B. 2018. Adoption of Conservation Agriculture Technologies by smallholder farmers in the semiarid region of Tunisia: resource constraints and partial adoption, *Journal of New Sciences, Sustainable Livestock Management* 6(1), 105–114.

Friedrich, T. 2005. Does no-till farming require more herbicides? *Outlooks on Pest Management* 16(4), 188–191.

Glover, D., Sumberg, J. and Andersson, J. A. 2016. The adoption problem: or why we still understand so little about technological change in African agriculture, *Outlook on Agriculture* 45(1), 3–6.

Gomez-Macpherson, H., Boulal, H., Mrabet, R. and Gonzalez, E. 2009. Rational and application of conservation agriculture for irrigated production in Southern Europe and North Africa, in 4th World Congress on Conservation Agriculture: Innovations for Improving Efficiency, Equity and Environment, Delhi, India, pp. 129–135.

González-Sánchez, E. J., Veroz-González, O., Caneway, G., Moreno-García, M., Kassam, A., Mkomwa, S., Ordoñez-Fernandez, R., Triviño-Tarradas, P. and Carbonell-Bojollo, R. 2019. Meta-analysis on carbon sequestration through conservation agriculture in Africa, *Soil and Tillage Research* 190, 22–30.

Guesmi, H., Ben Salem, H. and Moujahed, N. 2019. Integration crop-livestock under conservation agriculture system, *Journal of New Sciences, Agriculture and Biotechnology* 65(1), 4061–4065.

Hajer, G., Cyrine, D., Salah, B. Y., Mohamed, C., Sourour, A., Hichem, B. S. and Nizar, M. 2020. Wheat stubble from conventional or conservation agriculture grazed by ewes: biomass dynamics and animal performances, *Animal Nutrition and Feed Technology* 20(2), 187-200.

Hajjaj, B., Bouhache, M., Mrabet, R., Taleb, A. and Douaik, A. 2016a. Efficacité de quelques herbicides contre les mauvaises herbes du pois chiche et de la féverole conduits en semis direct, *Revue Marocaine des Sciences Agronomiques et Vétérinaires* 4(3), 37–47.

Hajjaj, B., Bouhache, M., Mrabet, R., Taleb, A. and Douaik, A. 2016b. Efficacité de quelques herbicides des céréales dans une culture de blé tendre conduit en semis direct, *Revue Marocaine des Sciences Agronomiques et Vétérinaires* 4(3), 48–56.

Hajjaj, B., Bouhache, M., Mrabet, R. and Taleb, A. 2018. Stock semencier des mauvaises herbes dans le sol et évolution de la flore de surface sous système de semis direct, *Revue Marocaine de Protection des Plantes* 12, 1–17.

Hamim, A., Essahat, A., El Othmani, S., Douiek, A., Mrabet, R., Duponnois, R. and Hafidi, M. 2012. Etude de l'effet de la rotation et du travail du sol sur la densité spécifique des champignons mycorhiziens arbusculaires, in International Congress: Microbial Biotechnology for Development (MICROBIOD 2), Marrakech, Morocco.

Harb, O. M., Abd El-Hay, G. H., Hager, M. A. and Abou El-Enin, M. M. 2015. Studies on conservation agriculture in Egypt, *Annals of Agricultural Sciences* 60(1), 105–112.

Harb, O. M., Abd El-Hay, G. H., Hager, M. A. and Abou El-Enin, M. M. 2016. Calibration and validation of DSSAT V.4.6.1, CERES and CROPGRO-models for simulating no-tillage in central Delta, Egypt, *Agrotechnol* 5, 143, doi:10.4172/2168-9881.1000143.

Hunt, J. R., Celestina, C. and Kirkegaard, J. A. 2020. The realities of climate change, conservation agriculture and soil carbon sequestration, *Global Change Biology* 26(6), 3188–3189.

Ibno-Namr, K. and Mrabet, R. 2004. Influence of agricultural management on chemical quality of a clay soil of semi-arid Morocco, *Journal of African Earth Sciences* 39(3–5), 485–489.

Ibrahim, M. and Lal, R. 2013. Climate change and land use in the WANA region with a specific reference to Morocco, in Sivakumar, M.V.K., Lal, R., Selvaraju, R. and Hamdan, I. (Eds), *Climate Change and Food Security in West Asia and North Africa*, New York: Springer, pp. 89–113.

IPCC 2018. Summary for policymakers. Global Warming of 1.5°C, in Masson-Delmotte, V., P. Zhai, H.-O. Pörtner, D. Roberts, J. Skea, P. R. Shukla, A. Pirani, W. Moufouma-Okia, C. Péan, R. Pidcock, S. Connors, J. B. R. Matthews, Y. Chen, X. Zhou, M. I. Gomis, E. Lonnoy, T. Maycock, M. Tignor, and T. Water (Eds), *Intergovernmental Panel on Climate Change (IPCC)*, Geneva, Switzerland: IPCC.

Jemai, I., Ben Aissa, N., Ben Guirat, S., Ben-Hammouda, M. and Gallali, T. 2012. On-farm assessment of tillage impact on the vertical distribution of soil organic carbon and structural soil properties in a semiarid region in Tunisia, *Journal of Environmental Management* 113, 488–494.

Jemai, I., Ben Aissa, N., Ben Guirat, S., Ben-Hammouda, M.. and Gallali, T. 2013. Impact of three and seven years of notillage on the soil water storage, in the plant root zone, under a dry subhumid Tunisian climate, *Soil and Tillage Research* 126, 26–33.

Jones, A., Breuning-Madsen, H., Brossard, M., Dampha, A., Dewitte, O., Hallett, S., Jones, R., Kilasara, M., Le Roux, P., Micheli, E., Montanarella, L., Spaargaren, O., Tahar, G., Thiombiano, L., Van Ranst, E., Yemefack, M. and Zougmore, R. (Eds). 2013. *Soil Atlas of Africa*, Luxembourg: European Commission, Publications Office of the European Union.

Kassam, A., Friedrich, T., Derpsch, R., Lahmar, R., Mrabet, R., Basch, G., González- Sánchez, E. J. and Serraj, R. 2012. Conservation Agriculture in the dry Mediterranean climate, *Field Crops Research* 132, 7–17.

Kassam, A., Derpsh, R. and Friedrich, T. 2020. Development of Conservation Agriculture systems globally, in Kassam, A. (Ed), *Advances in Conservation Agriculture Volume 1: Systems and Science*, Cambridge, UK: Burleigh Dodds Science Publishing, pp. 1-56.

Khaldi, R. and Saaidia, B. 2017. Analyse de la filière céréalière en Tunisie et identification des principaux points de dysfonctionnement à l'origine des pertes, *Final report of the FAO project Réduction des pertes et gaspillage alimentaire à travers le développement des chaînes de valeur pour la sécurité alimentaire en Egypte et en Tunisie*, 102 pp.

Khouri, N., Breisinger, C. and Eldidi, H. 2017. Can MENA reach the sustainable development goals? An overview of opportunities and challenges for food and nutrition security, in Mergosand, G. and Papanastassiou, M. (Eds), *Food Security and Sustainability*, Cham: Palgrave Macmillan, pp. 175–191.

Kribaa, M., Hallaire, V., Curmi, P. and Lahmar, R. 2001. Effects of various cultivation methods on the structure and hydraulic properties of soil in semi-arid climate, *Soil and Tillage Research* 60(1-2), 43-53.

Labad, R. and Hartani, T. 2016. Analyse des performances de quelques exploitations agricoles céréalières en semis direct dans la wilaya de Sétif, *Revue de l'Agriculture* 1, 78-81.

Labad, R., Hartani, T., Belguet, H., Bendada, H., Louahdi, N. and Taibi, M. 2018a. Evaluation de la biologie du sol sous l'effet du traitement chimique en semis direct dans une zone semi-aride de l'Algérie, *Agriculture Journal* 9(1), 46-55.

Labad, R., Hartani, T. and Shinde, G. U. 2018b. Optimum herbicide dose management in direct seeding for cereals production: case of semi-arid area of Algeria, *Journal of Agronomy* 17(2), 99-105.

Labad, R., Hartani, T. and Shinde, G. U. 2020. Effect of weed treatment on cereal yield in direct seeding: a challenge between soil pollution and seeds quality, *Indian Journal of Agricultural Research* 54(1), 101-106, doi:10.18805/IJARe.A-304.

Labbaci, T., Dugué, P., Kemoun, H. and Rollin, D. 2015. Innovation et action collective: le semis direct des cultures pluviales Au Moyen Sébou (Maroc), *Cahiers Agricultures* 24(2), 76-83.

Lagacherie, P., Álvaro-Fuentes, J., Annabi, M., Bernoux, M., Bouarfa, S., Douaoui, A., Grünberger, O., Hammani, A., Montanarella, L., Mrabet, R., Sabir, M. and Raclot, D. 2017. Managing Mediterranean soil resources under global changes: expected trends and mitigation strategies, *Regional Environmental Change* 18(3), 663-675, doi:10.1007/s10113-017-1239-9.

Lal, R. 2002. Carbon sequestration in dryland ecosystems of West Asia and North Africa, *Land Degradation and Development* 13(1), 45-59.

Lal, R. 2018. Promoting "4 Per Thousand" and "Adapting African Agriculture" by southsouth cooperation: conservation agriculture and sustainable intensification, *Soil & Tillage Research* 188, 27-34.

Lal, R. 2019. Promoting "4 Per Thousand" and "Adapting African Agriculture" by south-south cooperation: conservation agriculture and sustainable intensification, *Soil and Tillage Research* 188, 27-34.

Lal, R., Safriel, U. and Boer, B. 2012. *Zero Net Land Degradation: A New Sustainable Development Goal for Rio+ 20,* A Report Prepared for the Secretariat of the UNCCD.

Lalani, B., Dorward, P., Holloway, G. and Wauters, E. 2016. Smallholder farmers' motivations for using conservation agriculture and the roles of yield, labour and soil fertility in decision making, *Agricultural Systems* 146, 80-90.

Lalani, B., Dorward, P. and Holloway, G. 2017. Farm-level economic analysis - Is Conservation Agriculture helping the Poor?, *Ecological Economics* 141, 144-153.

Laouar, A., Djenadi, F. and Bendada, H. 2018. Weed management in conservation agriculture in Northern Algeria, in Second African Congress on Conservation Agriculture, Johannesburg, South Africa, pp. 115-118.

Lionello, P., Abrantes, F., Gacic, M., Planton, S., Trigo, R. and Ulbrich, U. 2014. The climate of the Mediterranean region: research progress and climate change impacts, *Regional Environmental Change* 14(5), 1679-1684.

Lipper, L., Thornton, P., Campbell, B. M., Baedeker, T., Braimoh, A., Bwalya, M., Caron, P., Cattaneo, A., Garrity, D., Henry, K., Hottle, R., Jackson, L., Jarvis, A., Kossam, F., Mann, W., McCarthy, N., Meybeck, A., Neufeldt, H., Remington, T., Sen, P. T., Sessa, R., Shula,

R., Tibu, A. and Torquebiau, E. F. 2014. Climate-smart agriculturefor food security, *Nature Climate Change* 4(12), 1068–1072.

Llewellyn, R. and Ouzman, J. 2019. Conservation agriculture in Australia: 30 years on, in Pratley, J. E. and Kirkegaard, J. A. (Eds), *Australian Agriculture in 2020: From Conservation to Automation*, Wagga Wagga: Agronomy Australia and Charles Sturt University, pp. 21-31.

Loss, S., Haddad, A., Khalil, Y., Alrijabo, A., Feindel, D. and Piggin, C. 2015. Evolution and adoption of conservation agriculture in the Middle East, in Farooq, M. and Siddique, K. H. M. (Eds), *Conservation Agriculture*, New York: Springer, pp. 197–224.

Magnan, N., Lybbert, T. J., Mrabet, R. and Fadlaoui, A. 2011. The quasi-option value of delayed input use under catastrophic drought risk: the case of no-till in Morocco, *American Journal of Agricultural Economics* 93(2), 498–504.

Magnan, N. 2015. Property rights enforcement and no-till adoption in crop-livestock Systems, *Agricultural Systems* 134, 76–83.

Mahgoub, F. 2014. *Current status of agriculture and future challenges in Sudan*, Uppsala, Sweden: Nordiska Afrikainstitutet, 98p. ISBN 978-91-7106-748-7.

Mbow, C., Rosenzweig, C., Barioni, L. G., Benton, H., Krishnapillai, M., Liwenga, E., Pradhan, P., Rivera-Ferre, M. G., Sapkota, T., Tubiello, F. N. and Xu, Y. 2019. Food security, in Shukla, P. R., Skea, J., Buendia, E. C., Masson-Delmotte, V., Pörtner, H.-O., Roberts, D. C., Zhai, P., Slade, R., Connors, S., van Diemen, R., Ferrat, M., Haughey, E., Luz, S., Neogi, S., Pathak, M., Petzold, J., Portugal Pereira, J., Vyas, P., Huntley, E., Kissick, K., Belkacemi, M. and Malley, J. (Eds), *Climate Change and Land*, IPCC special report on climate change, desertification, land degradation, sustainable land management, food security, and greenhouse gas fluxes in terrestrial ecosystems, Geneva, Switzerland: Intergovernmental Panel on Climate Change (IPCC).

Mekhlouf, A., Rouag, N., Boukhadra, R., Chenni, S., Fenni, M. and Makhlouf, M. 2013. Influence of cropping systems on soil properties in semi-arid conditions of Setif, Algeria, *Journal of Agricultural Science and Technology A* 3, 653–658.

M'hedhbi, K., Chouen, S. and Ben-Hammouda, M. 2003. A recent Tunisian experience with direct drilling, in II World Congress on Conservation Agriculture, Iguassu Falls, Parana, Brazil, pp. 132–135.

M'hedhbi, K., Ben-Hammouda, M., Letoumy, P., Nasr, K., Ali Hannachi, M., Chouen, S., Mahouachi, M. A., Jarrahi, T., Nasraoui, R., Zaouani, R. and Fakhfakh, M. M. 2004. Résultats agronomiques de production pour les semis directs et conventionnels, *Deuxièmes rencontres méditerranéennes sur le semis direct*, Tabarka, Tunisia, pp. 87–89.

Minasny, B., Malone, B. P., McBratney, A. B., Angers, D. A., Arrouays, D., Chambers, A., Chaplot, V., Chen, Z., Cheng, K., Das, B. S., Field, D. J., Gimona, A., Hedley, C. B., Hong, S. Y., Mandal, B., Marchant, B. P., Martin, M., McConkey, B. G., Mulder, V. L., O'Rourke, S., Richer-de-Forges, A. C., Odeh, I., Padarian, J., Paustian, K., Pan, G., Poggio, L., Savin, I., Stolbovoy, V., Stockmann, U., Sulaeman, Y., Tsui, C., Vågen, T., van Wesemael, B. and Winowiecki, L. 2017. Soil carbon 4 per mille, *Geoderma* 292, 59–86.

Montanarella, L., Pennock, D. J., McKenzie, N., Badraoui, M., Chude, V., Baptista, I., Mamo, T., Yemefack, M., Singh Aulakh, M., Yagi, K., Young Hong, S., Vijarnsorn, P., Zhang, G., Arrouays, D., Black, H., Krasilnikov, P., Sobocká, J., Alegre, J., Henriquez, C. R., de Lourdes Mendonça-Santos, M., Taboada, M., Espinosa-Victoria, D., AlShankiti, A., AlaviPanah, S. K., Elsheikh, E. A. E. M., Hempel, J., Camps Arbestain, M., Nachtergaele, F. and Vargas, R. 2016. World's soils are under threat, *SOIL* 2(1), 79–82.

Mouelhi, B., Slim, S., Arfaoui, S., Boussalmi, A. and Ben Jeddi, F. 2016. Effet du mode de semis et de la rotation culturale sur les paramètres de croissance et les composantes de rendement du blé dur (Triticum durum Desf.) variété « Karim », *Journal of New Sciences, Agriculture and Biotechnology* 28(11), 1638–1648.

Moujahed, N., Abidi, S., Ben Youssef, S., Darej, C., Chakroun, M. and Ben Salem, H. 2015. Effect of stocking rate on biomass variation and lamb performances for barley stubble in Tunisian semi-arid region and under conservation agriculture conditions, *African Journal of Agricultural Research* 10(50), 4584–4590.

Moussadek, R. and Mrabet, R. 2017. Carbon management and sequestration in dryland soils of Morocco: nexus approach, in Global Symposium on Soil Organic Carbon, Rome, Italy: Food and Agriculture Organization of the United Nations, pp. 497–500.

Moussadek, R., Mrabet, R., Dahan, R., Douaik, A., Verdoodt, A., Van Ranst, E. and Corbeels, M. 2011a. Effect of tillage practices on the soil carbon dioxide flux during fall and spring seasons in a Mediterranean vertisol, *Journal of Soil Science and Environmental Management* 2(11), 362–369.

Moussadek, R., Mrabet, R., Zante, P., Lamachere, L.-M., Pepin, Y., Le Bissonnais, Y., Ye, L., Verdoodt, A. and Van Ranst, E. 2011b. Influence du semis direct et des résidus de culture sur l'érosion hydrique d'un vertisol Méditerranéen, *Canadian Journal of Soil Science* 91(4), 627–635.

Moussadek, R., Mrabet, R., Dahan, R., Zouahri, A., El Mourid, M. and Van Ranst, E. V. 2014. Tillage system affects soil organic carbon storage and quality in Central Morocco, *Applied and Environmental Soil Science* 2014, 1–8, doi:10.1155/2014/654796.

Mrabet, R. 1997. *Crop Residue Management and Tillage Systems for Water Conservation in a Semiarid Area of Morocco*, PhD dissertation, Fort Collins, CO: Colorado State University, 220p.

Mrabet, R. 2002. Wheat yield and water use efficiency under contrasting residue and tillage management systems in a semiarid area of Morocco, *Experimental Agriculture* 38(2), 237–248.

Mrabet, R. 2000. Differential response of wheat to tillage management systems under continuous cropping in a semiarid area of Morocco, *Field Crops Research* 66(2), 165–174.

Mrabet, R. 2008a. Conservation agriculture in Morocco: a research review, in International Workshop on Conservation Agriculture for Sustainable Land Management to Improve the Livelihood of People in Dry Areas, Damascus, Syria: ACSAD & GTZ, pp. 183–208.

Mrabet, R. 2008b. *No-Tillage Systems for Sustainable Dryland Agriculture in Morocco*, Paris, France: Institut National de la Recherche Agronomique (INRA), 153pp.

Mrabet, R. 2008c. No-till practices in Morocco, in Goddard, T., Zoebisch, M., Gan, Y. T., Ellis, W., Watson, A. and Sombatpanit, S. (Eds), *No-Till Farming Systems*, Special Publication No. 3, Bangkok: World Association of Soil and Water Conservation, pp. 393–412.

Mrabet, R. 2011a. Climate change and carbon sequestration in the Mediterranean basin: contributions of no-tillage systems, *Options Méditerranéennes* 96, 165–184.

Mrabet, R. 2011b. Effects of residue management and cropping systems on wheat yield stability in a semiarid Mediterranean clay soil, *American Journal of Plant Sciences* 2(2), 202–216.

Mrabet, R. 2011c. No-tillage agriculture in West Asia & North Africa, in Tow, P. G., Cooper, I. M., Partridge, I. and C. J. Birch (Eds), *Rainfed Farming Systems*, Dordrecht Netherlands: Springer, pp. 1015–1042.

Mrabet, R. 2017. Chapitre 27: le semis direct sous couverture végétale, une voie de restauration durable de la productivité des sols semi-arides du Maroc, in Roose, E. (Ed), *Restauration de la productvité des sols tropicaux et méditerranéens. Contribution de l'agroécologie*, Paris, France: institut de recherche pour le développement, pp. 357–366.

Mrabet, R. 2018. Information and data collections for preparation of synthesis report of the general status of adoption of conservation agriculture (CA) in Africa towards 2ACCA: Case of Morocco, in Second African Congress on Conservation Agriculture, Johannesburg, South Africa, 9–12 October, 3p.

Mrabet, R., El-Brahli, A., Bessam, F. and Anibat, I. 2003. No-tillage technology: research review of impacts on soil quality and wheat production in semiarid Morocco, *Options Méditerranéennes* 60, 133–138.

Mrabet, R., Ibno-Namr, K., Bessam, F. and Saber, N. 2001a. Soil chemical quality changes and implications for fertilizer management after 11 years of no-tillage wheat production systems in semiarid Morocco, *Land Degradation and Development* 12(6), 505–517.

Mrabet, R., Saber, N., El-Brahli, A., Lahlou, S. and Bessam, F. 2001b. Total, particulate organic matter and structural stability of a Calcixeroll soil under different wheat rotations and tillage systems in a semiarid area of Morocco, *Soil and Tillage Research* 57(4), 225–235.

Mrabet, R., Lahlou, S., Le Bissonnais, Y. and Duval, O. 2004. Estimation de la stabilité structurale des sols semi-arides marocains : influence des techniques culturales simplifiées, *Bulletin du Réseau Erosion* 23(2), 405–415.

Mrabet, R., Essahat, E., Gomez-Macpherson, H., Falisse, A. and Alghoum, M. 2010. Coupling conservation agriculture and supplementary irrigation for improving yield and yield stability of wheat and chickpea in semiarid Morocco, in European Congress on Conservation Agriculture: Towards Agro-Enviromental, climate and energetic sustainability, Madrid, Spain, pp. 581–584.

Mrabet, R., Moussadek, R., Fadlaoui, A. and van Ranst, E. 2012. Conservation agriculture in dry areas of Morocco, *Field Crops Research* 132, 84–94.

Mrabet, R., Badraoui, M., Moussadek, R. and Soudi, B. 2017. Challenges of soil organic carbon sequestration in drylands, in Global Symposium on Soil Organic Carbon, Rome, Italy: Food and Agriculture Organization of the United Nations.

Mueller, N. D., Gerber, J. S., Johnston, M., Ray, D. K., Ramankutty, N. and Foley, J. A. 2012. Closing yield gaps through nutrient and water management, *Nature* 490(7419), 254–257.

Nefzaoui, A., Ketata, H. and El Mourid, M. 2012. Changes in North Africa production systems to meet climate uncertainty and new socio-economic scenarios with a focus on dryland areas, *Options Méditerranéennes* 102, 403–421.

OCP 2020. OCP: le semis direct, levier clé pour augmenter la résilience de l'agriculture face aux changements climatiques. Available at: https://www.agrimaroc.ma/ocp-semis-direct/.

Oulbachir, K., Zoubeidi, M., Kouadria, M. and Bouchenafa, N. 2014. Pratiques de conservation des agro systèmes en vue du développement durable, *Journal of Materials and Environmental Science* 5(Suppl.1), 2345–2349.

Peiretti, R. and Dumanski, J. 2014. The transformation of agriculture in Argentina through soil conservation, *International Soil and Water Conservation Research* 2(1), 14–20.

Pittelkow, C. M., Liang, X., Linquist, B. A., van Groenigen, K. J., Lee, J., van Gestel, N., Six, J., Venterea, R. T. and van Kessel, C. 2015. Productivity limits and potentials of the principles of conservation agriculture, *Nature*, 517, 365-368.

Prestele, R., Hirsch, A. L., Davin, E. L., Seneviratne, S. I. and Verburg, P. H. 2018. A spatially explicit representation of conservation agriculture for application in global change studies, *Global Change Biology* 24, 4038-4053.

Rahali, A., Makhlouf, M. and Benkherbache, N. 2011. Influence de l'itinéraire technique sur le stock semencier de mauvaises herbes de la zone semi-aride de Sétif, *Options Méditerranéennes* 96, 153-162.

Ramdani, A., Mrabet, R. and Gomez-Macpherson, H. 2010. Performance of Moroccan wheat cultivars under tillage systems and water regimes, *Journal of Agricultural Science and Technology A* 4(5), 106-111.

Rasheed, N. H., Al-Heeti, A. A., Burhan, H. O. and Abdulrazak, N. A. 2006. Experience of the Arab Authority for Agricultural Investment and Development in the adoption and dissemination of the zero tillage system to the traditional farmers in Sudan, *Journal of Agricultural Investment* 4, 6-13.

Rastoin, J. L. and Benabderrazik, E. H. 2014. *Céréales et oléoprotéagineux*, Construire la Méditerranée-IPEMED.

Raunet, M., Richard, J. F. and Rojat, D. 2004. Premiers résultats d'introduction du semis direct sous couvert et lutte antiérosive en Tunisie, in Roose, E., De Noni, G., Prat, C., Ganry, F. and Bourgeon, G. (Eds), *Gestion de la biomasse, érosion et séquestration du carbone. Séquestration du carbone et érosion des sols*, Paris, France: institut de recherche pour le développement, pp. 388-404.

Rekik, M., Frija, A., Idoudi, Z., López Ridaura, S., Louahdi, N., Dhehibi, B., Najjar, D., Rudiger, U., Bonaiuti, E., Becker, L., Ghallem, Z. D., Cheikh M'hamed, H., Devkota Wasti, M. and Rischkowsky, B. 2021. Use of Conservation Agriculture in Crop-Livestock Systems (CLCA) in the Drylands for Enhanced Water Use Efficiency, Soil Fertility and Productivity in NEN and LAC Countries - Progress Highlights: Year (3) - April 2020 to March 2021. Lebanon: International Center for Agricultural Research in the Dry Areas (ICARDA).

Requier-Desjardins, M. and Bied-Charreton, M. 2006. *Evaluation des coûts économiques et sociaux de la dégradation des terres et de la désertification en Afrique*, C3ED, Paris, France: AFD, 163 pp. Version révisée en mars 2009.

Richard, F. 2005. Le semis direct en Tunisie, La, *Gazette des SCV au Cirad*, 28 Nov. 2005.

Roe, S., Streck, C., Obersteiner, M., Frank, S., Griscom, B., Drouet, L., Fricko, O., Gusti, M., Harris, N., Hasegawa, T., Hausfather, Z., Havlík, P., House, J., Nabuurs, G., Popp, A., Sánchez, M. J. S., Sanderman, J., Smith, P., Stehfest, E. and Lawrence, D. 2019. Contribution of the land sector to a 1.5°C world, *Nature Climate Change* 9(11), 817-828.

Rouabhi, A., Dhehibi, B., Laouar, A., Houmoura, M. and Sebaoune, F. 2016. Adoption perspectives of direct seeding in the High Plains of Sétif-Algeria, *Journal of Agriculture and Environmental Sciences* 5, 53-64.

Rouabhi, A., Laouar, A., Mekhlouf, A. and Dhehibi, B. 2018. What are the factors affecting no-till adoption in the farming system of Sétif Province in Algeria? *Turkish Journal of Agriculture - Food Science and Technology* 6(6), 636-641.

Rouabhi, A., Laouar, A., Mekhlouk, A. and Dhehibi, B. 2019. Socioeconomic assessment of no-till in wheat cropping system: a case study in Algeria, *New Medit* 18(1), 53-64.

Ryan, J. 2011. Rainfed farming systems in the West Asia & North Africa (WANA) region, in Tow, P. G., Cooper, I. M., Partridge, I. and Birch, C. J. (Eds), *Rainfed Farming Systems*, Dordrecht Netherlands: Springer, pp. 65-393.

Ryan, J., de Pauw, E., Gomez, H. and Mrabet, R. 2006. Drylands of the Mediterranean zone: biophysical resources and cropping systems, in Peterson, G. A., Unger, P. W. and Payne, W. A. (Eds), *Dryland Agriculture*, 2nd edn., Madison, WI: American Society of Agronomy Monograph, vol. 23, pp. 577-624.

Sanz, M. J., de Vente, J., Chotte, J.-L., Bernoux, M., Kust, G., Ruiz, I., Almagro, M., Alloza, J.-A., Vallejo, R., Castillo, V., Hebel, A. and Akhtar-Schuster, M. 2017. Sustainable Land Management contribution to successful land-based climate change adaptation and mitigation, *A Report of the Science-Policy Interface*, Bonn, Germany: United Nations Convention to Combat Desertification (UNCCD).

Sanz-Cobeña, A., Lassaletta, L., Aguilera, E., del Prado, Ad, Garnier, J., Billen, G., Iglesias, A., Sánchez, B., Guardia, G., Abalos, D., Plaza-Bonilla, D., Puigdueta-Bartolomé, I., Moral, R., Galán, E., Arriaga, H., Merino, P., Infante-Amate, J., Meijide, A., Pardo, G., Álvaro-Fuentes, J., Gilsanz, C., Báez, D., Doltra, J., González-Ubierna, S., Cayuela, M. L., Menéndez, S., Díaz-Pinés, E., Le-Noë, J., Quemada, M., Estellés, F., Calvet, S., van Grinsven, H. J. M., Westhoek, H., Sanz, M. J., Gimeno, B. S., Vallejo, A. and Smith, P. 2017. Strategies for greenhouse gas emissions mitigation in Mediterranean agriculture: a review, *Agriculture, Ecosystems and Environment* 238, 5-24.

Sartori, M., Philippidis, G., Ferrari, E., Borrelli, P., Lugato, E., Montanarella, L. and Panagos, P. 2019. A linkage between the biophysical and the economic: assessing the global market impacts of soil erosion, *Land Use Policy* 86, 299-312.

Sassi, K., Abid, G., Melaouhia, M., Jabnouni, A., Helal, R., Cheikh M'hammed, H. and Ben Hammouda, M. 2012. Effect of direct and conventional drillings on grain yield and quality of durum wheat, *Egyptian Journal of Agricultural Research* 90(4), 347-353.

Schilling, J., Freier, K. P., Hertig, E. and Scheffran, J. 2012. Climate change, vulnerability and adaptation in North Africa with focus on Morocco, *Agriculture, Ecosystems and Environment* 156, 12-26.

Selvaraju, R. 2013.Implications of climate change for agriculture and food security in the Western Asia and Northern Africa region, in Sivakumar, M. V. K., Lal, R., Selvaraju, R. and Hamdan, I. (Eds), *Climate Change and Food Security in West Asia and North Africa*, New York: Springer, pp. 27-51.

Simpfendorfer, S., McKay, A. and Ophel-Keller, K. 2019. New approaches to crop disease management in conservation agriculture, in Pratley, J. and Kirkegaard, J. (Eds), *Australian Agriculture in 2020: From Conservation to Automation*, Wagga Wagga: Agronomy Australia and Charles Sturt University, pp. 173-188.

Sims, B., Corsi, S., Gbehounou, G., Kienzle, J., Taguchi, M. and Friedrich, T. 2018. Sustainable weed management for conservation agriculture: options for smallholder farmers, *Agriculture* 8(8), 0, doi:10.3390/agriculture8080118.

Sivakumar, M. V. K., Ruane, A. C. and Camacho, J. 2013. Climate change in the West Asia and North Africa region, in Sivakumar, M. V. K., Lal, R., Selvaraju, R. and Hamdan, I. (Eds), *Climate Change and Food Security in West Asia and North Africa*, New York: Springer, pp. 3-26.

Souissi, A., Annabi, M., Benyoussef, S., Chakroun, M. and Cheikh M'hamed, H. 2018a. Assessment of the use of conservation agriculture on durum wheat yield, water and nitrogen use efficiencies and soil health, *FARA Research Report*, 3(6), 1-13.

Souissi, A., Bahri, H., Cheikh M'hamed, H. and Annabi, M. 2018b. Une méta-analyse sur les effets de la fertilisation azotée sur le rendement et l'efficience de l'utilisation de l'azote chez le blé dur en Tunisie, *Annales de l'INRAT* 91, 62–85.

Stavi, I. and Lal, R. 2015. Achieving zero net land degradation: challenges and opportunities, *Journal of Arid Environments* 112, 44–51.

Sun, W., Canadell, J. G., Yu, L., Yu, L., Zhang, W., Smith, P., Fischer, T. and Huang, Y. 2020. Climate drives global soil carbon sequestration and crop yield changes under conservation agriculture, *Global Change Biology* 26(6), 3325–3335, doi:10.1111/gcb.15001.

Taibi, H. H. Y., Smail-Saadoun, N., Labidi, S., Abdellaoui, K., Makhlouf, M., Laouar, A., Benouaret, C., Rezki-Sekhi, L., Boukais, A. B. and Sahraoui, A. L. 2020. The Influence of no-till farming on durum wheat mycorrhization in a semi-arid region: a long-term field experiment, *Journal of Agricultural Science* 12(4), 77–96.

Tanji, A., El Gharras, O., Mayfield, A. and El Mourid, M. 2017a. On-farm evaluation of integrated weed management in no-till rainfed crops in semi-arid Morocco, *African Journal of Agricultural Research* 12(16), 1404–1410.

Tanji, A., El Gharras, O., Ouabbou, H. and Mladen, T. 2017b. Weed dynamics in no-till rainfed crops in Chaouia, Semi-Arid Morocco, *Journal of Agricultural Science* 9(11), 198–208.

The World Bank 2014a. *Réalisation de l'étude sur le semis direct dans le cadre du projet d'intégration du changement climatique dans la mise en œuvre du plan Maroc Vert (PICCPMV). Phase 1: Diagnostic analytique de la situation actuelle du système de semis direct (SSD)* (Document no: 11706-N711-13a), Washington, DC: The World Bank.

The World Bank 2014b. *Réalisation de l'étude sur le semis direct dans le cadre du projet d'intégration du changement climatique dans la mise en œuvre du plan Maroc Vert (PICCPMV). Phase 2 : Feuille de route pour le développement et promotion du semis direct au niveau des exploitations agricoles* (Document no: 943-N711-14b), Washington, DC: The World Bank.

Thomas, R. J. 2008. Opportunities to reduce the vulnerability of dryland farmers in central and West Asia and North Africa to climate change, *Agriculture, Ecosystems and Environment* 126(1–2), 36–45.

Tibbo, M. and van de Steeg, J. 2013. Climate change adaptation and mitigation options for the livestock sector in the Near East and North Africa, in Sivakumar, M. V. K., Lal, R., Selvaraju, R. and Hamdan, I. (Eds), *Climate Change and Food Security in West Asia and North Africa*, New York: Springer, pp. 269–280.

Touchan, R., Kherchouche, D., Oudjehih, B., Touchan, H., Slimani, S. and Meko, D. M. 2016. Dendroclimatology and wheat production in Algeria, *Journal of Arid Environments* 124, 102–110.

United Nations Economic and Social Commission for West Asia. 2019. *Status Report on the Implementation of Integrated Water Resources Management in the Arab Region: Progress on SDG indicator 6.5.1*. Beirut, Lebanon: United Nations Economic and Social Commission for West Asia.

United States Agency for International Development (USAID) 2016. Climate change risk in Sudan, in *Country Fact Sheet*, Washington, DC: United States Agency for International Development.

Vadon, B. and Marionneau, A. 2012. Le semis direct à l'essai (Maroc, Mali), *Afrique Agriculture* 2012, 34–35.

Vadon, B., Lamouchi, L., Elmay, S., Mahnane, S., Benaouda, H. and Elgharras, O. 2006. Organisations paysannes: un levier pour développer l'agriculture de conservation au Maghreb, *Options Méditerranéennes : Série A* 69, 87-99.

Zaghouane, O. and Makhlouf, M. 2011. Promoting no-till practices to stabilize durum wheat yields and sustain agricultural production in semi-arid regions of Algeria, in World Congress on Conservation Agriculture (WCCA), Brisbane, Australia.

Zaghouane, O., Abdellaoui, Z. and Houassine, D. 2006. Quelles perspectives pour l'agriculture de conservation dans les zones céréalières en conditions Algériennes ?, *Options Méditerranéennes: Série A* 69, 183–187.

Zaghouane, O., Abdellaoui, Z., and El Mourid, M. 2015. *L'agriculture de conservation: introduction et perspectives de développement du semis direct dans les systèmes de production céréaliers, une nouvelle expérience en Algérie*, ITGC-ICARDA-ACIAR. 209p.

Zdruli, P. 2014. Land resources of the Mediterranean: status, pressures, trends and impacts on regional future development, *Land Degradation and Development* 25(4), 373–384.

Zdruli, P., Kapur, S. and Çelik, I. 2011. Soils of the Mediterranean region, their characteristics, management and sustainable use, in Kapur, S., Eswaran, H. and Blum, W. E. H. (Eds), *Sustainable Land Management*, Berlin/Heidelberg: Springer, pp. 125–142.

Zouabi, O. and Peridy, N. 2015. Direct and indirect effects of climate on agriculture: an application of a spatial panel data analysis to Tunisia, *Climatic Change* 133(2), 301–320.

Zurqani, H. A., Mikhailova, E. A., Post, C. J., Schlautman, M. A. and Elhawej, A. R. 2019. A review of Libyan soil databases for use within an ecosystem services framework, *Land* 8(5), 82, doi:10.3390/land8050082.

Chapter 8

Conservation Agriculture in West and Central Africa

Lamourdia Thiombiano, Agriculture and Environmental Research Institute, Burkina Faso; Kalifa Coulibaly, Nazi BONI University, Burkina Faso; Saidi Mkomwa, African Conservation Tillage Network (ACT), Kenya; Kofi Boa, CA Center, Ghana; Robert B. Zougmoré, CCAFS/ICRISAT/WCA, Mali; Oumarou Balarabe, CIRAD, France; Traoré Salifou, Joseph Ki Zerbo University, Burkina Faso; Sylla Halimatu, Ministry of Agriculture, Burkina Faso

1 Introduction
2 Diversity of Conservation Agriculture practices within the region
3 Promoting Conservation Agriculture in the region
4 Conservation Agriculture adoption and spread in the region
5 The contribution of Conservation Agriculture to food production and security
6 The contribution of Conservation Agriculture to the environment and combatting climate change
7 Conclusions
8 Where to look for further information
9 References

1 Introduction

The Comprehensive Africa Agricultural Development Programme has been adopted by the African Union Commission to reform agriculture; its objectives are to minimize and eventually eliminate land degradation, improve yields, reduce burdens on smallholder farmers, improve productivity and profitability and enhance environmental conservation (NEPAD, 2003). In the West and Central Africa (WCA) region, agricultural production is mainly rain-fed and therefore sensitive to climatic variations and extreme weather events, particularly drought. Human pressure on land resources for tillage agriculture has led to the decline of humid and savannah forest and woodland vegetation cover as well as to soil and environmental degradation (Nana et al., 2014). Conservation Agriculture (CA) is increasingly seen as part of the solution to these challenges. The agroecological transition of agricultural systems is a strategic approach that could lead to more sustainable farming and food systems according to Cote et al. (2019). CA is

http://dx.doi.org/10.19103/AS.2021.0088.07

an alternative to the tillage-based green revolution in agriculture in improving household food security and sustainability (Nana et al., 2015; Kassam et al., 2017; Mkomwa and Kassam, 2021). CA can prevent soil loss, regenerate degraded land and improve soil productive capacity (Kassam et al., 2017; Mkomwa and Kassam, 2021). CA is now being promoted in 25 countries in Africa. In some countries such as South Africa, Malawi, Zambia, Ghana, Mozambique and Morocco, CA covers significant areas, benefitting farmers and contributing to food security.

The agro-ecological and socio-economic environment in WCA offers favourable conditions for CA practices and development. During the first Africa Congress on Conservation Agriculture held in Lusaka in March 2014, experts reported that the development of CA systems based on local technologies can achieve sustainable production intensification (Kassam et al., 2017). This was affirmed by the Second Africa Congress on Conservation Agriculture held in Johannesburg in October 2018 in the context of the Malabo Declaration and Agenda 2063 (Mkomwa and Kassam, 2021).

As stated by the Food and Agricultural Organisation of the United Nations (FAO, 2017), CA is based on the following three main principles:

- i continuous minimum mechanical soil disturbance (no-till seeding and weeding);
- ii permanent vegetative soil cover by plant biomass (of at least 30%);
- iii crop diversification through crop rotations and/or other methods such as cover crops.

As noted by Thiombiano and Malo (2009), the concept itself is not new in African smallholder farming. As an example, zero tillage and permanent soil cover have been in practice for centuries in the Congo Basin in Central Africa, while minimum soil disturbance combined with growing mixed crops and use of rotations are a common practice in the Sahel and dry savannah zones. However, the combined application of the three key principles of CA on a farm, supplemented by complementary production management practices, has been shown to have significant benefits in terms of yields, biodiversity preservation and delivery of ecosystem services. With current climate change threats, and the need to preserve natural resources, including soil, whilst improving yields, it is important to promote CA at various scales across all farming systems. The aim of this chapter is to examine the diversity of CA systems and practices within WCA, results in the field as well as options for upscaling and addressing climate change issues.

2 Diversity of Conservation Agriculture practices within the region

Soil and water conservation practices are part of a set of traditional technologies used over centuries in some parts of Africa. For example, no-tillage and minimum soil disturbance are age-old practices in the transition savannah zone of Ghana (Ghana Agricultural Sector Improvement Programme (GASIP), 2020). The three components of CA have been in use across semi-arid zones for a long time (Garrity, 2017) and have been developed during the postwar period. For example, the FAO (1993) reported the extensive research work on zero tillage and soil mulching done by Rattan Lal at the International Institute of Tropical Africa in the 1970s and 1980s.

Work on the modern version of CA within the WCA region began in 2001 led by the *Centre de Coopération Internationale en Recherche Agronomique pour le Développement* (

CIRAD), with the development of cropping systems using soil plant biomass cover in Northern Cameroon via the cotton company SODECOTON (*Société de Development du Coton from Cameroun*). In Burkina Faso, CA was introduced in 2001 by FAO working with national institutions (FAO, 2009), promoting integrated agricultural production systems (Réseau pour Organisation des Systémes - RODS) within its interdisciplinary actions (*Projet d'Appui aux Infrastructures Agricoles* - PAIA) approach. In 2006, cotton growing organisations such as SOCOMA (*Société Cotonnière du Gourma*) and GEOCOTON also promoted CA. The main constraint faced in these CA systems was weed infestation that significantly reduced yields. By 2015/16, farmers from at least 22 African countries had adopted CA systems. Among these countries, four belonged to WCA: Ghana, Burkina Faso, Senegal and Cameroon.

Research on CA in the WCA region began several decades ago. A key problem has been the gap in expectations between scientists focused on the long-term impact of CA on soil properties and sustainability of production systems, on the one hand, and the vision of farmers who were expecting immediate solutions to their need for food security, income and labour saving, on the other hand (Nana et al., 2014). This observation is also true for other regions. Kassam and Friedrich (2011) have also noted scientific research on CA often lags behind farmers' own discoveries, partly because CA comprises an interlinked set of practices that does not lend itself to easy scientific scrutiny through short-term research and reductionist approaches.

3 Promoting Conservation Agriculture in the region

There has been a gap between the commitment to CA and policy tools to promote it in practice. However, a few countries have broader initiatives to promote CA, including the following:

- The government in Burkina Faso is preparing a draft policy to boost smart agriculture and particularly agroecology which involves CA and related practices.

- In Ghana, the education and training system is promoting CA-related practices in the management of cocoa plantations, notably to address climate change issues and cocoa market needs.
- In Cameroon, attempts are being made by research organisations and the private sector to include CA practices in cotton-based systems as well as in tree-based plantations for more sustainable land management.
- The government in Gabon is encouraging conservation measures that preserve and protect natural resources using CA-related practices.

CA is similar to smallholder traditional practices and can be practised with lower input costs and labour requirements. It also has a huge impact on both production and ecological sustainability after only a few years of adoption. However, given its relatively recent adoption, education on CA is still only being offered occasionally in some colleges and universities as part of degree or diploma courses in agriculture and related fields. It is also taught within broader subjects such as sustainable production or sustainable development.

More recently, there has been a move to establishing CA-dedicated education institutes and training centres. For example, the Rwanda Institute for Conservation Agriculture (RICA, https://www.rica.rw/) was begun in 2017 and is the first institute in Africa offering a degree course specializing in CA. Other notable education and training efforts in WCA are those initiated by the African Conservation Tillage (ACT) Network under the Internation Fund for Agriculture Development - IFAD-financed Smallholder Conservation Agriculture Promotion Project (SCAP) in Western and Central Africa and the European Union-supported initiative, Agroecology-based Aggradation-Conservation Agriculture (ABACO): Targeting Innovations to Combat Soil Degradation and Food Insecurity in Semi-arid Africa. SCAP operated in Burkina Faso, Guinea and Niger from 2008 to 2011. ABACO targeted Burkina Faso in the WCA region from 2011 to 2014.

Following these initiatives, ACT is supporting the expansion and mainstreaming of formal CA education and training by supporting tertiary education curriculum development (Temu et al., 2018). It is also establishing a network of Conservation Agriculture Centres of Excellence (CA-CoEs) (ACT, 2017). One example of a WCA CA-CoE is the University of Nazi Boni (formerly University of Bobo Dioulasso) in Burkina Faso. Through the SCAP and the ABACO projects, the Universities of Ouagadougou and Nazi Boni in Burkina Faso have trained 19 students (7 BSc, 11 MSc and 1 PhD) in CA or related fields such as cropping systems, knowledge management and innovation networks (ACT, le Centre International pour la Recherche en Agroforesterie - ICRAF and CIRAD, 2012). These graduates are engaged in research, training or practice of CA in Burkina Faso, Guinea and Niger. Support for students included scholarships for studies in CA topics.

Over the past 20 years, several CA research and development projects in West Africa have generated new knowledge that has been shared using various platforms and publications. Notable platforms are the FAO-hosted Conservation Agriculture Community of Practice (CA-CoP) (http://www.fao.org/conservation-agriculture/resources/en/), the ACT library (http://www.act-africa.org/library.php?com=5) and the Africa CA Congresses Website (https://africacacongress.org/). The first and second Africa Congresses on CA have consolidated research experiences from across the continent summarised in condensed collections of papers (ACT, 2014, 2018) and in book publications (Kassam et al., 2017; Mkomwa and Kassam, 2021).

The Howard G. Buffett Foundation Centre for No-Till Agriculture (CNTA), in Amanchia, in Ghana (https://centrefornotill.org/#home), is dedicated to CA research and skills development, providing much-needed scientific evidence to inform curricula development. Equipped with farmer dormitories, land for practical learning and an irrigation system, the Centre trains farmers, Ministry of Agriculture staff and shares learning with hundreds of visitors to the centre every year. The Centre also offers on-farm technical support for the establishment of CA learning/demonstration fields which are supported by the GASIP and the Vibrant Villages Foundation (VVF) in the northern zone of the country. Currently, there are 66 permanent CA learning/demonstration plots established by GASIP and 3 plots established by VVF. Together, the 69 permanent learning plots have attracted over 3000 farmers who, in turn, are able to share the knowledge and skills acquired from the learning plots with their neighbouring farmers, following the farmer–farmer CA extension model.

The Deutsch Gesellschaft fuer Internationale Zusammenarbeit, Norwegian Agency for Development Cooperation, and the FAO, as well as other international and local organizations, have also supported CA promotion both through short-term technical cooperation projects and in establishing permanent CA training centres. A team of experts has been assembled to outline the curriculum for a CA course to be run at tertiary agricultural colleges which will then become a CA institute.

4 Conservation Agriculture adoption and spread in the region

The WCA region has been identified as an area where the potential benefits of CA are greatest and most likely to help farmers meet the challenges of sustainable intensification of production systems (Nana et al., 2014). There is a need for farmers to take up more sustainable, productive and profitable methods of production that do not damage the soil, land and environment (FAO, 2008).

The successful spread of CA in WCA requires that a number of constraints are overcome in a participatory manner. These constraints include the widespread perception amongst smallholder farmers in particular that tillage and/or slashing and burning techniques are indispensable for crop production. Soil cover is an essential part of the CA process. However, with food security as a major objective, many smallholder farmers are hesitant to invest scarce labour, land, seed and fertilizer in cover crops that do not result in something to eat or to sell immediately. In several places, there are other uses of crop residue such as for fuel, thatch and animal feed. Farmers also suffer from restricted access to relevant knowledge, inputs or credit.

There appeared growing interest in CA adoption within the WCA region, although this can be difficult to measure precisely. CA practices are being implemented in countries such as Benin, Burundi, Democratic Republic of the Congo, Guinea, Ivory Coast, Mali, Niger, Nigeria Rwanda and Senegal. Table 1 summarises what is known about the distribution of the estimated 245 510 ha under CA in WCA across seven countries, 95% of which is in Ghana.

4.1 Conservation Agriculture adoption in Burkina Faso

Research on the adoption of CA in Burkina Faso by Dabire et al. (2017) describes the promotion of CA systems by the government, FAO, CIRAD through research stations and farm demonstrations. Surveys conducted in the main intervention villages (Yilou, Sindri, Koumbia and Yé) show an average adoption rate of CA systems of 64% and the total area under CA practices estimated as 127 ha. The literature review on the Eastern region of Burkina Faso, which is the first region to have experienced large-scale and long-term CA interventions has revealed that the number of producers practicing could be around 400 for a total area of about 140 ha.

4.2 Conservation Agriculture adoption in Ghana

In response to concerns about soil degradation, by 1990, most farmers in the forest zone of Ghana had gone back to the use of no-till with mulch, a traditional practice that used to be very popular among cocoa farmers. This system was called 'proka' in the local Akan language.

According to Bonsu (2001), the earliest research on no-till in Ghana started in the late 1960s and was boosted by Mensah-Bonsu and Obeng (1979) who reported the beneficial effects of no-till on soil and water conservation. Other groups have continued this research into CA.

The Crops Research Institute (CRI), Soils Research Institute and the Savannah Agricultural Research Institute, part of the Council for Scientific and Industrial Research, together with various universities, have been involved in several CA-related research projects, including the following:

Table 1 CA practices promoted in the respective WCA countries, the extent of adoption, constraints and opportunities for scaling

Countries	Development organizations & zones	CA practices	Crops/value chains	Start year of CA project	Area under CA ha (2019)	Opportunities to scaling	Constraints
Burkina Faso	*Société Cotonnière du Gourma* in Eastern B/Faso		Cotton	2006	140	**High**. Cotton is still a priority crop	Low awareness
	Food and Agricultural Organisation and INERA in moist savannah - southwest Burkina Faso	Direct seeding (jab planters and animal-drawn seeders): cover crops & crop rotations	Crop-livestock integration, fodder and cover crops: (brachiaria, mucuna); other crops (cowpea, maize, cotton)	2002-2007	500	**High**. Lack of adequate supplies of good-quality livestock feed remains a challenge	High price of implements; weak local industry; inadequate credit for equipment; weak knowledge of comparative advantages
	African Conservation Tillage Network and MoA under Smallholder Conservation Agriculture Promotion Project IFAD-financed projects - central B/Faso	Zaï pits, cover crops and crop rotations	Cereals (millet, sorghum maize) and grain legumes	2009-2012	360		Low awareness
Burundi					200		

(*Continued*)

Table 1 (*Continued*)

Countries	Development organizations & zones	CA practices	Crops/value chains	Start year of CA project	Area under CA ha (2019)	Opportunities to scaling	Constraints
Cameroon	*Société de Development du Coton from Cameroun*				2000		
Congo DR	Catholic Relief Services & Howard G. Buffett Foundation in Maniema	Slash and mulch, dibble stick, cover crops, associations and rotations	Smallholders producing cassava, cereals, legumes and vegetables	2002	2060		
Ghana	Monsanto and Sasakawa Global 2000. National. Implemented by: Crops Research Institute (CRI), MOFA, University of Development Studies	Glyphosate-based no-tillage farming	Smallholders producing bananas, plantains, cereals, legumes and vegetables in forest, transition and Guinea savannah zones	1992–2005	226 352		High cost; limited supply of inputs; Limited awareness
	Sedentary Farming Systems Project. GTZ, Ministry of Food and Agriculture, Brong Ahafo Region	Cover crops, minimum tillage, integrated soil fertility management	Smallholders producing bananas, plantains, cereals, legumes and vegetables	1996–2004			
	Centre for No-Till Agriculture	Soil cover crops and seeds, mechanization services (including roller crimper) and crop rotations	Smallholders producing bananas, plantains, cereals, legumes and vegetables	2012		**High**: labour saving and precision of seed and fertilizer placement	

	MOFA, Research (CRI, Soils Research Institute, Savannah Agricultural Research Institute & Council for Scientific and Industrial Research-CRI), non-governmental organisation. National.	No-tillage, cover crops, crop rotations	Smallholders producing bananas, plantains, cereals, legumes and vegetables	2006 to date	
	West African Rice Company		Large and medium-scale	2018	500
	Technologies for African Agricultural Transformation in the Savannahs Initiative of the Savannah Agricultural Productivity Improvement Programme		Medium and large-scale	2018	8148
Niger	Indigenous in South East Niger (Maradi)	Farmer-managed natural regeneration of Faidherbia Albida	Millet, sorghum		5000
Rwanda	MoA-CIMMYT SIMLESA; Rwanda Institute for CA	No-tillage, cover crops, crop rotations, soil fertility management	Smallholder farmers producing cereals and grain legumes	2016	250
Estimated total area under CA in WCA (ha)					**245 510**

- Comparison of tillage systems involving bullock ploughing, tractor ploughing and manual ploughing using hoes.
- Control of spear grass (*Imperata cylindrica*) in the transition savannah zone using mucuna cover crops.
- Integration of multi-purpose cover crops into cropping systems.

CA practices have been promoted mostly by donor-funded projects within the Ministry of Food and Agriculture and the Department of Agriculture in the Ministry of Local Government, as well as by projects sponsored by non-governmental organizations.

Ekboir et al. (2002) described the introduction of the CRI-Sasakawa-Monsanto no-till package to farmers in the forest, transition and Guinea savannah zones in Ghana, estimating that no-till was used on 45 000 ha by 100 000 smallholder farmers. The CNTA (2019) estimated that the land area under smallholder CA in 2018 was 235 000 ha. Established plantation farms of cocoa, citrus, cashew, mango and oil palm are excluded from this estimate as those farms are traditionally managed as no-till farms. In a more recent study, GASIP (2020) found that approximately 87% of 1560 smallholder farmers interviewed in the Guinea savannah, Sudan savannah and transitional savannah agro-ecological zones of Ghana were practicing CA. This study found that CA practices offered smallholder farmers numerous opportunities including:

i reduction in soil erosion and maintenance of soil moisture, as reported by 61% of respondents;
ii long-term improvement in crop yields, as reported by 63% of the respondents;
iii improvement in household food security, as reported by 58% of respondents.

Large-scale and medium-scale CA farming started recently in Ghana. The West African Rice Company had 500 ha under CA at the end of 2020 and will be increasing this to 750 ha or 1000 ha in 2021. The Technologies for African Agricultural Transformation in the Savannahs Initiative of the Savannah Agricultural Productivity Improvement Programme had 76 farmers transforming a total land area of 8148 ha into CA at the end of 2020.

Over the past 25 years, farmers across the three major agro-ecological zones in Ghana (the dryland savannahs, forest-savannah transition and the forest) have been planting several food crops including bananas and plantains, cereals, legumes and vegetables under the CA system. About 30% of farmers planting ginger (a crop known traditionally planted under the full tillage system) have now moved to CA production.

The CNTA, established in Ghana's humid zone in 2012 and headed by Kofi Boa, is an NGO established to illustrate the benefits of CA to the general public and to offer CA training to farmers and extension officers. The Centre also specializes in the production of cover crop seeds for sale to farmers and has introduced mechanized CA services through its six service centres around the country. The mechanized CA service centres focus on the use of:

- the ripper or tine seeders (Fig. 1) to break soil surface crusts and hardpans;
- the roller crimper (Fig. 2) to manage vegetation cover without herbicides for no-till seeding, helping to form a carpet of biomass mulch on the soil;

Figure 1 Ripper.

Figure 2 Roller crimper.

- the no-till planter (Fig. 3) for planting cereals and legumes through the mulch.

These technologies save labour and make it easier to implement CA on a larger scale.

CNTA's experience is that mechanized CA is widely welcomed and needed by farmers but that use varies by region. Farmers in the northern savannah dryland areas prefer to rip their fields in the early years of implementing CA as most of their fields are bare and crusted with underlying hardpans from repeated ploughing and trampling by animals. Farmers in the more humid zone with higher biomass cover prefer the use of the roller crimper to manage no-till pre-sowing vegetation as a complete mulch cover. This is because they find it difficult to manage the heavy biomass without the use of fire. In between the dryland and the humid zones (i.e. transitional savannah areas), with some biomass availability and less crusted soils, farmers prefer to plant straight away with the no-till planter.

4.3 Conservation Agriculture adoption in Cameroon

The favourable climatic conditions and low population density in the sub-humid and humid zones of Cameroon offer favourable conditions for the production and conservation of biomass essential in CA. A review by Djamen et al. (2014) found that, in Cameroon and Ghana where CA is most developed, the most common practices were:

- cover cropping with a large diversity of species;

Figure 3 No-till planter.

- direct seeding under mulch provided by cover crops;
- cereal (maize or rice) grown in rotation with fodder crops or fallow enhanced by cover crops.

Higher rainfall is favourable for cultivation of a wide variety of cover crops such as *Brachiaria* sp., *Mucuna* sp., *Stylosanthes* sp., *Crotalaria* sp., *Lablab purpureus* and *Cajanus cajan*. In some cases, the rainfall pattern is bimodal, allowing two cropping seasons per year. It is possible to produce a large volume of crop biomass to ensure 100% soil cover. The low population density enables the implementation of fallow in rotations, including improved fallow using leguminous shrubs that can bolster nutrient supply through nitrogen fixation and nutrient cycling.

4.4 Conservation Agriculture adoption in other countries in the region

In several other countries of the region, there is growing interest in CA although there is shortage of data to document the situation. CA research and development is being conducted in countries such as Benin, Burundi, Democratic Republic of the Congo, Guinea, Ivory Coast, Mali, Niger, Nigeria, Rwanda and Senegal. Key developments are summarised below:

- Democratic Republic of the Congo: Under forest or/and in cassava and banana systems, soil cover is more permanent with minimum disturbance. This is combined with practices such as cover crops and rotations to produce effective CA production systems (Mulimbi et al., 2019). The area under CA is about 2060 ha.
- Rwanda: CA is increasingly recognized as a key foundation for climate-smart agriculture. The area of cropland under CA is about 250 ha. The presence of RICA should improve future adoption of CA systems.
- Burundi: As in Rwanda, the presence of RICA should promote CA adoption in Burundi, including annual cropping systems, annual cropping systems with trees and plantation systems such as coffee. The area of cropland under CA is about 200 ha.
- Mali: In Mali, farmers practise soil and water conservation practices such as zaï (traditional planting pits), half-moons (shallow pits to store runoff water) and contour stone bunds (to trap water). These practices are precursors of CA systems, together with other practices such as planting trees, grass strips, mulching, improved fallow, crop rotation, cereal-legume combined systems and regeneration of natural vegetation. These are all being promoted as elements of an integrated CA system.
- Senegal: Early CA practices adopted include mulching, rotations, legume-cereal associations, regeneration of agroforestry areas, prevention of slash and

burn practices during land clearing and livestock corralling on cropland. There has been promising research into manufacture and use of no-till seeders, for example, the animal-drawn Super-Eco seeder (Diakhaté et al., 2021).

- Niger: As in Mali, farmers have traditionally used soil and water conservation practices which predate CA. These include zaï and half-moon techniques, crop residue mulching, annual cropping with trees, intercropping, farmer-managed natural regeneration conservation practices. These practices can be grouped into different categories: Tree parks associated with annual cropping; assisted natural regeneration based on shrub pruning and green manure cover crops; and mulching, crop rotation and intercropping. These practices improve yields in millet- and sorghum-based cropping particularly in areas where rainfall is less than 600 mm. The area of CA cropland is about 5000 ha.
- Benin: CA adoption is being supported by design and fabrication of direct seeders (Dayou et al., 2021) Mechanized seeding with, for example, a working speed of 1.6 m per second will help farmers increase the area of direct sowing with vegetable cover. A recent example of a locally made two-row no-till seeder that can seed 0.9 ha per hour is described in the study of Dayou et al. (2021).

Table 1 gives estimated areas under CA practices in WCA in 2019. These data are gathered from expert estimates, reports and other publications.

5 The contribution of Conservation Agriculture to food production and security

It has been suggested that CA can improve farmer incomes and food security by increasing crop productivity while preserving soil health and the environment (Tittonell et al., 2012; Kassam et al., 2017; Mkomwa and Kassam, 2021). A key challenge is whether CA can meet the rising demand for food in WCA (Nana et al., 2015; Corbeels et al., 2020). Given the expected increase in population, a doubling of agricultural yields is required to meet future needs (van Ittersum et al., 2016). The large gap between potential and current yields points to the need for smart production systems to increase food production sustainably (Sanchez, 2015). Are substantial increases in crop yields possible with CA?

Crop yield increases in CA systems vary according to location and environmental conditions (Rodenburg et al., 2020). Depending on CA systems, crop types and environmental conditions, the effect on yields can be positive or negative (Nana et al., 2015). A key issue is that minimum soil disturbance on its own does not improve yields (Droppelmann et al., 2017). Many studies in different locations with mainly cereal crops have shown that no-tillage on its own is generally associated with lower yields compared to conventional tillage

(Rusinamhodzi et al., 2011; Mason et al., 2015; Buah et al., 2017; Naab et al., 2017). No-tillage cropping performs poorly in degraded soils of WCA semi-arid regions due to soil compaction impeding root development, water capture and retention. In severely degraded land (non-productive land) in the Sahel and Sudan zones of West Africa, soil compaction needs to be broken through sub-soiling and supplemented by soil and water conservation measures such as half-moon and zaï techniques to increase water infiltration and retention (Zougmoré et al., 2014; Nyamekye et al., 2018). Before efficient CA implementation is possible, an aggradation phase is essential for soil restoration (Lahmar et al., 2012).

No-tillage improves yields when it is associated with soil biomass cover and crop diversification. Managing soil biomass cover is a key CA practice in the agroecosystems of the Sahel and Sudan zones of West Africa due to low rainfall and soil productivity. Substantial improvements in crop yields have been consistently associated with retention of biomass from crops and crop diversification, especially in dry climates (Rusinamhodzi et al., 2011; Bayala et al., 2012). Crop residue cover improves yields by overcoming soil moisture and nutrient limitations, as seen in increase in crop yields after crop residue retention in drought and hot conditions (Thierfelder et al., 2017; Steward et al., 2018).

The combination of the three principles of CA (continuous minimum mechanical soil disturbance, permanent organic soil cover and crop diversification) is essential to improve soil quality and increase crop yields (Thierfelder et al., 2018). A certain level of residue is required to obtain a significant increase in crop yield, depending on soil conditions and climate. In the Sahel, mulching with millet stalks at a rate of 2 Mg ha^{-1} increased plant nutrient uptake and grain yield by 31% (Buerkert and Lamers, 1999). In the Sudano-Sahelian zone of West Africa, retention of more than 6 Mg ha^{-1} of biomass soil cover is required to obtain a substantial increase in yields (Nana et al., 2015). Adequate crop cover retention is critical in agropastoral systems to manage the competing use of crop residues to feed livestock (Giller et al., 2009; Ndah et al., 2014; Valbuena et al., 2015). In these mixed farming systems, a balanced and flexible integration of agropastoral practices can be essential during the initial years of transition to CA system, combining controlled grazing, residue retention and manure management, to allow equitable biomass allocation for competing uses.

Crop-livestock systems offer farmers a more diverse source of food and income (FAO, 2001a; Valbuena et al., 2015). Introducing trees and shrubs into cropping systems can increase the supply of biomass materials for mulching, reinforcing the positive effect of permanent plant cover on soil health and crop yields (Bayala et al., 2012; Garrity, 2017). The use of shrub- and tree-based ramial wood species resulted in equal or higher cereal yields as compared to conventional systems (Félix et al., 2018). The quality of residues, crop diversification and fertilization will influence crop performance. Poor-quality residues (low nutrient contents, high C/N ratio) lead to nutrient immobilization and limited crop nutrient uptake (Mupangwa et al., 2020). Crop diversification including N-fixing crops in association or in rotation seems essential to improve nutrient balances (Guébré et al., 2020a), soil function and crop yields (Rusinamhodzi et al., 2011).

Evidence of improved soil fertility and increased crop productivity with improved fallows have been observed across humid and sub-humid WCA. Improved fallow with *Callopogonium mucunoides* in northern Ghana, for example, significantly increased the yield of rice compared with natural fallows and chemical fertilizer treatments. Recent studies have also shown dry season livestock feeding from leaves of drought-tolerant tree/shrub legumes used in improved fallow systems provides good-quality green fodder and higher levels of crude protein and minerals to animals (Avornyo et al., 2018).

6 The contribution of Conservation Agriculture to the environment and combatting climate change

Good-quality CA systems have positive environmental impacts in terms of reduced soil erosion, improved water quality, enhanced biodiversity and reduced greenhouse gas (GHG) emissions, as discussed in Volume 2 of this series of books on CA. In the Sudano-Sahelian zone of West Africa, crop residue retention promotes soil macrofauna activity which enhances soil porosity and improves soil structure (Guébré et al., 2020b). It has also been reported that residue retention prevents nitrogen leaching by increasing NH_4-N availability over NO_3-N (Gnankambary et al., 2008; Guébré et al., 2020a). The amount of soil organic carbon (SOC) depends on the quantity of crop residues and their interaction with termite activity (Félix et al., 2018). In the West African Sahel where soil erosion is a serious problem, surface-mulched crop residues significantly reduced soil loss from erosion (Buerkert and Lamers, 1999). The use of residues for soil cover also reduces the need for damaging slash and burn practices.

CA has some potential environmental risks if the principles are not applied correctly. One issue is the reliance on herbicides in CA in Cameroon as discussed during the Symposium of Experts on Conservation Agriculture in WCA in 2005. However, CA systems do not generally use more herbicides than conventional tillage systems, and the use of agrochemicals in CA systems, both for crop protection and crop nutrition, reduces over time (Friedrich and Kassam, 2012; Goddard et al., 2021). More recently, research is showing that biological methods of integrated pest control are possible to manage weeds, insects and pathogens. A good example of this is the push-pull system of integrated pest management applied in CA systems (Khan et al., 2017).

Agricultural practices that increase organic inputs enable additional carbon storage. CA contributes to the reduction in GHG emission by reducing energy expenditure in the system and by sequestering carbon. CA therefore makes it possible to go from a situation of a net loss of soil organic matter and carbon emissions to a situation of gain in the form of carbon storage in the soil. According to the Intergovernmental Panel on Climate Change (2006), reducing or eliminating tillage increases the storage of SOC. Tillage destroys soil structure, thus contributing to the decomposition of organic matter in soil. Elimination of tillage would limit this decomposition and promote stabilization and enhancement of carbon in soil with associated environmental benefits.

Climate change is expected to lower the yields of major crops in the WCA region by 2050 (Jalloh et al., 2013). Smallholder farmers in the tropics and subtropics will be less able to cope with changes in climate because they have far fewer adaptation options than farmers in temperate zones (Brown and Funk, 2008). As shown in Volume 2, combining no-till practice with residue retention for soil cover and diversified cropping increases the potential for carbon sequestration by increasing plant biomass inputs to the soil, including directly from roots. Permanent soil cover with crop residues or mulch provides a constant source of fresh organic material, some of which is converted into stable carbon fractions that can remain in the soil for millennia. Increasing soil carbon only mitigates climate change if it represents an additional net transfer of carbon from atmosphere to land. In the case of retaining crop residues on the soil surface, the balance depends on the extent of alternate demands for the residue material. There is also the possibility of adding significant amounts of carbon to the soil through root biomass, and this requires designing effective multipurpose diversified cropping systems. Land uses and agricultural practices that increase organic inputs allow additional carbon storage. CA makes it possible to go from a net loss of organic matter and carbon emissions to a gain in the form of carbon storage in the soil (Mkomwa and Kassam, 2021). Two examples of such synergies for the WCA region are described in the following paragraph.

Various studies have analysed CA systems and practices from the Sahel-Soudan zones in terms of climate change adaptation and mitigation as well as increased agricultural resilience (Zougmoré et al., 2014; Zougmoré, 2018). They have demonstrated that zaï and half-moon practices of soil and water conservation can be considered climate-smart options. These practices contribute effectively to rehabilitation of degraded land, therefore improving the resilience of smallholder farming. Zaï and half-moon practices improve water storage in soil, increase decomposition and nutrient release, resulting in improved crop performance in semi-arid conditions, thus allowing farmers to better adapt to climate variability. Crop harvests of up to 1.5 Mg ha^{-1} can be reached in years of good rainfall, making it more attractive to farmers to restore degraded lands (Kaboré and Reij, 2004).

Bayala et al. (2011) have shown that zaï and half-moon techniques also favour regeneration of local tree and other plant species through seeds contained in manure used to fertilise plants, with regenerated trees improving soil carbon sequestration and crop production. Several studies have reported the re-establishment of over 20 herbaceous species and 15 woody species on formerly bare soil following two consecutive years of zaï soil management in the central part of Burkina Faso. The increase in vegetation cover in the Sahel-Sudan zones is associated with an increase in soil total carbon, especially in sandy soils.

The favourable conditions in humid and sub-humid WCA zones for cover cropping and improved fallow make these two practices an important part of a CA-based climate-smart production system. The improvement of water infiltration and reduction of evaporation, coupled with runoff and erosion control provided by soil cover and a no-tillage system, increase the stock of available water and the depth of wetting front, improving the stock of groundwater. There is also growing evidence that improved fallows reduce pests and diseases.

Although more field-level data from long-term experiments are required to assess regional GHG budgets in absolute terms, the integration of tree/shrub legumes into cropping systems does not seem to increase GHG emissions. Verchot et al. (2007) reported that improved fallows of *Inga edulis* and *Acacia mangium* on sandy soils did not appear to decrease the soil C sink or increase CO_2 and NO_2 emissions. A review by Partey et al. (2017) reported a significant increase in soil organic matter with planted fallows of *Tephrosia vogelii* in Cameroon, as well as *Tephrosia candida* and *Cajanus cajan* in Nigeria. An increase of soil C stocks in the top 5-cm soil depth by about 1.5 mg C ha^{-1} with a 2-year fallow with *Cajanus cajan* was reported in South-eastern Nigeria. A study in southern Ghana by MacCarthy et al. (2018) found that CO_2 emissions were higher in clay soil compared to sandy soil. In the clay soil environment, the highest average CO_2 emission was observed from cattle kraals (enclosures for livestock) (256.7 mg m^{-2} h^{-1}), followed by forests (146.0 mg m^{-2} h^{-1}) and rice paddy fields (140.6 mg m^{-2} h^{-1}). The lowest average emission was observed for maize-cropped land (112.0 mg m^{-2} h^{-1}). On sandy soil soils, the highest average CO_2 emission was observed for soya-cropped land (52.5 mg m^{-2} h^{-1}), followed by forests (47.4 mg m^{-2} h^{-1}) and woodlots (33.7 mg m^{-2} h^{-1}).

7 Conclusions

Progress in CA adoption and practices in WCA vary from country to country, depending on agroecological zones. Various traditional soil and water practices fit well within CA systems. CA systems have been found to promote biodiversity, diversify production, promote more efficient use of resources and preserve natural resources in all the ecological zones, from the Sahel-Sudan zones in the north to the sub-humid and forest zones in the south. Integrating traditional practices into CA systems could accelerate the uptake of CA in the region.

Although the potential exists, it is important to acknowledge the need for policy measures and actions to boost the adoption and spread of CA systems across the WCA region. Researcher could play a prominent role in providing the evidence to convert key stakeholders into supporting the spread of CA. An appropriate monitoring and evaluation system needs to be put in place in individual countries to track progress.

The future of CA in WCA also calls for sustained collaborative efforts to find local solutions and innovations - beyond the physical soil and water conservation structures of *zaï* pits, half-moons and stone bunds - to new soil carbon sequestering practices and regenerative management methods to maximise and sustain productivity benefits. Successful case studies of what can be achieved include (amongst others) the extremely dry conditions of Western Australia where CA is practised in conditions of 250 mm or less of rainfall a year (Jat et al., 2014), or the restoration of degraded Savannah and forest soils in the cerrados of Brazil to create productive agricultural land (Jat et al., 2014; Derpsch et al., 2010).

Successful examples of CA practice need to be promoted through appropriate communication channels. Key challenges include the following:

i robust advocacy and showcasing of CA success stories to policy makers at all levels to develop appropriate policy tools and regulations to facilitate upscaling of CA;
ii developing an appropriate combination of CA-related practices that are more attractive and profitable in the short term for farmers;
iii researching and demonstrating the links between CA, food security and nutrition to facilitate more investment, including mechanization;
iv ensuring that CA is fully part of the curricula for young farmers in training centres and for agronomy students in agricultural schools and universities;
v involving the private sector in the promotion of CA, moving from research trials on smallholders' farms to larger-scale mechanized farms which have more impact on the economy and value chains;
vi promoting CA as an agroecological system and a way to adapt to and mitigate climate change, for example, by providing data on its contribution to increased carbon stocks.

These and other measures will help CA practices to spread for the benefit of the whole region.

8 Where to look for further information

A general source of information on CA globally is the FAO website (http://www.fao.org/conservation-agriculture) and for Africa is the ACT website (http://www

.act-africa.org/), CIMMTY website (https://www.cimmyt.org/tag/conservation-agriculture/) and CIRAD website (https://www.cirad.fr/) and its open registry of CIRAD publications Agritrop (https://agritrop.cirad.fr/572218/). More information regarding CA in Africa can be found on websites of national and regional organizations including universities and in books and journals. Books include: Kassam et al. (2017) and Mkomwa and Kassam (2021) and Volumes 1 and 2 of this book.

9 References

ACT (2014). Book of condensed papers. First Africa Congress on Conservation Agriculture, held in Lusaka Zambia. Available at: http://www.africacacongress.org/presentations/book-of-condensed-papers.

ACT (2017). CA centres of excellence brochure. Available at: www.act-africa.org/lib.php?com=5&res_id=250.

ACT (2018). Making climate-smart agriculture real in Africa with conservation agriculture: supporting the Malabo declaration and Agenda 2063. Book of Condensed Papers. Second Africa Congress n Conservation Agriculture. Johannesburg South Africa. 9th–12th October 2018. Available at: https://www.google.com/search?client=firefox-b-d&q=1ACCA+condensed+papers+book.

ACT, ICRAF and CIRAD (2012). Smallholder CA Promotion (SCAP) in Western and Central Africa. Grant No. 873. ICRAF/IFAD Programme. SCAP Project - Final Report.

Avornyo, F. K., Zougmoré, R., Partey, S. and Tengan, K. (2018). Candidate fodder trees and shrubs for sustainable ruminant production in northern Ghana. *Livestock Research for Rural Development* 30, 154. Available at: http://www.lrrd.org/lrrd30/9/favor30154.html.

Bayala, J., Kalinganire, A., Tchoundjeu, Z., Sinclair, F. and Garrity, D. (2011). Conservation Agriculture with trees in the West African Sahel – a review. *ICRAF Occasional Paper No. 14*. World Agroforestry Centre, Nairobi.

Bayala, J., Sileshi, G. W., Coe, R., Kalinganire, A., Tchoundjeu, Z., Sinclair, F. and Garrity, D. (2012). Cereal yield response to conservation agriculture practices in drylands of West Africa: a quantitative synthesis. *Journal of Arid Environments* 78, 13–25.

Bonsu, M. (2001). A report on CA in Ghana. A Country Report Prepared for FAO Regional Office for Africa, Accra.

Brown, M. E. and Funk, C. C. (2008). Climate: food security under climate change. *Science* 319(5863), 580–581.

Buah, S. S. J., Ibrahim, H., Derigubah, M., Kuzie, M., Segtaa, J. V., Bayala, J., Zougmore, R. and Ouedraogo, M. (2017). Tillage and fertilizer effect on maize and soybean yields in the Guinea savannah zone of Ghana. *Agriculture and Food Security* 6(1), 17.

Buerkert, A. and Lamers, J. P. A. (1999). Soil erosion and deposition effects on surface characteristics and pearl millet growth in the West African Sahel. *Plant and Soil* 215(2), 239–253.

Centre for No-Till Agriculture (CNTA) (2019). Annual report.

Corbeels, M., Naudin, K., Whitbread, A. M., Kühne, R. and Letourmy, P. (2020). Limits of conservation agriculture to overcome low crop yields in sub-Saharan Africa. *Nature Food* 1(7), 447–454.

Cote, F.-X., Poirier-Magona, E., Perret, S., Rapidel, B., Roudier, P. and Thirion, M.-C., (Eds.) (2019). *The Agroecological Transition of Agricultural Systems in the Global South, Agricultures et Défis Du Monde Collection, AFD,CIRAD*. Editions Quae, Versailles, 360 p.

Dabire, D., Andrieu, N., Djamen, P., Coulibaly, K., Posthumus, H., Diallo, A. M., Karambiri, M., Douzet, J.-M.. and Triomphe, B. (2017). Operationalizing an innovation platform approach for community-based participatory research on conservation agriculture in Burkina Faso. *Experimental Agriculture* 53(3), 460–479.

Dayou, D. E., Zokpodo, B. L. K.,Montcho, M.and Kakai, G. L. R. (2021). Design, fabrication and assessment of direct seeder in Benin Republic. *AgricEngInt. CIGR Journal Open Access* 23(1), 138–147. Available at: http://www.cigrjournal.org.

Derpsch, R., Friedrich, T., Kassam, A. and Li, H. (2010). Current status of adoption of no-till farming in the world and some of its main benefits. *International Journal of Agricultural and Biological Engineering* 3(2), 1–25.

Diakhaté, D., Sarr, I., Júnior, D. S., Havard, M., Ralisch, R. and Fall, A. (2021). Evaluation of the technical capacity of artisans to fabricate the animal powered direct seeder super-eco in Senegal. In: Mkomwa, S. and Kassam, A. H. (Eds.) *Conservation Agriculture in Africa: Climate Smart Agricultural Development*. CABI Publishing, Wallingford, UK. pp. 217–223.

Djamen, N. P., Dugué, P., Mkomwa, S., Da, S. J. B., Essecofy, G., Bougoum, H., Zerbo, I., Ganou, S., Andrieu, N. and Douzet, J.-M. (2014). Conservation agriculture in West and Central Africa. In: Jat, A. R., Sahrawat, L. K. and Kassam, A. (Eds.). *Conservation Agriculture: Global Prospects and Challenges*. CAB International, UK. pp. 311–338.

Droppelmann, K. J., Snapp, S. S. and Waddington, S. R. (2017). Sustainable intensification options for smallholder maize-based farming systems in sub-Saharan Africa. *Food Security* 9(1), 133–150.

Ekboir, J., Boa, K. and Dankyi, A. A. (2002). *Impacts of No-Till Technologies in Ghana*. CIMMYT, Mexico D.F.

FAO (1993). Soil tillage in Africa: needs and challenges. FAO Soils Bulletin 69. Soil Resource Management and Conservation Services. Land and Water Development Division, Rome, Italy.

FAO (2001a). Conservation agriculture case studies in Latin America and Africa. Introduction. FAO Soils Bulletin No. 78. Food and Agriculture Organization, Rome.

FAO (2001b). *Mixed Crop-Livestock Farming*. Food and Agriculture Organization, Rome.

FAO (2008). *Investing in Sustainable Agricultural Intensification. The Role of Conservation Agriculture. A Framework for Africa*. Food and Agriculture Organization, Rome, Italy.

FAO (2009). Enhancing crop-livestock systems in Conservation Agriculture for sustainable production intensification: a farmer discovery process going to scale in Burkina Faso. *Integrated Crop Management* Vol. 7. Food and Agriculture Organization, Rome.

FAO (2017). *Conservation Agriculture*. Leaflet: http://www.fao.org/3/a-i7480e.pdf.

Félix, G. F., Clermont-Dauphin, C., Hien, E., Groot, J. C. J., Penche, A., Barthès, B. G., Manlay, R. J., Tittonell, P. and Cournac, L. (2018). Ramial wood amendments (Piliostigma reticulatum) mitigate degradation of tropical soils but do not replenish nutrient exports. *Land Degradation and Development* 29(8), 2694–2706.

Friedrich, T. and Kassam, A. (2012). No-till farming and the environment: do no-till systems require more chemicals? *Outlooks on Pest Management* 23(4), 153–157.

Garrity, D. (2017). How to make conservation agriculture evergreen. In: Kassam, A. H., Mkomwa, S. and Friedrich, T. (Eds), *Conservation Agriculture for Africa: Building Resilient Farming Systems in a Changing Climate*. CABI Publishing, Wallingford, UK.

GASIP (2020). Ghana agricultural sector improvement programme. Profiling and Characterization of Conservation Agriculture Practices and Adoption in Guinea Savannah, Sudan Savannah and Transitional Agro-Ecological Zones of Ghana.

Giller, K. E., Witter, E., Corbeels, M. and Tittonell, P. (2009). Conservation agriculture and smallholder farming in Africa: the heretics' view. *Field Crops Research* 114(1), 23-34.

Gnankambary, Z., Bayala, J., Malmer, A., Nyberg, G. and Hien, V. (2008). Decomposition and nutrient release from mixed plant litters of contrasting quality in an agroforestry parkland in the south-Sudanese zone of West Africa. *Nutrient Cycling in Agroecosystems* 82(1), 1-13.

Goddard, T., Kassam, A. and Mkomwa, S. (2021). Moving paradigms - conservation agriculture with alternative agronomics to minimize inputs. In: Mkomwa, S. and Kassam, A. (Eds), *Conservation Agriculture in Africa: Climate Smart Agricultural Development*. CABI Publishing, Wallingford, UK.

Guébré, D., Traoré, S., Hien, E., Somé, D., Bationo, B. A. and Wiesmeier, M. (2020a). Soil macrofaunal activity, microbial catabolic limitations and nutrient cycling in cropping systems amended with woody residues and nitrogen inputs. *Pedobiologia* 83, 150686.

Guébré, D., Traoré, S., Traoré, M., Mare, B. and Hien, E. (2020b). Conservation des sols en zone soudano-sahélienne : quelle est l'efficience des amendements ligneux? *Sciences Naturelles et Agronomie* 39, 199-220.

Jalloh, A., Faye, M. D., Roy-Macauley, H., Sérémé, P., Zougmoré, R., Thomas, T. S. and Nelson, G. C. (2013). An overview of West African agriculture and climate change. In: Jalloh, A., Nelson, G. C., Thomas, T. S., Zougmoré, R. and Roy-Macauley (Eds). *West African Agriculture and Climate Change: A Comprehensive Analysis*. International Food Policy Research Institute (IFPRI), Washington, DC. pp. 1-35.

Jat, R. A., Sahrawat, K. L. and Kassam, A. H. (Eds.) (2014). *Conservation Agriculture: Global Prospects and Challenges*. CABI Publishing, Wallingford.

Kabore, D. and Reij, C. (2004). The emergence and spreading of an improved traditional soil and water conservation practice in Burkina Faso. Environment and Production Technology Division, IFPRI, Washington, DC.

Kassam, A. and Friedrich, T. (2011). Mainstreaming CA: challenges to adoption, institutions and policy. 5th World Congress of CA, 25-29 September. Brisbane, Australia.

Kassam, A. H., Mkomwa, S. and Friedrich, T. (2017). *Conservation Agriculture for Africa: Building Resilient Farming Systems in a Changing Climate*. CABI Publishing, London. Wallingford, UK.

Khan, Z. R., Midega, C. A. O., Pittchar, J. O., Murage, A. and Pickett, J. (2017). Climate-smart push-pull: a conservation agriculture technology for food security and environmental sustainability in Africa. In: Kassam, A., Mkomwa, S. and Freidrich, T. (Eds), *Conservation Agriculture in Africa: Climate Smart Agricultural Development*. CABI Publishing, Wallingford, UK.

Lahmar, R., Bationo, B. A., Dan Lamso, N., Guero, Y. and Tittonell, P. (2012). Tailoring conservation agriculture technologies to West Africa semi-arid zones: building on traditional local practices for soil restoration. *Field Crops Research* 132, 158-167.

MacCarthy, D. S., Zougmoré, R. B., Akponikpè, P. B. I., Koomson, E., Savadogo, P. and Adiku, S. G. K. (2018). Assessment of greenhouse gas emissions from different land-use systems: a case study of CO2 in the southern zone of Ghana. *Applied and Environmental Soil Science* 2018, 1-12.

Mason, S. C., Ouattara, K., Taonda, S. J.-B., Palé, S., Sohoro, A. and Kaboré, D. (2015). Soil and cropping system research in semi-arid West Africa as related to the potential for conservation agriculture. *International Journal of Agricultural Sustainability* 13(2), 120-134.

Mensah-Bonsu and Obeng, H. G. (1979). Effects of cultural practices on soil erosion and maize production in the semi-deciduous rainforest-savanna transitional zones of Ghana. In: Greenland, D. J. and Lal, R. (Eds.) *Soil Physical Properties and Crop Production in the Humid Tropics*. John Wiley, Chichester. pp. 509-519.

Mkomwa, S. and Kassam, A. H. (Eds.) (2021). *Conservation Agriculture in Africa: Climate Smart Agricultural Development*. CABI Publishing, Wallingford, UK.

Mulimbi, W., Nalley, L., Dixon, B., Snell, H. and Huang, Q. (2019). Factors influencing adoption of conservation agriculture in the Democratic Republic of the Congo. *Journal of Agricultural and Applied Economics* 51(4), 622-645.

Mupangwa, W., Thierfelder, C., Cheesman, S., Nyagumbo, I., Muoni, T., Mhlanga, B., Mwila, M., Sida, T. S. and Ngwira, A. (2020). Effects of maize residue and mineral nitrogen applications on maize yield in conservation-agriculture-based cropping systems of Southern Africa. *Renewable Agriculture and Food Systems* 35(3), 322-335.

Naab, J. B., Mahama, G. Y., Yahaya, I. and Prasad, P. V. V. (2017). Conservation agriculture improves soil quality, crop yield, and incomes of smallholder farmers in North Western Ghana. *Frontiers in Plant Science* 8, 996.

Nana, P. D., Dugue, P., Mkomwa, S., Da Sanson, J. B., Essecofy, G., Bougoum, H., Zerbo, I., Ganou, S., Andrieu, N. and Douzet, J.-M. (2014). Conservation agriculture in West and Central Africa. In: Jat, R. A., Sahrawat, K. L. and Kassam, A. H. (Eds), *Conservation Agriculture: Global Prospects and Challenges*. CABI Publishing, Wallingford, UK. pp. 311-339.

Nana, P. D., Andrieu, N., Zerbo, I., Ouédraogo, Y. and Gal, P. (2015). Agriculture de conservation et performances des exploitations agricoles en Afrique de Œ Ouest. *Cahiers Agricultures* 24, 113-122.

Ndah, H. T., Schuler, J., Uthes, S., Zander, P., Traore, K., Gama, M. S., Nyagumbo, I., Triomphe, B., Sieber, S. and Corbeels, M. (2014). Adoption potential of conservation agriculture practices in sub-Saharan Africa: results from five case studies. *Environmental Management* 53(3), 620-635.

NEPAD (2003). *Comprehensive Africa Agriculture Development Programme (CAADP)*. NEPAD, Midrand, South Africa.

Nyamekye, C., Thiel, M., Schönbrodt-Stitt, S., Zoungrana, B. J.-B. and Amekudzi, L. K. (2018). Soil and water conservation in Burkina Faso, West Africa. *Sustainability* 10(9), 3182.

Partey, S. T., Zougmoré, R. B., Ouédraogo, M. and Thevathasan, N. V. (2017). Why promote improved fallows as a climate-smart agroforestry technology in sub-Saharan Africa? *Sustainability* 9(11), 1887. https://doi.org/10.3390/su9111887.

Rodenburg, J., Büchi, L. and Haggar, J. (2020). Adoption by adaptation: moving from conservation agriculture to conservation practices. *International Journal of Agricultural Sustainability*, 1-19. (Ahead of Print. Open Access.)

Rusinamhodzi, L., Corbeels, M., van Wijk, M. T., Rufino, M. C., Nyamangara, J. and Giller, K. E. (2011). A meta-analysis of long-term effects of conservation agriculture on maize grain yield under rain-fed conditions. *Agronomy for Sustainable Development* 31(4), 657-673.

Sanchez, P. A. (2015). En route to plentiful food production in Africa. *Nature Plants* 1, 14014.

Steward, P. R., Dougill, A. J., Thierfelder, C., Pittelkow, C. M., Stringer, L. C., Kudzala, M. and Shackelford, G. E. (2018). The adaptive capacity of maize-based conservation agriculture systems to climate stress in tropical and subtropical environments: a meta-regression of yields. *Agriculture, Ecosystems and Environment* 251, 194–202.

Temu, A. B., Kasolo, W., Mkomwa, S. and Kassam, A. (2018). Conservation agriculture: a curriculum guide for Africa. ACT and ANAFE. Nairobi, Kenya. Also Available at Conservation Agriculture: Conservation Agriculture: A Curriculum Guide for Africa.

Thierfelder, C., Baudron, F., Setimela, P., Nyagumbo, I., Mupangwa, W., Mhlanga, B., Lee, N. and Gerard, B. (2018). Complementary practices supporting conservation agriculture in southern Africa: a review. *Agronomy for Sustainable Development* 38, 1–22.

Thierfelder, C., Chivenge, P., Mupangwa, W., Rosenstock, T. S., Lamanna, C. and Eyre, J. X. (2017). How climate-smart is conservation agriculture (CA)? – its potential to deliver on adaptation, mitigation and productivity on smallholder farms in southern Africa. *Food Security* 9(3), 537–560.

Thiombiano, L. and Malo, M. (2009). *Scaling up Conservation Agriculture in Africa: Strategies and Approaches*. The Food and Agriculture Organization Sub Regional Office for Eastern, Africa. 30 p.

Tittonell, P., Scopel, E., Andrieu, N., Posthumus, H., Mapfumo, P., Corbeels, M., van Halsema, G. E., Lahmar, R., Lugandu, S., Rakotoarisoa, J., Mtambanengwe, F., Pound, B., Chikowo, R., Naudin, K., Triomphe, B. and Mkomwa, S. (2012). Agroecology-based aggradation-conservation agriculture (ABACO): targeting innovations to combat soil degradation and food insecurity in semi-arid Africa. *Field Crops Research* 132, 168–174. https://doi.org/10.1016/j.fcr.2011.12.011.

Valbuena, D., Tui, S. H.-K., Erenstein, O., Teufel, N., Duncan, A., Abdoulaye, T., Swain, B., Mekonnen, K., Germaine, I. and Gérard, B. (2015). Identifying determinants, pressures and trade-offs of crop residue use in mixed smallholder farms in Sub-Saharan Africa and South Asia. *Agricultural Systems* 134, 107–118.

van Ittersum, M. K., van Bussel, L. G. J., Wolf, J., Grassini, P., van Wart, J., Guilpart, N., Claessens, L., de Groot, H., Wiebe, K., Mason-D'Croz, D., Yang, H., Boogaard, H., van Oort, P. A. J., van Loon, M. P., Saito, K., Adimo, O., Adjei-Nsiah, S., Agali, A., Bala, A., Chikowo, R., Kaizzi, K., Kouressy, M., Makoi, J. H. J. R., Ouattara, K., Tesfaye, K. and Cassman, K. G. (2016). Can sub-Saharan Africa feed itself? *Proceedings of the National Academy of Sciences of the United States of America* 113(52), 14964–14969.

Verchot, L. V., Van Noordwijk, M., Kandji, S., Tomich, T., Ong, C., Albrecht, A., Mackensen, J., Bantilan, C., Anupama, K. V. and Palm, C. (2007). Climate change: linking adaptation and mitigation through agroforestry. *Mitigation and Adaptation Strategies for Global Change* 12(5), 901–918 https://doi.org/10.1007/s11027-007-9105-6.

Zougmoré, R., Jalloh, A. and Tioro, A. (2014). Climate-smart soil water and nutrient management options in semiarid West Africa: a review of evidence and analysis of stone bunds and zaï techniques. *Agriculture and Food Security* 3(1), 16.

Zougmoré, R. B. (Ed.) (2018). Promoting climate-smart agriculture Through water and nutrient interactions options in semi-arid West Africa: a review of evidence and empirical analysis. In: *Improving the Profitability, Sustainability and Efficiency of Nutrients Through Site Specific Fertilizer Recommendations in West Africa Agro-Ecosystems*. Springer, Cham. pp. 249–263.

Chapter 9

Conservation Agriculture in Eastern and Southern Africa

Saidi Mkomwa, African Conservation Tillage Network (ACT), Kenya; Sepo Marongwe, African Conservation Tillage Network (ACT), Zimbabwe; Raymond Nazare, Freelance Consultant, Zimbabwe; and Weldone Mutai, African Conservation Tillage Network (ACT), Kenya

1 Introduction
2 Background: Agriculture, livelihoods and climate change
3 Conservation Agriculture, Agenda 2063 and the Malabo Declaration
4 Conservation Agriculture as practised in Eastern and Southern Africa
5 Case studies on mainstreaming Conservation Agriculture in Tanzania and Zimbabwe
6 Conclusions and future prospects
7 Where to look for further information
8 Acknowledgements
9 References

1 Introduction

Smallholder farmers have been credited with providing up to 80% of the food consumed in a large part of the developing world, contributing significantly to poverty reduction and food security while creating the potential for smallholders to sustainably feed themselves (IFAD, 2013). Africa's impressive economic performance, sustained during the last decade, has inspired optimism about the realization of Africa's vision of a continent free from hunger and extreme forms of poverty.

However, according to De La O Campos et al. (2018), most of the extremely poor - about 80% - live in rural areas. Their incomes depend greatly on agricultural activities, either from work on their farms or agricultural wage employment. Increasing fragmentation of landholdings, coupled with reduced investment support and marginalization of small farms in economic and development policy, threatens this contribution, leaving many smallholders vulnerable to poverty. Smallholder farmers do also have a pivotal role as stewards of the environment and its ecosystems, capable of protecting or destroying services including soil fertility, fresh water delivery, pollination and pest control.

http://dx.doi.org/10.19103/AS.2021.0088.08

Looking ahead, the core question is whether today's agriculture and food systems are capable of meeting the needs of a global population and particularly that of Africa, which is expected to increase from 1.34 billion in 2020 to nearly 2.4 billion inhabitants by 2050 (UN, 2019). Can Africa achieve the required food production increases, even as the pressures on already scarce land and water resources and the negative impacts of climate change intensify? A further concern is therefore about whether Africa will be able to sustain growth in production intensification systems over the long term. This includes resiliency to recover from climatic or natural disaster shocks such as the coronavirus (COVID-19) pandemic, which has had a negative impact of between −2.1 and −5.1 on economic growth in 2020, from a modest 2.4% in 2019 in sub-Saharan Africa (World Bank, 2020).

Fortunately, the endorsement of the Comprehensive Africa Agriculture Development Programme (CAADP) by African Heads of State and Government in 2003 in Maputo, Mozambique, with the aim of transforming Africa's agriculture to ensure food and nutrition security in Africa, reduce poverty and create jobs, embraces a landmark vision to address the challenge. This was reinforced by the June 2014 African Heads of State and Government adoption of the Malabo Declaration (AUC, 2015) on Accelerated Agricultural Growth and Transformation for Shared Prosperity and Improved Livelihoods, with a wider mandate and more targeted approach to achieve the agricultural vision of shared prosperity and improved livelihoods for the continent. The specific goals and targets to be attained in 2025 include: recommitting to the principles of CAADP; enhancing investment finance in agriculture; ending hunger in Africa by 2025 (doubling productivity and halving the current level of post-harvest losses); halving poverty through improved agricultural growth and transportation; enhancing resilience in livelihoods and production systems to climate variability and other shocks; and commitment to mutual accountability for actions and results.

We can, without doubt, eliminate poverty and end hunger and malnutrition in Africa through massive agricultural transformations. It is possible for Africa to unlock its full agricultural potential - with about 300 days of sunshine each year, 65% of the world's remaining uncultivated arable land and cheap labour - to become a global powerhouse in food and agriculture, to feed not only its people but others in the world. The big question is, how to target and transform the millions of smallholders managing 1 ha farms to create wealth sustainably using their land and water resources.

The demand on agriculture to feed a larger and more urbanized population through global markets over the next 40 years will continue to grow, placing additional pressure on available land and other natural resources. Current tillage-based practises are undermining the ecological foundation of the global food system through overuse and by the effects of agricultural pollution, thereby enhancing degradation, reducing ecosystem capacity to generate sustainable yields and threatening to negatively impact food security and poverty reduction (Kassam et al., 2017; IFAD, 2013). Sustainable agricultural intensification can be the answer to enhanced food security, environmental protection and poverty reduction.

1.1 Agriculture in Eastern and Southern Africa

The Eastern and Southern Africa (ESA) region is home to some of the world's fastest-growing economies and to some of the poorest, displaying remarkably diverse levels of economic growth. Over 65% of the population of the region depends on farming (IFAD, 2021). It is the backbone of the sub regions' economies and plays a key role in their industrial development and trade. Unfortunately, this growth has not automatically led to a reduction in poverty. Instead, there has been an increase in the number of people living below the poverty line. This is partly because Africa has the largest proportion of young people, many of whom are unemployed, in the world. Yet, in recent years, there has been a decline in agricultural production triggered by drought and climate change.

Large areas of sub-Saharan African soils are affected by various types of degradation, including fertility decline. According to Gupta (2019), land degradation and declining soil fertility are a major problem. Inappropriate land management further increases loss of productivity of resource-poor farmers, in turn affecting their food security and livelihood. Weather and climate are important determinants of plant growth, water availability and soil replenishment and affect food security in the region. Evidence shows that Africa is warming all-year round and faster than the global average, and this is likely to continue (AMCEN, 2011). Predictions suggest a greater water stress, more frequent droughts and floods, greater variability in rainfall and increases in extreme weather events with many more negative consequences on agricultural productivity. This points to an overall increase in incidences of crop failure due to adverse weather, inevitably severely affecting the livelihoods of already vulnerable populations.

There is sufficient evidence that Conservation Agriculture (CA) offers a promising option, which, over time, can sustainably increase the productivity of smallholder production systems, as well as their profitability and resilience without compromising the environment (Thierfelder et al., 2015a; Kassam et al., 2017; Mkomwa and Kassam, 2021). CA systems are, however, not widely used in the ESA sub-region, partly because there is limited awareness, capacity, knowledge and policy support for the widespread use of the technologies.

This chapter aims to fill the CA knowledge gap through the search of a comprehensive answer to the following questions: is CA a promising alternative to ensure the sustainability of farming in ESA? What are the evidence, knowledge gaps and prospects for CA in ESA? The methodology used is literature review, lessons and results of several past and ongoing CA projects and programmes in which the authors have been involved in ESA. The chapter is organized into six main sections: overview of agriculture, livelihoods and climate change in ESA; CA, Agenda 2063 and the Malabo Declaration; CA practises, promotion and adoption; case studies from Tanzania and Zimbabwe; efforts being undertaken to change the status quo; conclusions and recommendations.

2 Background: Agriculture, livelihoods and climate change

While agriculture is the largest sector in the ESA region, employing 65% of Africa's labour force and accounting for over 30% of the region's GDP, economic growth has done little to reduce hunger in the region, where malnutrition levels are only five percentage points lower than in 1990 (IFAD, 2021). Maize, wheat, rice, millet, potatoes and cassava are the main agricultural trade commodities for the region, generating estimated annual trade revenues of USD50 billion.

The conventional tillage (CT)-based agriculture being practised has been accompanied by equally unprecedented degradation of agricultural land resources, biodiversity and the environment. This also includes degradation of many of the ecosystem functions and land-based societal services such as clean water; carbon sequestration, water and nutrient cycling; regulation of streams and rivers; control of erosion, runoff and floods; pollination; and control of insect pests and pathogens (MEA, 2005; Nkonya et al., 2016).

In modern as well as traditional agriculture, tillage is accompanied by maintenance of bare soils and low or no return of vegetative biomass to physically protect the soil surface and feed soil microorganisms. Tillage is also accompanied with cropping systems that are poor in diversity and therefore vulnerable to biotic and abiotic stresses. Under the dominant Green Revolution agriculture, intensive tillage is also accompanied by excessive application of fertilizers and pesticides, high use of fossil energy and heavy investment in mechanization for on-farm operation.

The demand to produce additional food is occurring within an environment of rising agricultural input costs and high levels of poverty and disease. In addition, this region constitutes part of the global regions where, predictably, the negative impacts of climate change will be most heavily felt. The Intergovernmental Panel on Climate Change (IPCC, 2018, 2019) projected that by the year 2050 crop losses in the region could reach 20% due to adverse temperatures. The region is also projected to experience higher frequencies of floods and droughts. Such impacts will affect an already vulnerable region which relies heavily on rain-fed agriculture.

Supportive climate-smart practises that complement land preparation activities include soil and water conservation structures and agroforestry. Some of the soil and water conservation structures, discussed in the next section, are indigenous and have been used for generations.

3 Conservation Agriculture, Agenda 2063 and the Malabo Declaration

3.1 Indigenous practises and soil and water conservation structures

Farmers in ESA have for generations been practising several traditional soil and water conservation techniques. Some of these techniques include the matengo pits *(Ngoro)* in the Mbinga district, Tanzania, and terracing in Ethiopia, Machakos Kenya and Lushoto, Tanzania. These techniques have shown immense benefits in terms of soil and moisture conservation for crops as well as fertility improvements (Temu and Bisanda, 1996). Under the *ngoro* system, for example, farm residues from the previous cropping season are arranged into square grids of around 1.5 m edge length and subsequently covered with soil dug up from the pits in the centre of the grids. The residues increase soil fertility while the pits retain rainwater, thus considerably increasing soil moisture.

Overall, these systems have succeeded in the control of erosion, maintenance of soil fertility and increase in productivity on the steep slopes of the adopting communities for more than a century. However, most of these traditional soil and water conservation practises being hand-hoe based are labour intensive, thus severely limiting the cropped land area and scaling.

3.2 Faidherbia albida

Faidherbia albida is an indigenous acacia tree that is widespread throughout the semi-arid regions of Africa. It is managed in fields as a 'fertilizer factory' tree which takes nitrogen out of the air and 'fixes' it into other forms available as plant parts (roots, stems and leaves), which are subsequently incorporated into the soil. Its pods and leaves are used as protein-rich livestock fodder and its wood is used for construction, as fuelwood and for charcoal. But what makes it special is its nitrogen-fixing properties and its unusual habit, known as 'reverse phenology'. It goes dormant and sheds its leaves during the rainy season. Its leaves regrow when the dry season begins. This makes it highly compatible with food crops because it does not compete with them for light, nutrients or water. On the contrary, annual crops in the vicinity of *Faidherbia* trees tend to exhibit improved performance and yield.

Farmers throughout the Sahelian region of Africa have been sustaining the trees in their sorghum and millet fields where it improved the performance of the crops planted nearby, as well as providing nutritious fodder for their livestock during the dry season from the leaves and pods. *Faidherbia* has long been an integral part of Sahelian agriculture, where farmers have nurtured and protected *Faidherbia* trees growing in their fields for centuries. Besides its prominence in the ESA countries of Ethiopia, Tanzania, Malawi and Zambia, the trees are a frequent component of the farming landscapes of Senegal, Mali, Burkina Faso, Niger, Chad, and Sudan. Interest in *Faidherbia* was further stimulated in the 1990s when farmers dramatically increased their efforts to regenerate and expand the tree populations on their farms, with a farmer-managed natural regeneration drive.

The *Faidherbia* tree has also been cultivated traditionally by farmers in various parts of Ethiopia, where it enhances cereal production up to 2800 m elevation in Tigray Province. It is equally valued by farmers in southern Africa, where it is estimated that about 500 000 farmers in Malawi and the southern highlands of Tanzania maintain *Faidherbia* trees in their maize fields to benefit their crops and livestock (Fig. 1).

The trees provide a natural form of fertilizer entirely free of charge through leaf fall at the beginning of the rains as the crops are planted. All that the trees require to thrive is sunshine during the dry season and sufficient moisture, which they obtain from their very deep root systems during the dry season after the crops are harvested.

Both Malawi and Zambia are encouraging farmers to establish *Faidherbia* trees in their maize farming systems, the aim being to double food production.

Figure 1 Sustained and productive maize farming system in an agroforest of *Faidherbia* trees in Madibira, Tanzania. *Source:* Saidi Mkomwa.

Agronomists in both of these countries' national programmes recommend that farmers establish 100 *Faidherbia* trees on each hectare of maize field. As the trees mature and develop a spreading canopy, they are gradually thinned down to about 25 trees per hectare. They are planted in a grid pattern to facilitate normal field crop operations, even with full mechanization. They may live to the ripe old age of 70–100 years, providing a lifetime of benefits for a very modest investment. The result is a sustained and productive maize farming system in an agroforest of *Faidherbia* trees.

3.3 Evolution of the modern Conservation Agriculture in ESA

In ESA, as in South America, commercial farmers and associated national and international agricultural research institutes caught the second wave of global interest in CA during the 1970s, encouraged by advances in the United States of America in the 1960s and the breath-taking increases in world oil prices. During the 1970s and 1980s, South African and Zimbabwean commercial farmers visited the United States of America and also launched research programmes on minimum tillage (Elwell, 1995). Zimbabwe's Agricultural Research Trust proved particularly influential among Zambian commercial farmers, who also sent farm delegations to the United States of America for study and established commercial contacts during the mid-1980s.

The predominantly hand-hoe CA package in Zambia and Zimbabwe represents a local variant of traditional minimum tillage technologies adopted in many parts of Africa (Reij et al., 2001; Haggblade et al., 2004). This involves the use of planting basins (shallow holes where the seed and fertilizers are placed), typically measuring about 30 cm long, 15 cm wide and 15 cm deep.

CA has been shown to be relevant and appropriate for small- and large-scale farmers at all levels of farm power and mechanization, from manually operated hand tools to equipment drawn by animals to operations performed by engine-powered machinery. However, its spread in Africa has been slower than desired compared to other continents for several reasons (Kassam et al., 2017; Mkomwa and Kassam, 2021).

3.4 The role of Conservation Agriculture in the Malabo Declaration and Agenda 2063

3.4.1 How do the Conservation Agriculture principles anchor benefits from technology practices?

3.4.1.1 Minimum mechanical soil disturbance

According to Kassam et al. (2017), the greatest weakness in the CT paradigm is exacerbated by the strident focus on yield as the main or only output of concern in an input-output production sense, not paying much attention to the ecological basis of high factor productivity (factor efficiencies) and agro-ecosystem resilience to biotic and abiotic stresses including climate variability and extreme events. Worst, even less attention is paid to soil health and ecosystem-related functions and services on-farm and from the landscape and agro-ecosystem as a whole.

Conservation Agriculture practises help to weather-proof production systems and build resilience in a number of ways, namely by rehabilitating and conserving natural resources, including reduction in soil erosion and compaction, and enhancing soil health; sequestering carbon; improving water cycling and water and nutrient use efficiency; timeliness in operation and improved cropping ratio (Kassam et al., 2017). All these aspects, along with genetically adapted and improved cultivars, operate together and synergistically in CA systems to weather-proof them and build and maintain system level resilience from ecological and socio-economic viewpoints. These are expressed in a number of ways such as improved and more stable yields and profit; reduced production inputs of seeds, water, nutrients, biocides and energy and greater farm output and return to investment (Rockstrom 2000; Kassam et al., 2009, 2013; Thierfelder et al., 2013, 2015a,b).

3.4.1.2 Soil cover with crop residues

The benefits of leaving crop residue on the soil surface through farm practises such as no till are numerous. In addition to reducing soil erosion, old crop residue will catch and retain moisture, mitigate soil temperature fluctuations during extremes and contribute to soil organic matter. These benefits are hard to measure and often lose out to short-term gains.

Data from No Till Farmer (2021) (https://www.no-tillfarmer.com/articles/10416-crop-residue-provides-many-benefits) indicates that, per tonne, wheat straw contains approximately 9 kg of nitrogen (N), 3 kg of phosphate (P_2O_5), 28 kg of potash (K_2O) and 2 kg of sulphur (S) valued at USD 12–14. Therefore, a 4 tonnes per hectare wheat crop will hold approximately USD 36–40 worth of nutrients. A maize crop stover with double the yield will have double the value of nutrients per hectare. These nutrients are lost when straw is removed from fields. The value of the nutrients does not include the cost of baling the residue or the intrinsic costs associated with the loss of residue and the role residue plays in protecting and building the soil.

However, the biggest cost of residue removal is the loss of carbon held in the residue. Carbon is one of the largest components of soil organic matter. The decomposition of old crop residue is part of the building process in soil organic matter. Higher levels of organic matter in soils correlate to improved soil structure, resilience, nutrient availability and water holding capacity. The carbon held in crop residue cannot be replaced through fertilizer applications like other nutrients. It has to be produced through the process of photosynthesis, residue decomposition and other biological processes, all of which take time. Soil quality isn't always easy to measure, and landowners and producers need to be compensated for practices that protect the soil, maintain its quality for long-term profit and our ability to feed ourselves.

3.4.1.3 Diversification of crops

Diversification of crops through rotations and associations has been reported to replenish soil fertility. This is particularly so when involved through sequences and/or intercropping with nitrogen-fixing legumes which transform atmospheric nitrogen through nitrogen-fixing bacteria in root nodules into biological forms of nitrogen compounds that are usable by plants. Increase in soil organic matter, also termed as 'recarbonization' due to CA, is the foundation for improved soil health resulting in numerous benefits including enabling crops to use the nutrients in the soil more effectively; helping to control weeds, diseases and pests by breaking their life cycles through the introduction of new crop species in the cropping system and reducing the risk of total crop failure in cases of drought and disease outbreaks.

3.4.2 Conservation Agriculture adaptations in ESA

Southern Africa has invented its own type of CA called 'planting basins', building on the indigenous Zai pits of West Africa and mechanization wisely, based on the predominantly hand-hoe tillage system. It is a compromise-reduced tillage system, losing some of the benefits acquired from not-till but benefiting from easy access and affordability. It is superior to the CT systems in many ways including precision farming - with optimized planting population, manure placement and timeliness of planting. The main drawbacks are that it is still labour intensive to dig the planting basins with the hand-hoe. The manageable farm size is thus restricted to much less than 1 ha.

4 Conservation Agriculture as practised in Eastern and Southern Africa

4.1 Adoption trends: lower and slower than desired/ anticipated adoption

Of the 205 million hectares under CA in the World in 2019/2021, Africa contributes 2.7 million hectares. The cropland under CA in Africa is 1.4%, about 70% of which is managed by large-scale commercial farmers, 1% by medium-scale farmers and the rest by smallholder farmers each with about 1 ha. In terms of numbers of farmers, the authors estimate that most (98%) of the CA farmers in Africa are smallholders. The number of countries in Africa opting to adopt CA as a core production component of climate-smart agriculture increased from 9 to 25 in the last decade (2008/09 to 2018/09), and likewise, the area under CA has also increased from 485 230 ha to 2 712 203 ha over the same period. This is most encouraging.

The huge benefits presented by CA to farmers, society and the environment create anticipations for faster and massive adoption of CA – way beyond the doubling of area in a decade. The targets set by the African Union under the Malabo Declaration Vision 25 × 25 (to have 25 million households practising CA by 2025) set the bar high, confirming the aspirations.

Eastern and Southern Africa has about 90% of the area under CA on the continent (2.4 million ha), with the average cropland under CA being 2.3%, with huge variations across countries (Table 1). South Africa has the largest area under CA (1 176 200 ha) followed by Zambia, Mozambique, Malawi and Zimbabwe. The lower than desired adoption rates are viewed as both an opportunity for anticipated massive scaling but also an opportunity to inform on the priority constraints which need to be addressed.

Ordinarily, assessment and comparison of adoption rates of CA require benchmarking with either comparable innovations/technologies or similar investments over a given timeframe. The usual rushed conclusions on the adoption rates of CA in ESA fail to recognize that CA is a system, comprised of sets of technologies, as opposed to a single technology or product. Even then, it would have added value to relate adoption of individual CA technologies with similar ones recently promoted in Africa – such as inorganic fertilizers, improved seeds, mechanization or soil conservation structures – before concluding. Such a co-relation should also justifiably reference the amount of investments and the duration of the promotions.

The magnitude and duration of investments in CA programmes is, in almost all cases, not stated when assessments on adoption rates are made. Highly localized short-term projects of say 3 to 4 years can successfully build capacity in CA for hundreds of farmers in isolated communities on selected CA technologies. However, it takes longer-term (6–10 year) programmes, along with entire value chains, to support smallholders transition from producing a surplus to adding value using say mechanization and efficiently utilizing saved time with enterprise diversification.

Leadership is needed at the pan-African level to coordinate the partnerships for the accurate documentation of resources that have been invested in the promotion of CA and gauge these as investments per farmer, per hectare or per country in a given period and compare these to adoption rates elsewhere. Our interim conclusion is that, with the exception of Zambia and South Africa,

Table 1 List of Eastern and Southern Africa countries that have adopted CA and cropland area (ha) under CA

	CA cropland area (ha)				Per cent of total cropland area
Country	2008/09	2013/14	2015/16	2018/19	2018/19
South Africa	368 000	368 000	439 000	1 176 200	9.48
Zambia	40 000	200 000	316 000	552 667	14.41
Mozambique	9000	152 000	289 000	289 000	3.08
Malawi		65 000	211 000	211 000	5.55
Zimbabwe	15 000	90 000	100 000	100 000	2.44
Kenya	33 100	33 100	33 100	33 100	0.52
Tanzania		25 000	32 600	32 600	0.21
Madagascar		6000	9000	9000	0.25
Uganda			7800	7800	0.09
Ethiopia				7500	0.04
DRC Congo				2060	0.02
Lesotho	130	2000	2000	2000	0.48
Namibia		340	340	800	0.10
Swaziland			1,300	800	0.42
Rwanda				250	0.02
Burundi				200	0.01
Total	**465 230**	**941 440**	**1 441 140**	**2 424 977**	**2.3**

Source: adapted from Kassam and Mkomwa (2021)

the investments to CA in Africa have been much lower than elsewhere in the world, and it is unjustifiable to expect dividends where no investments have been made.

4.2 Adoption by farmer typology

The majority of the CA farmers in Eastern Africa (EA) are smallholders managing 1 ha on average (Mkomwa et al., 2017). They constitute (99%) of the estimated total of 36 870 farmers in the sub-region. In terms of areas under CA, smallholders contribute some 22 748 ha, 36.6% of the nearly 62 000 ha in EA. There are close to 20 large-scale CA farmers in the EA sub-region. These have three to four decades of experience in CA, unlike their counterpart smallholders who have been engaged in CA only in the last 10 to 15 years. Large-scale farmers have transitioned from conventional inversion tillage driven by economic benefits on cost reductions with corresponding stabilized or increased yields. The major crops grown by smallholders are maize, sorghum and pulses, while the priorities for large-scale farmers are barley, sorghum, wheat, maize, beans, canola and safflower.

The situation in Southern Africa is the same as that in EA, with the exception of South Africa, where the use of CA is more prevalent in commercial than in smallholder agriculture. This can perhaps primarily be attributed to the very complex smallholder farming system and the factors influencing it, such as land tenure, lack of resources and infrastructure, etc., demanding a special holistic approach and skills to empower and adapt these farmers and their systems to CA (Midgley et al., 2015).

4.3 Delayed potential to contribute to the Malabo, The Africa We Want Vision 2063 and the UN SDGs.

The commitments under the African Union Malabo Declaration include (i) recommitment to the principles and values of the Comprehensive Africa Agriculture Development Programme (CAADP) process, (ii) enhance investment finance in agriculture, (iii) ending hunger by 2025, (iv) halving poverty, by 2025, through inclusive agricultural growth and transformation, (v) boosting intra-African trade in agricultural commodities and services (vi) enhancing resilience on livelihoods and production systems to climate variability and other related risks and (vii) mutual accountability to actions and results.

The impact of CA on food security (increase productivity), climate change resilience and mitigation of climate change risks will be realized when millions of the smallholder farmers in the region practise technologies based on the CA system. Attainment of the set targets under the Malabo Declaration Vision 25 × 25 requires more speed and investments in the implementation process.

The adoption of CA-based climate-smart agriculture (CSA) in Africa by 25 million smallholder and medium-scale farming households will permanently transform them into more productive and resilient livelihood status with improved soil health as the catalyser. Public and private sector institutional support will be required for the targeted value chains to sustain the mainstreaming processes involved. Key outcomes of the transformation and CA systems optimization include doubling of yields of staples even in the semi-arid systems with similar or reduced production costs; improvement in mother and child nutrition from the surplus food production and crop diversification with nutrient rich legumes; more equitable inputs markets creating rewarding employments in agribusiness and services; better rural employment created for mechanization service providers and processors; significant contribution to greenhouse gas emission reductions and carbon sequestration from reduced fuel consumption for agricultural production, carbon sequestration from no-till and reduced use of nitrogen fertilizers offset by biological nitrogen fixation with leguminous cover crops.

Women in the ESA region generally fare worse than their male counterparts on most social and economic indicators. Their unequal status is shaped by the interlocking factors of poverty and discriminatory treatment in the family and public life. Various studies indicate that CA has a reduced labour requirement as compared to conventional practices, thus making CA attractive to women who constitute a large proportion of small-scale farmers and farm labour force. Women engaged in CA have increased involvement in decision-making at household level, spanning from agricultural practises and crop use to household expenditure. Social status is heightened among women engaged in CA because of improved reliable production and income, with a larger proportion reporting self-confidence and an elevated status in their community and within their extended families.

At a higher level, adoption of CA at the magnitude of 25 million households will enable the African Union to attain the ambitions of Malabo Declaration goals and the SDGs.

4.4 Current adoption trends by value chains (maize, livestock, horticulture, rice, irrigated fields)

Although CA was primarily developed for the sustainable production of grain and other field crops (Midgley et al., 2015), the underlying principles of sound integrated management of natural resources for long-term sustainability of production equally apply to farming in the livestock and horticultural sub-sectors, with context-specific adjustments.

Most of the area under CA in ESA is cultivated with maize and wheat, with sparse inclusion of legumes such as soya beans as a rotational crop in large-scale farming or pigeon peas, beans and *Lablab purpureus* as intercrops in smallholder systems. Rice, the third most important staple in Africa, is not produced under CA systems, in either rain-fed or irrigated systems. CA in horticulture is a lucrative business opportunity, to serve steady and increasing demands for healthy foods, including the needs of the growing middle class in cities. In fruit production, all three of the principles are applicable to and should be adapted to each situation. Perennial crops such as orchards, for example, with a productive life-span of at least 25–30 years, benefit tremendously with the use of cover crop species in the rows, apart from minimum or no disturbance of the soil and permanent soil cover. However, the horticulture value chain is hardly addressed in CA programmes.

The same is true with the livestock value chain, where according to Mloza-Banda (2016), literature is replete with sharply divided and contradictory ideas of crop–livestock integration specifically versus CA on 35% of Africa's territory comprising savannah grassland deemed suitable for cultivation of crops and livestock rearing. When appropriately stocked and managed, livestock on the open veld or that under stall-feeding schemes have much to contribute to sustainable agriculture. Blignaut et al (2015) note that feedlot cattle are more profitable under CA than the conventional production system, mainly due to faster and more efficient growth. Livestock herds and flocks provide an invaluable safety net to poor households and poor women, contributing to their diverse livelihoods portfolios. Yet, CA is faulted for creating new hurdles where farmers must address trade-offs between using crop residues/cover crops to enhance soil properties for improving production of staple crops or to feed livestock in-field, to cope with fodder deficits during the long dry seasons (Andriarimalala et al., 2013).

To scale the use of CA, an understanding of the current agricultural value chains, their functioning, regulatory framework and constraints is essential and this raises a key question: What are the main shortfalls and deterrents in agricultural value chains, and why might CA be faced with challenges to feed into these existing structures, through which it could create a more inclusive and sustainable farming system for long-term food security?

4.5 Reasons contributing to the lower-than-desired adoption of Conservation Agriculture in ESA

The adoption of CA in Africa benchmarked with similar successfully adopted agricultural technologies/innovations with comparable investments and time frames is not considered low; however, it was lower than expectations. This is in consideration of the immense benefits being derived from the fewer than expected adopters. The reasons contributing to the lower-than-desired adoption rates, as described above, are summarized as follows:

1 continued promotion and development support of tillage-based agricultural systems by national and international, public and private institutions and sector industries;
2 weak policies, regulatory frameworks and institutional arrangements to support the promotion and mainstreaming of CA;
3 inadequate skills and competencies among farmers and other CA practitioners;
4 inability to disaggregate the opportunistic CA approaches for smallholder commercial versus smallholder subsistence farmers; this is compounded by the challenge that smallholder CA in Africa is synonymous to smallholder farming;
5 inadequate linkages of farmers to equitable markets to harness the market pull forces while overemphasizing on the CA technology (push);
6 poor availability and access to the required CA equipment, machinery and inputs;
7 inadequate innovation and enterprise diversification;
8 inadequate coordination and capacity development for CA.

5 Case studies on mainstreaming Conservation Agriculture in Tanzania and Zimbabwe

Mainstreaming is defined by Kassam and Mkomwa (2021) as the agricultural development context to the establishment of an ongoing national development process aimed at making the application of CA by farmers and their communities the new normal and for all relevant institutions and stakeholders to support this transformation. Such a change requires that the change process also involves aligning and mobilizing the support of all the key institutional stakeholders in the public, private and civil sectors to work together for this common objective.

This section analyses policy instruments from Tanzania and Zimbabwe dedicated to fostering science and technology and innovation in CA and the governance of its innovation system. The section explores the implications of current policies and governance on innovative performance by looking at the policies the governments have used to promote adoption and mainstreaming of CA to encourage the transition to make climate-smart agriculture real in Africa and their main limitations.

5.1 Tanzania's innovation policy on Conservation Agriculture: institutional reform and learning curve

A summary of the analysis on Tanzania's innovation policy on CA, the institutional reform and learning curve is presented schematically in Fig. 2.

The CA policy reform and development evolved in four main phases (1980–2003, 2004–2010, 2011–2014 and 2015–2016) marked by the periods at which strategic decisions, policy shifts and investments were taken. The reform of CA policy has taken an incremental approach, characterized by a progressively deeper understanding of policies, expected to lead to systemic transformation and institutional innovation.

The CA developments in the first phase (1980–2003) were on incubation and early experimentation targeting smallholder farmers. The promoters were non-governmental organizations (NGOs) supporting CA in ESA as part of their diverse objectives. These NGOs were not specialists in CA, which typically contained 10 to 20% of their portfolio. These NGOs were supported under programmes such as The Sasakawa Global 2000 in the 1990s, the 1990–1996 FAO Projects in EA, the 1996–1999 GTZ (Now GIZ) programmes through agricultural research institutes (ARI) Selian and Uyole and the 2000–2003 World Bank and ministry of agriculture soil fertility initiative implemented by ARI Mlingano, Uyole and Selian.

The second phase (2004–2010) was on early adoption under smallholder farming. It was supported by local NGOs but now with an interest in international agencies programmes with a varied CA agenda, such as agroforestry and pest control. The international agencies include FAO support for the Conservation Agriculture for Sustainable Agricultural Rural Development (CA-SARD), the Regional Land Management Unit based at the World Agroforestry Centre (RELMA in ICRAF); Swedish International Development Agency and the World Agroforestry Centre (ICRAF) on Conservation Agriculture with Trees (CAWT).

The third phase (2011–2014) saw deepening and scaling-up under smallholder farming, with the entry of more NGOs, international development programmes and agencies. The ability of early experimentation and incubation efforts in generation of food surplus and addressing food security attracted involvement of national governments and national agricultural research institutes. Some of the programmes included a consortium of north–south research networks led by ACT and funded by the European Union under the agroecology-based aggradation conservation agriculture (ABACO), the Australian Centre for International Agricultural Research (ACIAR) funded, Farm mechanization and Conservation Agriculture for Sustainable Intensification (FACASI) and the Alliance for a Green Revolution in Africa (AGRA) programmes with ACT. Efforts to bring on board the private sector to supply demanded equipment were only marginally successful. They expressed inadequate demand volumes to justify investments while farmers on the other hand lamented low quality.

The fourth phase, which is still on going from 2015, has seen the entry of medium- and large-scale farmers, the local manufacturing industry and increasing participation of the national agricultural research systems (NARS),

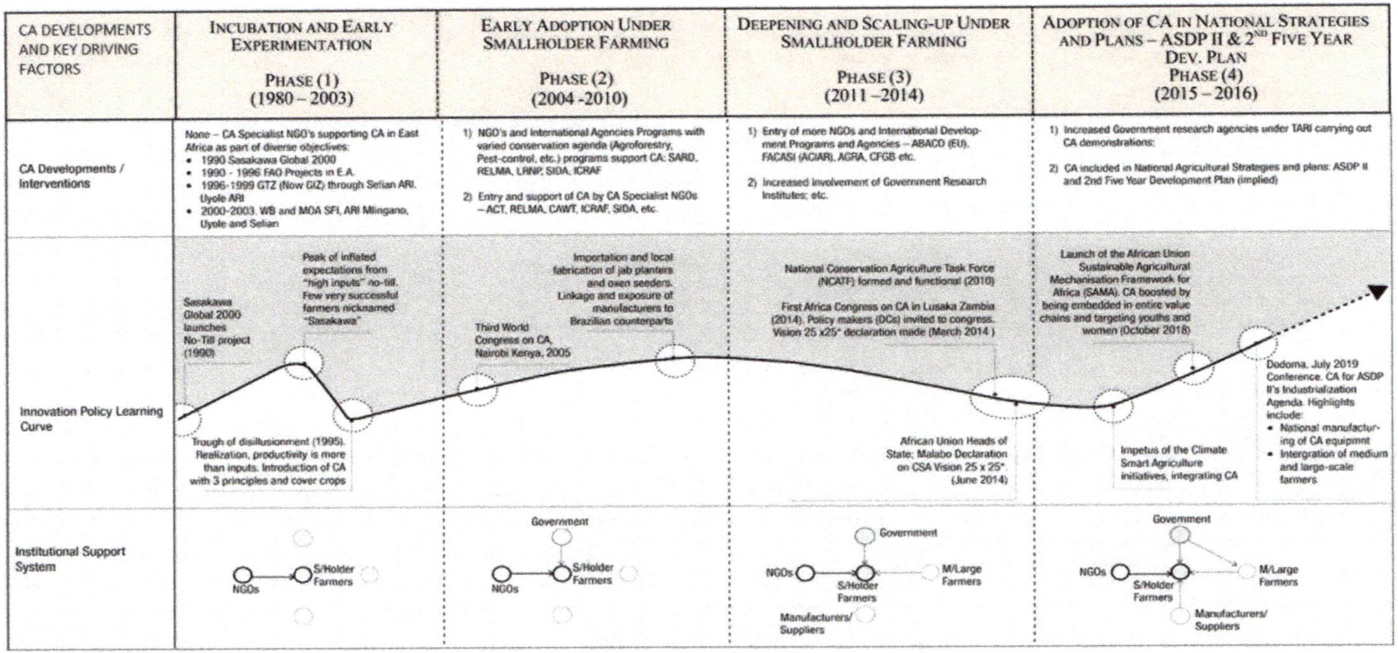

Figure 2 Tanzania's innovation policy on Conservation Agriculture: Institutional Reform and Learning Curve. Source: Mkomwa and Makungu (2019). ABACO: Agroecology based Aggradation Conservation Agriculture; CFGB: Canadian Foodgrains Bank; RELMA: Regional Land Management Unit; ARI: Agricultural Research Institute; FACASI: Farm Mechanization and CA for Sustainable Intensification; SFI: Soil Fertility Initiative; CIDA: Canadian International Development Agency; MOA: Ministry of Agriculture; Sida: Swedish International Development Cooperation Agency.

CA DEVELOPMENTS, KEY DRIVING FACTORS INCUBATION AND EARLY EXPERIMENTATION	INCUBATION AND EARLY EXPERIMENTATION PHASE (1) (1980 – 2003)	EARLY ADOPTION UNDER SMALLHOLDER FARMING PHASE (2) (2004 -2010)	DEEPENING AND SCALING-UP UNDER SMALLHOLDER FARMING PHASE 3 (2011 –2014)	ADOPTION OF CA IN NATIONAL STRATEGIES AND PLANS – ASDP II & 2ND FIVE YEAR DEVELOPMENT PLAN PHASE 4 (2015 – 2016)
Drivers and Value Propositions	Support mainly aimed as livelihood and poverty alleviation to targeted smallholder farm households	Support expands to other livelihood and conservation methods involving government research agencies	Entry of new drivers for CA adoption: i). Agricultural productivity and profitability (entry of local machinery manufacturers); and, ii). Need for land conservation	Broadening of perceived CA benefits – entry of: i). Lead-farmer extension; and ii). Medium-scale farmer service providers with 2W & 4W tractors
Adoption Barriers	1. Low awareness of CA 2. High labour requirements at CA introduction stage 3. Lack of mechanization support	1. Low awareness of CA 2. High labour requirements in at CA introduction stage 3. Lack of mechanization support	1. Lack of support mechanism to local manufacturers 2. "Newly introduced" machinery and equipment not well performing (Learning going-on)	1. Support mechanism for service providers not in-place; 2. Low operator knowledge 3. Low demand for service
Key Success Factors	Increased productivity and resilience against climatic factors	Increased knowledge of CA benefits both at international and local context attracting more NGO's and International organizations	Visible "long-term" benefits from early adopters Increased availability and access of CA mechanization machinery	Increased access to mechanization inputs from mechanization service providers and local manufacturers reducing labour requirements
Policy Environment Support	International agenda on poverty reduction focusing on livelihood of poor farmers in environmentally challenging areas	International agenda expanded to include land conservation	International agenda promoting success stories from other countries (Brazil, Zambia, etc.) on benefits and innovations (Brazilian manufacturers, etc.)	National policies giving new focus on agriculture and its linkage with industrial growth where CA is identified as input for modernization of agriculture in ASDP II and indirectly in 2nd Five Year Development Plan
Funding Instruments	NGOs	NGOs and International agencies	NGOs, International Agencies support; Start of Government support	Expanded NGOs and International Agencies support; Increased Government commitment

Figure 2 *(Continued)*

particularly the Tanzania Agricultural Research Institute (TARI) and the academia (Sokoine University of Agriculture and the Nelson Mandela African Institution of Science and Technology). CA has been recognized and included in national agricultural strategies and plans, namely the agricultural sector development programme (ASDP II) and also in the second 5-year development plan. Some NGOs and regional organizations, such as the African Conservation Tillage Network (ACT), restructured their mandates to include policy advocacy as exemplified by their role in the scaling-up of CA in EA project, in Ethiopia, Kenya and Tanzania, in partnership with the Canadian Foodgrains Bank (CFGB).

5.1.1 Innovation policy learning curve

The innovation policy learning curve is informed by the significant investments and deductions from their outcomes under the different phases. These included inflated expectations from 'high inputs' no-till under the Sasakawa Global 2000 programme, followed a few years later with a low trough of disillusionment with the realization that productivity is more than inputs. This 'justified' the introduction of CA with three principles and cover crops.

Hosting of the third world congress on CA in EA, in Nairobi, Kenya, in 2005, generated learnings that influenced policy decisions, including establishment of the FAO supported CA-SARD project in Kenya and Tanzania and the substantive policy support to CA by the Government of Zambia that has continued to date. Under the CA-SARD project, efforts were made for importation and local fabrication of jab planters and animal drawn seeders under business linkages and exposure of Eastern African manufacturers to Brazilian counterparts.

Another important landmark was the First Africa Congress on CA in Lusaka Zambia (2014), with the Vision 25 × 25 Declaration. Building on the Lusaka Declaration Vision 2025, the African Union Heads of State made the Malabo Declaration on CSA Vision 25 × 25 in June 2014. Tanzania was represented by policy makers (including a District Commissioner), who also learned from the field visits and participated in drafting the Lusaka Declaration.

The second Africa Congress on CA held in Johannesburg in October 2018 created opportunities for decision makers from Tanzania (directors in the ministry of agriculture, researchers, academia and parliamentarians) to comprehend the role of CA beyond food security to contributing to Tanzania's industrialization agenda. The interests culminated into the stakeholders' conference in the capital city Dodoma in July 2019. This was followed by another presentation to the special parliamentary committee on agriculture, livestock and water in January 2020. In both events, emphasis was on operationalizing the funding commitments for ASDP II, which included a sizeable amount for CA. It was also emphasized to link CA to Tanzania's industrialization agenda as a means to create demand for nationally manufactured CA equipment as well as integration of medium- and large-scale farmers.

5.1.2 Drivers and value propositions

The value propositions and drivers for CA adoption from this case study can be summarized as follows:

- In phase 1, support is mainly aimed at livelihood improvement and poverty alleviation to targeted smallholder farm households.
- In phase 2, support is expanded to other livelihood and conservation methods involving agenda government research agencies.
- Phase 3 has seen the entry of new drivers for CA adoption. These are: (i) agricultural productivity and profitability, with the entry of local machinery manufacturers; and (ii) the need for land conservation.
- Phase 4 has broadened the perceived CA benefits, with the entry of lead-farmer extension and medium-scale farmer service providers with more reliable sources of farm power (two-wheel and four-wheel tractors).

5.1.3 Key success factors

- Phase 1: increased productivity and resilience against climatic factors;
- Phase 2: increased knowledge of CA benefits both at international and local context attracting more NGOs and international organizations;
- Phase 3: visible long-term benefits from early adopters and increased availability and access of CA mechanization machinery;
- Phase 4: increased access to mechanization inputs from mechanization service providers and local manufacturers reducing labour requirements.

5.1.4 Policy environment support (including funding)

- International agenda on poverty reduction focusing on the livelihood of poor farmers in environmentally challenging areas. Funding through NGOs.
- Expansion of International agenda to include land conservation. Funding through NGOs and international agencies.
- International agenda promoting success stories from other countries (Brazil, Zambia, etc.) on benefits and innovations (Brazilian manufacturers, etc.). Funding from NGOs and international agencies, and start of government support.
- National policies giving new focus on agriculture, and its linkage with industrial growth where CA being identified as an input for modernization of agriculture in ASDP II and indirectly in the Second Five Year Development

Plan. Financing from expanded NGOs and international agencies support; increased private sector and government commitment.

Public research organizations are increasingly demanded to commercialize research results and bridge the gap between research and development and industrial activity. The large-scale commercial farmers in ESA, who are doing much better than the smallholders, have their own channels for knowledge and advice different from public research. Support for smallholders is mainly on inputs (improved seeds and inorganic fertilizers) but limited to natural resource management.

It is in this context that ACT established the network of Conservation Agriculture Centres of Excellence (CA-CoEs) to create platforms for research and training on CA innovations that are responsive to the needs of smallholders and which actively tap into the huge pool of technical expertise at public research institutions and universities.

5.2 Commercial and non-commercial approaches for promotion of Conservation Agriculture in Zimbabwe

5.2.1 Public and civil society promotion of Conservation Agriculture

CA in Zimbabwe has been promoted intensively by NGOs and civil society, focusing mainly on impacting household food security (Mazvimavi et al., 2010). Due to intensive technical and input support, number of farmers using manual CA practises increased rapidly between 2005 and 2012 (Fig. 3). A larger proportion of these were, however, receiving a basic input package to support them to apply CA principles on parts of their farms which created a high level of dependency (FAO, 2008). During the same period, coordination through the formation of a Conservation Agriculture Task Force supported by the FAO and the Government brought together numerous NGO partners and international organizations promoting CA with various funding sources but mainly from the European Union (EU), European Commission's Humanitarian Aid and Civil Protection department (ECHO) and the United Kingdom's Department for International Development (DFID) (FAO, 2011). The coordination platform facilitated effective documentation, reduced support over-laps within districts and provided information-sharing and learning platforms. Many research activities were also commissioned and supported through the Task Force. The main aspects of this approach used by most of the NGOs in Zimbabwe included the following:

1 Beneficiaries were mostly selected on the basis of their vulnerability (FAO, 2011). This approach created a stigma among many farming

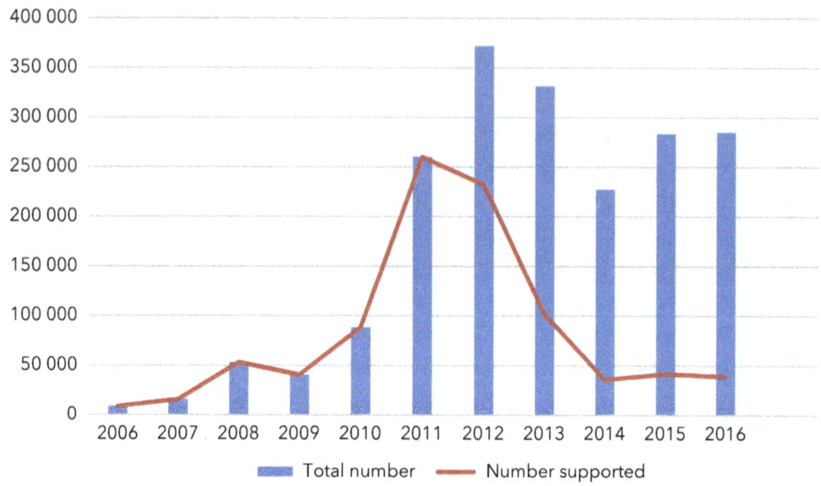

Figure 3 Conservation Agriculture adoption trends 20062016.Source: Zimbabwe CATF presentation at the Regional CA Dialogue, 24-28 /09/2019, Lusaka, Zambia.

communities as the manual CA systems were viewed as meant for the poorest households, who had very few assets and had no livestock, which is the main source of draft power in the smallholder farming sector. Consequently, the more resource-endowed farmers distanced themselves from the concept.

2 Farmers received inputs on the basis of having dug planting basins and the programme became just a free source of expensive inputs for some farmers who lacked the financial capacity to purchase their own inputs.
3 Training initially focused on the areas of operation for donor funded organizations, focusing solely on identified beneficiaries and their field officers, until it was internalized and mainstreamed into routine government programmes as a viable and sustainable option for farmers struggling with low productivity and high production costs.
4 To ensure donor programmes meet their targets, very prescriptive approaches were used as even those farmers who are able to expand their areas seem to stick to the prescribed small pieces of land.
5 The FAO, through the CA Task Force, also supported the training of government officers, which proved to be a very sustainable approach as many of whom went on to promote the CA concept in their areas of operation. According to statistics from Agritex in 2015, more than 50% of field extension officers were trained on CA principles.

The above approaches resulted in an exponential increase in the number of farmers applying basic CA practises. As the level of input support from NGOs declined after 2012, a significant proportion of the farmers who were applying the CA concept in order to access seed and fertilizer dropped the concept. However, there were still a sizeable number of farmers who continued to apply CA principles even beyond the input support (Fig. 3). The major lesson learnt is that the learning curve for CA technologies promoted from the inputs perspective is long but eventually yielding results. Furthermore, reasonable estimates need to be made on the proportion of farmers likely to adopt the innovation systems, in this case 70–80% of those reached. This is not a small proportion of adoption compared to other innovations of technologies. It is also prudent to compare the CA adoption rates with comparable benchmarks.

Some researchers classify, erroneously, farmers who are practising CA but still learning and receiving such free inputs as adopters. When such farmers drop out of practising CA due to withdrawing the inputs incentives, they are classified as dis-adopters. This is again an error, since they were not adopters yet in the first instance.

The academia has been active and loud, unfortunately, not in innovating and providing the much needed solutions to problems facing the smallholders but mostly in evaluating its non-performance. The misreporting by the academia and researchers has been substantively covered by Rolf Derpsch et al., 2014 'Why do we need to standardize no-tillage research?' ESA needs a lot more research for development in CA to address the numerous challenges faced by medium and smallholder farmers. However, the innovations and evaluations need to be on CA which qualifies for scientific comparisons with comparable technologies, methodologies and definitions.

According to the 2017 National Crop and Livestock Assessment, the number of smallholders practicing CA is 255 410 farmers in an area of 84 000 ha, which is 2.4% of the total cropland area. This proportion is very low, indicating the very high potential for scaling CA and optimizing its impact on national food security (Mazvimavi et al., 2010).

To increase and sustain the adoption of CA and enhance its impact on National Food Security, approaches for its promotion need to take into account the many lessons from the past. For many smallholder farmers where cereal crops were predominantly promoted under CA to enhance food security, the uncertainties in the cereal markets have found farmers who achieve increased production become stranded with their produce, especially when they lack adequate and quality storage facilities as it is the case in most instances (MLAWRR, 2020). Seed access for leguminous crops that are promoted in rotation in CA fields has also been a challenge due to the underdeveloped seed markets or the crops. Facilitating linkages to markets and identifying and addressing the bottlenecks in both the cereal and legume value chains would enhance the uptake of CA technologies.

5.2.2 Private sector promotion of Conservation Agriculture as a more efficient means of production

To significantly increase the area sown to CA crops in the large-scale farming sector, commercially oriented approaches such as re-branding CA as a more efficient means of production are likely to enhance uptake. The reduction in production costs by commercial farmers as a result of reduced tractor runs when no-till planters are used is a good selling point.

5.2.2.1 Production cost reduction

The shift from CT to CA systems presents significant cost savings in crop establishment operation costs. Table 2 presents a case in which all crop establishment operations are based on hired equipment and use standard rates currently followed in Zimbabwe.

CA systems have the potential to drastically reduce machinery costs for crop establishment operations by at least 60%. This is a significant cost saving with potential to contribute to the motivation to shift from CT to CA systems assuming the crop yield levels are maintained.

5.2.2.2 Improved timeliness of operations

Crop establishment operations in CT systems have traditionally combined ploughing, discing, planting and herbicide spraying. The standard work rates with an 80Hp tractor, three-disc plough, 3 m wide disc harrow, a four-row planter and a 9 m wide boom sprayer are given in Table 3. The assumed crop is maize.

The equation for total area (At) planned for a maize crop with this kit over total available time (Tt) is given by:

$$At = Tt\left(\left(1/WRplough\right)+\left(1/WRdisc\right)+\left(1/WRplant\right)+\left(1/WRspray\right)\right)$$

Table 2 Cost savings in crop establishment operations in Zimbabwe

Operation	Cost/ha (US$)	Area targeted for crop establishment (ha)	Conventional tillage system: crop establishment machinery costs on 54 ha (US$)	Conservation Agriculture system: crop establishment machinery costs on 54 ha (US$)
Ploughing	110	54	5940	0
Discing	60	54	3240	0
Planting	55	54	2970	2970
Spraying	30	54	1620	1620
			13 770	4590

where:

At is the total area on which the crop can be established within the given total time.
Tt is the total time available for crop establishment operations.
WRplough is the ploughing work rate (d/ha).
WRdisc is the discing work rate (d/ha).
WRplant is the planting work rate (d/ha).
WRspray is the spraying work rate (d/ha).

Using data in Table 3,

$$At = 26\left(\left(1/0.33\right)+\left(1/0.07\right)+\left(1/0.08\right)+\left(1/0.026\right)\right)$$
$$At = 54\ Ha$$

Working with a 26-day/1-month crop establishment window, the kit specified in Table 3 will, based on the equation, establish a total of 54 ha of maize.

A shift to CA will allow for elimination of the ploughing and discing operations and adoption of a no-till planter while retaining the boom sprayer. The resulting kit is specified in Table 4.

The new equation for total area (At) the kit can cover in a given total time (Tt) is expressed as:

$$At = Tt\left(\left(1/WRplant\right)+\left(1/WRspray\right)\right)$$
$$At = 26\left(\left(1/0.08\right)+\left(1/0.026\right)\right)$$
$$At = 260\ ha$$

The shift from CT to CA resulting in the elimination of the two very slow ploughing and discing operations results in the area that can be established under maize increasing from 54 ha to 260 ha with a lower cost machinery kit, that is, lesser use of plough and disc harrow.

Table 3 Data for estimating improved timeliness associated with the shift from CT to CA

Operation	Implement specifications	Working speed (m/s)	Work rate ha/8 h/day	Work rate Days/ha
Ploughing	Three-disc plough	1.5	3	0.33
Discing	3 m wide	2	14	0.07
Planting	Four row conventional planter with an inter row spacing of 0.9 m	2	12	0.08
Spraying	9 m wide boom sprayer	2	38	0.026

Table 4 Planning data for CA operations

Operation	Implement specifications	Working speed (m/s)	Work rate ha/8h/day	Work rate Days/ha
Planting	Four row no-till planter with an inter row spacing of 0.9 m	2	12	0.08
Spraying	9 m wide boom sprayer	2	38	0.026

Alternatively, the CA kit will establish the maize crop on the same 54 ha as the conventional kit but in just 5.4 days. This simple calculation indicates a five-fold increase in timeliness of crop establishment operations resulting from the shift from CT to CA systems for the maize crop.

A good entry point will be targeting farmers doing maize-wheat rotations. Many have delayed their wheat planting activities due to delays in maturity of long season maize varieties and the need to carry out land preparation (ploughing, harrowing) before wheat planting. The delayed planting negatively affects wheat yield and may also result in damage by early rains as maturity of the wheat crop may occur after the onset of the rainy season.

The close to 20 years of promoting CA in Zimbabwe has resulted in many lessons that can be utilized to increase and enhance sustainable impact of CA at the national level. There is a need to scale and increase the area put to CA, which has remained lower than potential despite the high numbers of farmers. This is mainly due to the small average area per farmer as a result of the nearly exclusive promotion of manual CA options that have been perceived to be labour intensive. The promotion of small mechanization and related technologies, which has made big strides, is yet to have an impact on area planted due to limited access to appropriate equipment resulting from the length of time required for machinery development processes and lack of confidence in the use of mechanical planters by smallholder farmers.

Although the Conservation Agriculture Framework 2010–2015 highlights the need to use market-oriented strategies for promoting CA, very little has been achieved due to the overall harsh economic environment that prevailed in the country for some time as well as due to the absence of an implementation strategy for the framework. Identification and addressing value chain bottlenecks along the CA production chain, including appropriate mechanization and legume crop value chains that are an integral part of CA systems, are critical. The medium-scale farmers with larger cropping areas at their disposal need to be targeted with parallel effort to facilitate access to finance for appropriate equipment as well as continued partnerships with the private sector and other value chain players for local research and development of small mechanization. This will ensure sustainable access and uptake of CA. Large-scale commercial farms, where appropriate no-till planters already exist, should be targeted to demonstrate benefits in time saving, cost saving and related efficiency in crop establishment that is achieved by the precision no-till equipment. The Government's *pfumvudza* CA programme, which targeted 1.6 million farmers by linking an existing government input support programme to the application of CA principles, has contributed immensely to creating awareness of CA across different farming sectors. The programme has, in the 2020/2021 season, attained maize yields averaging 5.28 metric tons per hectare and contributed to an overall 1.066 million tonnes, which is 40% of the 2020/21 total maize production. Training and support for the farmers to ensure the uptake is sustained will be critical moving forward.

6 Conclusions and future prospects

CA works faster and optimally when all three interlinked principles are used simultaneously to build the soil health basis for climate-proofing agriculture. Attainment of the Malabo Declaration and the Africa We Want Agenda 2063 with smallholders managing 1 ha as the majority requires mainstreaming of CA into public and private sector agendas, strategies and programmes. Mainstreaming should eventually attract investments from the private sector for farming and related agri-food systems as de-risked investment options, while national governments and the civil sector develop the institutions for an enabling environment and quality assurance.

The millions of smallholder farmers require support for coordination as food producers and stewards of the environment to enable their collective pooling of production inputs and produce, negotiation for equitable produce markets and advocacy for incentives for public and environmental services rendered. As a low-cost innovation that smallholders can relate to, CA empowers smallholders to produce a surplus at minimal cost, which subsequently enables them to penetrate markets and commercialize their farming systems without debt burden.

The largest opportunity for Africa lies in the answer to the question of how to reach and transform the millions of smallholders managing 1 ha farms to sustainably support livelihoods using their land and water resources. As a way of life or a social safety net for those unable to be employed in the formal sector, national governments have a duty to compensate such communities for the CA-based public goods and environmental services rendered at national and global levels. The payment modalities need to be structured to incentivize farmers to practise and adopt CSA. It is therefore imperative to transition CA from the short period and uncoordinated NGO operated projects to mainstream national strategies and programmes.

As a thriving business, even for the smallholders managing 1 ha, agriculture requires prioritization for greater investments from the private sector (including farmers), national governments and development partners.

This in-turn requires scientific evidence that the technologies, innovations and practises being fronted for scaling are economically profitable, socially and environmentally sustainable. Such empirical evidence needs to originate from real-size business operated model farms and Centres of Excellence, proving to investors that the CA-based climate-smart agriculture innovations are more resilient to climate change and entail less drudgery, thus appealing to women and youth and less risky as well as worthwhile investment options to build greater resilience for African economies. Successful CA-based business cases should support mobilization of recognition, institutional support and leverage additional funding from governments and international development institutions to support CA programme expansion.

Guided by the SDGs and the climate-smart agriculture vision of triple gains in sustainable agricultural productivity and income increases, strengthening resilience to climate change and variability and reduction of greenhouse gas emissions, CA is continually transforming to less use of agrochemicals towards uncertified 'organic CA'. Advances in machinery innovations, including in electrical and mechanical weed management, supported by digitalization are making the vision realizable.

Now that CA has been perfected for the cereals and legumes value chains production aspects, the scope should be widened to include other priority value chains (such as rice, livestock and horticulture) and at the same time drive for compliance with CA for all operations carried out along the entire agri-food chain for these and the current value chains. The role of mechanization, research and development, capacity development, networking and information sharing will become apparent.

In the era of information overload, it is essential to synthesize and package CA knowledge for specific target groups to achieve intended behavioural changes. The role of CA-based knowledge and innovation systems and information sharing and communication has become even more important to learn from past mistakes and to take advantage of the latest innovations from elsewhere in the world. In the twenty-first century, digitalization needs to be integrated for greater efficacy and speed in knowledge, information sharing and communication.

Going forward, the key is development of group-specific strategies to motivate key players such as farmers, private sector companies and national governments to drive adoption and practise of the CA concept, be it as individual entities or as partners.

7 Where to look for further information

A general source of information on CA globally is the FAO website (http://www.fao.org/conservation-agriculture) and for Africa is the ACT website (http://www

.act-africa.org/), CIMMYT website (https://www.cimmyt.org/tag/conservation-agriculture/) and CIRAD website (https://www.cirad.fr/) and its open registry of CIRAD publications Agritrop (https://agritrop.cirad.fr/572218/). More information regarding CA in Africa can be found on websites of national and regional organizations, including universities, and in books and journals. Books include: Kassam et al. (2017) and Mkomwa and Kassam (2021) and Volumes 1 and 2 of this book.

8 Acknowledgements

The authors acknowledge the support of the Norwegian Development Agency (NORAD) for the financial support to the African Conservation Tillage Network (ACT) to undertake several CA-based projects in ESA.

9 References

AMCEN (2011). *African Ministerial Conference on the Environment (AMCEN): Addressing Climate Change Challenges in Africa; A Practical Guide Towards Sustainable Development*. Addis Ababa.

Andriarimalala, J. H., Rakotozandriny, J. N., Andriamandroso, A. L. H., Penot, E., Naudin, K., Dugué, P., Tillard, E., Decruyenaere, V. and Salgado, P. (2013). Creating synergies between conservation agriculture and cattle production in crop-livestock farms: a study case in the Lake Alaotra Region of Madagascar. *Experimental Agriculture* 49(3): 352–365.

AUC (2015). *African Union Commission Agenda 2063: The Africa We Want*. African Union Commission, Midrand South Africa. Also Available at: https://au.int/Agenda2063/popular_version. ISBN: 978-92-95104-23-5.

Blignaut, J., Knot, J., Smith, H., Nkambule, N., Crookes, D., Saki, A., Drimie, S., Midgley, S., de Wit, M., von Loeper, W. and Strauss, J. (2015). Promoting and advancing the uptake of sustainable, regenerative, conservation agricultural practices in South Africa with a specific focus on dryland maize and extensive beef production. *Asset Research, booklet Nr 2. Pretoria: ASSET Research*.

CATF (2019). *Zimbabwe Conservation Agriculture Task Force*. Presentation at the Regional CA Dialogue, Lusaka, Zambia, 24–28 September 2019. Harare.

De la, O., Campos, A. P., Villani, C., Davis, B. and Takagi, M. (2018). *Ending Extreme Poverty in Rural Areas: Sustaining Livelihoods to Leave No One Behind*. Food and Agriculture Organization, Rome. Available at http://www.fao.org/3/CA1908EN/ca1908en.pdf.

Derpsch, R., Franzluebbers, A. J., Duiker, S. W., Reicosky, D. C., Koeller, K., Friedrich, T., Sturny, W. G., Sá, J. C. M. and Weiss, K. (2014). Why do we need to standardize no-tillage research? *Soil and Tillage Research* 137: 16–22. ISSN 0167-1987, doi:10.1016/j.still.2013.10.002.

Elwell, H. (1995). *An Assessment of the Performance of Minimum Tillage Technologies in Zimbabwe in Terms of Drought Risk Alleviation, Yields and Cost-Effectiveness*. World Bank, Harare.

FAO (2008). *The Conservation Agriculture Toolbox for Zimbabwe*. Food and Agriculture Organization, Rome. Available at: http://www.fao.org/conservation-agriculture/en/. Accessed 25 June 2021.

FAO (2011). *Socio-Economic Analysis of Conservation Agriculture in Southern Africa*. Network Paper #. Johannesburg.

Gupta, G. S. (2019). Land degradation and challenges of food security. *Review of European Studies* 11(1). Canadian Center of Science and Education. ISSN 1918-7173.

Haggblade, S., Tembo, G. and Donovan, C. (2004). *Household Level Financial Incentives to Adoption of Conservation Agricultural Technologies in Africa*. Working paper no. 9. Food Security Research Project Lusaka, Zambia. February 2004.

IFAD (2013). *Smallholders, Food Security, and the Environment*. International Fund for Agricultural Development, Rome. Available at: www.ifad.org/pub/viewpoint/index.htm. Accessed 20 January 2013. Accessed June 2021.

IFAD (2021). *IFAD East and Southern Africa*. Available at: https://www.ifad.org/en/web/operations/regions/esa. Accessed June 2021.

IPCC (2018). *Special Report: Global Warming of 1.5°C*. Summary for Policymakers. 24 pp. Available at: https://www.ipcc.ch/sr15/chapter/spm/.

IPCC (2019). *Climate Change and Land: An IPCC Special Report on Climate Change, Desertification, Land Degradation Sustainable Land Management, Food Security, and Greenhouse Gas Fluxes in Terrestrial Ecosystems*. Summary for Policymakers. Available at: https://www.ipcc.ch/report/srccl/.

Kassam, A., Basch, G., Friedrich, T., Shaxson, F., Goddard, T., Amado, T., Crabtree, B., Hongwen, L., Mello, I., Pisante, M. and Mkomwa, S. (2013). Sustainable soil management is more than what and how crops are grown. In: Lal, R. and Stewart, R. A. (Eds.) *Principles of Soil Management in Agro-ecosystems. 2013. Advances in Soil Science*. 337–399. Boca Raton, Florida, CRC Press.

Kassam, A., Friedrich, T., Shaxson, F. and Pretty, J. (2009). The spread of conservation: agriculture justification, sustainability and uptake. *International Journal of Agricultural Sustainability* 7(4): 293–320.

Kassam, A. H. and Mkomwa, S. (2021). Mainstreaming of the conservation agriculture paradigm in Africa. In: Mkomwa, S. and Kassam, A. (Eds). *Conservation Agriculture in Africa: Climate Smart Agricultural Development*. CAB International, Wallingford, UK.

Kassam, A. H., Mkomwa, S. and Friedrich, T. (2017). Weather-proofing agriculture with conservation agriculture. In: Kassam, A., Mkomwa, S. and Friedrich, T. (Eds.) *Conservation Agriculture for Africa: Building Resilient Farming Systems in a Changing Climate*. Chapter 2, pp. 10–29. CAB International, Wallingford, UK. ISBN: 9781780645681. DOI: 10.1079/9781780645681.0000.

Mazvimavi, K., Ndlovu, P. V., Nyathi, P. and Minde, I. J. (2010). *Conservation Agriculture Practices and Adoption by Smallholder Farmers in Zimbabwe*. A poster presentation at the joint 3rd AAAE and 48th AEASA Conference, Cape Town South Africa, September 19–23, 2010.

MEA (2005). *Ecosystems and Human Well-Being: Synthesis*. Millennium Ecosystem Assessment. Island Press, Washington, DC.

Midgley, S., Drimie, S., von Loeper, W., Mudavanhu, S., Blignaut, J., Knot, J., Smith, H. and de Wit, M. (2015). *Implications for a Future Agrarian Structure in South*

Africa Based on Conservation and Sustainable Agriculture: Alignment of a Multi-Institutional and Multi-Policy Landscape. Asset Research, booklet Nr 1. Pretoria: ASSET Research.

Mkomwa, S. and Kassam, A. (Eds.) (2021). *Conservation Agriculture in Africa: Climate Smart Agricultural Development*. CAB International, Wallingford, UK.

Mkomwa, S., Lugandu, S., Kuria, P. and Mutai, W. (2017). Empowering smallholder farmers with profitable and sustainable farming using Conservation Agriculture - the case of East Africa. In: Kassam, A., Mkomwa, S. and Friedrich, T. (Eds.) *Conservation Agriculture for Africa: Building Resilient Farming Systems in a Changing Climate*. Chapter 4, pp. 41. CAB International, Wallingford, UK. ISBN: 9781780645681. DOI: 10.1079/9781780645681.0041.

Mkomwa, S. and Makungu, P. (2019). *Tanzania's Innovation Policy on Conservation Agriculture: Institutional Reform and Learning Curve*. ACT, Nairobi.

MLAWRR (2020). *Ministry of Lands, Agriculture, Water and Rural Resettlement, Zimbabwe. Postharvest Crop and Livestock Assessment Report* (vol. 2020). Harare, Zimbabwe.

Mloza-Banda, H. (2016). Streamlining conservation agriculture into private and public sector systems: policy dimension. Article published in the ACT. *Newsletter*. Accessed July 2016. Available at: https://www.google.com/url?sa=t&rct=j&q=&esrc=s&source=web&cd=&ved=2ahUKEwjnyOjHuL_xAhXmzlUKHT41AF0QFjAAegQIBBAD&url=http%3A%2F%2Fwww.act-africa.org%2Flib.php%3Fcom%3D5%26res_id%3D208&usg=AOvVaw1ATlCEhGTiicGY-XQ1MH66.

Nkonya, E., Mirzabaev, A. and von Braun, J. (Eds.) (2016). *Economics of Land Degradation and Improvement - A Global Assessment for Sustainable Development*. IFPRI and ZEF Springer Open. 695 pp. Available at: http://link.springer.com/book/10.1007/978-3- 319-19168-3/page/1.

No Till Farmer Magazine (2021). Available at: https://www.no-tillfarmer.com/articles/10416-crop-residue-provides-many-benefits. Accessed June 2021.

Reij, C., Scoones, I. and Toulmin, C. (2001), Chapter 1. Sustaining the soil: indigenous soil and water conservation in Africa. In: Reij, C., Scoones, I. and Toulmin, C. (Eds.) 1996. *Sustaining the Soil: Indigenous Soil and Water Conservation in Africa*. Earthscan Publications, London.

Rockstrom, J. (2000). Water resources management in smallholder farms in Eastern and Southern Africa: an overview. *Physics and Chemistry of the Earth, Part B* 25(3): 275–283.

Temu, A. E. M. and Bisanda, S. (1996). Pit cultivation in the Matengo Highlands of Tanzania. In: Reij, C., Scoones, I. and Toulimn, C. (Eds.) *Sustaining the Soil - Indigenous Soil and Water Conservation in Africa*. International Institute for Environment and Development.

Thierfelder, C., Chisui, J. L., Gama, M., Cheeseman, S., Jere, Z. D., Bunderson, W. T., Eash, N. S., Ngwira, A. and Rusinamhodzi, L. (2013). Maize-based conservation agriculture systems in Malawi. Long-term trends in productivity. *Field Crops Research* 142: 147–157.

Thierfelder, C., Bunderson, W. T., Jere, Z. D., Mutenje, M. and Ngwira, A. (2015a). Development of Conservation Agriculture (CA) systems in Malawi: lessons learned from 2005–2014. *Experimental Agriculture*. doi:10.1017/S0014479715000265.

Thierfelder, C., Rusinamhodzi, L., Ngwira, A. R., Mupangwa, W., Nyagumbo, I., Kassie, G. T. and Cairns, J. E. (2015b). Conservation Agriculture in Southern Africa: advances in knowledge. *Renewable Agriculture and Food Systems* 30(4): 328–348. doi:10.1017/S1742170515000332.

United Nations, Department of Economic and Social Affairs, Population Division (2019). *World Population Prospects 2019: Highlights* (ST/ESA/SER.A/423).

World Bank. (2020). *The World Bank in Africa*. Available at: https://www.worldbank.org/en/region/afr/overview. Accessed June 2021.

Chapter 10

From theory to practice: key lessons in the adoption of Conservation Agriculture in South Africa

H. J. Smith, ASSET Research, South Africa; G. Trytsman, Independent Researcher, South Africa; A. A. Nel, Independent Agronomist, South Africa; J. A. Strauss, Western Cape Department of Agriculture, South Africa; E. Kruger, Mahlathini Development Foundation, South Africa; R. K. Mampholo, National LandCare Program Coordinator, Department of Agriculture, Land Reform and Rural Development, South Africa; J. N. Van Coller, Conservation Agriculture Farmer - Libanon Farm, South Africa; H. Otto, Conservation Agriculture Farmer - Korannafontein Farm, South Africa; J. G. Steyn, Conservation Agriculture Farmer - Humanskraal Farm, South Africa; I. D. Dreyer, Conservation Agriculture Farmer - Skulpspruit Farm, South Africa; D. Slabbert, Conservation Agriculture Farmer - Van Rooyenswoning Farm, South Africa; R. Findlay, No-Till Club of KwaZulu-Natal, South Africa; E. Zunckel, Conservation Agriculture Farmer - Zunckels Farm, South Africa; and L. Visser, Agriculture Research Council – Small Grains, South Africa

1 Introduction
2 The status of Conservation Agriculture adoption and spread in South Africa
3 Challenges in Conservation Agriculture adoption and spread in South Africa
4 Appropriate theories supporting Conservation Agriculture innovation
5 Farmers as intuitive innovators
6 The spontaneous adoption of Conservation Agriculture in a South African context
7 Engaging with Conservation Agriculture farmer innovators
8 Conclusion and future trends
9 References

1 Introduction

Most indicators show that South African agriculture is generally causing and facing a downwards spiral of environmental and economic degradation,

http://dx.doi.org/10.19103/AS.2021.0088.09

specifically in the grain sector (Borrelli et al., 2017). Impacts of especially tillage-based cropping systems are severe, not only through soil erosion and fertility loss (Le Roux et al., 2008) but also through a conspicuous number of off-site effects (e.g. sedimentation, siltation and eutrophication of waterways or enhanced flooding). In addition, the impact of bare (fallow) soil and the loss of soil organic carbon (SOC) on climate remain an increasing threat.

The South African Department of Environmental Affairs (DEA, 2016) stated that there is no evidence to suggest the slowing of anthropogenic climate change trends in South Africa since 2000. Temperature trends consistent with anthropogenic warming continue to be seen across South Africa, with the strongest trends in the west and east, but less so in the central interior. Average rainfall trends are ambiguous, but evidence for longer dry spells, fewer rainy days and more intense rainfall events is mounting. According to Engelbrecht et al. (2015) the 'likely occurrence of more frequent heatwaves, El Niño events and more frequent droughts' suggests that dryland agriculture will increasingly require more climate-resilient systems to avoid possible catastrophic impacts as we move deeper into the twenty-first century. Model projections suggested significant warming and rainfall change for South Africa over several decades, even under strong international mitigation scenarios. Local warming of several degrees by mid-century is virtually certain, and rainfall variability is very likely to increase, but the direction and amount of rainfall change cannot yet be projected with confidence (DEA, 2016).

Rising input and overhead costs and sliding real prices of commodities (e.g. maize) put the economic viability and profitability of cropping systems under severe pressure resulting in many crop producers being affected by the cost squeeze and having financial difficulties. A study done by De Wit et al. (2015) confirms this trend but shows a very large monetary benefit of adopting Conservation Agriculture (CA) in South Africa. The outcomes of this study demonstrated that the transition from tillage systems to CA systems has the potential of not only reducing costs, increasing yields, increasing net farm income but it also has ecological benefits. This is through lower greenhouse gas (GHG) emissions, lower input use and carbon sequestration. Crop farmers should, therefore, be encouraged to adopt CA systems to improve the profitability of their farms.

Consequently, as an effective strategy to stop and reverse this trend, the adoption and implementation of CA principles and practices have been identified as the way forward by most relevant global and local policies, programs and institutions (De Wit et al., 2015; WWF South Africa, 2018). South Africa has, in fact, a fairly favorable national policy environment supporting the promotion of CA among farmers and other actors. What is more, in the mostly semi-arid climate of South Africa, is water limiting crop yield, where, according to Pittelkow et al. (2015), no-till grain yields often equal or exceed yields of conventional tillage practices.

In sections 24 and 27 of the South African Constitution as well as in their Bill of Rights, provision is made for an enabling environment that addresses the right of citizens to a secure environment that is conducive to sustainable human enterprise and food security (Parliament of the Republic of South Africa, 1996). The South African National Development Plan (National Planning Commission, 2011) on the rural economy recommends:

> To deal with the consequences of industrialised agriculture . . . demands that serious attention is paid to advances in ecological approaches to sustainable agriculture. This includes greater attention to alternative energy, improvement of soil quality, minimum tillage and other forms of conservation farming.
>
> *(National Planning Commission, 2011)*

The Strategic Plan for 2011–2015 of the Department of Agriculture, Forestry and Fisheries (DAFF) and the Agricultural policy action plan (APAP) states that an agroecological approach to agriculture will be the key area of focus. Other sectoral legislation, such as the Conservation of Agricultural Resources Act, 43 of 1983, entrenches sustainable use of natural resources. The strategic plan of the DAFF also places sustainable natural resource management at the center of its strategic goals, which is supported by the Conservation Agriculture Policy (DAFF, 2017).

Since around 2000, especially from 2010, numerous local studies and papers have shown that CA systems are more sustainable, being able to foster synergies between agricultural production and conservation enhancing rural livelihoods. These studies highlighted the fact that the success of CA depends on the application and adaptation of all the principles of CA in any specific local context, which requires a (holistic) systems approach with farmers as key innovators (De Wit et al., 2015; Swanepoel et al., 2017; Smith et al., 2017; Kruger and Gilles 2014). This array of local studies joins a large pool of international studies and farmers showing that CA is gaining acceptance in many parts of the world as a proven sustainable and cost-effective alternative production system to conventional farming systems (Kassam et al., 2015, 2017, 2018; Kassam, 2020).

Progress has been made with CA adoption in certain areas of the country. In 2020, the growth in the use of no-till planters, the high adoption of sufficient soil cover, minimal soil disturbance, appropriate crop rotations, plus the integration of cover crops and livestock, have led to an increased percentage of commercial farmers who adopted CA in South Africa.

However, to accelerate the adoption and adaptation of CA in certain commercial farming and in all the smallholder regions of South Africa, a number of underlying challenges and gaps in awareness, knowledge, institutions and capacity need to be effectively and innovatively addressed. Fortunately, from about the 1970s, a few key theories and concepts emerged, leading to the implementation of a growing number of successful initiatives and approaches aiming to understand, engage with, support, strengthen, sustain and/or scale-out so-called 'farmer-driven processes', which were key to the success of the world-wide spread of CA and which provided a solid framework for CA innovation in other countries, including South Africa.

2 The status of Conservation Agriculture adoption and spread in South Africa

South Africa has a fairly favorable policy environment for the promotion of CA, and there are indications that the country is in the early stages of a genuine agricultural revolution toward a more sustainable crop-livestock industry. In 2015/2016, South Africa had the highest area (under CA in Africa (439 000 ha), followed by Zambia (316 000 ha), Mozambique (289 000 ha), Zimbabwe (100 000 ha) and Malawi (65 000 ha). By 2015, the implementation of CA in South Africa has primarily been done by large-scale commercial farmers with a small percentage of smallholders using CA (Kassam et al., 2018).

In 2015, the first formal CA adoption survey was done by Findlater et al. (2019), and in 2021, a second formal CA adoption study was conducted by the Food and Agriculture Organization of the United Nations (FAO, 2021). The latter study found that CA is applied on 1 607 081 ha comprising 25% of the total area under commercial annual crop-livestock systems in South Africa. This is a significant increase from the 2015/2016 figure, by far the biggest in Africa and 12th on the list of all countries in the world. Table 1 shows the distribution of CA per Province (in ha) under commercial farming and the percentage it comprises of the total area cultivated by annual crop-livestock systems in South Africa as assessed in the FAO (2021) study, compared with results from Findlater et al. (2019).

According to the FAO (2021) study, CA areas under semi-commercial and smallholder systems occupied a very small percentage of the total area under CA (13 556 ha) in the 2020/2021 season (see Table 2). These farmers face various unique challenges with a general lack of resources to carry out their farming and even more so for starting new practices, such as CA. A continuous, concentrated, and improved effort is necessary to support and empower this category of farmers with CA as it will be crucial for their sustainability.

There are an estimated 3 million smallholder farmers in South Africa, of whom around 72% fall within a non-commercial category consisting primarily of unemployed women who rely on social grants (89%), who farm for household food purposes on small plots (0.1–1 ha), the ones with very low household incomes (~R2000/month), low productivity (maize yields between 0.1 t/ha and 2 t/ha), and with negligible external support. A further 23% are considered semi-commercial, as they produce for both household consumption and sale and are slightly better resourced. Commercial smallholders make up the remaining 5% and are often supported through employment in the family (Cousins, 2015).

Table 1 Summary of the distribution of CA in South Africa by 2015 and 2021 (from FAO, 2021)

Province	Total annual crop area (ha)[1]	Adoption of CA: Ha		Adoption of CA: %	
		2015[3]	2021[2]	2015[3]	2021[2]
Eastern Cape	160 307	0	3194	0%	2%
Free State	2 196 986	175 759	73 520	8%	3%
Gauteng	173 435	116 202	57 649	67%	33%
KwaZulu-Natal	164 620	82 310	62 956	50%	38%
Limpopo	255 866	63 967	68 834	25%	27%
Mpumalanga	850 484	314 679	205 598	37%	24%
North West	890 437	142 470	330 464	16%	37%
Northern Cape[4]	69 498	7645	0	11%	0%
Western Cape	1 569 277	564 940	804 866	36%	51%
Total	**6 330 910**	**1 467 971**	**1 607 081**	**23%**	**25%**

[1] Data provided by GEOTERRAIMAGE.
[2] Data collected in FAO (2021) study.
[3] Results adapted from Findlater *et al.* (2019).
[4] Total area under irrigated double-cropping conservation tillage system; does not qualify for CA

Table 2 CA areas under semi-commercial and smallholder systems per Province, South Africa

Province	Area under smallholder CA (ha) during the 2020/2021 season
Eastern Cape	344.73
Free State	1137.65
Gauteng	59.37
KwaZulu Natal	196.47
Limpopo	1179.69
Mpumalanga	7516.25
North West	1593.50
Northern Cape	000.00
Western Cape	1528.56
Total	**13 556.22**

The majority of smallholder farmers in South Africa live in scattered communal tenure communities, more often than not in agriculturally marginal production areas. They suffer under the yoke of extreme poverty, highly degraded natural environments and are highly vulnerable to the effects of climate change. Agricultural production is central to rural livelihoods and food production and is undertaken as a mixed farming system approach that typically includes vegetable production in small household gardens, field cropping in dryland fields between 0.1 ha and 2 ha, and livestock rearing

- mostly cattle, goats, and sheep in village-based commonages (De Wit et al., 2015).

A number of projects were launched from around 1997 to promote CA among smallholders in South Africa of which most were done under the umbrella of the National LandCare Program (NLP), the Department of Rural Development and Land Reform's CA Technologies (CAT's) program and since 2013, the Maize Trust-funded CA Farmer Innovation Programme (FIP) for smallholders. The latter initiative was conceptualized and implemented through Mahlathini Development Foundation (MDF), which pioneered the use of agricultural innovation systems as a methodological approach for the promotion of an appropriate smallholder CA farming system, as well as awareness-raising and adaptive research into specific elements of this system (Kruger and Smith, 2019). This approach takes cognizance of the complexity of introducing CA into a farming system, including working with smallholder farmers as partners in the knowledge co-creation process through on-farm research and experiential learning, as well as embedding the process into the existing socio-political environments and economic value chains. A range of CA smallholder projects implemented an innovation systems process that incorporates many of these elements, which were very successful in horizontal scaling as well as developing locally appropriate CA practices (Kruger et al., 2021). Feedback from all these initiatives indicated that there is a gradual yearly increase in the number of participants practicing CA, despite adverse weather conditions and the many constraints smallholder farmers face.

The FAO (2021) study found that CA farmer pioneers and innovators, together with their structures, played a key role in the adoption and spread of CA in South Africa. Other factors that played a key role are local CA equipment manufacturers promoting equipment as part of the whole CA system, local study groups and awareness events, local research initiatives where farmers, researchers and other key stakeholders collaborate, international success stories and cross-visits, international pioneer CA farmers, prominent international and local CA scientist, local service providers, and agribusiness.

The FAO (2021) study recommended that the existing CA farmer-led structures (e.g., clubs and groups) that successfully helped to spread the adoption of CA in the various regions in South Africa, should be supported and used as ideal platforms to implement, strengthen, and scale out/up CA initiatives to other farmers, stakeholders, and regions.

3 Challenges in Conservation Agriculture adoption and spread in South Africa

To accelerate the adoption and adaptation of CA in South Africa, specifically in certain regions under commercial crop farming in the Highveld summer rainfall region (Free State, North West, Mpumalanga and Gauteng Provinces) and smallholders in areas like the eastern sea-board (KwaZulu-Natal and Eastern Cape Provinces), Limpopo and Mpumalanga, the following underlying challenges need to be effectively and innovatively addressed:

- Lack of awareness, knowledge and understanding of environmental problems, such as soil erosion, loss of biodiversity and climate change.
- Lack of knowledge and understanding of the nature and function of natural resources (ecology), especially soils.
- Ignorance of appropriate solutions to address these issues (e.g. CA principles and practices).
- Unique and challenging environmental conditions, specifically a combination of sandy soils and a semi-arid climate in the western cropping areas of the Highveld. Large parts of this area, such as the north-western Free State, are dominated by aeolian sandy soils on which rainfed crop production is practiced. These fine sandy soils are up to several meters deep and are eutrophic. The sandy nature of these soils enables the effective capturing of rainwater due to their very high infiltration rates. These soils have substantial water storage capacity, provided the soil is deep enough and underlain with an impermeable deeper clay (plinthic) horizon on which a shallow hanging water table develops which are available for crop consumption. However, these soils have two very negative characteristics, namely their extreme susceptibility to wind erosion and their extreme vulnerability to subsurface soil compaction by in-field vehicular traffic and naturally by drying and wetting cycles. To address the latter problem, most farmers on these soils have adopted a deep tillage practice of up to 1 m deep, the so-called rip-on-row (RoR) system, to alleviate the compaction. In the higher rainfall eastern parts of the Highveld, many soils have extreme soil infertility and high soil acidity. These natural problems are furthermore all underlain by low SOC and soil health levels across all arable soils due to decades of tillage and poor return of crop biomass.
- Resistance by individuals and institutions in various spheres of the agricultural sector, such as agri-business, academics, researchers and farmers, of which some have vested interests in agribusinesses (e.g. the sale of implements, fertilizers, etc.) or are influenced by individuals in the academic, farming or agribusiness circles.
- A lack of appropriate research and extension support due to diminishing capacity and poor training at the tertiary level. 'Working with farmers may call for skills that not everyone has' (Bentley, 1994), plus the need for new knowledge around CA systems and ecology has not been adequately met by tertiary training institutions. Although all universities in South Africa with agriculture departments have started to include CA in their training

syllabus, especially at the Masters and PhD level, there is still not a steady release of graduates into the workforce having appropriate capacity in both natural (i.e. CA) and social sciences required for a systems research approach. On a school level, there are a few positive movements toward CA training, such as the newly established Agricultural Academy linked to the Reitz High School in the north-eastern Free State Province, which is busy to develop and provide accredited training curriculums on CA and precision farming for grade 10 to 12 pupils.

- Inappropriate research and development (R&D) approaches. Most researchers and extension officers still follow linear R&D approaches with low effectiveness and poor impact on the farmer level. This factor in combination with the diminishing capacity of South African agricultural research institutions, to a large extent, mean that farmers are not supported effectively and/or that they mostly depend on the services of agribusiness representatives for soil and agronomic advice. Most of these support services are also ill-equipped with CA knowledge and experience, and, therefore, do not prioritize CA as a preferred alternative. According to Hoffmann et al. (2007), 'It needs to be acknowledged that, although the agricultural sciences have long been dominated by the positivist (linear) worldview, different disciplines, particularly the social sciences, have already enriched the epistemological debate within the academy. The sociology of scientific knowledge, for instance, has looked at the practice of science and come to the conclusion that the elements contributing to scientific knowledge underlie influences similar to the products of any other way of knowing, e.g. rules of thumb, improvization, tacit knowledge and craft skills, face-to-face interaction, and discipline or culture-specific circumstances'. Success stories of appropriate farmer–research interaction and approaches during the evolution of CA in South Africa are found in the Western Cape and KwaZulu-Natal Provinces, which is clearly reflected by the much higher adoption rates in these areas.

4 Appropriate theories supporting Conservation Agriculture innovation

In all the original cases of CA adoption, it was the farmers' organization and innovation that generated knowledge that not only eventually led to wider adoption but also mobilizing public, private and civil sector support (Kassam et al., 2014, 2018). Although research institutions were involved in nearly all cases of the emergence of CA in the world, at least at some stage in the process (Goddard et al., 2020), their effectiveness and real impact on the farmer level were questioned (Bentley, 1994; Braun and Duveskog, 2008; Holt-Gimenez, 2006) because of the perceived and familiar 'challenges and pitfalls relative to traditional agronomic research' (Snapp et al., 2019). Despite many challenges, there were institutions, such as 'land grant' universities in the United States of America, solitary and innovative scientists or NGOs and the like, who were convinced about the need and potential of creating or facilitating the necessary linkages and processes between researchers and farmers or research institutions and the industry to increase the uptake of CA systems (Smith, 2006; Kassam et al., 2014; Goddard et al., 2020). This has helped formal agricultural research to 'justify' their role as it '. . . can help overcome the limitations of informal innovation; scientists have received substantial formal education and have highly sophisticated research equipment and methods at their disposal. Not only do they have more time and money for doing research in a professional way, they also have better access to written material and exchange methods and results with colleagues worldwide' (Hoffmann et al., 2007).

Despite criticism and resistance from research institutions, in particular, the recognition of farmers as key innovators in agriculture research and technology development, however, was a critical milestone that led to a true transformation in this field, including CA (Coughenour and Chamla, 2000; Kassam et al., 2018). Already in the early stages, scientists working in this field came to the conclusion that 'farmers have valuable knowledge, that they do agricultural research on their own, and that scientists could work with them to improve agrarian R&D' (Bentley, 1994). The development of farmer participatory research (FPR) has been a difficult journey to most scientists exposed to it, as it *inter alia* showed that 'initiating and sustaining farmer participation beyond initial, diagnostic stages was more difficult than researchers had anticipated. In most cases, the participation of farmers in FPR was relatively passive' (Bentley et al., 1994). The main issues often raised were 'questions regarding whom should be asked to participate, in which phases of the research and extension process and in what ways' (Snapp et al., 2019). The relatively few success stories, as mentioned previously, were mainly due to a lack of leadership, capacity and ability to create participatory project teams and co-design appropriate R&D projects and activities (Merrill-Sands, 1991, as cited by Bentley, 1994). However, from about the 1970s, a few key theories and concepts emerged, firstly, motivated by the poor impact that linear approaches had on the transformation of complex whole-farming systems and rural livelihoods and secondly, to underpin this spontaneous, intuitive farmer innovation processes. From an emerging, but solid theoretical basis, a growing number of successful initiatives and approaches were conceptualized and implemented aiming to understand, engage with, support, strengthen, sustain and/or scale-out these farmer-driven processes, which provided a solid framework for CA innovation (Freire, 1973; Chambers and Jiggins, 1987; Holt-Gimenez, 2006; Kassam et al., 2014; Snapp et al., 2019).

The following are seen as some of the key philosophies:

Systems thinking (Checkland, 1981; Anandajayasekeram and Rukuni, 1989; Capra and Luisi, 2014): This involves a new kind of thinking - thinking in terms of relationships, patterns and context. In this field, scientists realized that a living system - an organism, ecosystem or social system - is an integrated whole system whose properties cannot be reduced to those of smaller parts. It shifted the (agricultural) R&D process from a linear, technology transfer mode, where ideas are formulated, solutions are designed and the process is controlled by the researchers, to a facilitation mode (action research or co-learning) approach, where boundaries and conditions are negotiated among actors, where action and reflection become part of an on-going cycle. It supports the theory of *Autopoiesis* or self-production, which is a process generated through a closed organization of production processes such that the same organization of processes is regenerated through the interactions of its own products (components) (Zeleny, 1981). From a biophysical premise, there is an emphasis on the adoption of a system-based CA, which includes no-till farming in conjunction with residue mulching, cover cropping, integrated nutrient and pest management, complex rotations and integration of crops with trees and livestock, also termed as 'regenerative agriculture' (Lal, 2020a,b).

Experiential (Kolb, 1984), ***discovery, action and adult non-formal learning and action research*** (Lewin, 1946; Fell, 1996; Argyris, 1983; Röling and Wagemakers, 1998): These theories are similar in practice and share a commitment to pursue action and research through a cyclic process, alternating action with critical reflection. They are usually *participative* and *responsive* to the situation.

Transformative Learning (Mezirow, 1978): This addresses the need to move to joint learning, aiming to create frames of mind that are more inclusive, discriminating and integrative of experience. To do so, transformative learning facilitators seek to foster individual experience, critical reflection, dialogue, holistic learning and authentic relationships taking into account the context that influences the process (Taylor, 2009; as cited by Probst et al., 2019). Concepts of transformative learning have been applied in various fields and regions, including sub-Saharan African agriculture (Duveskog et al., 2011; Friis-Hansen and Duveskog, 2012).

Communicative action (Habermas, 1981) and ***Tree of knowledge*** (Maturana and Varela, 1987): These theories include the central insights that cognition and knowledge are not a representation of the world 'out there', but rather a 'bringing forth of the world through the process of living itself' by action, communication, storytelling, dialogue, reflection and last, but not least, love, in solving wicked problems (Conklin, 2005). Communicative action requires a discourse that (1) includes all stakeholders without exception; (2) 'guarantees all participants equal opportunity to contribute to argumentation'; and (3) establishes rules that ensure the right to (1) and (2) (Habermas, 1990). Communicative action is thus a process based on the 'improbable conditions' (Habermas, 1990) to facilitate change and learning, which has become mainstream in agricultural research and practice for development, mainly in the form of *innovation platforms* (Probst et al., 2019; Misiko, 2017).

Social learning (Bandura, 1977): 'People learn by watching other people. We can learn from anyone – teachers, parents, siblings, peers, co-workers, YouTube influencers, athletes, and even celebrities. We observe their behavior, and we mimic that behavior. In short, we do what they do'.

These theories influenced a family of agricultural R&D movements and approaches, which were developed and implemented over a number of decades.

Farmer Field School (FFS) (Braun and Duveskog, 2008), Campesino a Campesino (Holt-Gimenez, 2006), Australian Landcare (Robins, 2018; Love, 2012) and Innovation Systems and Platforms (Misiko, 2017; Smith et al., 2017; Kruger et al., 2021), which were employed in many parts of the world, were able to quite successfully facilitate research, learning and adaptation all together in comprehensive, coherent and sometimes massive scale implementation processes (the FFS movement is a good example of the latter where the Global Survey of 2005 (data presented in EcoPort http://ecoport.org/ep) and a rough estimate by Braun and Duveskog (2008) suggested that by 2008 10–20 million farmers have graduated from FFS globally in 87 countries). The key elements in the implementation approach are awareness, reflection, facilitation, on-farm experimentation, season-long training, learning-by-doing, participatory monitoring and evaluation (PM&E) and social learning (the facilitation of sharing and learning in groups) (Smith, 2006; Love, 2012; Robins, 2018; Kruger and Gilles 2014; Kruger et al., 2021; Woodhill and Robins 1998).

As alluded to above, in a number of cases, the support from researchers and extension staff came much later (years or even decades after farmers started with an innovation, such as CA), but in many cases, it was ineffective or absent, which necessitated farmers to carry on by themselves without the involvement of support services (Bentley, 1994). According to Kassam et al. (2018), farmers who do become aware of and seriously interested in adopting CA develop local solutions to all these barriers in adoption. Some of these cases have been reported for smallholder (Holt-Gimenez, 2006) and large-scale farms (Coughenour and Chamla, 2000; Awada et al., 2014; Bellotti and Rochecouste, 2014; De Freitas and Landers, 2014; Peiretti and Dumanski, 2014) in all continents showing that farmer-led transformation of agricultural production systems based on CA has been a key driver. Although the FAO provided, with their AQUASTAT database, a platform for the publication of adoption data (Kassam et al., 2018), it does not give details on the mechanisms and approaches behind this adoption; in other words, it is not clear from this data source if it was solely a farmer-driven approach or a collaborative effort, for example. Other initiatives, such as the Global Partnership Program 'PROmoting Local INNOVAtion' (PROLINNOVA), have been collecting inventories of local innovations related to ecology-oriented agriculture and natural resource management (PROLINNOVA, 2003). What is known from these sources is that smallholders are very innovative but more reliant on external support or facilitation in a farmer participatory development process (Holt-Gimenez, 2006; Braun and Duveskog, 2008; Grabowski, 2015), while large-scale farmers are usually better organized and equipped to start and lead on-farm development initiatives (Coughenour and Chamla, 2000; Knowler and Bradshaw, 2007). Bentley (1994) commented that in his experience, scientists working with commercial large-scale FPR are much more effective and successful than smallholder FPR because they usually have had the same culture and, therefore, better or easier relationships; 'the relationship of agricultural scientists to small farmers may be like music critics to solo folk artists, but in large-scale, commercial agriculture scientists and growers play in the same orchestra'.

5 Farmers as intuitive innovators

It is a demonstrated fact that farmers and other rural people also invent technology on their own (Gamser et al., 1990, as quoted by Bentley, 1994). Over the last few decades, many books and articles have been published that draw attention to farmer experimentation and local innovation (Hoffmann et al., 2007; Kassam et al., 2014). Any farmer's daily efforts to adapt, survive and prosper produce numerous innovations. In fact, the entire agricultural revolution, which started about 10 000 years ago, has been a prime example of farmer-led innovation and invention (Hoffmann et al., 2007; Diamond, 2017). Diamond (2017) emphasized the fact that 'over entire continents and other large areas containing hundreds of competing (agricultural) societies, some societies will be more open to innovation, and some will be more resistant'. He concluded that the more innovative societies were more successful than societies resisting innovation. According to Hoffmann et al. (2007), indigenous technical knowledge, ecological 'wisdom' and traditional farming practices were first studied and documented by cultural anthropologists and geographers, long before the first programmatic statements on 'FPR' were made in public agricultural research. Anthropologists initially paid little attention to the inventions and experimentation done by small-scale farmers, and Johnson (1972; as quoted by Hoffmann et al., 2007) was the first anthropologist to document how farmers routinely experiment with their cropping systems. Snir et al. (2015) offered the first evidence that farmer innovation through trial plant cultivation began far earlier, some 23 000 years ago. 'While full-scale agriculture did not develop until much later, our study shows that trial cultivation began far earlier than previously believed and gives us reason to rethink our ancestors' capabilities', said Sternberg, one of the co-authors, to *American Friends of Tel Aviv University* (2015). 'Those early ancestors were cleverer and more skilled than we knew'. The study determined that early humans there had gathered over 140 species of plants. These included 13 known weeds mixed with edible cereals, such as wild emmer, wild barley and wild oats. From his experiences in FPR, Bentley (1994) concluded that (1) most of the important agricultural inventions bear witness to the creativity and intellect of illiterate villagers; (2) at least in some cases, farmers do better when left alone than when outsiders interact with them; and (3) the few successful eco-development projects he visited were ones without scientist participation.

The examples listed above illustrated that through multiple 'experiences' over long periods of time, farmers have developed 'tacit knowledge' (vs explicit knowledge), which is described as the intuitive understanding that people have and that cannot easily be separated from them and the way that they behave. Tacit knowledge is hard to express and communicate, but 'it is deeply rooted in action, involvement and commitment within a specific context'; it is 'a continuous activity of knowing' and is sometimes described as 'just the way things are done around here' (Parminter and Neild, 2013). Tacit knowledge serves as a true source of innovation, provides dynamic responses to context-specific problems, serves as a basis for developing, interpreting and applying explicit knowledge and provides a mechanism for transferring best practices. If an environment is more innovation-oriented, tacit knowledge becomes increasingly important to the generation of new knowledge and to meet emerging environmental trends and challenges. Therefore, effective exploitation and management of tacit knowledge have become critical (Hoffmann et al., 2007).

Farmers are the only persons able to enact tacit knowledge in their own contexts guiding them to adapt continuously and intuitively. According to Hoffmann et al. (2007), this tacit knowledge, often used intuitively and unconsciously, includes not only technical elements (concrete know-how, crafts, and skills) but also 'mental models' (schemata, paradigms, perspectives, beliefs and viewpoints) that help individuals perceive and define their world. It is, therefore, claimed that the great majority of knowledge capital is tacit, which is embedded in people's experiences, expertise, know-how, skills, techniques, insights, judgments, actions or behaviors. Based on the definition of Dreyfus and Dreyfus (1986; as cited by Hoffmann et al., 2007), 'experienced farmers must clearly be seen as experts who have developed a deep situational understanding of their environment and their profession. A considerable part of their performance is intuitive and non-reflective and entails a large body of tacit knowledge'.

According to Hoffmann et al. (2007), 'externalization (articulation or description) holds the key to knowledge creation, because it creates new, explicit concepts from tacit knowledge that can be passed on as information and used by others'. They argued that the externalization of the tacit knowledge of expert farmers (e.g. pioneer CA farmers) is a field of activity neglected by agricultural researchers. In this process farmers, like children in their development stages, 'proceed like (little) scientists, testing and revising their hypotheses in the light of new evidence'.

Another way how farmers have been sharing tacit knowledge and experiences (making it explicit) was to document (write) their farmer innovation journeys and stories, which has been done by many farmers over the years. More recently, during the advancement of the CA movement under the regenerative agriculture initiative, several prominent farmer innovators published their journeys on how they have learned and adopted CA practices on their farms, essentially through a farmer-led innovation process (Salatin, 2017; Berry, 2015; Brown, 2018; Massy, 2015). These documentaries underline the fact that farmers are primarily the key innovators in agriculture production systems, especially with complex systems like CA, and are in most cases totally reliant on their own abilities and innovation capacity to experiment and adapt CA principles in their contexts. This is despite the great efforts and intentions of the fairly high number of successful farmer-participatory R&D initiatives performed by external change agents (e.g. researchers and extension officers), institutions, programs, NGOs, etc. (as mentioned earlier) but reaching and transforming only a fairly small percentage of farmers. In the very end, it is the farmer's own responsibility to adopt and adapt new ideas and technologies in their unique reality, if they want to survive and prosper. Only they are in a position to 'just do it'.

6 The spontaneous adoption of Conservation Agriculture in a South African context

On the Highveld summer rainfall area of South Africa, a remarkable story emerged in the crop farming sector, which underlined the capacity of pioneer farmer innovators and their influence on the transformation of the industry, where a significant number of commercial crop farmers spontaneously reached unexpected high levels of CA innovation, adaptation and adoption in North West, Mpumalanga, Gauteng, KwaZulu-Natal and Limpopo Provinces as shown in Table 1 (FAO, 2021;Findlater et al., 2019; Strauss et al., 2021).

According to Strauss et al. (2021), a percentage of commercial farmers in the Highveld summer rainfall area of South Africa (see Table 1) converted their farming systems to CA. However, in contrast with the first no-till pioneer farmers in the KZN and Western Cape, who started in partnership with researchers and manufacturers from the 1980s to about 2000 (Goddard et al., 2020; ARC,

2014; Goddard et al., 2020; Western Cape Department of Agriculture, 2017), the Highveld farmers' journey roughly started around 2000, but they did it without any support or influence of formal research studies in their regions. The Highveld farmers' journey in adapting to holistic CA systems in their own contexts, initially led by a smaller group of CA or 'no-till' pioneers, in many cases partnering with CA equipment manufacturers, was triggered by an intense awareness (observation) of their dire environmental and socio-economic situation, prompting them to seek for solutions. Their answers and sources of advice and guidance, which is where they actually discovered CA as a potential solution, mostly came from other CA farmers (local and global), followed by the internet/social media and other local awareness events, such as farmers' days and conferences. 'All these sources of information guided farmers to get started in their own situation and to transform their farms to CA, in some cases, working with farmer groups, manufactures and technical advisors, through a trial-and-error adaptation process' (Strauss et al., 2021). Expert pioneer CA farmers and their networks (i.e. their experiences, insights and expertise or tacit knowledge) were driving this process, sharing with and influencing others through direct observation (demonstrations), interaction, socialization (social learning), storytelling (dialogue), learning by doing and experimentation.

From around 2000, a few formal researcher-managed (RM) trials were established on specific locations to investigate and improve the development of CA in the Highveld area (Nel, 2017; Swanepoel et al., 2017; Swanepoel, 2018; Visser, 2020). These trials were very effective in building formal research capacity and producing numerous formal research outputs, such as peer-reviewed articles, reports, presentations and academic theses. For example, an on-station, RM trial was established on the Zeekoegat site of the Agricultural Research Council (ARC) Roodeplaat campus outside Pretoria from 2008 to 2013, which produced 11 peer-reviewed papers, 7 technical reports, 4 theses (3 MSc's and 1 PhD), 18 presentations, 4 popular articles, 4 information days, 3 awareness material publications and 5 TV/Radio talks (Swanepoel, 2018). However, Swanepoel et al. (2017) pointed out that the number of CA initiatives in the Highveld area were still very low, and furthermore, according to the chairman of the national CA forum in South Africa and one of the prominent CA farmers on the Highveld (Cobus van Coller, CA farmer, Viljoenskroon, Free State, personal communication), these formal RM trials, whether they were on-farm or on-station, had very little influence on farmers' innovation levels and adoption. 'Yet, by 2020, a rising number of successful commercial CA crop farmers were found in all conditions on the Highveld – from higher rainfall regions with clay soils to hot and dry regions with very sandy soils. Obviously, the latter situation is more challenging to adapt CA successfully (as described earlier) leading to a much lower adoption in those regions' (Strauss et al., 2021). Although not purely CA focused, the Sandy Soils Development Committee research project was launched in this sandy soil area of the northwest Free State (Beukes et al., 2018, 2019), aiming to work closely with farmers to improve the sustainability of cropping systems in a very unique environment with deep sandy soils and a semi-arid climate.

In 2013, the CA Farmer Innovation Programme (FIP), funded by a local funding body for research on maize-based systems, that is, The Maize Trust (MT) and implemented through Grain SA, a local grain farmers' association, was launched with the task to coordinate and facilitate CA research under commercial, semi-commercial and smallholder grain farmers in different summer rainfall regions in South Africa (Smith and Visser, 2014). This program's main purpose was specifically to respond to a need or request from CA farmers on the Highveld to make CA research more relevant since they haven't received any real or direct benefits from the CA research projects running in this area up till that moment (as indicated earlier). Therefore, the program's terms of reference were to engage with, support and strengthen the activities of CA farmers on the ground and to reach out to as many other farmers as possible. The philosophy and design of the program's approach and the implementation of the projects and their activities had to be unique and appropriate to fit the situation created and demanded by the farmers; it also had to join, learn from and support successful local initiatives (such as the Western Cape CA movement and the KZN No-till Club) and government CA programs such as the National LandCare Program, as best as possible. Essentially, the message from different sources was to start to work with the farmers, build on what they were doing, from where they were. The advice from one of the original creators of the FFS approach (Dr. Tonie Putter, *EcoPort*, Berlin, Germany, personal communication) in using a 3R approach serves as strong guidance in this respect: 'Recognise farmer competence, Respect it and Reward it by learning about/from it'. He reminded us that 'there is always infinite knowledge outside you (e.g. in the farming community) compared to what you (think) you know'.

7 Engaging with Conservation Agriculture farmer innovators

The extensive global and local experiences on CA R&D (some described above) informed the CA FIP concept development and approach. Central to this concept is that complex, multi-component technologies such as CA require the development of innovation systems (IS) to adapt technologies to local conditions, which critically place the farmer innovator in the central role. The IS approach (for both commercial and smallholder farmers) served as an umbrella for a family of theories and approaches (introduced previously), to facilitate and coordinate awareness, research and learning all together in a single implementation process. The following key strategic objectives (KSOs) were proposed during its launch in 2013 (Smith and Visser, 2014). These KSOs and themes were open for revision on a continuous basis, namely: (1) awareness and access to information; (2) on-farm research; (3) incentive and market-based mechanisms; and (4) education and training. KSOs involved in project implementation were awareness and on-farm research on which almost all the resources were spent during the first 7 years of implementation.

The selection of suitable projects was done according to a range of criteria created around the core philosophy of the CA FIP, which is discussed in more detail below as part of the governance process. To select a potential project study area in a key grain-producing region on the Highveld, the presence of an active CA group(s) (farmers are usually organized into structures such as learning or study groups, associations, clubs or cooperatives, of which some focus their efforts on CA) was a prerequisite. After a CA group was identified in a priority area, such as the North West Province, they were approached to start a more formal CA project on a voluntary basis, which was followed by a participatory project proposal development process with the farmers and key stakeholders. After a project was submitted and approved, a project team, or also called an innovation platform (IP), was created to launch and implement the project, starting with a participatory planning event from where an iterative adaptive management process followed during a period of at least 4 years. For smallholders, the process was much more intense, which required dedicated resources and support around facilitation, inputs (for small trial plots), infrastructure (no-till planters, sprayers), training and experimentation. Kruger et al. (2021) described how this model was successfully used for implementation, networking and up-scaling CA practices and activities in smallholder farming in South Africa; it was also been promoted nationally as a strategic approach to smallholder adaptation and mitigation programming.

> Introduction of CA and associated climate-smart agricultural practices within an innovation system approach and using farmer-level experimentation and learning groups as the primary learning and social empowerment processes have created a sustainable and expanding farming alternative for smallholders that is improving their resilience to climate change substantially.
>
> Through a knowledge co-creation process, smallholder farmers in the program have adapted and incorporated a wide range of practices into their farming system, including minimum tillage, close spacing, improved varieties, judicious use of fertilizer, pesticides and herbicides, crop diversification, intercropping and crop rotation as well as fodder production and livestock integration. They have organized themselves into learning groups, local savings and loan associations, water committees, farmer centers and cooperatives, and in so doing, they have created innovation platforms for local value chain development. They have built on-going relationships with other smallholders, NGOs, academic institutions, government extension services and agribusiness suppliers and have promoted CA tirelessly within their local communities and social networks.
>
> *(Kruger et al., 2021)*

7.1 Channeling Conservation Agriculture funding

One of the first actions of the CA FIP was to create a 'structure' to govern the roll-out of CA funds allocated by the MT in a fair and transparent manner. Part of that action was to engage and involve a range of stakeholders in the core structure created, which was a 'CA forum' or national CA innovation platform involving a wide stakeholder group in sharing and assisting with CA FIP governance. A range of governance matters was handled in this forum, such as evaluation criteria and protocol, communication, monitoring, evaluation and coordination, of which the evaluation of new proposals and progress reports was perhaps the most important - see evaluation criteria for proposals in Box 1.

Box 1 Evaluation Criteria for CA Research Proposals

1 The extent to which on-farm, farmer-centered (research and development) approaches have been included.
2 Level of farmer/community participation, support and/or networking.
3 Addressing key principles and practices of CA.
4 Proof of systematic planning and preparation of proposal (technical content, quality and rationale).
5 Clear and viable projections and assumptions on bio-physical, economic and social sustainability and impact, including potential for adoption.
6 Co-funding from and collaboration with other institutions and organizations.
7 Demonstrated capacity, suitability and ability of the organization to deliver the project.
8 Feasibility and realism of achieving project objectives through proper implementation and management.

It is important to understand that these structures were 'independent' and primarily served the goal of the funder (The Maize Trust), which was to appropriately channel and manage funds allocated toward CA research with all interested parties. In this respect and on request from the MT, the CA forum formed a CA evaluation panel, comprising a range of mostly local and also international experts and CA farmers to assist in the evaluation of new CA project proposals, as well as annual progress reports. Appropriate evaluation criteria were developed for this purpose, aligned with the CA FIP concept (see Box 1). Most of the CA forum members are also members of the CA panel.

7.2 Conservation Agriculture project interventions: was it sufficient and efficient?

Three commercial farming projects were implemented under the banner of the CA-FIP between 2013 and 2020, together with three smallholder projects. Kruger et al. (2021) described the CA smallholder program implemented in South Africa in detail, and no further information is provided on that quite successful and comprehensive process in this chapter. Table 3 reflects on the salient features, activities and results of each of the commercial CA FIP projects. In the North West Province, a project was implemented in collaboration with the Ottosdal No-till Club and investigated commercial CA practices under semi-arid, sandy soil conditions, as described by (Smith et al., 2019a). Another two projects were implemented in two commercial farming study areas in the northeastern Free State. These projects were implemented in collaboration with the Rimland (Reitz) and Ascent (Vrede) study groups (Smith et al., 2019b).

The main goal of these projects was to research, develop and adapt CA systems in a range of local contexts in those regions, by working with and empowering farmers to adopt CA in their own realities. In this process, relevant research questions on all aspects of the CA system were raised by the project teams, which were consequently tested in a range of collaboratively managed trials (CMTs) linked to or complemented by a participatory monitoring and evaluation (PM&E) system feeding into social learning and critical reflection in the IP's for continuous improvement and adaptation.

At the annual planning events before the start of every growing season, the implementation of the different work packages and activities approved in the project business plan was coordinated. The key activity planned was on-farm experimentation, which was undertaken by a number of volunteering farmers in a group. On-farm experimentation, as described by Biggs (1987), Selener (1998), Smith (2006) and Smith et al. (2021), was employed as a key methodology in the CA FIP as it created a platform for further awareness and joint learning. A distinction was made between non-participatory RM, consultative researcher-led, collaboratively managed (CM) and farmer-managed (FM) trials, with increasing participation and farmer-level impact toward FM trials.

CM trials, done mostly in partnerships between farmers and researchers, were initially viewed to be the most suitable to address key research questions and were used extensively in all commercial CA FIP projects; the Ottosdal project had seen the peak of this effort with around 80 CMTs done between 2013 and 2020, of which key results were described by Smith et al. (2019; 2021) and Haarhoff and Swanepoel (2020), the latter as part of a PhD study. FMTs have been rigorously used, mostly 'under the radar' at individual farms by all categories of farmers in this area as an 'inherent, intuitive part of a farmers' innovativeness and resilience to adapt their practices (to CA), which

Table 3 The salient features of the three commercial CA projects implemented under the CA FIP

Salient features, activities and results	Ottosdal, North West	Vrede, Free State	Reitz, Free State
Climate	The climate is semi-arid (aridity index = 0.37), with mean annual summer precipitation from 500 mm to 620 mm. Dry spells often occur during December, January and February, which can seriously affect crops, as it usually coincides with the flowering and grain filling growth stages. A large proportion of rainfall events occur in the afternoon as thunder storms with a high intensity, causing runoff.	MAR around 700 mm, concentrated in the summer months of October to March with very little precipitation in the winter. Frosts are severe, usually from May to September, where minimum temperatures below −10°C are often recorded in winter. Ranges of mean daily maximum temperature in January and mean daily minimum temperature in July is from 25.5°C to 27.5°C and from 0.8°C to −2.8°C, respectively.	
Geology	Mainly sandstone and other sedimentary rocks	Mainly mudstone and sandstone of the Adelaide Formation, Beaufort Group with, in the north-east, shale of the Volksrust formation, Ecca Group. Dolerite intrusions occur frequently.	Underlain mainly by mudstone and sandstone of the Tarkastad Formation, Beaufort Group.
Soils	Deep, freely drained sandy to sandy loam soils, with about 50% soils having a water impermeable deep, plinthic sub-soil horison, are predominant in this area with a low water holding capacity and low organic matter content.	The soils are mainly dark, swelling clays (Vertisols) along with duplex soils (sandy, often bleached topsoil abruptly overlying gleyed clay).	The soils here are mainly grey and yellow, sandy loam to sandy clay loam, moderately deep soils with grey, mottled plinthic subsoils. Duplex soils, as well as shallow, rocky soils, are also present.
Land use (% cropland: natural pastures)	80:20	20:80	50:50
Dominant cash crops and rotations	Maize and sunflower	Maize and soya bean	Maize and soya bean
Major natural resource problems	Drought and soil erosion	Soil erosion and soil fertility	Soil erosion and soil fertility
Other problems	Low profit margins and production risk	Low profit margins and weeds	Low profit margins
Description of farmer group	Ottosdal No-till Club with specific focus on CA; strong group of CA farmers with good leadership	Ascent study group not with specific focus on CA; only one strong CA farmer	Rimland study group with major focus on CA; young group of farmers with good leadership
Other key stakeholders	Input suppliers, NWK and Senwes cooperatives	Input suppliers, VKB cooperative	Input suppliers, VKB cooperative
Cumulative number of CMT's from 2013 to 2020	80	12	12

Major topics, or research questions of CMT's	Cover crops and livestock integration, crop rotations, green fallow systems, crop density (plant population × narrower row width), intercropping, tine vs disc no-till planters, soil fertility and fertilizer rates, weed management, conventional vs CA crop systems and maize cultivar evaluation under high density CA systems	Crop density (higher plant population × narrower row width); soil fertility and fertilizer rates	Crop rotations, crop density (higher plant population × narrower row width)
Key indicators used at CMT's	Soil health (various physical, chemical and biological parameters); soil water balance; yields (grain and biomass); soil cover; plant roots	Soil health (various physical, chemical and biological parameters); soil water balance; yields (grain and biomass)	Soil health (various physical, chemical and biological parameters); soil water balance; yields (grain and biomass)
Number of monitoring points (ecotopes)	None	10	10
Description of ecotopes	NA	Site 1: • Sandy loam to loamy sandy soils with clay from 14% to16% in the topsoil Site 2: • Sandy to sandy loam soils, with 4% to 14% clay in the topsoil Site 3: • ('Turf soils') – sand clay to sand clay loam, clay between 30% and 40% in the topsoil	Site 1: • Sandy loam • Sandy • Loamy sand • Natural pasture Site 2: • Sandy field with high potential or yields • Sandy field with lower potential or poor yields Natural pasture conventionally grazed, moving over to UHDG
Treatments or practices at monitoring points	None	Conventional vs CA crop systems, the latter including cover crops and livestock integration, crop rotations, green fallow systems, higher crop density, intercropping; natural pastures (grazing)	Conventional vs CA crop systems, the latter including cover crops and livestock integration, crop rotations, green fallow systems, higher crop density, intercropping; different grazing systems on natural pastures
Key indicators measured at monitoring points	None	Soil health (various physical, chemical and biological parameters); soil water balance; yields (grain and biomass); Nematodes (as soil health indicators); dung beetles (as indicators of biodiversity)	Soil health (various physical, chemical and biological parameters); soil water balance; yields (grain and biomass); Nematodes (as soil health indicators); dung beetles (as indicators of biodiversity); plant survey (biodiversity and impact of grazing systems)

(*Continued*)

Table 3 (*Continued*)

Salient features, activities and results	Ottosdal, North West	Vrede, Free State	Reitz, Free State
Key project results or outcomes	• strong CA innovation platform led by farmers; • strong partnerships with key scientists coordinating and implementing project activities, especially CMTs; • 80 CMTs on relevant topics; • major annual CA conferences reaching many non-participating farmers, plus two formal feedback sessions; • various field visits to CMTs and CA farmers; • cross-visits (outreach) to new semi-commercial farmers from previously disadvantaged communities in the region; • awareness materials, including a number of popular and scientific publications and videos; • a growing positive footprint among the greater farming community in the North West Province; • one PhD study.	• existing farmers' study group serving as CA innovation platform; • strong partnerships between 1x CA farmer and key scientists coordinating and implementing project activities, especially monitoring of selected ecotopes; • the fixed monitoring points at the ecotopes provided a good alternative or complimentary tool for CMTs; the method helped to analyze real farming practices or treatments without taking much extra effort from farmers; • major annual CA farmers' days reaching many non-participating farmers, plus 1 formal feedback session; • various informal field visits to CMT's and CA farmer; • awareness materials, including a number of popular and scientific publications and videos; • a growing positive footprint among the greater farming community in the region; • one Master's study.	• existing farmers' study group serving as CA innovation platform; • fairly strong partnerships between CA farmers and key scientists coordinating and implementing project activities, especially monitoring of selected ecotopes; • major annual CA conferences reaching many non-participating farmers, plus one formal feedback session; • various informal field visits to CMTs and CA farmers; • awareness materials, including a number of popular and scientific publications and videos; • a growing positive footprint among the greater farming community in the region; • one Master's study.

Reflection on activities and results	• A strong and passionate CA local farmer leadership group was the main factor behind project successes. • CMTs and the annual conference were the activities with biggest impacts so far. • Excellent relationships with core scientists were key. • Data from CMT's made critical breakthroughs in CA system understanding and development.	• A strong local CA farmer leadership group was absent; • A model CA farm and the annual farmers' days were the activities with the widest impacts. • Excellent relationships between CA farmer and core scientists, as well as field workers from the local cooperative (VKB). • Data from ecotopes and on-farm discussions made critical breakthroughs in CA system understanding and development on one farm.	• A strong local CA farmer leadership group was present. • A few model CA farms and the annual conference were the activities with the widest impacts. • Good relationships between CA farmer and core scientists, as well as field workers from the local cooperative (VKB). • Individual farmer innovation made critical breakthroughs in CA system understanding and development on CA farms. • The creation of the Reitz Agricultural Academy at the local High School developing a CA and precision farming curriculum for grade 10–12 learners was very positive.
General cross-cutting reflections	• The CA FIP governance process was effective in channeling funding towards CA projects with a greater potential for farmer-level impact. • Is it possible and necessary to link the more formal research (CMTs) with the informal research (FMTs), or tacit with implicit knowledge (in order to increase learning and adaptation) and what would be the most effective way? • There is a need to improve and expand the monitoring system at the ecotopes as it shows tremendous potential. • Too little engagement, feedback and discussion from the general farming community, outside formal project activities. • There was too little engagement with and support of the frontline CA farmer adopters and non-adopters, or information on that, outside formal project activities. • A few farmers in the region started to adopt CA, or change their systems; lots of perceptions still exist and very little is known about their systems, experiences and stories. Better communication, feedback and reflection were needed. • Many different perceptions still existed among farmers and very little is known about their systems, experiences and stories.		

happens spontaneously, every day, rather than as the result of some external motivation' (Hansson, 2019). Very little data was formally collected from these FM trials at commercial farms, partially due to an unawareness of the untapped value of these farmer innovations and also due to a lack of ideas or tools to include these types of trials into the main innovation or research process. However, by acknowledging and including FM experimentation we 'need to become more open to farmers' informal experiments and more systematically seek farmers' assessment of technology and the reasons they give for their assessment. Particularly those aspects that seem irrational, odd or inexplicable to a researcher at first glance may give valuable hints for further research' (Hoffmann et al., 2007).

In the Vrede and Reitz projects in the north-eastern Free State, the monitoring of FM experimentation or treatments (their actual practices) on land areas with homogeneous soil, slope and climate, also called 'ecotopes', were employed later in the project cycle as a complementary research tool to fill this gap (Smith et al., 2019b), especially after the ending of CM trials due to the inability of participating farmers to commit to these trials at that point (which will probably happen more often than not). This monitoring framework included an annual soil assessment on referenced GPS points within selected farmer fields as a key component to monitor the effect of CA and conventional tillage (CT) practices on soil health. As a reference, the natural pasture (veld) adjacent to these crop fields was also sampled.

On individual farms of 'non-participating' CA farmers, the informal FM 'trial-and-error' process continued, but no formal data capturing took place around that. It was assumed that experiences were frequently shared between farmers in informal meetings at different occasions, in isolated corners of the farming community, between neighbors or in smaller group gatherings. More formal feedback and reflection meetings took place in the project context during annual or biannual planning events, farmers' days and conferences.

Smith et al. (2019b) and Smith et al. (2021) described the progress and results of these three commercial projects in detail. Table 3 below summarizes and reflects on the salient features, activities and results of the three projects where CA interventions were made, as a way to create new ideas for the way ahead.

8 Conclusion and future trends

By 2020, the arable land under CA in South Africa was an estimated 1 607 081 million hectares, which showed a steady increase in adoption to 25% of primarily large-scale commercial grain farmers. Although it comprises a small fraction of the total adoption figure, there was also a gradual yearly increase in the number of smallholders practicing CA, despite the many constraints smallholders face. Despite a number of challenges that still appear on many levels, this positive increase in CA adoption in South Africa could be ascribed to an effective collaborative effort between farmers and other key stakeholders who laid a solid foundation in certain provinces (i.e. the Western Cape and KwaZulu-Natal), the leadership of pioneer CA farmers, the implementation of successful participatory systems research approaches supporting commercial and smallholder farmers in CA innovation, as well as a conducive policy environment and government support (e.g. through the LandCare program).

To effectively engage with and support local CA innovation, the cultivation of a conducive environment for communication (dialogue) and constructive engagement with and between key innovators, or pioneer CA farmers in a specific area, is necessary. Experience in South Africa showed that this is something that develops during the life of a project. If relations and dialogue are honest and respectful, through skillful process design and facilitation, farmers will open up to full participation.

Change agents (e.g. researchers or extension officers) who have the opportunity and responsibility to work with farmers should make an effort to facilitate and enhance activities that firstly, help to understand these farmer experts and their contexts and secondly, convert or articulate (externalize) their tacit knowledge effectively and efficiently into explicit information or knowledge, and feeding it into further formal collaborative research processes. In this regard, participatory systems research approaches should be implemented and/or enhanced where farmers are recognized, rewarded and utilized as key innovators in a long-term iterative spiral of on-farm research, learning and adaptation.

Tools are needed to extract tacit knowledge (e.g. from pioneer CA farmers) and make it explicit for transfer to other parts of an industry (i.e. other farmers, researchers, etc.), such as those developed by Nonaka (1994) and Semeon and Garfield (2019). These tools or pathways are iterative and spiral processes of socialization, externalization, combination and internalization and correspond broadly with the participatory innovation systems research approach implemented by the CA FIP. The following major phases in this pathway are complementary to what CA pioneer farmers, as well as the CA FIP, were doing: (1) creating time and space for exchange; (2) dialogue and collective action; (3) networking and linking explicit knowledge; and (4) learning by doing and experimentation.

CA is primarily a social innovation, frequently leading to a misperception of the function and potential impacts, and consequently, a misuse of natural science research methods among some scientists involved in CA research. Commonly, a predominant natural scientific approach is taken in most CA research initiatives, which reveals the

> 'psycho-statistical mindset common to (natural) scientists, based on the assumed objectivity of the researcher, and focussing on hypothesis testing, cause-effect relationships, consensus-seeking, and generalization. However, humans are not organisms like seeds or plants, with variables that can be investigated scientifically. Humans are idiosyncratic individuals whose beliefs, attitudes, and behaviors are influenced by a complex web of factors that are fundamentally affected by people's life experiences. These factors are typically referred to and investigated as "variables" in scientific research. The sheer number and nature of the human factors involved resulting in "confounding variables" in natural scientific investigations - variables that cannot be controlled or accounted for and are usually left out or simplified. Conclusions from such studies are thus often fundamentally flawed. Research involving human participants is more appropriately conducted as social science investigations'.
>
> *(Sanders, 2020)*

A much more preferred research approach with CA projects aiming to adopt CA systems in local farming contexts should rather build on a conceptual framework based on a thorough knowledge of the pertinent R&D theories as well as the experiences in the CA FIP described in this chapter. Such a framework would fundamentally change the approach for the study, the design, implementation and facilitation of the research, the expected outcomes, the interpretation of the findings and the way it is reported back. Participatory innovation systems approaches create an environment where the mutual advantages, strengths and roles of each main discipline and actor are balanced in CA research (i.e. social and qualitative vs quantitative research, farmers vs researchers), which should primarily be on-farm, involving people as key actors in the system.

It is important to introduce research trials, or experimentation, that suit the real purpose and context of the CA initiative, where all the major stakeholders, especially the farmers, can benefit from. If the purpose is to develop the capacity (human resource) of natural scientists to complete their academic studies and publish scientific articles, or to screen/test specific technologies, a pure statistical laid-out CM or RM trial is the most effective. However, care should be taken to train these students right inside the 'real' CA context of the on-farm CA system (compared to a fictitious and highly controlled environment of an on-station or on-farm RM trial). There is also a justifiable purpose and function for CM and even RM trials (mother trials) where and when (1) the project team collectively identifies (agree on) a real need for it, and (2) where there are sufficient and appropriate resources to conduct the trial (especially capacity and funding). Both farmer volunteers willing to provide land and implements for more sophisticated, better controlled and bigger trials, as well as researchers with the right mindset and experience should be available. Typical topics to be investigated are, for example, screening of different systems, crop species or cultivars, testing new and scarce implements made available by the manufacturer and testing of new soil amelioration products (such as biochar, inoculants, etc.). These types of trials are also suitable to be used for awareness events, such as farmer's days, cross visits and conferences.

A broader definition of 'experimentation' must be adopted to accommodate farmers' 'experiments'. According to Okali et al. (1994; as cited by Hoffmann et al., 2007) an 'experiment' can also occur without being predetermined and without the treatments being especially created, as long as there is an initial observation of the conditions followed by observation and monitoring of the results and effect. This is precisely how CA farmers on the Highveld informally experimented with CA principles in the development of CA systems in their own contexts. Farmers used this intuitive experimentation process to adopt and adapt CA principles through a continuous process of experiential learning.

Involving a large number of farmers through FM trials (single replicates of subsets of technologies chosen by farmers, also called 'baby trials') systematically linked to CM trials (also called 'mother trials'), CA technologies could be evaluated in a replicated manner within-site while supporting joint learning among farmers, extension staff and researchers (Snapp et al., 2019).

CA research 'should put more emphasis on monitoring and making sense of what different farmers are doing' (Hoffmann et al., 2007) in their CA 'experiments' and capturing their experiences (stories) behind their actions. Their improvements and adaptations, as well as their challenges and pitfalls experienced should be monitored, relayed back and assessed by all actors in order to crystallize, document and disseminate the respective principles or lessons that can be generalized. Furthermore, the wide variation across the Highveld, for example, in farmers' cropping practices provides an opportunity for on-farm monitoring and simultaneously engaging farmers in the research and learning process. This type of monitoring framework could be used to understand how different tillage practices (treatments) interact with prominent site-specific environmental conditions (i.e. dominant ecotopes) to determine CA outcomes such as crop yield, profit and soil health (as applied by Snapp et al., 2019). This approach makes it possible to capitalize on the diversity of cropping practices used by many farmers, as well as their local experiences, knowledge and interest within a relevant environmental context (i.e. a specific agroecological zone or ecotope).

M&E of on-farm trials and analysis of the data collected (social, economic and biophysical) is part of the research process. Appropriate indicators should be measured and monitored to track changes and influence decisions on CA system adaptation. Subject matter specialists (researchers) should assist with technical monitoring, data management, analyses, interpretation and reporting where possible and necessary, especially with CMTs. Farmer observations, experiences and reflections, as part of PM&E, are essential in FMTs.

A novel opportunity for M&E systems is to serve as a conduit to effective incentive mechanisms and niche markets for CA farmers. For example, a set or index of indicators could be used to assess the level and impact of CA implementation indicating whether the CA practice, its impact on the environment and people involved are empowered and the product quality improves. This M&E 'score' could be linked to a system-wide risk mitigation approach (including short-term insurance, where the premiums decline as the robustness or health of the land or agricultural system improves as measured by the M&E system), market access (better demand and prices for higher levels of CA implementation), etc. This idea is based on the theoretical concept of an easement payment (Blignaut, 2019), which is currently implemented by two venture capital companies recently established in South Africa, collectively 'Restore Africa Funds' or 'RAFFs' to: (1) facilitate and manage investments in the restoration of natural capital especially in conservation and regenerative (climate smart) agriculture under the banner of *heal the land, heal the people*, in Africa; (2) achieved through private and commercial ventures in partnership with local and global development assistance, public sector contributions and R&D (see https://www.restore-africa.com/home).

Opportunities and platforms for dialogue, storytelling, reflection and social learning should be frequent and robust, such as project, Landcare, study or learning group meetings. Personal stories providing a glimpse into what people experienced in their lives have long been an important part of evaluations. 'Stories can be powerful tools for helping people understand old and new situations, and their attendant risks and opportunities' (Franz, 2016). As explained by Sole and Wilson (2002), 'the tacit, experience-based knowledge that comes up more easily in stories, can be more important in problem-solving than information coming through more formal options'. According to Maturana and Varela (1987), 'not knowing what makes up our world of experience, which is the closest world to us, is a crying shame. There are many things to be ashamed of in the world, but this ignorance is one of the worst'. This they summed up as 'All doing is knowing, and all knowing is doing'. Furthermore, stories can memorably convey meaning and relevance as opposed to facts and statistics alone (Dahlstrom, 2014).

9 References

American Friends of Tel Aviv University (2015). First evidence of farming in Mideast 23,000 years ago: evidence of earliest small-scale agricultural cultivation, *ScienceDaily* [Online] Available at: www.sciencedaily.com/releases/2015/07/150722144709.htm (accessed 20 July 2020).

Anandajayasekeram, P. and Rukuni, M. (1989). Institutionalisation of on-farm research with Farming Systems Perspective in eastern and southern African region: achievements and future directions, *Zimbabwe Journal of Agricultural Research* 29, 67–82.

ARC (2014). Assessing the impact of conservation agriculture practices on wheat production in the Western Cape, *Report for ARC-Small Grain Institute and the*

Western Cape Department of Agriculture. [Online] Available at: https://www.greenagri.org.za/assets/documents-/SmartAgri/Other-Resources/ARC-2014-Conservation-Agriculture-report.pdf (accessed 5 September 2020).

Argyris, C. (1983). Action science and intervention, *Journal of Applied Behavioral Science* 19(2), 115–135.

Awada, L., Lindwall, C. W. and Sonntag, B. (2014). The development and adoption of conservation tillage systems on the Canadian Prairies, *International Soil and Water Conservation Research* 2(1), 47–65.

Bandura, A. (1977). *Social Learning Theory*. New York: General Learning Press.

Bellotti, B. and Rochecouste, J. F. (2014). The development of Conservation Agriculture in Australia - farmers as innovators, *International Soil and Water Conservation Research* 2(1), 21–34.

Bentley, J. W. (1994). Facts, fantasies, and failures of farmer participatory research. *Agriculture and Human Values* 11(2-3), 140–150.

Bentley, J. W., Rodriguez, G. and Gonzalez, A. (1994). Science and people: honduran campesinos and natural pest control inventions. *Agriculture and Human Values* 11, 178–182.

Berry, W. (2015). *The Unsettling of America: Culture & Agriculture*. Berkeley, CA: Counterpoint Press.

Beukes, D. J., Nel, A. A., Trytsman, G., Steenkamp, S., Rhode, O. H. J., Abrahams, A. M., Van Staden, P., Marx, F. and Van Zyl, B. (2018, 2019). Investigating the impacts of conservation agriculture practices on soil health as key to sustainable dry land maize production systems on semi-arid sandy soils with water tables in the north western Free State. Annual Progress Report to the Maize Trust.

Biggs, S. (1987). Proposed methodology for analysing farmer participation in the ISNAR OFCOR study, *Agricultural Administration (Research and Extension) Network Newsletter No. 17*. London: ODI.

Blignaut, J. N. (2019). Making investments in natural capital count, *Ecosystem Services* 37, 1–4.

Borrelli, P., Robinson, D. A., Fleischer, L. R., Lugato, E., Ballabio., C., Alewell, C., Meusburger, K., Modugno, S., Schütt, B., Ferro, V., Bagarello, V., Van Oost, K. V., Montanarella, L. and Panagos, P. (2017). An assessment of the global impact of 21st century land use change on soil erosion, *Nature Communications* 8(1), 2013.

Braun, A. and Duveskog, D. (2008). The farmer field school approach - history, global assessment and success stories, Background Paper for the IFAD Rural Poverty Report 2011.

Brown, G. (2018). *Dirt to Soil - One Family's Journey into Regenerative Agriculture*. London, UK: Chelsea Green Publishing.

Capra, F. and Luisi, P. L. (2014). *The Systems View of Life: A Unifying Vision*. Cambridge: Cambridge University Press.

Chambers, R. and Jiggins, J. (1987). Agricultural research for resource-poor farmers. Part I: Transfer-of-technology and farming systems research. Part II: A parsimonious paradigm, *Agricultural Administration and Extension* 27: 35–52 (Part I) and 27: 109–128 (Part II).

Checkland, P. B. (1981). *Systems Thinking, Systems Practice*. Chichester: Wiley.

Conklin, J. (2005). *Dialogue Mapping: Building Shared Understanding of Wicked Problems*. Chichester: Wiley, CogNexus Institute. http://www.cognexus.org.

Coughenour, M. and Chamla, S. (2000). *Conservation Tillage and Cropping Innovation: Constructing the New Culture of Agriculture*. Iowa: University Press.

Cousins, B. (2015). Through a glass darkly: towards agrarian reform in South Africa. In: Cousins, B. and Walker, C. (Eds) *Land Divided, Land Restored. Land Reform in South Africa for the 21st Century*. Auckland Park, Pretoria: Jacana, pp. 250–269.

DAFF (2017). Draft Conservation Agriculture Policy. [Online] Available at: https://cer.org.za/wp-content/uploads/2018/01/Draft-Conservation-Agriculture-Policy.pdf (accessed 17 August 2020).

Dahlstrom, M. F. (2014). Using narratives and storytelling to communicate science with non-expert audiences, *Proceedings of the National Academy of Sciences* 111(4), 13614–13620.

DEA (2016). Climate change trends, risks, impacts and vulnerabilities. South Africa's 1st Annual Climate Change Report, Theme C. Available at: https://www.environment.gov.za/sites/default/files/reports/themeC_vulnerabilities_risks.pdf.

De Freitas, P. L. and Landers, J. N. (2014). The transformation of agriculture in Brazil through development and adoption of zero Tillage Conservation Agriculture, *International Soil and Water Conservation Research* 2(1), 35–46.

De Wit, M. P., Blignaut, J. N., Knot, J., Midgley, S., Drimie, S., Crookes, D. J. and Nkambule, N. P. (2015). Sustainable farming as a viable option for enhanced food and nutritional security and a sustainable productive resource base. Synthesis report. Green Economy Research Report, Green Fund, Development Bank of Southern Africa, Midrand.

Diamond, J. (2017). *Guns, Germs and Steel: A Short History of Everybody for the Last 13,000 Years*. London: Vintage Book Company.

Duveskog, D., Friis-Hansen, E. and Taylor, E. W. (2011). Farmer field schools in Rural Kenya: A transformative learning experience, *Journal of Development Studies* 47(10), 1529–1544.

Engelbrecht, F., Adegoke, J., Bopape, M.-J., Naidoo, M., Garland, R., Thatcher, M., McGregor, J., Katzfey, J., Werner, M., Ichoku, C. and Gatebe, C. (2015). Projections of rapidly rising surface temperatures over Africa under low mitigation, *Environmental Research Letters* 10(8), 085004.

FAO (2021). *An assessment of the adoption of conservation agriculture in annual crop-livestock systems in South Africa*.

Fell, R. F. (1996). *Application of Learning Theory to Extension. IPM Tools for Implementation Course Manual*. Brisbane, Australia: Cooperative Research Center for Tropical Pest Management, University of Queensland.

Findlater, K. M., Kandlikar, M. and Satterfield, T. (2019). Misunderstanding conservation agriculture: challenges in promoting, monitoring and evaluating sustainable farming, *Environmental Science and Policy* 100, 47–54.

Franz, N. (2016). *The Extension Storyteller: Using Stories to Enhance Meaning and Catalyse Change*, *Journal of Extension* 54(3), Article # 3TOT1.

Freire P. (1973). *Education for critical consciousness*. New York: Continuum.

Friis-Hansen, E. and Duveskog, D. (2012). The empowerment route to well-being: an analysis of farmer field schools in East Africa, *World Development* 40(2), 414–427.

Goddard, T., Basch, G., Derpsch, R., Hongwen, L., Jin, H., Karabayev, M., Kassam, A., Moriya, K., Peiretti, R. and Smith, H. J. (2020). Institutional and policy support for Conservation Agriculture uptake. In: Kassam, A. (Ed) *Advances in Conservation Agriculture Volume 1: Systems and Science*. Cambridge, UK: Burleigh Dodds Science Publishing. (ISBN: 978 1 78676 264 1; www.bdspublishing.com).

Grabowski, P. P. (2015). Promotion and adoption of conservation agriculture in Mozambique and Zambia. PhD Thesis. Michigan State University.

Haarhoff, S. J. and Swanepoel, P. A. (2020). Narrow rows and high maize plant population improve water use and grain yield under conservation agriculture, *Agronomy Journal* 112(2), 921-931.

Habermas, J. (1981). *Theory of Communicative Action, Volume One: Reason and the Rationalization of Society*. Boston, MA: Beacon Press. ISBN 978-0-8070-1507-0.

Habermas, J. (1990). *Moral Consciousness and Communicative Action*. Cambridge, MA: The MIT Press. ISBN 0-262-58118-3.

Hansson, S. O. (2019). 'Farmers' experiments and scientific methodology, *European Journal for Philosophy of Science* 9(3), Article number 32.

Hoffmann, V., Probst, K. and Christinck, A. (2007). Farmers and researchers: how can collaborative advantages be created in participatory research and technology development?, *Agriculture and Human Values* 24(3), 355–368.

Holt-Gimenez, E. (2006). *Campesino a Campesino: Voices from Latin America's Farmer to Farmer Movement for Sustainable Agriculture*. Oakland, CA: Food First Books.

Kassam, A. (Ed) (2020). *Advances in Conservation Agriculture. Volume 1 – Systems and Science, Volume 2 – Practice and Benefits*. Cambridge: Burleigh Dodds.

Kassam, A., Friedrich, T., Shaxson, F., Bartz, H., Mello, I., Kienzle, J. and Pretty, J. (2014). The spread of Conservation Agriculture: policy and institutional support for adoption and uptake, *Field Actions Science Reports* (vol. 7) [Online] Available at: http://factsreports.revues.org/3720 (accessed 15 July 2020).

Kassam, A., Friedrich, T., Derpsch, R. and Kienzle, J. (2015) Overview of the worldwide spread of conservation agriculture, *Field Actions Science Reports* [Online], Vol. 8 . Available at: http://factsreports.revues.org/3966 (online since 26 September 2015, connection on 15 October 2015).

Kassam, A., Mkomwa, S. and Friedrich, T. (Eds) (2017). *Conservation Agriculture for Africa: Building Resilient Farming Systems in a Changing Climate*. Wallingford: CABI Publishing.

Kassam, A., Friedrich, T. and Derpsch, R. (2018). Global spread of Conservation Agriculture, *International Journal of Environmental Studies* 76(1), 29–51. DOI: 10.1080/00207233.2018.1494927.

Knowler, D. and Bradshaw, B. (2007). 'Farmers' adoption of conservation agriculture: a review and synthesis of recent research, *Food Policy* 32(1), 25–48.

Kolb, D. A. (1984). *Experiential Learning: Experience as the Source of Learning and Development*. Englewood Cliffs, NJ: Prentice-Hall.

Kruger, E. and Gilles, J. (2014). A review of participatory agricultural research and development in South Africa and KwaZulu-Natal. [Online] Available at: www.mahlathini.org (accessed 5 March 2020).

Kruger, E. and Smith, H. J. (2019). *Conservation Agriculture Farmer Innovation Programme* (CA FIP) for smallholders, Grain SA. *Farmer Centred Innovation in Conservation Agriculture in Upper Catchment Areas of the Drakensberg, KwaZulu-Natal, Pretoria: the Maize Trust Annual Progress Report*.

Kruger, E., Smith, H. J., Ngcobo, P., Dlamini, M. and Mathebula, T. (2021). CA innovation systems build climate resilience for smallholder farmers in South Africa. In: *Conservation Agriculture in Africa: Climate Smart Agricultural Development*. Wallingford, UK: CAB International.

Lal, R. (2020a). Managing soils for negative feedback to climate change and positive impact on food and nutritional security, *Soil Science and Plant Nutrition* 66(1), 1-9, DOI: 10.1080/00380768.2020.1718548.

Lal, R. (2020b). Regenerative agriculture for food and climate, *Journal of Soil and Water Conservation* 75(5), 123A-124A. DOI: 10.2489/jswc.2020.0620A.

Le Roux, J. J., Morgenthal, T. L., Malherbe, J., Pretorius, D. J. and Sumner, P. D. (2008). Water erosion prediction at a national scale for South Africa, *Water SA* 34(3), 305-314.

Lewin, K. (1946). *Action research* and minority problems, *Journal of Social Issues* 2(4), 34-46.

Love, C. (2012). *Evolution of Landcare in Australia: In the Context of Australian Government Natural Resource Management Policy and Programs (Perspectives from the Former National Landcare Facilitator)*. Australian Landcare Council Secretariat.

Massy, C. (2015). *Call of the Reed Warbler: A New Agriculture, A New Earth*. London: Chelsea Green Publishing.

Maturana, H. R. and Varela, F. J. (1987). *The Tree of Knowledge*. Boston: Shambhala.

Mezirow, J. (1978). Perspective transformation, *Adult Education Quarterly* 28(2), 100-110. DOI: 10.1177/074171367802800202. ISSN 0741-7136.

Misiko, M. (2017). Extending conservation agriculture benefits through Innovation Platforms. In: Kassam, H., Mkomwa, S. and Friedrich, T. (Eds) *Conservation Agriculture for Africa: Building Resilient Farming Systems in a Changing Climate*. Wallingford, UK: CAB International.

National Planning Commission (2011). National Development Plan. [Online] Available at: https://www.salga.org.za/SALGA%20Municipal%20Managers%20Forum/Documents/National%20Development%20Plan%202030.pdf (accessed 17 Aug 2020).

Nel, A. (2017). Evaluation of conservation agriculture principles on two soil types on the Highveld. Final progress report to The Maize Trust.

Nonaka, I. (1994). A dynamic theory of organizational knowledge creation, *Organization Science* 5(1), 14-37.

Parliament of the Republic of South Africa (1996). The constitution of the Republic of South Africa. [Online] Available at: https://www.justice.gov.za/legislation/constitution/SAConstitution-web-eng.pdf (accessed 16 August 2020).

Parminter, T. and Neild, J. (2013). Just knowing: tacit knowledge, *Extension Farming Systems Journal* 2, 1-12.

Peiretti, R. and Dumanski, J. (2014). The transformation of agriculture in Argentina through soil conservation, *International Soil and Water Conservation Research* 2(1), 14-20.

Pittelkow, C. M., Linquist, B. A., Lundy, M. E., Liang, X., van Groenigen, K. J., Lee, J., van Gestel, N., Six, J., Venterea, R. T. and van Kessel, C. (2015). When does no-till yield more? A global meta-analysis, *Field Crops Research* 183, 156-168.

Probst, L., Ndah, H. T., Rodrigues, P., Basch, G., Coulibaly, K. and Schuler, J. (2019). From adoption potential to transformative learning around conservation agriculture, *The Journal of Agricultural Education and Extension* 25(1), 25-45. DOI: 10.1080/1389224X.2018.1520733.

PROLINNOVA (2003). *Progress with PROLINNOVA. Launching the Participatory Design of a Global Partnership Programme*. Leusden, The Netherlands: PROLINNOVA.

Robins, L. (2018). More than 30 years of 'Landcare' in Australia: five phases of development from 'childhood' to 'mid-life'- crisis or renewal?, *Australasian Journal of Environmental Management* 25(4), 385-397, DOI: 10.1080/14486563.2018.1487342.

Röling, N. G. and Wagemakers, M. A. E. (1998). *Facilitating Sustainable Agriculture: Participatory Learning and Adaptive Management in Times of Environmental Uncertainty*. Cambridge: Cambridge University Press.

Salatin, J. (2017). *The Marvelous Pigness of Pigs: Respecting and Caring for All God's Creation*. New York: Hachette Book Group.

Sanders, M. (2020). Research into human cognition, attitudes, and beliefs requires a social sciences approach, *South African Journal of Sciences* 116 (special issue), Art. #8588, 4 pages. DOI: 10.17159/sajs.2020/8588.

Selener, D. (1998). *Participatory Action Research and Social Change*. Quito, Ecuador: Global Action Publications, ISBN 9978-95-130-X.

Semeon, G. and Garfield, M. (2019). Framework for externalization of tacit knowledge in participatory agricultural research in Ethiopia: the case of Farmers Research Group (FRG). Proceedings of the 52nd Hawaii International Conference on System Sciences. [Online] Available at: http://hdl.handle.net/10125/59970 (accessed 5 September 2020).

Smith, H. J. (2006). Development of a systems model facilitating action research with resource-poor farmers for sustainable management of natural resources. PhD Thesis. University of Free State.

Smith, H. J. and Visser, M. (2014). A farmer-centred innovation systems approach to stimulate adoption of Conservation Agriculture in South Africa, *Poster presented at the World Congress on Conservation Agriculture*, 23–25 June 2014, Winnipeg, Canada.

Smith, H. J., Kruger, E., Knot, J. and Blignaut, J. (2017). Conservation agriculture in South Africa: lessons from case studies. In: Kassam, A. H., Mkomwa, S. and Friedrich, T. (Eds) *Conservation Agriculture for Africa: Building Resilient Farming Systems in a Changing Climate*. Wallingford, UK: CAB International.

Smith, H. J., Nel, A. A. and Trytsman, G. (2019a). Farmer innovations in Conservation Agriculture (CA) systems for sustainable crop intensification in semi-arid, sandy soil conditions, north west province. Annual Progress Reports for The Maize Trust.

Smith, H. J., Visser, L., Trytsman, G., Jankielsohn, A., Du Preez, G., Fourie, D., Loggenberg, A. and Van Zyl, J. (2019b). The promotion of conservation agriculture in the north-eastern Free State - phase 1 (two study areas). In collaboration with Riemland (Reitz) and Ascent (Vrede) study groups. Annual Progress Report for The Maize Trust.

Smith, H. J., Trytsman, G. and Nel, A. A. (2021). On-farm experimentation for scaling-out conservation agriculture using an innovation systems approach in the North West Province, South Africa. In: Mkomwa, S. and Kassam, A. (Eds). *Conservation Agriculture in Africa: Climate Smart Agricultural Development*. Wallingford, UK: CAB International.

Snapp, S. S., DeDecker, J. and Davis, A. S. (2019). Farmer participatory research advances sustainable agriculture: lessons from Michigan and Malawi, *Agronomy Journal* 111(6), 2681–2691.

Snir, A., Nadel, D., Groman-Yaroslavski, I., Melamed, Y., Sternberg, M., Bar-Yosef, O. and Weiss, E. (2015). The origin of cultivation and proto-weeds, long before Neolithic farming, *PLoS ONE* 10, e0131422. DOI: 10.1371/journal.pone.0131422.

Sole, D. and Wilson, D. (2002). Storytelling in organizations: the power and traps of using story to share knowledge in organizations. Harvard Learning Innovations Laboratory. Presidents and Fellows of Harvard College, Cambridge, USA. [Online] Available at:

http://providersedge.com/docs/km_articles/Storytelling_in_Organizations.pdf (accessed 20 April 2020).

Strauss, J. A., Swanepoel, P. A., Laker, M. C. and Smith, H. J. (2021). CA (no-till based) in rainfed annual crop production in South Africa, *South African Journal of Plant and Soil*. In Press.

Swanepoel, C. M. (2018). Assessment of the potential of conservation agriculture management practices to sequester organic carbon in South African soils. PhD thesis, University of Pretoria.

Swanepoel, C. M., Swanepoel, L. and Smith, H. J. (2017). A review of conservation agriculture research in South Africa, *SA Journal of Plant and Soil* DOI: 10.1080/02571862.2017.1390615. (accessed 21 April 2020).

Visser, L. (2020). The long-term influence of soil cultivation methods on soils under wheat production in the Eastern Free State. Agricultural Research Council - Small Grain Report number P15000046. Annual Report, Winter Cereal Trust.

Western Cape Department of Agriculture (2017). Promoting climate smart intensification of winter grain production while improving soil health in the Western Cape, case study #2 Conservation Agriculture. [Online] Available at: https://www.greenagri.org.za/assets/documents-/SmartAgri/Case-Studies/2.-Case-Study-CA-FINAL.pdf (accessed 5 September 2020).

Woodhill, J. and Robins, L. (1998). *Participatory Evaluation for Landcare and Catchment Groups: A Guide for Facilitation*.Greening Australia, Yarralumla.

WWF South Africa (2018). Investing in soil. WWF business case on emerging climate-smart business opportunities [Online] Available at: https://www.wwf.org.za/our_research/publications/?26481/Investing-in-soil (accessed 17 August 2020).

Zeleny, M. (1981). *Autopoiesis: A Theory of Living Organization*. New York: North-Holland.

Chapter 11

Conservation Agriculture in West Asia

Isam Bashour, Roula Bachour, Nicolas Haddad and Razan Dbaibo, American University of Beirut, Lebanon; Kassem Jouni, World Food Programme, Lebanon; Faten Adada, FAO Regional Office for the Near East and North Africa, Egypt; Yahya Shakhatreh, Yahya Bani-Kalaf, Abedaraheem Bawaliz, Iyad Musallam, Faddel Ismael, Mahmud Huwaian and Nabeel Bani Hani, National Agricultural Research Center, Jordan; Mina Devkota, International Center for Agricultural Research in the Dry Areas (ICARDA), Morocco; Atef Haddad and Yaser Musa, Aga Khan Foundation, Syria; Rabea Al Hayek, Agricultural Research Station, Syria; Abdulsattar Asmair Alrijabo, University of Mosul and Local Leader of the Conservation Agriculture Programme in Nineveh Province and North of Iraq, Iraq; Irfan Gultekin, Bahri Dagdas International Agricultural Research Institute, Turkey; Mohammad Esmaeil Asadi, Crop, Soil and Conservation Agriculture Systems, Golestan Agricultural and Natural Resources Research and Education Center, Iran; and Amir Kassam, University of Reading, UK

1 Introduction
2 Lebanon
3 Jordan
4 Syria
5 Iraq
6 Turkey
7 Iran
8 Conclusion
9 Where to look for further information
10 References

1 Introduction

The West Asia region has a semi-arid to arid Mediterranean climate with irregular rainfall mainly during the winter season. The average rainfall in the region is around 500 mm, but most of the cropped lands may receive rainfall in the range of 300–900 mm, varying from year to year, with rainfall distribution also varying considerably within years. Climate change is predicted to cause a reduction in rainfall and an increase in temperature, which will exacerbate the existing water shortages in the region. Crop production in the West

http://dx.doi.org/10.19103/AS.2021.0088.10

Asia region is primarily wheat in wetter areas and barley in drier areas with leguminous crops such as lentils, *vicia* beans, vetches and chickpeas grown in rotation.

The concept of Conservation Agriculture (CA) was introduced to the region and promoted by European Union and international agencies. CA embraces crop production systems, which are based on minimal soil disturbances, retention of crop biomass and stubble on the soil surface and diversified cropping involving rotations and/or associations. This contrasts with conventional tillage (CT) agriculture with operations that plough the land and invert the soil, thus burying surface crop biomass or removing it for livestock while maintaining poor crop diversification.

The soils of agricultural regions in West Asia are experiencing degradation, erosion and reduction in fertility that is reflected in declining yields. The results of field research in the region for the past 10 years or more, comparing CA to CT methods, show that CA is environmentally and economically more beneficial than the conventional methods.

In this chapter, the results of field research and extension programmes in countries of the West Asia region are presented. The countries covered are Lebanon, Syria, Jordan, Iraq, Turkey and Iran. These countries have been active in promoting CA in various forms over the past two decades. Some governmental inputs in Iran, Syria and Turkey are more organized, effective and consistent, whereas others are less so. Arab Centre for the Studies of Arid Zones and Dry Lands (ACSAD), International Center for Agricultural Research in Dry Areas (ICARDA) and Food and Agriculture Organization of the UN (FAO) along with local institutions, bilateral and multilateral donor agencies, and NGOs have been providing support to promote CA in the region, and there has been some significant success.

The CA research has been conducted on rain-fed and irrigated agriculture on various soil types, which are farmed by people of varying educational backgrounds. Land management in the West Asia region is being encouraged to shift from the current degrading conventional methods of farming to the no-till systems. The results have generally verified that the no-till method of farming is more beneficial to farmers and the environment, when there is effective weed management, because it rebuilds soil structure and allows crops to use rainfall more effectively, resulting in greater agricultural productivity.

2 Lebanon

The location of Lebanon on the eastern Mediterranean and its mountainous geomorphology gives it a range of climatic conditions from subtropical coastal, the typical Mediterranean at low elevations, cold snowy weather at higher elevations, reaching a semi-arid climate in the inland plain (the Bekaa valley). Lebanon, like many countries with the Mediterranean climate, benefits from a wet season extending from November to April and a dry season from May to October. The average annual rainfall on the Lebanese coast varies between 700 mm and 1000 mm and increases towards the North. Mount Lebanon forms a barrier against the rain movement and the annual precipitations can reach more than 1400 mm (the majority of which is snow). Rainfall decreases rapidly in the eastern slope of Mount Lebanon to reach about 600 mm. Annual rainfall in the Bekaa valley varies between 800 mm (southern Bekaa) and 200 mm (extreme north-east of the valley). As for the Anti-Lebanon Mountains, rainfall is around 600 mm and increases up to more than 1000 mm in Jabal el Cheikh. The total volume of water as rainfall is estimated at 8600 million m^3 (MCM) per year and 2200 MCM of snow.

Agriculture is an important sector in Lebanon with about 1/3 of the total area being arable lands (360000 hectares) (MOA, 2010). The largest agricultural area is in the Bekaa (42%) followed by northern Lebanon and Akkar valley with 26%, 22% in Southern Lebanon and 9% in Mount Lebanon. Thirty-one per cent of the total agricultural production is fruit trees, 23% is olives, 20% is cereals, 17% is vegetables, and the remaining 9% is industrial crops, like tobacco, grape vineyards, and others (MOA, 2010). Climate change is having a large effect on agriculture in Lebanon. Olive trees are generally resistant to rapid climate swings but are still vulnerable to longer-term climate change. Cereal yields are more vulnerable to decreased rainfall. Higher temperatures are also a factor for cereal production. Just a 1°C rise in temperature leads to a 13% loss in yields (Tsusaka and Otsuka, 2013). Legumes are also detrimentally impacted by heat and drought (World Bank, 2012).

In addition to climate change, excessive tillage of agricultural soils in Lebanon combined with overgrazing lead to poor soil aggregate structure, low organic matter and exposure of soils to erosion, causing land degradation and desertification (Montgomery, 2007; Lahmar and Ruellan, 2007; Bashour et al., 2016). The traditional farming practices in Lebanon rely on intensive tillage and removal of crop residues by grazing or burning. These practices decreased soil fertility, and hence, reduced the productivity of the soils (Lal et al., 2007; Bashour et al., 2016). CA is an effective system to restore soil structure, fertility and health, and reduce soil degradation (Kassam et al., 2010; Busari et al., 2015). CA adoption in Lebanon has been occurring at a slow pace, with around 1200 ha of cropland (less than 1% of total cropland), and remaining mainly under perennial systems with fruit trees, olive groves and vineyards. This section presents the results of CA field trials in Lebanon while highlighting the institutional support of the Ministry of Agriculture and other institutions and the major reasons why CA has not spread more widely.

2.1 Methodology

2.1.1 Study area

To promote CA in Lebanon, collaboration among the American University of Beirut (AUB), the Lebanese Agricultural Research Institution (LARI), the German Corporation for International Cooperation (GIZ) and ACSAD started in 2007 with a series of field trials. The field trials were conducted at the Agricultural Research and Education Center (AREC) of AUB, LARI in Tal Amara and in farmers' lands in the Bekaa and North Lebanon, which continued for several subsequent years with GIZ funding. At a later stage, agricultural service providers and input suppliers became involved as well. GIZ supported the work by providing two Fankauser multi-planters, from Brazil (Fig. 1), that are able to sow cereals and pastures and perform precision planting of large grains. These seeders have the ability to plant small seeds from alfalfa, to vetch with minimum distance between rows for row crops, to big seeds that require large spacing (60–80 cm), such as maize. The same planters were also provided by GIZ to research institutes AREC and LARI who tried various types of seeders, such as Ozdoken (Turkey) and small Syrian brands made in local workshops. The work was conducted in collaboration with service providers to be able to reach a large number of farmers.

In the first season of 2007/2008, two pioneer farmers participated in addition to the experiments at AREC and LARI, totalling an area of 4 ha. The number and acreage increased in the season of 2008/2009 to 60 farmers in 13 locations, covering 562 hectares. In the 2009/2010 season and after the promotion to olive farmers, the areas under CA increased to about 1200 ha. The field experiments and trials were conducted across the major agricultural regions in the country, including rain-fed wheat, barley and a barley-vetch mixture, irrigated corn, alfalfa, within olive trees, apple orchards and vineyards.

Figure 1 Fankhauser direct seeder used to plant barley at AREC. *Source:* Photo by N. Haddad, 2009.

2.1.2 Experimental design and field management

At both experimental sites (AREC and LARI) and the first farmers who adopted CA, the fields were divided into two halves, one cultivated with CA and the other with CT.

During the 2008–2009 growing season, two trials were conducted at AREC and LARI to evaluate the relevance of CA compared with CT in increasing the biomass yield, income and net revenue for barley and vetch, as individual crops and as vetch/barley mixtures. The total area allocated for each individual trial was 4 ha at AREC and 10 ha at LARI. At both research stations, all the plots were sown at the same time after the first germinating rains (October-November), using the Fankhauser direct seeder for the CA treatment and the conventional method of cultivation for the CT treatment (Fig. 2). At AREC, the trials followed a split-plot design, with tillage practices as main plots and the vetch/barley mixture as subplots. The trial at LARI followed a randomized complete block design with three replications.

2.1.3 Data collection

The biomass yield of the individual barley and vetch crops and the vetch-barley mixture were recorded in the middle of May in the AREC and LARI trials and at farmers' fields. Soil samples were collected and analysed for nutrient content and organic matter. Soil moisture was also monitored throughout the season at the experimental plots as well as earthworm count. Weed infestation was also monitored at AREC.

On the other hand, economic parameters and costs were calculated to effectively compare the profitability under CA compared with CT. The variable costs included all inputs and operations for tillage, sowing, fertilizer application, weed control, harvesting and labour.

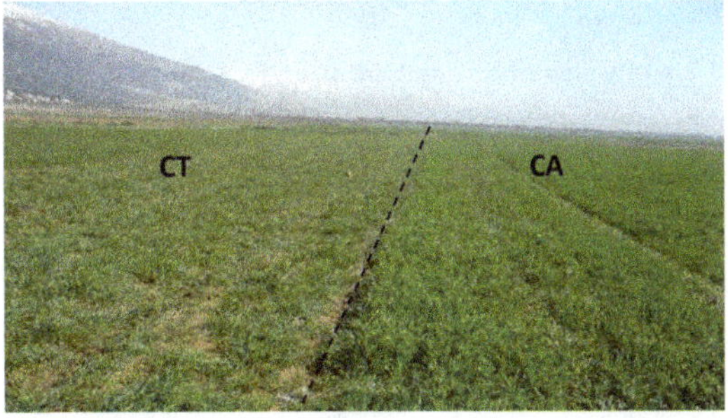

Figure 2 Barley-vetch mixture after germination in CT vs. CA plots. *Source:* Photo by K. Jouni, 2010.

2.2 Results and discussion

2.2.1 Impact on yield

The results of the experimental trials in Lebanon showed that the harvested green barley/vetch mixture yields (at early seed formation stage) were 26.5% higher in CA compared to CT at AREC, while no significant ($P < 0.05$) difference was shown at LARI. The vetch yield was significantly ($P < 0.05$) higher under CA compared to CT in the two sites (AREC and LARI). The same was noticed for the green fresh barley yields as well (Table 1).

In addition to the experimental sites, the harvested green yield of barley/vetch mixture was also recorded for the same season at a farmer's plot in Ammiq (Bekaa), where CA resulted in 41.2 t ha^{-1} and CT was 34 t ha^{-1} encouraging the farmers to be early adopters of CA in Lebanon after seeing more than a 21% increase in yield.

2.2.2 Impact on soil moisture

It was observed that the no-till practices allowed the soils to hold moisture longer than CT practices, and the mulch from plant residues acted as a preventive layer for evaporation of soil moisture and decreased the soil surface temperature. Soil samples were taken from different fields at AREC during the dry season of the year and they showed an increase of 2% in soil moisture close to harvest (Fig. 3). In farmers' language, that equated to 30 water truck loads of water per hectare (120 m^3 ha^{-1}). Similar results were observed at LARI with a 2 to 3% increase in soil moisture.

2.2.3 Impact on soil quality

Two years after starting the experiment, several soil samples were collected at AREC's experimental trial for the 2008/2009 season, and the average results

Table 1 Harvested green yields of different crops under CA and CT at two sites in Lebanon during the 2008–2009 growing season

	Green yields (t ha^{-1})					
	Barley/Vetch		Vetch		Barley	
Farming system	AREC	LARI	AREC	LARI	AREC	LARI
CA	45.0^a	38.1^a	25.3^a	28.5^a	38^a	50^a
CT	34.0^b	38.7^a	17.5^b	24.0^a	33^b	43^b
Yield increase (%)	26.5	−1.6	44.6	18.8	15.2	16.3

Note: Within the same column, values denoted by the same letter are not significantly different at $P < 0.05$.
^{a,b} Data with different notations indicate that there is a presence of significance at 5%.

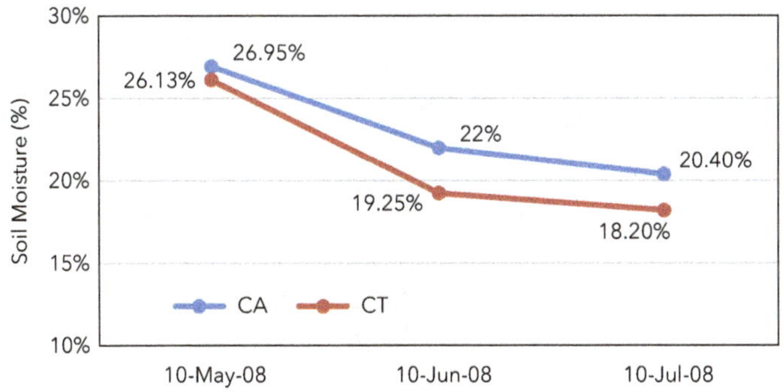

Figure 3 Impact of CA on soil moisture (season 2008/2009 at AREC) (Bashour et al., 2016).

are presented in Table 2. The analysis showed an increase in organic matter, potassium, and phosphorous in the CA fields compared to the CT fields. The consecutive CA practices enabled a build up of organic material with an increase of 0.7% compared to CT, an increase of 51 ppm in potassium, and around 4 ppm in phosphorous in the topsoil layer (0-20 cm). The increase in the values of organic matter, most probably, was due to residual non-decomposed roots or straw and the increase in P and K most probably was due to residual fertilizers accumulation on topsoil. Random sampling was also done at farmers' plots, and the soil analysis showed a small increase or no significant difference in these soil parameters after only 1 or 2 years of adopting CA farming systems.

2.2.4 Impact on weed infestation and earthworms

Most farmers claim that their ploughing activities are mainly to control weeds and improve soil aeration. During the early years, CA fields showed less weed infestation than conventionally ploughed fields due to the use of glyphosate before planting and broadleaf herbicide at the tillering stage (Table 3); the plant residues help inhibit the growth of weeds at the beginning. However, after several years, around 5 years, an increase in the perennial weed population of the CA plot was observed.

An earthworm count was done at AREC after the second season and on average, there were 5.7 earthworms m^{-2} in CA plots compared to 1 earthworm

Table 2 Results of soil analysis at AREC at two depths

Farming system	OM (%)		k mg kg^{-1} (ppm)		P mg kg^{-1} (ppm)	
	(0-20 cm)	(20-40 cm)	(0-20 cm)	(20-40 cm)	(0-20 cm)	(20-40 cm)
CA	4.64%	3.98%	634	554	22.9	17.9
CT	3.90%	3.90%	581	584	18.3	17.6

Table 3 Comparison of weed infestation at Ammiq farmer plot after 2 years (season 2008/2009)

	Weed count, per m^2			
Farming system	*Papava* spp.	*Raphanus* spp.	*Anthemis* spp.	Others
CA	1	5	3.5	2
CT	4	14.5	11	4.5

Table 4 Average net revenue for barley and barley/vetch mixture CA and CT in Lebanon

	Barley		Barley/Vetch mixture	
	CA	CT	CA	CT
Production cost (US$ ha^{-1})	870	1130	800	1150
Income (US$ ha^{-1})	1940	1890	2,250	2040
Net Revenue (US$ ha^{-1})	1070	760	1450	890

m^{-2} in CT plots. This is an indication of improved soil aeration and biological activity, which, in the long run, improves soil structure and water infiltration. Earthworms also help in incorporating plant residues into the soil and in improving soil health.

2.2.5 Impact on economics

The economic parameters and costs of production for barley and barley/vetch mixture for the 2008/2009 season were calculated by a simple gross margin and benefit-cost ratio to compare the profitability under CA compared with CT. The results in Table 4 show that farmers may obtain higher net revenues when applying CA, namely about US \$ 310 ha^{-1} for barley and US \$ 560 ha^{-1} for barley-vetch mixture.

The variable costs including all inputs and operations for tillage, sowing, fertilizer application, weed control, harvesting, and others (including labour) are shown in Table 5 for barley. It is obvious that farmers can increase net revenue by at least US \$ 260 ha^{-1} by adopting CA. The use of ZT seeders can produce savings in the cost of ploughing and seedbed preparation, and reduced seed inputs.

The results show that the application of CA in different agricultural areas in Lebanon gave similar results to those which were obtained from CA in other countries in the region and elsewhere. The adoption of CA led to the reduction in the cost of fuel, labour and machinery required for land preparation Also, it led to an increase in crop yield after 3 to 4 years of adopting the CA system. When discussing with farmers why they did not adopt CA at a faster and larger scale, it was apparent that soil and water conservation and regeneration of soil health were not the farmers' main concerns but rather the economic savings

Table 5 Cost of operations for barley in the Bekaa valley for CA and CT systems

	Cost (US$ ha^{-1})	
Operation	CA	CT
Ploughing	0	100
Seedbed preparation	0	100
Planting	100	100
Seeds	200	300
Fertilization	250	250
Herbicide	100	60
Harvest	120	120
Others	100	100
Total	**870**	**1130**

and higher returns. The lack of knowledge about CA practices and systems as well as the absence of affordable CA seeders are discouraging farmers from giving up ploughing and initiating the transformation to CA systems.

The results of the CA promotional programme were more successful for orchards than for field crop production. The introduction of hairy vetch, an annual legume, as a cover crop between olive and other trees was successful, and now is practiced in many apple and olive orchards (Fig. 4), in addition to several vineyards. This success under orchard systems was partly attributed to well-designed institutional support, which consisted of technical seminars on CA principles at academic institutions and agriculture technical schools, building capacity within extension officers, development of educational material and field trials, coupled with the distribution of vetch seeds to farmers. In collaboration with AUB, LARI, FAO, GIZ and the Lebanese Ministry of Agriculture, the promotional programme supported the establishment of a national committee aiming to promote sustainable crop intensification and scale-up CA success stories in Lebanon.

Figure 4 Vetch under olive trees. *Source:* Photo by K. Jouni, 2013.

2.3 Conclusion

Although the adoption of CA in Lebanon is developing at a slower pace than some other countries, it is progressing, albeit gradually from 0 ha in 2007 to 1200 ha in 2015/16. The potential benefits that farmers may be able to achieve through the CA systems will encourage more farmers to adopt the CA practices and systems in the future. The lower cost of production and higher yields and income are the main motivating factors. The main challenges remain the lack of available CA machinery across the country, weed control, and management of crop residues, especially overgrazing. The education of farmers to understand the practices of CA and its advantages is necessary to promote this technique in Lebanon. All these challenges may be eased once decision-makers realize that CT-based cropping practices are no longer sustainable and cannot be relied upon to satisfy the ever-growing need for food security and ecosystem stability. Therefore, the continuous support of institutional organizations, the involvement of ministries of agriculture, NGOs, international and national research bodies, and funding agencies is essential to increase the adoption of CA in our region.

3 Jordan

Jordan is centrally located in the eastern Mediterranean and it's bordered with Syria from the North, Saudi Arabia from the South and southeast, Iraq from the East, and Palestine from the West. The area of the country is about 89 300 km^2, over 80% of which are semi-arid and arid.

The climate in Jordan varies from one region to another. Western parts of Jordan have a Mediterranean climate, characterized by a hot dry summer and mild wet winter and extreme variability in rainfall within and among years. Mild summers and cold winters characterize the climate in the highlands of Jordan. Aqaba and Jordan valley regions are very similar in their temperature pattern to the subtropical climate. The steppe and steppe desert regions have continental climates with large amplitudes of temperature.

Rainfall mainly occurs in the period between November and March. The annual average rainfall in the steppe desert is only 30–100 mm, while it exceeds 600 mm in very small areas of the highlands. The average annual rainfall volume in Jordan is 8.5 billion m^3. About 12% of this volume is available for use in springs and runoff of groundwater, while the remaining 88% is lost through evaporation.

The total population of the kingdom was 5 480 000 in 2004 and reached 10 420 000 in 2019 (DOS, 2019). Jordan has witnessed many waves of compulsory migrations that have significantly contributed to the abnormal increase in population, most of which were due to the influx of over 700 000 Iraqis and about 1.5 million Syrian refugees due to the current political unrest in both countries. Rapid population increase in the narrow strip within the main cities of Amman, Zarqa, Irbid and Mafraq has placed unprecedented demands on natural and water resources, as it has on the infrastructure and the economy in general. The total demand of water resources is approaching one billion cubic meters per year, which approximates the limit of Jordan's renewable and economically developable water resources.

The contribution of agriculture to the gross domestic product (GDP) reached $1.5 billion in 2016 (3.8% of GDP). However, when agribusiness activities were included, agriculture contributed 29% to the GDP. Due to scarcity of water and low rainfall, about 380 000 ha are cultivated and only 17% of this area is irrigated, which accounts for less than 0.1 ha per capita.

The prevailing agricultural production systems in Jordan are fragile and unsustainable and may not remain economically viable in view of the steadily increasing input costs. Scarce rain and irrigation water are not sufficient for agricultural needs because of high evaporation losses and CT, frequent, deep ploughing and lack of soil cover. The grazing or burning of crop residues combined with intensive tillage has resulted in low soil organic matter content (less than 1%), destruction of soil structure, and a decrease in water infiltration and storage capacity.

One of the deepest and most direct impacts of climate change over the next years will be on agriculture and food production systems. Quantitative assessments show that climate change will adversely affect food security, and small farmers in poor countries will likely be affected (Brown and Funk, 2008; Schmidhuber and Tubiello, 2007). The impacts of increased maximum and minimum temperatures and decreased annual precipitation in three regions in Jordan were reported by Shakhatreh (2010). The Intergovernmental Panel on Climate Change (IPCC, 2007) reported that global mean surface temperature is projected to increase in a range of 1.8°C to 4.0°C by 2100. Therefore, in drier regions, all climate models predict an increase in evapotranspiration and lower soil moisture levels, as drought and heat waves increase, especially in the Mediterranean region.

Despite the problems described above, there are still opportunities to stabilize or even increase crop production and enhance food security in Jordan. Many studies have shown that the gap between farmers' yields and those obtained with improved management packages is high, and productivity can be significantly improved and sustained with improved varieties and crop management technologies. These practices could have a major impact if widely disseminated. Among these opportunities is the adoption of CA or zero tillage (ZT) systems. CA is widely recognized as an important and viable concept for sustainable agriculture due to its comprehensive benefits in economic, environmental and social sustainability (Derpsch and Friedrich, 2009; Kassam et al., 2010, 2018; FAO, 2012a; Hermans et al., 2020).

This report presents the results of CA/ZT field trials and demonstrations in Jordan while highlighting the institutional support from the Ministry of Agriculture and other institutions and the major reasons why CA was modestly disseminated in Jordan.

3.1 Methodology

3.1.1 Target areas

The real work to promote CA in Jordan was during the period (2007–2016) in collaboration with different institutions and they were GIZ, ACSAD, Arab Authority for the Agricultural Investment and Development (AAAID), FAO, ICARDA and the National Agricultural Research Center (NARC). Farmers' field demonstrations were conducted in Irbid, Madaba and Karak governorates, and field trials were conducted at the Maru experimental research station of (NARC) in the Irbid governorate of Jordan (32°36′20.16N and 32°54′3.95E with an elevation 590 m above sea level and 412 mm long term rainfall).

3.1.2 Wheat and vetch farmer demonstrations (2009–2011)

The work and activities during these two seasons were performed in collaboration with the FAO and NARC through the 'Sustainable agricultural production through promotion of Conservation Agriculture in Jordan' project, funded by the FAO. Field demonstrations in farmers' fields for applying CA on wheat-based systems were conducted in three regions (north, middle and south) in Jordan during the 2009–2010 and 2010–2011 growing seasons in order to compare biological yield, grain yield and water use efficiency under the CA system to those in the CT system. Fifteen farmer fields' demonstrations were chosen for the implementation of CA and CT in the wheat-based system in the north (Irbid), middle (Madaba) and south (Al-Karak) governorates. The FAO supported providing three Vence Tudo zero-till seeders imported from Brazil (Fig. 5). These seeders are suitable for small and mid-size farms and to plant small seeds as well as large seeds.

Before starting the implementation of the field demonstrations, several inception and awareness workshops were held in order to explain the principles and benefits of CA.

During the first season (2009–2010), five wheat demonstrations were conducted in each governorate and planted during the first 2 weeks of December 2009. Each field was divided into two parts. One part was for CA and the other part was for CT with two tillage operations using a chisel plough. Five random samples (3 m^2), each from both CA and CT, were taken for measuring biological, grain yields and water use efficiency. In the second season, the same

Figure 5 Vince Tudo zero till seeder used to plant wheat, barley and vetch at NARC and in farmers field demonstrations. *Source:* Photo by Y. Shakhatreh, 2010.

procedures were followed and the 15 demonstration plots were planted with vetch. The area of each demonstration ranged between 2 and 5 ha.

Results showed that wheat biological yield, grain yield and water use efficiency under the CA system were significantly higher than those of the CT practices. The overall mean of wheat biological yield, grain yield and water use efficiency under CA was 5222 kg ha^{-1}, 1331 kg ha^{-1} and 0.399 kg m^{-3}, respectively. Whereas, the mean was 4076 kg ha^{-1}, 1071 kg ha^{-1} and 0.32 kg m^{-3}, respectively, under CT. The same trend was found for vetch production, where vetch's biological, grain yield and water use efficiency under CA superseded those under CT by 22%, 25% and 27%. In addition, the increase of net return due to the application of CA was \$ 20 ha^{-1} (14 JDs) compared to that of CT on wheat crop.

3.1.3 Barley and vetch farmer demonstrations (2011-2014)

These activities were performed as a collaboration between ICARDA and NARC and sponsored by IFAD under the project 'improving the food security and climate change adaptability of livestock producers using the rainfed barley-based system in Iraq and Jordan'. The location of the target areas is shown in Fig. 6. Target areas include two clusters of villages that represent the main barley-livestock agricultural activities in the Al-Karak governorate. A group of researchers and extension agents visited the target areas and interviewed farmers to identify their profiles during the first of October 2013. The site was selected in the first year of the project, and new farmers were added to participate in the project. The total number of demonstrations was 29 and the total area was 94.5 ha.

3.2 Methodologies and data collection

The soil moisture content at three depths (0-15 cm, 15-30 cm and 30-45 cm) was monitored before planting and after harvesting for both conventional and zero-till demonstrations. Twenty-nine farmer fields' demonstrations were chosen for the implementation of ZT and CT practices on barley, *Vicia sativa* (Bekia) and *Vicia ervilia* (Kersana) crops in Al Karak governorate. Each field was divided into two parts: ZT without any kind of tilling and CT with two tillage passes using a chisel plough. Barley ZT practices included two improved drought-tolerant varieties (Muta' and Adrouh), in addition to the check variety (RUM), while in case of CT, only RUM was used. Five random samples (5 m^2/plot) were taken for measuring biological and grain yields.

Since the locations of the implemented demonstrations were scattered across the governorate and two clusters were pooled (Fig. 6), we chose Rabbah weather station to represent climate data (rainfall), (Fig. 7).

The three seasons (2011-2012, 2012-2013 and 2013-2014) differed in rainfall and its distribution. During 2011-2012, there was a delay in the first effective rain until late December, and the rain stopped in late March. Whereas during 2013-2104, good rainfall was received in December, but the rainfall ended in February. These situations impacted negatively on the growth of crops, especially vetch varieties, forcing farmers to graze it early.

3.3 Crop performance

In general, the biological growth of crops grown under ZT was higher than CT (Fig. 8).

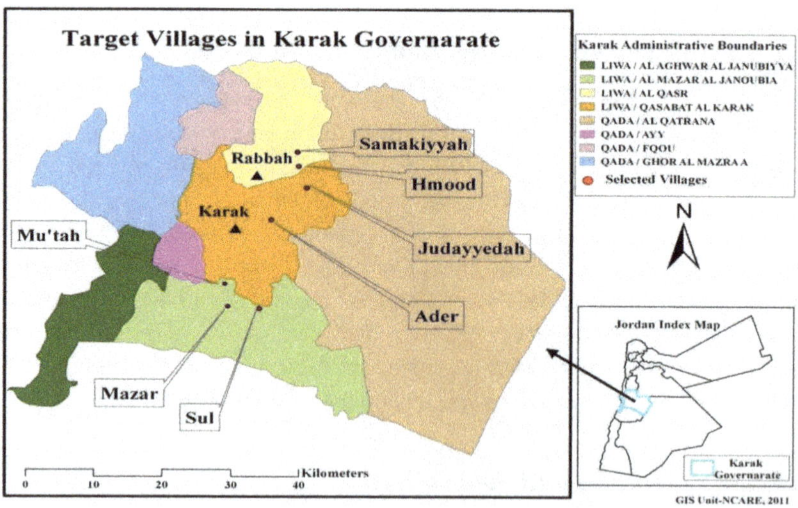

Figure 6 Distribution of CA farmers across Al-Karak. *Source:* Photo by GIS Unit-NCARE, 2011.

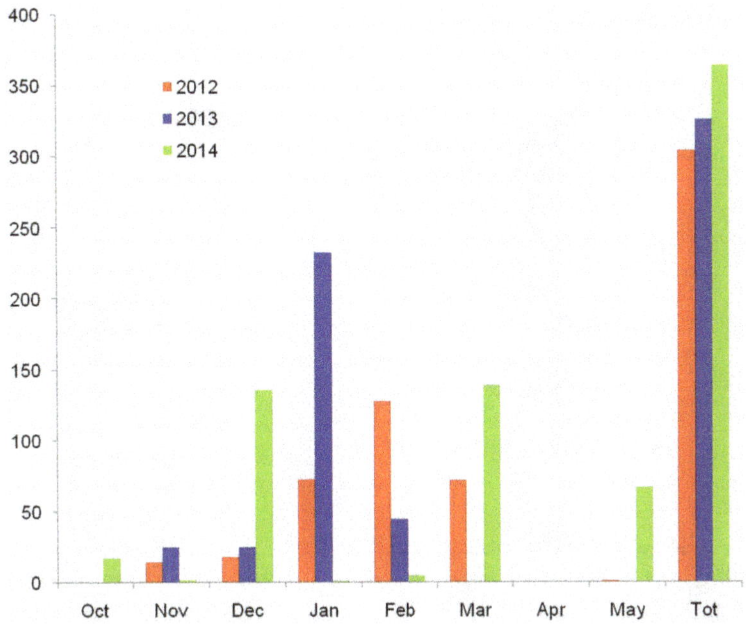

Figure 7 Monthly rainfall (mm) distribution in Rabbah station during the three seasons 2011-2012, 2012-2013 and 2013-2014 (Y. Shakhatreh, 2014).

3.3.1 Wheat and vetch demonstration (2017-2018)

These activities were implemented with the collaboration of ICARDA and NARC through the 'enhancing food security in Arab countries' and sponsored by different donors and implemented in three districts in the Northern part of Jordan (Irbid governorate). Seven wheat demonstrations were conducted in farmers' fields in the Irbid governorate.

Tables 6 and 7 summarize the obtained yields of wheat applying ZT vs. CT planting methods and also for vetch grown in north Jordan.

3.3.2 Experimental field trials

The below research work is a collaboration between the University of Nottingham (UoN) and NARC through an EU project 'adaptation to climate change of the Mediterranean agricultural system' (ACLIMAS).

With the aim to increase grain and dry matter yields under rain-fed agriculture, a field experiment was conducted at Maru Agricultural Station (Irbid, Jordan) in the 2012/13 growing season to investigate the performance of two wheat genotypes growing under ZT and CT during three sowing dates.

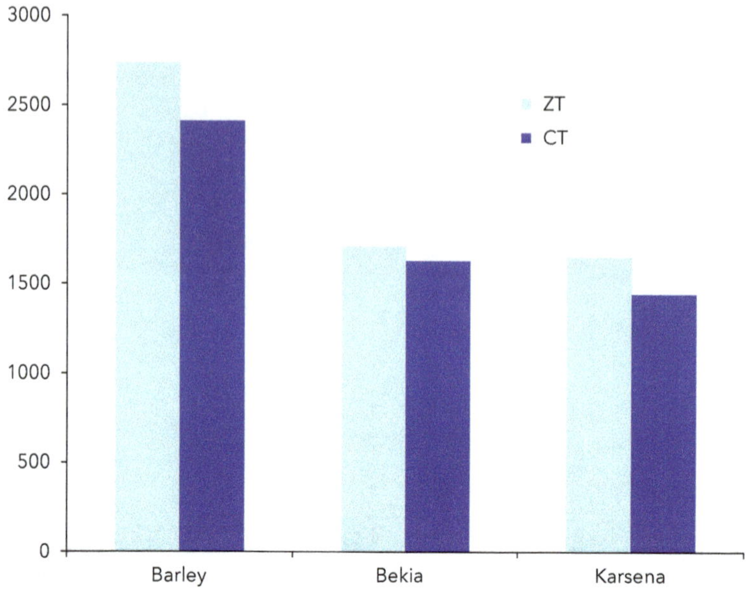

Figure 8 Average biological yield (kg ha^{-1}) of crops grown under ZT and CT in Karak, south part of Jordan during 2012, 2013 and 2014 growing seasons (Y. Shakhatreh, 2014).

3.3.2.1 Experimental site and design

A field experiment was implemented at the Maru experimental research station of the National Center for Agricultural Research and Extension Center (NARC) in Irbid, elevation 590 m above sea level and 412 mm long-term rainfall. The soil type was heavy clay soil and the available water was 180 mm m^{-1}. The experimental design was a split-split block design with three sowing dates (sub-sub plot) and 2 weeks interval: (i) early, planted on 28 November 2012, (ii) mid, planted on 16 December 2012 and (iii) late, planted on 2 January 2013 using two durum wheat cultivars (sub plot): (i) ACSAD 65 and (ii) Hourani grown under two tillage systems (plot): (i) ZT and (ii) CT with three replicates. The Zadoks scale was used to determine wheat developmental stages (Zadoks et al., 1974). The sub-sub plot size was 50 m^2.

3.4 Results

Temperatures for the 2012/13 growing season were typical of the Irbid region. Although rain started relatively late in the growing season and there was a very dry March and May in 2013, the total rainfall was 358.2 mm, with average rainfall values for the region as shown in Fig. 9.

Table 6 Wheat grain yield (GYLD), biological yield (BYLD), and profitability under zero tillage (ZT) and conventional tillage (CT) systems applied in farmers' fields in northern Jordan during the 2017/2018 season

			BYLD (t ha^{-1})			GYLD (t ha^{-1})			
Number	Farmer Name	Variety	ZT	CT	Increase % BYLD	ZT	CT	Increase % GYLD	[a]Profitability (US$ ha^{-1})
1	Mahmoud Subhi Bwaneh	Deir alla6	7.11	6.25	13.7	1.95	1.74	12.1	400.80
2	Jaser Al Quran	Deir alla6	8.52	6.76	26.0	2.84	2.13	33.3	810.80
3	Ibrahim Shtewi Rwaq	Deir alla6	8.14	6.08	33.9	2.75	2.05	34.1	898.00
4	Fareed Suleiman Alshoha	Acsad1103	9.37	8.00	17.1	2.91	2.33	24.9	657.40
5	Ahmed Hussein Bataeneh	Acsad1103	9.88	8.59	15.0	3.00	2.41	24.5	636.20
6	Mohamed Mahmoud Jaradat	Acsad1187	7.50	6.43	16.6	2.53	2.23	13.4	489.00
7	Amar abbabneh	Cham1	6.22	5.52	12.7	1.71	1.58	8.2	330.40
	Average		8.11	6.80	19.3	2.53	2.07	21.5	603.23

[a] This value was calculated depending on money saved as a result of decreasing the cost of land preparation plus the increase in farmer's income due to grain and straw yield increases. This was based on a price of wheat grain of US$580 t^{-1}, of wheat straw of US$ 300 t^{-1}, and the cost of two tillage operations of US$ 84 ha^{-1}.

Table 7 Vetch grain yield (GYLD), biological yield (BYLD) for zero tillage (ZT), conventional tillage (CT) systems applied in farmers' fields and profitability in northern Jordan during the 2017/2018 season

		BYLD (t ha^{-1})			GYLD (t ha^{-1})			
Number	Farmer name	ZT	CT	Increase % BYLD	ZT	CT	Increase % GYLD	[a]Profitability (US$ ha^{-1})
1	Montaha Natheer Sweed	3.32	2.96	12.2	1.21	1.01	19.8	327.60
2	Mohamed Kher Ababneh	3.01	2.89	4.1	0.97	0.88	10.2	170.10
3	Falehyouesf Rfaai	3.45	2.80	23.2	1.11	0.85	30.6	502.60
4	Salehmahmood Ababneh	3.27	3.00	9.0	0.92	0.78	17.9	264.60
5	Huseentalal Altaweel	3.00	2.81	6.7	0.78	0.70	11.4	207.20
6	Ali Suleiman Deifallah	3.73	3.44	8.4	1.05	0.98	7.1	261.10
7	Wedadmohamed Ubedat	3.47	3.12	11.2	0.96	0.90	6.7	292.60
8	Mohamed Abdalazeez Wardat	4.14	4.00	3.5	1.45	1.23	17.9	208.60
9	Anees Ismael Taweel	4.71	4.36	8.0	1.52	1.33	14.3	319.90
10	Khalil Khasawneh	3.32	3.11	6.7	0.87	0.81	7.4	214.20
11	Esam Mohamed Alnaeem	3.59	3.21	11.8	1.05	0.94	11.7	319.90
Average		**3.55**	**3.25**	**9.5**	**1.08**	**0.95**	**14.1**	**280.76**

[a] This value was calculated depending on money saved as a result of decreasing the cost of land preparation plus the increase in farmer's income due to grain and straw yield increase. This was based on a price of vetch grain of US$ 770 t^{-1}, of vetch straw US$ 560 t^{-1}, and cost of two tillage operations of US$ 84 ha^{-1}.

There was an apparent trend for ZT to increase above-ground dry matter (AGDM) under both mid and late sowing dates compared to CT (AGDM). However, the increase in yield slightly decreased under early planting. As for harvest index (HI), ZT significantly ($P < 0.05$) decreased the overall HI by 6%, but there were no significant differences among cultivars and sowing dates (Table 8).

It was found that ACSAD 65 variety gave a higher grain yield under ZT than under CT, but this trend was not observed for the Hourani wheat variety (Table 9).

3.5 Conclusion

Increasing and sustaining agricultural production of wheat-based systems under climate change impact is a challenge in Jordan. This raised the question: Shall we change the existing CT systems and crop agricultural practices?

The above results show that ZT could be a helpful adaptive measure for climate change impact. ZT led to increased GY, AGDM, plant height, grain number and fertile shoot number when compared with CT, and hence the profitability.

Although the adoption of CA in Jordan is developing at a slower pace than in other countries in the region, with a current area of around 2000 ha, the potential of benefits that farmers may be able to achieve through the CA systems should encourage them to adopt CA practices and systems in the future. The main challenges are (1) lack of available and accessible CA machinery, (2) weed control, (3) crop residue management and 4) lack of awareness. These challenges could be overcome by convincing decision-makers about

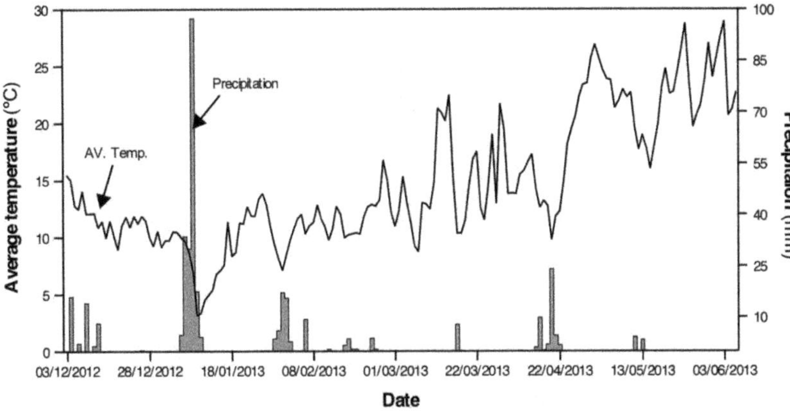

Figure 9 Daily average temperature and precipitation in 2012–13 at Maru Irbid, Jordan (Shakhatreh, Y. 2013).

Table 8 Above ground dry matter (AGDM) at harvest, grain yield (GY) and harvest index (HI) for two durum wheat genotypes (ACSAD 65 and Hourani) grown under CT and ZT and at three sowing dates 28 November 2012 (early), 16 December 2012 (mid) and 2 January 2013 (late)

		AGDM harvest (g m^{-2})		GY (t ha^{-1})		HI	
Genotype	Sowing date	CT	ZT	CT	ZT	CT	ZT
ACSAD65	Early	661	554	2.86	2.34	0.43	0.43
	Mid	423	544	2.03	2.46	0.48	0.46
	Late	392	441	1.76	1.70	0.45	0.39
Mean		503		2.19		0.44	
Hourani	Early	605	537	2.31	2.09	0.38	0.38
	Mid	419	565	1.81	2.20	0.44	0.39
	Late	330	389	1.44	1.57	0.44	0.40
Mean		474		1.90		0.40	
$SED_{tillage}$ (Df)		47.7^{ns} (2)		0.181^{ns} (2)		0.0047^{a} (2)	
$SED_{genotype}$ (Df)		12.7^{ns} (4)		0.090^{a} (4)		0.0173^{ns} (4)	
SED_{sowing} (Df)		27.8^{c} (16)		0.139^{c} (16)		0.0139^{ns} (16)	
$SED_{tillage*sowing}$ (Df)		57.5^{b} (4.13)		0.242^{b} (5.91)		0.0167^{ns} (17.83)	

[a] for $P \leq 0.05$, [b] for $P \leq 0.01$, [c] for $P \leq 0.001$ and ns for a non-significant result for the ANOVA test.

Table 9 Plant height (PLH), grain number (GrNo) and fertile shoot number (FeNo) for two durum wheat genotypes (ACSAD 65 and Hourani) grown under CT and ZT and at three different sowing dates 28 November 2012 (early), 16 December 2012 (mid) and 2 January 2013 (late)

			PLH (cm)		GrNo (m^{-2})		FeNo (m^{-2})	
Genotype		Sowing Date	CT	ZT	CT	ZT	CT	ZT
ACSAD 65		Early	80.1	86.6	9309	10444	233	236
		Mid	72.6	82.3	7757	8385	181	235
		Late	72.1	80.6	7032	8034	175	159
	Mean		79.1		8494		203	
Hourani		Early	90.0	96.9	7760	7642	264	247
		Mid	87.4	91.7	6834	7469	195	234
		Late	82.4	79.1	5963	5830	193	188
	Mean		87.9		6916		220	
	$SED_{tillage}$ (Df)		2.59^{ns} (2)		740.6^{ns} (2)		17.7^{ns} (2)	
	$SED_{genotype}$ (Df)		3.16^{a} (4)		789.2^{ns} (4)		21.5^{ns} (4)	
	SED_{sowing} (Df)		2.51^{b} (16)		765.7^{a} (16)		14.9^{b} (16)	
	$SED_{tillage*sowing}$ (Df)		3.89^{ns} (8.44)		1153.4^{ns} (9.38)		24.6^{ns} (6.82)	

[a] for $P \le 0.05$, [b] for $P \le 0.01$ and ns for a non-significant result for the ANOVA test.

the importance of replacing existing CT-based cropping practices with more sustainable farming systems based on the CA principles. Therefore, real support of institutional organizations, the involvement of policy makers, NGOs, international and national research bodies, and funding agencies is essential to improve the adoption of CA in Jordan. Farmer incentives (e.g. subsidized ZT seeders) and training will help promote the adoption of CA.

4 Syria

Crop production in West Asian countries is characterized by frequent soil tillage, grazing, burning or harvesting of the crop stubbles, and delayed seeding to allow for weed control by frequent soil tillage (Haddad et al., 2013). Soil degradation is a great threat to crop production and food security for rural livelihood in West Asian countries. CA practices, such as minimum soil disturbance, biomass soil cover management, and diversified crop rotation, have advantages over CT by reducing costs, increasing water use efficiency, reducing run-off and soil erosion, and increasing soil organic matter without compromising yield.

In this context, ICARDA undertook research and development activities on CA for sustainable crop production in the region since 1984 (Fig. 10) (Haddad et al., 2013). This section highlights the constraints, opportunities and elements of CA systems and practices.

4.1 Suitability of Conservation Agriculture in different contexts

Since its establishment in 1977, ICARDA has incorporated a strong farming system component into its research with its multidisciplinary component focusing on the drier, less productive, and more-risky agro-ecological zones

Figure 10 First CA adopter farmers holding with zero tilled (with good growth) and conventional tilled (poor growth) barley plants at Hama, Syria (left) and lentil plants at Aleppo, Syria (right). *Source:* Photo by A. Haddad, ICARDA, 2010.

in West Asia, North Africa (WANA) regions. Its applied and adaptive research programme focused on different scales in key target areas that match local conditions in collaboration with national, regional and international research institutions. In the last 3-4 decades, ICARDA has carried out its research activities with a special focus on CA practices; however, early research activities were mainly concerned with no-till or direct seeding, with less importance on crop rotation and residue management, which are more complex system changes. A comprehensive CA, research and development, work started with ICARDA in Syria and Iraq in 2005 with funds from the Australian Centre for International Agricultural Research (ACIAR) and Australian expertise, which was later expanded to some West Asian countries, that is, Syria, Jordan, Iraq, Lebanon, and many WANA countries through different long-term (2005–2015) and short-term projects.

ICARDA's research and development programme has focused on generating scientific evidence from CA systems through many short and long-term research trials, and on-farm verification of the research outcomes under the local condition in different aspects. In this context, Loss et al. (2015a) found that there were appreciable benefits of ZT systems, which enabled early sowing, more efficient use of rainfall and yield increases, even in the absence of crop biomass retention or diverse rotations. This became an important starting point for adopting no-till technology in integrated crop-livestock systems in Syria and Iraq. In medium and low rainfall areas, no-tillage sowing for cereal crops as early as possible, either immediately after the first effective autumn rains or even into dry soil before the first rains (in marginal land), was shown to maximize water-use efficiency and grain yields in the longer term (Piggin et al., 2011, 2015; Loss et al., 2015a). Similarly, the adoption of ZT helped to improve production efficiency and reduce the production risk for wheat in Syria (El-Shater et al., 2020).

Varietal evaluation of major crops on yield performance between CA and CT identified no significant interaction between tillage and crop variety, suggesting that the existing crop varieties grown by farmers could be used under CA systems without yield penalty (Piggin et al., 2011). Other CA aspects were also evaluated, for example, yield performance of different crops (Bashour et al., 2016; Piggin et al., 2015; Loss et al., 2015), effect on soil quality (Sommer et al., 2011), economic benefits (El-Shater et al., 2020; ICARDA, 2012), water use efficiency (Bashour et al., 2016), reduction of soil erosion, weed infestation and disease pressure on different crop genotypes (Ahmed et al., 2012). The outcomes of the site-specific evidence were communicated through numerous publications (Ahmed et al., 2012; El-Shater et al., 2020; Haddad et al., 2013; Kumar et al., 2011; Piggin et al., 2015; Sommer et al., 2011; Yigezu et al., 2018), a practical manual (Loss et al., 2015), and workshops and training courses.

4.2 Availability of low-cost zero tillage seeders

Together with verification of the technology, special attention was paid to the establishment of a commercial supply chain for the machinery required for the successful adoption of CA in the region (Haddad et al., 2013). During the time of initiation of CA in the region, the commercially available ZT seeders, mostly made in wealthy countries, were sophisticated, disc-type, and expensive, which placed them out of reach of smallholders in developing countries. In response to this situation, ICARDA coordinated an effort, in collaboration with different national, international and private sectors, to make simple, low-cost seeders in small local workshops in Syria, Iraq and Jordan (Loss et al., 2015). With this effort, by the middle of 2014, there were more than five different brands of low-cost no-till seeders originating from Syria, (Figs 11 and 12) and later, after the civil conflict, from Iraq and Jordan. Also, the effort was dedicated to the establishment of local workshops, which provide repair and maintenance services for ZT seeders and thereby creating local youth employment (Loss et al., 2015).

The local Syrian ZT seeders were designed to match local farming systems and farmer's budget, costing US$5–10 000 (Loss et al., 2015). These seeders were characterized by being:

- Light in weight (300–800 kg).
- Appropriate for farm size (2–4 m active width).
- Suited to available tractors (40–80 hp).
- Tine-type seeders, which are generally lighter, simpler, cheaper, and better suited to a wide range of soil types compared to disc-type seeders.
- Easily maintained and repaired with locally available parts.

To assess the effectiveness of these locally produced ZT seeders, their performance was compared with imported seeders (Fig. 13). All local seeders performed as well as the imported seeders for all major crops, that is, wheat, barley, lentil, and chickpea (ICARDA, 2012).

ICARDA also helped the local seeder manufacturers to convert the existing CT seeder into ZT seeders in different countries, that is, Syria, Iraq and Jordan. This drastically reduced the cost of seeders (US$500–1000) compared to manufactured or imported seeders. Feedback from the farmers who bought the adapted seeders was provided to manufacturers (Fig. 14), which helped to gradually improve the performance of the seeders. The local manufacturing of ZT seeders has been an important factor in driving adoption, as imported seeders are expensive and too large for small-scale farmers in Syria and the region (Jalili et al., 2011; Piggin et al., 2011).

Figure 11 Low cost no-till tine seeders manufactured in Syria. These dual hopper seeders are 2.3 m to 3.8 m wide, with price range of US$ 5–10 k. *Source:* Photo by A. Haddad, ICARDA, 2009.

Figure 12 Local workshop preparing no-till seeders for shipment to North Africa, Qabbaseen-Syria, 2011 (left). New local workshop preparing no-till seeders for AKF and UNDP, Salamiyeh-Syria, 2019 (right). *Source:* Photo by A. Haddad and ICARDA, 2011; AKF, 2019.

Figure 13 Locally made tines, simple and easy to replace, badly worn knife openers (left) with a characteristic 'bullnose' rounding at the leading edge will not penetrate the soil as effectively as a new opener (right). *Source:* Photo by J. Desbiolles, 2014.

Figure 14 Key farmers and all seeder manufacturers comparing different no-till seeders to evaluate their seeding performance in a field day organized by ICARDA at Tel Hadya station, Syria. *Source:* Photo by ICARDA, 2010.

4.3 Capacity development for scaling Conservation Agriculture adoption

Capacity development of different industry sectors and awareness campaigns about CA technology through different means are the foundation for future research, participatory extension, and dissemination of technology at different scales (Yigezu et al., 2018). In this regard, ICARDA, in collaboration with national, regional, international and private sector partners supported the training of several scientists, extension specialists, machinery suppliers and farmers at different levels. For example, ICARDA followed participatory extension principles by establishing local CA farmer working groups to evaluate the CA system, where each working group was provided with a simple and affordable locally manufactured no-till seeder, (Haddad et al., 2015a). Similarly, in different CA-related projects, ICARDA facilitated the development of the capacity of more than 700 Iraqi and 800 Syrian scientists, extension specialists, machinery suppliers, and farmers through different means, that is, short-term training courses, research fellowships, and on-the-job training (ICARDA, 2012). This provided the foundation for future research and extension of CA in the country.

Field days were conducted extensively, allowing a large number of diverse groups to observe, learn and discuss the new technology with different farmers and stakeholders (Fig. 15). Field days at sowing time were particularly popular, giving participants the opportunity to observe and assess no-till planters in action (Loss et al., 2015a).

Based on scientific evidence from field trials and learning from participatory extension programmes, ICARDA also developed principles for productive and sustainable cropping systems with conservation tillage (ICARDA, 2012). In 2015, ICARDA published a comprehensive technical manual in English and Arabic 'The Practical Implementation of Conservation Agriculture in the Middle East' (Loss et al., 2015). This publication was designed to help innovative farmers, machinery manufacturers, extension specialists and researchers learn more about CA in the Middle East, especially the practicalities of implementing the various CA principles in the field.

4.4 Conservation Agriculture in Syria (2011-2020)

Before the outbreak of civil unrest in Syria in 2011, the integrated research, development and extension efforts of ICARDA and national and international partners achieved a widespread application of CA with a relatively rapid adoption rate. Since then, however, the ongoing Syrian conflict has severely restricted agriculture, as farmers lost their assets or were displaced, and most funds and subsidies were reallocated from agricultural development into humanitarian relief activities. Despite the absence of governmental institutions and support from development agencies, many farmers kept practicing CA, benefiting from input savings and more productive farming systems, especially during times of severe fuel scarcity. However, machinery manufacturing was also disrupted and no-till local seeders became unavailable or unaffordable due to disconnected country areas and/or migration of local expertise. There have been modest efforts to restore the agricultural sector and rural livelihoods, and

Figure 15 Different stakeholders assessing no-till planter during field day in Amman-Jordan, 2015 (left) and in Salamiyeh-Syria, 2010 (right). *Source:* Photo by A. Haddad, ICARDA, 2015.

in line with this, ICARDA, Aga Khan Foundation (AKF), World Food Programme (W F P) and UNDP have recently supported small CA development projects.

4.4.1 Research and development

Very limited research was conducted during the conflict period, as many of the agricultural research stations were destroyed or not functioning. However, post-graduate student research continued, where seven different thesis topics were carried out since 2015 (M. Alzoubi, personal communication). A long-term study to assess the impact of CA on orchard management was established in 2015, which showed improvements in soil physical and chemical properties due to CA as well as the reduction of the spread of two olive diseases, Peacock Eyespot (*Piloceae oleaginea*) and Verticilium Wilt (*Fusarium dali)* (Barhoum B. and Haddad A. 2015–2020, personal communication).

4.4.2 No-till seeder manufacturing

Three of the original workshops kept operating during the conflict, however, they were not able to supply seeders because of the disconnected communication and travel between districts within the country. The AKF supported local workshops to supply seeders in the middle areas of Syria in 2015, and the UNDP funded a project in 2019 to restore local no-till seeder manufacturing based on ICARDA expertise. These funds helped the development of three new workshops, through support and training to produce no-till seeders and local workshops. The workshops produced 24 no-till seeders (2019–2020) and made them available to the farmer community for a reasonable price of US$ 1500–2000 each.

4.4.3 Conclusion

4.4.3.1 Extension for wider adoption

Recently attempts are being made to resume CA promotion campaigns in the country. National and international development agencies have already started individually (and hopefully collectively) to accelerate CA adoption in Syria, although the current situation requires a different dissemination strategy. These new strategies include considering disconnected areas, encouraging and supporting new local workshops to produce affordable no-till seeders and, unlike the past, assisting small farmers with other inputs, such as quality seeds, fertilizers and herbicides. The national Syrian government programme has recently prioritized CA systems and practices based on documented benefits and FAO recommendations. A national CA workshop, organized by AKF and

MoA, was held in December 2019, where representatives of FAO, UNDP, AKF, ACSAD, national programme and Ministry of Agriculture attended. The workshop recommended accelerating the adoption and spread of CA based on past local experience and international funding. The workshop also decided to establish a national coordination mechanism to support this acceleration. Currently, the CA area in Syria is estimated to be about 30 000 ha, similar levels as in 2012.

5 Iraq

Nineveh province is known as the 'bread basket of Iraq' because it is the most important province for cereal production. The total area of Nineveh province is 32 308 km^2, about 1.64 million ha. The agricultural lands in Nineveh province are described according to the annual rain rates of three regions, high rainfall area (HRA, more than 500 mm), moderate rainfall area (MRA, 300–500 mm), and low rainfall area (LRA, less than 300 mm). Rain-fed agriculture constitutes about =95% of agricultural land and only about 5% is given supplementary irrigation.

5.1 Farming systems

5.1.1 Rain-fed farming systems

The prevailing crop in the LRA region is barley due to its drought tolerance. Two cropping systems are followed in the LRA region: barle-fallow-barley rotation and continuous barley. The first cropping system during the fallow year includes four steps: (1) wait for the first effective rain (more than 30 mm) to germinate the weeds, then implement the winter tillage (first tillage) to eliminate the weeds. (2) perform spring tillage to eliminate the weeds again (second tillage). (3) perform autumn tillage to soften the soil (third tillage). (4) sow the barley with a disc seeder. For the continuous barley cropping, sowing is done either before or after the first effective rain (Alrijabo, 2012a).

The prevailing crops in the MRA region are 60–80% bread wheat and 20–40% barley. Cropping systems in the MRA region are mainly barley-fallow-barley, continuous barley, wheat-fallow-wheat or continuous wheat. No more than 5% of this area applies a wheat-legume-wheat rotation. The steps in the cropping systems are similar to LRA (Alrijabo, 2014).

The prevailing crops in the HRA region are 80% bread wheat and 20% durum wheat. Cropping systems in this region are mainly continuous bread wheat, continuous durum wheat and not more than 10% wheat-legume-wheat. Steps in the cropping system are to (1) wait for the first effective rainfall, (2) wait for the emergence of weeds, (3) till with a disc plough and (4) sow with a modern seeder (Al-Juburi and Alrijabo, 2012).

5.1.2 Farming system with supplementary irrigation

The prevailing crops are 80% bread wheat and 20% durum wheat. Cropping systems under supplementary irrigation are continuous bread wheat, continuous durum wheat and 10% wheat-legume-wheat. Sometimes, vegetables are grown if there is enough water. Steps in the cropping system are (1) irrigation of the field as a substitute for the first effective rain, (2) ploughing the field for weed control and (3) crop sowing with a modern seeder (Alrijabo, 2012b).

5.2 Risks facing agriculture in rain-fed areas

1. Climate change in LRA and MRA regions increased drought in these two regions, which represent more than 80% of the area of cultivated land in Nineveh.
2. Desertification, dust storms and sand inundation of rural homes of many villagers, which forced the residents of 15 villages in the Tal Abata district in Nineveh to migrate to other villages.
3. Farmers insisting on using multiple tillage to eliminate weeds has led to severe soil degradation, erosion and dust storms impacting villages and nearby cities Fig. 16).
4. High fuel consumption (32 L ha^{-1}), which in return increases the cost of production and produces carbon dioxide and other greenhouse gases contributing to climate change and air pollution (Fig. 17).
5. The decrease in rainfall accompanied by high temperatures led to a decrease in wheat cultivation and to an increase in barley. If this trend continues cultivation of the land will become uneconomical and will stop.

Thus, according to the previous review, the Iraqi government planned and implemented a CA programme in Nineveh province that was administered by

Figure 16 Rural homes covered by sand dune due to wind erosion. *Source:* Photo by J.K. Shallal, 2009.

Figure 17 Air pollution caused by intensive tillage of dry soils. *Source:* Photo by A. Alrijabo, 2011.

the Iraqi Ministry of Agriculture in cooperation with the University of Mosul, with financial support from the ACIAR and technical representation from Australian universities and the ICARDA. The programme activities started in 2005 and ended in 2013. The programme expanded, due to its success, to include three additional provinces: Kirkuk, Salahuddin and Anbar, where the programme included human, economic and environmental development as given below.

5.3 Human and knowledge development

Human development activities included enhancing the skills of farmers and researchers to change their convictions about old planting methods and adopting the new ZT planting method through the implementation of field observations in farmers' fields under different environmental conditions in Nineveh and the other provinces.

CA programme started with testing the no-till planting method using an Indian-made seeder with a width of only 2 m and without press wheels. However, this was light weighted and not very effective. More suitable ZT seeders with a wider working width (3 m and 4 m) with press wheels were introduced from Syria and later Jordan, in addition to a small number of locally made seeders. The programme succeeded in supporting the local manufacturing of ZT seeders in cooperation with local manufacturers and pioneer farmers (Jalili et al., 2011). The programme also converted 20 traditional seeders owned by farmers to ZT seeders (Saad Abdul-Jabbar et al., 2013).

The direct implementation was then made in the farmers' fields after the farmers were introduced to the new technology's benefits and application mechanisms. No-till seeding demonstrations were conducted in 35 farmers' fields in Nineveh, 5 farmers' fields in Anbar, 4 farmers' fields in Salah al-Din and 4 farmers' fields in Kirkuk.

The staffs of the provinces of Kirkuk, Anbar and Salah al-Din were trained inside and outside Iraq to become field leaders and agricultural extension

Figure 18 Training courses for field staff working in the CA programme: (a) & (b) Anbar province, (c) Kirkuk province, (d) Salah Aldin province. *Source:* Photo by A. Alrijabo, 2012.

agents for the newly introduced technology (Fig. 18; Darweesh et al., 2013). The programme provided more than 60 specialist training courses and field days for farmers inside and outside Iraq.

The programme also achieved a high level of scientific activity in research and postgraduate studies, which were implemented at the University of Mosul and in the Nineveh Research Department. More than 125 research papers, Masters and PhD theses were produced.

5.4 Economical and environmental development

The economic success of CA is due to the fact that this system seeks to reduce inputs (seed rates, fuel, hours of work, equipment maintenance and the amount of water used for supplemental irrigation) while increasing production most of the time. Hence, in our opinion, the CA programme also succeeded in boosting farmers' income and in expanding the spread of ZT planting practice. Figure 19 shows the progress in the ZT areas (ha) in Nineveh from the first season 2005–2006 to the eighth season 2012–2013 (Loss et al., 2013). In 2008, 12 000 ha were seeded with ZT drills in CA cropping systems (Fig. 19).

5.5 Conclusion

From the research conducted on CA in Iraq, we can summarize the benefits of CA systems as follow:

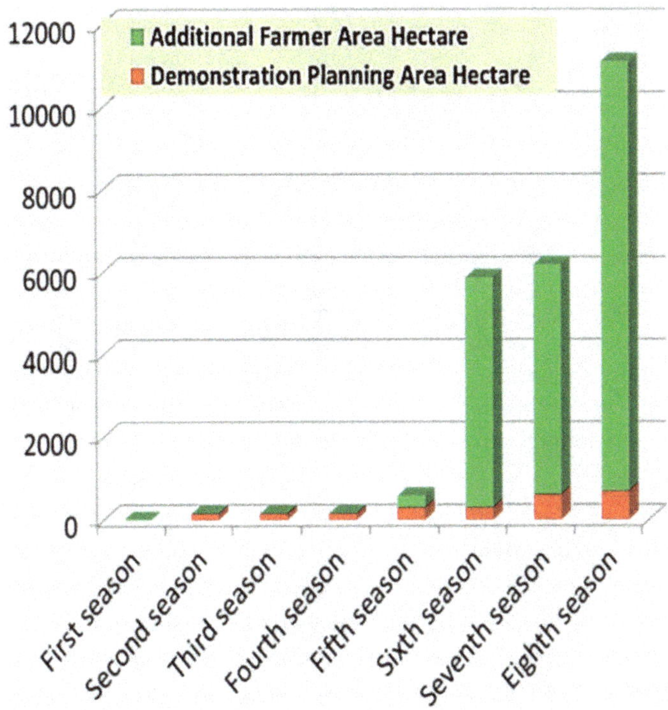

Figure 19 Area under CA system in Iraq (green) and under on-farm field demonstrations (red) from 2006 to 2014 (Alrijabo, 2014).

1 Reduced soil erosion: Leaving the crop residues of the previous harvest on the soil surface reduced or prevented soil erosion and led to building up the level of organic matter in the soil as well.
2 Flexible sowing time. Farmers no longer had to wait for the first effective rain to start ploughing (as in CT) but rather were able to start ZT sowing immediately after the first rains.
3 Improved soil water retention: By reducing surface water runoff and evaporation, the retained water in the soil profile increased. Under supplementary irrigation, CA achieved higher water productivity benefits compared to conventional agriculture.
4 Improved crop establishment: The press wheels, which were added to the no-till seeder, had a great benefit in capturing more of the rainwater, as well as pressing the soil over the seed line and increasing the contact between grains and soil, thus enhancing grain germination and crop growth. ZT sowing improved crop growth with rapid emergence in the field compared to CT and enabled farmers to obtain higher yields with lower costs.

5 Reduced seed inputs: The recommended sowing rate of wheat was reduced from 160 kg ha^{-1} in CT to 100 kg ha^{-1} in no-till seeding.
6 Reduce fuel use: ZT seeding consumed less fuel (8 L ha^{-1}) compared to CT (32 L ha^{-1}) with a reduction in the machinery maintenance and labour costs.
7 Continuous crop production: By applying ZT, farmers were able to eliminate fallow phases and adopt continuous cropping, which greatly enhanced their total income (Alrijabo and Al-Ajrawi, 2013).

Five years after the ACIAR programme was concluded, CA continues to have an impact in Iraq and beyond. Despite the ongoing civil conflict, Iraqi researchers, manufacturers and farmers are continuing to advance the development of ZT systems. ZT seeders are being put to use again, with an estimated 12 000 ha planted using no-till seeding in the 2018 growing season.

6 Turkey

Agriculture has been playing a leading role in the history of Turkey as the largest employer and a major contributor to the national GDP. With the aim of increasing production and national income by modernizing agricultural production techniques, research studies were initiated as early as the mid-1920s. Almost all studies from this period concluded that soil tillage was the most important component of productivity. These studies indicated that soil tillage is a necessary practice, but the number of operations, timing and type of machinery depends on the conditions (Gerek, 1987). Yield responses from tilling were reported by Numan Kıraç in the Central Anatolia region in the 1930s, and subsequent similar research results formed the basis of the tillage mindset. Later on, the extension services were established to contribute to the widespread dissemination of this knowledge. Farmers have adopted tillage throughout the country, albeit slowly. This is because farming was a family business, passed on from parents to children and grandchildren. On the other hand, farmers who cultivated their lands wanted their ploughed soils to be deep, smooth and loose. To achieve this, they performed multiple passes of tillage operation.

Excessive tillage has now led to major problems with soil degradation and lower productivity. While the tillage costs were compensated at the beginning with increased yields, its economic return over time disappeared. The question now is how to reduce the costs of tillage, mainly fuel use. Increasing costs and results of studies investigating the options for alternative tillage methods enabled the spread of equipment, such as disc harrow, tandem disc harrow and chisel plow. During the same period, the use of a cultivator, which was considered as secondary tillage equipment, became more common. However, these tillage applications usually differed in their intensity of soil disturbance depending on the specific equipment used. Additionally, stubble burning has become widespread throughout the country to further reduce the cost of tillage (Fig. 20). Moreover, CT cropping reduces soil organic matter content and increases the risk of soil erosion (Bayat, 1993).

6.1 The search for possible solutions

To overcome these problems, several studies were initiated in the 1970s into direct sowing systems based on the elimination of tillage (Yeşilsoy, 1972). Pioneer researchers, who conducted the first CA studies, found encouraging cost reductions, but yields were reduced under direct seeding compared to tilled treatments (Sungur et al., 1994; Kamburoğlu, 2001; Gültekin et al., 2011). In the years that followed, environmental gains of CA systems across the world were becoming clearer. Soil organic matter (SOM) played an important role in the work conducted on sustainable agriculture because 69% of the country's soil was found to be low in organic matter due to excessive soil tillage (Okur et al., 2003). Some studies conducted during this period showed that SOM content under ZT was always higher than any other method, particularly at a depth of 0 cm to 5 cm (Bayhan et al., 2006). The same study also illustrated that soil moisture levels varied based on the tillage method used; the highest soil moisture content was always under the no-till system. This finding in dryland farming systems, which represented the majority of the country's agriculture, showed that more soil moisture can be stored in the soil with the CA system and yields could be increased.

The effects of soil tillage on soil physical properties were studied. Aggregate stability was found to be affected negatively by soil tillage. The soil under no-till or lands with reduced-tillage management had a better aggregate structure than the common tillage management (Özpınar and Çay, 2006; Öztürkmen

Figure 20 Poor management for some agricultural areas in Turkey showing intensively tilled topsoil, burnt crop biomass and bare soil surface. *Source:* Photo by İ. Gültekin, 2010.

and Kavdir, 2012). Soil bulk density and penetration resistance were also found to increase under no-tillage (Kolay and Öztürkmen, 2007; Barut et al., 2007). Furthermore, researchers found that no-tillage methods increased soil carbon sequestration rates and reduced CO_2 emissions (Akbolat et al., 2007). The CT treatments improved the soil hydraulic conductivity temporarily (Çelik, 2011).

According to long-term experimental results, CA was a feasible method but results also showed that diversified crop rotations were absolutely necessary for superior performance (Gültekin et al., 2013). During this study process, local agricultural machinery manufacturers began the production of ZT drills for farmers because locally produced machines have an advantage over imported machines in terms of price and size (Fig. 21).

6.2 Implementation of Conservation Agriculture

Turkey has 37.71 million hectares of agricultural land, 38.75 million hectares of meadows and pasture, 9.34% perennial crops and 51.91 % (19.56 million hectares) of arable land. Of this arable land, 33.22 % (6.5 million hectares) is irrigated, and 1.2% is planted twice a year. Due to the lack of and irregularity of the rainfall, almost 3.52 million hectares of arable land under winter cereal-based systems are left for fallow (TSAI, 2020) that are planted with wheat and barley, and in the irrigated areas, sugar beet, cotton, maize, sunflower, pulses are the most commonly grown crops. Excessive tillage is practiced under all of these different production systems throughout the country. The current CT systems involve the use of a mouldboard plough as the primary tillage tool followed by multiple trips using secondary tillage tools like disks, harrows and field cultivators. The main purpose of tillage is to increase crop yields, but it is quite expensive because of the fuel cost and its effect on land degradation and soil erosion.

Figure 21 Two examples of locally manufactured no-till seeders for grain and row crops in Turkey, Bora produced by Altınöz Ltd. Co. and Milenyum, produced by Özdöken A.Ş. respectively. *Source:* Photo by Z. Altınöz, 2011; İ. Gültekin, 2010.

The Ministry of Agriculture and Forestry is mainly responsible for policy formulation. Increased agricultural production, food security, conservation and improvement of agricultural land and the use of environmentally friendly production techniques are the primary aims of the Turkish National Strategic Plan. Results of the research showed that CA methods can be successfully applied in different parts of Turkey. In 2011, farmers were supported through subsidies to apply CA within a pilot project. The programme was designed to prevent and reduce water and wind erosion, desertification, salinity and soil contamination. Soil erosion is one of the most urgent environmental problems in the country. The programme provided subsidy and technical support (Fig. 22) to the farmers participating in the environmentally based agricultural land protection (CATAK) programme. This programme has been expanding and continuing up till now.

In Turkey, CA adoption levels are still in the initial phase. Today, the area under the CA system is not known accurately because there is no specific information about CA in the archives of the Turkish Statistical Institute or Ministry of Agriculture and Forestry (MAF); however, unofficial estimates of agricultural lands under CA are close to 100 000 ha. Most of this area is in the drylands, where winter cereals, such as wheat and barley, are produced in crop rotation with chickpea, lentil, safflower and Hungarian vetch (Fig. 23). It is possible to find the same farmers growing continuous wheat. For farmers, profit is the most important factor when considering the adoption of CA systems and practices, such as ZT seeding. As far as the ZT yield advantage is concerned, there has been some disagreement in the perceived yield benefits of ZT farming practices because of the perception that ZT systems perform well only under rain-fed conditions.

Although there is a positive return from CA implementation for farmers and there is also support available, it is possible to say that the uptake of CA has reached the desired level and is continuing at a desired rate of adoption. Lack

Figure 22 Demonstrating no-till drilling on CA fields in Turkey. *Source*: Photo by İ. Gültekin, 2019.

Figure 23 Samples of farmers' barley fields after: (a) dry beans, (b) sunflower, (c) & (d) maize grown under CA in Turkey. *Source:* Photo by İ. Gültekin, 2019; S. Gültekin, 2019; İ. Gültekin, 2018.

of information and awareness about the CA system seems to be a problem. It is emphasized that a large mass of people is not aware of the future benefits of CA. At the level of policymakers, the system has not received the attention that it deserves. Even among technical staff and farmers, there has not been enough emphasis on agricultural education and publications. Therefore, poor CA implementation, coordination and planning have made it impossible for local examples to become mainstream.

6.3 Conclusion

Turkey has a high potential for the adoption of the CA system. In the last few years, a number of projects have addressed the issue, especially taking into consideration the need to mitigate and adapt to climate change, whilst improving the sustainability of agriculture through the diffusion and adoption of low-carbon technologies with win-win benefits for production efficiencies and the environment.

In essence, CA is concerned with profitable and sustainable agricultural production - that is maximizing farm profits in the long-term by optimizing agricultural production while conserving inputs (labour, fuel, seed, fertilizer, pesticides, etc.) and minimizing or mitigating any impacts on natural resources (soil, water and air). In most parts of the world, CA concepts are primarily applied to crop production, but the underlying philosophies can and should also be applied to the whole farming system, including animal production, especially where crop and livestock production are highly integrated, as in many parts of Turkey.

The adoption of CA principles in locally adapted production systems for the growth of annual and perennial crops, pastures and forages, together with good quality seeds and optimally integrated nutrient, water and pest management, would achieve environmental and economic benefits in Turkey by:

- Providing similar or even higher yields through improvements in soil structure, organic matter and overall soil fertility.
- Lowering production costs through reduced inputs of energy, labour, machinery, fertilizers, water, and pesticides, thus raising related productivity and efficiency.
- Mitigating CO_2 emissions through reduced fuel consumption and sequestration of atmospheric carbon into soil organic matter, reducing N_2O and CH_4 emissions through reduced use of mineral nitrogen and improved soil drainage.
- Reducing runoff and wind/water erosion through better soil aggregate stability.
- Improving water infiltration rate and protective cover of the soil by crops and/or crop residues.
- Diminishing off-site damage of infrastructure and pollution of water bodies through less runoff with a much-reduced sediment load.
- Enhancing in-field and off-site biodiversity through the absence of destructive soil disturbance, protective soil shelter and less off-site transport of contaminants.
- Encouraging diversity of rural landscapes through enhanced crop and species diversity and cover crops.
- Maintaining less favoured rural areas under production through the adoption of economically and environmentally viable production methods.

7 Iran

The population of Iran in 2016 was 80.2 million inhabitants. Iran has an area of 1 648 195 km^2 (Malekloo, 2018). The northern part of Iran is covered by the lush lowland Caspian Hyrcanian mixed forests, near the southern shores of the Caspian Sea. The eastern part consists mostly of the desert basin. The only large plains are found along the coast of the Caspian Sea and at the northern end of the Persian Gulf, where the country borders the mouth of the Arvand River. Smaller, discontinuous plains are found along the remaining coast of the Persian Gulf, the Strait of Hormuz and the Gulf of Oman.

About 52 % of the country consists of mountains and deserts, and some 16% of the country has an elevation of more than 2000 m above sea level. The central or interior plateau covers over 50% of the country. It is partly covered by a remarkable salt swamp (Kavir) and partly by areas of loose sand or stones with stretches of better land near the foothills of the surrounding mountains.

Iran's climate is diverse, ranging from arid and semi-arid to subtropical regions along the Caspian coast and the northern forests. For example, temperature extremes can reach 54°C in arid areas and down to −30°C at high altitudes. Approximately, 88% of Iran is located in arid and semi-arid regions (Fig. 24). Water shortages are compounded by the unequal distribution of rainfall. The mean annual precipitation during the period 1980–2018 was 250 mm for the entire country (Malekloo, 2018). Near the Caspian Sea, rainfall averages about 1280 mm per year, but the central plateau and the lowlands to the South seldom exceeds 100–200 mm, far below the minimum requirement for dryland farming. Moreover, the annual distribution of precipitation is diverse in terms of timing and location but early drought and terminal drought are common in unirrigated winter crops. The temperature is the second major constraint that affects rain-fed agricultural production. The rate of evaporation per annum ranges from 700 mm to 4200 mm.

Iran lies within the West Asian diversity centre, where crops of global importance originated and includes a diversity of farming systems prevailing in Central and West Asia and North Africa (CWANA) region and in the non-tropical drylands. The diversity of climate in Iran provides an opportunity for producing diversified horticultural and field crops.

7.1 Agriculture in Iran

In 2005, 18.1 million ha of land was cultivated, of which 10 million ha were rain-fed and 8.1 million ha were irrigated. Major crops grown include wheat, barley, rice, chickpea, cotton, sugar beet, vegetables, fruits and nuts. The western and north-western portions of the country have the most fertile soils and 3% of the total land area is used for grazing and small fodder production. Most of the grazing is done on mostly semi-dry rangeland in mountain areas and on areas surrounding the large deserts ('Dasht's') of Central Iran.

In 2007 the GDP was $270.9 billion. Agriculture accounted for around 9% of GDP, while in 1992, it accounted for 23% (Frenken, 2009). The economically active population is about 27.6 million (2005), of which 70% is male and 30% is female. In agriculture, 6.7 million inhabitants are economically active, of which 56% are male and 44% are female. From 1960 to 1993, the number of farming units increased from 1.8 million units to 2.8 million units, with the average area per unit decreasing from just over 6 ha to less than 5.5 ha. More than 80% of these farming units have a total size of less than 10 ha, and even these 10 ha are on average scattered over five different locations. About 5% of the agricultural land is used by cooperatives. Usually, each cooperative has eight members

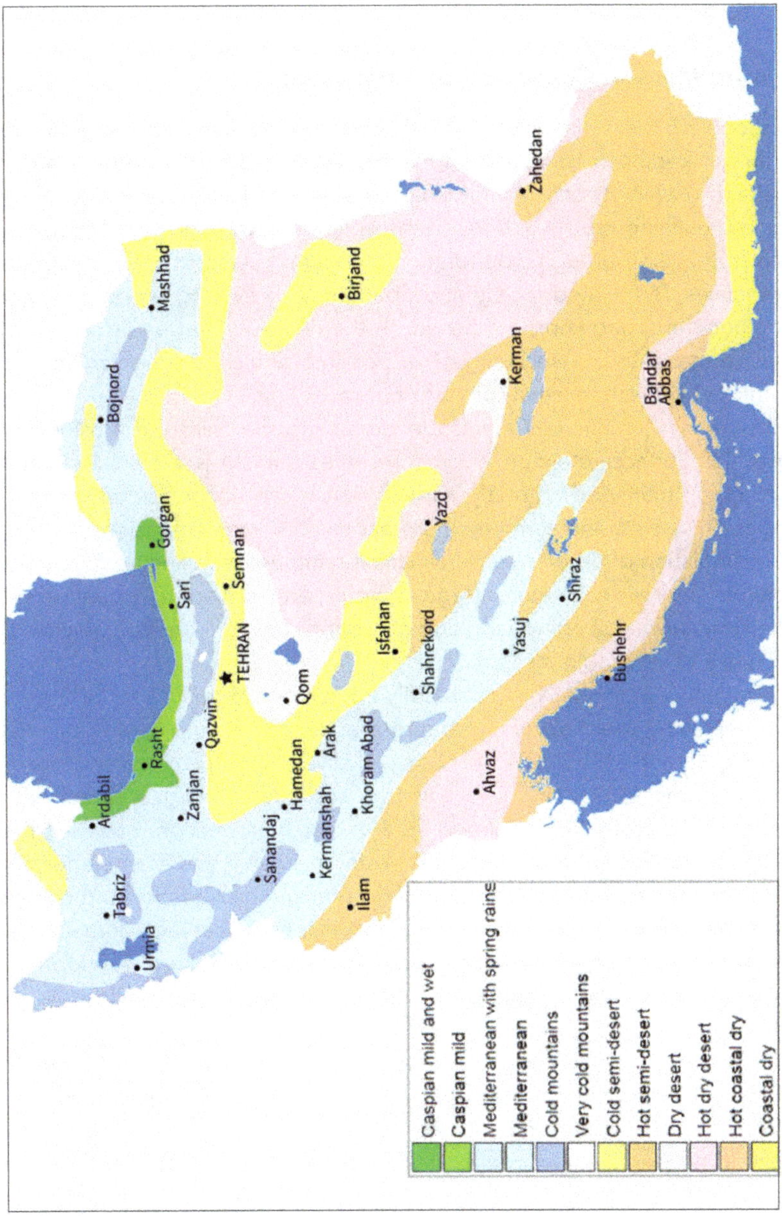

Figure 24 Climate map of Iran showing the location of province capitals (Vaghefi et al., 2019).

with an average size of 40 ha. Commercial companies cover around 14% of the agricultural land, mostly located in Khuzestan province in the southwest of the country.

7.2 Importance of Conservation Agriculture

CA is one of the most important components of sustainable agriculture. The three interlinked principles of CA are no or minimum mechanical soil disturbance, which means no-till seeding and weeding, maintaining a soil cover by retaining crop residues on the soil and diversification of cropping systems by using thorough crop rotation of legumes (Kassam et al., 2018). The CA practices, formulated locally and adapted to prevailing biophysical and socio-economic conditions, are based on the necessity that undisturbed soil is kept permanently covered as much as possible, and crops are sown through the mulch cover with minimal soil disturbance.

The many benefits of CA include protecting the soil from erosion and saving fuel. Surface retention of crop residues (mulch) is a crucial element in CA. When managed correctly, surface mulch can buffer soil temperature changes (in winters and summers), protect the soil against erosion, promote the accumulation of organic carbon, facilitate nutrient recycling, reduce weed infestation, improve soil water storage capacity, reduce evapotranspiration and hence slow down soil salinization and provide niches for beneficial microbes, soil fauna and flora to flourish.

Agriculture alone has grown steadily from about 7% to 10% of GDP since 2010, employing over 16% of the country's labour force. Iran's varying climates and geography have made it the centre of diversity of many major crops of global importance, including wheat, which constitutes the largest contribution to grain yield, accounting for almost 70% of cereal production. However, that growth was fuelled in large measure by unsustainable and environmentally damaging policies. The sector now suffers from land degradation and climate change. Meanwhile, total grain imports remain high to meet local demand and maintain strategic stockpiles. A decrease of support in agricultural development in recent years may also be attributed to sanctions and increased inflation since 2011.

Iran's agriculture sector can be credited to the country's push towards food self-sufficiency, part of the country's agenda since the Islamic Revolution of 1979. In 1999, the government initiated a self-sufficiency strategy for wheat by increasing production through input subsidies and the adoption of new methods and technologies, such as improved crop varieties, mechanization and farmer training.

Iran's Ministry of Agriculture (MoA) is currently investing in sustainable solutions to agriculture with the goal of increasing self-reliance for major agricultural commodities, such as wheat in the next 5 to 10 years. This includes plans to spread CA-based crop management practices across three million hectares of farmland. CA-based crop management technologies and practices may ensure a more sustainable and equitable food security using less water and other inputs, lowering costs of sustainable production, while creating healthier soils, the natural resource that all people depend on.

7.3 Promotion of Conservation Agriculture

It is a keystone of Iran's MoA current strategy to conserve water and make farming more sustainable. These practices reduce costs for farmers – especially by saving fuel for the soil tillage – improve soil quality, reduce soil erosion and improve biological activity, all while increasing agricultural productivity, especially by increasing resilience to drought (Asadi, 2017). Building knowledge of CA practices at a local level is critical to change agriculture in Iran. Investing in farmers' learning process for the adoption of CA will be required too.

Since 2008, Iran's MoA has been importing large ZT seeders from South America, Italy and Germany, which have been helping farmers improve soil and soil moisture conservation. (Figs 25 and 26). But these machines need big tractors with more than 110 horsepower (HP). Most farmers have access to tractors with only 75 HP.

In Iran, CA was first introduced in 2007 to four provinces: Golestan, Khouzestan, Fars and Khorasan in an area of 150 hectares and was later expanded to other provinces. A case study to promote CA in Golestan province is described in the following section (Asadi, 2015, 2018). In 2014, a comprehensive organization office of CA under the control and supervision of the Deputy of Agronomy in the MoA in Tehran was established. The office of CA was responsible for the development of macroeconomic policies in support of CA's technology in interaction with intra and inter-organizational sectors, including the design, development, approval and notification of programmes,

Figure 25 Gaspardo no-till planter in Golestan Province. *Source:* Photo by Asadi, 2015.

Figure 26 Amazon no-till planter in Golestan Province. *Source:* Photo by Asadi, 2015.

projects, guidelines and other related activities across the country. The CA technical committee was responsible for the operating procedures of the programmes and projects, development or revision of operating instructions, identification of equipment list and allocation of funds for each project. This office also trained two CA specialists for each province as local experts. Their duties were to train farmers and other experts at the provincial level. Most of these experts were mechanization experts. At the provincial level, the CA technical committee tracked issues related to the implementation of programmes using provincial and national guidelines, identified provincial priorities and took responsibility for the implementation, extension and education of farmers. They also compared the results of research projects under field conditions within the framework of CA-based applied research and extension model farms, aggregated the results and evaluated the implementation of the plan. According to this plan, such extension hubs were established in all provinces.

In addition, MoA trained some CA researchers and experts at CIMMYT in Mexico and at Borlaug Institute for South Asia (BISA) in Ludhiana, India. The MoA sent specialists in agriculture to Mexico and India in 2008 to undertake advanced courses in CA. Furthermore, the trained personnel trained farmers on CA in Iran and developed technical guidelines for irrigated and rain-fed lands. Every year, the MoA provincial organizations have implemented more than 1000 extension workshops, training courses and other programmes for the development of CA in collaboration with technical and scientific staff of FAO, CIMMYT and ICARDA (Figs 27 and 28).

In 2015/16, around 150 000 ha in Iran were under CA (Kassam et al., 2018) and in 2020, the CA area increased to 300 000 ha. Although the government has promoted CA in a systematic way, the rate of farmers' adoption of CA was considered low because of poor access to affordable ZT seeders. Recently, seeders and other CA equipment are being made locally. Boukan Kaveh

Sazeh-Kesht Company is one of the agricultural machinery factories in Iran that has successfully developed and manufactured CA machines.

The following section describes the CA systems and their practices that are being managed in Golestan province, the northern part of Iran. Golestan was the second highest producer of wheat in the country during the period 2016–2020.

7.4 Conservation Agriculture in Golestan province

Wheat and barley are grown in Golestan province on more than 300 000 ha. After the harvest of wheat and barley, some farmers burn the residue, as part of their practice of land preparation for the second crop production with soybean,

Figure 27 Workshop participants included farmers, agricultural and extension experts, Dr. Ken Sayre, Dr. Kamali (Principal scientist CIMMYT, Iran) and Dr. Asadi. *Source:* Photo by Asadi, 2015.

Figure 28 A senior CA scientist (Dr. Asadi) speaking on long-term adaptation to CA-based crop management in West Azerbaijan Bukan city. *Source:* Photo by Rorett, K. 2016.

maize or sorghum. These tillage practices cause soil disturbance, compaction and deterioration. Consequently, in many areas, conventional agriculture has led to a decline in crop yields and profitability. In the past, farmers burnt crop residues in 120 000 ha of their lands after the harvesting of wheat and barley, but in 2016, due to the promotion of CA activities, this practice declined to less than 5000 ha (Fig. 29). It is anticipated that in the next few years, the burning of crop residues will stop altogether.

In the 2015–16 season, CA promotion activities included working directly with farmers, TV programmes and field workshops related to the promotion of CA systems and practices (Fig. 30). Consequently, during the winter of 2019 and in the summer of the same year, almost 30% of soybean (a total of 50 000 ha) was directly sown into wheat residues without removing or burning the residues.

Unploughed fields, where crop residues are retained, are better at capturing and holding moisture and thus can boost yield potentials. Also, farmers have a short time in the autumn to sow barley, canola and wheat, therefore, they have to take advantage of the Mediterranean seasonal rains during the winter period. If tillage can be avoided, farmers can plant earlier and keep crop residues on the soil surface for protection from erosion. The potential fuel savings and the more efficient use of water are the main points of interest to farmers.

Figure 31 shows sowing a model soybean CA farm of 2 ha. In 2016, the soybean crop was sown into the wheat stubble after harvest. The ZT seeder used on this farm was Aske 2200, produced by Sazehkesht Company in Bukan city in West Azerbaijan province, Iran (http://sazehkesht.com/en).

Figure 32 shows the very good establishment of soybean after ZT sowing without any irrigation (9 July 2016). Water savings in the CA cropping system

Figure 29 Burning of wheat residue after harvest in Golestan province in 2016. *Source:* Photo by Asadi, 2016.

Figure 30 TV programme recording for broadcasting on provincial TV regarding the disadvantages of burning crop residue in Golestan province in 2015. *Source:* Photo by Asadi, 2015.

from reduced soil evaporation has been an essential element in the Golestan region, especially for the second crop, such as soybean, corn, sorghum and cotton. Figure 33 represents full soil coverage of soybean farm without any irrigation (July 26, 2016).

There are about 30% savings in water consumption with CA system compared with conventional cultivation methods (CT and pre-sowing flood irrigation). This difference came from the presence of wheat stubble versus bare soil surface. The soybean crop was harvested in November 2016 producing 3.6 t ha^{-1} grain yield. The average grain yield of soybean in this province is about 2 t ha^{-1}. The farmer and other similar regional farmers were trained by CA experts.

Figure 31 Direct soybean sowing after wheat harvest using Aske 2200 no-till drill seeder in Golestan province, 7 June 2016. *Source:* Photo by Asadi, 2016.

Figure 32 Establishment of soybean 32 days after sowing in Golestan province, 9 July 2016. *Source:* Photo by Asadi, 2016.

Figure 33 Full ground coverage of the field with soybean on 26 July 2016. *Source:* Photo by Asadi, 2016.

7.5 Technology for Conservation Agriculture development

Starting in 2010, several separate initiatives arose across the country, which aimed to modify the tillage practices, employed to protect the soil, water and environment and enhance the profitability of agribusinesses. This encouraged several private manufacturers to produce ZT seeders, minimum tillage equipment and raised-bed planters. The largest companies are Sazehkesht Kaveh (Bukan city), Hamadan Machine barzegar (Hamadan city), Hamedan Machine Zeraaat (Hamadan city), Tarashkadeh (Karaj city) and Taka (Arak). They produced all kinds of ZT planters, both tine and disc openers.

Recently some of the manufacturers produced ZT planters suitable for raised beds. In this method, crops can be planted on preformed or existing

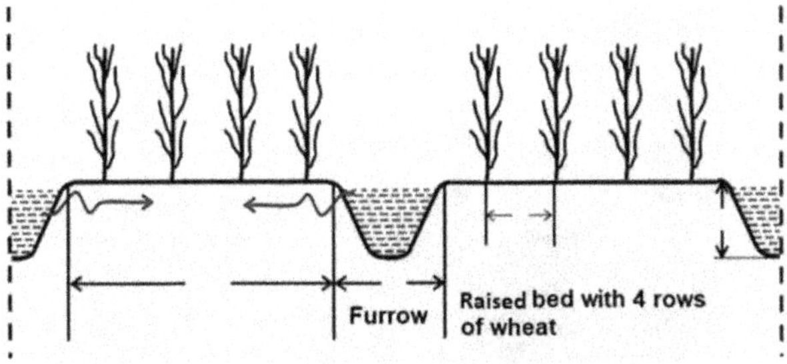

Figure 34 Permanent raised bed illustration. *Source:* Photo by Asadi et al., 2020.

beds of any length, 10-50 m long and 15–30 cm high, with two (or 4) defined planting rows per bed depending on the crop types - four rows for wheat, two rows for soybean and one row for cotton. The furrows between the beds supply the irrigation water and provide drainage and access to labourers for hand operations, such as weeding, and to the wheels of machinery for weeding, fertilizer and harvest operations (Fig. 34). The beds are retained and maintained on a continuous basis for successive crops without tillage. Machinery wheels avoid the beds and compaction is restricted to the furrows only (Figs 35–37).

The Tarashkadeh Company designed a pneumatic raised-bed planter, which is suitable for crops such as wheat in water shortage areas and was offered to farmers in 2019. This machine can also plant cotton seeds through mulch cover (Fig. 38). Tarashkadeh planters have been sent to most provinces of the country. In 2010, Tarashkadeh produced a no-till planter that was cheaper, lighter and can be worked with 80 HP tractors.

7.6 Main challenges for promotion of Conservation Agriculture

Researchers, experts, policymakers and farmers in Iran are faced with challenges in promoting CA. Some of these challenges may be related to technical, social and environmental issues, but most of them are related to mindset, knowledge and behavioural changes.

The adoption of CA technology is influenced by the level of farmers' education, quality of extension services, type and farm size and crops and livestock being produced and integrated. Also, training courses and programmes are needed for extension personnel, in cooperation with international organizations, such as the FAO, CIMMYT and ICARDA, to build public awareness of the modern CA systems and technologies. Most importantly, CA involves a change in the mindset, and without this change, farmers will easily resort back to tillage-intensive systems when any problem occurs (Asadi, 2017). This has been the case in all other countries where CA has been spreading, and we believe that Iran will be no different (Karabayev et al., 2012). However, there has been clear evidence that the CA systems work well in

Figure 35 Raised bed planter of Machine Barzegar Hamadan being used for wheat, cotton and soybean sowing in Golestan province in 2016-2020. *Source:* Photo by Asadi, 2020.

Figure 36 Wheat at final stage of growth on permanent raised bed farm in Golestan province in 2017. *Source:* Photo by Asadi, 2020.

Figure 37 Cotton at flowering stage on permanent raised bed, the fifth year in Golestan province (16 Aug 2020). *Source:* Photo by Asadi, 2020.

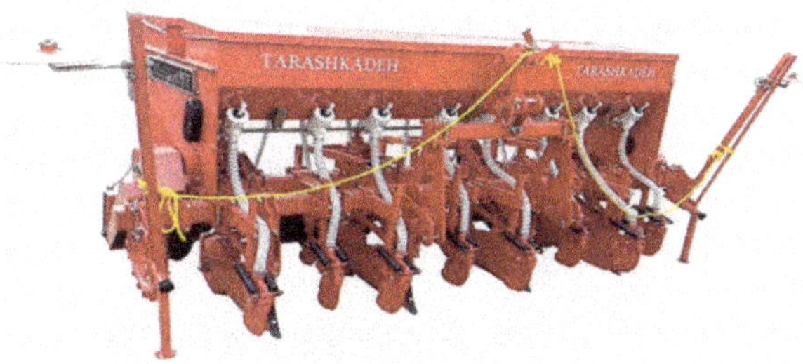

Figure 38 Pneumatic raised bed planter of Tarashkadeh company. *Source:* Photo by Asadi, 2020.

the region and are important for halting and reversing the marked, albeit slow, soil and land degradation in the region.

For agriculture mechanization development, the government must support more research and development opportunities and increase the financial support for the industry of CA machinery across the country.

8 Conclusion

Field experiments and farmers' trials as well as some longer-term research is taking roots in the region, particularly in Lebanon, Syria, Jordan, Iraq, Turkey and Iran comparing CA systems to CT systems. The results of these field trials generally show the following:

- CA reduces soil water evaporation and increases water availability.
- CA increases organic matter in the soil and restores soil fertility.
- CA reduces soil erosion.
- CA reduces labour and machinery usage.
- Since this method of land management is relatively new to the region, governmental support, more research, active extension programmes and financial subsidy to the local farmers are needed.

CA now covers some 446 400 ha in the region spread as 1200 ha in Lebanon, 200 ha in Jordan, 30 000 ha in Syria, 15 000 ha in Iraq, Turkey 100 000 ha in Turkey and 300 000 ha in Iran. More adoption is expected in the coming years as knowledge, experience and the availability of equipment spreads. Already, training on CA to farmers and CA courses in colleges and universities are expanding, particularly in Lebanon, Syria, Iran and Turkey with locally manufactured no-till seeders becoming more accessible. It is possible that more nationally or sub-nationally coordinated CA extension efforts will be organized in the coming years as is already happening in Iran and Syria. Also, increasingly, perennial systems, such as fruit orchards, olive groves and vineyards, are being transformed into CA systems.

In light of the information compiled in this chapter regarding the diverse status of CA-related activities and uptake in the region, the following is a list of what is needed to strengthen the support needed in the countries of the region.

1 Systematic efforts for the promotion of CA are less frequent in West Asia region than desirable. In order to transform agriculture, innovative thinking and action are needed. Also, everything possible must be done to involve farmers in adopting this thinking. It requires a change in everyone's mindset (tradition, prejudice) to be able to produce more with less.
2 Development of workshops, farmer's fields schools (FFS) and field days for farmers.
3 Establishment of long-term demonstration sites (including on farms) side by side comparing CA systems and CT systems.
4 Development of regional and national strategies, action plans, and coordination platforms.
5 Development of manuals and guidelines, policy and regulations for each region based on climate and geographic conditions.
6 Legalizing the new crop management techniques and introducing them into national curriculum and extension services
7 Establishment of favourable environment to support modern crop management techniques, including the development of CA associations and organizations, development of relevant technologies and input supply through commercial networks.
8 Promotion of incentives as payment for application of environment-friendly methods of land use and community services.

The above needs and concerns must be overcome not only by farmers but also by agricultural researchers, extension experts, university professors, politicians and all stakeholders involved in the farming industry if a greater adoption and regional mainstreaming are to be achieved.

9 Where to look for further information

Kassam, A., Friedrich, T., Shaxson, F. and Pretty, J. (2009). The spread of conservation agriculture: justification, sustainability and uptake, *International Journal of Agricultural Sustainability*, 7:4, 292–320, DOI: 10.3763/ijas.2009.0477.

Bobby, I., Stewart, A., Asfary, F., Belloum, A., Steiner, K. and Friedrich, T. (2008). *Conservation Agriculture for Sustainable Land Management to*

Improve the Livelihood of People in Dry Areas. Damascus, Syria. ACSAD and GTZ.

Communication Team ICARDA (13/11/2013). *Principles of Conservation Agriculture in Northern Iraq*. Beirut, Lebanon: International Center for Agricultural Research in the Dry Areas (ICARDA).

Cornell University College of Agriculture and Life Sciences. (2015). *Advantages of Conservation Agriculture*. Retreived from: http://conservationagriculture.mannlib.cornell.edu/gfx/Interim Files/CALS_2line_blk.gif

Mutua, J., Muriuki, J., Gachie, P., Bourne, M. and Capis, J. (2014). *Conservation Agriculture with Trees: Principles and Practice: A Simplified Guide for Extension Staff and Farmers*. Nairobi, Kenya: World Agroforestry Centre (ICRAF).

Corsi, S. and Muminjanov, H. (2019). *Conservation Agriculture: Training Guide for Extension Agents and Farmers in Eastern Europe and Central Asia*. Rome, Italy: FAO.

10 References

Ahmed, S., Piggin, C., Haddad, A., Kumar, S., Khalil, Y. and Geletu, B. 2012. Nematode and fungal diseases of food legumes under conservation cropping systems in northern Syria. *Soil and Tillage Research* 121: 68–73.

Akbolat, D., Ekinci, K., Uysal, S. and Onursal, E. 2007. Farklı Toprak İşleme Sistemlerinde Topraktan CO2 Çıkışının Orta Vadeli Ölçüm Ile Saptanması,2. *Koruyucu Toprak İşleme ve Doğrudan Ekim Çalıştayı, S*: 134–148, İzmir.

Al-Juburi, N. H. and Alrijabo, A. A. 2012. A study of growth, yield and its components traits of new varieties of wheat and barley crops grown in different environmental regions in Nineveh Province. *Mesopotamia Journal of Agriculture* 40 (Suppl. 1): 94–104.

Alrijabo, A. A. 2012a. Effect of the new farming system zero Tillage in growth, yield and its components of barley crop in low rainfall area in Ninevah province. Iraqi. *Journal for Soil Science* 12(1): 178–188.

Alrijabo, A. A. 2012b. Effect of the new farming system - Zero Tillage in growth, yield and its components of bread wheat and durum wheat crops under supplementary irrigation in Ninevah province. Minia International Conference for Agriculture and Irrigation in the Nile Basin Countries. El-Minia, Egypt, pp. 576–585.

Alrijabo, A. A. 2014. Effect of the new farming system zero Tillage in growth, yield and its components of bread wheat, durum wheat and barley crops in moderate rainfall area in Ninevah province. *Mesopotamia. Journal of Agriculture* 42(1): 242–254.

Alrijabo, A. A. and Al-Ajrawi, A. 2013. The effect of cultivation methods, row spacing and the addition of polymer jell in growth, yield and its components of wheat crop Triticum aestivum L. Kirkuk University. The Second International Scientific Conference for Agricultural Research. October 30–31, 2013.

Asadi, M. E. 2015. Promoting conservation agriculture in Golestan, Iran. 2015. *Informa (CIMMYT Newsletter)* 1829: 3.

Asadi, M. E. 2017. *Healthy Soils with Conservation Agriculture Systems*. Nouruzi Publication, Gorgan, Iran. 135 p. (In Farsi with English summary).

Barut, Z. B., Çelik, İ., Özgüven, F. and ve Ortaş, İ. 2007. Killi Topraklarda Farklı Toprak İşleme ve Ekim Sistemlerinin Toprağın Bazı Özelliklerine Etkisi. 2. Koruyucu Toprak İşleme ve Doğrudan Ekim Çalıştayı. 13 Haziran 2007, s:117-131. İzmir, Turkey.

Bashour, I., Al-Ouda, A., Kassam, A., Bachour, R., Jouni, K., Hansmann, B. and Estephan, C. 2016. An overview of Conservation Agriculture in the dry Mediterranean environments with a special focus on Syria and Lebanon. *AIMS Agriculture and Food* 1(1): 67-84. https://doi.org/10.3934/agrfood.2016.1.67.

Bayat, A. İ. A. 1993. Toprak işleme sistemlerinin organik madde içeriği bakımından incelenmesi ve Konya' nın Kadınhanı ve Ilgın ilçelerindeki uygulamaları. Proceedings. 5. Uluslararası Tarımsal Mekanizasyon ve Enerji Kongresi. Kuşadası.

Bayhan, Y., Kayisoglu, B., Gonulol, E., Yalcin, H. and Sungur, N. 2006. Possibilities of direct drilling and reduced tillage in second crop silage corn. *Soil and Tillage Research* 88(1-2): 1-7.

Brown, M. E. and Funk, C. C. 2008. Climate: food security under climate change. *Science* 319(5863): 580-581.

Busari, M. A., Kukal, S. S., Kaur, A., Bhatt, R. and Dulazi, A. A. 2015. Conservation tillage impacts on soil, crop and the environment. *International Soil and Water Conservation Research* 3(2): 119-129.

Darweesh, N. H., Alrijabo, A. A. and Abdullah, H. A. 2013. The effect of tillage system in yield and its components of bread wheat in Salahaldin province-north of Iraq. Fourth Agriculture Scientific Conference, Anbar University.

Department of Statisitcs 2019. Table 3: Estimated population of the kingdom by governorate urban/ rural 2004-2019. http://jorinfo.dos.gov.jo/Databank/pxweb/en /Demographi_Statistics/-/Table3.px/.

Derpsch, R. and Friedrich, T. 2009. Development and current status of no-till adoption in the world. Proceedings on CD, 18th Triennial Conference of the International Soil Tillage Research Organization (ISTRO), Izmir, Turkey, June 15-19, 2009.

El-Shater, T., Mugera, A. and Yigezu, Y. A. 2020. Implications of adoption of zero tillage (ZT) on productive efficiency and production risk of wheat production. *Sustainability* 12(9): 3640.

Frenken, K. 2009. Irrigation in the Middle East region in figures AQUASTAT survey - 2008. *FAO Water Reports* 34.

Gerek, R. 1987. İç Anadolu'da nadaslı ziraat sistemi, nadası kaldırma veya nadas oranını azaltma imkanları. Sayfa 9-15. Türkiye Tahıl Sempozyumu. 6-9 Ekim 1987, Bursa.

Gültekin, İ., Arısoy, R. Z., Kaya, Y., Partigöç, F., Taner, A., Aksoyak, Ş. and Gültekin, S. 2013. Doğrudan Ekim Yönteminin Ülkemizde Dünü, Bugünü ve Geleceği, Bahri Dağdaş Uluslararası Tarımsal Araştırma Enstitüsünün Çalışmaları. 10. Tarla Bitkileri kongresi 10-13 Eylül 2013, cilt 1, s.753-759, KONYA.

Gültekin, İ., Arısoy, R. Z., Taner, A., Kaya, Y., Patigöç, F. andAksoyak, Ş. 2011. Comparison of different soil tillage systems, under several crop rotations in wheat production at Central Anatolian Plateu in Turkey. 5th World Congress of Conservation Agriculture Incorporating 3rd Farming Systems Desing Conference, September 26-29, 2011, Brisbane Avustralia.

Haddad, N., Piggin, C., Haddad, A. and Khalil, Y. 2013. Conservation agriculture in West Asia. In: *Conservation Agriculture: Global Prospects and Challenges* Jat, R. A.,

Sahrawat, K. L., Kassam, A. H. (Eds). Amman, Jordan:CABI, p. 248. http://doi.org/10.1079/9781780642598.0248.

Hermans, T. D. G., Whitfield, S., Dougill, A. J. and Thierfelder, C. 2020. Bridging the disciplinary gap in conservation agriculture research, in Malawi: a review. *Agronomy for Sustainable Development* 40(1): 3. https://doi.org/10.1007/s13593-020-0608-9.

ICARDA 2012. Conservation agriculture: opportunities for intensified farming and environmental conservation in dry areas, Research to Action 2. ICARDA.

Intergovernmental Panel on Climate Change (IPCC) 2007. *Impacts, Adaptation and Vulnerability, Contribution of Working Group II to Fourth Assessment Report of the Intergovernmental Panel on Climate Change*. Cambridge, UK: Cambridge Univ Press.

Jalili, S., Fathi, G., Fathi, Y., Al-Rijabo, A. S., Piggin, C. and Desbiolles, J. 2011. Farmer innovation: seeder fabrication and uptake of zero-tillage in Iraq. Proceedings of the 5th World Congress on Conservation Agriculture, pp. 26–29, 306–308.

Kamburoğlu, İ. 2001. Kırklareli Kuru Tarım Koşullarında Buğday Tarımında Toprak İşlemesiz, Azaltılmış Toprak İşlemeli Ve Geleneksel Toprak İşlemeli Sistemlerin Toprağın Rutubet Değişimine Ve Ürün Verimine Etkisi. Trakya Toprak ve Su Kaynakları Simpozyumu. Mayıs 24–27, 2001, Kırklareli.

Karabayev, M., Wall, P., Sayre, K. and Morgounov, A. 2012. *Conservation Agriculture Adoption in Kazakhstan: History, Status and Outlooks*. CIMMYT Report.

Kassam, A. H., Friedrich, T. and Derpsch, R. 2010. Conservation agriculture in the 21st century: A paradigm of sustainable agriculture. Proceedings of the European Congress on Conservation Agriculture, Madrid, Spain, October 2010.

Kassam, A., Friedrich, T. and Derpsch, R. 2018. Global spread of conservation agriculture. *International Journal of Environmental Studies* 76(1): 29–51. https://doi.org/10.1080/00207233.2018.1494927.

Kolay, B. and ve Öztürkmen, A.R. 2007. Diyarbakır Koşullarında II. Ürün Soya Tarımında Farklı Toprak İşleme Yöntemlerinin Bazı Toprak Özelliklerine Etkisi. 2.Koruyucu Toprak İşleme ve Doğrudan Ekim Çalıştayı. 13 Haziran 2007, s: 55–67. İzmir.

Kumar, S., Singh, R. G., Piggin, C., Haddad, A., Ahmad, S. and Kumar, R. 2011. No till lentil: A profitable option in dry areas. Res. Bull. GRAIN Legum. No. 56 – April.

Lahmar, R. and Ruellan, A. 2007. Dégradation des sols et stratégies coopératives en méditerranée La pression sur les ressources naturelles et les stratégies de développement durable. *Cahiers Agricultures* 16(4): 318–323.

Lal, R., Reicosky, D. C. and Hanson, J. D. 2007. Evolution of the plow over 10,000 years and the rationale for no-till farming. *Soil and Tillage Research* 93(1): 1–12.

Loss, S., Feindel, D., Haddad, A., Khalil, Y., Alrijabo, A. and Piggin, C. 2013. Adoption of conservation cropping in dryland regions of northern Iraq and Syria. Proc. of the Int. Conference on Development of Drylands, Beijing China. Mar. 2013.

Loss, S., Haddad, A., Desbiolles, J., Cicek, H., Khalil, Y. and Piggin, C. M. 2015a. The practical implementation of conservation agriculture in the Middle east. International Center for Agricultural Research in the Dry Areas (ICARDA).

Loss, S., Haddad, A., Khalil, Y., Alrijabo, A., Feindel, D. and Piggin, C. 2015b. Evolution and adoption of conservation agriculture in the Middle East. In: *Conservation Agriculture* Farooq, M. and Siddique, K. (Eds). Cham: Springer, pp. 197–224. https://doi.org/10.1007/978-3-319-11620-4_9.

Malekloo, A. 2018. *A Review of the Iranian Water Crisis and Proposed Countermeasures*. Northern Cyprus: Middle East Technical University Northern Cyprus Campus.

Ministry of Agriculture (MOA) 2010. *The Development Plan of Cereals and Legumes in Lebanon*. Ministry of Agriculture (MOA), Lebanon.

Montgomery, D. 2007. *Dirt: The Erosion of Civilizations*. University of California Press, Berkeley, Los Angeles.

Okur, B., Okur, N. and Anaç, D. 2003. Tarım Topraklarında organik Maddenin sürdürülebilirliği. Koruyucu Toprak İşleme ve Doğrudan Ekim Çalıştayı, E.Ü. Ziraat Fakültesi, Tarım Makinaları Bölümü, Bornova, İzmir.

Özpınar, S. and Çay, A. 2006. Effect of different tillage systems on the quality and crop productivity of a clay-loam soil in semi-arid North-Western Turkey. *Soil and Tillage Research* 88(1-2): 95–106.

Öztürkmen, A. R. and Kavdir, Y. 2012. Comparison of some quality properties of soils Around land-mined areas and adjacent agricultural fields. *Environmental Monitoring and Assessment* 184(3): 1633–1643.

Piggin, C., Haddad, A. and Khalil, Y. 2011. Development and promotion of zero-tillage in Iraq and Syria. Proceedings of the 5th World Congress on Conservation Agriculture, Brisbane, Australia, pp. 26–29.

Piggin, C., Haddad, A., Khalil, Y., Loss, S. and Pala, M. 2015. Effects of tillage and time of sowing on bread wheat, chickpea, barley and lentil grown in rotation in rainfed systems in Syria. *Field Crops Research* 173: 57–67.

Rorett, K. 2016. Iran tackles environmental challenges with new approach to cropping systems. *Informatica (CIMMYT Newsletter)* 1968: 6.

Saad Abdul-Jabbar, A., Alrijabo, A. A. and Husham, A. 2013. Assessed the performance of the modified Caspardo seeder to work under zero tillage system in wheat crop cultivation. The Second International Scientific Conference for Agricultural Research, University of Kirkuk.

Schmidhuber, J. and Tubiello, F. N. 2007. Global food security under climate change. *Pans* 104: 19703–19708.

Shakhatreh, Y. 2010. Trend analysis for rainfall and temperatures in three locations in Jordan. Food Security and Climate Change in Dry Areas, 1–4 February 2010. Amman Jordan.

Shakhatreh, Y., Carvalho, P., Foulkes, J., Musallam, I., Ismail, F., Khalaf, Y. and Bani Hani, N. 2014. Improving wheat agricultural practices to cope with climate change effects in Jordan. Proceedings in: Water, environment and agriculture: Challenge for sustainable development. Lamaddalena, N., Todorovic, M. and Pereira L. S. (Eds). 10–14 September 2013, Bari-Italy.

Sommer, R., Ryan, J., Masri, S., Singh, M. and Diekmann, J. 2011. Effect of shallow tillage, moldboard plowing, straw management and compost addition on soil organic matter and nitrogen in a dryland barley/wheat-vetch rotation. *Soil and Tillage Research* 115–116: 39–46.

Sungur, N., Ulusoy, E. and Yalçın, H. 1994. Ege Bölgesi Koşullarında Buğday ve İkinci Ürün Mısır Elde Etmede Mekanizasyon Olanakları. *Tarımsal Mekanizasyon 15. Ulusal Kongresi Bildiri Kitabı, s*:582–591. Antalya.

Tsusaka, T. and Otsuka, K. 2013. The changes in the effects of temperature and rainfall on cereal cropyields in Sub-Saharan Africa: a country level panel data study,1989 to 2004. *Environmental Economics* 4(2).

Turkish Statistical Institute (TSAI) 2020. http://www.tuik.gov.tr/PreTablo.do?alt_id =1001.

Vaghefi, S. A., Keykhai, M., Jahanbakhshi, F., Sheikholeslami, J., Ahmadi, A., Yang, H. and Abbaspour, K. C. 2019. The future of extreme climate in Iran. *Scientific Reports* 9(1): 1464.

World Bank 2012. Droughts and agriculture in Lebanon causes, consequences, and risk management. Report. Washington, DC.

Yeşilsoy, Ş. 1972. Winter wheat summer-fallow research iin Central Anatolia first International Winter Wheat Conference-Proceedings, Ankara June 5-10, 1972.

Yigezu, Y. A., Mugera, A., El-Shater, T., Aw-Hassan, A., Piggin, C., Haddad, A., Khalil, Y. and Loss, S. 2018. Enhancing adoption of agricultural technologies requiring high initial investment among smallholders. *Technological Forecasting and Social Change* 134: 199-206.

Zadoks, J. C., Chang, T. T. and Konzak, C. F. 1974. A decimal code for the growth stages of cereals. *Weed Research* 14(6): 415-421.

Chapter 12

Adoption of Conservation Agriculture in Central Asia

Aziz Nurbekov, Tashkent State Agrarian University, Uzbekistan; Alisher Mirzabaev, ZEF-University of Bonn, Germany; Muratbek Karabayev, CIMMYT, Kazakhstan; Nurali Asozoda, President Tajik Academy of Agricultural Science, Tajikistan; Nasriddin Khalilov, Samarkand Institute of Veterinary Medicine, Uzbekistan; and Dossymbek Sydyk, South-Western Research Institute of Livestock and Crop Production, Kazakhstan

1 Introduction
2 Research and Conservation Agriculture systems in Central Asia
3 Adoption of Conservation Agriculture in Central Asia
4 What is next: challenges and opportunities
5 Conclusion
6 Where to look for further information
7 References

1 Introduction

Agriculture is an important sector of the economy in Central Asian countries because it serves as a major source of employment and income for the rural population. During the past 30 years, the agricultural sector has been experiencing gradual shifts from a centralised administrative management system to more market-oriented economies across the region, with varying degrees of liberalisation of the agricultural sector. The agricultural potential also differs considerably between countries based on the geography, climatic zones and combined soil–climate conditions.

The most important factor limiting agricultural production in the Central Asian region is the scarcity and poor use of water resources. Global warming of 2°C has been projected to increase the gap between precipitation and crop water demands in some parts of Central Asia, particularly in northern Kazakhstan and the Fergana valley (Li et al., 2020). The decrease of water level in the Aral Sea is attributed to expanding irrigation systems – the unsustainable management of irrigation and drainage systems have brought about serious

http://dx.doi.org/10.19103/AS.2021.0088.11

ecological damage, with large areas becoming unsuitable for agriculture due to increased salinity. In Uzbekistan, the Aral Sea crisis has led to increasing water shortages and the loss of huge agricultural areas due to salinity. Farmers have limited experience in water conservation practices and in utilising crops that are tolerant to salt and drought, despite the introduction of resource-conserving water and soil management practices and drought- and salt-tolerant crops, on pilot scales.

Therefore, in Central Asia, due attention needs to be given in the coming years to the introduction of new resources conserving crop management practices, including crop diversification and crop residue retention. A number of projects have been implemented recently with the assistance of international organizations (Food and Agriculture Organization (FAO), International Centre Agricultural Research in the Dry Areas (ICARDA), CIMMYT[1], ZEF[2]-University of Bonn, National Agricultural Research Systems (NARS) etc.) and within the framework of testing new crop management practices, such as Conservation Agriculture (CA), for all agricultural crops and cropping systems. These research activities have shown that CA can help achieve sustainable growth in crop productivity and improve soil fertility and productivity. CA practices include no-till (NT), crop residue retention and crop rotation. Irrigation efficiency is addressed through improved land levelling; drainage; water-saving technologies including furrow, drip and sprinkler irrigation; improved control of irrigation water; crop irrigation scheduling using deficit irrigation practices; and saline soil leaching, all with the primary objective of utilising the scarce water resources in an efficient and sustainable manner. CA practices allow workload to be reduced (by up to 50%, Hickmann, 2006), save fuel (by up to 70%, Nurbekov et al., 2016), and reduce equipment needs (by up to 50%, Paroda, 2007). Sustainable crop production systems can be achieved with NT methods and appropriate crop rotations. In addition to the quantitative benefits of CA practices, the reduced leaching of soil nutrients and farm chemicals, together with the reduced soil erosion, lead to a significant improvement of the water quality in watersheds and to the sequestration of atmospheric carbon, thereby lowering the carbon footprints (de Moraes Sá et al., 2013; Hassan et al., 2015).

The up-to-date, scientific and practical experience of adopting CA systems and practices in Central Asia has been limited compared to other regions of the world. There are limited reliable data on the process of CA adoption, spread and uptake, except for Kazakhstan. The purpose of this chapter is to contribute to filling these knowledge gaps by discussing present trends and future prospects for CA in the region.

1 CIMMYT - International Maize and Wheat Improvement Center
2 ZEF - Center for Development Research

1.1 Crop production

After the dissolution of the former Soviet Union, the agriculture sector of Central Asia has declined, with a sharp reduction in the production of agricultural and horticultural crops. Wheat and cotton production are predominant in most Central Asian countries, for example in Turkmenistan and Uzbekistan, where cotton exports are the main source of foreign currency in the agricultural sector. The main agricultural areas in the southern warmer zones are under cotton, whereas the northern cooler zones are under spring wheat and irrigated summer crops. The changes in the area allocated to cotton and cereals over the period from 1992 to 2017 are presented in Table 1. The overall area under cotton, wheat, rice and barley decreased from 24.29 M ha in 1992 to 19.54 M ha in 2017 (−19.56%), whereas production decreased from 37.00 M MT to 34.11 M MT (−7.81%).

On average, crop production now varies by approximately one-third from one year to the next (World Bank, 2015). Apart from the domestic food security problem, this creates an international market risk, given that 70% of Kazakhstan's grain is exported (World Bank, 2015). The demand for food grains is expected to continue to grow in the Central Asian countries. The fastest growth in demand is for rice, followed by fodder crops for livestock.

The monoculture of wheat and cotton production prior to independence neglected the development of other crops in the cropping systems, especially alternative crops such as vegetables (melon, watermelon, cucumber, tomato etc.) and fruit trees. The predominance of cotton and cereals has led to certain problems in the agriculture sector. To enable diversification of the agricultural sector, government policies need to be put in place to promote a wider range of agricultural, horticultural and fodder crops. The diversification and expansion of the agricultural sector will enhance domestic food markets and improve the living standards of farming populations in Central Asia.

1.2 Climate and soils

The climate in Central Asia is continental, with a high level of spatial and intertemporal variability in temperatures and precipitation, and mostly arid and semi-arid moisture regimes, with long and hot summers and short and cold winters. Average annual precipitation is concentrated during the winter and spring months and is approximately 270 mm but ranges from 600 mm to 800 mm in the mountainous zones and from 80 mm to150 mm in the desert and arid regions. The mean winter temperatures throughout the region during the last century ranged between −25°C and +7°C, while the mean summer temperatures have ranged between +2°C and +31°C (Mirzabaev, 2013).

According to the United Nations Educational, Scientific and Cultural Organization (UNESCO) classification, there are 21 agroclimatic zones in

Table 1 Area, production and productivity of agricultural crops in Central Asian countries, 1992 and 2017

		Kazakhstan		Kyrgyzstan		Tajikistan		Turkmenistan		Uzbekistan		Total		
Crops	Item/unit	1992	2017	1992	2017	1992	2017	1992	2017	1992	2017	1992	2017	% change
Cotton	Area (m ha)	0.11	0.14	0.02	0.021	0.29	0.17	0.57	0.55	1.67	1.17	2.66	2.05	−22.90
	Prod (m t)	0.23	0.33	0.04	0.07	0.53	0.39	1.18	1.01	3.86	3.79	5.84	5.59	−4.45
	Yield (t ha^{-1})	2.10	2.41	1.90	3.18	1.80	2.24	2.10	2.02	2.30	3.21	2.04	2.61	28.00
Wheat	Area (m ha)	13.72	11.91	0.25	0.25	0.18	0.28	0.20	0.90	0.63	1.40	14.98	14.74	−1.60
	Prod (m t)	18.29	14.80	0.68	0.60	0.17	0.89	0.38	1.00	0.96	6.08	20.48	23.37	14.10
	Yield (t ha^{-1})	1.30	1.24	2.70	2.41	0.90	3.14	1.90	1.10	1.50	4.31	1.66	2.44	46.90
Rice	Area (m ha)	0.12	0.10	0.05	0.01	0.01	0.01	0.03	0.14	0.18	0.07	0.39	0.33	−2.94
	Prod (m t)	0.47	0.49	0.07	0.04	0.02	0.10	0.06	0.13	0.54	0.40	1.16	1.16	0.00
	Yield (t ha^{-1})	4.00	4.68	1.50	3.56	2.00	7.80	2.30	0.94	3.00	5.54	2.56	4.50	25.00
Barley	Area (m ha)	5.63	2.06	0.26	0.19	0.06	0.08	0.06	0.01	0.30	0.10	6.31	2.44	−61.49
	Prod (m t)	8.51	3.30	0.62	0.42	0.04	0.15	0.13	0.01	0.29	0.13	9.59	4.01	−58.19
	Yield (t ha^{-1})	1.50	1.59	2.40	2.20	0.70	1.89	2.10	1.43	0.90	1.34	1.52	1.69	11.18
Total for all crops	**Area (m ha)**	**19.58**	**14.21**	**0.58**	**0.47**	**0.54**	**0.54**	**0.86**	**1.59**	**2.78**	**2.74**	**24.34**	**19.55**	**−19.56**
	Prod (m t)	**27.50**	**18.92**	**1.41**	**1.13**	**0.76**	**1.52**	**1.75**	**2.15**	**5.65**	**10.40**	**37.07**	**34.12**	**−7.81**

Source: Statistical committee/Agency of Kazakhstan (https://stat.gov.kz/), Kyrgyzstan (http://www.stat.kg/), Tajikistan (https://www.stat.tj), Turkmenistan (http://www.stat.gov.tm/) and Uzbekistan (https://stat.uz/), 2018.

Central Asia (Table 2). Estimates show that 68% of the total area is covered by only two agroclimatic zones: (i) semi-arid, with cold winter and warm summer; and (ii) arid, with cold winter and warm summer. The remaining 30% of the area is covered by ten zones, and other zones cover the remaining small territories (De Pauw, 2008; Gupta et al., 2009).

Uzbekistan is subdivided into ten agroclimatic zones. The vegetative period (when the temperature is higher than 10°C) makes up 190 days in the northern part and 240 days in the southern part of the country.

In the high-altitude zones of Tajikistan, there is a subzone with precipitation up to 1000 mm a year, including dry farming land, which includes the southern slopes of Hissar mountains, all of Karategin–Darvaz, and the foothills of the Kulyab group of areas.

The agroclimatic zones of Turkmenistan and the southern part of Kazakhstan have characteristics similar to those of Uzbekistan. The agroclimatic zones of northern Kazakhstan, however, sharply differ from other territories of Central Asia. The northern part of Kazakhstan is a forest–steppe zone with chernozem soil and with annual precipitation of 350–400 mm. Moving south, the climate becomes more arid and the chernozem soils are replaced by chestnut soils. During the winter period, 70–100 mm of precipitation falls as both snow and rainfall, representing approximately one-third of the annual total precipitation.

Kyrgyzstan is subdivided into four agroclimatic zones: northern, northwest Kyrgyzstan; northeast Kyrgyzstan (Issyk Kul hollow); Internal Tien Shan; and Southwest Kyrgyzstan.

In summary, the agroclimatic zones of Central Asia allow the production of a wide range of crops. There are huge opportunities for the diversification of agricultural crops under CA systems and practices. The current crop production systems are as follows:

- Spring cereals, legume and oil crops, widely grown in rainfed conditions in the northern and central parts of the region, mostly in Kazakhstan and in some regions of northern Kyrgyzstan;
- Rainfed winter cereal and legume crops, mainly in Uzbekistan, Kyrgyzstan and Tajikistan and the southern part Kazakhstan;
- Irrigated agriculture in Kyrgyzstan, Tajikistan, Uzbekistan and the southern part of Kazakhstan.

1.3 Land degradation

Land degradation and desertification are common all over the Central Asian region. Nachtergaele et al. (2011) confirm that the Central Asian drylands are very prone to land degradation - in particular water and soil erosion - due to low vegetation cover. In Kazakhstan, 66% of the land area is affected by land

Table 2 Agro climatic zones of Central Asia derived from the underlying data

Agroclimatic zone	Description	Aridity index 1	Temp. range coldest month (°C)	Temp. range warmest month (°C)	% of Total
SA-K-W	Semi-arid, cold winter, warm summer	0.20-0.50	<=0	20-30	37.9
A-K-W	Arid, cold winter, warm summer	0.03-0.20	<=0	20-30	30.8
SA-K-M	Semi-arid, cold winter,	0.20-0.50	<=0°C	10-20°C	6.6
SH-K-M	Sub-humid, cold winter,	0.50-0.75	<=0°C	10-20°C	5.9
A-C-W	Arid, cool winter, warm summer	0.03-0.20	0-10°C	20-30°C	4.9
A-C-VW	Arid, cool winter, very warm summer	0.03-0.20	0-10°C	>30°C	2.9
PH-K-C	Per-humid, cold winter, cool summer	>1.00	<=0°C	0-10°C	2.0
H-K-M	Humid, cold winter, mild summer	0.75-1.00	<=0°C	10-20°C	1.6
SA-C-W	Semi-arid, cool winter, warm summer	0.20-0.50	0-10°C	20-30°C	1.5
SH-K-W	Sub-humid, cold winter, warm summer	0.50-0.75	<=0°C	20-30°C	1.4
A-K-VW	Arid, cold winter, very warm summer	0.03-0.20	<=0°C	>30°C	1.2
PH-K-M	Per-humid, cold winter	>1.00	<=0°C	10-20°C	1.2
SH-K-C	Sub-humid, cold winter, cool summer	0.50-0.75	<=0°C	0-10°C	0.5
SA-K-C	Semi-arid, cold winter, cool summer	0.20-0.50	<=0°C	0-10°C	0.5
H-K-C	Humid, cold winter, cool summer	0.75-1.00	<=0°C	0-10°C	0.5
H-K-W	Humid, cold winter, warm summer	0.75-1.00	<=0°C	20-30°C	0.2
SH-C-W	Sub-humid, cold winter, warm summer	0.50-0.75	0-10°C	20-30°C	0.1
A-K-M	Arid, cold winter, mild summer	0.03-0.20	<=0°C	10-20°C	0.1
PH-K-K	Per-humid, cold winter, cold summer	>1.00	<=0°C	<=0°C	0.1
PH-K-W	Per-humid, cold winter, warm summer	>1.00	<=0°C	20-30°C	0.0
A-K-C	Arid, cold winter, cool summer	0.03-0.20	C	0-10°C	0.0

(Source: De Pauw, 2008)

degradation, while in Turkmenistan and Uzbekistan this figure is as high as 80% (Fig. 1). Between 40% and 80% of irrigated lands in the region are salt-affected and/or waterlogged (Qadir et al., 2008; Toderich et al., 2013). Erosion affects over 88% of the arable land in Kyrgyzstan and 97% of the agricultural land in Tajikistan.

Soil degradation is due to continuous monocropping production of cotton and rice, with intensive tillage, in the irrigated conditions of Uzbekistan, Turkmenistan, Tajikistan and the southern part of Kazakhstan, while rainfed agriculture of the region is also challenged by problems of soil degradation due to intensively tilled wheat monocropping and fallow. The introduction of conservation tillage in the mid-1950s reduced wind erosion in Northern Kazakhstan (Barayev, 2008). Poor irrigation management practices have led to increased salt-affected areas in the region (ELD Initiative, 2015; Pender et al., 2010). Inefficient management of croplands has led to significant decreases in soil fertility in the region, and the low soil fertility now has a negative effect on the productivity of agricultural crops in the region.

The current agricultural production practices in Central Asian countries are based on intensive crop management and full tillage in irrigated conditions, while in rainfed conditions conservation tillage dominates, mostly in Northern Kazakhstan. The system has led to increased demand for agricultural inputs (fuel, fertilizer and pesticides), increased labour, and higher production costs. Prior to 1990, pollution by fertilisers and pesticides was the main cause of soil degradation. The highest level of soil pollution is in Uzbekistan, where 54% of the surveyed land was polluted by pesticides. Drainage water from agricultural fields,

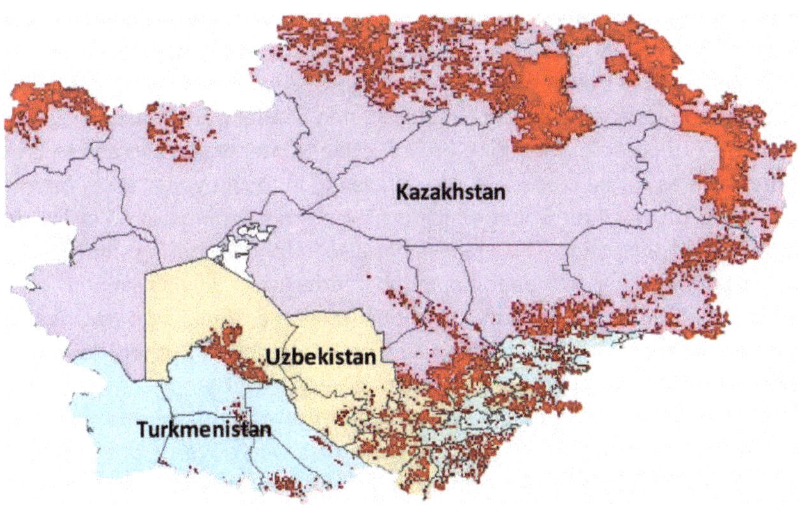

Figure 1 Land degradation hotspots in Central Asia (in red), a negative change in NDVI between 1982 and 1984 and 2006. Source: Mirzabaev et al. (2016).

dispersion of pesticides into the atmosphere, and direct application of chemicals for seaweed control (NPDOOS, 1999) were the main sources of pollution.

Land degradation is the main problem in agricultural production systems and is expected to continue in the near future, with the traditional agricultural practices in Central Asia. Moreover, climate change is projected to exacerbate soil degradation problems in the region (Ma et al., 2020). Hence, there is a need to study and adopt new sustainable land management systems and practices including CA systems and practices.

2 Research and Conservation Agriculture systems in Central Asia

2.1 Crop rotation and double cropping under Conservation Agriculture

2.1.1 Crop rotation

Vladykina et al. (2006) and Borin (2003) have reported that in a three-course rotation (wheat 50–67%, legume and fallow 50–33%), the humus content reduced between 21% and 26% in the upper layer of the soil with traditional tillage agriculture, whereas Nurbekov (2018) has shown that humus content increased with short-term crop rotation under CA practices. The wheat–legume rotation is a good practice in the irrigated conditions of Uzbekistan and Tajikistan (Nurbekov, 2018). This short-term crop rotation can improve land use productivity and efficiency in the irrigated areas, which are extremely important in Central Asian countries. Suleymenov and Akshalov (2009) have reported that the wheat–legume rotation under CA will decrease soil bulk density and increase water infiltration and moisture storage. Wheat provides abundant residues which decompose slowly, while legumes improve soil fertility and generally do not leave large amounts of residues, facilitating seeder performance.

Earlier, because of their long growth cycle, winter wheat and cotton crops were grown as single crops in a year. However, in recent years, it has become possible to grow winter wheat, which is planted in standing cotton stalks after the harvest of raw cotton. This practice has also helped in sparing some land for growing other crops, thus enabling much-needed crop diversification.

Paroda (2007) proposed that field pea, chickpea, lentil and buckwheat are possible options for inclusion into existing crop rotations in northern Kazakhstan to increase agricultural production in the country.

As already mentioned, winter wheat and cotton are currently the main crops in the region, receiving all policy support from the local government. There is an urgent need for crop diversification to include grain legumes (soybean, common bean, chickpea, lentil and field pea), oilseed crops (groundnut, safflower and sunflower) and short-duration cereals (buckwheat and maize). Short-duration grain legumes also provide protein-rich food for human consumption and improve water use efficiency. Similarly, other oilseed crops like groundnut, safflower, sunflower etc. can be included in crop rotations to help meet the ever-increasing demand for vegetable oil for human consumption. Maize can be used as a dual-purpose crop for both human consumption and as animal feed and fodder.

2.1.2 Double cropping

The main climatic factors that determine the possibility of double cropping and crop diversification are the duration of the warm period after the harvest of winter cereals, the amount of precipitation, temperature and daylength. The activity of endogenous nitrogen-fixing bacteria living in the roots of legumes is closely related to the life of the host plant and depends on crop and water management, right seeding and the rate and date of fertiliser application (Baranov and Kochegura, 2005).

In recent years, in the irrigated conditions of Central Asia, in order to maximise the accumulation of kinetic energy from the sun by green plants and make efficient use of irrigated lands, double cropping (Nurbekov, 2018) with catch and intermediate crops (Ostanakulov and Khalilov, 2017) is being implemented in two ways. The first method is growing early-maturing high-yielding crops by building agrocenoses that allow maximum use of Photosynthetically Active Radiation (PAR) by green plants. The second method is to cultivate the crop during the hot season by occupying the crop fields with intermediate crops (Ostanakulov and Khalilov, 2017).

Good results have been obtained with the inclusion of mung bean in the wheat–cotton cropping system, by sowing mung bean after the wheat and before the cotton crop (Nurbekov et al., 2016). A number of studies have been conducted on systems with two or three crops per year on irrigated lands in Uzbekistan (Atabayeva, 2004; Iminov and Khalikov, 2005), and these have reported increased land use efficiency and gross production using no or minimum tillage.

In Tajikistan, many crops used for double cropping such as maize and mung bean, followed by common bean, soybean, vegetables, buckwheat, millet, tobacco, groundnut or sesame, have been grown after the winter wheat harvest and are widely adopted by small farmers. Since 1991, Turkmenistan began focusing on grain self-sufficiency, and wheat has become the second-most important crop after cotton (Paroda, 2007).

CA practices will reduce the risks of pest and disease build-up compared to long-term cereal monocropping sequences in the field (Bukhoriev and To'khtayev, 2012), while planned crop rotation will increase soil fertility and system productivity and profitability (Abdurakhimov et al., 2018).

The introduction of food legumes and other alternative crops through double cropping in the cotton-wheat rotation is another important activity carried out with great success in irrigated agriculture in Central Asia. Food legumes such as mung bean, soybean and common bean can be successfully grown under irrigated conditions, providing grain yields of 1.5–2 t/ha (Muhayyo, 2009). Nurbekov (2018) argues that there is a definite advantage of double cropping as compared to leaving the field fallow after the harvest of winter wheat (Fig. 2). Apart from food legumes, maize, sesame, melon, carrots etc. also have good possibilities to enhance crop diversification in Central Asia. Rice-wheat is a good option in the Aral Sea Basin area, which is surrounded by Kazakhstan, Turkmenistan and Uzbekistan. Rice is salt tolerant, therefore growing rice can help reduce the salinity problem without requiring any additional amount of water for leaching of salts. Devkota et al. (2015a,b) showed that the application of CA practices including dry seeding of rice could reduce irrigation water requirement by 60% compared to the alternate wet and dry irrigation.

Early-maturing varieties of soybean can be used as a catch crop in the irrigated conditions of Uzbekistan to increase land use efficiency and fix atmospheric nitrogen in the root zone (Gabilov, 2011; Umarova and Abitov, 2016). Sidikov and Husenov (2008) recommend planting winter wheat between September 15 and October 15 using minimal tillage in the standing cotton crop, in order to obtain higher yields and to sow the second crop on time after the winter wheat harvest in June.

In Central Asia, there are favourable soil and climatic conditions for producing two to three crops per year. Earlier long-term investigations on cotton did not anticipate producing two crops per year. Today, the winter wheat area under irrigation is about 3 M ha, of which 1.5 M ha is suitable for growing second crops after the winter wheat harvest using NT seeding.

2.2 Alley cropping with Conservation Agriculture practices

Many farmers grow different fruit trees on their farm and grow alfalfa, wheat, barley and legume crops in the alleyways between widely spaced rows of fruit trees (T.A. Bukhoriev Dushanbe, Tajikistan, 2020, pers. comms). Alley cropping is used strategically to improve forage production, mitigate drought, serve as a windbreak, capture windblown sediments in dry areas and reduce the need for fertiliser. Additionally, the soil surface between the rows of trees remains protected from the water erosion generated by the direct impact of raindrops. In addition to these benefits, it is an important practice in cultivated areas as it also offers long-term stability to soils.

The alley cropping technology has been adopted to improve livestock feeding options in Uzbekistan. The productivity of forage crops (pea-triticale,

Figure 2 Permanent ground cover in Khujand, Tajikistan, 2016.

barley) and fodder shrubs in an alley cropping system was evaluated to develop alternative, low-cost feed options (single species of forages or mixes) to improve the quantity and quality of feed supply and to assist farmers to optimise their livestock management in CA systems (Nurbekov et al., 2019).

The species used for alley cropping in irrigated conditions of Central Asia include cereals, legumes, vegetables and melons, while in rainfed conditions safflower, chickpea etc. Nowadays, the agri-environmental benefits of NT and alley cropping in tree and shrub plantations are widely recognised, whereas the unspecific practices under the umbrella of minimum tillage do not allow its acceptance for soil conservation anymore. Therefore, there is a need to use only NT crops in alley cropping system in the region.

Alley cropping in rainfed conditions can diversify farm income, increase crop production, decrease soil erosion and provide protection and conservation benefits to the crops (Turok et al., 2013). The introduction of alley cropping in the mountain regions of Tajikistan increased with triticale+forage pea as an alternative in order to increase forage production and decrease soil erosion (T. A. Bukhoriev, 2016, personal communication).

Jannatul Ferdush et al. (2019), in a field experiment on alley cropping, reported that the alley cropping system had a positive effect on soil properties in the top layer, including soil pH, organic carbon, total N, available P, available S, exchangeable K, Ca and carbon exchange capacity (CEC).

The crop seeded in the alleys must not grow too tall, so as to avoid hindering the transit of machinery through the plot; must grow rapidly to cover the soil quickly; must have shallow roots so as to not excessively explore the soil profile and extract water that can be used for the tree crop and should not host crop pests. Proper crop management of the vegetative cover is fundamental to optimise water utilisation and to maintain potential productivity of the tree crop, especially in dry climates like Central Asia where the long-term annual precipitation is low and seasonal. Therefore, the most important factors to be taken into account in the management of alley cropping will be determined by the soil water balance – firstly, the method of mowing to be used; and secondly, the timing of mowing of the cover, in order to reduce to a minimum necessary the consumption of water at times of maximum crop need, as well as to avoid competition for nutrients.

2.3 Crop productivity under Conservation Agriculture practices

A few research experiments were done on winter wheat seeding rate under permanent bed NT planting system in the irrigated conditions of South-Kazakhstan, but all research has been done using locally made bed planters. Using permanent raised beds, yields in the cotton–wheat system in Uzbekistan did not differ significantly from those under the traditional tillage system (Suleymenov et al., 2006, 2004). However, the savings in energy use and labour yielded a higher profit.

Atakulov et al. (2014) reported that yields of winter wheat, soybean and corn were higher by up to 10% under the permanent NT bed planting systems compared to the traditional tillage planting system. The introduction of resource-saving technology in the irrigated conditions of the Jambil province of Kazakhstan will increase irrigated land use efficiency and the productivity of agricultural crops and decrease the cost of agricultural inputs throughout the region.

Nurbekov et al. (2018) report that NT wheat planted less than 1 cm deep may result in uneven germination due to seed exposure or dry soil conditions. Shallow-planted wheat is also more susceptible to soil heaving due to freezing. NT wheat planted more than 4 cm deep may result in seedling death due to pre-mature leaf opening or poor tiller development and winter survival (Nurbekov et al., 2018). Grain yields tended to be the greatest with the highest seeding rates compared to yields with the lowest seeding rates. Uniform seed placement and seeding depth are important in promoting crop health in the fall, while using the NT method. In 2008, low yields resulted from severe low temperatures and the absence of snow cover during the winter, which posed a severe limitation for winter wheat growth and development (Nurbekov et al., 2018). The responses to seeding depth under NT and traditional till were significant, resulting in an increased number of tillers and greater grain yield.

Kienzler (2010) reported that N-160 kg/ha had the highest yield in Khorezm province of Uzbekistan. These results confirmed those of Frederick and Camberato (1995) and Nurbekov (2018). Different tillage methods and fertiliser applications positively affect the thousand-kernel weight, the number of tillers per plant and the spike weight. Fertiliser rate and placement are two factors that determine winter wheat productivity in the region under both the conventional and the CA systems, and fertiliser use efficiency increases while fertilizer rate decreases in the CA system (Nurbekov et al., 2012).

Sydyk et al. (2008, 2018) reported that furrow irrigation at 900 m^3/ha^{-1} and 1200 m^3/ha^{-1} is more profitable than conventional flood irrigation for growing maize on permanent raised beds. At these rates of furrow irrigation, the yields of maize were higher than under float irrigation and saved considerable water (up to 30%), while the input costs were similar.

Figure 3 shows winter wheat growing successfully on permanent beds in Uzbekistan. Nurbekov (2018) reported that winter wheat grain yield was significantly higher under the NT system compared to the yield under the conventional system. Grain yield from verified permanent bed planting was the highest compared to conventional planting across all varieties and years, while Košutić et al. (2005) reported that the grain yield of winter wheat among conventional, conservation tillage and NT treatments did not differ significantly.

Several authors compared traditional tillage, NT and minimal tillage methods, and found that the NT method can reduce field work (including the number of field operations and labour), improve soil structure (Sydyk et al., 2008; Tynybaeva et al., 2017) and increase crop productivity (Wall et al., 2007). An FAO report (2012) reported that NT systems had a greater soil moisture-holding capacity during the 2012 drought year in Kazakhstan, and this led to a higher yield compared with the conventional and minimum tillage systems. Kireyev (2004) reported that NT wheat crop had 10% to 15% higher yield compared to conventional and minimum tillage treatments in Almaty region of Kazakhstan. It is well known that a longer vegetation period has a positive effect on the productivity of agricultural crops, which may have been the cause for the low yields of winter wheat under controlled tillage treatment.

Frost resistance in wheat depends on the biological properties, severity of low temperatures during the winter period, crop growth stages, fertiliser rates and seeding dates, and frost tolerance also depends on soil condition and snow cover (Dvurechenskiy, 2010; Landers et al., 2003). Keldiyorova and Xozhaev (2004) reported that the timing of seeding in winter wheat and residue cover can also be important factors in managing frost damage under NT technology. Crop residue cover of 10–20 cm deep does not impede the NT seeding operations. It keeps the snow in place well and promotes the accumulation of winter moisture for future crops, preserving the winter crops

Figure 3 Double cropped mungbean after winter wheat harvest in Karshi, Uzbekistan, 2017.

from freezing. Maize crop residue cover of 30–60 cm deep creates excellent conditions for snow retention (Barayev, 2008).

Whaley et al. (2004) reported that wheat might be at risk of damage by low temperatures at all stages of crop development, and Nurbekov (2018) reported that in spring 2008, low temperature negatively affected the wheat stand without crop residue cover, while with crop residue, the wheat stand was better. Cold hardening may occur in the crop over winter, a process whereby compounds, usually water-soluble carbohydrates, accumulate to protect cell membranes from disruption in very cold temperatures (Suzuki and Nass, 1988). In Karakalpakstan conditions, NT winter wheat was only slightly injured by cold during the tillering stage, while conventional wheat was injured much more with symptoms such as leaf chlorosis, burning of leaf tips, silage odour, or blue cast to fields, brownish tillering nodes and root and plant death (Nurbekov, 2018).

2.4 Crop residue and salinity management under Conservation Agriculture

2.4.1 Crop residue

CA practices, combined with maintaining permanent raised beds, have been shown to reduce soil salinity by 32% in the top 10 cm of soil and by 22% in the top 90 cm of soil in the cotton-wheat-maize rotation system in Khorezm, Uzbekistan (Devkota et al., 2015a,b). In Kazakhstan, loss of the organic content of the soil is widespread, which, combined with extreme climatic events, drives extreme instability in the food supply. As an example, a drought in northern Kazakhstan in 2012 affected 1.1 M ha of grain crops, causing a 53% decrease in grain crop production (UNEP, 2016).

Dridiger et al. (2016) argue that crop residue retention under NT technology accumulates 30–35% more moisture in the soil compared to the situation without crop residue under traditional technology, and it also increases crop yield (15%) (Hickmann, 2006; Banik and Sharma, 2008). The retention of crop residues in NT systems also reduced evaporation (30%) and increased water infiltration (15%) into the soil (Nurbekov, 2018). This effect was more pronounced in a dry climate where residues are slower to degrade (Asoyev et al., 2015). Maize provides abundant residues, whereas cotton does not (Qilichev and Khalilov, 2008).

Due to the minimisation of field operations in the crop rotation and the practice of leaving crop residues in the field, soil physical properties and biological activity improve (Dvurechenskiy, 2010). The NT method, with crop residue mulching on the surface of the soil using the chopped straw and stubble and minimised tillage, helps to preserve and increase moisture reserves in the soil (Khaitov and Allanov, 2014; Singh et al., 2014). With crop residue retention, greater soil health and productive capacity and lower cost of production lead to higher crop yields and factor productivities (Gonzalez et al., 2018).

Soil degradation from monocropping of cotton, wheat and to a lesser extent rice cultivation continues unabated in the region (or in particular countries or areas). In order to reduce soil degradation and improve soil structure in Central Asia, NT and the retention and use of crop residues must be encouraged. Crop residue also makes a very beneficial contribution to the nutrient status and fertility of soils, particularly where fertiliser is in short supply.

2.4.2 Salinity management under Conservation Agriculture

Many piedmont basins of Central Asia have little natural drainage. Irrigation of such basins contributes to the formation of shallow ground water, with the subsequent increase in secondary soil salinity. Therefore, salinity is becoming a major problem in irrigated agriculture in Central Asia, which is increasing year after year and thereby adversely affecting crop yields (Nkonya et al., 2016). We have to tackle this problem scientifically by adopting the best sustainable land management practices so that younger generations can also use the available land resources more effectively. There are different options by which the salinity problem can be managed; these include the use of salt-tolerant varieties, CA, and rice-wheat cropping system (Gonzalez et al., 2018). Rice is comparatively a more salt-tolerant crop, and when it is grown after wheat, it provides continuous land cover, thus reducing salt uptake during the summer period (Bakhitbay Aybergenov, Nukus, Uzbekistan, 2020, pers. comms). In saline areas, instead of leaching and then growing any crop, it is better to grow rice, as leaching will automatically be done and farmers would get an additional rice crop. The winter wheat variety 'Dostlik' can be successfully grown in low- to medium-salinity areas (Paroda, 2007) using NT methods (Nurbekov, 2008).

Soil mulching, through the reduction of physical evaporation from the bare soil surface, assists in reducing the upward movement of salts from shallow saline groundwater to the crop root zone through the capillary rise. Tursunov (2009) reports that the soil salinity level in the soil profile had an increasing trend from the bottom of the furrow to the crest of permanent raised beds. Nurbekov (2008) reported that soil mulching decreased the salt content in 0–10 cm soil layer by up to 1.6 to 4 times where different mulches were applied in order to cover the soil surface in the salt-affected areas. CA has already provided evidence in the field that due to the reduced evaporation, salts remain in the deeper soil layers. However, to achieve the objective of less salt movement in the soil profile, soil cover should be more than 60%, but to avoid evaporation from the soil (which would imply salt ascension), the optimum would be to cover 100% of the soil surface with crop residues.

The productivity of wheat in general is low in the Aral Sea basin of Central Asia, due to water scarcity and salinity (Nurbekov, 2018). Egamberdiev (2007) reported that under the NT method, soil salinity decreased compared to conventional tillage in the Khorezm province of Uzbekistan.

Farmers are benefiting from CA technology in salt-affected areas of Central Asia. We believe that in the long run, the NT practice – with the retention of crop residues – will help in lowering the salinity levels due to the combined effects of reduced evaporation and recycling of organic matter.

2.5 Economics of Conservation Agriculture

The social and economic benefits of CA can help address the interlinked problems of food insecurity and the degradation of natural resources under a changing climate in the region. Although economic information on NT practices is not extensively available for Central Asian countries, recent experiences with NT systems point to various opportunities to increase farmers' incomes and livelihoods and strengthen local and national economies.

To illustrate this, Nurbekov et al. (2016) assessed the economic impact of conventional versus NT agriculture. NT can be recommended as a more appropriate method in increasing the profit from winter wheat in Tajikistan. Nurbekov (2018) studied the economics of tillage methods in winter wheat cultivation under irrigated conditions in the Gissar Valley of Tajikistan and reported that profit (US$651 per ha) was highest under NT wheat, compared to US$309 per ha and US$435 per ha in control and minimum-till (MT) treatments respectively. Thus, adopting NT practices in the irrigated areas of Tajikistan can improve the livelihoods of farmers.

Similarly, NT was found to be 30% more profitable and efficient than traditional tillage practices for growing maize in southern Kazakhstan (Sydyk et al., 2018). With the expected continuation in the rise of fuel prices, the relative profitability of NT direct seeding compared to the traditional tillage practice will only increase over time. Moreover, experiments conducted by ICARDA in Kazakhstan showed that the NT and MT cultivation of maize provided greater savings and net benefits for producers compared to traditional tillage (ICARDA-CAC, 2013). The net benefits from NT maize and MT maize were US$736.5 per ha and US$641.9 per ha respectively, which were 123.7% and 107.8% higher than conventional till (CT) maize with US$595.3 per ha. The benefit-cost ratio varied from 45.9% in CT to 57.3% in NT (ICARDA-CAC, 2013). These results show that NT forage maize crop will help livestock producers to have access to low-cost forage resources and thus improve the productivity and efficiency of livestock production in Kazakhstan, and perhaps in other Central Asian countries as well. A knowledge of maize growth and development is also essential to use cultural practices more efficiently and obtain higher yields and profits.

Furthermore, Asoyev et al. (2015) reported a comparison of the returns over variable costs for the average yield of comparable management practices for both traditional-tilled and permanent bed-planted winter wheat. The average returns over variable costs were US$842 per ha and US$873 per ha higher for the bed-planted maize versus US$224 per h and US$189 per ha for the traditional-till-planted maize. Permanent bed planting produced the highest net benefit and net revenue. This is of critical importance in Tajikistan, where agricultural input prices are steadily increasing year on year while the prices of agricultural products are decreasing in the local markets.

Some other results showed that the sowing of new varieties of winter wheat at the beginning of October under NT methods will provide the highest gross income, net profit and the highest level of profitability compared to late planting treatments (Khalilova, 2019).

A cost-benefit analysis of double-cropped soybean and kidney bean showed that the highest profitability rate was observed with early planting treatments under minimum tillage, while the lowest profitability rate was recorded with late planting treatments under conventional tillage (Qilichev and Khalilov, 2008; Khalilova, 2019).

Economic modelling by Boboev et al. (2019) showed that the application of CA practices was profitable even in degraded lands where conventional practices were economically unviable. Moreover, diversified cropping systems with CA practices in the cotton–wheat–maize cropping system had higher profitability over the cotton monoculture under conventional practices (Boboev et al., 2019). CA can also maintain ecosystem functions of soils in croplands and thus help to avoid losses of ecosystem services with the cropland expansion that is occurring in Central Asia (e.g. see Li et al., 2019, on croplands and their ecosystem values).

3 Adoption of Conservation Agriculture in Central Asia

3.1 Adoption status of Conservation Agriculture

Conservation agriculture (CA) is becoming more known in Central Asia and several good quality publications and practical guides are available (Nurbekov, 2008; Nurbekov and Ziyadullaev, 2012; Nurbekov et al., 2012, 2013, 2016). The characteristics of locally adapted CA production systems, together with rational and responsible use of external inputs, can optimise crop yields, farm income, competitiveness and biodiversity, while minimising some of the negative ecological impacts associated with intensive tillage farming.

CA based on minimum soil disturbance techniques is recommended in order to reduce tractor and machinery requirements per hectare, conserve soil moisture through reduced cultivation and improve soil structure and yields. Field grading and levelling both need to be performed much more effectively in order to improve the irrigation efficiency with CA systems in the region.

CA is a systems approach to farming that has the potential to reverse the trend of soil degradation by eliminating tillage and managing residues using diversified cropping systems. It is a proven approach to sustainable production in all regions of the world and is gaining greater acceptance throughout Central Asia. The region could benefit from international scientific and empirical experiences in NT, conservation and minimum tillage systems.

In 2018, CA practices in Central Asia were spread over to 3.195 M ha (Table 3), of which about 3.0 M ha are in Kazakhstan, with 0.195 M ha in the other countries of the region. Also, an area of more than 11.41 M ha of cropland is under minimum tillage in the region, with the greatest area located in Kazakhstan. This area is gradually being transformed to the NT system. In addition, there are NT perennial fruit trees and vineyards with permanent cover crops on about 0.70 M ha, of which the largest area (0.37 M ha) is in Uzbekistan. With the exception of Kazakhstan, the adoption of CA practices in the region has been slow compared to other regions in the world, especially in the irrigated lands.

Central Asian countries have traditionally practiced alley cropping and groundcover maintenance principally in irrigated areas where farmers grow field crops in orchards (Fig. 4). Gonzalez-Sanchez et al. (2015) and Kassam et al. (2020) have reported on agricultural practices within perennial CA systems. A minimum permanent groundcover of at least 30% must be met in CA perennial systems. Considering this, we can also include the permanent fruit orchards under CA and the vineyard plantation areas in Table 3. The total area of permanent CA systems of about 0.70 M ha across Central Asia makes the total area of CA systems in the region equal to some 3.79 M ha.

Since 2008, the government of Kazakhstan has been subsidising farmers who adopt CA-based technologies. In this regard, policy support was deemed to be a crucial factor for up-scaling CA practices in other countries of the region. The main issues to be addressed in the future are weed control, crop rotations, fertilisation strategies, training and awareness. The existing financial support system fails to reward smaller land users for maintaining the ecological integrity of agricultural and forest landscapes. For example, farms between 100 ha and 300 ha make, on average, an income of US$280 per ha from wheat production and do not receive any subsidies for adopting sustainable agricultural practices, while farms with over 15 700 ha receive a subsidy of US$220 per ha in addition to crop production income. Considering these subsidies in context, large-scale farms account for only 18% of the total number of agricultural producers while small- and medium-sized farm holders represent 82% of the producers but have no, or limited, access to government subsidies. Small community-based family farms that have, in many cases, less than 150–350 ha are not eligible for subsidies at all. This farmer support scheme sends a perverse signal and motivates farmers to simply increase farm area to be eligible for subsidies. In the crop sector, the government's support scheme is extremely biased towards wheat production. Moreover, wheat production activities are focused on increasing the total production, without proper consideration of ecological sustainability. This, in turn, leads to the proliferation of large-scale monoculture crop production.

Kyrgyzstan adopted the 'Law on protection of soil productivity on agricultural lands' in 2012. The law regulates soil conservation, soil fertility, conservation of the soil quality and protection from degradation and other negative impacts associated with land management, related to the ownership, use and/or disposal of agricultural lands. In Kyrgyzstan, many organisations, research institutions and international organisations support sustainable land management practices including CA practices. However, the adoption of CA practices such as NT, crop residue retention and crop diversification, remains limited. This is because there is limited access to seeds and inputs, a lack of access to long-term credit and inadequate training and extension support available for farmers to transform their systems.

Currently, the government of Uzbekistan is putting in place new agricultural policies. One of the five priority strategies during 2017–2025 focuses on the diversification and intensification of agricultural production and expansion of agricultural exports. The Government set out a longer-term strategy to

Table 3 Area of agricultural crop land under different tillage methods in Central Asia[a]. '000 ha (2002–2018)

Country	Technology and crops	2002	2005	2010	2015	2018	Operations
Kazakhstan	Conventional	19 000	15 200	11 600	7700	5600	Multiple tillage methods with blades and sweeps (mostly monocropping system)
	Minimum tillage	0	3400	6200	9300	10 500	Direct seeding with V-shaped openers or with narrow chisels regularly combined with shallow harrowing/ cultivation (few diversifications)
	No-till perennial crops	130	140	150	150	160	Perennial fruit trees and plantation of grapevines with vegetative cover and no-till
	No-till cropland	0	400	1200	2000	3000	Direct seeding with narrow chisels or double-disc openers (diversification)
	No-till cropland and perennial crops	130	540	1350	2150	3160	Perennial fruit trees and plantation of grapevines with vegetative cover and no-till
Kyrgyzstan	Conventional	1150	1160	1150	1200	1180	Multiple tillage methods with blades and sweeps (mostly monocropping system)
	Minimum tillage	0	0	0	30	60	Direct seeding with chisels with shallow harrowing/cultivation and planting wheat on standing cotton, (few diversification)
	No-till perennial crops	50	50	50	60	50	Perennial fruit trees and plantation of grapevines with vegetative cover and no-till
	No-till cropland	0	0	0	70	60	Direct seeding with narrow chisels or double-disc openers (diversification)
	No-till cropland and perennial crops	50	50	50	130	110	Perennial fruit trees and plantation of grapevines with vegetative cover and no-till

Tajikistan	Conventional	900	900	840	870	640	Multiple tillage methods with blades and sweeps (mostly monocropping system)
	Minimum tillage	30	30	50	80	200	Direct seeding with chisels combined harrowing/cultivation and planting wheat on standing cotton (diversification)
	No-till perennial crops	89	100	110	120	120	Perennial fruit trees and plantation of grapevines with vegetative cover and no-till
	No-till cropland	0	0	0	02	5	Direct seeding with narrow chisels or double-disc openers (diversification)
	No-till cropland and perennial crops	89	100	110	120	125	Perennial fruit trees and plantation of grapevines with vegetative cover and no-till
Uzbekistan	Conventional	3500	3410	3410	3310	3210	Multiple tillage methods with blades and sweeps (mostly monocropping system)
	Minimum tillage	450	540	620	630	650	Direct seeding chisels regularly combined harrowing/cultivation and planting wheat on standing cotton (diversification)
	No-till perennial crops	320	340	310	360	370	Perennial fruit trees and plantation of grapevines with vegetative cover and no-till
	No-till cropland	0	0	0	30	120	No-till including alley cropping
	No-till cropland and perennial crops	320	340	310	390	490	Perennial fruit trees and plantation of grapevines with vegetative cover and no-till

Source: Statistical committee/Agency of Kazakhstan (https://stat.gov.kz/), Kyrgyzstan (http://www.stat.kg/), Tajikistan (https://www.stat.tj), Turkmenistan (http://www.stat.gov.tm/) and Uzbekistan (https://stat.uz/), 2018.
[a]Data for Turkmenistan are not available.

diversify and intensify crop production in the country in the Presidential Decree 'Measures for Agricultural Reform and Promotion from 2016 to 2020' (#2460 of 29 December 2015). Under this decree, approximately 170 000 ha of cotton and 50 000 ha of wheat-producing areas will be diverted to the cultivation of potatoes, vegetables, intensive orchards, fodder, oil and other crops, between 2016 and 2020. The Government of Uzbekistan adopted Presidential Decree # 5742, dated 17 June, 2019, on measures for the efficient use of land and water resources in agriculture. As per the decree, land users have the right to independently organise land use plans involving NT methods. This creates new opportunities for pursuing crop diversification and the production of winter forage crops such as barley, rye, maize, sorghum and Sudan grass that will strengthen the livestock sector. There is also an opportunity to introduce forage pea and pearl millet crops under NT to the existing crop production systems. The above-mentioned decrees can accelerate the wider adoption of CA in the country through crop diversification and NT methods.

Tajikistan developed the National Strategy on Agrarian Policy in 2008 for increased social protection measures and safety nets, with a strong focus on agricultural development and food security. Tajikistan adopted the 'Law on soil protection' in 2009. The law defines the main principles of the state policy, legal bases of activities of state bodies and physical and legal entities, with the aim of promoting the rational and careful use of soils, preservation of soil quality and fertility and the protection of soils from negative practices. The law regulates the complex relations associated with the protection of soils. In Tajikistan, some basic principles of CA were introduced through ICARDA, the International Fund for Agricultural Development (IFAD) and FAO projects. NT and permanent-bed planting systems have been tested for food and feed crops such as wheat, barley, safflower, sunflower, mung bean, soybean, kidney bean, maize, forage pea, common vetch and sainfoin in Tajikistan. These projects addressed the problems of soil degradation, reduced soil fertility and increased crop diseases through the introduction of CA-based technologies including NT, cereal-legume rotations and integrated crop management.

Turkmenistan has adopted new programmes on agricultural development in order to support resource-conserving technologies, including the 'National programme of social economic development to 2030'. One of the priorities highlighted in this programme is the development of agriculture. The programme emphasises improvement in the efficiency of agricultural production through the use of modern resource-conserving technologies for land management and the use of drought-resistant and high-yielding varieties of the main crops, while minimising water consumption. The priority issues in the area of protection of land resources of Turkmenistan are as follows: a) combatting soil salinisation; b) combatting land desertification and degradation of pasture areas and c) combatting wind and water erosion. The programme

Figure 4 Permanent bed planted winter wheat at milk stage seeded after corn in Karshi, Uzbekistan, 2014.

takes into account measures aimed at resource-conserving technologies including minimum soil disturbance through the application of no-tillage, minimum tillage and crop diversification systems.

The adoption of CA practices in the region should be carried out with the close integration of crop breeders, farmers, technologists, machine builders and scientists. Only the creation of favourable institutional conditions for the adoption and spread of NT methods and the intensification of the economic activities of agricultural enterprises will allow sustainable crop production systems to thrive in the region.

3.2 Prospects for crop diversification under Conservation Agriculture

There are a number of factors that determine the prospects of diversification of agricultural crops under CA in the countries of Central Asia.

Natural factors: First, the limitation in water resources and the frequent droughts (Aitekeyeva et al., 2020) necessitate the production of crops that require less water - cotton is not such a crop. This is of special economic importance to Uzbekistan and Turkmenistan, as in these countries, more than 90% of the water resources come from outside their territory, highlighting the importance of transboundary management of water resources (Yu et al., 2019). Second, the level of soil fertility in the countries of Central Asia has declined to the extent that further monoculture production without diversification and sustainable soil management can lead to dire consequences. The authorities of the countries of the region have recently realised this situation, and increasing attempts are being made to limit monoculture production of cotton and wheat. The natural climatic conditions of Central Asia allow the production of various kinds of crops and, in some cases, the production of two or even three crops per year, which occasionally leads to saturation in the domestic markets. In some irrigated areas (precisely, the Southern part of Uzbekistan and Tajikistan), three crops per year can be grown under CA practices. At the same time, in other countries of the Commonwealth of Independent States (CIS), there exists a growing demand for agricultural products, particularly grapes, melons, early vegetables and onions, which offer more profit than cotton and require less labour and inputs.

Socio-economic factors: In recent years, strong price fluctuations and price increases in food commodities and retail food products have been observed globally. At the same time, fluctuations in the price of cotton fibre have been within normal limits. Therefore, priority will be given to producing agricultural crops demonstrating price increases. With the growing population (population growth in the Central Asian countries is relatively high), the demand for more diverse and higher quality foodstuffs is also expanding. Given the situation of increasing world prices of food commodities, their imports by the Central Asian countries are decreasing. As a result, there is a need to increase domestic production to satisfy the local demand for food products. In particular, animal products (eggs, meat, butter, various oils and others) and the traditional and non-conventional agricultural products which can be cultivated in Central Asia (different kinds of tropical fruits, olive oil, etc.) are important, and it will be necessary to increase the production of forage crops, legume crops and vegetables using CA practices in the region.

The increasing migration of the agricultural population to cities is observed in the region. Though this does not strongly influence the production of agricultural products now, there will come a time when there will be an increasing shortage of labour in agriculture, in particular in cotton production, and in some regions of Central Asia, this trend can already be observed. This will gradually force agricultural producers to change crops. Most likely this tendency will be observed principally in those regions where there are large industrial cities, and therefore, steady growth in demand for agricultural products (vegetables, fruits, potatoes, meat and oilseeds). Filling this demand will require sustainable perennial cropping system diversification, moving away from the unsustainable production of wheat and other annual crops. This can include mixed cropping systems, rangeland seeding and re-seeding, NT seeding of legumes (alfalfa, sainfoin, sorghum, Sudan grass and ryegrass) between trees as alleys, as well as integrated pest and disease management (e.g. entomophagous biological plant protection), snow retention, CA practices, salt-tolerant and drought-tolerant fodder systems and enhanced food and seed storage capacities.

4 What is next: challenges and opportunities

Despite the progress of CA and soil conservation technologies in the region, there are several persistent technological, institutional and policy-related challenges for the wider adoption and spread of CA systems and practices in the region (Kienzler et al., 2012). The Central Asian governments and their institutions and international research organisations should join the efforts to address the following objectives in the field with CA:

- Weed control (infestation management): This remains one of the major problems of CA management and needs to be addressed through further scientific research. It is important to develop efficient integrated methods to combat weeds, especially for the early stages of the introduction of NT technologies. Selection of herbicide types, identification of efficient combinations and concentrations and application schedule depending on the weed species present are the main objectives of this research, which is expected to provide recommendations on the most efficient and cost-effective measures to combat weeds under CA conditions. Experience shows that after a while, CA cropping systems help to significantly reduce the level of weed infestation. However, as indicated above, in the initial stages of CA introduction, weed control is a serious technical and management task.
- Challenges in Kazakhstan: North - Readjust incentive policies to improve the quality of CA and include smaller farmers; South - provide similar support for the uptake of CA for the diversified, irrigated and smaller-scale cropping systems (i.e. permanent-bed planting technologies).
- Challenges in Kyrgyzstan: Crop diversification by farmers in small-sized areas; knowledge about CA at all levels; availability of equipment; access to medium- to long-term credit at affordable interest rates.
- Challenges in Tajikistan: Practical farmers' experience in pilot areas with CA - proof of concept; knowledge about CA at all levels; availability of equipment; accessible and affordable credit.

- Challenges in Turkmenistan: Knowledge about CA systems among extension and technical staff; availability of equipment; permission for kolxozs[3] to decide on diversified crop rotations.
- Challenges in Uzbekistan: small-scale farm adoption; knowledge about CA systems among extension and technical staff; availability of equipment; permission for farmers to decide on diversified crop rotations.

5 Conclusion

Timely policy support is critical for any breakthrough in agriculture and for the adoption of new and appropriate technology. The introduction of such technologies creates important new opportunities and directions in agriculture for farmers and all stakeholders. In the former Soviet Union, technology transfer followed a top-down approach and was implemented through centralised systems. Hence, once policy makers decided on a technology to be transferred, large-scale adoption and dissemination were never a problem, and technical knowledge and inputs were simultaneously made available. Farm workers had practically no role in decision making and merely followed instructions. During the transition from a centralised to a decentralised system of independent farmers, changing technology transfer mechanisms from the top-down to a participatory approach is indeed a challenge for all stakeholders: the governments, research institutions, farmers' associations and farmers themselves, non-governmental organisations and the private sector. To date, mechanisms to link key stakeholders involved in technology development and transfer do not exist. Cognisant of poor extension services, the role of NARS and civil society organisations linking 'scientists to farmers' through on-farm demonstrations, field visits, travelling workshops, participatory research, etc. becomes vital. Similarly, training and education to support the transfer of new knowledge and learning are equally important.

Government approval must be relied on to support technology uptake and farmer innovation at scale. The Governments of Central Asia should recognise CA and permanent-bed planting as a necessary part of future sustainable and profitable agriculture. Regional networks of CA stakeholders would help to create an enabling environment for change at the national and regional levels. A regional CA website has already been created, but countries need to actively participate to provide inputs. The close collaboration of all stakeholders at all levels for the introduction and up-scaling of CA is important, particularly for farmers, farmers' communities and associations, NGOs, the machinery industry, universities, government offices and institutions, including the higher levels of the relevant ministries. Multiplication of prototypes of less costly machinery and

3 Kolxoz - former collective farms during the Soviet time, which still exist in Turkmenistan

implements is another area that requires government support. As an example, raised-bed planters and NT drills, flame weeders and similar tools have to be manufactured and popularised. Their import is costly, and hence, promoting their manufacture in the region is critical. Promotion of these CA machines would require appropriate subsidies and Government intervention.

Alley cropping is another production system where CA practices can be introduced in Central Asia, under both irrigated and rainfed conditions. There are many orchards and vineyards which could be intensified by sowing field crops, cover crops and/or pastures between the rows. Application of this production system using the NT method will help to create permanent vegetation cover and improve soil fertility and the environment. There is a large potential to intensify and protect the land under orchards and vineyards, especially in sloping areas where soil erosion is widespread.

Cost-benefit analyses of improved technologies to assess the economic viability of useful options for better livelihood of farming communities in Central Asian countries have been reviewed in this chapter. The application of permanent-bed planting for winter wheat, mung bean, corn, sorghum and other crops have been found to be useful for increasing farmers' income and improving their livelihoods.

It is very clear that soil and water management are very important for sustainable agriculture in Central Asia. Much work has been done to improve soil- and water-related aspects. However, it has been observed that public policy support for the uptake of new efficient technologies such as CA is required, in addition to technology transfer mechanisms for small farmers, so that they are empowered to take full benefit and can increase their productivity and profitability. In this context, international agricultural research centres have been able to work with national scientists for generating CA practices, especially suited to small and marginal farmers in Kyrgyzstan, Tajikistan, Turkmenistan and Uzbekistan, while in Kazakhstan, large farmers have been the principal adopters of CA. These technologies need to be promoted further for large-scale adoption and desired impact on all farmers' fields.

6 Where to look for further information

The sources for information on the research of Conservation Agriculture in Central Asia have been growing rapidly over the last two decades. These sources could be divided into three categories: 1) scientifically published peer-reviewed papers, 2) specific project reports, brochures and pamphlets, 3) websites and blogs on CA development and news from Central Asia. The references provided in this chapter can serve as valuable sources for further information on CA development in the region.

In addition, the following websites cover CA news, events and major development relevant for the region: the website of the FAO (FAO Regional Office for Europe and Central Asia (http://www.fao.org/europe/central-asia/en/), and the ICARDA regional programme for Central Asia and the Caucasus (http://cac-program.org/).

More information on adoption of Conservation Agriculture in Central Asia can be found at the following websites:

- http://agroprod.kg/.
- http://minagri.gov.tm/.
- http://www.fao.org/.
- http://www.stat.gov.tm/.
- http://www.stat.kg/.
- https://moa.tj/en.
- https://stat.gov.kz/.
- https://stat.uz/.
- https://www.agro.uz/.
- https://www.cimmyt.org/.
- https://www.gov.kz/memleket/entities/moa?lang=en.
- https://www.icarda.org/.
- https://www.stat.tj.
- www.eld-initiative.org/.

7 References

Abdurakhimov, M., Bobomirzayev, P. B. and Khalilova, L. N. (2018). Sugoriladigan yerlarda kuzgi bugdoy va takroriy ekinlarni uygunlashgan holda yetishtirish texnologiyasi. Tavsiyanoma. "F.Nasimov" press, Samarkand, Uzbekistan (in Uzbek language).

Aitekeyeva, N., Li, X., Guo, H., Wu, W., Shirazi, Z., Ilyas, S., Yegizbayeva, A. and Hategekimana, Y. (2020). Drought risk assessment in cultivated areas of Central Asia using MODIS time-series data. *Water* 12(6). DOI: 10.3390/w12061738.

Asoyev, N. A., Ibragimov, N. I., Nurbekov, A. I., Yatimov, B. N. and Kholmatov, A. (2015). Winter wheat productivity under different tillage methods. *J. Tajik Acad. Sci.* 1/1(156), 197–201 (in Russian).

Atabayeva, H. N. (2004). *A Soybean Crop*. Uzbekistan National Encyclopedia, Tashkent (in Uzbek lang).

Atakulov, T., Ospanbaev, Z. and Alkenov, Y. (2014). Permanent raised beds using; efficiency of direct seeding in the south-east region of Kazakhstan. *J. Lif. Sci.* 11(11), 554–557.

Banik, P. and Sharma, R. C. (2008). Effect of integrated nutrient management with mulching on rice (*Oryza sativa*) -based cropping system in rainfed Eastern plateau area. *Indian J. Agric. Sci.* 78, 240–243.

Baranov, B. F. and Kochegura, A. B. (2005). Soya biologiya i technologiya vozdelyvaniya,– Krasnodar. Sovetskaya Kuban press (in Russian).

Barayev, A. I. (2008). *Selected Articles* (vol. III). Gylym Press, Almaty, Kazakhstan (in Russian).

Boboev, H., Djanibekov, U., Bekchanov, M., Lamers, J. P. A. and Toderich, K. (2019). Feasibility of conservation agriculture in the Amu Darya River Lowlands, Central Asia. *Int. J. Agric. Sustain.* 17(1), 60–77.

Borin, A. A. (2003). Obrabotka pochvy pod zernovyye v sevooborote. *J. Zemledeliye Mosc.* 4, 14–15.

Bukhoriev, T. A. and To'khtayev, M. O. (2012). Fotometricheskie parametry ozimoy pshenytsy a zavisimosti ot srokov seva. *J. Agrarnaya Nauk.* 10, 10–11.

de Moraes Sá, J. C., Séguy, L., Tivet, F., Lal, R., Bouzinac, S., Borszowskei, P. R., Briedis, C., dos Santos, J. B., da Cruz Hartman, D., Bertoloni, C. G., Rosa, J. and Friedrich, T. (2013). Carbon depletion by plowing and its restoration by no-till cropping systems in oxisols of subtropical and tropical agro-ecoregions in Brazil. *J. Land Degrad. Develop.* 26(6), 531-543. DOI: 10.1002/ldr.2218.

De Pauw, E. (2008). Principal biomes of Central Asia. In: *Climate Change and Terrestrial Carbon Sequestration in Central Asia* Lal, et al. (Eds). pp. 3–24. London, UK.

Devkota, K. P., Hoogenboom, G., Boote, K. J., Singh, U., Lamers, J. P. A., Devkota, M. and Vlek, P. L. G. (2015a). Simulating the impact of water saving irrigation and conservation agriculture practices for rice–wheat systems in the irrigated semi-arid drylands of Central Asia. *J. Agric. For. Meteorol.* 214–215, 266–280.

Devkota, M., Martius, C., Gupta, R. K., Devkota, K. P., McDonald, A. J. and Lamers, J. P. A. (2015b). Managing soil salinity with permanent bed planting in irrigated production systems in Central Asia. *J. Agric. Ecosyst. Environ.* 202, 90–97. DOI: 10.1016/j.agee.2014.12.006.

Dridiger, V. K., Likhodievskaya, S. A., Kashchaev, E. A., Stukalov, R. S. and Yu, P. (2016). `Urozhaynost' i ekonomicheskaya effektivnost' sel'skokhozyaystvennykh kul'tur v sevooborote v zavisimosti ot tekhnologii vozdelyvaniya i udobreniy. *Mikhaylovsk. J. Agro. and For. Manag.* 5(I), 32–36 (in Russian).

Dvurechenskiy, V. I. (2010). Vozdelivaniye Zernovykh Kultur na Osnove Novoy Vlagosberegayuschey Tekhnologii i Soveremennoy Tekhniki. Izdatelskiy Dom, Kazakhstan (in Russian).

Egamberdiev, O. (2007). Changes of soil characteristics under the influence of resource saving and soil protective technologies within the irrigated meadow alluvial soil of the Khorezm region. Ph.D. Thesis, Tashkent, Uzbekistan (in Uzbek lang).

ELD Initiative (2015). The value of land: prosperous lands and positive rewards through sustainable land management. Available at: www.eld-initiative.org (Accessed on 21.08.20).

FAO (2012). Advancement and impact of conservation agriculture/no-till technology adoption in Kazakhstan [Internet]. *FAO Information Note*. Available at: http://www.fao.org/tempref/AG/agp/ca/CA_CoP_March13/FAO_Info_note_on_CA_in_Kazakhstan.pdf (Accessed 13.09.20).

Ferdush, J., Meftahul Karim, Md., Noor, I. J., Jui, S. A., Ahamed, T. and Saha, S. R. (2019). Impact of alley cropping system on soil fertility. *Int. J. Adv. Geosci.* 7(2), 173-178. DOI: 10.14419/ijag.v7i2.29942.

Frederick, J. R. and Camberato, J. J. (1995). Water and nitrogen effects on winter wheat in the southeastern coastal plain: I. Grain yields and kernel traits. *Agron. J.* 87(3), 521–526.

Gabilov, M. A. (2011). Posledyestviya mineralnykh udobreniy pri vyrashivaniya ozimoy pshenitsy. *J. Zernoviye Kult.* 1, 11–13.

Gonzalez-Sanchez, E. J., Nurbekov, A., Muminjanov, H. and Kassam, A. (2018). *National Strategy for the Promotion of Conservation Agriculture in Uzbekistan*. FAO-Uzbekistan Press.

Gonzalez-Sanchez, E. J., Veroz-González, O., Blanco-Roldan, G. L., Marquez-Garcia, F. and Carbonell-Bojollo, R. (2015). A renewed view of conservation agriculture and its evolution over the last decade in Spain. *Soil Till. Res.* 146 (PB), 204–212.

Gupta, R., Kienzler, K., Martius, C., Mirzabaev, A., Oweis, T., Pauw, E., Qadir, M., Shideed, K., Sommer, R., Thomas, R. J., Sayre, K., Carli, C., Saparov, A., Bekenov, M., Sanginov, S., Nepesov, M. and Ikramov, R. (2009). Research prospectus: a vision for sustainable land management research in Central Asia. *ICARDA Central Asia and Caucasus Program. Sustainable Agriculture in Central Asia and the Caucasus* Series 1. CGIAR-PFU, Tashkent. Uzbekistan.

Hassan, A., Ijaz, S. S., Lal, R., Ali, S., Hussain, Q., Ansar, M., Khattak, R. H. and Baloch, M. S. (2015). Depth distribution of soil organic carbon fractions in relation to tillage and cropping sequences in some dry lands of Punjab, Pakistan. *Land Degrad. Develop.* 27(4), 1175–1185 DOI: 10.1002/ldr.2345.

Hickmann, S. (2006). Conservation agriculture in northern Kazakhstan and Mongolia. FAO, Agricultural and Food Engineering Working Document 4. Available at: http://www.fao.org/3/a-j8349e.pdf (Accessed on 08.09.2020).

ICARDA-CAC (2013). Available at: http://www.cac-program.org/publications/sections/8 (Accessed on 12.7.2021).

Iminov, A. A. and Khalikov, B. M. (2005). Takroriy ekinlarni tuproqdagi oziqa moddalr miqdoriga ta'siri. In: *O'zbekiston tuproqshunoslar va agrokimyogarlar jamiyati 5-qurultoyi maqolalar to'plami*. pp. 257–258. Tashkent, Uzbekistan.

Kassam, A., Friedrich, T. and Derpsch, R. (2019). Global spread of conservation agriculture. *Int. J. Environ. Stud.* 76(1), 29–51, DOI: 10.1080/00207233.2018.1494927.

Kassam, A., Gonzalez-Sanchez, E. J., Cheak, S. C., Rahman, Z. A., Roman-Valquez, J., Mrquez-Garcia, F., Carbonell-Bojollo, R., Veroz-Gonzalez, V. and Garrity, D. (2020). Managing perennial conservation agriculture systems: orchards, plantations and agroforestry. In: *Advances in Conservation Agriculture Volume 1: Systems and Science* Kassam, A. (Ed). pp. 327–358. Burleigh Dodds, Cambridge, UK.

Keldiyorova, Kh. Kh. and Xozhaev, J. H. (2004). Zarafshon Vodiysida Har Xil Muddatlarda Ekilgan Bugdoy Navlarining Iqtisodiy Samaradorligi. Toshkent, Uzbekistan. In *O'zbekistonda bug'doy selektsiyasi, urug'chiligi va yetishtirish tekhnologiyasiga bagishlangan birinchi milliy konferentsiya materaillari*. pp. 216–220.

Khaitov, B. and Allanov, Kh. (2014). Crop rotation with no-till methods in cotton production of Uzbekistan. *Eurasian J. Soil Sci. (EJSS)* 3(1), 28–32. DOI: 10.18393/ejss.52631.

Khalilova, L. N. (2019). Zarafshon *vodnogo khozyaystva suriridigan kergi* buzadoy va takroriy ekinlar yetishtirishni ilmiy asoslash. In *Dissertatsiya*, pp. 187.Tashkent, Uzbekistan.

Kienzler, K. M. (2010). Improving the nitrogen use efficiency and crop quality in the Khorezm region, Uzbekistan. In: *Ecology and Development Series* (Bd. 72). Vlek, P. L. G., Denich, M., Martius, C., Manschadi, A. and Bogardi, J. (Eds), ZEF Bonn.

Kienzler, K. M., Lamers, J. P. A., McDonald, A., Mirzabaev, A., Ibragimov, N., Egamberdiev, O., Ruzibaev, E. and Akramkhanov, A. (2012). Conservation agriculture in Central Asia–what do we know and where do we go from here? *Field Crops Res.* 132, 95–105.

Kireyev, A. K. (2004). Minimalizatsiya obrabotki zyabi pod yarovyye zernovyye kul'tury na neobespechennoy bogare. *J. Vestnik s.-kh. nauki Kazakhstana* 5, 25-28 (in Russian).

Košutić, S., Filipović, D., Gospodarić, Z., Husnjak, S., Kovačev, I. and Čopec, K. (2005). Effects of different soil tillage systems on yield of maize, winter wheat and soybean on albic luvisol in north-west slavonia. *J. Centr. Eur. Agri.* 6(3), 241-248.

Kyrgyzstan Statistical Agency (2018). Available at: http://www.stat.kg/ (Accessed on June 2019).

Landers, J. N., Sant'Anna De Barros, G., Rocha, M. T., Manfrinato, W. A. and Weiss, J. (2003). Environmental impacts of zero tillage in Brazil–A first approximation. In: *Conservation Agriculture* García-Torres, L., Benites, J., Martínez-Vilela, A. and Holgado-Cabrera, A. (Eds). Springer, Dordrecht.

Li, J., Chen, H., Zhang, C. and Pan, T. (2019). Variations in ecosystem service value in response to land use/land cover changes in Central Asia from 1995-2035. *PeerJ* 7, e7665. DOI: 10.7717/peerj.7665.

Li, Z., Fang, G., Chen, Y., Duan, W. and Mukanov, Y. (2020). Agricultural water demands in Central Asia under 1.5°C and 2.0°C global warming. *Agric. Water Manag.* 231, DOI: 10.1016/j.agwat.2020.106020.

Ma, X., Zhu, J., Yan, W. and Zhao, C. (2020). Assessment of soil conservation services of four river basins in Central Asia under global warming scenarios. *Geoderma* 375, DOI: 10.1016/j.geoderma.2020.114533.

Mirzabaev, A. (2013). Climate volatility and change in Central Asia: economic impacts and adaptation. Doctoral thesis at Agricultural Faculty, University of Bonn. urn:nbn:de:hbz:5n-3238.

Mirzabaev, A., Goedecke, J., Dubovyk, O., Djanibekov, U., Bao, Le, Q., Le, Q. B. and Aw-Hassan, A. (2016). Economics of land degradation in Central Asia. In *Economics of Land Degradation and Improvement - A Global Assessment for Sustainable Development* Nkonya, et al. (Eds). pp. 261-290. Springer International Publishing. DOI: 10.1007/978-3-319-19168-3.

Muhayyo, N. (2009). Priyomy vozdelivaniya pojnivnogo masha (aziatskiy fasol) v usloviyakh Centralnogo Tajikistana. Thesis of dissertation. 122, Dushanbe, Tajikistan.

Nachtergaele, F. O., Petri, M., Biancalani, R., van Lynden, G., van Velthuizen, H. and Bolise, M. (2011). Global land degradation information system (GLADIS v 1.0). LADA Technical Report #17. An information base for land degradation at global level. FAO, Rome, Italy.

Nkonya, E., Mirzabaev, A. and von Braun, J. (2016). Economics of land degradation and improvement–a global assessment for sustainable development. DOI: 10.1007/978-3-319-19168-3.

NPDOOS (1999). *Natsional'nye plan deistvii po okhrane okruzhayushchei sredy Rossiiskoi Federatsii na 1999-2001 gody*. Goskomekologiya, Moscow, Russia.

Nurbekov, A., Akramkhanov, A., Lamers, J. K., Friedrich, A., Gupta, T., Muminjanov, R., Karabayev, H., Sydyk, M., Turok, D. J. and Bekenov, M. (2013). Conservation agriculture in Central Asia: past and future. In: *Conservation Agriculture: Global Prospects and Challenges* Jat, R. A., Sahrawat, K. L. and Kassam, A. (Eds). pp. 223-248. CABI Publishing, Wallingford.

Nurbekov, A., Kassam, A., Sydyk, D., Ziyadullaev, Z., Jumshudov, I., Muminjanov, H., Feindel, D. and Turok, J. (2016). *Practice of Conservation Agriculture in Azerbaijan, Kazakhstan and Uzbekistan*. FAO. Available at: http://www.fao.org/3/a-i5694e.pdf (Accessed on 12.7.2021).

Nurbekov, A., Suleymenov, M., Friedrich, T., Taher, F., Ikramov, R. and Nurjanov, N. (2012). Effect of tillage methods on productivity of winter wheat in the Aral Sea Basin of Uzbekistan. J. Ar. L. Stu. 22, 255–258.

Nurbekov, A. and Ziyadullaev, Z. (2012). Adoption of no-till technology for sustainable agriculture in the drought-affected regions of Karakalpakistan. International Conference on improving soil fertility through adoption of resource saving technologies in agriculture. Tashkent, Uzbekistan. pp. 101–105.

Nurbekov, A. I. (2008). *Manual on Conservation Agriculture Practices in Uzbekistan*. Tashkent, Uzbekistan.

Nurbekov, A. I. (2018). Cultivation of agricultural crops under conservation agriculture in irrigated conditions Central Asia. Tashkent, Uzbekistan. Thesis of doctoral dissertation. Tashkent, Uzbekistan.

Nurbekov, A. I., Kassam, A., Muminjanov, H. and Nurbekova, R. A. (2018). Effect of seeding depth on winter survival and yield of wheat under conservation agriculture in Uzbekistan. Second Afr. Congress on Conservation Agriculture. Johannesburg, South Africa.

Nurbekov, A. I., Khalilova, L. N., Isroilov, B. and Xalilov, U. M. (2019). Alley cropping system under conservation agriculture practices in Tajikistan. *J. Euro Sci. Rev.* 11-12, 48–51.

Ostanakulov, T. E. and Khalilov, N. Kh. (2017). *Takroriy Ekinlar Farovonlik Manbai*. Samarkand, Uzbekistan (in Uzbek lang).

Paroda, R. (2007). Addressing the challenges for sustainable agriculture in central Asia. In: *Climate Change and Terrestrial Carbon Sequestration in Central Asia* Lal, et al. (Eds). pp. 109-127. London, UK.

Pender, J., Mirzabaev, A. and Kato, E. (2010). Economic analysis of sustainable land management options in Central Asia. *Central Asian Countries Initiative for Land Management Multicountry Partnership Framework Support Project*. Asian Development Bank.

Qadir, M., Tubeileh, A., Akhtar, J., Larbi, A., Minhas, P. S. and Khan, M. A. (2008). Productivity enhancement of salt-affected environments through crop diversification. *J. Land Deg. Dev.* 19(4), 429–453.

Qilichev, A. H. and Khalilov, N. (2008). Go'za qator oralariga ekilgan kuzgi bugdoy hosidorligi va don sifati. J. AGRO-ILM, 163–168 (in Uzbek).

Sidikov, R. and Husenov, Sh. (2008). Kuzgi gallanii sifatli ekish. *J. O'zbekiston Qishloq Xo'jaligi* 9, 8–9 (in Uzbek lang).

Singh, A., Rajneesh, K. and Kang, J. S. (2014). Tillage system, crop residues and nitrogen to improve the productivity of direct seeded rice and transplanted rice. *J. Cur. Agri. Res.* 2(1), 14–29. DOI: 10.12944/CARJ.2.1.03.

Statistical Committee/Agency of Kazakhstan (2018). Available at: https://stat.gov.kz/ (Accessed on March 2020).

Stewart, B. A. (2007). Water conservation and water use efficiency in drylands. In: *Proceedings of the International Workshop on Conservation Agriculture for Sustainable Land Management to Improve the Livelihood of People in Dry Areas* Stewart, B., Fares Asfary, A., Belloum, A., Steiner, K. and Friedrich, T. (Eds). pp. 57–66. ACSAD and GTZ, Damascus, Syria.

Suleymenov, M. K. and Akshalov, K. A. (2009). Vzaimodeystviye s tochki zreniya razvitiya i urovnya agrotekhniki kul'turnogo razvitiya. *Ekonomicheskaya i sotsial'naya Politika v Oblasti Resursosberezheniya dlya Intensifikatsii Proizvodstva*

Produktsii. Mezhdunarodnaya konferentsiya. Astana-Shortandy. Kazakhstan, pp. 252–258.

Suleymenov, M. K., Akhmetov, K. A., Kaskarbayev, J. A., Khasanova, F., Kireyev, A., et al. (2004). Developments in tillage and cropping systems in central Asia. In: *Agriculture in Central Asia: Research for Development* Ryan, J., Vlek, P. L. G. and Paroda, R. (Eds). pp. 188–211. International Center For Agricultural Research In The Dry Areas, Aleppo, Syria.

Suleymenov, M. K., Pala, M., Paroda, R., Akhshalov, K. F., Martynova, L. I. and Medeubayev, R. (2006). New technologies for Central Asia. *Caravan* 23, 19–22.

Suzuki, M. and Nass, H. G. (1988). Fructan in winter wheat, triticale, and fall rye cultivars of varying cold hardiness. *Can. J. Bot.* 66(9), 1723–1728.

Sydyk, D. A., Karabalayeva, A. D. and Nurbekov, A. I. (2018). Rezhim orosheniya ozimoy pshenitsy pri grebne-borozdkovom sposobe poseva. International Conference, Tashkent, Uzbekistan. pp. 415–418.

Sydyk, D. S., Jarasov, Sh., Sydykov, M. A., Isabekov, B. and Babakhodjaev, A. T. (2008). *Rekomendatsii Po Resursosberegauyshey Technologii Vozdelyvaniya Zernovykh Kolosovykh Kultur v Usloviyakh Bogarnogo i Oroshaemogo Zemledeliya Yujnogo Kazakhstana*. Jebe-Disayin, Shymkent, Kazakhstan (in Russian).

Tajikistan Statistical Agency (2018). Available at: https://www.stat.tj (Accessed in May 2019).

Toderich, K. N., Shuyskaya, E. V., Taha, F., Matsuo, N., Ismail, S., Aralova, D. B. and Radjabov, T. B. (2013). Integrating agroforestry and pastures for soil salinity management in dryland ecosystems in Aral Sea Basin. In: *Development in Soil Salinity Assessment and Reclamation: Innovative Thinking and Use of Marginal Soil and Water Resources in Irrigated Agriculture* Shahid, S. A., et al. (Eds). Springler Science+Business Media, Dordrecht. DOI: 10.1007/978-94-007-5684-7_38.

Turkmenistan Statistical Committee (2018). Available at: http://www.stat.gov.tm/ (Accessed in May 2019).

Turok, J., Bekmuratov, M., Akramov, K., Carli, C., Christmann, S., Glazirina, M., Jumaboev, K., Karimov, A., Kazbekov, J., Khalikulov, Z., Mavlyanova, R., Nishanov, N., Nurbekov, A., Saidov, S., Sharma, R., Toderich, K., Turdieva, M. and Yuldashev, T. (2013). Agricultural research collaboration in Tajikistan. Working Paper 14. International Center for Agricultural Research in the Dry Areas (ICARDA). Available at: https://hdl .handle.net/20.500.11766/7901.

Tursunov, M. (2009). Potential of conservation agriculture for irrigated cotton and winter wheat production in Khorezm, Aral Sea Basin. PhD dissertation, ZEF/Rheinische Friedrich-Wilhelms-Universität Bonn, Germany.

Tynybaeva, K., Saparov, A. and Dzhalankuzov, T. (2017). Development of receptions purposeful management of soil fertility in Northern Kazakhstan. *IJAAEE* 4(1), DOI: 10.15242/IJAAEE.ER0117002.

Umarova, N. S. and Abitov, I. I. (2016). *Kultura Soi v Pojnivnykh Posevakh*. Tashkent Institute of Irrigation and Agricultural Mechanization Engineers Press.

UNEP (2016). Republic of Kazakhstan–technology needs assessment for adaptation to climate change.

Uzbekistan Statistical Agency (2018). Available at: https://stat.uz/ (Accessed in May 2020).

Vladykina, N. I., Lentochkina, L. A. and Lentochkina, A. M. (2006). Produktivnost sevooborota v zavisimosti ot system osnovnoy obrabotki pochvy i vneseniya udobreniy. *J. Zernovoye khozyastvo* 6, 13–14 (in Russian).

Wall, P. C., Yushenko, N., Karabayev, M., Morgounov, A. and Akramhanov, A. (2007). Conservation agriculture in the steppes of northern Kazakhstan: the potential for adoption and carbon sequestration. In: *Climate Change and Terrestrial Carbon Sequestration in Central Asia* Lal, R., et al. (Eds). pp. 333–348. Taylor and Francis, New York.

Whaley, J. M., Kirby, E. J. M., Spink, J. H., Foulkes, M. J. and Sparkes, D. L. (2004). Frost damage to winter wheat in the UK: the effect of plant population density. *Eur. J. Agron.* 21(1), 105–115.

World Bank (2015). Agricultural risks study on Kazakhstan and Central Asia. Available at: https://www.worldbank.org/en/about/annual-report-2015 (Accessed on 12.7.2021).

Yu, Y., Pi, Y., Yu, X., Ta, Z., Sun, L., Disse, M., Zeng, F., Li, Y., Chen, X. and Yu, R. (2019). Climate change, water resources and sustainable development in the arid and semi-arid lands of Central Asia in the past 30 years. *J. Arid Land* 11(1), 1–14. DOI: 10.1007/s40333-018-0073-3.

Chapter 13

Conservation Agriculture in Eurasia

Hafiz Muminjanov, Food and Agriculture Organization of the United Nations (FAO), Italy; Tatiana Semenova, World Food Programme, Kyrgyzstan; Amir Kassam, University of Reading, UK; Theodor Friedrich, Retired Food and Agriculture Organization of the United Nations (FAO) Expert on Conservation Agriculture, Germany; Avetik Nersisyan, International Plant Protection Convention, Italy; Turi Fileccia, formerly Food and Agriculture Organization of the United Nations (FAO), Italy; Gagik Mkrtchyan, Armenian Technology Group Foundation, Armenia; Armen Zakaryan, Food and Agriculture Organization of the United Nations (FAO), Armenia; Imran Jumshudov and Yagub Guliyev, National Agrarian Research Centre, Azerbaijan; Alena Abarava, Lyakhovichi State Agricultural College, Belarus; Allan Pineda Burgos and Demna Martsvaladze, Food and Agriculture Organization of the United Nations (FAO), Georgia; Muratbek Karabayev, formerly International Maize and Wheat Improvement Center (CIMMYT), Kazakhstan; Gulnaz Kaseeva, Public Organization "AgroLead", Kyrgyzstan; Omurbek Mambetov and Cholpon Alibakieva, Food and Agriculture Organization of the United Nations (FAO), Kyrgyzstan; Boris Boincean, Research Institute of Field Crops, Moldova; Mihail Rurac, State Agricultural University, Moldova; Vadim Bandurin, Association of the Supporters of No-till Planting, Russia; Luidmila Orlova, National Movement for Conservation Agriculture, Russia; Alexandr Toigildin, Ulyanovsk State Agricultural University, Russia; Khalil Safin, Bashkortostan State Agricultural University, Russia; Muhammadi Muminov, Jamshed Sangynov and Shavkat Juraev, Cooperative of Agricultural Extension Specialists "Sarob", Tajikistan; Askarsho Zevarshoev, Public Organization "Eco Consultancy", Tajikistan; Nicola Kosolap, Ukraine Agrarian University, Ukraine; Oksana Davis and Oleksandr Zhuravel, Food and Agriculture Organization of the United Nations (FAO), Ukraine; Michail Draganchuk, YouTube channel "No-tiller", Ukraine; Alim Pulatov, Tashkent Institute of Irrigation and Agricultural Mechanization Engineers, Uzbekistan; Aziz Nurbekov, Tashkent State Agrarian University, Uzbekistan; Josef Kienzle and Sandra Corsi, Food and Agriculture Organization of the United Nations (FAO), Italy; and Emilio J. Gonzalez-Sanchez, Universidad de Córdoba, Spain

1 Introduction
2 Conservation Agriculture in Eastern Europe
3 Conservation Agriculture in the Caucasus
4 Conservation Agriculture in Central Asia
5 Conclusion
6 Where to look for further information
7 References

http://dx.doi.org/10.19103/AS.2021.0088.17

1 Introduction

Conservation Agriculture (CA) plays an important role in the development of agricultural production in the Eurasian region that covers almost all countries of the former Soviet Union. These countries gained independence in 1991; they have a common history and similar production, economic and agricultural development systems, but rather diverse climatic conditions and natural resources (Fig. 1). There are still a number of obstacles with management systems related to lack of support to agriculture development and consequences of climate change at the regional and national levels.

The region covers three geographic zones - Eastern Europe, the Caucasus and Central Asia - with a total territory of 21 695 052 km^2. Total population of the countries is 292 440 607, with annual growth of 1 356 078 people (0.46%). The area of agricultural land is 567 162 600 ha, including 201 659 400 ha of cropland. The area of forests is 851 448 900 ha (http://www.fao.org/faostat) (Table 1). Most of the agricultural land in the northern latitudes of Russia and Kazakhstan is under rainfed farming systems; the crops are also located on irrigated zones in Central Asia, the Caucasus, Moldova and Ukraine. The main field crops cultivated in the region are cereals (wheat, barley, maize and rye), perennial grasses and fodder legumes (alfalfa), pulses, industrial crops (rapeseed, cotton), etc. Three countries of the region - Russia, Kazakhstan and Ukraine - are major grain exporters with annual contributions of exports, respectively, in 43 965 626 tonnes

Table 1 Land distribution by countries in the region

	Country area, km^2	Agricultural land, '000 ha	Cropland, '000 ha	Forest area, '000 ha
Armenia	29 743	1677.3	445.6	328.5
Azerbaijan	86 600	4779.5	1631.0	1131.8
Belarus	207 600	8532.7	5712.0	8767.6
Georgia	69 700	2371.8	311.0	2822.4
Kazakhstan	2 724 900	216 036.5	29 748.4	3454.7
Kyrgyzstan	199 951	10 541.3	1288.0	1315.4
Moldova	33 846	2257.1	1681.7	386.5
Russia	17 098 250	215 494.0	121 649.0	815 311.6
Tajikistan	143 100	4727.7	720.2	423.8
Turkmenistan	48 810	33 838.0	1910.0	4127.0
Ukraine	603 628	41 329.0	32 498.0	9690.0
Uzbekistan	448 924	25 577.7	4064.5	3689.7
TOTAL	**21 695 052**	**567 162.6**	**201 659.4**	**851 448.9**

Source: http://www.fao.org/faostat.

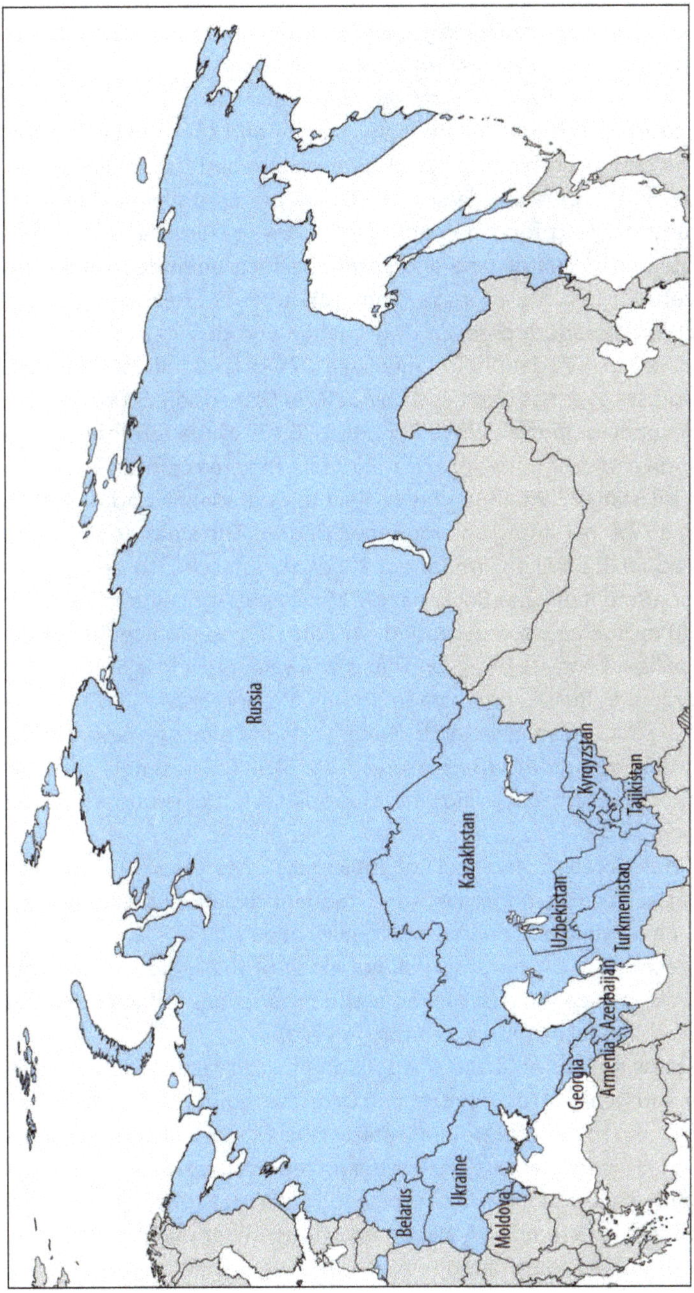

Figure 1 Map of the Eurasia region.

(USD10 543 554), 6 198 354 tonnes (USD2 336 962) and 16 373 389 million tonnes (USD3 794 356).

The region has contrasting soils and climatic conditions: the Eastern Europe part is famous for its fertile black soils and relatively mild and cold humid climates. However, soils in Central Asia and the Caucasus are less fertile and more degraded; the climate is dry, steppe and sub-tropical agro-climatic zones. The global problems of land degradation and desertification are also very acute in the Eurasian region, and they are intensifying due to climate change and growing food demand. Particularly, desertification is one of the main problems in Central Asia and severely affects countries in the Caucasus and Eastern Europe. On average, from 20% to 90% of all land in the region is subject to different degrees of degradation and desertification (Kovalevich, 2010; FAO, 2016a). To date, in Turkmenistan, 37% of land suffers from salinization and the forest cover has decreased by 32%. In Uzbekistan, 50% of all irrigated land is desertified. In the Kyrgyz Republic, 70.6% of the territory is subject to degradation and desertification; over 70 years, the area of forests decreased by 35%. In Tajikistan, 97.9% of land is subject to degradation and desertification, including 88.7% in a strong and moderate degree. The area of desertification in Russia is about 1.2 million km^2 (about 7% of the total country area). According to the Department of Land Reclamation, Ministry of Agriculture, Russia, in 2015 about 130 million ha were degraded, and the degradation continues at a rate of 1.5–2 million hectares per year. This is equivalent to a loss of 3.2–3.9 million tons of grains. In Russia, on average about 65% of cropland, 28% of hayfields and 50% of pastures are affected by water and wind erosion, periodic droughts, dry winds and dust storms (Bagdasaryan, 2015). In Belarus more than 4 million ha of land, including 2.6 million ha of cropland (65%) is under erosion risks (Kovalevich, 2010).

The deterioration and lack of main resources – soil degradation and depletion, lack of irrigation water – and frequent climatic and natural anomalies lead to a loss of yield and decrease its quality of output. This explains the urgent need for scaling a CA system in all countries of the region to preserve and protect soils, reduce the cost of labour and production, save natural resources, generate additional income and improve yield.

To assess the current state of applied farming practices, to determine the strengths and weaknesses of promoting CA in the region, an analytical study was conducted, and the results are presented in this Chapter. With this purpose, the available statistical data were analysed, and reports on the status and results of the implementation of the projects promoting CA at the regional and national levels as well as the regulatory documents (laws, government resolutions, etc.), national strategies, programmes and other documents, including detailed information on the state of land resources, degradation, restoration and conservation measures were reviewed. Additional information was collected through a survey of the agricultural and extension specialists conducted by e-mails and personal telephone calls. The status of CA adoption and promotion was discussed during an online Regional Expert Consultation conducted by FAO and European Conservation Agriculture Federation (ECAF) on 29 October 2020 for the preparation for the 8th World Conservation Agriculture Congress (https://8wcca.org) to be held during 20–24 June 2021 in Bern, Switzerland. More than 27 representatives from 12 countries actively participated in the discussion that was a significant contribution to a comprehensive analysis of the current situation and the perspectives on promoting CA in the Eurasia region.

2 Conservation Agriculture in Eastern Europe

Agriculture with tillage-based farming systems has a long history in Central Asia and the Caucasus. In Eastern Europe farming developed mainly through opening up the land by cutting and burning the forests that required hard work and a large number of workers. Therefore, early farming in forests presupposed the existence of cooperative associations - communities. In the steppe zone, the initial farming system was based on shifting cultivation and after the first harvest, the plot was left fallow for several years. However, due to the increase in population and rising demand for animal feed, the three-field crop rotation systems were introduced and the time for fallow was reduced. In parallel, soil science began to develop in Russia with internationally recognized researchers like M. V. Lomonosov, V. R. Williams, V. V. Dokuchaev and others. Indeed, several researchers consider soil science to be of Russian origin because Vasiliy Dokuchaev is known as the founder of soil science. He was the first scientist who carried out soil studies, developed a soil classification system and investigated the influence of external conditions on soil life. Moreover, most of the current soil scientists confirm that they are still developing Dokuchaev's ideas.

Back in the 19th century, the famous Russian agronomist Ivan Ovsinsky developed, tested and demonstrated a 'new farming system' in different agro-ecological zones, including the Far East and the dry zones in Ukraine for many years (Ovsinsky, 1909). He managed to stop ploughing, obtained high yield of wheat in a severe drought and maintained soil fertility. Instead of ploughing, he recommended cultivating the soil to a depth of 5 cm to create a loose surface layer that protects soil moisture and soil layers further below from evaporation. These techniques to overcome drought were further investigated and improved by academician N. M. Tulaykov in the Volga region (1911, 1932), Professor M. Z. Zhuravlev in Western Siberia (1932) and other Russian and foreign scientists (Rychkov, 2018).

Studies on economic and environmental efficiency and the local application of reduced tillage were carried out in the USSR (Union of Soviet Socialistic Republics) in the 1950s. In this period, during the cultivation of virgin lands in Siberia and Kazakhstan, huge areas of fertile lands were affected by severe wind erosion. Therefore, an interest in soil-saving technologies increased in the 1960s by the evidence-based research work devoted to the practical application of resource-saving technologies carried out by Professor Terentiy Maltsev and academician Alexander Barayev (Rychkov, 2018).

T. Maltsev (2020) is the founder of modern resource-saving and soil-protective farming in the Trans-Urals. The main components of Maltsev's system includes application of grain-fallow-grass crop rotation with short rotation schemes and a high proportion of fallow; a soil cultivation system, in which deep mouldboard ploughing alternates with shallow surface tillage; usage of a special field equipment and machines; selection of sowing dates depending on the crops in the rotation and local varieties of spring wheat, adapted to the climatic conditions of the region, etc.

A. Baraev has developed the soil protection system based on the concept proposed by T. Maltsev and the farming practices in Canada. He proved the need to replace ploughing with flat-cut cultivation combined with preserving stubble and crop residue on the soil surface. Moreover, he proposed the grain-fallow crop rotation schemes with a short rotation cycle (3–5 years) instead of grain-grass crops with a long period (8–10 years). This soil-protective system was applied in North Kazakhstan and Western Siberia, which prevented ecological catastrophe in the 1960s due to the rapid development of wind erosion from ploughing of virgin lands. For these purposes, a set of special field equipment and crop management technologies were developed.

Despite having a rich history of soil preserving and resource-saving research work in the region, the recent adoption and promotion of CA has been successful, thanks to the exchange of experiences and study visits between the farmers and the researchers of Canada, the USA and South America. Moreover, the CA farmers from Argentina and Brazil as well as experts from World Bank, FAO, International Maize and Wheat Improvement Center (CIMMYT) and International Center for Agricultural Research in the Dry Areas (ICARDA) as well as local NGOs significantly contributed to adoption and promotion of CA as well as modification and production of suitable field equipment.

2.1 Belarus

The total area of Belarus is 207 600 km^2, including 8 532 700 ha of agricultural land. The forests occupy 8 767 600 ha (FAO Country profiles: Belarus, 2016c; FAOSTAT, 2019).

Agriculture is an important sector of the national economy of Belarus, which, together with forestry and hunting, provides about 8% of the country's GDP, 17.1% of investments in fixed assets and about 10% of employment. The total area of agricultural land is over 8.5 million ha, or more than 40% of the country's territory. There are 0.9 ha of agricultural land per capita, including 0.6 ha of cropland. The agricultural sector employs more than 330 000 people or 8% of the total employed population, and 23% of the population living in rural areas.

The agriculture of Belarus is located in the zone of 'risky farming', and climatic fluctuations can lead to changes in its gross product by 15–20%. Drought causes the greatest damage to agricultural production. In dry years, the yield of cereals and legumes can decrease by 10–20%, and in years of extremely severe drought, which today occurs once every 100 years, the yield can decrease by 30–40%. The uneven snow cover and sharp fluctuations in temperature, cause freezing or rotting of winter crops.

Soil types in the regions of Belarus show large differences: there are light sandy and loamy sandy soils in the Brest and Gomel regions, and heavy clay-loamy and loam soils in Mogilev and Vitebsk. Peat soils occupy from 5% to 10% of the Polesye region. The light soils are exposed more to erosion, but rich with minor-nutrient elements heavy soils have unsatisfactory physical structures that decrease soil fertility and yield.

In recent years, there has been a strong fluctuation in precipitation during the growing season. In 2015, weather was dry during the growing period of sugar beets, potatoes and several grain crops that eventually led to a decrease in crop yield. Water and wind erosion is also a serious problem. Moreover, crop management is still far from being effective. This does not allow increasing the crop yield and is the main reason for the high cost of production. Therefore, the development and introduction of efficient resource-saving technologies, including CA/no-till practices, is one of the most important issues of agriculture in Belarus. The main elements of new technologies should be the improvement of crop rotations, minimising soil disturbance and the combination with the application of fertilizer and plant protection practices. Optimization of these agricultural practices for specific soil and climatic conditions, combined with the development and introduction of high yielding and adapted crop varieties, is expected to reduce the cost of inputs by 1.4–2.0 times with a significant increase in the productivity of cropland (Grib, 2002; Gusakov, 2002; Bulavin et al., 2010).

The climatic conditions of Belarus with an average of 600–700 mm of precipitation, mean temperature during the growing season of 19°C and high humidity are favourable for the development of diseases, pests and weeds. Therefore, researchers are still sceptical and assume that CA cannot be applicable in all regions of the country. They consider, among other factors that a high level of weed seeds in the soil in the absence of mouldboard ploughing will lead to increased application of herbicides. Such misperceptions are existing in the country because no large-scale CA/no-till studies have been carried out in Belarus. Moreover, there is lack of information exchange and experience sharing. The researchers in Belarus might not be well aware that climatic conditions in parts of Argentina and Brazil are also humid and favourable for developing pests and diseases. However, CA adoption and spread is progressing well, and farmers are satisfied with the outcomes.

There is also a lack of specialized field equipment for no-till planting. The researchers of the Scientific and Practical Center of the National Academy of Sciences of Belarus assessed the impact of non-inversion full tillage and minimum tillage and concluded that soil compaction, temporary excessive soil moisture, as well as low soil fertility, may hinder the introduction of CA in the farmers' field. Thus, they also recommend that before the introduction of no-till systems, fields should be cleared from perennial weeds, fields should be evenly levelled and plough pans and soil compaction should be removed. They also do not ignore the human factor: crop management needs to be carried out timely and in high quality. Moreover, they realize that establishing CA systems will take time; transition can take several years before yields and costs stabilize at a fully acceptable level.

Field experiments carried out by the Scientific and Practical Center indicate that for many field crops, minimising tillage can be beneficial. For example, on the fertile light loamy soil of 'Zhodino' Farm (Smolevichi district, Minsk region) with traditional ploughing and application of glyphosate and nitrogen fertilizers (N120), the yield of winter triticale was 5.5 t ha^{-1}, but under the direct sowing it was 5.37 t ha^{-1}, i.e. only 103 kg ha^{-1} less. At the same time, the insignificant yield shortfall was compensated by a significant reduction in the cost of fuel, lubricant and labour, which provided an additional economic return of USD16.06 per ha.

Based on soil studies, the estimated area under minimum tillage (100% surface area) and direct sowing in Belarus is 800 000–900 000 ha, including 640 000 ha for winter rye and triticale, 100 000 ha for spring cereals and 60 000 ha for maize. Direct sowing is also applicable for cultivating forage grasses and cover crops. Considering the positive results, researchers are working on developing soil-protective technologies and adaptive machines for reducing soil disturbance and transforming farming systems towards CA, including the pneumatic no-till seeder S-9. The current area of land under CA is estimated to be about 400 000 ha.

Farmers in Belarus are interested in adopting and promoting the CA system. They are enhancing their knowledge and obtaining the updated information from colleagues in the neighbouring countries such as Russia and Ukraine, participating in the regional capacity building events and farmer exchange visits organized by the FAO and other institutional development activities. However, up to date, no government or donor-funded technical support projects or large-scale testing and promotion of CA have been implemented in the country.

2.2 Moldova

Agriculture plays an important role in the economy of Moldova and accounts for more than 12–13% of GDP. Agricultural land occupies 2 257 100 ha, or 75.2% of the total area, including cropland – 1 681 700 ha (53.1%), orchards and vineyards – 370 700 ha (10.9%), meadows and pastures – 376 000 ha (11.1%), forests and shrubs 422 900 ha, or 12.5%. In Moldova, as in other countries, there is a decrease in cropland per capita, which currently is 0.407 ha (FAO Country Profiles: Republic of Moldova, 2016g; FAOSTAT, 2019).

Agricultural production and processing account for about 50% of export earnings. Currently, over 390 380 individual agricultural producers own more than 40.7% of the total land area. The majority of the population lives in rural areas, about 30% is employed in the agricultural sector. The salary of workers in the agricultural sector is one of the lowest in the country, and villages are rapidly emptying due to migration to cities and abroad.

The climate of Moldova is temperate continental and mild; winters are short with little snow; summers are long and warm with low precipitation. The negative impact of climate change, mostly drought is increasing in all regions of the country. The intensive soil bed preparation with mouldboard plough has accelerated the desertification process, surface and linear erosion, land subsidence and landslides that cause loss of biodiversity and environmental instability. Particularly, the agricultural lands in the centre and the south of the country are significantly eroded and degraded. Annual losses of fertile soil on all agricultural lands due to erosion amount to 26 million tonnes, including humus – 700 thousand tonnes, nitrogen – 50 thousand tonnes, phosphorus – 34 thousand tonnes and potassium – 597 thousand tonnes. Consequently, these losses cause a significant decline in crop yields.

Therefore, since 2009 the government of Moldova began supporting the initiatives of testing and adopting soil conservation technologies. In this framework, the government has been subsidising 30% cost of no-till drills. At the initial stage, different approaches as no-till, strip-till, vertical-till, ridge-till and minimum or reduced-till were demonstrated and promoted. The awareness was raised on the importance of adopting and promoting CA through conducting national workshops, the establishment of field demonstrations, developing and publishing guidelines and textbooks to change the farming system. There were several projects supporting the promotion of CA funded by FAO, Global Environment Facility (GEF), World Bank (WB), International Fund for Agricultural Development (IFAD) and other donors implemented in Moldova. Farmer Field Schools was an efficient modality for training large numbers of farmers to promote CA.

During 2014–2017, the WB project provided grants in the amount of USD 1.9 million for sustainable soil management. In this framework, 165 farms received financial support for the procurement of no-till drills, mowers, sprayers, harvesters and other field equipment. As the result, the area under soil preservation technologies increased up to 52 119 ha. Most of these farms continue and promote CA. For instance, farm 'Panfil Ion' grows maize, soybean, wheat, barley and sunflower on 1000 ha with no-till technology. This allowed reducing the fuel cost by two to three times and fertilizer application by 20–25%. The savings allowed the farmer to procure a combine harvester. The farm 'Agroelit' started applying minimum-till and now moving more on no-till and diversified crop rotation for growing rapeseed, soybean, winter wheat and barley on 3911 ha. Farm 'Khiliucanul' fully converted to minimum and no-till since 2013 and grows wheat, maize, barley, rapeseed and soybean on over 1000 ha. The transition to the CA system allowed reducing the number of field operations, saving fuel, time, labour and at the same time improving the soil. The farmer also uses soil mulching: all plant residues remain on the soil surface after harvesting field crops. Farm 'Agro-Popuros' is successfully applying no-till and s on 1740 ha since 2014. One-third of crop area (over 1000 ha) in the farm 'Girlea Andrei Pavel' is under minimum and no-till (Figs 2 and 3).

Transition from traditional to CA system is ongoing in Moldova. Although at the beginning there was a lack of adequate understanding of concepts and

Figure 2 No-till planting of sunflower in the farm of V. Chernii in 2000 (Photo: M. Rurac).

Figure 3 No-till planting on cover crop (Photo: Dorin Cebanu).

principles, some positive results were achieved. Several knowledge products and informational materials on CA have been published to raise awareness. There were also some perceived risks to farmers during transformation from tillage agriculture to CA because not everybody was supporting the CA concept. Some researchers, agricultural experts and practitioners presented the concept and shared personal opinions in mass media which led to several debates on the pros and cons of CA. Overall, this had a positive impact, but it also jeopardized the wider adoption of CA by farmers. Despite this, there were progressive farmers who adopted CA with its fundamental principles, and today they are successful and happy with the transition to CA. This encourages more farmers who need support for scaling up the successful experiences on the adoption of CA.

Wider adoption of CA also requires educating and training future agricultural experts. For this purpose, an MSc course on CA was established in the State Agrarian University of Moldova, and the curriculum has been developed. The International Conferences held in Moldova in 2012; 2014; 2018 and 2019 with the participation of experts from different countries allowed to learn and further promote CA in the farmers' fields. Moreover, Amir Kassam (CA expert) and Tony Reynolds (CA farmer) shared global and their own knowledge and experiences on the promotion and adoption of CA during the workshops and field visits in Moldova through online webinars. In 2019, 'Springer' published the book 'Farming the Black Earth: Sustainable and Climate-Smart management of Chernozem Soils' by Boris Boincean and David Dent that summaries the research and extension work carried out in the country on soil conservation.

According to Kassam et al. (2019), the CA area in Moldova in 2015/16 was 60 000 ha which had increased to 180 000 ha in 2018/19.

2.3 Russia

Russia is the largest country in the region with diverse environment and rich natural resources. The total area of Russia is 17.10 million km^2, including 215.5 million ha of agricultural land and over 815 million ha of forests (FAO Country Profiles: Russia, 2016h; FAOSTAT, 2019).

Agriculture in Russia is a large sector of the national economy with a share in GDP of 4.5% (2016). The total population of Russia is over 144 million (2018) and about 9% is employed in agriculture (2015). The volume of agricultural production in Russia, in 2017, amounted to 5.7 trillion rubles (about USD 100 billion). The leading industry is crop production, which accounts for 54% of the agricultural production; the share of livestock is 46%. The structure of the agricultural production by types of farms: agricultural organizations - 53%, household plots - 35% and small-scale (family) farmers - 13%. The value of gross agricultural output amounted to 5.11 trillion rubles in 2018.

Russia is a major exporter of agricultural products. In particular, in terms of wheat exports, Russia ranks 1st among the countries of the world. The total volume of exports of food and agricultural raw materials from Russia reached its highest value in 2018 – USD 25 billion.

Like in other countries in the region, farmers began testing, adopting and promoting CA in Russia many years ago. However, there is little fundamental research work being done on the subject. In the past two decades, Russian farmers have become increasingly interested in CA (no-till technology) for the production of cereals, legumes and oilseeds in many regions (mainly in the forest-steppe, steppe and dry-steppe zones of the European territory of Russia, the North Caucasus, the Volga region, the Urals and Trans-Urals and Western Siberia).

However, there are no official statistics on the area of CA in Russia. According to rough estimates of foreign and domestic experts, the area under minimal, strip-till and no-till in Russia is about 15 million ha. However, currently, CA (with three key interlinked principles) is implemented approximately on about 6 million ha, and it is growing from year to year at an ever-increasing pace. In the Rostov region alone farmers practice CA on an area of more than 700 000 ha (V. Bandurin – personal communication). In the Belgorod region, within the framework of the regional programme for organic agriculture, by 2015, over 20% of cropland (365 000 ha) was converted to no-till systems. There is a successful experience in the use of CA in the Republic of Bashkortostan, Stavropol, Samara, Saratov and other regions (Safin, 2021).

A large number of farmers express a desire to adopt no-till system, but not knowing the theoretical basics and not having practical experience lead to mistakes and miscalculations during the implementation. Accordingly, the effectiveness of the technology decreases followed by their refusal to adopt and promote CA. This is part of the regional problem, not only Russia.

Adopting a new technology without regulatory support and scientifically grounded adaptation to local conditions often leads to negative results. Supporters of CA systems and practices from agricultural producers and the scientific community recognize the need for comprehensive systemic studies for adaptation of CA systems to local soil and climatic conditions. At the same time, in the development concepts of the agro-industrial complex of some regions of Russia, in the national reports and in regional recommendations for the intensification of crop production, the need and prospects of introducing no-till are recognized.

Some of the key driving forces for adopting CA in Russia are the recurrent drought, erosion and dust storms that are a consequence of global climate change towards warming. Decrease in precipitation is observed in arid climatic zones almost every year. The impact of anthropogenic factors on climate change is also significant and includes tillage-based agriculture leading to soil erosion and degradation. Moreover, reducing the number of forest belts (windbreakers), aging of trees, cutting down and lack of work for planting new ones are increasing wind erosion. Along with this, excessive livestock, especially sheep, has an extremely negative effect on the vegetation cover of pastures and meadows, leading to erosion and land degradation. The sheep knocks out all the vegetation necessary for protecting soil health, including the stubble and biomass from cereals which, with their root system and standing stalks, ensure the binding of the upper soil layer, preventing water and wind erosion.

There is no prohibition on the ploughing of land in the protective virgin zones near ponds and on slopes. Therefore, farmers still practice bare fallow that not only leads to an increase in dust storms but also facilitates soil erosion followed by loss of organic matter, destruction of soil mechanical structure, loss of capacities for the accumulation, retention of moisture, etc. A man-made imbalance in the agricultural landscape through excessive ploughing of soils and leaving extensive area of bare fallow lands without a protective cover leads to an increase in wind speed that facilitates dust storms that have been occurring more frequently during the last few years (Fig. 4).

The research institutes of Russia have been working on soil conservation studies and technologies, developing manuals and guidelines for farmers. For instance, in the 1990s the Don Zonal Institute developed methodological manuals on improving soil fertility and agricultural landscape farming systems for soil and climatic zones, etc. However, the outputs of the research work and the recommendations are not always implemented or practiced by farmers. In addition to that, during recent years, agricultural research in Russia experienced a significant reduction in financial support. Moreover, the Research Institutes do not benefit from the incentives provided to farmers for procurement of field equipment. As a result, the technical capacity of researchers and the research work is not optimal in relation to CA research. Therefore, the progressive

Figure 4 Dust storm in Rostov region, September 2020 (Photo: https://glavagronom.ru).

farmers and the associations are focusing on testing different CA-related techniques and conducting applied research work in the fields. The outputs of farmers' work are usually shared during the winter conferences organised by the four active CA Associations. The farmers are more interested in testing different crops, varieties, seed and fertilizer rates, timing and type of application, management of weeds and volunteer plants, etc. under no-till system. They also experiment with different crop rotation schemes, crop residue and stubble management, growing pure and mixed cover crops, and discuss the usefulness of living roots to soil health and function, application of liquid nitrogen fertilizer, timing, depth and type of phosphorus fertilizers in CA systems. Unfortunately, in many farms soil is over cultivated by mouldboard and disc ploughs that lead to the loss of the top fertile layer of soil and to pollution of water. Burning the crop residue and grasses is still practiced in some farms. Therefore, the government is considering making changes in the agricultural development programme, which would prohibit the provision of subsidies to agricultural producers who allow burning crop residues and pastures. In 2020, the Russian government adopted new amendments to the fire legislation imposing more severe punishment on farmers for fire accidents in their fields.

Also, there are no crop rotations being practiced, and there is a high degree of weed infestation including from perennial grasses. At least half of the farmers do not have the necessary knowledge and appropriate agricultural equipment to test and adopt CA with no-till, crop residue retention and appropriate crop rotation. Therefore, these issues should attract greater attention by government to protect soil resources and environment for future generations and for supporting the intensification of sustainable agricultural production.

It should be noted that the lack of scientists of the necessary specialties, and the absence of necessary research, aggravates the situation with both soil conservation and climate change mitigation and adaptation. Soil conservation has been pushed to the limit. Several research institutes have been working on soil conservation issues. As a result of a lack of specialists, and, in general, knowledge, there is currently over-cultivation of the soil, especially with disc tools. The soil cover and resilience are destroyed to produce a predominance of the dust-like fraction, increasingly being washout into rivers - siltation and transport of fertile layer of soil (humus) over long distances. There are practically no crop rotations that suffer from infestation of perennial grasses when they are combined with the practice of soil protection technologies (no-till with the retention of plant residues on the soil surface). However, at least half of the farmers do not have the necessary knowledge and appropriate agricultural equipment to practice CA and therefore this should attract government's attention as an opportunity to improve agricultural sustainability.

At the early stages of promoting CA in Russia, there was a lack of suitable field equipment. Today, more than ten factories are designing and producing no-till drills in Russia in cooperation with foreign companies. Therefore, in principle, the availability of field machines should not be a problem to advance the CA. However, there is currently no government support, subsidies or incentives available to help implement and promote CA adoption. On the other hand, in some regions, the government is interested and pays attention to promoting CA where there is clear evidence about its ability to cope with climate change and recurrent droughts. For instance, the introduction of CA in the Samara region, through the regional programme 'Improving grain production in the Samara region using resource and moisture saving technologies' was implemented from 1998 to 2015. Under this programme, 257 no-till drills were procured for the farmers. By 2006, the area under resource-saving technologies in the Samara region increased to 600 000 ha.

Farmers in Russia are recognising CA as an ecosystem solution for many problems, including recurrent droughts occurring due to climate change. Therefore, more farmers are converting to the no-till-based farming system. For instance, in the Altay region, every tenth farmer is practicing CA. In 2020, the region faced a severe drought too. The snow started melting earlier and daily temperatures sharply increased by mid-April, which is the time of planting. Farmers were in a hurry to sow wheat to avoid losing soil moisture. The dry weather persisted almost throughout the season – with short and unproductive rains. This provided a good opportunity to define the strengths and weaknesses of traditional technologies and the benefits offered by the CA system. As a result, no-till practice has proven itself at its best level: in several CA farms, the yield was much higher compared to farms with a traditional farming system. For instance, on the farm 'Virt' in the Tselinny district, the yield of wheat variety Buran reached 8.92 t ha^{-1}. A whole range of factors facilitated this achievement: continuous improvement and correct application of CA system and practices, selection of suitable crops for rotation and selection of high-quality seeds of superior and well-adapted varieties. However, no such results were observed among farmers using traditional soil tillage. Therefore, the areas with no-till are expected to grow further (Fig. 5).

The large commercial companies marketing CA equipment and plant protection chemicals (e.g. Agro-Soyuz, AgroMir, August, etc.) also put significant efforts into the adoption and promotion of CA in Russia and other countries in the region.

Public movements to promote CA technologies in Russia originated in 2003, when the National Fund for the Development of Conservation Agriculture (NFDCA) was established with the participation of several leading agricultural research institutes and with the support of the Ministry of Agriculture of Russia. Ms. Luidmila Orlova was elected as the President of the NFDCA. In 2008, the NFDCA was transformed into the National Movement for Conservation Agriculture (NMCA). The non-profit partnership was created to disseminate

Figure 5 Soybean, sunflower and barley planted in no-till system in the pilot farm in Pokhvistnev District, Samara region (Photo: Luidmila Orlova).

advanced agricultural practices based on CA and participate in the formulation and implementation of the agricultural policy in Russia.

The NMCA has united the farmers, business representatives and researchers, to foster their further collaboration in spreading theoretical and practical knowledge on CA, it regularly organizes seminars and conferences. For example, since 2016, together with the German-Russian forum 'St. Petersburg Dialogue' the NMCA has been organising an annual conference 'Climate, Soil Health and Agriculture Technologies' supported by the Russian Ministry of Agriculture, Russian Association of Specialized Machinery and Equipment

Manufactures, Association of the Engineering Industry (Germany, VMDA), Russian Grain Union, business and research institutions. The conference has become a platform for an open dialogue of the industry, business, research and policymakers on the challenges and prospects for further promotion of CA in Russia (Fig. 6).

The NMCA is actively cooperating with the lead Agricultural Research Institutes and the Universities, including the Eurasian Center for Food Security in Moscow State University, regional agrochemical stations and branches of the Russian Agricultural Center to stimulate the research and build up the evidence basis to adapt the CA. NMCA is also developing joint projects not only to prove the benefits of the three interlinked principles of CA but also to study the integration of CA with other good agricultural practices and modern technologies, including precision agriculture, air seeding, drone applications, biological control of pests and diseases, etc.

Thanks to the awareness raised, the Russian Government recognized the potential benefits of CA for the national economy and the environment. Russia has ratified the international climate conventions and agreements and improved the national legislation on carbon sequestration. In this framework, the national legislation is to be adopted to motivate farmers to practice CA through special incentives. These measures will provide real opportunities to the farmers and significantly contribute to the further promotion of CA in Russia.

The demand for agronomists with the CA understanding and knowledge is increasing and that has served as a driving force for creating specialization courses on CA in some agricultural Universities in Russia. Moreover, the Central Agricultural Research Library has opened a special window on the CA on its website (http://www.cnshb.ru/news/vex_pzt.shtm) where the relevant literature is posted. CA farmers are establishing associations to adopt and promote the promising system. For instance, the Association of the No-till Supporters is established in the Rostov region. The Association is actively engaged in the adaptation and promotion of the no-till system throughout the Russian

Figure 6 Annual conference 'Climate, Soil Health and Agriculture Technologies' with the field day and the demonstration of the CA machinery (Photo: Luidmila Orlova).

Federation. The Association closely collaborates with the No-till Associations and farmers in the USA, Argentina, Sweden and other countries. In cooperation with the research institutes, the Association carries out the R&D activities and has established large-scale demonstrations. It promotes CA systems adapted to local conditions and participatory and evidence-based decision-making approaches with special attention to different climatic zones and farmers' practices. The Association also promotes suitable equipment for CA and provides services to the farmers. Capacity development is a priority area for the Association. Therefore, it hosts training events, Field Days, workshops and conferences, and organizes farmer exchange visits to share experiences and knowledge. The Association organizes the winter conferences to raise awareness and share the knowledge and information on CA among the farmers and other beneficiaries. During 4-5 February 2021 the Association organized the 9th Winter Conference with participation of over 200 farmers, researchers and extension specialists from different regions of Russia, Ukraine and Kazakhstan. The conference was also translated online, which allowed more people to benefit from the presentations and discussions. The CA farmers from Argentina and the USA also made presentations and shared their experiences. The exhibition of CA machinery was also organized in the hall next to the meeting room (Fig. 7).

The experts from the Eurasia Center and the NMCA in 2019 surveyed ten agricultural enterprises in the Samara and Rostov regions, practicing no-till direct sowing in various soil and climatic conditions. The survey outputs demonstrated that the main objective of the transition to no-till technology is to reduce fuel cost (because the fuel price is rising), increase the profitability and efficiency

Figure 7 Participants of the 9th Winter Conference of the No-till Supporters, Rostov-na-Donu, Russia, 4-5 February 2021 (Photo: Association of the No-till Supporters).

of agricultural production and preserve soil fertility. The agricultural producers recognized that application of CA principles improves the efficient use of precipitation, restores soil fertility, reduces water and wind erosion and increases product quality. The farm areas of the surveyed agricultural producers range from 650 ha (Olimp LLC, Rostov Region) to 85 000 ha (Zerno Zhizn LLC, Samara Region). The introduction of CA (direct seeding) in the surveyed farms covered the period 2008–2014 which is in line with the CA transition requirements - that is the practice should be applied for at least 5 years to assess the impact.

The main cultivated crops in the surveyed farms were cereals (winter and spring wheat, barley, triticale, corn, sorghum), pulses (peas, chickpeas, lentils) and oilseeds (soybean, sunflower, mustard, rapeseed, flax). The crop rotations included perennial grasses, including perennial legumes (clover, alfalfa, sainfoin and vetch). The share of legumes and perennial legumes in the crop rotations of the surveyed farms ranged from 10% to 25%, and, according to agricultural producers, is constantly increasing. Agricultural producers reported that the yield of these crops under the no-till system is not low, and in some cases even exceeds the crop yield under traditional technology in similar soil and climatic conditions. However, farmers concluded that agro-climatic conditions remain the main factor limiting the yield in the middle and the south of Russia regardless of any crop management technology. Even under the condition of strict adherence to crop management measures (seed rate, fertilisation, pest ad disease control, etc.), the yield varied greatly, primarily depending on the amount of precipitation.

Agricultural producers use both locally produced field equipment (Eurotechnika JSC, Sterlitamak Machine-Building Company CJSC, NAIR LLC, etc.) as well as imported machinery (Gherardi, Semeato, John Deere, etc.). The high cost of the equipment is noted as one of the problems when switching to no-till direct seeding. Moreover, the main difficulties that agricultural producers face when switching to the no-till system are the lack of scientific justification for adapting technology to local climatic conditions. Difficulties arise with the selection of crop rotations, seeds and hybrids for various soil and climatic conditions, optimal sowing dates, plant protection products, technical characteristics of machinery and equipment. Agricultural producers gain knowledge mainly using foreign publications from the Internet, study trips to Canada, Argentina, Brazil, through the exchange of experience with colleagues, attending conferences, training seminars organized by NMCA and manufacturers of equipment for direct seeding and plant protection products.

2.4 Ukraine

The total territory of Ukraine is 603 628 km^2. The area of agricultural land is 41.4 million ha, or 70% of the entire territory of the country, and the area of cropland is 32.5 million ha, or 78.4% of all agricultural land (FAO Country Profiles: Ukraine, 2016k; FAOSTAT, 2019). The agricultural land per capita is 0.9 ha, including 0.7 ha of cropland (the average in Europe is 0.44 and 0.25 ha, respectively). The area of chernozems (black soils) in Ukraine ranges from 15.6 to 17.4 million ha, or about 8% of world reserves.

Agriculture is an important sector of the economy of Ukraine. Irrigated lands are mostly located in Kherson, Zaporozhe and Odesa oblasts and have particular value in the arid zone of Ukraine. However, due to lack of water, not all potentially suitable land for production is irrigated. For instance, in 2019 only 0.49 million ha out of a total 1.77 million ha of the potentially suitable area was irrigated. Rehabilitation of the irrigation system requires significant investment (up to 2 billion USD). The on-farm irrigation infrastructure for 926 000 ha is degraded. On 660 000 ha the automatic (sprinkler) water system is missing, on 361 000 ha the irrigation system is lost and on 122 000 ha on-farm the water pumping system does not function properly.

The country is the largest exporter of sunflower oil and one of the largest grain exporters. The main crops are grain and forage crops, including wheat, maize, barley, sunflower, sugar beet, tobacco, legumes, fruits and vegetables. However, the area under fodder crops significantly reduced from 12 million ha in the 1990s to 1.8 million ha in 2017, or almost 85%. The average yield of crops is: winter wheat - 4.17 t ha^{-1}, barley - 3.48 t ha^{-1}, buckwheat - 1.33 t ha^{-1}, millet - 1.81 t ha^{-1}, sunflower - 2.51 t ha^{-1}, soybean - 2.35 t ha^{-1} and maize - 7.14 t ha^{-1}.

Climate change is the key driving force behind CA advancement as crop losses increase significantly each year. This is especially important for crop production in the dry zones in Ukraine. Therefore, some large-scale and small-scale farmers in Ukraine widely practice integrated crop-livestock production systems, resource-saving technologies and diversified crop rotations.

CA has a long history in the region, dating back to the time of the Russian Tsar, who made efforts to promote soil and resource-saving technologies in Ukraine. Previously, the lack of no-till planters was the main problem in promoting CA, but there are now enough manufacturers and suppliers on the market. Thanks to the access of Ukraine to the sea, the new and second-hand no-till equipment are imported from Europe and South America. At present, an area of some 900 000 ha is under CA systems.

Currently, there is a lack of special research and development work focused on the adoption of CA as a system or its elements to the specific soil and climatic conditions of different regions of Ukraine. Therefore, the farmers search for solutions and the large companies and holdings are creating separate scientific structures to address these issues. There is also a lack of farmers' associations, which would allow pooling resources to address the issues of a successful accelerated transition to CA, knowledge and experience. There is a lack of trained personnel for the development of a roadmap for the transition to CA and consulting support. Today, this is done to the greatest extent by representatives of large agrochemical (pesticides and fertilisers), machinery (seeders, harvesters, tillage equipment) and seed companies.

The interpretation of the CA system is quite diverse in Ukraine (Fileccia, 2014). The term, its content and essence are not clear to some farmers, researchers and farmers; therefore, it is used by a very limited number of people. The subject of CA is not taught in agricultural universities. They clearly understand and know that the farmers and specialists are working in zero tillage, strip tillage, organic farming, but not in CA. All of this is viewed as separate farming systems and their supporters do not interact with each other.

Farmers in general are interested in adopting and promoting CA, as it helps to cope with the climate change. There is a need for promoting CA/no-till under sandy soils in Polesye and drip-irrigation (sub-surface) in the Steppe zone of Ukraine. Various organisations (FAO, USAID and private companies, e.g. Agro-Soyuz) were able to change the mind-set of farmers. In Soviet times, soil protection technologies were promoted through the command-administrated system and pressure. However, this did not produce good results because the farmers were not sure about the outputs. Only through conducting demonstration trials, training and practicing over several years, farmers realized the benefits of CA, especially in terms of improving soil health, fertility and increased yields (Figs 8 and 9).

Currently, the FAO is implementing a GEF-funded project that supports the adoption and promotion of CA in Ukraine. Moreover, with the support of the project, demonstration fields on improving water use efficiency through the application of surface and subsoil drip-irrigation under the CA system were established. It is a very good approach to increase water use efficiency, especially when many beneficiaries are complaining about the lack of water for irrigation and precipitation that may force them to stop farming. However, the replication of the demonstrated technology to other farms in the region requires additional financial resources. The project team closely collaborates with

Figure 8 Farmers in Ukraine are widely practicing CA in large scales (Photo: H. Muminjanov).

Figure 9 Some farmers use large-scale field machinery for CA (Photo: H.Muminjanov).

several farms (e.g. 'Zorya-Yug' in Chaplinka district with 2800 ha under irrigation and Agrofirma 'Dodola' in Nova Kakhovka district with 7000 ha under rainfed) that have fully converted their cropping system into CA. These farms support the initiatives on raising awareness and promotion of CA through training of farmers and neighbouring farms. The main crops in the rotation under CA are winter wheat and barley, rapeseed, sunflower, maize and cover crops. Some of the CA farms have already sold out all types of field equipment for soil bed preparation and investing huge funds (through bank credits) for procurement of a large-scale modern no-till drill and combined harvesters. The only concern is that the farms are applying glyphosate as the main herbicide. However, the farm managers confirm that in the fields with the proper application of crop rotation and a thicker layer of mulch, there is a significant reduction in weeds' pressure observed and fewer herbicides are applied. They also noted the improvement of soil health and development of soil biota, especially increased population of earthworms in the fields with the application of CA principles (Fig. 10).

During recent years (especially in 2019/20 cropping season), the mice were damaging crops in CA fields. The farmers are using chemicals to control

Figure 10 Farmers realised that the standing stubbles protect crop field from sand and the mulch increases the soil moisture (Photo: H. Muminjanov).

mice, although rearing owls could be an excellent environmentally friendly approach. However, it could be applicable only in the fields surrounded by the shelterbelts. Unfortunately, in many places, the trees are being cut down for firewood, particularly for heating the greenhouses. Therefore, there is a huge risk on the sustainability of the approach on establishing shelterbelts unless the relevant legislation is adopted, and land ownership defined. Another concern for the sustainability of CA in Ukraine is the ongoing land reform that might not consider soil quality and CA fields as important.

There are also several initiatives on improving the national capabilities on the application of CA in Ukraine. In parallel to FAO and other development partners, farmers are leading in the organization of capacity building activities. For instance, farmer and blogger Mr. Michail Draganchuk during recent years has been organising training for beginners and the regional workshops named 'No-till Lab' in Kiev. The training and workshops are organized fully on a paid basis and farmers attending the events need to cover the related cost. In 2020, over 400 interested farmers and researchers from 15 countries attended the event. The workshop demonstrates the rising interest of farmers in CA that is a very positive sign. Parallel to the workshop, an exhibition of no-till drills, drones, sprayers and other field equipment is organized where a number of national and foreign companies from Argentina, Finland and Russia demonstrate their products (Fig. 11).

Figure 11 'No-till Lab' attracts interested CA farmers; is useful to demonstrate modern no-till equipment and other useful tools and inputs (Photo: ' No-till Lab-2020').

Figure 12 Farmer and blogger M. Draganchuk in no-till field (Photo: www.glavagronom.ru).

Mr. M. Draganchuk has also established a channel 'No-tiller' on YouTube through that he shares no-till practices, field visits and video lessons. (https://www.youtube.com/user/MDraganchuk) (Fig. 12).

Considering the low capacity of the national partners (Research Institute of Soil Science and Agro-chemistry, etc.) the Ukrainian Soil Partnership Association was established that enables an environment for the collection of data from different institutions to provide a platform for a national soil monitoring system. There is also opening for new specializations and further education and training of agronomists (BSc and MSc) on CA that requires developing curricula, textbooks and guidelines. There is a need for conducting training programmes and field workshops for farmers on established long-term CA fields.

Further promotion of CA in Ukraine as in other countries requires political interest and support. In this regard, farmers are advocating for the establishment of the national association on CA that can defend the farmers' interest, especially during the upcoming reforms. Further, the detailed assessment of the status of CA in Ukraine can serve as a basis for the formulation of the National Strategy on CA. Such a document could be a road map for both the Government of Ukraine and the farmers to move towards successful adoption and promotion of a sustainable crop production system.

3 Conservation Agriculture in the Caucasus

There are three countries - Armenia, Azerbaijan and Georgia located in the Caucasus. These countries have many similarities in their institutional systems in agriculture from the past Soviet era. However, there are also differences in farming and cropping systems, depending on soil and climatic conditions of the agricultural regions.

3.1 *Armenia*

Armenia is a mountainous country with a total area of 29 743 km^2. Hills and mountains cover over 90% of the country. The area of agricultural land is 1.7 million ha, of which 1 million ha account for pastures and 0.5 million ha for cropland (FAO Country Profiles: Armenia, 2016a; FAOSTAT, 2019).

The agricultural sector remains essential for the economy of the country as being the main source of economic activity in rural areas and a significant contributor to GDP, accounting for 14.9% of it (2017). The plant production sector accounts for 52%, and livestock husbandry - 48% of the total production. The share of livestock in the structure of gross production is growing. More than a third of the population lives in rural areas. Agriculture employs about 36.6% (2017) of the working population of whom almost 56% are female farmers. The informal employment in agriculture is substantial which leads to reduced access to social protection schemes and coupled with the limited access to land and other agricultural assets leaves the rural population, particularly women, in a vulnerable situation, thus contributing to poverty.

In the 1990s, Armenia implemented one of the most comprehensive land reform programmes in the former Soviet Union. As a result, around 340 000 private farms were created, mostly smallholders, and left with limited access to resources and inputs (water for irrigation, quality seed and planting material), finances, knowledge (on sustainable and good food production practices) and services (e.g. extension and advisory, machinery and equipment, information and communication technologies, etc.).

A large number of small farms with fragmented land holdings dominate the farming structure in Armenia. There are about 320 000 family farms, which contribute more than 97% of total agricultural output, comprise 99.86% of all active agricultural holdings. About 45% of the farms have three or more land plots and 20% of farms have five or more plots. Smallholders constitute the vast majority of farmers in the country, representing around 95% of all farms and produce around 97% of gross agricultural product.

Cereals and leguminous crops occupy more than half of the cropping area. The main cropping system is wheat based and wheat accounts for 30% of all cultivated area. The share of melons and gourds is 8%, vegetables and potatoes - 4% each. At the same time, the share of cereals, including wheat is decreasing, although the country relies heavily on grain imports, especially cereals.

Armenia is prone to natural disasters, and this trend is expected to worsen under climate change. These natural hazards pose serious threats to food security, rural livelihoods and agricultural development in the country, due to both crop losses and, the spread of plant and animal diseases. While there are limited possibilities to prevent these disruptive events, good practices exist, which can help increase the resilience of rural communities in risk-prone areas that may help these communities better mitigate the impact of these natural hazards and facilitate an early recovery after the shock.

There are several negative trends affecting agriculture in Armenia, including inefficient farm structure and production systems, land degradation, land fragmentation, aging rural population, migration, heavy dependency on water and irrigation facilities, lack of state support and inappropriate land management practices, which result in low yield and production. There is also a lack of policy, legal framework and available tools for sustainable agricultural production. Due to intensive tillage-based mono-cropping systems, the land in Armenia is severely degraded. Therefore, the adoption and promotion of CA are very important for controlling degradation and improving agricultural productivity, economic, environmental and social benefits for farmers and society.

In response to the challenges in the agricultural and rural sectors, the Government of Armenia has adopted the 'Sustainable Agriculture and Rural Development Strategy of the Republic of Armenia 2010–2020' (SADS) and Action Plan (AP), which emphasizes both short-term (food security and productivity increase) and long-term goals (sustainable growth and competitiveness). The latter is to be achieved through sustainable production intensification, natural disaster risk management, strengthened extension and advisory services and farmer cooperation, land consolidation, farmer risk management, insurances, etc.

The first attempts at testing and demonstrating CA in Armenia were made under the USAID support by local NGO 'ATGF'(Armenian Technology Group Foundation). For this purpose, no-till drills were imported, demonstration plots were established and training was conducted. However, the CA did not find any follow-up support from the Government and, therefore, was not promoted further due to several factors, including the lack of suitable field equipment, inability to raise the interest of farmers to produce more wheat to sustain food security, lack of incentives, etc.

Therefore, the EU provided the complimentary technical support to the Ministry of Agriculture of Armenia on this initiative under the project 'European Neighbourhood Programme for Agriculture and Rural Development (ENPARD)'. The overall objective of the project was to support the Government of Armenia in ensuring efficient and sustainable agriculture and contributing to better conditions in rural areas. In this framework, several potentially good agricultural technologies and/or practices, including CA (no-till practice), use of drought-tolerant varieties, anti-hail nets, drip-irrigation systems and integrated pest management (IPM) were identified and promoted (FAO, 2017). Moreover, the project supported the establishment of five demonstration fields on CA, organization of workshops/field days on promoting CA, including integrated soil management and enhancing the capacities of relevant stakeholders. Considering that the main share of cereal production is concentrated in the north part of Armenia, the Shirak and Lori regions have been selected for the demonstration of the CA system. Selected fields within rainfed farming systems were located on the sloppy landscape, which was ideal to demonstrate the main advantages of the no-till system (Figs 13 and 14).

One of the lessons learned by the project team was the importance of continued no-till and crop rotation for improving soil fertility and increasing yield. In the beginning, in most of the demonstration plots, wheat yield under no-till was somewhat lower or on the same level as in the plots with conventional tillage. However, in the permanent demonstration plot established in the Arapi village in the Shirak region with continued no-till for 5 years, the wheat yield was the highest (Fig. 15).

The FAO continues supporting the government in the promotion and encouragement of CA in Armenia. In particular, in the scope of the project 'Enhancing Human Security and Building a Resilient Society in the Disadvantaged Communities of Armenia', three more no-till demonstration fields in the Shirak region have been established. The main aim of the current intervention is not

Figure 13 The no-till drill used for establishing the demonstration trials by ATG (Photo: H. Muminjanov).

Figure 14 No-till planted wheat and Field Day in Shirak region (Photo: A. Zakaryan).

Figure 15 The ENPARD project team and the farmers growing no-till wheat are happy with the achievements: there are no yield differences between no-till and ploughed wheat fields in the first year (Photo: H. Muminjanov).

only the promotion of modern and effective soil management practices but also the demonstration of the importance of continued application of the CA system for improving soil health and increasing productivity.

Despite the obstacles that tend to slow down the development and expansion of CA in Armenia (e.g. lack of proper machinery, fragmented lands, etc.), the area of no-till is continuously rising. According to the data provided by the ATGF, about 300 ha are already under the CA system in Armenia.

3.2 Azerbaijan

The total territory of the Republic of Azerbaijan is 86 600 km^2, of which 52.3% is agricultural land, and 12% – forest. There is 1.631 million ha of cropland of which 1.350 million ha or 67.6% is irrigated. The total population of Azerbaijan is over 11 million of which 50% live in rural areas (FAO Country profiles: Azerbaijan, 2016b; FAOSTAT, 2019).

In Azerbaijan, agriculture is the third most important source of income for the national economy after the oil and construction industries. Agricultural landholdings are 99.8% privately owned; more than 66.8% of them are family farms, 32.8% are engaged in subsistence farming, and 0.2% are owned by agricultural enterprises. A long-term strategy that is based on environmentally friendly and climate-smart agriculture, research and development activities aimed at exploring technologies adapted to local conditions and supported by training and participatory agricultural extension can facilitate the intensification process and allow Azerbaijan to achieve increased productivity in agriculture and related industries.

According to the national statistics, there are 0.2 ha of cropland and 0.6 ha of farmland per capita in the republic. The largest share in the structure of farmland falls on pastures – 51.6%, cropland – 30.2%, perennial plantations – 12.8%. Hayfields account for 5.4%. Almost one-third of the land area is unsuitable for agriculture. The most common type of erosion is water erosion. The economic damage from erosion is significant.

The main factors causing soil degradation are mechanical disturbances, erosion, compaction, salinization, alkalization and leaching, all leading to the deterioration of soil properties and health. Anthropogenic factors, including tillage-based farming, have a significant impact on soil degradation. In addition, chemical pollution from excessive use of mineral fertilisers and pesticides also contributes to soil degradation.

Water and wind erosion in the republic are most acutely manifested in the form of washouts, erosion, blowing away of soil particles, ravines and landslides. According to the available data, the area of eroded land is 3 685 000 ha or 42.5% of the territory of the republic, ranging from 10% to 84% in the individual natural and economic regions. Erosion affects 33.7% of cropland, 68% of summer pastures, 15% of hayfields, 16% of orchards, 24% of vineyards and 26% of forests.

Grain crops and food legumes occupy over 1 million ha of which wheat is sown to 670 000 ha. The yield of these crops is increasing and on average was 3.25 t ha^{-1} in 2019. However, it is still low and does not fully meet the rising demand. Therefore, the Government of Azerbaijan is aiming to increase the yield to 4 t ha^{-1}.

In this framework, it is considered that the sustainable increase of crop yields is closely linked with the application of resource-saving technologies to restore soil fertility to improve production. The outputs of the FAO regional project (2011–2013) 'Conservation Agriculture for Irrigated Areas in Azerbaijan, Kazakhstan, Turkmenistan and Uzbekistan' funded by the FAO-Turkey Partnership Programme demonstrated the advantages of CA in the cropping systems that include wheat, legumes, maize, sorghum and sunflower. The large-scale field demonstrations showed that the application of CA has reduced soil erosion and increased its organic matter. Moreover, with CA there was a reduction in the application of inputs (water, seeds, fuels and lubricants), in the cost of production and an increase in profit. This led to increased interest in the adoption of the CA system.

Further, the Government of Azerbaijan attempted in cooperation with the donors and the development partners to promote CA in farmers' fields. The FAO, ADB, WB, the EU and ICARDA supported the initiatives and carried out several projects in Azerbaijan. The local NGO 'Agro Information Center' was also actively engaged in training the farmers, developing manuals and guidelines and disseminating information (Fig. 16).

Drought and irrigation water scarcity due to climate change observed in recent years is serving as a driving force for the promotion of resource-saving practices, including CA. Under irrigated conditions, CA involving bed planting is developing and secures the establishment of permanent raised beds to obtain high crop yields with reduced production cost. The bed planting is also attracting cotton-growing companies. In 2020, irrigated cotton with no-till bed

Figure 16 No-till planting of wheat under rainfed conditions in Azerbaijan (Photo: Guliyev in 2020).

planting was produced on 7 ha in the Beylagan region: the yield was 6 t ha^{-1}, almost two times greater than the country average (Fig. 17).

Researchers and agricultural specialists promoting CA in Azerbaijan observe the difficulty of going directly from the traditional system to no-till as a key element of the CA system. Therefore, they propose a step-by-step transformation of the traditional tillage-based farming system to minimising tillage, with bed planting and then to no-till with bed planting.

Currently, crop cultivation under the CA system is rapidly spreading among individual farmers and companies in Azerbaijan. Minimum soil tillage is practiced by 'Agro Dairy LLC' in 10000 ha, by ' Garabagh Grain LLC' in 5000 ha, by 'Kroun KO LLC' in 1000 ha, by ' Azertohum LLC' in 1000 ha, and by the farmer enterprise 'Hajhi Jhamalkhan LLC' in 500 ha. According to 2020 data, individual farmers practice minimum soil tillage on over 20000 ha and the total area under the CA system is around 37500 ha.

CA will be continued to be further promoted in Azerbaijan in the future. However, the main obstacle in irrigated lands is the lack of field machinery, particularly no-till planters, tine rippers and combine harvesters.

Figure 17 Field demonstrations to promote CA to the farmers' fields (Photo: Guliyev, 2020).

3.3 Georgia

Georgia is a small country with a total area of 69 700 km^2. About 43% of the total area is used for agricultural purposes; forests, other natural resources and urban settlements cover the remaining 57%. Approximately 25% of agricultural land is annual cropland, while 9% is under permanent crops. Pastures and grasslands occupy 65% (FAO Country Profiles: Georgia, 2016d; FAOSTAT, 2019).

Georgia is a country with a complex landscape, climate and a variety of soil conditions. Livestock husbandry and crop production are traditionally well developed. There are 13 agro-ecological zones in Georgia and each zone is focused on a specific product or farming activity, e.g. viticulture, horticulture, fodder production, livestock, suburban agriculture, tea, grain and tobacco production. As per Vavilov's theory on the centers of diversity and origin of crop species, Transcaucasia is the center of origin for wheat. There are about 12 species and 30 subspecies of wheat in the region, of which two - Makha and Zanduri, are found only in Georgia.

Land resources in Georgia are limited. Only 15% of the country's territory is cultivated, and 70% are forests, shrubs, meadows and pastures. Due to climatic and landscape conditions, as well as unsustainable farming practices, more than a third of agricultural land is prone to degradation, erosion, pollution and soil damage. In most regions water is abundant year-round and of very good quality for agriculture use; however, not all of it is found where it is most needed. Roughly 4% of agricultural land is subject to desertification. Overgrazing, unsustainable forest management, loss of forest cover and unplanned urban development are major drivers of land degradation in Georgia.

According to official statistics, the country is 70% self-sufficient in vegetables and only 8% in wheat. Wheat yields (Table 2), for example, are lower than in other countries, e.g. neighbouring Azerbaijan.

The Ministry of Environment Protection and Agriculture (MEPA) of Georgia and the non-governmental organization Regional Environmental Center for the Caucasus (REC Caucasus) have been working on a project to introduce crop rotation practices in the Shida Kartli and Kakheti regions of central and eastern Georgia since autumn 2016. The project is called 'Generating Economic and

Table 2 Trends on wheat production in Georgia, 2015-2018

Year	2015	2016	2017	2018
Area, '000 ha	49.3	50.1	44.8	43.6
Total production, '000 t	125.6	126.6	97.9	107.1
Yield, t ha^{-1}	2.6	2.6	2.2	2.5

Source: National Statistics Office of Georgia, 2021.

Environmental Benefits from Sustainable Land Management for Vulnerable Rural Communities in Georgia' ('Greenlands' GE). Funded by the Global Environment Facility and implemented by the United Nations Environment Program (UNEP), it aims to promote sustainable land management. This is a continuation of the 2016–2019 project 'Applying Landscape and Sustainable Land Management (L-SLM) for Mitigating Land Degradation and Contributing to Poverty Reduction in Rural Areas'. About 100 farms will take part in the 'Greenlands' pilot project. Ultimately, the project plans to reach 10 000 farmers to support practicing crop rotation. More than 500 ha of wheat is already being grown under crop rotation; peas, corn, beans and buckwheat are used to rotate. The previous project showed that crop rotation nearly doubled the wheat yield per hectare. Pilot projects to promote sustainable land management practices are planned until 2022 and will cover 20 000 ha. The goal is to increase soil productivity and achieve food security through crop rotation in the municipalities of Gori, Kareli, Kvareli and Sagarejo. Crop rotation and mixed crops increase yields and reduce soil erosion by improving soil nutrition. By planting multiple crops, farmers could make the most of the land while reducing the risk of crop failure. The project relied heavily on local knowledge and local inputs, partly addressing the problem of seed import. The Ministry of Agriculture under this project promotes local methods of obtaining sustainable agricultural products.

REC Caucasus has been working also in the establishment of windbreakers. Most of the windbreakers from the Soviet Union times have been destroyed in several locations in the country, but more evidently in locations like Dedoplistskaro municipality in the east, where climate change has impacted harshly by reducing precipitation in the last decades. Reduced precipitation, destruction of crop residues by burning, mono-cropping and lack of protection by windbreakers are all part of the annual scenery in this municipality that is one of the most important for wheat production in the country.

GIZ (German Corporation for International Cooperation) made combined efforts with REC Caucasus to establish windbreakers in Dedoplistskaro, but the indifference of the population did not allow for much success. In Gori, Kvareli and Kareli a total of 3646 ha of agricultural land will be under protection by windbreakers through planting saplings along main roads under IFAD-funded projects.

The projects 'Generating Economic and Environmental Benefits from Sustainable Land Management for Vulnerable Rural Communities of Georgia', and GEF-funded 'Applying Landscape and Sustainable Land Management (L-SLM) for Mitigating Land Degradation and Contributing to Poverty Reduction in Rural Areas' contributed to the establishment of wind breakers on 3.3 km and 12.4 km, respectively.

There are other zones in Georgia where wheat production is important and where precipitation is almost null that cannot support wheat. Very few pivot irrigation systems have been established in the last decade, mostly with foreign investment and some government support. Arid lands in Gardabaani would have enough water transported through channels to irrigate the valley, but there are no clear regulations/enforcement for water distribution and a significant portion is wasted.

GIZ also promoted crop rotation in Dedoplistskaro municipality and initiated the promotion of no-till planting of wheat and other small grains. The lack of no-till drills was an impediment to the dissemination of the technology during the intervention. Another challenge identified was the lack of modern harvesters that could do an even distribution of cereal straw on the fields to have a relatively even layer of mulch cover. Farmers then recurred to burn the wheat stubbles to facilitate land preparation and destroy rodents (Fig. 18).

The FAO, under the activities of the European Union Neighbourhood Program for Agriculture and Rural Development (3rd phase), started promoting CA in Dedoplistskaro in 2019 and 2020. The first trials/demos were established with spring wheat (not common in Georgia), oats and barley. The spring of 2020 was particularly dry and the wheat yield of 1.0 t ha^{-1} or equal to the yield harvested from the traditionally ploughed plots which was a good start (Fig. 19).

Farmers from different municipalities were invited to field days on promoting CA. By autumn 2020, 41 farmers from five different municipalities tried no-till in their farms in total of 190 ha. Initial results give hope that CA will help farmers in Georgia alleviate the impact of climate change. More farmers from the same five municipalities and other two other municipalities have expressed interest in trying out the no-till practice. In parallel, the FAO project is also promoting no-tillage in a maize production system in dry parts of Lagodekhi municipality in the east. The adoption of CA has started from 2 ha of trial in 2020 to 41 ha in the spring of 2021.

4 Conservation Agriculture in Central Asia

Agriculture in Central Asia rapidly developed after the 1960s with the opening of virgin lands in North Kazakhstan and irrigated land in the south, for cotton cultivation. In parallel, the infrastructure, field machinery and services were

Figure 18 First trials on no-till planting by the FAO in Dedoplistskaro (Photo: FAO).

Figure 19 Testing, demonstration and promotion of CA under the FAO project (Photo: D. Martsvaladze).

strengthened. During 1960–1990 the area of irrigated land in all Central Asia countries expanded by three times. As the result, agricultural production significantly increased, and livelihood of people improved. However, there were severe catastrophic consequences not anticipated while cultivating extensive areas based on an unsustainable approach for agricultural development. Thousands of hectares of newly cultivated virgin land were severely eroded. Dust and sand bowls were covering cities and settlements. Also, wind transported the intensively tilled topsoil to longer distances, from Kazakhstan to Mongolia and China. Intensive use of water for cotton cultivation significantly reduced water flow to the Aral Sea that slowly began to dry out: in 1960 the sea area was 60 900 km^2 and now it is 8600 km^2; salt storms became more frequent, and a huge gene pool of plants and animals was lost.

In response to these challenges, the soil protection technologies to replace the use of mouldboard ploughs and to promote the use of tine rippers were developed, tested and adopted in dryland farming systems. The governments of the countries also implemented efforts to diversify cropping systems. In the north, in the drylands, mono-cropping of wheat was diversified with pulses and other pasture plant species, including grasses. The area under cotton was significantly reduced in the south to promote cereals, vegetables and orchards. Moreover, water-saving technologies were adopted on many farms. In these initiatives, several development agencies and research centres, in close cooperation with the governments of the countries in Central Asia, tested and promoted CA systems and practices in the farmers' fields through participatory programmes.

The FAO Sub-Regional Office for Central Asia (FAO-SEC) in cooperation with CIMMYT, ICARDA and the national counterparts in 2012 conducted a study on the status of CA in Central Asia. The study outputs were presented and discussed in the Regional Workshop on 'Save and Grow': Promotion of Conservation Agriculture and Modern Plant Protection Methods held on 4-6 December 2012 in Antalya, Turkey. Considering the study outputs and the contributions made by the workshop participants, the study report and the policy recommendations for promotion of CA were formulated and published (FAO, 2012). This document served as a guideline for further promotion of CA in Central Asia through demonstration, adoption, promotion and developing the national strategies.

The FAO and the Ministry of Agriculture of the Republic of Uzbekistan organized the International conference on 'Strategies for the promotion of Conservation Agriculture in Central Asia' during 5–7 September 2018 in Tashkent, Uzbekistan (FAO, 2020). The Conference: (a) reviewed the status of CA in the countries of the region, (b) analysed and identified opportunities for and constraints to adoption and spread of CA, including the diversification of cropping systems and improved crop management practices; and (c) defined the opportunities and strategies for upscaling the adoption of CA at the regional level. Over 170 participants attended the International Conference, including about 60 international experts from Austria, Azerbaijan, Brazil, China, Georgia, Iran, Kazakhstan, Kyrgyzstan, Moldova, Portugal, Russia, Spain, Tajikistan, Turkey and the United Kingdom. Representatives of international organizations, ministries, universities, research institutes and private companies joined the International Conference, and more than 40 farmers from Central Asia attended the event. About 500 students of the Tashkent Institute of Irrigation and Agricultural Mechanization Engineers joined the opening session, learning the key principles of CA and the deliverables it offers to Central Asia.

Positive experiences of more than ten countries were presented at the Conference, constituting the living proof that the adaptation and adoption of CA by farmers in a diverse range of local agro-ecologies and conditions is not only feasible but also was occurring. Most of the experiences presented for CA in Central Asia showed higher agricultural production at a reduced cost with CA systems compared to conventional tillage production systems. Therefore, these economic benefits are the basis for ensuring stable incomes for farmers and for sustainable production of sufficient food, biological raw materials and domestic energy for a growing population, while demonstrating significant potential as a strategy for poverty alleviation. Some of the key benefits for Central Asia presented at the Conference included the control of soil erosion and run-off, improved water management and salinity control, climate change mitigation and adaptation, improved biodiversity and better productivity and income for farmers. The mechanisation challenges posed by no-tillage practice in CA were also addressed, especially during the machinery and equipment demonstration session in the field. A common concern in the region of the integration of feed demand by livestock while implementing the permanent soil cover principle of CA was also addressed at the Conference, including during the field day. The International Conference called upon stakeholders of Central Asia countries to conceive and endorse appropriate long-term strategies to promote the adoption and spread of CA, and to further develop this sustainable agriculture system with effective support from public, private and civil sector institutions and stakeholders including research, education and service providers. Policies in favour of CA, and National Development Programs based on the current regional and international knowledge and experiences were recommended for the region (FAO, 2020). CA is one of the priority components under the current regional 'Integrated natural resources management in drought-prone and salt-affected agricultural production landscapes in Central Asia and Turkey - CACILM-2' (2017–2022) project of FAO that is supported by GEF and implemented in Kazakhstan, Kyrgyzstan, Tajikistan and Uzbekistan.

4.1 Kazakhstan

Kazakhstan is the largest country in Central Asia (and second in Eurasia, after Russia) with a total territory of 2 724 902 km². The total land area is 2 699 700 km², including 216 036.5 km² of agricultural land and 34 546.8 km² of forest (FAO Country Profiles: Kazakhstan, 2016e; FAOSTAT, 2019). The population of the country is 18 809 210. In terms of area, Kazakhstan ranks ninth in the world; in terms of cropland per capita, Kazakhstan ranks second in the world.

The agro-industrial complex is one of the important sectors of the economy, which forms the food and economic security of the country, as well as the labour and settlement potential of rural areas. Forty-three per cent of the population lives in rural areas and agriculture remains the main activity for the majority of the population. Kazakhstan is the largest grain-producing country in the world. Wheat is the main crop and cultivated on over 12 million ha, out of over 20 million ha of cropland. Kazakhstan is one of the largest exporters of grain and flour. The majority of crop fields, especially under cereals are located in the northern part of the country under rainfed conditions. Maize, rice, buckwheat, rapeseed, soybeans, oats, cotton, sugar beets, many vegetables and fruits are also cultivated under irrigation in South Kazakhstan.

Cultivation of virgin land facilitated severe soil degradations in the steppe regions of North Kazakhstan in the 1960s. Taking it into account, research work on developing soil protection and crop management systems as well as anti-erosion equipment was initiated. In this context, special crop rotation schemes with reduced cycle and fallow and use of tine rippers instead of ploughs were recommended. There were also attempts to promote no-till technology. However, the revolutionary progress in sustainable soil management was made through the joint efforts of the FAO, CIMMYT, World Bank, government of Kazakhstan, National Agricultural Research System (NARS) and farmers on adoption and promotion of CA. Initially, the Technical Cooperation (TCP) project was implemented jointly by the FAO and CIMMYT. The outputs of the TCP project were up-scaled further by the WB-funded investment project (Fileccia, 2013). Moreover, the promotion of CA-based systems has become an official state policy in agriculture in Kazakhstan and in 2008, the government started subsidising CA farmers. The incentives facilitated the further application of no-till and diversification of wheat-based cropping systems. As the result, the area under CA-based practices has increased from 0 ha in 2002 to more than 3 million ha in 2020. With this progress, Kazakhstan was ranked 7th in the world in terms of CA area in 2010. The adoption of CA systems increased income and food security and the yield gain amounted to satisfying the food requirement of additional about 5 million people annually. The output of this progress also positively contributed to agriculture, responding to climate change mitigation: currently Kazakhstan contributes to the annual carbon sequestration in the soil of about 1.8 million tonnes of CO_2 (Fig. 20).

Farmers recognised the positive impact of CA to mitigate drought and climate change. One of the companies (cooperatives) successfully practicing CA is 'August' that has several farms in Kazakhstan and Russia. Before introducing no-till, the company studied the experiences of farmers in Brazil, Argentina, Canada, who were successfully implementing CA for decades. In 2020, there was a severe drought in Zhambyl rayon (Northern Kazakhstan Oblast) that seriously affected crops. However, the wheat fields of farm 'Ak-Zher 2010' (part of 'August') that had adopted CA for the last 4 years were not much affected: the average grain yield was 2.7 t ha^{-1} that is about twice as much as in neighbouring farms that used regular ploughing. Based on this result, farmers conclude that CA is a universal system that works for any region where the conventional farming system is practiced. They also became aware that mulch covering the soil surface plays an important role in protecting the crops against hot conditions: it prevents soil from heating up by reflecting more sunshine, slows down water loss through evaporation and preserves soil moisture.

Figure 20 Conservation Agriculture for rainfed areas of Northern Kazakhstan. No-tillage, direct seeded spring wheat (Photo: M. Karabayev).

Despite the difficulties due to lack of experience and expertise, there are many opportunities to accelerate the uptake and spread of CA systems in Kazakhstan. Support from the government on promotion of CA is still available. For instance, the state supports testing and promoting CA and other related technologies through the Research Institute of Crop Husbandry and Farming, organization of Field Days, workshops, conferences and study visits to the countries with advanced agriculture production.

Farmers practicing CA in Kazakhstan also realized that yield increases gradually after the introduction of the CA system. However, there is a need for patience to accept little yield advantage in the first 2–3 years for the sake of significant increase after that. For instance, the yield of spring wheat in 'Ak-Zher 2010' farm in 2018 was 2.3 t ha^{-1} (with 336 mm of precipitation from March to August), and a year later it was more than 2.6 t ha^{-1} (200 mm during a similar period). And, despite the drought in the summer of 2020 (96 mm in 3 months against 144 mm in the same period of 2019), the yield again increased.

CA system in Kazakhstan was mainly introduced in the wheat-based cropping system in the drylands. However, FAO, ICARDA, CIMMYT demonstrated the successful application of CA under irrigation in the South of Kazakhstan. Currently, FAO is implementing the regional CACILM-2 project funded by GEF on sustainable soil management that has a strong component in demonstrating and improving national capabilities on CA (Fig. 21).

Figure 21 Conservation Agriculture for irrigated areas of Southern Kazakhstan. No-tillage, direct seeded winter wheat on permanent beds (Photo: M. Karabayev).

Despite good progress in adoption of CA in Kazakhstan, there are still a number of issues not allowing faster uptake of CA in the farmers' fields, including problems related to weed control, crop rotation, lack of suitable field equipment, lack of information on crop management under the CA system, lack of trained experts, etc. There is a need for further research on weed control that should allow the use of different options for weed management depending on the conditions. Diversification of the cropping and production system in Kazakhstan and the whole of Central Asia remains a challenge to the successful promotion of CA. Therefore, the improved crop rotations should be adaptable, affordable, functional and economically feasible. There is also a need for developing and marketing appropriate field equipment suitable for small-scale and mid-size farms. Another bottleneck for adoption of CA is lack of information and knowledge on CA. The agricultural and extension services still provide recommendations based on traditional crop management technologies, particularly on irrigation and the fertiliser rate and timing. Therefore, there are needs for modern research and evidence-based recommendations on balanced and optimal application of water and nutrients under CA systems. However, the state support for agricultural research and development is declining, which has been aggravated even more by the COVID-19 pandemic. Moreover, the technical capacity of agricultural and extension experts, researchers and farmers on CA needs to be strengthened through trainings, exchange visits, developing and publishing manuals, guidelines, etc.

During the last 20 years, the international research centres (particularly CIMMYT and ICARDA) and donors (WB, USAID, etc.) have very actively supported the adoption and promotion of CA in Kazakhstan. However, recently this support is decreasing sharply. In this regard, the government should focus on increasing funding and extending the cooperation with the international research centres and the donors to support further development of sustainable food production in the country. Only continued support from all interested parties (governments, NARS, CGIAR (Consultative Group for International Agricultural Research), donors, stakeholders, farmers associations, etc.) can and will accelerate progress towards sustainable agriculture and nature resources management in Kazakhstan and the whole of Central Asia.

Further acceleration of the uptake and spread of CA in the region requires continuous practice of the systems. Farmers, extension and agricultural specialists, researchers and policymakers should have patience to see the fuller manifestation of benefits of CA after several years. This is because, changing the mind-set and transforming from the traditional tillage approach to CA-based systems requires time for better understanding and for the soil to regenerate its health and function. It is obvious, that adopting a new approach and new technologies is always challenging. Experiences of farmers in Kazakhstan demonstrate that in the first years of introducing CA, the cost for herbicides may go up temporarily. Additional investment would be needed to procure suitable seeders, sprayers and harvesters. There will be difficulties with handling crop residue and direct planting. In most cases, it has been easier to adopt CA in larger farms compare to smaller ones that require more attention and different equipment to suit the smaller scale of operation (Fig. 22).

CA is a complex technology and calls for changes in a number of practical components of the existing traditional systems of agriculture. It is necessary to change two basic approaches: the approach to soil tillage and the approach of linear knowledge flow.

Many agricultural research and extension systems are based on a linear model of knowledge flow, with new knowledge and information being developed in research organizations which is passed on to agricultural extension agents, who in turn pass it on to farmers. While this model may be applied to simple technology, it does not always effectively work with a complex set of system-based practices, especially when research institutes do not have the capacity to develop functional system packages of multiple

Figure 22 Field days, seminars, trainings and public awareness - the main approach to promote and spread CA (Photo: M. Karabayev).

technological components for all farmer situations. Innovative systems based on complex technologies are needed for adaptation, system development process and promotion. Innovative systems (platforms) are based on networks of multiple agents, including farmers-innovators and decision-makers, all utilising their own knowledge, external information and policy support to help overcome problems and develop functional systems for local farming conditions and farmer circumstances. Successful adaptation and promotion of CA in Kazakhstan can be considered as an example of this approach.

4.2 Kyrgyzstan

Kyrgyzstan is a mountainous country with a total area of 199 951 km^2. The total agricultural land area is 10.54 million ha and the forest area is 1.31 million ha (FAO Country Profiles: Kyrgyzstan, 2016f; FAOSTAT, 2019). Agriculture in Kyrgyzstan is one of the leading sectors of the national economy: the share of agriculture to GDP is about 24%. The total population of Kyrgyzstan is 6 586 600 (2020). More than 65% of people live in rural areas, and 14% are employed in agriculture. Therefore, agriculture in Kyrgyzstan has not only economic but also significant social and political importance.

Agricultural reforms for the transition from the centrally planned economy to market relations in Kyrgyzstan started in the 1990s. At the initial stage, 576 collective and state farms were restructured into small farms to develop private property in agriculture. Despite the predominant number of private farms, today different forms of agricultural entities are functioning in the agriculture of Kyrgyzstan. Currently, there are 453 000 agricultural entities, including 342 100 (75%) of private (farmers') farms. Besides, there are 726 600 household plots that are contributing to agriculture and food production.

The total area of pastures and hayfields is 9.2 million ha. There are 1.29 million ha of cropland. Irrigated land occupies 1.035 million ha and provides more than 90% of agricultural production. Pasture resources and natural hayfields occupy more than 40% of the total territory of the country or about 90% of all agricultural land. The pastures are the exclusive state property, and the local governments are responsible for control of pasture management.

The main crops grown by farmers are cereals (wheat, barley, maize, rice), fodder crops (sainfoin, alfalfa), industrial crops (cotton, sunflower), melons (watermelons, melons) and vegetables (potato, tomatoes, onions, etc.). On average, annual cereals occupy 56% of cropland, fodder crops - 23%, vegetables - 11%, industrial crops - 10%. On average, there is 0.19 ha of cropland per capita in Kyrgyzstan. The crop yield remains low due to improper crop management, lack of inputs and soil degradation. For instance, in 2019, the average yield of cereals was 3.2 t ha^{-1}, wheat - 2.37 t ha^{-1}, barley - 2.27 t ha^{-1} and maize 66 t ha^{-1}. However, food demand is rising in parallel to population increase and improving living standards. Therefore, there are several challenges to produce more food with efficient utilisation of the available resources.

Lands in Kyrgyzstan are severely degraded both under rainfed and irrigated farming systems and degradation has significantly accelerated under the influence of climate change. About 90% of agricultural land is recognized as degraded and prone to desertification processes. Due to the long-term absence of amelioration measures, cleaning of drainage systems and raising water table the area of secondary soil salinization sharply increased (FAO, 2016a). The area of such land is about 5 million ha or 45.7% of all agricultural land, including 700 000 ha of cropland. There are also sandy and rocky soils that are not suitable for agriculture.

More than half of the cropland is subject to water and wind erosion. Particularly, flash flood irrigation is still practiced by farmers in the northern region which leads to significant loss of fertile soil. Moreover, this technology also facilitates water pollution and loss. There are also huge water losses due to the unsatisfactory condition of the irrigation systems: about 25% of the water is lost during transportation. The government recognizes that the degradation of agricultural land poses a significant threat to food security; therefore, they are addressed in the key strategic documents, including the 'National Strategy for Sustainable Development for 2013-2017' and 'National Development Strategy of the Kyrgyz Republic for 2018–2040'. Kyrgyzstan is a member of the Eurasian Soil Partnership and consolidates the efforts towards conservation of soil as an important and limited natural resource (Figs 23 and 24).

Mono-cropping of cereals in drylands and intensive furrow irrigation also facilitate soil erosion and degradation. The risks intensify due to climate change

Figure 23 Example of flood irrigation that facilitates soil erosion and degradation followed by water pollution and loss (Photo: A. Asanaliev).

Figure 24 Soil preparation for bed planting (Photo: S. Corsi).

followed by continued climatic stresses, especially drought followed by water stress. Therefore, the introduction and promotion of CA could significantly contribute not only to sustainable food production but also to the conservation of natural resources and saving inputs, particularly protection of soil and efficient utilization of water for irrigation.

Several years back, farmers began to realize the importance of soil conservation, the introduction of innovative technologies and compliance with agricultural practices. Soil scientists and agronomists began to raise questions about soil degradation and preservation of fertility. Local NGOs also began to take an active part in the dissemination of knowledge with the support of international organizations, as the main development partners.

Thus, there were several attempts on testing, demonstrating and promoting resource-saving technologies in the farmers' fields in Kyrgyzstan. At the initial stage, CIMMYT and GIZ demonstrated the advantages of permanent no-till bed planting to improve water use efficiency for wheat production in the Chuy Valley in 2003. A series of research and development experiments on bed planting was carried out at the experimental stations and farm level (Pojarskiy, 2008). Further, Sida (Swedish International Cooperation Development Agency) funded a project that supported the successful experiments on bed planting systems that served as a clear example of the need to transform from flood ('wild') irrigation to no-till bed planting with and furrow irrigation. The outputs are also very useful for promoting CA on permanent raised-bed systems. There were also experiments on covering the seedbed with plastic but that was not supported by farmers due to the cost and negative impact of plastics on the environment.

Like in other countries, companies producing field machine were also supporting the promotion of CA in Kyrgyzstan. For instance, the JSC 'Eurasia Group' as the official representative of John Deere is marketing different field equipment, including no-till drills and tools for minimum soil tillage during the last 10 years (Fig. 25).

Figure 25 Minimum-till is considered as the first step for transition to no-till based CA system, although it makes the soil vulnerable to wind erosion (Photo: AgroLead).

The development partners and the donors are actively supporting the government's initiatives on the sustainable development of agricultural production. In this framework, several projects with components on CA were successfully implemented through FAO, GEF, GIZ, GCF (Green Climate Fund), UNDP (United Nations Development Programme) , Aga Khan Foundation, ICARDA, etc. Civil society organizations were very active in supporting these projects in testing, demonstrating and promoting CA. For instance, NGOs 'AgroLead' and 'Chuy-Tals' RAS were key implementing partners in adopting and promoting CA in the farmers' fields in the Chuy Valley, in Issyk-Kul and South Kyrgyzstan.

The FAO project also supported the government in the formulation of a national strategy on further adoption and promotion of CA. One of the farmers pioneering CA in Chui valley was Tasin Izatov from Issyk-Ata Rayon who has 60 ha of cropland and over 50 head of cattle. Farmer Abdrakunov Dostuk from Jety-Oguz Rayon promoted CA on 60 ha for wheat and barley seed production, where seed crops were rotated with perennial grass pasture.

Several season-long trainings for the agricultural and extension specialists, researchers and farmers were conducted, and the training and extension materials were developed and published in the Russian and Kyrgyz languages. Establishment and running of the Farmer Field Schools (FFS) have been tried for testing and promoting CA through training a big number of male and female farmers. During 2017–2018, under the FAO project funded by GEF 'Sustainable management of mountainous forest and land resources under climate change conditions', 176 FFS were established for training farmers on sustainable crop and soil management. In this project, CA was a key practical topic. In total 2493 farmers, including 573 women participated in the trainings. Moreover, 2276 farmers, including 611 women received on-the-job training on CA (Figs 26–28).

Figure 26 FFS trainings and consultations on no-till planting (Photo: C. Alibakieva).

Figure 27 Abdrakunov Dostuk, Farmer (Photo: O. Mambetov).

Figure 28 Farmers, extension specialists and students were trained on CA (Photo: AgroLead).

During 2019-2020 'AgroLead' under the EU-funded project trained over 250 farmers and promoted CA on 700 ha drylands in Jalalabad oblast. Moreover, the FAO project provided support in developing the curricula for the MSc and BSc students in agronomy in the Kyrgyz National Agricultural University. The 60 h curricula with two credits aimed to master the elements of the CA system, including no-till, suitable equipment, crop rotation and diversification of wheat-based cropping systems, plant residues management, mulching, cover crops, etc.

The technical support and development projects paved the way for the introduction and adoption of CA by farmers in Kyrgyzstan. As the result, today the area under CA in Kyrgyzstan has grown to 60 000 ha. Farmers practicing CA realized that it has the potential to sustainably increasing productivity, reduce production costs, restore degraded land, minimize soil erosion and improve ecosystem services. They are assured that CA is an ecosystem approach that not only can improve economic and ecological crop production and land use, but also is a key approach to sustainable intensification and facilitates to produce 'more yield with fewer resources'. However, the CA has not yet become a widespread production approach among farmers. There is a need to carry out more research and extension work, train qualified CA specialists, develop evidence-based recommendations and guides, etc. The government of Kyrgyzstan should also consider promoting CA through the adoption and implementation of relevant policies and strategies, providing incentives to the importers of CA field equipment and to CA farmers.

4.3 Tajikistan

Tajikistan is a mountainous and agricultural country with a total territory of 143 100 km^2. Mountains and hills occupy about 93% of the country. The total area of agricultural land is 4.72 million ha, including 720 200 ha of cropland. Almost 423 800 ha is under forest (FAO Country Profiles: Tajikistan, 2016i; FAOSTAT, 2019). Overall, the agricultural sector contributes 19.8% to the national GDP and 61% of the total employment. Tajikistan is the leading country for dependency on remittances (40% of GDP, 2017) in the globe.

The total population of Tajikistan is 9.54 million (2020), of which more than 60% is below the age of 16 years. Approximately 25% of the population is urban and >75% is rural. The growth rate of the population is high, with 2.1% of the rural population increasing at a rate higher than the urban population.

Irrigated land is used intensively, producing two harvests of wheat and vegetables a year, 5-8 cuts of alfalfa, and cover crops are grown and intercropping is practiced. The development of agriculture, and primary cotton production in Tajikistan, is closely connected with extensive use of pesticides for controlling insect pests, diseases and weeds.

The climate of Tajikistan is continental with hot summers, mild winters, from semi-arid to cold climate in the Pamir mountains. In valleys and on the river floodplains, the climate is hot and dry in the summer and mild and warm with some precipitation in the winter. About 95% of crop production is from irrigated land. Rainfed agriculture in the arid area is not guaranteed because of low precipitation. Rainfed land is used to grow mainly cereals and for pastures. Cereal yields on irrigated areas are two to three times higher than on rainfed land. Temperature conditions allow for the cultivation of a wide variety of crops, such as cereals, legumes, vegetables, orchards (apricots, grapes, apples, pears, pomegranates, figs, walnuts, pistachio, peaches, cherries, plums, cucurbitaceous, citrus), watermelons, melons, pumpkins, non-food crops (cotton, including fine fiber, tobacco, geranium) and medicinal plants.

Cotton and aluminium are the main export products. Marketing of agricultural products helps to raise 30–35% of the national income from the export and 35–38% from taxes and levies. Thus, agriculture has great potential for stable economic growth, poverty alleviation and food security. After independence, there were significant changes in the cropping patterns and production of agricultural products. The government revised agricultural policy and oriented it toward meeting people's demands for food products and reinforcing the economy. The government's concern was to increase cotton production that brings hard currency to the national economy and not to reduce the production of food products. The private sector, including rural and urban households, is firstly concerned with increasing food production which brings income. Due to that, only grain and potato production increased per capita, 185.3% and 207.0% respectively, but the production of other food products per capita sharply declined.

Unsustainable agricultural practices and climate change followed by melting glaciers, degradation of land, droughts and sand and dust storms significantly affected agricultural production in Tajikistan. Due to these complex factors, out of the 14 000 glaciers of Tajikistan, which are of vital importance for the entire region, more than 1000 glaciers have disappeared over the past 30 years. The most concerning land degradation issue in Tajikistan is soil erosion. There is an issue with land tenure, including lack of control on soil management and quality during the rental period. Farmers can rent plots from other farmers and usually treat the soil carelessly because their main objective is to obtain the highest yield in the short run without paying much attention to soil quality aspects. The Country Environmental Assessment (2008) done by the World Bank, reports that – on State Land Management Committee data – over 82% of all lands are eroded. However, the exact extent and degree of land degradation in Tajikistan is unknown. Allegedly, most of the lands are either medium, strongly or very strongly eroded, while only a small portion is considered weakly eroded or not eroded. Over time, salinity and waterlogging seriously affected an extensive area of irrigated lands due to poor infrastructure and inadequate water management practices.

Agricultural development initiatives and associated unsustainable cropping practices play a significant role in the erosive processes of sloping lands. The erosive processes have been especially active in the foothill regions. The hilly parts of cotton fields suffer from a deficient supply of water, while in those places where there are depressions, crops suffer from an oversupply of water. Both situations negatively affect the productivity of cotton fields as well as that of other crops. Under rainfed agriculture, slopes have been widely cultivated, ignoring the observance of anti-erosion measures. Therefore, there is a need for the adoption and promotion of CA to stop erosion, improve soil health and restore the landscapes.

Taking the above into account, development partners and donors have been undertaking actions to assist the government in piloting CA under rainfed and irrigated farming systems. One of the first initiatives was taken by CIMMYT and (Agency for Technical Cooperation and Development (ACTED), International NGO with Headquarters in Paris, France) together with the Tajik Soil Science Institute involving the demonstration of no-till seeding of wheat into standing cotton in Vakhsh Valley (South Tajikistan) in 2002 (Fig. 29).

In 2002–2004 CIMMYT and GIZ supported the promotion of no-till bed planting in cotton-wheat-based cropping system, ensuring that the permanent beds remain for further no-till planting in CA systems combined with diversified crop rotations and covering the soil with the crop residue and stubble (Rashidov, 2009). In 2004–2006, the WB took the initiative to support a project with a CA component that was implemented by the FAO and the Tajik Soil Science Institute. The project supported testing no-till drills, demonstration of CA practices and strengthening the national capabilities on sustainable soil management (Fig. 30).

The local NGOs as Mountain Society Development Support Programme (MSDSP), supported by Agha Khan Foundation), Cooperative of Agricultural Consultants 'Sarob', 'Neksigol' and ATAC Agriculture Training and Advisory

Figure 29 Demonstration of no-till drill and jab-planter (Photo: FAO, 2006).

Figure 30 No-till planting for rehabilitation of pasture under IFAD funded project (Photo: H. Muminjanov).

Centre (ATAC) have been actively involved in the implementation of the CA projects. 'Sarob' provides services to no-till farmers and MSDSP has lately been promoting CA in high altitude areas through provision of two-wheeled tractors equipped with no-till seeders.

Through these projects, no-till drills have been procured and delivered, the demonstration plots established, Training for Trainers (ToT) organized, Field Days and a season-long training have been conducted for farmers and rural households. Moreover, training manuals and guidelines for agricultural and extension specialists and farmers have been developed and published. The curricula and textbooks on CA for the students at Tajik Agricultural University have also been developed (FAO, 2015).

The CA approach has proven validity in Tajikistan's rainfed area. The principal issue for scaling is with the availability of appropriate seeders and with the need to capture the economies of scale for establishing CA systems. Sustainability in rainfed areas increases when access to tractors and seeders is provided through service providers (Shavkat Juraev, personal communication). Problematic areas relate to acquiring the skills for weed, residues and crop rotation management in CA systems. Currently, it is estimated that the CA approach has been adopted over about 6000 ha. 'Sarob' is actively providing advice and technical support to the CA farmers. It also serves as a service provider for no-till planting and marketing relevant field equipment (Fig. 31).

The Regional Workshop on Sustainable Crop Production and Conservation Agriculture was conducted in Dushanbe, Tajikistan during 15–17 October 2015 with the participation of over 100 researchers, agricultural and extension specialists and farmers. Based on the resolution of the regional workshop, a Regional Alliance on CA was established. The FAO assisted in the development of a website in Russian (www.caincentralasia.com) that serves as a platform for the Regional Alliance to share knowledge and information. Many training materials, books, conference proceedings and presentations are posted on the website. Currently, 'Sarob' is hosting and maintaining the website.

Figure 31 No-till planting of winter wheat after harvesting maize (a) and planting pulses right after harvesting winter wheat (b) (Photo: H. Muminjanov).

Similar to other countries in the region, there are some bottlenecks that would need to be overcome to achieve wider adoption and spread of CA in Tajikistan. The most important issue is changing the mind-set of policy and decision-makers, researchers, agricultural and extension specialists, farmers and the population in general. Further support is required for conducting research and education work, advocacy and raising awareness through mass media on the benefits of CA. There is a need for strengthening the extension system, in general, to promote sustainable practices, including CA, among the farmers. Lack of relevant CA field equipment could be solved through importation and assembling locally, and eventually local production. In this regard, FAO can facilitate the initiatives coming from the entrepreneurs and private investors to link them with the lead CA machinery producing companies.

Results of adoption of CA in terms of soil fertility improvement, yield increases, and costs of production reduction are effective. Therefore, FAO supported the Ministry of Agriculture in the formulation of the National Strategy for the promotion of CA in Tajikistan that requires adoption and implementation. Moreover, related policies and actions should be harmonized for the successful promotion of CA in the farmers' fields.

For successful introduction and spread of CA in Tajikistan, it is necessary to improve the legal and normative framework on regulation of the crop production sector. To improve the professional level of CA specialists it is necessary to address the issue of their practical training in countries that have successfully adopted CA systems.

There is a need to develop curricula, special courses or separate training modules on introduction of CA for students of higher education institutions and secondary special institutions of biological, geographic, environmental, agricultural and economic profiles. State support is needed for strengthening the capacity of national extension services on the application of innovation based on scientific and research results in agriculture, and specific sources of funding for this support should be identified.

4.4 Turkmenistan

Turkmenistan is a country with a total area of 48 810 km^2. The famous Kara Kum desert covers 80% of the country and about 12% of the area is covered by water and non-soil formations (talus, rocks and precipices). The total population is over 5.85 million, of which 50% is rural. Agriculture is a key sector of the national economy. Although its contribution to the GDP is around 12.7%, over 30% of the economically active population is involved in agriculture (FAO Country Profiles: Turkmenistan, 2016j; FAOSTAT, 2019).

The area under the agricultural land is 33 million ha or 72% of the total area. However, only 4% of the total land area, or 1 910 000 ha, is cultivated. The total area under annual crops is 1 850 000 ha and 60 000 ha is occupied by perennial orchards (vineyards, pistachio nuts, figs and olives). Two of the largest agricultural crops are cotton, most of which is produced for export, and wheat, which is domestically consumed.

The government initiated a land reform in 1994 with further reorganization of large-scale farms (kolkhozes and sovkhozes). In this reform, land was distributed to farmers under a lease contract of 99 years. The government's priority is to ensure food self-sufficiency by focusing on key food products. Wheat and rice are the main traditional crops cultivated to ensure food security. These crops are closely correlated to large-scale irrigation schemes. Individual rural entrepreneurs produce 99% of potatoes, 69% of watermelons, 24% of grape production, 82.5% of meat, 96% of milk and 93% of eggs. The population's food demands are met by large-scale irrigation schemes, small-scale production and importation.

Climate in Turkmenistan is distinctly continental and arid. Average annual precipitation is about 191 mm, ranging from less than 80 mm in the northeast to 300 mm in the Kopetdag mountains in the southwest. The summer is very hot and dry. In July, average temperatures exceed 30°C throughout the country. Therefore, agriculture depends entirely on irrigation. However, due to unsustainable management of natural resources for agricultural production, 37% of land area suffers from salinization and the forest cover has decreased by 32%.

Government is supporting research and development activities on soil and water conservation. There were several research programmes on testing, demonstration and promotion of application of liquid mineral fertilizer and deep tine ripper to establish mole or aeration drainage to improve water, air and temperature balance on saline and heavy soils for increasing soil fertility and crop productivity. Financial support is also provided for land rehabilitation and planting forests and green zones around the cities (Danatarov, 2011).

The first attempt on testing and adoption of CA in Turkmenistan was made by the FAO in 2011-2013 under the project 'Conservation Agriculture in Irrigated Regions of Azerbaijan, Kazakhstan, Turkmenistan and Uzbekistan'. However, the government did not sign the project document and no activities were carried out in Turkmenistan. Currently, the CACILM-2 project funded by GEF and implemented by the FAO in cooperation with the Government of Turkmenistan is aiming to promote CA. The project will introduce no-till planters to demonstrate and train researchers, farmers, agricultural and extension specialists on CA systems.

4.5 Uzbekistan

The total area of Uzbekistan is 448 924 km² and 34 036 800 people live in the country. The agricultural sector has generated about 23% of Uzbekistan's GDP on average for the past 3 years. It is the largest contributor among all sectors of the economy. Uzbekistan's agriculture is the main source of income and livelihood for nearly 60% of people living in rural areas. The total area of agricultural land is almost 26 million ha and irrigated cropland accounts for 4.4 million ha (FAO Country Profiles: Uzbekistan, 2016l; FAOSTAT, 2019).

The climate of the country is continental; the northern part belongs to the temperate zone, and the southern part, to the dry subtropics. The arid (deserts) climate covers over 60% of the territory. Therefore, agriculture is vulnerable to frequent droughts and is highly dependent on water availability. However, irrigated area occupies only 15% of the agricultural, but provides more than 95% of all agricultural production. Erosion is the main issue for most of the provinces. Thus, the eroded area in Uzbekistan is 1.772 million ha, or 40% of all the cropland. Of this, 712 900 ha of land is subjected to water erosion, 300 000 ha to wind erosion and 50 000 ha to gully erosion.

Uzbekistan was the major cotton-growing region in the Soviet Union, accounting for 61% of total production. The country was the 4th largest producer of cotton in the world and the 3rd largest cotton exporter. However, the rapid expansion of the area for growing cotton and inefficient irrigation systems triggered the drying out of the Aral Sea followed by severe ecological catastrophes. After the disintegration of the Soviet Union, the government policy moved towards the expansion of the wheat area to meet local food needs. As a result, the wheat area expanded from 230 000 ha in 1991 to 1.35 million ha in 2019 with a similar decline in area under cotton. Wheat production increased substantially, from 386 000 tonnes in 1991 to over 6.9 million tonnes in 2019. Today cotton and wheat are the country's main crops occupying over 80% of irrigated land. However, the government has adopted a long-term programme of crop diversification encouraging farmers to diversify from cotton production towards higher-value crops such as fruits and vegetables.

Cotton, with vegetables and fruits, is the country's principal exports. The main grain crops of Uzbekistan are wheat, barley, corn and rice, which are grown in intensively irrigated oases. Farmers also grow sesame, onion, flax and tobacco. Fresh and dried fruits are consumed domestically and are also exported.

Irrigated agriculture consumes 88% of all water resources. Furrow irrigation is practiced on 99.9% of the area. Two major land problems in the country are salinity and waterlogging. Salinity generally increases towards the Aral Sea in the Amudarya and Syrdarya basins. Further, in Karakalpakistan, about 95% of the land is either highly or very highly saline. A reduction in the quantity and quality of water delivered to farms and poor maintenance of drainage canals are considered the main causes of increased salinisation. As a result, despite a slight increase in the yields of main crops (e.g. wheat - 5.8-5.9 t ha^{-1}, cotton 2.5-2.7 t ha^{-1}, vegetables 27-30 t ha^{-1}), they remain relatively low compared to countries with developed irrigated agriculture. One of the major reasons for lower yield is the inefficient use of water. Consequently, water issues are a priority area for the government and most donor agencies supporting agricultural development and natural resource management in Uzbekistan.

After achieving independence in 1991 and transition to a market economy, large inefficient 'shirkat' (company) farms were transformed into many small private farms. The lack of proper management of irrigation systems and drainage and the dismantling of former state-operated large-scale irrigation systems have led to resource degradation and inefficiencies in water use and distribution. Appropriate water management and field irrigation practices for maximising water use efficiency using different sources of water (rainfall, run-off and alternative irrigation water sources) on small farm holdings, together with better management of inputs such as nutrients and appropriate soil conservation practices will result in improved productivity (FAO, 2016). Therefore, high input use technologies will not be suitable for these small farms and there is a need to look for appropriate and realistic approaches to enhance cropping intensity and achieve diversification in agriculture. Moreover, due to the restructuring of large farms, today there are almost 92 554 private farms, and improving the technical capacity of farmers on the sustainable crop, water and land management remains a top national priority (MoA, Uzbekistan, 2019).

Therefore, CA is considered by several development partners and government as a proven approach to improving soil health and structure, and control salinity and promote its uptake through the development and technical support projects. Consequently, CA will contribute to the development of agriculture in Uzbekistan that has several constraints associated with it such as limited potential and diversity in cropping, use of intensive tillage methods, poor soil fertility management t, inefficient on-farm water management and low product quality. Thus, most donor-funded projects with CA as a component have demonstrated the possibility of desired changes in agricultural practices that can improve agricultural production in the irrigated areas of Uzbekistan (Fig. 32).

One of the first projects implemented by FAO to adopt and promote CA was the TCP/UZB/3001 'Enhanced Productivity of Cotton-Wheat Systems through the Adoption of Conservation Agriculture Practices' (2004–2006). The overall objective of the project was to change inappropriate agricultural practices that lead to environmental degradation toward producing food more sustainably. There was a second project TCP/UZB/2903 (A) 'Sustainable Agriculture Practices in the Drought-affected Region of Karakalpakstan' (2004-2007) implemented in cooperation with ICARDA. The overall objective of the project was to demonstrate alternative, profitable and more sustainable forms of agricultural production methods such as efficient water and soil conservation practices and CA for small farmers in Karakalpakstan, where water is very scarce, valuable and a diminishing resource.

The technical assistance aimed at increasing the productivity of cotton and wheat by diversifying crop rotation through the introduction of CA. These projects facilitated the promotion of no-till planting of wheat and cotton in Uzbekistan. It also supported the farmers' initiatives on planting wheat into standing cotton crop. This was because there were delays with planting winter

Figure 32 Demonstration of no-till drill and diversification of cropping systems (Photo: FAO).

wheat due to extended harvesting of cotton until later autumn - early winter. After sowing of wheat, the cotton stalks might be removed manually from the field later on. As a result, the annual area under the direct planting of winter wheat into standing cotton by 2008 reached 600 000 ha (FAO, 2012). Currently, the area under CA is some 120 000 ha. Now with well-proven research findings, timely planting of winter wheat using no-till seeders in standing cotton has shown promising results (FAO, 2012). In Uzbekistan, no-till cultivation has been tried in the case of wheat, sunflower, mung bean, sesame and sorghum and the results had been quite encouraging. The lodging of sunflower was much less under no-till planting as compared to the traditional planting, which is an important factor for higher yield. Similarly, raised-bed planting of winter wheat provides higher yields than conventional planting and also saves seed (about 50%) and water (about 30%). Planting sorghum on beds also saves water up to 32%, which is an important factor in Uzbekistan.

Following that, several collaborative research and development projects have been implemented to promote CA in Uzbekistan. The projects are demonstrating appropriate management techniques for rehabilitation and improvement of salt-affected and gypsiferous irrigated lands to support food security in the country. Some of the studies and guidelines produced by these projects serve as useful reference materials for others. No-till and raised-bed planting technologies tested in Karakalpakstan and Tashkent provinces proved technically and economically suitable for local conditions; and can provide similar or higher crop yields while saving considerable production resources and costs including fuel, seeds and labour. Before introducing CA practices, seeding equipment was adapted and calibrated in Uzbekistan (Egamberdiev, 2007). As a first step, seedbed preparation and planting/seeding was tested in North-western Uzbekistan. The modifications in an imported Indian no-till seeder (Happy Seeder) included the introduction of a seeding-depth regulator, appropriate soil openers for planting into the hard and mulched soil, the seeding blade that is now suitable for various crops, and adoption of the row distance regulator. The modified seeder became suitable for planting cotton and wheat on permanent beds. Experience and results gathered with the introduction of CA were positive for the adoption of sustainable farming systems in Karakalpakstan and Tashkent were further disseminated to other regions of Uzbekistan in the frameworks of different projects funded by FAO, FAO-Turkey Partnership Programme, GEF, ADB, GEF-Small Grant Program in Uzbekistan, UNDP and WB.

The research activities on crop management including the CA practices are carried out by the following institutions: Kashkadarya Research Institute for Breeding and Seed Production of Cereals; Uzbek Research Institute of Cereal and Legume Crops and its branch in Gallaaral; Uzbek Research Institute of Rice and its branches; Uzbek Research Institute of Plant Industry; Samarkand

Institute of Agriculture; Tashkent State Agrarian University; Tashkent Institute of Irrigation and Agricultural Mechanization Engineers, etc. The above-mentioned research institutes and universities implement different projects on land management and water-saving technologies with the elements of CA.

Considering the needs on developing the national capabilities, the FAO projects supported the development and publishing of several manuals and guidelines on CA for extension specialists and farmers in Russian and Uzbek. Tashkent State Agricultural University and Tashkent Institute of Irrigation and Agricultural Mechanization Engineers adopted curricula on CA for MSc students and developing relevant textbooks and manuals. There are some civil society organisations supporting adoption and promotion of CA in Uzbekistan. For instance, the local NGO 'Khorezm Rural Advisory Support Service - KRASS' (http://www.krass.uz) has been implementing projects on CA and bed planting funded by the German Federal Ministry for Economic Cooperation and Development.

The government of Uzbekistan realized that the CA system contributes to sustainable management of land resources in a changing climate that is badly affected by land degradation and desertification. Therefore, a grant programme was established to support research groups addressing different aspects of CA systems and practices. Further, together with the FAO, the government has formulated the National Strategy for promotion of CA in Uzbekistan and initiated the organisation of the International Conference 'Strategies for the Promotion of Conservation Agriculture in Central Asia' held during 5–7 September 2018 in Tashkent, Uzbekistan (FAO, 2020). The draft National Strategy was presented and discussed at the special session of the International Conference. The document must be approved by government before it can be implemented. Timely policy support is critical for the adoption of CA. Farmers need adequate support for seeds or planting materials of new crops, varieties and inputs (fertilizers, pesticides, etc.) for crop diversification. Farmers also need access to credit for buying various inputs and machinery, which should be available to them at affordable low-interest rates through rural banks. Multiplication of prototypes of less costly machinery and implements in a new area such as CA requires Government support. As an example, raised-bed planters and no-till drills, flame weeders or similar tools have to be manufactured and popularized. Their import is costly and hence promoting their manufacture in Uzbekistan is critical. Similarly, policy support for use of water-saving technologies such as sprinklers and sub-surface drip-irrigation systems are urgently needed in the national interest. Promotion of these technologies would require appropriate subsidies and government intervention, besides ongoing efforts on land privatisation, crop diversification and capital investment in agriculture for linking farmers to markets.

5 Conclusion

The chapter allowed reviewing some history of agriculture development in the Eurasia region. It also helped update the status and the perspectives on the wider promotion of CA in the future. The region has very contrasting and diverse soils and climate characteristics. It is obvious that extensive use of land and water resources through application of unsustainable technologies has had severe negative impact on agricultural productivity and on environment degradation, soil erosion and biodiversity loss. Therefore, the countries are convinced that the CA approach in combination with other good agronomic practices can help to rehabilitate, maintain and improve soil health and fertility, increase crop productivity and farmers' income. Ecosystem services that can be harnessed with CA and their contribution toward combating climate change are clearly defined in relation to improving water quality and carbon sequestration.

It is obvious that the unsustainable approach in agricultural development or the concept of 'production at any cost' applied in the region since the 1950s has led to substantial loss of natural resources and degradation of the environment. The famous black soil in Russia and Ukraine is degraded. Shelterbelts are significantly reduced and no longer can protect the soil from wind erosion. The ploughing of virgin land in Siberia and North Kazakhstan facilitated severe erosion, leading to sand and dust storms. The area under irrigation in all Central Asian countries was expanded three-fold with the aim of increasing cotton production. As a result, the Aral Sea has been drying out, causing biodiversity loss and salt storms. There were national research programmes that focused on developing and adoption of soil conservation technologies. The outputs of these research activities, particularly promotion of tine rippers instead of mouldboard ploughs significantly contributed to the protection of soils in the region. However, after the countries gained independence in 1991, the financial support for agricultural research significantly shrunk and activities on soil conservation studies were reduced. Therefore, special attention needs to be paid to research programmes supporting the adoption and promotion of sustainable approaches to soil management, including CA.

Integrated production, diversified technologies and crop rotation exist in most countries and farmers are practicing them on large- and small-scale farms. In this regard, the key driving force for promoting CA is to address the impact of climate change due to which annual yield losses are increasing significantly. This is especially important in the dry agro-ecological zones.

Initially, the main challenge for promoting CA was the lack of no-till drills; however, there are enough producers and suppliers in the market now. Farmers are motivated to adapt and promote CA as it helps to cope with the effects of climate change. Only through testing and practicing for several years that the farmers are able to realize the full benefits of CA, especially on improving soil health, followed by fertility and increasing the crop yields. Rising the input prices is another driving force for adopting CA as it significantly reduces the amount of input application. Currently, the total area of CA in the region is about 10–11 million ha.

Considering the above, the national governments in the region are focusing more on the promotion of soil protection approaches. UN and donor agencies support the governments' initiatives on adoption and wider promotion of CA through implementation of different projects on demonstration, capacity development and policy formulation, particularly in Ukraine, the Caucasus and Central Asia. In this framework, FAO actively cooperates with the governments, the WB, IFAD, GEF, GIZ, UNDP and other agencies. There are civil society organizations that are actively involved in the testing, adoption and promotion of CA in the region. For instance, 'ATGF' in 'Armenia', 'Agro Information Centre' in Azerbaijan, 'AgroLead' in Kyrgyzstan, 'Sarob' in Tajikistan and 'KRASS' in Uzbekistan strongly support development projects promoting CA in farmers' fields. Visiting experts from South America, the USA, Canada, Australia, Europe and the United Kingdom have shared their knowledge and experiences that have facilitated greater adoption of CA by farmers. These joint efforts have enabled to change the mind-sets of farmers. However, many researchers and policy makers are not fully convinced about the benefits of CA.

Indeed, a systematic approach is needed for successful adoption of CA in the future as many farmers appear not to be interested in improving soil fertility and productivity through CA but aim at immediate and short-term economic benefits. Moreover, most countries have adopted laws on the protection of soil fertility, but they are rarely implemented by farmers. This is because, in all countries, renting of agricultural land is commonly practiced by farmers and there is no control on soil quality management. Therefore, in most cases the farmers who have rented the land plots are not interested in soil conservation but aim for maximum yield and profit. As a result, land degradation is common, followed by soil and water pollution due to high application rate of mineral fertilizers and pesticides.

Currently, CA is mostly applied in rainfed farming systems; however, it is also suitable for irrigated fields. Obviously, the promotion of CA may require step-by-step transformation of traditional tillage-based farming system to minimum tillage, bed planting and then to no-till. It would be difficult to transfer directly in all cases from traditional systems to no-till-based systems. The key driving force for transformation is the consequences of climate change, especially the recurrent droughts and the severe sandstorms that are taking place at greater frequencies in some countries of the region during recent years.

Further adoption and wider promotion of CA requires training of young professionals. However, CA is not included in training curricula, and there is a lack of agronomists with good knowledge about CA. Only in some Agricultural

Universities are training courses on CA included in the curricula for MSc students. Some students from the region are studying in the universities abroad such as in Japan, Canada, Germany and the United Kingdom on sustainable soil management, mechanisation and CA.

Wider promotion of CA also requires committed state support at scale that, unfortunately, is missing in most of the countries. Some countries have developed national strategies for promotion of CA; however, they are either not adopted or not implemented. Therefore, the governments should focus more on creating an enabling environment and providing incentives to CA farmers, which is being practiced in many other countries and has yielded positive results. For instance, in Russia the insurance companies refuse to compensate damages on CA fields, so regulations need to be improved. The State Programmes should support initiatives that would promote CA focused on CO_2 sequestration, improving soil health and contributing to climate change mitigation. Government should also improve the cooperation and coordination between research, universities and farmers. Moreover, it should motivate the promotion of systematic research programmes and support the development of agricultural extension systems with clear vision and understanding of CA. Only the well-trained experts should carry out training of farmers to minimise risk of failures, which can lead to loss of trust. It would help farmers to acquire a positive long-term perspective on soil fertility and agree to return a portion of the crop residue as mulch. FAO has developed a sustainable soil management programme as well as a drylands development programme that could be used by the countries of the region for strengthening their support for CA in the future.

In general, CA is moving to the farmers' fields in the region, but it will take time and patience to establish it at scale. There is a strong need for building trust and changing mind-set with practical knowledge and demonstrations that could be driven forward increasingly by farmers themselves. This calls for establishing farmer networks and associations to capture economies of scale and generate momentum and farmer innovations.

6 Where to look for further information

Further information on CA could be obtained from the FAO website on CA: http://www.fao.org/conservation-agriculture/en/ and the website of the Regional Alliance on CA (in Russian): http://www.caincentralasia.org/.

Moreover, the following publications are useful to obtain further information:

- Corsi, S. and Muminjanov, H. 2020. Conservation agriculture: training guide for extension agents and farmers in Eastern Europe and Central Asia. FAO. 134 p.

- FAO. 2012. *Conservation Agriculture in Central Asia: Status, Policy, Institutional Support, and Strategic Framework for Its Promotion*. FAO. Ankara. 57 p.
- FAO. 2016. Practice of conservation agriculture in Azerbaijan, Kazakhstan and Uzbekistan. FAO. Ankara. 84 p.
- FAO. 2020. Strategies for the promotion of Conservation Agriculture in Central Asia Proceedings of the International Conference, 5-7 September 2018, Tashkent, Uzbekistan. Tashkent. https://doi.org/10.4060/ca8659en.
- Fileccia, T. 2009. Conservation agriculture and food security in Kazakhstan. Working Paper, FAO Investment Centre Division. Rome: FAO.

7 References

Bagdasaryan, A. (2015). Деградация на миллиарды: в России истощены свыше 60% сельхозугодий Армина Багдасарян / Агроинвестор, 9 ноября 2015. Available at: https://www.agroinvestor.ru/technologies/article/22499-degradatsiya-na-milliardy-v-rossii-istoshcheny-svyshe-60-selkhozugodiy/.

Boincean, B. and Dent, D. (2019). Farming the black earth. In: *Sustainable and Climate-Smart Management of Chernozem Soils*. Springer.

Bulavin, L. A., Bulavina, T. M. and Nilova, O. V. (2010). *Булавин Л.А., Булавина Т.М., Нилова О.В. Проблема ресурсосбережения в земледелии и особенности ее решения в условиях Беларуси*.

Corsi, S. and Muminjanov, H. (2020). *Conservation Agriculture: Training Guide for Extension Agents and Farmers in Eastern Europe and Central Asia*. Food and Agriculture Organization. 134 p.

Danatarov, A. (2011). Данатаров, Агахан. Совершенствование ресурсосберегающих и почвозащитных технологий и технических средств обработки почвы в условиях Туркменистана / Агахан Данатаров, С. Ч. Ашыров. - // Молодой ученый. - 2011. - № 12 (35). - Т. 2.- С. 207-210.

Egamberdiev, O. J. (2007). Dynamics of irrigated alluvial meadow soil properties under the influence of resource saving and soil protective technologies in the Khorezm region. PhD dissertation. National University of Uzbekistan.

FAO (n.d.). CA website. Available at: www.fao.org/ag/ca.

FAO (2011). *Save and Grow: A Policymaker's Guide to the Sustainable Intensification of Smallholder Crop Production*. Food and Agriculture Organization: Rome. Food and Agriculture Organization (2012). 98 p.

FAO (2012). *Conservation Agriculture in Central Asia: Status, Policy, Institutional Support, and Strategic Framework for Its Promotion*. Food and Agriculture Organization: Ankara. 57 p.

FAO (2015). *The Terminal Report of the TCP/TAJ/3405: Support to Adoption and Promotion of Modern Crop Management Practices*.

FAO (2016a). Country Profiles: Armenia. Available at: http://www.fao.org/countryprofiles/index/en/?iso3=ARM.

FAO (2016b). Country Profiles: Azerbaijan. Available at: http://www.fao.org/countryprofiles/index/en/?iso3=AZE.

FAO (2016c). Country Profiles: Belarus. Available at: http://www.fao.org/countryprofiles/index/en/?iso3=BLR.
FAO (2016d). Country Profiles: Georgia. Available at: http://www.fao.org/countryprofiles/index/en/?iso3=GEO.
FAO (2016e). Country Profiles: Kazakhstan. Available at: http://www.fao.org/countryprofiles/index/en/?iso3=kaz.
FAO (2016f). Country Profiles: Kyrgyzstan. Available at: http://www.fao.org/countryprofiles/index/en/?iso3=KGZ.
FAO (2016g). Country Profiles: Republic of Moldova. Available at: http://www.fao.org/countryprofiles/index/en/?iso3=MDA.
FAO (2016h). Country Profiles: Russia. Available at: http://www.fao.org/countryprofiles/index/en/?iso3=RUS.
FAO (2016i). Country Profiles: Tajikistan. Available at: http://www.fao.org/countryprofiles/index/en/?iso3=TJK.
FAO (2016j). Country Profiles: Turkmenistan. Available at: http://www.fao.org/countryprofiles/index/en/?iso3=TKM.
FAO (2016k). Country Profiles: Ukraine. Available at: http://www.fao.org/countryprofiles/index/en/?iso3=UKR.
FAO (2016l). Country Profiles: Uzbekistan. Available at: http://www.fao.org/countryprofiles/index/en/?iso3=UZB.
FAO (2016m). *Practice of Conservation Agriculture in Azerbaijan, Kazakhstan and Uzbekistan*. Food and Agriculture Organization: Ankara. 84 p.
FAO (2016n). *Land Resources and Food Security of Central Asia and Southern Caucasus*. P. Krasilnikov, M. Konyushkova and R. Vargas (Ed.). FAO. 434 p.
FAO (2017). Report of the Project "European Neighbourhood Programme for Agriculture and Rural Development (ENPARD)".
FAO (2020). Strategies for the promotion of Conservation Agriculture in Central Asia. Proceedings of the International Conference, 5-7 September 2018, Tashkent. Available at: https://doi.org/10.4060/ca8659en.
Fileccia, T. (2009). *Importance of Zero-Tillage with High Stubble to Trap Snow and Increase Wheat Yields in Northern Kazakhstan*. FAO Investment Centre Country Notes.
Fileccia, T. (2013). *Advancement and Impact of Conservation Agriculture/No-Till Technology Adoption in Kazakhstan*. FAO Investment Centre Country Notes. Available at: http://www.eastagri.org/publications/pub_docs/Info%20note_Print.pdf
Fileccia, T. (2014). *Ukraine: Soil Fertility to Strengthen Climate Resilience - Preliminary Assessment of the Potential Benefits of Conservation Agriculture*. Joint FAO–World Bank Publication. Available at: http://www.fao.org/3/a-i3905e.pdf.
Grib, S. I. (2002). Гриб, С.И. Стратегия интенсификации адаптивного растениеводства/С.И. Гриб, М.М. Севернев, И.М. Богдевич // Сельскохозяйственный вестник. - 2002. - №5. - С.4 - 6. 7.
Gusakov, V. (2002). Гусаков, В. Основные предпосылки адаптивной интенсификации с/х производства / В. Гусаков, М. Севернев, С. Гриб // Агроэкономика. - 2002. - №6. - С.3 - 6. Available at: https://rep.polessu.by/bitstream/123456789/121/1/5.pdf.
Kassam A., Friedrich T. and Derpsch R. (2019). Global spread of Conservation Agriculture, International Journal of Environmental Studies, 76:1, 29-51, DOI: 10.1080/00207233.2018.1494927

Kovalevich, A. (2010). Ковалевич А. Опустынивание и деградация земель в странах СНГ //Лесное и охотничье хозяйство. Раздел: Экология. № 2 (2010). – С.17-22.

Maltsev, T. (2020). Т.С. Мальцев – родоначальник современного ресурсосберегающего почвозащитного земледелия Зауралья. Информационное агентство "Светич". Сайт о сельском хозяйстве. Электронный ресурс. Available at: http://svetich.info/publikacii/opyt-mirovogo-zemledelija/t-s-malcev-rodonachalnik-sovremennogo-re.html.

MoA, Uzbekistan 2019. Annual Report.

Ovsinsky, I. (1909). Овсинский И.А. Новая система земледелия. Москва.

Pojarskiy, V. (2008). Пожарской В.Г. Продуктивность озимой пшеницы при возделывании на гребнях в условиях Чуйской долины Кыргызской республики. Автореф. канд. с.-х. н. Бишкек-2008. 18 с.

Pulatov, A., Choudhary, A. and Akramkhanov, A. 2001. Status of conservation tillage practices in Uzbekistan. In: Gill, M. A. (Ed.), *International Workshop on Conservation Agriculture for Food Security and Environment Protection in Rice-Wheat Cropping Systems*. Lahore, Pakistan

Rashidov, Q. (2009). Продуктивность перспективных сортов пшеницы при гребневом посеве в условиях Центрального Таджикистана./ Автореферат диссертации на соискание ученой степени кандидата сельскохозяйственных наук. Душанбе.

Rychkov, A. (2018). Рычков А.В. Сибирские саратники академика А.И. Бараева. 2018 – С. 142–147.

Safin, Kh. (2021). Сафин Х. М. Сберегающие технологии земледелия No-till и Strip-till: теория и практика внедрения. Уфа, Мир печати, 2021. 204 с.

Журнал (2019). Ресурсосберегающее земледелие 4(44) /2019. Электронный Ресурс. Available at: https://agriecomission.com/base/pochvozashchitnoe-i-resursosberegaushchee-zemledelie-v-rossii-i-v-mire.

Сельское хозяйство Украины. Available at: https://ru.wikipedia.org/wiki/%D0%A1%D0%B5%D0%BB%D1%8C%D1%81%D0%BA%D0%BE%D0%B5_%D1%85%D0%BE%D0%B7%D1%8F%D0%B9%D1%81%D1%82%D0%B2%D0%BE_%D0%A3%D0%BA%D1%80%D0%B0%D0%B8%D0%BD%D1%8B.

ГлавАгроном (n.d.). No-Till с топ-блогером: опыт применения технологии от Михаила Драганчука. Available at: https://glavagronom.ru/articles/No-till-s-top-blogerom-opyt-primeneniya-tehnologii-ot-Mihaila-Draganchuka.

Chapter 14

Conservation Agriculture in South Asia

Yashpal Singh Saharawat, International Fertilizer Development Centre (IFDC), India; Mushtaq Gill, SACAN Services, Pakistan; Mahesh Gathala, CIMMYT, Bangladesh; Tika Bahadur Karki, Nepal Agricultural Research Council, Nepal; D. B. T. Wijeratne, Food and Agriculture Organization of the United Nations (FAO), Sri Lanka; Sayed Samiullah Hakimi, Kabul University, Afghanistan; Neelam Chaudhary, Directorate of Plant Protection and Quarantine, India; Md. Enamul Haque, Murdoch University, Australia; Richard W. Bell, Murdoch University, Australia; C. M. Parihar, Indian Agricultural Research Institute, India; Harisankar Nayak, Indian Agricultural Research Institute, India; Rajbir Singh, ICAR-Agricultural Technology Application Research Institute, India; R. K. Malik, CIMMYT, India; Upendra Singh, International Fertilizer Development Centre, USA; Raj Paroda, Trust for Advancement of Agricultural Sciences, India; and Amir Kassam, University of Reading, UK

1 Introduction
2 History and status of Conservation Agriculture in South Asia
3 Crop productivity and income in Conservation Agriculture
4 Soil health in Conservation Agriculture
5 Water productivity under Conservation Agriculture
6 Climate change mitigation and adaptation in Conservation Agriculture
7 Nutrient management in Conservation Agriculture
8 Weed management in Conservation Agriculture
9 Further upscaling of Conservation Agriculture in South Asia
10 Where to look for further information
11 References

1 Introduction

South Asia (SA) is the southern region of Asia, defined in both geographical and ethnocultural terms. The region consists of Afghanistan, Bangladesh, Bhutan, India, Nepal, Pakistan, Sri Lanka and the Maldives. It is home to 1.7 billion people and has achieved high economic growth in the past six decades but still houses around 42% of the world's poorest population, 21% of the malnourished and 41% of underweight children globally (Akhtar, 2016). The combined population of Pakistan, India, Nepal and Bangladesh is

http://dx.doi.org/10.19103/AS.2021.0088.12

expected to reach 2.4 billion by 2050, demanding an almost 50% increase in food production (https://www.eastasiaforum.org/2013/04/06/demographic-dividends-in-south-asia-a-window-of-opportunity/). The food security challenge is further exacerbated by climate-related natural disasters that have already cost around US$2.97 trillion, affecting more than 4.03 billion people between 2000 and 2019 (UNDRR report, 2020, https://www.undrr.org/publication/undrr-annual-report-2020). SA agriculture therefore faces a trio of challenges: to increase production to meet the food demands of a growing human population with a lower environmental footprint; restore and preserve natural resources; and mitigate or adapt to climatic change (Jat et al., 2020a). At the same time, per capita availability of agricultural land and water is declining significantly.

The intensive agriculture practices of the 'Green Revolution' were successful in achieving production goals in the past but simultaneously led to degradation of natural resources, reduced total factor productivity, depletion of water resources and endangered future agricultural production (Kumar et al., 2018). Tillage-based practices have deteriorated soil health by removing crop residues and depleting soil organic matter, degrading the physical structure of the soil, enhancing soil erosion, promoting excessive use of mineral fertilizers and pesticides, reducing water infiltration, using inputs (water, energy, fertilizer and labour) inefficiently and making agriculture a significant source of greenhouse gas emissions and global warming (Parihar et al., 2018a). Furthermore, in Conventional Agriculture (CA), the open burning of crop residues causes atmospheric pollution through the emission of greenhouse gases (GHG), toxic smoke, soot and loss of valuable plant nutrients, which along with inadequate crop rotation and diversity have worsened the agricultural development situation in SA (Yadvinder-Singh and Sidhu, 2014; NAAS, 2017). Intensive and frequent use of tillage with heavy machinery destroys soil structure, compacts the soil, enhances the release of sequestered soil organic carbon (SOC) and releases GHG into the environment from the combustion of fossil fuel. Large yield gaps exist, particularly 'management yield gaps' ranging from 14% to 47%, 18% to 70% and 36% to 77% in wheat, rice and maize respectively, the region's three staple food crops (Jat et al., 2011).

To overcome the current multi-faceted problems of SA's agriculture, there is an urgent need for new science-based sustainable management approaches that are resource-efficient, more productive and environmental friendly. In the past two decades, CA has emerged as an alternative to inefficient tillage-based conventional agriculture across the region. CA is an ecosystem approach to regenerative sustainable agriculture and land management that helps manage agroecosystems in terms of improved and sustained productivity, increased profits and food security while preserving and enhancing the resource base and the environment. It is based on the context-specific application of three interlinked principles - continuous minimal soil disturbance, permanent soil mulch (crop biomass and cover crops) and diversification of the cropping system, along with other complementary agricultural production and land management practices (Kassam et al., 2019). Sustainability is one of the major concerns facing the region's agriculture, and CA constitutes a comprehensive approach for moving in that direction. In the past decade, CA-based resource conservation technologies (RCTs) have shown enhanced soil health, better input use efficiency, profitability while promoting conjunctive use of organics (avoiding residue burning), reduced environmental pollution and timely planting in the South Asia region (Ladha et al., 2015).

2 History and status of Conservation Agriculture in South Asia

The history of CA in SA is inextricably linked with wheat production constraints in the rice-wheat system, a dominant cropping system in the Indo-Gangetic Plains (IGP) (Ladha et al., 2003b). Byerlee et al. (1984) reported various problems with wheat production, including late planting, poor plant stand, weed problems, reduced seed replacement rate and poor management (fertilizer and water) in Pakistan Punjab. Randhawa et al. (1981) reported late sowing of wheat as one of the major constraints with a yield decline of about 1% per day after the optimal date. Identifying late planting of wheat as the major problem due to late rice harvesting and prolonged intensive tillage practices for land preparation to sow wheat, zero till (ZT) was recommended as a potential solution. However, ZT was attempted in the IGP in the late 1970s but the technology missed the mark for several reasons. The ZT technology was reintroduced in the early 1980s in Pakistan by using a seed drill imported from New Zealand for the first time in Pakistan and India, but it did not perform well in the first season. In 1996, when the ZT practice was again attempted with local versions of ZT drills from India and Pakistan, it performed better than conventional agriculture. Real-time demonstrations and evaluations of the technology in farmers' fields in a cost-effective manner aided its faster adoption. From 2002 onwards, the rice-wheat consortium (RWC) of the Consultative Group on International Agricultural Research (CGIAR), via its collaborators including the United States Agency for International Development (USAID), introduced and promoted second-generation ZT seeders. Today second-generation CA planters are available, which can be operated with four- or two-wheel tractors and are being adopted at different scales in different countries based on agroecology and socioeconomic conditions. The initiation and spread of CA technology vary among the different SA countries and therefore documentation of skill, knowledge and success stories is an important tool for mutual learning and progress. Historical milestones for CA in South Asia are given in Table 1. The development of a common platform for CA in South Asia can facilitate networking and its further adoption and spread.

2.1 Rice-Wheat Consortium (RWC): a stepping stone for Conservation Agriculture in South Asia

The RWC was formed in 1994 as an ecoregional programme (EP) of the CGIAR with strong support from national partners in India, Pakistan, Bangladesh and Nepal (Table 1). As an EP, the consortium has a special kind of research network to address natural resource management (NRM) issues in a sustainable manner and problems related to agricultural production and within a geographically defined area. The success of the RWC can be attributed to initiatives by the National Agricultural Research System (NARS), CGIAR institutes and other partners, having defined roles and commitment to make it successful. Initially, RWC encouraged tillage and crop establishment methods with a focus on providing a more favourable window for wheat sowing. When the ZT technology was introduced, this was further extended to other RCTs. Although the RCTs were adopted by a large number of farmers in SA countries, they could not cope with the negative effect of traditional agricultural practices, such as loss of SOC associated with high intensity of inversion tillage, in-situ crop residue burning, soil degradation and associated problems, given the level of diversity and endowments prevailing among the farmers in SA.

In SA, the terms 'conservation agriculture' and 'resource conservation technologies' (RCTs) are used interchangeably as if their meanings are similar, but in reality, they differ greatly. RCTs refer to those practices that enhance resource or input-use efficiency (Hobbs et al., 2008). The RCTs in long-term experiments across SA apply minimum soil disturbance and crop residue retention to the dry season crop only and continue with full tillage and puddling of soils for conventional rice transplanting. The limitation of these experiments is that the benefits of RCT/CA take 2 to 5 years of continuous practice for changes in soil properties (Salahin et al., 2021; Alam et al., 2018) and weed ecology (Hossain et al., 2021a,b) to become evident. Therefore, results from the long-term experiments and long-term practice of CA in farmers' fields are most influential in drawing conclusions about its benefits (Bell et al., 2019; Gathala et al., 2021). Overall, CA as a potential alternate land-use system reconciles economic viability, enhances social balance, restores soil health, improves environmental conservation and mitigates climate change and was propagated by the RWC to improve long-term smallholders' living conditions and to alleviate poverty. There are some challenges in large-scale adoption of CA in SA, such as changing farmers' mindsets, weed control, residue management, nitrogen management, availability of ZT planters, socio-economic issues, climate change, irrigation, policy interventions and non-incentives for residue retention/carbon credit. The history and status of CA per country are presented in the following sections.

2.2 India

In India, various state agricultural universities (SAUs) trialled ZT in the 1970s, but their efforts failed due to technical problems including insufficient planting equipment and a lack of solutions to manage weed infestation. In the 1990s, CIMMYT introduced the inverted-T openers in India, which were originally developed in New Zealand by Aitchison industry. In 1991, the first prototype of the Indian ZT seed drill was developed at GB Pant University of Agriculture and Technology, Pantnagar. In 1992–93, a collaborative programme for further development and commercialization of ZT was initiated, involving small-scale industries in the Punjab state of India. Following the significant investment of resources and several rounds of design modifications, the first ZT seed drill was made available for field-testing within 12 months. From 1995, RWC, in collaboration with the National Agricultural Research and Extension System (NARES), provided support for farmer participatory research to accelerate the pace of ZT technology adoption in the rice-wheat systems. The DFID-supported project acquired several ZT drills and donated them to the CCS Haryana Agricultural University for farmer participatory research, providing a much-needed push for this programme. In the late 1990s, Monsanto supported no-till research at the universities but soon reduced its activities when it perceived that the potential for glyphosate marketing was small. In 1997, with further refining of the machines based on feedback received from scientists and farmers, private manufacturers supplied over 150 improved ZT drills to SAUs and the Indian Council for Agricultural Research (ICAR) institutions located in different states. The combined efforts of NARES, private manufacturers, RWC and CGIAR institutes resulted in widespread adoption of ZT in rice-wheat systems during the 1990s in India. Haryana and Punjab were the leading adopters of the ZT technology as it ensured timely planting of wheat, reduced the cost of tillage and increased the wheat yield.

The ZT, along with other resource-efficient technologies such as laser land levelling, bed planting, leaf colour charts etc. which have the potential to save fuel, improve plot-level water productivity and nutrient use efficiency, are considered as RCTs. Currently, the RCTs are being trialled on an area of more than 20 million ha in India alone (VK Singh personal communication during RCTs Vaibhav summit 2020). The RCT practices will be referred to as being CA only when the RCTs are combined with the three interrelated CA principles. The distinction is important because some RCTs, while adoptable in the short term, maybe unsustainable in the longer term. For example, the use of ZT without

Table 1 Historical milestones in Conservation Agriculture in South Asia

Timeframe	Constraint/Problem/Events	Innovation/Technology	Reference
1980–1990	**Transformation from disciplinary to commodity to systems research based on farmer-defined problems and issues**		
1981–1985	• Diagnostic surveys identify wheat production problems – late planting, poor stand, weeds, old varieties and poor management by farmers		Randhawa et al. (1981), Byerlee et al. (1984), Beebe (1985)
1986–1990	• Intensive tillage after rice, main reason for late planting of wheat • Soil physical constraints due to puddling in rice – adverse impact on wheat yields • Diagnostic surveys highlight natural resource degradation (salinization, groundwater contamination), pest build-up and soil nutrient depletion in the rice-wheat cropping system	• Introduction of zero-till (ZT) drill in Pakistan from New Zealand in 1984–1985 by Peter Hobbs from CIMMYT • Successful research and demonstration of ZT drill in 1985–1988 in Pakistan • Shipment of four Aitcheson drills from New Zealand to India in 1988 by Peter Hobbs from CIMMYT. Establishment of Project Directorate for Cropping System Research (PDCSR) in 1989 in India	Byerlee et al. (1986), Khan et al. (1986), Gupta (1989), Flinn and Khokhar (1989), Khuhro (1989), Harrington and Hobbs (2009)
1990–2000	**Consortium to Ecoregional Approach**		
1990–1995	• Isoproturon herbicide resistant *Phalaris minor* weed problem reported and flagged in wheat by R K Malik in 1992–93 under ACIAR project • FAO 'Regional Expert Consultation on sustainability of rice-wheat production system in different agroecological settings in Asia' highlighted in analysing rice-wheat productivity trends and emphasized in rice-wheat research for agroecological zones in 1993 • RWC, along with NARES and CGIAR, identify six research themes to overcome current constraints of rice-wheat system: productivity and sustainability trends, improved crop establishment methods, sustaining soil fertilizer, water management, IPM, and enabling policies for sustainable resource management in 1994	• Descon (Lahore-based implement manufacturer) develops first prototype of local multicrop ZT seedmatic drill • First prototype of Indian ZT drill by GB Pant University of Agriculture and Technology in 1991 'Pantnagar Drill' • National Agro-Industries Punjab India develops own 'T-openers' in 1993 and initiates localized development of ZT drills • First low cost ZT drill prototype developed by NARC Islamabad Pakistan in 1993. • Formation of a Regional Steering Committee and Facilitation Unit (RSC) in 1994 through World Bank meeting of NARES, IRRI and CIMMYT leadership – leading towards RWC along with donor pledge for an Eco-regional programme in New Delhi • RWC for IGP first meeting held in IDRC office as a formal launch of RWC on 19–20 May 1994, attended by NARES, CGIAR and donor leaders	Aslam et al. (1991), Singh and Paroda (1993), Ahmad et al. (1994), Malik and Singh (1995), Malik et al. (1998, 2000), Mehla et al. (2000), Harrington and Hobbs (2009)

1995–2000	• RWC facilitation unit operational in 1995 with Inder Abrol as a permanent facilitator • *Phalaris minor* reported as a major problem in wheat crop in South Asia in 1995–96 under Australian project on *Phalaris minor* resistance operated by HAU, Hisar • Mechanization of small holder farms a major problem in Nepal and Bangladesh • RTCC meeting of 1994 and 1995 prioritizes RWC research agenda on nutrient management, water management, integrated pest management and tillage and crop establishment methods • Visit by RWC team to China in 1995 – introduction to surface seeding and Chinese two-wheel handheld tractor • Long-term soil fertility experiments (LTSFE) highlight decline in rice yields over time and stagnation in wheat yields in 1996, also decline in SOM in chemical fertilizer treatments • Decline in total factor productivity over time reported from long-term experiments in Nepal and India • Decline in ground water table reported from North-western IGP in the late 1990s • Project on herbicide resistance and integrated management in India and Australia implemented 1997–2002. Beginning of accelerated ZT tillage adoption	• ZT farmer participatory research initiation on large scale in rice-wheat system from 1995 onwards • Farmer participatory demonstration of ZT drills on large scale in Haryana through DFID project by CCS HAU and CIMMYT collaboration from 1997 onwards • ZT seeding of wheat helps to reduce *Phalaris minor* problem in wheat and to reduce production cost in upper IGP during 1995–2000. • Chinese two-wheel hand tractor with ZT attachment testing for wheat sowing initiated in Nepal and Bangladesh during 1995 onwards – DFID project • Parachute planting of rice seedlings introduced in Pakistan from 1995 onwards • Strip tillage system developed in Bangladesh in the late 1990s by RWC • ZT drills refined several times based upon farmer and scientist feedback by National Agro-industries, ASS foundry etc • Diversification through legumes introduced by ICRISAT in 1996 through ADB gran • Over 150 ZT drills supplied to NARES for farmers' participatory research and demonstrations in 1997 • Introduction of leaf colour charts (LCC) for nitrogen management by IRRI in the late 1990s • Initiation of two-wheel tractor for ZT wheat in Bangladesh and Nepal in 2000 • ZT wheat exceeds 10 000 ha in upper IGP, and custom hiring of machinery becomes popular in the region	Malik and Singh (1995), Abrol et al. (1996), Abrol et al. (2000), Meisner et al. (1998), Hobbs and Morris (1996), Adhikari et al. (1999), Vincet and Quirke (2002), Ladha et al. (2003a), Tripathi et al. (2006), Malik et al. (2002a), Malik (2005), Harrington and Hobbs (2009)

(Continued)

Table 1 (*Continued*)

Timeframe	Constraint/Problem/Events	Innovation/Technology	Reference
2000–2009	**Zero Tillage (ZT) to Resource Conserving Technologies (RCTs) to Conservation Agriculture (CA)**		
2000–2005	• Acceleration of Resource Conservation Technologies ICAR project implemented from 2001 to 2005. Expansion in the areas in EIGP • Identical problems of rice-wheat system identified in the region (Pakistan, India, Nepal and Bangladesh) • Emphasis put on crop-based research rather than cropping system-based research in rice-wheat system • Low water use efficiency depleting ground water table in upper IGP • RWC, along with NARES, initiates research on nutrient management, water management, IPM, nematodes, weeds etc. and develops Resource Conserving technologies (RCTs) • Cornell University project on soil health funded by USAID in RWC countries except for India • World Bank loan to India through NATP enhances the RCTs in IGP. • RWC awarded 'Excellence in Science Award' by CGIAR in 2000, enhancing its profile and credibility • King Baudouin Award to RWC by CGIAR in 2004 for a unique consortium for Government and International research institutions benefiting around 250 000 farm households	• Cropping systems approach with stronger farmer participatory research adopted for RCTs evaluation • Initiation of Resource Conserving Technologies (RCTs) through integrated systems research over six sites in SA, including improved tillage and crop establishment, new varieties in ZT system, site-specific nutrient management, improved irrigation method and diversification jointly by NARES, CIMMYT, IRRI, IWMI, ICRISAT and CIP under ADB funding • Agri-machinery manufacturer visit to Australia in 2002 through ACIAR • Residue management prototype development project with PAU and ACIAR – Happy seeder • Wider usage of IT and GIS for RCTs dissemination and its impact analysis	RWC (2000), Gupta et al. (2000a), Ahmad et al. (2001), Franke et al. (2001), Vincent and Quirke (2002), Malik et al. (2005), Thakur (2005), Laxmi et al. (2007), Franke et al. (2007), Seth et al., 2003

2005–2009	• Farmer-led and contributory research on CA helps to change the perspective of NRM research for development • Tillage project commissioned in India by the ICAR and the Laser and water management project in Pakistan • IRRI/RWC ADB funded project on scaling out RCTs initiated in 2005 • Throwback residue system of Happy Seeder modified to in-situ residue management by Turbo Seeder through RWC • Turbo seeder initiates the double zero till system in rice-wheat system • Shifting of CGIAR focus to challenge programmes • Shifting of RWC Facilitation unit from CIMMYT to IRRI • Integrated crop management using RCTs introduced in Eastern-IGP including unpuddled transplanting of rice/zero-till transplanting of rice, bed planting for rice and wheat • Amar Agro develops two-wheel 5 HP tractor as well as 20HP narrow wheel tractor for zero till • CA machinery pool system developed in the SA as a new model of custom hiring • Inclusion of social science research on RCTs and CA from 2005 onwards • Laser land levelling introduced in India by Joe Rickman from IRRI and locally manufactured by Bery Industries Karnal and promoted by PDFSR and SVBPUAT Modipuram	• By 2006–07, ZT area expands to around 3 M ha in SA through NARES and CG funded research • RCTs' scope expanded to laser land levelling, lining of canals, surface seeding wheat, bed planting, direct seeding of rice, LCC and intercropping from 2000 onwards • Happy seeder for residue management in rice-wheat system developed under PAU-ACIAR project and initiated conservation agriculture in SA in 2007 • Happy Seeder modified to Turbo seeder to suit local rice-wheat system by RWC and Dashmesh Industries India • Turbo seeder machine more compact, lightweight, with depth control wheel and ensured in-situ residue management with less air pollution/dust • Laser land levelling gains importance in India after success in Pakistan from 2007 onwards • Emphasis on stakeholder capacity development on CA, travelling seminars and cross-site visits • RWC and NARES publish packages of practices for DSR to ensure double no-till system in SA during 2006

(Continued)

Table 1 (*Continued*)

Timeframe	Constraint/Problem/Events	Innovation/Technology	Reference
2009–2020	**Conservation Agriculture to Climate Resilient Agriculture**		
2009–2015	• Fourth World Conference on Conservation Agriculture, 4–7 February 2009 in New Delhi • Bill and Melinda Gates Foundation Funding to address cereal systems constraints in South Asia • Changes in front and rear wheels of combines by RWC in collaboration with Sardar Hari Singh of Nabha. • Significant changes in combine harvesters to fit in CA systems • Dismantling of the Rice-Wheat Consortium • Statistical analytical tools for testing significance of adaptive and farmer participatory field trials • Gupta & Prasad from IASRI-ICAR publish first book on adaptive research trials' statistical analysis • Initiation of 'Advance Course on Conservation Agriculture' by CIMMYT 2010 • Shifting of CA focus from upper IGP to lower IGP • Abandonment of RWC to CSISA • ICAR launches CRP on conservation across different parts of India • ACIAR initiates CASI project in Lower IGP	• Adopted resolution on CA for sustainability • Need to start mission mode on CA • Initiation of Cereal System Initiative for South Asia (CSISA) Program by IRRI, CIMMYT, ILRI and IFPRI, along with NARES for four countries • Hub concept of public-private and international centres collaboration for CA • Redesign of the axles of the front and rear wheels of combines aligned at 134 cm spacing to reduce combine compaction in CA • Design change allows controlled traffic movement • Maize combining blade attached to wheat combines • HS Sidhu fits the straw spreader and a sensor for alerting the driver to any interruption in residue throws in the combine back • Dr VK Gupta and Dr R Prasad from IASRI visit farmer fields and develop methodology for statistical analysis of the data from farmers' fields for easy publishing of results • 149 researchers from nine countries train in eight courses • South Asia platform for CA finished • Research and demonstration of CA extended in different cropping systems and agro-ecologies by ICAR-CRP programme • CASI extends CA in new areas of acid soils and covers new cropping systems • CASI emphasizes on small-scale machinery	

2016–2021	• CSISA emphasizes working in lower IGP • Fourth International Agronomy Congress (IAC) November 2016 • TAAS, ACIAR, CIMMYT, NARS (India, Pakistan, Nepal, Bangladesh) meeting on CASI for Sustainable Agriculture (CASI). • Human health problems due to rice residue burning problem recognized • PAU designs new lucky seeder • Private sector supports wide scale adoption of CA through Corporate Social Responsibility (CSR) funds • CA support base extended among different stakeholders (CSR, NGOs, private sector, CGIAR, ICAR, etc) • World Bank funds new project to Punjab Government on saving water through CA and converts residues to electricity • IFDC initiates research work on improving NUE in CA-based systems • Indigrow-CIMMYT-Punjab Government initiates carbon trading through CA in Punjab	• CA extended outside the rice-wheat system • Large-scale adoption of small-scale CA machinery in Bangladesh, Nepal and eastern India • Fourth IAC emphasis on small farm mechanization and the need for a shift in research focus from crop oriented to farming system oriented; focus on innovation in CA • Advocated Neutral platform on CASI • Government provides new policies and funding to extend turbo Happy Seeder in rice-wheat system of Punjab and Haryana • Multi-crop planter in residue retained condition promoted in Punjab and Haryana • Ecosystem-based incentive system for farmers through CA initiated in Punjab, India

residue retention and without suitable rotations, under some circumstances, can be more harmful to agroecosystem productivity and resource quality than a continuation of conventional practices (Thind et al., 2019). The development, testing and refinement of resource-conserving practices are based on the innovative low-cost seed-cum-fertilizer drill (costing US$400–500), which can plant crops with minimal soil disturbance.

With the advancement of CA technology, second-generation no-till drills and seeders were developed, particularly in India, in general for seeding into anchored stubble and loose residue retained soil. During the past few years, a major emphasis has been put on developing CA-based machinery, keeping in mind the needs of small farmers. Sidhu et al. (2015) were successful in developing a no-till seeder, popularly known as 'Happy Seeder', which turned wheat or any other grain crop into heavy loose residue retained on the soil surface. Combine harvesters used in SA led to an accumulation of straw in the rear, which hindered the proper working of the Happy Seeder. An attachment called the Punjab Agriculture University (PAU) Super SMS (straw management system) was developed, which can chop and evenly spread the straw coming out of the rear of a combine harvester (Mahal et al., 2019). In recent times, amid the labour shortage due to the Coronavirus outbreak, PAU has developed a new direct seeding of rice (DSR) machine called the 'Lucky Seeder' to address the issue during the rice transplanting season. Rice is direct seeded with a tractor-operated Lucky seed drill (LSD), which also sprays herbicide simultaneously.

Currently, efforts have focused on expanding CA in rain-fed areas and outside IGP among smallholder farmers. ICAR has ambitiously launched the Consortia-Research Platform on CA (CRP on CA) to adapt and mainstream available best-bet CA technologies across India by covering dominant crops under diverse ecologies, i.e. from a rain-fed to an irrigated system. Recently, the International Fertilizer Development Centre (IFDC) has initiated research work on efficient nutrient management under CA-based cropping systems in different states of India in collaboration with NARES. It is estimated that CA is presently being practised by more than one million farmers in India in an area of around two million hectares (Gupta et al., 2021). Others have indicated a figure of 3.5 million hectares of CA in the IGP in India (Paroda, 2018). No official CA data are available as there is no CA policy to date.

2.3 Pakistan

CIMMYT introduced the inverted-T openers in Pakistan in the 1990s, which were originally developed in New Zealand by the Aitchison industry. Descon Private Limited developed the first ZT drill prototypes in 1990–1991 in Pakistan. During the same season, prototypes were introduced for field testing among farmer groups, but the programme of farmer testing was not entirely successful.

The drills were too heavy to be used with local small tractors, tines were easily broken and frequent blockages of loose straw were reported (Aslam et al., 1991). In 1993, the Farm Machinery Institute, National Agriculture Research Council, Islamabad developed a low-cost drill that was successful in different locations of Punjab and the Sindh province of Pakistan. M/s. Greenland Engineers Daska, a local machinery manufacturer, started fabrication of such drills on a commercial scale, which were subsequently modified according to requirements. At the same time, popular RCTs among the farmers were laser land levelling, ZT, raised-bed planting and crop residue management. RWC and NARES enhanced the promotion and adoption of RCTs among farmers, including soil and water testing, the establishment of a machinery pool for equipment rentals and a village seed bank for provision of certified seed to smallholders. To effectively introduce, validate and promote RCTs, close collaboration among various partners was the key. A win-win partnership was developed among farmers, agricultural implement manufacturers, researchers, extensionists, service providers and private seed and fertilizer companies.

A second key milestone for the adoption of RCTs was via the New Zealand Ministry of Foreign Affairs and Trade funded project 'NZOD/ADAF' launched in April 2000. The project was implemented in irrigated and rain-fed areas of Lahore, Okara and Chakwal districts of Pakistan in 2001–2002. Under this project, ZT wheat was adopted across an area of 442 ha, bed planting on 58 ha and laser land levelling on 183 ha. Human resource development remained an integral component of the project wherein about 2000 farmers/professionals were trained in different RCT interventions. Permanent raised bed (PRB) was a game-changing RCT that saved water and energy, provided timeliness of operations, enhanced yield and soil fertility, reduced soil erosion, facilitated mechanized farming, improved agriculture sustainability and thus improved food security. It was helpful in diversifying crops in the saline and waterlogged rice crop zones of Hafizabad, Sheikhupura and Gujranwala, where growing other non-aquatic crops was not possible due to standing monsoon rainwater, leading to rice being grown as a compulsion rather than choice.

In the changing climatic scenario, with predicted intense and short duration rains, PRB is an effective technology due to larger surface and groundwater storage capacity. Second-generation planters such as Happy Seeder were replicated from India via the Australian Council of Agricultural Research (ACIAR) and Cereal System Initiative for South Asia (CSISA) during 2009–2020. These efforts need to be replicated in other cropping systems including rain-fed, arid, semi-arid and mountainous regions within and outside IGP. The CA expansion in Pakistan is confined to rice-wheat cropping systems through ZT wheat and DSR using the Happy Seeder and covers around 1.3 million ha. The current data is compiled through contacts with manufacturers, provincial agriculture departments and federal Ministry of Food Security and Research. No data on

CA areas from the national bureau of statistics are available as there is no CA policy to date in Pakistan.

2.4 Bangladesh

Research and development on CA in Bangladesh have focused on smallholders (<0.5 ha farm size and small fields of about 1000 m^2) and the two-wheel tractor (12 to 16 hp) as the primary power source for operating seeders (Krupnik et al., 2013; Haque et al., 2018b). The two-wheel tractor-operated ZT seeder and strip-till technology were first introduced by RWC/CIMMYT-Bangladesh in wheat fields during the late 1990s (Hossain and Haque, 2015). Subsequently in 2002, the Wheat Research Centre (WRC) of Bangladesh Agricultural Research Institute (BARI) developed the toolbar-attached raised-bed planter, which was successfully evaluated for other cereal and pulse crops. In 2003–2004, the Food and Agriculture Organization (FAO) initiated CA-based tillage technology to intensify wheat-based cropping systems in Bangladesh. The WRC developed a two-wheel tractor-attached pull-type ZT drill. The FAO, RWC and CIMMYT played a key role in promoting CA technology in farmers' fields during the early 2000. From 2005 to 2007, IRRI-ADB put special emphasis on CA DSR and wheat in Northern Bangladesh. Initial CA work focused on the cool and dry (Rabi) season crops such as wheat, maize, lentil, mustard, chickpea and demonstrated in research plots substantial yield gains (e.g. Bell et al., 2018; Rashid et al., 2019; Vance et al., 2014a) and cost savings (e.g. Miah et al., 2019). From 2006 to 2009, ACIAR, in collaboration with NARS, emphasized improving the quality of the ZT drill and strip-till drills for better adoption by farmers. In 2006 only, the farm machinery division of BARI developed a rotary-type bed planter with a single seed box. While machinery for seeding crops into crop residues with minimum soil disturbance is well developed for four-wheel tractors, there has been limited development of such planters for two-wheel tractors (Johansen et al., 2012). But most cropping systems in Bangladesh involve one or two rice crops per year, and developing a system of CA that is suitable for wetland rice has been a major challenge (Bell et al., 2019). For wetland rice, direct seeding and non-puddled transplanting have been developed to establish rice with minimal soil disturbance and crop residue retention (Haque et al., 2016b; Gathala et al., 2021).

Machinery development for mechanized planting into standing crop residue has therefore been the focus of CA research and development in Bangladesh. Nevertheless, the two-wheel tractor planters can only reliably sow into 40–50% standing rice and wheat residue (Haque et al., 2018a). Considering the ease of bed formation, the WRC-BARI-Cornell University (2010–2014) Food for Progress project further improved the rotary type of bed planter with the introduction of an inclined plate seed metering system, separate seed and fertilizer box, split power transmission chain and set-up of the operator's seat for easy movement and greater adoption in the farmer's field. Adaptations of imported Chinese seeders are available in Bangladesh for mechanized strip planting (Krupnik et al., 2013), but most of this machinery is used for full tillage rather than minimum soil disturbance planting (Haque et al., 2018a). The VMP is a planter developed in Bangladesh for operation on the two-wheel tractor (Haque et al., 2016b, 2018a). It can sow the seed and place fertilizer in rows using strip planting and zero tillage modes and sow into 40 cm high cereal residues (equivalent to 5 t of cereal residue/ha). Following a decade of research and development to improve seed metering, seed delivery and cover, for a range of seed sizes, and with continuous or spaced seed placement, the VMP is now in a machinery commercialization phase with a private sector manufacturer.

The CSISA and ACIAR-SRFSI projects are further expanding these tillage technologies in the Southern and Central parts of Bangladesh. Moreover, leading NGOs such as RDRS, IDE, CARITAS and PROVA are also expanding CA-based technology in the farmer's field. Current adoption levels of CA are not available in the country, but in 2015–2016, an area of around 1500 ha was under CA (Kassam et al., 2019). However, CA literature in Bangladesh is still dominated by experiments reporting on a single crop established using minimum soil disturbance and residue retention in a field previously under conventional tillage. While the Government of Bangladesh has implemented a series of machinery subsidy schemes, no special consideration is given to CA planters. The scale out of CA is ready for investment in widespread demonstrations of the CA planters on farmers' fields with a diverse range of crops. Initial pilot testing of the market suggests that a mixture of individual farmers and small entrepreneurs operating planting businesses will be attracted to the purchase of planters such as the VMP, but targeted incentives to decrease the net purchase price will accelerate the scale out. Business models for the service providers appear to be highly profitable when >20 ha per year is planted, but for an individual operator, 8–9 ha ensures a breakeven position (Miah et al., 2019).

2.5 Nepal

The ZT was first introduced in the wheat crop of the Terai region of Nepal in the late 1990s through RWC in order to reduce yield losses due to late planting of wheat (Giri, 1998). Over time, other RCTs such as bed planting and laser land levelling, along with CA, were introduced in Nepal. The NARC, in close collaboration with CIMMYT, initiated ZT maize in central Terai and the western mid-hills of Gulmi and Palpa in 2010. Many studies of the rice-wheat and rice-maize system in Terai and maize-based system in the hills have been carried out by CIMMYT to evaluate the effectiveness of RCTs. Upland and lowland maize under CA-based practices were found to be superior to conventional practices. For successful adoption of DSR, many efforts have been made by NARC, CIMMYT's Cereal System Initiatives for South Asia (CSISA) and IRRI and local agriculture-based NGOs such as LIBIRD, Pokhara and FORWARD, Chitwan. The CSISA was first initiated in 2009 by IRRI and CIMMYT in the central Terai region, with a focus on the rice-wheat system to promote widespread adoption of RCTs to increase productivity with less water, labour and input costs, and efforts were further extended to Western Terai of Nepal. Through the 'Sustainable and Resilient Farming Systems Intensification' (SRFSI) project (2014–2020), CIMMYT conducted research and promoted CA-based Sustainable Intensification (CASI) and efficient water management practices to increase crop productivity and resilience in the Eastern Terai of Nepal.

To date, CA-based practices have been tested and verified at the farm level, but the rate of upscaling is very poor. A dozen of postgraduates have been granted scholarships by CIMMYT for MScs and PhDs in CA from Nepalese universities. The Agriculture and Forestry University in Nepal has run a course on CA in both graduate and post-graduate studies since 2011. The momentum has not been as expected in Nepal compared to other parts of the globe. However, of the total rice-growing area of 1 497 744 ha, wheat area of 703 992 ha and maize area of 956 447 ha, an estimated 800 ha are covered under DSR, 500 ha under ZT maize and 1000 ha under ZT wheat in Nepal. In the hills, legumes are planted as relay or intercropping with maize on an area of about 30% of the total maize growing area. Of the total fruit growing area of 120 028 ha, about 15% ha is under legumes mulching. Mulching has become common in vegetables, and ginger and turmeric are 100% mulched. Garlic in western Terai is commonly grown under a no-till field with anchored rice residue after rice harvesting. Almost all of the cardamom growing area of 1555 ha is under the agroforestry system. Similarly, of the total vegetables grown under tunnels, about 30% are mulched with polythene. Currently, the International Fertilizer Development Centre (IFDC) is evaluating efficient nutrient management technology under CA-based cropping systems in collaboration with CIMMYT through a USAID-funded project. The uptake of CA is limited in Nepal despite tremendous potential. The key constraints are the fixed mindset of farmers and extension for intensive tillage, lack of knowledge, expertise and inputs (especially equipment and machinery), inadequate financial resources and infrastructure and poor policy support (Karki and Shrestha, 2014). The current area of CA in Nepal is estimated at 1000 hectares.

2.6 Sri Lanka

Traditionally, shifting cultivation was the predominant practice in Sri Lanka but, due to population growth, farmers started to settle in one place and practiced subsistence agriculture. This form of cultivation was not for large-scale commercial use but for limited small-scale farming to meet family needs. Since farmers had to settle at a particular location, the land was continuously used for cultivation, where land degradation, particularly soil erosion, was unavoidable, thus productivity declined. The Department of Agriculture launched research studies in 'Conservation Agriculture' way back in the early 1980s with the intention of incorporating such practices into farmers' fields to form a stabilized rain-fed cultivation system. The CA research programme commenced activities in 1980, led by Ray Wijewardena at Maha-illuppallama in collaboration with a Nigerian institution (Weerakoon, 1983).

CA was tried elsewhere in the dry zone of Sri Lanka to address problems associated with rain-fed upland farming. In Sri Lanka, CA includes diversified components - a combination of agriculture with forestry through mixed cropping, integration of perennials, green manuring, biological N_2 fixation, use of live crop or crop residue mulching, integration of botanical pesticides, integrated pest management, integration of crops with livestock, supplementary mineral fertilizer and use of low-cost energy-conserving tools. During the subsequent two decades, many field experiments were conducted and most produced encouraging results.

CA is a very promising approach, and a wealth of research findings and information are available based on two decades of research in the 1980s and 1990s. However, it must be stated that, although the research information is available, farmers did not fully accept the technology, probably because the programme was research-biased with no extension arm to upscale the technology. Moreover, social problems have not been adequately addressed. The Department of Agriculture is presently the authority for the implementation of the Soil Conservation Act of Sri Lanka. The Natural Resources Management Centre (NRMC) is responsible for the implementation of the Soil Conservation Act. The NRMC is launching an annual soil conservation programme in collaboration with Provincial Departments of Agriculture on 1000–1200 ha of farmers' fields supported by technical and financial assistance. In this programme, mechanical soil conservation, along with biological and agronomic measures, are being promoted. Lessons learned during conservation farming research programmes need to be incorporated into the above programme, especially for the rain-fed uplands in the dry zone.

In 2019, the FAO's 'Save and Grow' and 'Climate-Smart Crop and Mechanization Systems Scaling Up' projects started to introduce CA again in collaboration with the Department of Agriculture. The approach will create opportunities for profit by way of using CA combined with sustainable mechanization services and high-quality inputs (seed varieties and selected fertilizers). The projects operate in the dry zone of Sri Lanka, targeting small- and medium-scale farmers, who cultivate lowland paddies under minor irrigation systems and maize-based crops in rain-fed conditions. The project promotes a switch from maize monoculture to maize produced in rotation with groundnut and a 'sandwich crop' between the two main crops.

To control weeds in the off-season and to get additional income, sun hemp and sesame are used as sandwich cover crops. Crop residues are maintained. This production system is being promoted in farmers' fields and in the climate-smart agriculture village that is run by researchers of the Department of Agriculture. As the main objective of the project was to provide the necessary guidance for agricultural mechanization, which is an essential component of CA, several activities were carried out as well. Mechanization is not common among farmers due to the high cost of agricultural machinery, lack of machinery availability and lack of knowledge of mechanization practices. The usual farmer practice is to use only tractors for land preparation and harvesting machineries to harvest. The remaining operations are done manually. Most of the time, female farmers are involved in operations such as collecting, threshing, drying and haulage operations. The FAO and the Department of Agriculture have worked together to develop a programme to rent agricultural machinery at affordable prices, and the construction of the building was done by the department at one of the most suitable locations in the dry zone. The necessary machinery was selected, procured and provided by the project. The technical contribution of the engineers and extension officers of the department was invaluable in terms of providing mechanized training and demonstrations. Under the machinery hub concept, machinery was introduced for paddy cultivation, maize, groundnut, finger millet and other supplementary food crops grown in the dry zone. A television programme was also aired to promote the practice among farmers. More information is available at: http://www.fao.org/in-action/save-grow-climate-smart/our-approach/en/. The area of CA is about 50 ha (Sandra Corsi and Priyadharshanee Herath, FAO, 21 June 2021).

2.7 Afghanistan

In war-torn Afghanistan, agriculture is the backbone of the country's GDP and source of livelihood for more than 74% of the population. The country has made significant progress in the agriculture sector during the past two decades, but its sustainability and food security are still challenged by multiple intermingled factors, including limited access to technology, lack of trained human resources and insufficient policy support. Farming conditions are more traditional, with more than 50% of agriculture production from dryland conditions, which could be attributed to a lack of resources, ageing scientific institutions and unskilled human resources in the agricultural production system, which is still very traditional.

During 2008-2009, ZT technology was first introduced in Afghanistan through International Fertilizer Development Centre under the USDA-funded project. The ZT trials were conducted in the Nangarhar province of Afghanistan in rice crops, along with different nutrient management options. The ZT technology performed better than traditional rice cultivation. The IFDC organized the first CA study tour of Afghanistan stakeholders at RWC and Punjab Agricultural University (PAU) and other sites in Haryana and Punjab in India during 2009. During 2011-2019, CIMMYT and ICARDA organized several in-country and out-country trainings on CA for different stakeholders and introduced the ZT drill and Happy Seeder. ZT and laser levelling technologies have been adopted on a small scale in the country.

2.8 Conservation Agriculture in rain-fed areas: an emerging avenue in South Asia

Rain-fed farming soil health and moisture conservation practices directly influence the extent of potential crop production (Singh et al., 2015b). Healthier soil is more resilient to climate extreme events such as excess rainfall or drought, high or low temperatures etc. The adoption of CA is an emerging dimension that improves the SOC stock in the soil by recycling crop residues into the soil system. The competitive need for biomass for fodder must be taken care of in rainfall-deficit conditions. CA systems improve yields in rain-fed conditions on alfisols, which are shallow, often with compacted subsoil horizon and susceptible to hard setting. Traditionally, the area under degraded alfisols of Southern India is a mono-cropped area with maize being cultivated during the rainy season, but with adoption of CA, there is the potential to produce a post-rainy legume crop (e.g. horse gram, *Macrotyloma uniflorum*).

To harness the advantages of CA systems in SA's semi-arid tropics, it is essential to retain crop residues on the surface as mulch, which during the early years of transformation to CA can be a major challenge in rain-fed regions of South Asia due to the competing demands of crop residues for livestock. The potential for CA exists in rain-fed cropping systems involving crops such as maize, pigeon pea, mung bean, castor, cotton and sunflower, where crop residues are not used as feed. Ensuring good seed germination and crop stand establishment are major challenges to be addressed by CA management. Being the only source of water, interactions between rainwater conservation and CA must be studied in an integrated manner. Sharif et al. (2017) reported that studies on CA conducted to date in dryland areas of Pakistan have been short term and CA has the potential to improve soil structural stability by enhancing organic matter content, providing equal yield and economic benefit by reducing input cost. There is great potential for CA in dryland areas of Afghanistan, Pakistan and India as the climate, soils and crop diversity are conducive to adoption of CA systems.

3 Crop productivity and income in Conservation Agriculture

Long-term studies in SA show that CA has the potential to improve sustainability, profitability and input use efficiency in both irrigated and rain-fed regions. A meta-analysis by Jat et al. (2020b) showed that CA has the potential for 11.1% yield enhancement in on-station trials compared to 4.6% in on-farm CT. In EIGP, the yield advantage of 19.4% has been reported from ZT compared to CT in wheat as reported in the survey conducted in Bihar (Keil et al., 2015). The results from a long-term experiment in upper IGP covering India and Pakistan showed higher system productivity in CA-based rice-wheat/maize-wheat-mungbean systems compared to CT-based rice-wheat system. This might be due to the legume effect (integration of mungbean crop into the cropping system) and better thermal regime with reduced terminal heat stress in wheat crop (Gupta et al., 2010), higher soil water availability (Gathala et al., 2013; Jat et al., 2018; Sharma et al., 2019a) and regulation of soil moisture retention and release, thus avoiding extreme moisture conditions such as waterlogging and soil dryness in the absence of rainfall (Parihar et al., 2019a) and improved soil health, including increased plant nutrient availability (Singh et al., 2016; Choudhary et al., 2018a,b). In addition, it was found that the quality of wheat grain was improved under ZT compared to CT. For example, wheat grain grown under ZT had a higher protein, grain hardness and chapatti (Indian bread) score from all four rotations (rice-wheat, sugarcane-wheat, pearl millet-wheat and cluster bean wheat) than CT (Coventry et al., 2011).

In Pakistan, the use of village seed-bank-supplied good-quality seed alone increased wheat yield by 297 kg ha^{-1}. Laser levelling adopted on an area over 2080 ha improved seed germination by 15–25% and reduced irrigation water use by 20–30% and labour for irrigation by 30%. The ZT wheat planting saved 60 L of diesel fuel per hectare and produced an additional grain yield of 141 kg ha^{-1}. Bed planting of wheat increased yield by 566 kg ha^{-1} and reduced irrigation water use by 33%. With the integration and adoption of these technologies in pilot villages, farmers earned an estimated additional income of US$3 million in 2005–2008.

A diverse range of CA research has been carried out in Bangladesh, and much of it was reported in two international conference proceedings (Vance et al., 2014b; Haque et al., 2017a). In addition, regional studies on CA in the Eastern Gangetic Plain have included sites in Bangladesh (Gathala et al., 2021). Based on over a decade of research, the best practices for CA in rice-based cropping systems have recently been assembled in a manual for operators, extension officers and researchers (Haque et al., 2018a). The manual addresses site selection including the important issues of pre-existing weeds and the height of crop stubble from the previous season, seed quality, types of planters including their merits and limitations, crop management guidelines, weed control, best practice for a range of crop species, as well as the overall benefits of adopting CA in Bangladesh. For dry season crops, there are generally significant and immediate increases in crop yield when switching to CA (Bell et al., 2018; Salahin et al., 2021). With rice, there may be no significant effects on yield for 2–5 years. However, even when there are no increases in crop yield, as in the case of non-puddled transplanted rice, profit increases are generally (Haque et al., 2016a) even on farmers' fields (Haque and Bell 2019) but not on sandy loam soils (Chaki et al., 2021). Mechanized non-puddled transplanting of rice has been reported to increase crop yield and profit (Hossen et al., 2018).

The larger yield response of winter crops (wheat and maize) to crop residue retention and ZT (wheat only) might be associated with the higher conserved soil moisture, improvement in physical properties of the soil, increased SOC and moisture-dependent plant nutrient accessibility (Laik et al., 2014; Gathala et al., 2017; Nandan et al., 2019). In the rice-maize-mungbean system in Bangladesh, Rashid et al. (2019) reported that maize yields were significantly higher under the permanent bed (PB) method than in the conventional till method, but no differences were observed in mungbean and rice yields. Haque et al. (2018b) in Bangladesh reported a yield increase of 28% for lentils, 19% for mustard and 6% for wheat by farmers who adopted CA planting. Profits increased by 47% for lentil, 55% for maize, 46% for mustard and 76% for wheat due to the adoption of CA. In studies that recorded improvements in soil health under CA management systems, significant increases in yield have been obtained.

On a cropping system basis, the highest yield enhancement was observed in CA-based maize-wheat systems (18.6%) (Jat et al., 2020b), followed by rice-wheat (5.1%) and rice-maize (3.6%). The CA systems provided higher profitability due to reduced labour cost, reduced input cost for land preparation and lower water requirements, and partly from increased yield in irrigated systems (Jat et al., 2014; Kumar et al., 2018; Devkota et al., 2019). Increased yields due to retention of crop residues in established CA systems are attributed to (a) conservation of soil moisture and nutrients; (b) improved soil water infiltration; (c) improved soil biological activities and nutrient cycling; (d) better weed control; (e) improved soil quality through increased soil organic matter content; and (f) regulation of soil temperature, thereby minimizing high-temperature effects during the wheat grain filling period.

4 Soil health in Conservation Agriculture

Soil organic carbon (SOC) is one of the most important indices of soil health and sustainability in agriculture. Soils of semi-arid and arid regions of SA are generally low in OC content. Carbon sequestration in soil is an important strategy to improve soil quality, mitigate climate change and improve soil quality associated with higher crop productivity in CA (Parihar et al., 2020). Potentially one-third of the C emitted in current fossil fuel use could be offset by implementing CA globally in the next decade. A significant build-up in SOC in 0–20 cm under CA (ZT+ residue) compared to the CT system was recorded by Paudel et al. (2014). In Nepal, ZT+ residue-retained plots sequestrated 0.91g kg^{-1} $year^{-1}$ SOC, which was 23% higher than CT after five cycles of rice-wheat system. Several other studies from SA also recorded significant increases in SOC contents in CA treatments over CT (Choudhary et al., 2018a,b; Bera et al., 2018; Sharma et al., 2019b; Bhattacharyya et al., 2013; Singh et al., 2016; Jat et al., 2014, 2019; Thind et al., 2019; Yadvinder-Singh et al., 2009, 2014). From long-term on-farm trials maintained since ACIAR project in 1997, it was found that soil health of ZT plots was superior to CT as studied by Singh et al. (2015a). Data show that the carbon stock in surface 0.4 m soil depth increased by 19.0%, 34.7% and 38.8% over CT in 15 years in sandy loam, loam and clay loam soil, respectively.

Ghimire et al. (2012) concluded that the R-W system would serve as a greater sink of SOC under the ZT system with plant residues retained than with the CT system without residues. Sharma et al. (2019b) reported the highest carbon management index (CMI) values under the CA-based rice-wheat system, which provided better options for C sequestration in soils than conventional rice-wheat systems. CA improves the physical properties of soil, such as structure, aggregate stability, infiltration rate and plant available water capacity, as well as biological quality. In medium-term CA-based experiments, Parihar et al. (2018a) recorded a 34% and 21% increase in SOC under CA in maize-based cropping systems compared to CT systems, respectively. Based on research in Bangladesh, Bell et al. (2019) reported that CA practices decreased soil bulk density and increased soil porosity. The release of polysaccharide compounds during the decomposition of crop residue acts as a cementing agent and has a crucial role in micro- and macro-aggregate formation (Six et al., 2002; Bhattacharyya et al., 2013).

CA-based agricultural practices significantly influence infiltration rate and soil water retention. The lower infiltration in conventional tilling as compared to ZT is due to a breakdown of aggregates and formation of surface seal by the impact of raindrops, increased compaction and reduction in pore volume of the surface soil. CA leads to the formation of stable soil aggregates and ultimately results in less soil erosion. The majority of the reports from SA showed a decrease in bulk density under CA over CT in different cropping systems and soil types (Jat et al., 2018b; Choudhary et al., 2018a,b, Parihar et al., 2020). Crop residue cover in CA buffers the soil temperature regime in the surface layer by lowering the maximum soil temperature (often between 2°C and 8°C) and increasing the minimum soil temperature (by 1–3°C) compared to CT (Singh et al., 2016). The decreased soil temperatures in residue mulched wheat plots at grain filling stage resulted in a positive effect on reducing canopy temperature (by 1-3°C) and ultimately decreased the adverse impact of terminal heat on the yield (Balwinder-Singh et al., 2011; Yadvinder-Singh et al., 2014).

In Sri Lanka, CA, crop residues and live mulch are used to control weeds and protect soils. Studies revealed that the application of sun hemp mulch at the rate of 1.5–3.5 t/ha can reduce soil loss to 1.8–3.5 mt ha^{-1} $year^{-1}$ (Weerakoon, 1983). Investigations into the effects of tree components on upland soil fertility in the dry zone revealed improvements in bulk density, organic matter content, exchangeable K, Ca and Mg, N and earthworm activities in soils in Sri Lanka. Suppression of grass weeds and dominance of broad-leaved species were also observed between Gliricidia rows. However, available P in soil was depleted with continuous cropping on the land (Handawala and Kendaragama, 1991). The inclusion of a grass component on the tree row has become necessary in order to improve alley cropping as a grass-tree alley system, which would be more beneficial than tree alley cropping (Dharmasena, 1996).

SOC improvements in CA systems can have a significant effect on plant nutrient availability, due to both changes in the quantity of nutrients available and their distribution in the soil profile. The continuous addition of large quantities of crop residues in CA is associated with the addition of plant nutrients in residues to the soil, resulting in increased storage and availability of macronutrients (Jat et al., 2013, 2019; Bera et al., 2018; Nandan et al., 2019) and micronutrients (Gupta et al., 2007; Nandan et al., 2018; Rashid et al., 2019). Soil microorganisms appear to be excellent indicators of soil health as they respond quickly to changes in the soil ecosystem. Jat et al. (2019) recorded 35% and 54% higher microbial biomass carbon and nitrogen in CA-based rice-wheat-mungbean system and 50% and 110% in the CA-maize-wheat-mungbean system, respectively, than in CT-rice-wheat from a 7-year-old field experiment. The CA approach improves soil health, decreases costs of production and increases crop profitability.

5 Water productivity under Conservation Agriculture

CA-based cropping systems require 30–71% less irrigation water and use 24–47% less energy compared to CT-rice-wheat (Kumar et al., 2018). The CA-based maize-wheat-mungbean system improved productivity by 16% and profitability by 50%, using 71% less irrigation water and 47% energy input compared to the conventional rice-wheat system/farmers' practice (Kumar et al., 2018). In a cotton-wheat system, Choudhary et al. (2016) reported that mean system water productivity and energy output were 131% and 21% higher with permanent broad bed (PBB)-mungbean, respectively, compared with CT. Surface retention of crop residues reduced evaporative loss and contributed to moisture conservation, thus requiring less irrigation water than in the CT system with no residues (Jat et al., 2016). Mustard (*Brassica juncea* L.) can be an economically alternative crop to wheat in both irrigated and rain-fed agroecologies in SA, while grown under the CA-based system.

Devkota et al. (2019) evaluated the performance of ZT wheat in the Western Terai region of Nepal. ZT wheat produced a similar or higher grain yield with lower total production cost by $70 ha^{-1} and higher water productivity by 30% and net profit by US$115 ha^{-1} as compared to CT in rain-fed or limited irrigation areas. From a study conducted in Pakistan, Nawaz et al. (2017) concluded that DSR followed by ZT wheat was the best RCT to sustain the productivity of the rice-wheat system and improve net profit and soil properties. In conclusion, CA-based systems generally enhance crop productivity, water productivity and profitability, while reducing irrigation water and energy requirements compared to CT systems. Therefore, CA should be out-scaled to improve soil and environmental quality in SA.

6 Climate change mitigation and adaptation in Conservation Agriculture

Projections state that almost half of the IGP in SA – the major food basket of the SA region – may become inappropriate for wheat production by 2050 due to heat stress as a result of higher temperatures associated with GHG emissions (Ortiz-Monasterio et al., 2010). CA is reported to reduce GHG emissions and promote C sequestration in soil and play an important role in contributing to mitigation and adaptation/resilience to climate change for food security (Nandan et al., 2019). In medium- and long-term CA-based experiments, the CA components had higher SOC sequestration potential and the sequestered SOC has greater stability against higher temperatures as evident from twice less Q_{10} value as compared to CT (Parihar et al., 2019b). According to Ashraf et al. (2015), improved production practices under CA improve SOC and N storage while reducing soil erosion and N leaching from the soil profile without significantly altering crop yields and has GHG mitigation potential as compared to CT. Parihar et al. (2018b) estimated annual N_2O emissions using gas chamber methods and concluded that GWP per year or per hectare, associated with nitrate-N emission, was less in an intensive CA-based system as compared to CT. Arshad et al. (2017) suggested that agronomic management such as ZT with residue retention will help alleviate the period of critically high heat during flowering in wheat, which can advance sowing dates while conserving soil moisture in SA.

Carbon sequestration under CA helps to mitigate GHG emissions and increase soil productivity and avoid further environmental damage from the unsustainable use of intensive tillage systems. CA helps reduce CO_2 emissions via energy savings through less fuel consumption and reduction of the mineralization of soil organic matter, leading to a markedly less negative effect on climate change. Improved soil nutrient availability in CA systems lowers external N and other fertilizer requirements, which means less potential for N_2O emissions and energy saving from the manufacture of fertilizers. CA systems also mitigate climate change by enhancing soil moisture conservation, thereby helping save electricity costs and reduction in GHG associated with energy production consumed in irrigation. A study by Ladha et al. (2015) showed that adoption of CA practices in rice can reduce total global warming potential (GWP) for rice by 23% annually. Diversifying rice with maize has the potential to reduce GWP by 38%. On a system basis, CA-based rice-wheat and maize-wheat systems have the potential to reduce GWP by 16–26% compared with the CT-RW system. A significant reduction in GWP in CA-based maize-wheat system suggests that in areas where diversification of rice-wheat system is feasible, there is great scope for mitigation of GHG emissions. Soil organic carbon (Alam et al., 2018) and total nitrogen (Alam et al., 2020) concentration (0–10 cm) increases of up to 65% were reported after 5 years of continuous CA relative to the current practice of intensive tillage and <20 % crop residue retention. According to Alam et al. (2020), a 26% increase in soil organic carbon stock, equivalent to 5.1 t ha^{-1} after 5 years, was achieved with CA practices. Continuous strip planting plus increased crop residue retention decreased life cycle greenhouse gas emissions by 35%, equivalent to a savings of 0.25 t emissions per metric tonne of rice production relative to current practices (Alam et al., 2019a,b). Another study from Northwest India shows that the CA-based rice-wheat system reduced CH_4 emissions and resulted in the lowest GWP (CO_2-eq ha^{-1} $year^{-1}$) compared with CT systems (Jat et al., 2020a).

7 Nutrient management in Conservation Agriculture

Nutrient management is an important aspect of CA for enhancing crop productivity and its adoption by farmers. The conventional nutrient recommendations are not necessarily a valid basis for fertilizer recommendations in CA. Nutrient management can become more complex with CA due to higher residue levels at the soil surface, which makes the surface broadcasting and band placement application of fertilizer less efficient. In addition, deciding the timing and amount of fertilizer application must include the nutrient release pattern from crop residue and the soil pool. Thus, nutrient management should focus on local site-specific conditions, not only to achieve the full benefit of available resources but also to overcome crop yield barriers. Site-specific nutrient management (SSNM) captures the spatial and temporal variability in soil fertility and provides an approach to 'feeding' crops with all the required nutrients based on crop needs and thus improving the crop yield. As the concept of CA is a relatively recent introduction in SA, only a few studies are available on nutrient management in different crops and cropping systems. In CA, organic mulch will improve the soil moisture regime and nutrient dynamics that in turn will influence nutrient response and economic profitability. Yadvinder-Singh et al. (2009) found that rice straw mulch containing 40 kg N ha^{-1} released only a small fraction of N during the life span of the wheat crop (~150 days). With such small amounts of N released from the residue, significant savings in fertilizer N in wheat are unlikely in the first 2 to 3 years. CA in most cases improves the availability of P and K in the surface soil layer due to reduced mixing of fertilizer and surface retention of crop residues. The reduced mixing of P and K fertilizers with the soil in CA is expected to lower P-fixation (Gupta et al., 2007).

In CA production systems, innovations are needed in no-till seeding equipment to allow separate seed and fertilizer band placement. Major technologies focused on the adoption of new diagnostic tools for nutrient management under CA include the use of a leaf colour chart, Green Seeker and Nutrient Expert® (NE®) Software tool (a fertilizer decision support tool based on SSNM principles) and use of fertigation through micro-irrigation. As applicable under the conventional system, NUE under CA could be improved through the '4Rs' nutrient stewardship (right rate, right source, right timing and right place) and precision nutrient management practices that are specific to field soil, climate and crop management. Improvements in soil organic matter content and nutrient availability will have a significant bearing on nutrient management under CA (Kakraliya et al., 2018; Parihar et al., 2016). Parihar et al. (2016) reported that a combination of ZT/PB and SSNM increased maize-wheat productivity by 23% and 41% compared with CT recommended fertilizers and farmer's practices, respectively. Nayak (2019) and Nayak et al. (2019) addressed the right placement issues under long-term CA and concluded that point placement of second urea split enhances the maize grain yield, energy output and profitability as compared to CT. The layering of CA-based management practices with precision nutrient prescriptions using SSNM-based decision support tools offers a new management paradigm for scaling up the CA-based system.

8 Weed management in Conservation Agriculture

Traditional intensive tillage practices eliminate early crop weed competition, whereas under CA, weed control with tillage is not possible. The adoption of the full CA package of practices (CA with all three principles) modifies the soil environment, while some favour weed control, while others exaggerate weed proliferation. In the no-till CA environment, about 60–90% of the weed seed lies within 5 cm of the soil profile, as seeds move through the cracks and by fauna (Bàrberi et al., 2001). The chances of higher weed seed germination during the initial years of CA adoption are greater. Gradually, the soil seed bank depletes due to less soil disturbance/inversion (deeper weed seed does not come to the surface), increased predation due to more accessibility, reduced viability due to greater fluctuating environment at surface soil layers, etc. (Nichols et al., 2015).

Several field experiments and modelling-based findings suggest, if carefully managed, the higher weed germination during the initial year of CA adoption provides an opportunity for long-term weed control. Preventing weed establishment is more crucial under CA than the conventional till (CT) system. The less disturbed CA system favours perennial weeds, whereas intensive tilling under CT favours the dominancy of annual broad-leaved weeds. Due to no tillage, the reproductive structures of perennial weeds are neither buried too deep to avoid emergence nor come to the surface for desiccation (Sharma and Singh, 2014). Along with tillage, the nature/type of crop rotation dictates the nature of weed communities. The residue retention under CA decreases or delays the weed seed germination by moderating daily soil temperatures (weed seed requires certain threshold temperatures for germination), allelopathic effects seem to encourage seed decay (residue maintains a moist environment) and by acting as a physical barrier (avoids light penetration and seedling emergence), but the timing, form and application methods of herbicides must be chosen wisely as surface-retained residue intercepts the applied herbicide to the tune of 15–80% (Chauhan et al., 2012). The granular form of pre-emergence herbicide may be more effective under CA systems.

The principle of crop rotation under CA was historically advised as a component of integrated weed management. Crop-specific weeds are effectively controlled by efficient crop rotation, as well as the rotation of an allopathic crop, a weed-resistant crop and smothering crop, which favours less weed germination. Although all three principles of CA favour reduced weed growth, the transitioning phase to CA during the initial 4 to 10 years is important, and perennial weeds should be effectively controlled using glyphosate or other non-selective herbicides. Therefore, a greater amount of herbicides might be needed during the initial years in CA. Strip planting and crop residue retention for 2–5 years continuously has been reported to decrease the size but increase the diversity of the weed seed bank, particularly the dominance of perennial species, under intensive rice-based crop rotations in Bangladesh (Hossain et al., 2021a,b). In recent years, several low-dose, high-potency, selective, post-emergence herbicides and their mixtures have become available in India and South Asia and can effectively manage weeds in crops like rice and wheat grown in sequence under CA. Herbicide use must be complemented by other measures such as fertilizer placement, planting dates, microbial weed control, implementing 4+ year rotation, etc. Weed control is more challenging when CA is adopted partially, where the benefits of synergism between the three principles are not achieved. In northwest India, reduced *Phalaris* infestation has been reported under zero-till-wheat. CA involves the implementation of many skilful techniques such as the use of appropriate machineries, herbicides, proper crop rotation, soil and residue management. No doubt CA has multiple advantages, but its adoption is still not widespread with the full package of practices.

9 Further upscaling of Conservation Agriculture in South Asia

The total cropland area under CA globally in 2015/16 was estimated to be about 180 M ha in more than 75 countries (Kassam et al., 2019). The area under CA cropland is expanding worldwide at an average rate of around 10.5 M ha year^{-1} since 2008–2009. In SA across India, Pakistan, Bangladesh and Nepal, in the R-W system in the IGP, the estimated adoption of ZT wheat was about 5 M ha area, which has increased since. The adoption of various CA systems has been reported to be over 4 M ha (Paroda, 2018; Kassam et al., 2019). The success of CA spread in SA is attributed to a multi-stakeholder approach and low-priced drills. The success of CA drills from within the region has created demand for CA in other countries and regions. According to Devkota et al. (2019), while the benefits are clear, coordinated efforts are required to overcome technology-scaling bottlenecks that have kept adoption rates of these technologies at a nascent state in SA. The double no-till CA system of rice and wheat has not yet become mainstream in South Asia. The main barriers to the adoption of CA practices could be due to competing uses for stover/straw as forage, particularly under rain-fed situations, the large-scale practice of open residue burning in the IGP, lack of suitable machinery for no-till seeding of rice, lack of skilled manpower and scarcity of trained service providers (SPs) with the right equipment. This is partly because policy environments are not favourable and poor understanding of the CA practices by farmers due to the prevailing tillage mindset. These challenges relate to development, standardization and adoption of farm machinery for seeding with minimum soil disturbance and developing crop harvesters with effective residue management systems. Bell et al. (2019) concluded that CA in Bangladesh is feasible and a profitable venture for smallholder farmers and that the local service provider business is profitable for ZT planting. However, suitable planters should be made available at an affordable price. Several manufacturers were reluctant to invest without evidence of demand. Low-quality machinery will hamper efforts to develop farmer confidence in CA.

The lack of subsidies and incentives in a context of high poverty in rural areas does not create a favourable environment for CA adoption. Farmer participatory on-farm research to evaluate/refine the technology in the initial years, followed by large-scale demonstration in subsequent years is needed. Due to the wide range of agroecological conditions ranging from arid, semi-arid to sub-humid zones in SA, it is important for the promotion and development of CA to continue to identify entry points for its adoption and spread. There is a need to think about the problems faced at the implementation level and devise a strategy involving all concerned stakeholders. The potential limitations for future wide-scale adoption of CA include the high initial cost of no-till seeders, inability of ZT seeders to handle large amounts of loose straw left on the surface after combined harvesting of the previous crop and a lack of scale-appropriate and locally adapted machinery for smallholders in some locations (Sapkota et al., 2015; Mishra et al., 2021). Promotion of farm mechanization for CA through custom hiring centres (CHCs), private entrepreneurs and farmers' organizations does benefit small and marginal farmers, and this strategy must be expanded. Many medium- and large-scale farmers who purchase CA machinery in turn also become service providers to smallholder farmers. These farmers have the opportunity to supplement their farm income with the purchase of CA machinery for use on their own farms as well as neighbouring fields. They should be encouraged and supported to acquire CA machinery, leading to a successful business model for large-scale adoption of CA by all types of farmers. Agencies like Krishi Vigyan Kendras (KVKs) in India and extension wings of universities need to increase their involvement in introducing farmers to the custom hiring adoption of CA systems.

Policy barriers that constrain the adoption of CA technologies include issues such as farmers' financial capacity, short timeframes for economical use of ZT seeders, a serious lack of farmer training opportunities and resources and poor enforcement of the crop residue burning ban. Success stories and technical discussions generated by the growing spread of CA technology, as told by farmers and others, could be garnered to make government department heads, policymakers, institutional leaders and others aware of the benefits and desirability of backing initiatives to help farmers adopt CA systems. Training and capacity development are the core areas for out-scaling of CA. Consideration of extension approaches such as the 'Lead Farmer Approach' should also be made as a way to mitigate extension shortages at the local level. Targeting key policymakers, advisors and engaging governments/decision makers' policy in events related to CA can also create public awareness and knowledge. There is also a need for more private sector investment and government funds to refine and develop farm equipment and machinery to promote CA and complementary precision agriculture practices.

10 Where to look for further information

Volumes 1 and 2 of this book have more information related to CA in SA. The main global website for more information is the FAO at: http://www.fao.org/conservation-agriculture. Information is also available at the websites of regional CA organizations and they are: http://www.act-africa.org/ for Africa; https://ecaf.org/ for Europe and http://www.caa-ap.org/ for Asia-Pacific. More information is available at the national level in several countries that have CA organizations. The CIMMYT website (https://www.cimmyt.org/tag/conservation-agriculture/) also has information about CA in SA and generally. Research papers on CA from SA countries can be found in international journals.

11 References

Abrol, I. P., Bronson, K. F., Duxbury, J. M. and Gupta, R. K. 1996. Long-term soil fertility experiments in rice-wheat cropping systems: proceedings of a workshop. Rice-Wheat Consortium Research Series No. 1. New Delhi, India.

Abrol, I. P., Bronson, K. F., Duxbury, J. M. and Gupta, R. K. 2000. Long-term soil fertility experiments in rice-wheat cropping systems. Rice-Wheat Consortium Research Series No. 6. New Delhi, India.

Adhikari, C., Bronson, K. F., Panuallah, G. M., Regmi, A. P., Saha, P. K., Dobermann, A., Olk, D. C., Hobbs, P. R. and Pasuquin, E. 1999. On-farm soil N supply and N nutrition in the rice-wheat system of Nepal and Bangladesh. *Field Crops Res.* 64(3), 273–286.

Ahmad, M., Zaidi, M. A. and Khan, A. S. 1994. Development and adoption of no-till technology for wheat sowing. *Agric. Mech. Asia Afr. Lat. Am. (AMA), JPN* 25(4), 24–28.

Ahmad, N., Malik, R. K., Aroa, R. and Mehla, R. S. 2001. Adoption of resource conservation technologies by farmers in Haryana - a success story. Conservation Agriculture for Food Security and Environment Protection in Rice-Wheat Cropping Systems, Lahore, Pakistan, 17–19, February 6–9.

Akhtar, S. 2016. Malnutrition in South Asia—a critical reappraisal. *Crit. Rev. Food Sci. Nutr.* 56(14), 2320–2330. 10.1080/10408398.2013.832143.

Alam, Md. K., Bell, R. W. and Biswas, W. K. 2019a. Decreasing the carbon footprint of an intensive rice-based cropping system using conservation agriculture on the Eastern-Gangetic Plains. *J. Cleaner Prod.* 218, 259–272.

Alam, Md. K., Bell, R. W. and Biswas, W. K. 2019b. Increases in soil sequestered carbon under conservation agriculture cropping decrease the estimated greenhouse gas emissions of wetland rice using life cycle assessment. *J. Cleaner Prod.* 224, 72–87.

Alam, Md. K., Bell, R. W., Haque, M. E., Islam, M. A. and Kader, M. A. 2020. Soil nitrogen storage and availability to crops are increased by conservation agriculture practices in rice-based cropping systems in the Eastern Gangetic Plains. *Field Crops Res.* 250, 107764.

Alam, Md. K., Bell, R. W., Haque, M. E. and Kader, M. A. 2018. Carbon cycling under conservation agriculture practices in rice-based cropping systems in the Eastern Gangetic Plains (EGP). *Soil Till. Res.* 183, 28–41.

Arshad, M., Amjath-Babu, T. S., Krupnik, T. J., Aravindakshan, S., Abbas, A., Kächele, H. and Müller, K. 2017. Climate variability and yield risk in South Asia's rice-wheat systems: emerging evidence from Pakistan. *Paddy Water Environ.* 15(2), 249–261.

Ashraf, S., Iftikhar, M., Ali Khan, G., Javed Khan, M. A. and Raza, H. 2015. Climate change mitigation through conservation agriculture in Pakistan: critical analysis. *Sci. Int. (Lahore)* 27, 1519–1522.

Aslam, M. 2016. Agricultural productivity current scenario, constraints and future prospects in Pakistan. *Sarhad J. Agric.* 32(4), 289–303. 10.17582/journal.sja/2016.32.4.289.303.

Aslam, M., Ahmed, M., Hashmi, N. I., Chatha, M. Q. and Vander, V. M. 1991. *Zero Tillage Wheat Pilot Production Programme for the Punjab Rice-Wheat System*. Palestinian American Research Council, Islamabad.

Bàrberi, P., Bonari, E., Mazzoncini, M., García-Torres, L., Benites, J. and Martínez-Vilela, A. (Vol. 2: offered contributions, XUL) 2001. Weed density and composition in winter wheat as influenced by tillage systems: conservation agriculture, a worldwide challenge. Proceedings of the First World Congress on Conservation Agriculture, Madrid, Spain, 451–455, October 1–5.

Beebe, J. 1985. Rapid rural appraisal: the critical first step in a farming systems approach to re- search. Farming Systems Support Project. Networking Paper No. 5. Washington, DC.

Bell, R., Haque, M., Jahiruddin, M., Rahman, M., Begum, M., Miah, M., Islam, M., Hossen, M., Salahin, N., Zahan, T., Hossain, M., Alam, M. and Mahmud, M. 2019. Conservation agriculture for rice-based intensive cropping by smallholders in the Eastern Gangetic Plain. *Agriculture* 9(1), 1–17. 10.3390/agriculture9010005.

Bell, R. W., Haque, M. E., Johansen, C., Vance, W., Kabir, M. E., Musa, M. A., Mia, M. N. N., Neogi, M. G. and Islam, M. A. 2018. Mechanised minimum tillage establishment and yield of diverse crops in paddy fields using a two-wheel tractor-mounted planter suitable for smallholder cropping. *Exp. Agric.* 54(5), 755–773.

Bera, T., Sharma, S., Thind, H. S., Singh, Y. and Jat, M. L. 2018. Changes in soil biochemical at different wheat growth stages under conservation-based sustainable intensification of rice-wheat system. *J. Integr. Agric.* 17(8), 1871–1880.

Bhattacharyya, R., Das, T. K., Pramanik, P., Ganeshan, V., Saad, A. A. and Sharma, A. R. 2013. Impacts of conservation agriculture on soil aggregation and aggregate-associated N under irrigated agroecosystem of Indo-Gangetic Plains. *Nutr. Cycl. Agroecosyst.* 96(2–3), 185–202.

Byerlee, D., Hobbs, P., Khan, B. R., Majid, A., Akhtar, M. R. and Hashmi, N. I. 1986. Increasing wheat productivity in the context of Pakistan's irrigated cropping systems: a view from farmers' fields. PARC and CIMMYT Paper 86-87.

Byerlee, D., Sheikh, A. D., Aslam, M. and Hobbs, P. R. 1984. Wheat in the rice-based farming system of the Punjab: Implications for research and extension. NARC/CIMMYT Reports Series No. 4. Islamabad: NARC/CIMMYT (National Agricultural Research Center [Pakistan]/Centro Internacional de Majoramiento de Maize y Trigo [Mexico].

Chaki, A. K., Gaydon, D. S., Dalal, R. C., Bellotti, W. D., Gathala, M. K., Hossain, A., Siddquie, N. A. and Menzies, N. W. 2021. Puddled and zero-till unpuddled transplanted rice are each best suited to different environments – an example from two diverse locations in the Eastern Gangetic Plains of Bangladesh. *Field Crops Res.* 262, 108031.

Chauhan, B. S., Singh, R. G. and Mahajan, G. 2012. Ecology and management of weeds under conservation agriculture: a review. *Crop Prot.* 38, 57–65.

Choudhary, K. M., Jat, H. S., Nandal, D. P., Bishnoi, D. K., Sutaliya, J. M., Choudhary, M., Singh, Y., Sharma, P. C. and Jat, M. L. 2018a. Evaluating alternatives to rice-wheat system in western Indo-Gangetic Plains: crop yields, water productivity and economic profitability. *Field Crops Res.* 218, 1–10.

Choudhary, M., Datta, A., Jat, H. S., Yadav, A. K., Gathala, M. K., Sapkota, T. B., Das, A. K., Sharma, P. C., Jat, M. L., Singh, R. and Ladha, J. K. 2018b. Changes in soil biology under conservation agriculture based sustainable intensification of cereal systems in Indo-Gangetic Plains. *Geoderma* 313, 193–204.

Choudhary, R., Singh, P., Sidhu, H. S., Nandal, D. P., Jat, H. S., Singh, Y. and Jat, M. L. 2016. Evaluation of tillage and crop establishment methods integrated with relay seeding of wheat and mungbean for sustainable intensification of cotton-wheat system in South Asia. *Field Crops Res.* 199, 31–41.

Coventry, D. R., Singh, R. S., Yadav, A., Gupta, R. K., Gill, S. C., Chhokar, R. S., Kumar, V., Sharma, R. K., Kumar, A., Mehta, A., Kleemann, S. G. L. and Cummins, J. A. 2011. Effect of tillage and nutrient management on wheat productivity and quality in Haryana, India. *Field Crops Res.* 123(3), 234–240.

Devkota, M., Devkota, K. P., Acharya, S. and McDonald, A. J. 2019. Increasing profitability, yields and yield stability through sustainable crop establishment practices in the rice-wheat systems of Nepal. *Agric. Syst.* 173, 414–423.

Dharmasena, P. B. 1996. Grass-tree hedge rows: an improvement to alley cropping. Proc. 7th Regional Workshop on Multipurpose Trees (Ed. Gunasena, H. P. M.). University of Peradeniya, Sri Lanka, 69–76.

Erenstein, O. 2009. Zero tillage in the rice-wheat systems of the indo-gangetic plains. A Review of Impacts and Sustainability Implications. IFPRI Discussion Paper 00916.

Erenstein, O., Farooq, U., Malik, R. K. and Sharif, M. 2008. On-farm impacts of zero tillage wheat in South Asia's rice-wheat systems. *Field Crops Res.* 105(3), 240-252.

Flinn, J. C. and Khokhar, B. B. 1989. Temporal determinants of the productivity of rice-wheat cropping systems. *Agric. Syst.* 30(3), 217–233.

Franke, A. C., McRoberts, N., Marshall, G., Malik, R. K., Singh, S., Nehra, A. S. and Gill, G. S. 2001. The contribution of zero tillage for the management of *Phalaris minor* (Retz.) in the Indian rice-wheat system. Proc. Brighton Crop Prot. Conf.-Weeds 2, 901-906.

Franke, A. C., Singh, S., Mcroberts, N., Nehra, A. S., Godara, S., Malik, R. K. and Marshall, G. 2007. Phalaris minor seedbank studies: longevity, seedling emergence and seed production as affected by tillage regime. *Weed Res.* 47(1), 73–83.

Gathala, M. K., Jat, M. L., Saharawat,Y. S., Sharma, S. K. and Singh, Y. 2017. Physical and chemical properties of a sandy loam soil under irrigated rice-wheat sequence in the Indo-Gangetic plains of South Asia. *J. Ecosys. Ecograph.* 7, 1-12. 10.4172/2157-7625.S7-002.

Gathala, M. K., Kumar, V., Sharma, P. C., Saharawat, Y. S., Jat, H. S., Singh, M., Kumar, A., Jat, M. L., Humphreys, E., Sharma, D. K., Sharma, S. and Ladha, J. K. 2013. Optimizing intensive cereal-based cropping systems addressing current and future drivers of agricultural change in the north-western Indo-Gangetic Plains of India. *Agric. Ecosys Environ.* 177, 85–97.

Gathala, M. K., Laing, A. M., Tiwari, T. P., Timsina, J., Rola-Rubzen, F., Islam, S., Maharjan, S., Brown, P. R., Das, K. K., Pradhan, K., Chowdhury, A. K., Kumar, R., Datt, R., Anwar, M., Hossain, S., Kumar, U., Adhikari, S., Magar, D. B. T., Sapkota, B. K., Shrestha, H. K., Islam, R., Rashid, M., Hossain, I., Hossain, A., Brown, B. and Gerard, B. 2021. Improving smallholder farmers' gross margins and labor-use efficiency across a range of cropping systems in the Eastern Gangetic Plains. *World Dev.* 138, 105266.

Ghimire, R., Adhikari, K. R., Chen, Z. S., Shah, S. C. and Dahal, K. R. 2012. Soil organic carbon sequestration as affected by tillage, crop residue, and nitrogen application in rice-wheat rotation system. *Paddy Water Environ.* 10(2), 95–102.

Giri, G. S. 1998. Review on the research results of no-till (surface/relay) wheat planting in Nepal. Paper Presented at the winter Crops Workshop held at NWRP/RARS, Bhairahawa, Nepal, October 28–30.

Gupta, N., Pradhan, P., Jain, A. and Patel, N. 2021. Sustainable agriculture in India 2021: what we know and how to scale up. Council of Energy, Environment and Water (CEEW), New Delhi, 30. https://www.ceew.in/sites/default/files/CEEW-FOLU-Sustainable-Agriculture-in-India-2021-20Apr21.pdf.

Gupta, R., Gopal, R., Jat, M. L., Jat, R. K. and Sidhu, H. S. 2010. Wheat productivity in Indo-Gangetic Plains of India during 2010: terminal heat effects and mitigation strategies. Conservation Agriculture Newsletter May issue 14: 1–3, 9. Professional Alliance for Conservation Agriculture, New Delhi, India.

Gupta, R. K., Hobbs, P. R., Salim, M., Chowdhary, N. H. and Bhuiyan, S. I. 2000a. Study of research and extension issues in the Sichuan Province of China for farm-level impact on the productivity of the rice-wheat system. Traveling Seminar Report 2. Rice-Wheat Consortium for the Indo-Gangetic Plains, IARI Campus, Pusa, New Delhi, India.

Gupta, R. K., Hobbs, P. R., Salim, M., Malik, R. K., Varma, M. R., Pokharel, T. P., Thakur, T. C. and Tripathi, J. 2000b. Research and extension issues for farm-level impact on the productivity of the rice-wheat systems in the Indo-Gangetic Plains of India and Pakistan. Traveling Seminar Report Series 1. Rice-Wheat Consortium for the Indo-Gangetic Plains, New Delhi, India.

Gupta, R. K. and Seth, A. 2007. A review of resource conserving technologies for sustainable management of the rice-wheat cropping systems of the Indo-Gangetic Plains (IGP). *Crop Prot.* 26(3), 436–447.

Gupta, R. K., Singh, Y., Ladha, J. K., Singh, B., Singh, J., Singh, G. and Pathak, H. 2007. Yield and phosphorus transformations in a rice-wheat system with crop residue and phosphorus management. *Soil Sci. Soc. Am. J.* 71(5), 1500–1507.

Gupta, R. P. 1989. Soil physical constraints to wheat crop growth on puddled soil. Presented at the ICAR-IRRI-CIMMYT Collaborative Research Project Meeting, May 22–23.

Handawala, J. and Kendaragama, K. M. A. 1991. Effect pf Gliricidia sepium on upland soil fertility in the dry zone of Sri Lanka. Proc. of 2nd Regional Workshop on Multipurpose Tree Species (Ed. Gunasena, H. P. M.), 156–160.

Haque, M. E. and Bell, R. W. 2019. Partially mechanized non-puddled rice establishment: on-farm performance and farmers' perceptions. *Plant Prodn. Sci.* 22(1), 23–45.

Haque, M. E., Bell, R. W., Islam, M. A. and Rahman, M. A. 2016a. Minimum tillage unpuddled transplanting: an alternative crop establishment strategy for rice in conservation agriculture cropping systems. *Field Crops Res.* 185, 31–39.

Haque, M. E., Bell, R. W., Kassam, A. and Mia, M. N. N. 2016b. Versatile strip seed drill: a 2-wheel tractor-based option for smallholders to implement conservation agriculture in Asia and Africa. *Environments* 3(4), 1–13. 10.3390/environments3010001.

Haque, M. E., Bell, R. W., Islam, A. K. M. S., Sayre, K. D. and Hossain, M. M. 2017a. An innovative Versatile Multi-crop Planter for crop establishment using two-wheel tractors. *Agric. Mech. Asia Afr. Lat. Am.* 48, 33–37.

Haque, M. E., Bell, R. W. and Vance, W. H. (Eds.) 2017b. Proceedings of the 2nd Conference on Conservation Agriculture for Smallholders in Asia and Africa. Mymensingh, Bangladesh. Published as an E-book, 201, February 14–16. http://researchrepository.murdoch.edu.au/id/eprint/36818/.

Haque, M. E., Bell, R. W., Jahiruddin, M., Hossain, M. M., Rahman, M. M., Begum, M., Hossen, M. A., Salahin, N., Zahan, T. and Hossain, M. M. 2018a. *Manual for Smallholders' Conservation Agriculture in Rice-Based Systems*. Murdoch University, Perth, Australia, 108.

Haque, M. E., Bell, R. W., Jahiruddin, M., Hossain, M. M., Rahman, M. M., Begum, M., Hossen, M. A., Salahin, N., Zahan, T., Hossain, M. M., Hashem, A., Islam, M. A., Vance, W. H., Hossain, M. I., Esdaile, R. J. and Kabir, M. E. 2018b. Manual for smallholders' conservation agriculture in rice-based systems. Murdoch University, 108 (in English), 233 (in Bangla) http://researchrepository.murdoch.edu.au/id/eprint/41693.

Harrington, L. W., Fujisaka, M. L., Morris, P. R., Hobbs, H. C., Sharma, R. P., Singh, M., Chaudhary, K. and Dhiman, S. D. 1993. *Wheat and Rice in Karnal and Kurukshetra Districts, Haryana, India; Farmers' Practices, Problems and Agenda for Action*. CCS Haryana Agricultural University, Indian Council of Agricultural Research, Centro International de Mejoramiento de Maízy Trigo and International Rice Research Institute, Mexico, DF, 44p.

Harrington, L. W. and Hobbs, P. R. 2009. The rice-wheat consortium and the Asian Development Bank: a history. In: *Integrated Crop and Resource Management in the Rice-Wheat System of South Asia* Ladha, J. K., Yadvinder, S., Erenstein, O. and Hardy, B. (Eds). International Rice Research Institute, Los Baños. 1–69.

Hobbs, P. and Morris, M. 1996. Meeting South Asia's future food requirements from rice-wheat cropping systems: priority issues facing researchers in the post-green revolution era. NRG Paper 96-01. CIMMYT, Mexico, DF, 46.

Hobbs, P. R., Gupta, R., Jat, R. K. and Malik, R. K. 2017. Conservation agriculture in the Indo-Gangetic plains of India: past, present and future. *Exp. Agric.* 55(2), 339–357 10.1017/S0014479717000424[Opens in a new window].

Hobbs, P. R., Sayre, K. and Gupta, R. 2008. The role of conservation agriculture in sustainable agriculture. *Philos. Trans. R. Soc. Lond. B Biol. Sci.* 363(1491), 543 -555.

Hossain, I. and Haque, M. A. 2015. Status of conservation agriculture based tillage technology for crop production in Bangladesh. *Bangladesh J. Agric. Res.* 40(2), 235-248. 10.3329/bjar.v40i2.24561.

Hossain, M. M., Begum, M., Hashem, A., Rahman, M. M., Ahmed, S., Hassan, M. M., Javed, T., Shabbir, R., Hadifa, A., El Sabagh, A. and Bell, R.W. 2021a. Strip tillage and crop residue retention decrease the size but increase the diversity of the weed seed bank under intensive rice-based crop rotations in Bangladesh. *Agronomy* 11, 1164. 10.3390/agronomy11061164.

Hossain, M. M., Begum, M., Hashem, A., Rahman, M. M., Haque, M. E. and Bell, R. W. 2021b. Continuous practice of conservation agriculture for 3–5 years reduces soil weed seedbanks size in intensive rice-based cropping patterns in Bangladesh. *Agriculture* 11(9), 2021 10.3390/agriculture11090895.

Hossen, M. A., Hossain, M. M., Haque, M. E. and Bell, R. W. 2018. Transplanting into non-puddled soils with a small-scale mechanical transplanter reduced fuel, labour and irrigation water requirements for rice (*Oryza sativa* L.) establishment and increased yield. *Field Crops Res.* 225, 141–151 10.1016/j.fcr.2018.06.009.

Jaipal, S., Malik, R. K., Yadav, A. and Gupta, R. 2005. IPM issues in zero-tillage system in rice-wheat cropping sequence. Technical Bulletin (8). CCS Haryana Agricultural University, India, 32.

Jat, H. S., Choudhary, K. M., Nandal, D. P., Yadav, A. K., Poonia, T., Singh, Y., Sharma, P. C. and Jat, M. L. 2020a. Conservation agriculture-based sustainable intensification of cereal systems leads to energy conservation, higher productivity and farm profitability. *Environ. Manag.* 65(6), 774–786. 10.1007/s00267-020-01273-w.

Jat, H. S., Choudhary, M., Datta, A., Yadav, A. K., Meena, M. D., Devi, R., Gathala, M. K., Jat, M. L., McDonald, A. and Sharma, P. C. 2020b. Temporal changes in soil microbial properties and nutrient dynamics under climate smart agriculture practices. *Soil Tillage Res.* 199, 104595.

Jat, H. S., Datta, A., Sharma, P. C., Kumar, V., Yadav, A. K., Choudhary, M., Choudhary, V., Gathala, M. K., Sharma, D. K., Jat, M. L., Yaduvanshi, N. P. S., Singh, G. and McDonald, A. 2018. Assessing soil properties and nutrient availability under conservation agriculture practices in a reclaimed sodic soil in cereal-based systems of North-West India. *Arch. Acker. Pflanzenbau Bodenkd* 64(4), 531–545.

Jat, M. L., Dagar, J. C., Sapkota, T. B., Singh, Y., Govaerts, B., Ridaura, S. L., Saharawat, Y. S., Sharma, R. K., Tetarwal, J. P., Jat, R. K., Hobbs, H. and Stirling, C. 2016. Climate change and agriculture: adaptation strategies and mitigation opportunities for food security in South Asia and Latin America. *Adv. Agron.* 137, 127–235.

Jat, M. L., Gathala, M. K., Saharawat, Y. S., Ladha, J. K. and Singh, Y. 2019. Conservation agriculture in intensive rice-wheat rotation of western Indo-Gangetic Plains: effect on crop physiology, yield, water productivity and economic profitability. *Int. J. Environ. Sci. Nat. Res.* 18(2), 1–15.

Jat, M. L., Gathaha, M. K., Sahrawat, Y. S., Tetawal, J. P., Gupta, R. and Singh, Y. 2013. Double no-till and permanent raised beds in maize-wheat rotation of northwestern Indo-Gangetic plains of India: effects on crop yields, water productivity, profitability and soil physical properties. *Field Crops Res.* 149, 291–299.

Jat, M. L., Saharawat, Y. S. and Gupta, R. 2011. Conservation agriculture in cereal systems of South Asia: nutrient management perspective. *Karnataka J. Agril. Sci.* 24(1), 100–105.

Jat, R. K., Sapkota, T. B., Singh, R. G., Jat, M. L., Kumar, M. and Gupta, R. K. 2014. Seven years of conservation agriculture in a rice-wheat rotation of eastern Gangetic Plains of South Asia: yield trends and economic profitability. *Field Crops Res.* 164, 199–210.

Johansen, C. J., Haque, M. E., Bell, R. W., Thierfelder, C. and Esdaile, R. J. 2012. Conservation agriculture for small holder rainfed farming: opportunities and constraints of new mechanized seeding systems. *Field Crops Res.* 132, 18–32. 10.1016/j.fcr.2011.11.026.

Kakraliya, S., Jat, H., Singh, I., Sapkota, T., Singh, L., Sutaliya, J., Sharma, P., Jat, R., Lopez-Ridaura, S. and Jat, M. 2018. Performance of portfolios of climate smart agriculture practices in a rice-wheat system of western Indo-Gangetic plains. *Agril. Water Manag.* 202, 122–133.

Karki, T. B. and Shrestha, J. 2014. Conservation agriculture: significance, challenges and opportunities in Nepal. *Adv. Plants Agric. Res.* 1(5), 186–188.

Kassam, A., Friedrich, T. and Derpsch, R. 2019. Global spread of conservation agriculture. *Intl J. Environ. Stud.* 11(1&2), 159–170.

Keil, A., D'Souza, A. and McDonald, A. 2015. Zero-tillage as a pathway for sustainable wheat intensification in the eastern Indo-Gangetic Plains: does it work in farmers' fields? *Food Sec.* 7(5), 983-1001 10.1007/s12571-015-0492-3.

Keil, A., Mitra, A., McDonald, A. and Malik, R. K. 2020. Zero-tillage wheat provides stable yield and economic benefits under diverse growing season climates in the eastern Indo-Gangetic Plains. *Int. J. Agric. Sus.* 18(6), 567-593. 10.1080/14735903.2020.1794490.

Khan, B. R., Khan, B. M., Razzaq, A., Munir, M., Aslam, M., Ahmed, S., Hashmi, N. I. and Hobbs, P. R. 1986. Effect of different tillage implements on the yield of wheat. *Pak. J. Agric. Res.* 7(3), 141-149.

Khuhro, G. A. 1989. Rice-wheat system of Sindh. Prepared for the Planning Meeting on Raising and Sustaining Productivity and Profitability of Rice-Wheat Systems in South Asia, Bangkok, December 4-6.

Krupnik, T. J., Valle, S. S., Hossain, I., Gathala, M. K. and Justice, S. E. 2013. *Made in Bangladesh: Scale-Appropriate Machinery for Agricultural Resource Conservation in the Context of Small-Holder Farming*. International Maize and Wheat Improvement Center, Mexico, DF.

Kumar, V., Jat, H. S., Sharma, P. C., Singh, B., Gathala, M. K., Malik, R. K., Kamboj, B. R., Yadav, A. K., Ladha, J. K., Raman, A., Sharma, D. K. and McDonald, A. 2018. Can productivity and profitability be enhanced in intensively managed cereal systems while reducing the environmental footprint of production? Assessing sustainable intensification options in the breadbasket of India? *Agric. Ecosyst. Environ.* 252, 132-147.

Kumar, V., Singh, S., Chhokar, R. S., Malik, R. K., Brainard, D. C. and Ladha, J. K. 2013. Weed management strategies to reduce herbicide use in zero-till rice-wheat cropping systems of the Indo-Gangetic Plains. *Weed Technol.* 27(1), 241-254.

Ladha, J. K., Dawe, D., Pathak, H., Padre, A. T., Yadav, R. L., Singh, B., Singh, Y., Singh, P., Kundu, A. L., Sakal, R., Ram, N., Regmi, A. P., Gami, S. K., Bhandari, A. L., Amin, R., Yadav, C. R., Bhattarai, E. M., Das, S., Aggarwal, H. P., Gupta, R. K. and Hobbs, P. R. 2003a. How extensive are yield declines in long-term rice-wheat experiments in Asia? *Field Crops Res.* 81, 159-180.

Ladha, J. K., Hill, J., Gupta, R. K., Duxbury, J. and Buresh, R. J. (Eds.) 2003b. Improving the productivity and sustainability of rice-wheat systems: issues and impact. ASA, Spec. Publ. 65. Chapter 1:1-25. ASA, Madison, WI.

Ladha, J. K., Hill, J. E., Duxbury, J. D., Gupta, R. K. and Buresh, R. J. (Eds.) 2003c. Improving the productivity and sustainability of rice-wheat systems: issues and impact. American Society of Agronomy Special Publication 65. ASA, CSSA, SSSA, Madison, WI, 211.

Ladha, J. K., Rao, A., Raman, A. K., Tirol-Padre, D. A., Gathala, M., Kumar, V., Saharawat, Y. S., Sharma, S., Piepho, H. P., Alam, M. M., Liak, R., Rajendran, R., Reddy, C. K., Parsad, R., Sharma, P. C., Singh, S. K., Saha, A. and Noor, S. 2015. Agronomic improvements can make future cereal systems in South Asia far more productive and result in a lower environmental footprint. *Glob. Change Biol.* 22, 1054-1074.

Laik, R., Sharma, S., Idris, M., Singh, A. K., Singh, S. S., Bhatt, B. P., Saharawat, Y. S., Humphreys, E. and Ladha, J. K. 2014. Integration of conservation agriculture with best management practices for improving system performance of the rice-wheat rotation in the eastern Indo-Gangetic Plains of India. *Agric. Ecosyst. Environ.* 195, 68-82.

Laxmi, V., Erenstein, O. and Gupta, R. K. 2007. *Impact of Zero-Tillage in India's Rice-Wheat Systems*. CIMMYT, Mexico, DF.

Mahal, J. S., Manes, G. S., Singh, A., Kaur, S. and Singh, M. 2019. Complementing solutions and practices for managing rice straw and their impact in the state of Punjab. *Agric. Res. J.* 56, 588–593.

Malik, R. K. 2005. *Success Story of No-Tillage in India*. Unpublished memoir.

Malik, R. K., Gill, G. and Hobbs, P. R. 1998 (English). Herbicide resistance - a major issue for sustaining wheat productivity in rice-wheat cropping systems in the indo-gangetic plains. Rice Wheat Consortium Paper Series 3, 1998 (English). Rice-Wheat Consortium for Indo-Gangetic Plains, New Delhi, India, 36.

Malik, R. K., Gupta, R. K., Singh, C. M., Yadav, A., Brar, S. S., Thakur, T. C., Singh, S. S., Singh, A. K., Singh, R. and Sinha, R. K. (Eds.) 2005. Project workshop on accelerating the adoption of resource conservation technologies in rice-wheat systems of the Indo-Gangetic Plains. Directorate of Extension Education. CCS Haryana Agricultural University, Haryana, India, June 1-2.

Malik, R. K., Singh, B. K. and Mehla, R. S. 2000. Proceedings of the international workshop on developing an action programme for farm-level impact in rice wheat systems of Indo-Gangetic plains. Rice-Wheat Consortium Paper Series 14. Rice-Wheat Consortium for the Indo-Gangetic Plains, New Delhi, India, 31–45, September 25-29.

Malik, R. K. and Singh, S. 1995. Littleseed canarygrass resistance to isoproturon in India. *Weed Technol.* 9(3), 419–425.

Malik, R. K., Yadav, A., Singh, S., Malik, R. S., Balyan, R. S. and Banga, R. S. 2002a. *Herbicide Resistance Management and Evolution of Zero Tillage: A Success Story*. CCS Haryana Agricultural University, Hisar, India.

Malik, R. K., Yadav, A., Singh, S., Malik, R. S., Balyan, R. S., Jaipal, S., Hobbs, P. R., Gill, G., Singh, S., Gupta, R. K. and Bellinder, R. 2002b. Herbicide resistance management and evolution of zero-tillage - a success story. Research Bulletin - 2002. CCS HAU, Hisar, 1–43.

Malik, R. K., Yadav, A., Gill, G. S., Sardana, P., Gupta, R. K. and Piggin, P. 2004a. Evolution and acceleration of no-till farming in rice-wheat cropping system of the Indo-Gangetic Plains. Proceedings for the 4th International Crop Science Congress, Brisbane, Australia, 26 September–1 October. www.cropscience.org.au.

Malik, R. K., Yadav, A., Singh, S., Sardana, P. K., Hobbs, P. R. and Gupta, R. K. 2004b. *No-Tillage Farming in the Rice-Wheat Cropping Systems in India In Book:* Sustainable Agriculture and the International Rice-Wheat System. CRC Press. eBook ISBN9780429216343.

Mehla, R. S., Verma, J. K., Gupta, R. K. and Hobbs, P. R. 2000. Stagnation in the productivity of wheat in the Indo-Gangetic Plains: zero-till-seed-cum-fertilizer drill as an integrated solution. Rice-Wheat Consortium Paper Series 8. Rice-Wheat Consortium for the Indo-Gangetic Plains, New Delhi.

Meisner, C., Hobbs, P. R., Badaruddin, M., Razzaque, M. A., Giri, G. S. and Justice, S. 1998. *Mechanical Revolution among Small Landholders of South Asia: The Growing Use of Chinese Hand Tractors*. CIMMYT, Mexico, DF.

Miah, M. M., Haque, M. E. and Bell, R. W. 2019. Impact of multi-crop planter business on service providers' livelihood improvement in some selected areas of Bangladesh. *Bangladesh J. Agric. Res.* 44(3), 409–426.

Mishra, J. S., Poonia, S. P., Kumar, R., Dubey, R., Kumar, V., Mondal, S., Dwivedi, S. K., Rao, K. K., Kumar, R., Tamta, M., Verma, M., Saurabh, K., Kumar, S., Bhatt, B. P., Malik, R. K., McDonald, A. and Bhaskar, S. 2021. An impact of agronomic practices of sustainable rice-wheat crop intensification on food security, economic adaptability, and environmental mitigation across eastern Indo-Gangetic Plains. *Field Crops Res.* 267, 108164.

NAAS 2017. Innovative viable solution to rice residue burning in rice-wheat cropping system through concurrent use of super straw management system-fitted combines and turbo happy seeder. (Eds. Singh, Y., Jat, M. L., Sidhu, H. S., Singh, P. and Varma, A.). *Policy Brief No. 2*, National Academy of Agricultural Sciences, New Delhi, 16.

Nandan, R., Singh, V., Singh, S. S., Kumar, V., Hazra, K. K., Nath, C. P., Poonia, S. P. and Malik, R. K. 2018. Comparative assessment of the relative proportion of weed morphology, diversity, and growth under new generation tillage and crop establishment techniques in rice-based cropping systems. *Crop Prot.* 111, 23–32.

Nandan, R., Singh, V., Singh, S. S., Kumar, V., Hazra, K. K., Nath, C. P., Poonia, S. P., Malik, R. K., Bhattacharyya, R. and McDonald, A. 2019. Impact of conservation tillage in rice-based cropping systems on soil aggregation, carbon pools and nutrients. *Geoderma* 340, 104–114.

Nawaz, A., Farooq, M., Lal, R., Rehman, A. and Hafeez-ur-Rehman 2017. Comparison of conventional and conservation rice-wheat systems in Punjab, Pakistan. *Soil Till. Res.* 169, 35–43.

Nayak, H. S. 2019. Effect of nitrogen placement methods and crop establishment options on nitrogen use-efficiency and productivity of maize-wheat system under long term conservation agriculture. Dissertation, Division of Agronomy, ICAR- IARI, New Delhi.

Nayak, H. S., Parihar, C. M., Jat, S. L., Nain, L., Mandal, B. N., Singh, V., Garnaik, S., Muduli, L. and Sahu, S. 2019. Effect of nitrogen point placement on energetic and soil enzymatic activities on long-term conservation agriculture-based maize (*Zea mays*) - wheat (*Triticum aestivum*) system of western Indo-Gangetic plains. *Indian J. Agric. Sci.* 89(12), 2102–2106.

Nichols, V., Verhulst, N., Cox, R. and Govaerts, B. 2015. Weed dynamics and conservation agriculture principles: a review. *Field Crops Res.* 183, 56–68.

Ortiz-Monasterio, I., Wassman, R., Govaerts, B., Hosen, Y., Nobuko, K. and Verhulst, N. 2010. Greenhouse gas mitigation in the main cereal systems: rice, wheat and maize. In: *CABI Climate Change Series (Volume 1): Climate Change and Crop Production* Reynolds, M. (Ed.). CABI Publishing, Wallingford. 151–176.

Parihar, C. M., Jat, S. L., Singh, A. K., Datta, A., Parihar, M. D., Varghese, E., Bandyopadhyay, K. K., Nayak, H. S., Kuri, B. R. and Jat, M. L. 2018a. Changes in carbon pools and biological activities of a sandy loam soil under medium-term conservation agriculture and diversified cropping systems. *Eur. J. Soil Sci.* 69(5), 902–912.

Parihar, C. M., Nayak, H. S., Rai, V. K., Jat, S. L., Parihar, N., Aggarwal, P. and Mishra, A. K. 2019a. Soil water dynamics, water productivity and radiation use efficiency of maize under multi-year conservation agriculture during contrasting rainfall events. *Field Crops Res.* 241, 107570.

Parihar, C. M., Parihar, M. D., Sapkota, T. B., Nanwal, R. K., Singh, A. K., Jat, S. L., Nayak, H. S., Mahala, D. M., Singh, L. K., Kakraliya, S. K., Stirling, C. M. and Jat, M. L. 2018b. Long-term impact of conservation agriculture and diversified maize rotations on carbon pools and stocks, mineral nitrogen fractions and nitrous oxide fluxes in inceptisol of India. *Sci. Total Environ.* 640-641, 1382–1392.

Parihar, C. M., Singh, A. K., Jat, S. L., Dey, A., Nayak, H. S., Mandal, B. N., Saharawat, Y. S., Jat, M. L. and Yadav, O. P. 2020. Soil quality and carbon sequestration under conservation agriculture with balanced nutrition in intensive cereal-based system. *Soil Till. Res.* 202, 104653.

Parihar, C. M., Singh, A. K., Jat, S. L., Ghosh, A., Dey, A., Nayak, H. S., Parihar, M. D., Mahala, D. M., Yadav, R. K., Rai, V., Satayanaryana, T. and Jat, M. L. 2019b. Dependence of temperature sensitivity of soil organic carbon decomposition on nutrient management options under conservation agriculture in a sub-tropical inceptisol. *Soil Till. Res.* 190, 50–60.

Parihar, C. M., Yadav, M. R., Jat, S. L., Singh, A. K., Kumar, B., Pooniya, V., Pradhan, S., Verma, R. K., Jat, M. L., Jat, R. K., Parihar, M. D., Nayak, H. S. and Saharawat, Y. S. 2018c. Long-term conservation agriculture and intensified cropping systems: effect on growth, yield, water and energy-use efficiency of maize in Northwestern India. *Pedosphere* 28(6), 952–963. 10.1016/S1002-0160(17)60468-5.

Parihar, C. M., Yadav, M. R., Jat, S. L., Singh, A. K., Kumar, B., Pradhan, S., Chakraborty, D., Jat, M. L., Jat, R. K., Saharawat, Y. S. and Yadav, O. P. 2016. Long term effect of conservation agriculture in maize rotations on total organic carbon, physical and biological properties of a sandy loam soil in north-western Indo-Gangetic Plains. *Soil Till. Res.* 161, 116–128.

Paroda, R. S. 2018. Strategy paper for doubling farmer income. Publication 60. Trust for Advancement of Agricultural Sciences (TAAS), New Delhi, India, 27. https://www.taas.in/documents/pub-sp-11.pdf.

Paudel, M., Kumar Sah, S., McDonald, A. and Kumar Chaudhary, N. 2014. Soil organic carbon sequestration in rice-wheat system under conservation and conventional agriculture in western Chitwan, Nepal. *World J. Agric. Res.* 2(6A), 1–5.

Raghu, P. T., Aravindakshan, S., Rossi, F., Krishna, V., Baksh, E. and Miah, A. A. 2016. *A Biophysical and Socioeconomic Characterization of the Cereal Production Systems of Northwest Bangladesh. Cereal Systems Initiative for South Asia Project, Phase III.* CIMMYT, Dhaka, Bangladesh.

Randhawa, A. S., Dhillon, S. S. and Singh, D. 1981. Productivity of wheat varieties, as influenced by the time of sowing. *J. Res. Punjab Agric. Univ.* 18, 227–233.

Rashid, M. H., Timsina, J., Islam, N. and Islam, S. 2019. Tillage and residue management effects on productivity, profitability and soil properties of a rice-maize-mungbean system in Bangladesh. *J. Crop Improv.* 33(5), 683–710, 10.1080/15427528.2019.1661056.

RWC (Rice-Wheat Consortium for the Indo-Gangetic Plains) 2000. Minutes of the Kathmandu Meeting for ADB Project Planning. Hotel Annapurna, Nepal, November 24.

Salahin, N., Jahiruddin, M., Islam, M. R., Alam, M. K., Haque, M. E., Ahmed, S., Baazeem, A., Hadifa, A., EL Sabagh, A. and Bell, R. W. 2021 Establishment of crops under minimal soil disturbance and crop residue retention in rice-based cropping system: yield advantage, soil health improvement, and economic benefit. *Land* 10(6), 581. 10.3390/land10060581.

Samal, S. K., Rao, K. K., Poonia, S. P., Kumar, R., Mishra, J. S., Prakash, V., Mondal, S., Dwivedi, S. K., Bhatt, B. P., Naik, S. K., Choubey, A. K., Kumar, V., Malik, R. K. and Mc Donald, A. 2017. Evaluation of long-term conservation agriculture and crop intensification in rice-wheat rotation of Indo-Gangetic Plains of South Asia: carbon dynamics and productivity. *Eur. J. Agron.* 90, 198–208 10.1016/j.eja.2017.08.006.

Sapkota, T. B., Jat, M. L., Shankar, V., Singh, L. K., Rai, M., Grewal, M. S. and Stirling, C. M. 2015. Tillage, residue and nitrogen management effects on methane and nitrous oxide emission from rice-wheat system of Indian northwest Indo-Gangetic Plains. *J. Integr. Environ. Sci.* 12(sup1), 31–46.

Seth, A., Fischer, K. S., Anderson, J. M. and Jha, D. 2003. *The Rice-Wheat Consortium: An Institutional Innovation in International Agricultural Research on Rice-Wheat Cropping Systems of the Indo-Gangetic Plains*. The Review Panel Report, RWC, New Delhi, India.

Sharif, M., Ijaz, S. S., Ansar, Md., Latif, R., Hassan, A. and Nasir, Md. 2017. Conservation agriculture: research status, opportunities and challenges in dryland areas of Pakistan. *J. Bio. Env. Sci.* 11, 102–112.

Sharma, A. R. and Singh, V. P. 2014. Integrated weed management in conservation agriculture systems. *Indian J. Weed Sci.* 46(1), 23–30.

Sharma, S., Thind, H. S., Singh, Y., Sidhu, H. S., Jat, M. L. and Parihar, C. M. 2019a. Effects of crop residue retention on soil carbon pools after 6 years of rice-wheat cropping system. *Environ. Earth Sci.* 78(10), 296–301.

Sharma, S., Vashisht, M., Singh, Y. and Thind, H. S. 2019b. Soil carbon pools and enzyme activities in aggregate size fractions after seven years of conservation agriculture in a rice-wheat system. *Crop Pasture Sci.* 70(6), 473–485. 10.1071/CP19013.

Sidhu, H. S., Singh, M., Singh, Y., Blackwell, J., Lohan, S. K., Humphreys, E., Jat, M. L., Singh, V. and Singh, S. 2015. Development and evaluation of the Turbo happy seeder for sowing wheat into heavy rice residues in NW India. *Field Crops Res.* 184, 201–212.

Singh, A., Phogat, V. K., Dahiya, R. and Batra, S. D. 2015a. Impact of long-term zero till wheat on soil physical properties and wheat productivity under rice-wheat cropping system. *Soil Till. Res.* 140, 8–105.

Singh, A. P., Bhullar, A. P., Yadav, R. and Chowdhury, T. 2015b. Weed management in zero-till wheat. *Indian J. Weed Sci.* 47, 233–239.

Singh, B., Humphreys, E., Eberbach, P. L., Katupitiya, A., Singh, Y. and Kukal, S. S. 2011. Growth, yield and water productivity of zero till wheat as affected by rice straw mulch and irrigation schedule. *Field Crops Res.* 121(2), 209–225.

Singh, M., Kumar, P., Solanki, I. S., McDonald, A., Kumar, A., Poonia, S., Kumar, V., Kumar, A., Singh, D. K., Singh, B., Singh, S. and Malik, R. K. 2020. Intercomparison of crop establishment methods for improving yield and profitability in the rice-wheat systems of Eastern India. *Field Crops Res.* 250, 10776 10.1016/j.fcr.2020.107776.

Singh, R. B. and Paroda, R. S. 1993. Sustainability and productivity of rice-wheat system in Asia-Pacific region: research and technology development needs. Paper presented at Regional Expert Consultation on the Sustainability of the Rice-Wheat Production System in Different Agro-ecological Settings in Asia, Bangko, July 6-9.

Singh, V. K., Singh, Y., Dwivedi, B. S., Singh, S. K., Majumdar, K., Jat, M. L., Mishra, R. P. and Rani, M. 2016. Soil physical properties, yield trends and economics after five years of conservation agriculture-based rice-maize system in north-western India. *Soil Till. Res.* 155, 133–148.

Six, J., Conant, R. T., Paul, E. A. and Paustian, K. 2002. Stabilization mechanisms of soil organic matter: implications for C-saturation of soils. *Plant Soil* 241(2), 155–176.

Thakur, T. C. 2005. Personal information from T.C. Thakur, Professor of farm machinery and power engineering, Pantnagar University, as reported via email in April 2005 to Dr. Raj Gupta, facilitator of the Rice-Wheat Consortium for the Indo-Gangetic Plains.

Thind, H. S., Sharma, S., Singh, Y. and Sidhu, H. S. 2019. Rice-wheat productivity and profitability with residue, tillage and green manure management. *Nutr. Cycl. Agroecosystems* 113(2), 113–125.

Tripathi, J., Adhikary, C., Lauren, J. G., Duxbury, J. M. and Hobbs, P. R. 2006. Assessment of farmer adoption of surface seeded wheat in the Nepal Terai. Rice-Wheat Consortium Paper Series No. 19. Rice-Wheat Consortium for the Indo-Gangetic Plains, New Delhi, India, 1-50.

Vance, W. H., Bell, R. W. and Haque, M. E. (Eds.) 2014a. Proceedings of the Conference on Conservation Agriculture for Smallholders in Asia and Africa. Mymensingh, Bangladesh. Published as an E-book, 213, December 7-11. http://researchrepository.murdoch.edu.au/26081/.

Vance, W. H., Bell, R. W., Johansen, C. J., Haque, M. E., Musa, A. M., Shahidullah, A. K. M. and Mia, M. N. N. 2014b. One-pass mechanised row-sowing of rainfed winter chickpea minimises soil water constraints - an example for small holder farms in north-west Bangladesh. *Crop Pasture Sci.* 65(7), 602-613 10.1071/CP13331.

Vincent, D. and Quirke, D. 2002. Controlling Phalaris minor in the Indian rice-wheat belt. ACIAR Impact Assessment Series No. 18. ACIAR, Canberra, Australia, 1-35.

Weerakoon, W. L. 1983. Conservation farming research program at Mahailluppallama, IN Conservation Farming in the Commonwealth. Papers from a Training Workshop, Sri Lanka, 140, January 17-25.

Yadav, A., Malik, R. K., Bansal, N. K., Gupta, R. K., Singh, S. and Hobbs, P. R. 2002. Manual for using zero-till seed-cum-fertilizer drill and zero-till drill-cum-bed planter. Rice Wheat Consortium Technical Bulletin Series 4, 2002 - (English).

Yadav, A., Malik, R. K., Chauhan, B. S. and Bellinder, R. 2001. Temperature moderation - an important factor for higher grain yield of wheat under zero tillage. *Haryana J. Agron* 17(1&2), 18-21.

Yadvinder-Singh and Sidhu, H. S. 2014. Management of cereal crop residues for sustainable rice-wheat production system in the Indo-Gangetic plains of India. *Proc. Ind. Natl Sci. Acad.* 80(1), 95-114.

Yadvinder-Singh, Gupta, R. K., Gurpreet-Singh, Jagmohan-Singh, Sidhu, H. S. and Bijay-Singh. 2009. Nitrogen and residue management effects on agronomic productivity and nitrogen use efficiency in rice-wheat system in Indian Punjab. *Nutr. Cycl. Agroecosyst.* 84(2), 141--154.

Yadvinder-Singh, Thind, H. S., and Sidhu, H. S. 2014. Management options for rice residues for sustainable productivity of rice-wheat cropping system. *J. Res. Punjab Agric. Univ.* 51, 239-245.

Chapter 15

Conservation Agriculture in Southeast Asia

Yuji Niino, FAO Regional Office for Asia and the Pacific, Thailand; Victor B. Ella, University of The Philippines Los Baños, The Philippines; Florent Tivet, Centre de Coopération Internationale en Recherche Agronomique pour le Développement (CIRAD), Agroécologie et Intensification Durable des Cultures Annuelles (AIDA), University of Montpellier, France and General Directorate of Agriculture (GDA), Department of Agricultural Land Resources Management (DALRM), Cambodia; Vira Leng, General Directorate of Agriculture (GDA), Department of Agricultural Land Resources Management (DALRM), Cambodia; Hoa Tran Quoc, Centre de Coopération Internationale en Recherche Agronomique pour le Développement (CIRAD), University of Montpellier, France, and Department of Agricultural Land Management (DALaM), Laos; Pascal Lienhard, Centre de Coopération Internationale en Recherche Agronomique pour le Développement (CIRAD), Agroécologie et Intensification Durable des Cultures Annuelles (AIDA), University of Montpellier, France and Northern Mountainous Agriculture and Forestry Science Institute (NOMAFSI), Vietnam; Pham thi Sen, Northern Mountainous Agriculture and Forestry Science Institute (NOMAFSI), Vietnam; Thatheva Saphangthong, Department of Agricultural Land Management (DALaM), Laos; Vang Seng, General Directorate of Agriculture (GDA), Department of Agricultural Land Resources Management (DALRM), Cambodia; Lyda Hok, Royal University of Agriculture, Cambodia; Manuel Reyes, Kansas State University, USA; Stéphane Boulakia, Centre de Coopération Internationale en Recherche Agronomique pour le Développement (CIRAD), Agroécologie et Intensification Durable des Cultures Annuelles (AIDA), University of Montpellier, France; and Sim Choon Cheak, Sime Darby Plantation Research Sdn. Bhd., Malaysia

1 Introduction
2 Current status of Conservation Agriculture in Southeast Asia
3 Regional networking
4 Conclusion
5 Where to look for further information
6 References

1 Introduction

Southeast Asia (Fig. 1) comprises a diverse range of countries with varying levels of development and endowments (Table 1). Over recent decades, the region

http://dx.doi.org/10.19103/AS.2021.0088.13

Figure 1 The Southeast Region (dark gray shading).

as a whole has undergone significant development. Structural changes in many of its economies have led to significant gains, and the rise of 'Factory Asia' has placed Southeast Asia in a central role in a varied mix of manufacturing global value chains. With this, the region has experienced strong growth in Gross Domestic Product (GDP). Meanwhile, the regional population growth rate has been slowing down in line with socio-economic development and currently the growth rate is close to 1.3% per year (OECD/FAO, 2017).

The countries in the region made remarkable progress in improving food security. The proportion of undernourished was around 31% in the early 1990s, but the proportion had fallen below 10% by 2014–16. There were significant structural changes in the agriculture sector in the region as its share in GDP and employment declined, and as those workers were absorbed in other sectors of the economy with however large disparities between countries. Rice cultivation is the main agricultural production activity in the region, but the contribution of rice to total gross agricultural production value fell from around 40% in the early 1990s to 30% in 2013, due to dietary shifts and the development of other commodities (OECD/FAO, 2017). Farm size in the region is relatively small (less than 1 ha) and is reducing due to further fragmentation, which could have an impact on future increase in productivity, except in a few cases where farm size is increasing due to land consolidation (FAO, 2015). To compensate for labor shortages, mechanization has accelerated, but its rate has not kept pace with the declining labor force, maybe because of a shortage of appropriate machinery for small field scales compared to other regions with a greater proportion of larger farm size. There is also some form of national

Table 1 Contextual indicators for selected countries in Southeast Asia, 2015

	GDP per capita	Population (millions)	Rural population (%)	Total land area (km^2)	Agricultural land (km^2)	Agricultural land per capita (ha)	Freshwater resources (billion m^3)	Freshwater withdrawals agriculture (billion m^3)	Freshwater resources per capita ('000 m^3)
Cambodia	1 159	15.6	79.3	176 520	54 550	0.36	120.6	2.1	7.9
Indonesia	3 346	257.6	46.3	1 811 570	570 000	0.22	2019.0	92.8	7.9
Lao PDR	1 818	6.8	61.4	230 800	23 690	0.35	190.4	3.2	28.5
Malaysia	9 768	30.3	25.3	328 550	78 390	0.26	1003.0	2.5	33.5
Myanmar	1 161	53.9	65.9	653 080	126 450	0.24	580.0	29.6	10.9
Philippines	2 904	100.7	55.6	298 170	124 400	0.13	429.0	67.1	4.8
Thailand	5 815	68.0	49.6	510 890	221 100	0.33	224.5	51.8	3.3
Vietnam	2 111	91.7	66.4	310 070	108 737	0.12	359.4	77.7	4.0

self-sufficiency or subsistence policies for rural communities that can have an impact on preventing land consolidation and productivity improvement, thus causing further land fragmentation and consequently food insecurity.

With these dynamic changes in the region, extensive deforestation and unsustainable agricultural practices, together with inadequate soil conservation, cultivation of steep slopes, overgrazing and rapid urban and industrial growth have had a devastating impact on land resources. Land degradation processes of particular concern in the region include soil erosion, compaction, acidification, declining soil organic matter, weed infestation, soil fertility depletion and biological degradation (FAO, 2009). To minimize soil degradation and build a foundation for a functioning ecosystem, Conservation Agriculture (CA) is considered to be the entry point in making intensification sustainable, as well as in rehabilitating degraded agroecosystems. It is relatively new and adopted in a limited area in the Southeast Asian region, but awareness of it and the area under practice has been growing steadily over the past decade, and it is being recognized for its potential, particularly under the increasing climate risks.

2 Current status of Conservation Agriculture in Southeast Asia

2.1 Greater Mekong Sub-region (GMS) countries: Vietnam, Laos and Cambodia

2.1.1 History of Conservation Agriculture initiatives and actors in the GMS region

The first initiatives related to Conservation Agriculture (CA) in the GMS started in the early 2000s (Table 2) to combat deforestation and land degradation, notably soil erosion and soil fertility depletion. The French Development Agency (AFD) was the first donor to support national authorities through action-research projects, and the Centre de Coopération Internationale en Recherche Agronomique pour le Développement (CIRAD) was the first international institution that supported national research institutes on CA. All these projects benefitted from the scientific and technical expertise of Lucien Séguy on CA systems design and management (Séguy et al., 1998).

Nowadays, good competences and expertise in CA management, research and capacity building can be found at national levels, in various governmental agencies, for example, within the General Directorate of Agriculture (MAFF-GDA), Department of Agricultural Land Resources Management (GDA/DALRM), Department of Agricultural Engineering (GDA/DAEng), the Royal University of Agriculture (RUA) and the Center of Excellence on Sustainable Agricultural Intensification and Nutrition (CE SAIN), the University of Battambang (UBB), the Institute of Technology (ITC) in Cambodia, within the Department of

Agricultural Land Management (MAF-DALaM) in Laos, and within the Northern Mountainous Agriculture and Forestry Science Institute (MARD-NOMAFSI) in Vietnam.

2.1.2 Main Conservation Agriculture farming systems design and out-scaling strategies tested

The design of CA farming systems has evolved in line with the rapid transformation of agricultural systems that occurred in the GMS sub-region, notably in the Uplands. Within the past two decades, agricultural land has expanded at the expense of the forest and fallows (Ingalls et al., 2018; Kong et al., 2019). Farmers have rapidly shifted from extensive cropping systems based on shifting cultivation and long fallow periods to more intensive and permanent mono-cropping systems (Castella, 2012). Soil fertility has decreased due to organic matter mineralization and erosion, which adversely affects the technical performance of agricultural production systems. Rice is the cornerstone of farming systems in the sub-region (Castella, 2012). Maize and cassava cultivation has boomed over the past two decades (Ingalls et al., 2018), explaining why these three crops have been and are still at the center of CA systems design during that period.

Six main technical changes can be observed in CA systems in the recent development (Fig. 2):

- *Minimizing soil disturbance with no-tillage and covering the soil with mulch:* This was the prevailing entry point in the early 2000s to stop soil erosion and combat plow-induced land degradation. Crop residue management (Laos), and/or mulching (Vietnam) were seen as a first step toward more diversified and complex CA cropping systems in a step-by-step approach.
- *Crop diversification:* This was promoted at the same time with farmers. Various maize- and cassava-based systems incorporating pulses, forage crops, cover crops and alternative cash crops cultivated in association, relay, and/or rotations with the main crops were tested in stations and on-field, with farmers following an iterative approach (Husson et al., 2016); in Cambodia, no-tillage and agricultural systems diversification using cover crops were jointly promoted from the start (Boulakia et al., 2010; Kong et al., 2016).
- *Mechanical control of cover crops:* In line with the rising concerns related to herbicide misuse and overuse in the region, the mechanical control of cover crops has rapidly emerged as a high-ranking priority in CA systems development; annual grasses (e.g. finger millet, pearl millet, sorghum) were preferred to perennial grasses (e.g. ruzi grass) as cover crops in annual-based cropping systems to reduce herbicide use; *Stylosanthes*

Table 2 Main projects and initiatives in the recent past related to soil conservation and conservation agriculture in Cambodia, Laos, and Vietnam

Location		Period	Project	Intervention type	Main crops/activities	Donor	Partners
Cambodia	Kampong Cham	2004–2008	PHF (rubber for smallholder project), crop diversification component	R&D	Rubber and diversification with annual upland crops	AFD	GDR, CIRAD
	Kampong Cham, Battambang	2008–2013	PADAC (Support Project for the Development of Cambodian Agriculture)	R&D	Upland crops and lowland rice	AFD	GDA-DALRM, CIRAD
	Battambang, Siem Reap	2010–2014	SANREM CRSP (Sustainable Agriculture and Natural Resources Management Collaborative Research Support Program)	R&D	Upland crops: maize, cassava and vegetable	USAID	Virginia Tech, North Carolina A&T, RUA, GDA, CIRAD
	Kampong Thom	2011–2014	Alternatives to irrigated rice-based cropping systems	R&D	Irrigated rice	AFD	GDA/DALRM, CIRAD
	Kampong Cham & Thom, Battambang	2015–2019	Action-Research on Conservation Agriculture within the framework of the ACTAE project (Supporting the Agroecological transition in SEA)	R&D	Upland crops, lowland rice, various R&D activities	AFD	GDA-DALRM, RUA/CE SAIN, ITC, CIRAD
	National	2016–2020	SIIL (Sustainable Intensification Innovation Lab) and support to Bos Khnor CA Research Station, network of Technology Parks of RUA/CE SAIN	Research, Education	Research and Education: various research and development activities	USAID, SIIL	RUA/CE SAIN, US University (Kansas, Illinois, Pennsylvania), GDA, UBB, CIRAD
	Kampong Cham, Battambang	2016–2019	EISOFUN: Ecological Intensification and Soil Ecosystem Functioning	Research	Soil ecosystem services	CCCA	GDA, RUA, ITC, CIRAD
	National	2018–2021	Conservation Agriculture Service with a Fee (CASF) project	Development	Private sector engagement	USAID Cambodia	RUA/CE SAIN, Kansas State Univ., Swisscontact, GDA/DALRM, CIRAD
	National	2020–now	CASIC (Cambodia Conservation Agriculture Sustainable Intensification Consortium), Ministry of Agriculture, Forestry and Fisheries	Development	Policy dialogue, public - private partnership	MAFF Cambodia	Swisscontact, GDA, RUA/CE SAIN, Kansas Univ., CIRAD
	Battambang, Preah Vihear, Kandal	2020–2023	WAT4CAM (Water Resources Management and Agroecological Transition for Cambodia), agricultural research component	R&D	Rice, crop diversification, water management	AFD	MAFF, GDA/DALRM & CIRAD

Laos	Sayabouri	2000–2002	PRODESSA (Project for the Development of the South of Sayabouri Province), agricultural research component	R&D	Maize	AFD	NAFRI, CIRAD
	Sayabouri, Xieng Khouang	2003–2009	PRONAE (National Program in Agroecology), PASS (Development project for the South of Sayabouri Province)	R&D	Maize, upland rice, forages	AFD	NAFRI, CIRAD
	National	2007–2011	PROSA (Sector-based program in agroecology)	Development	Scaling-up strategy	AFD	MAF, CIRAD
	Northern Laos	2014–2019	EFICAS (Eco-friendly Intensification and Climate-resilient Agricultural Systems)	R&D	Participatory LU Planning	AFD, EU	DALaM, CIRAD
Vietnam	Bac Kan	1998–2004	SAM (Mountainous Agrarian Systems Project)	R&D	Rice, maize, cassava	French Government	VAAS, CIRAD, IRD; IRRI
	Pleiku	1999–2004	ADP (Agricultural diversification Project)	R&D	Rubber	WB	NIR, CIRAD
	Son la, Yen Bai	2008–2012	ADAM (Support to CA Extension in Mountainous Areas of Vietnam)	R&D	Maize, Tea	AFD	NOMAFSI, CIRAD
	Son la, Yen Bai	2009–2013	Improved Market Engagement for Sustainable Upland Production Systems in the North West Highlands of Vietnam	R&D	Maize	ACIAR	NOMAFSI, UQ
	Yen Bai, Son La, Dien Bien	2010–2013	Unfavorable rice environment management in the Northern mountainous region of Vietnam: 2009–2013 activities by NOMAFSI	R&D	Rainfed rice	IFAD, IRRI	NOMAFSI, IRRI/CURE
	North Vietnam	2011–now	Extension of minimum-tillage for potato production	Development	Potato	Gov. of VN	Provincial DARD
	Yen Bai	2017–2019	CGIAR Research Program on Climate Change, Agriculture and Food Security (CCAFS)	R&D	Cassava, grass hedgerow	CGIAR/ CCAFS	CIAT, IRRI & local partners
	Phu To, Son la, Yen Bai, Lao Cai	2019–2021	Extension of sustainable cassava-based and maize-based systems in the Northern Mountainous region of Vietnam	Development	Maize, cassava, grass hedgerow	Gov. of VN	NOMAFSI
Regional		2009–now	CANSEA (Conservation Agriculture network in South East Asia) turned into ASEA (Agroecology for South East Asia) in 2019	Research, Education	Diverse	AFD, CIRAD	CIRAD

Figure 2 Main evolutions in CA-based cropping system design during the past two decades.

guianensis, a perennial legume forage used in many intercropping systems, was successfully controlled simply using rollers with cutting knives and discs (Boulakia et al., 2013).

- *CA vegetable production:* In Cambodia, CA vegetable production along with drip irrigation has been implemented since 2009 within the SANREM project (Edralin et al., 2017) and later on by the Sustainable Intensification Innovation Lab (RUA/CE SAIN, Kansas State University) (Table 1).
- *Identifying the best cover crop mixtures:* More recently, research on CA has been focusing on identifying the best cover crop mixtures according to ecologies, agricultural systems, and expected ecosystem services, assuming that these mixtures support increased productivity and stability as compared to a single cover crop (Wortman et al., 2012). Pearl millet, sorghum, *Crotalaria juncea*, *Crotalaria ochroleuca*, cowpea and *Stylosanthes guianensis* are the main cover crops used at the moment in CA systems experiments in Cambodia and Laos for their agronomic and ecological performances. Support for capacity development of the local institutions for multiplication of those seeds has also been provided.
- *Sowing in green standing vegetation:* This is currently being practiced in Laos and Cambodia to increase the flexibility of agricultural operations, reduce production costs, and improve water, nutrients, and weed control management as compared to sowing in a dead mulch. Green-sowing is also a pathway toward organic CA cropping systems in which cover crops are terminated mechanically using roller crimpers with cutting discs or elliptical bars.

Apart from technical innovations in CA systems development, major changes in the actors involved in the innovation process and in out-scaling strategies are also visible:

- *Favoring farmers' access to CA-specific agricultural equipment and/ or services:* In a context of increasing labor scarcity in agriculture, CA systems must enable labor productivity improvement without introducing bottlenecks in the working calendar (Chan et al., 2018); seed broadcasters, roller-crimpers, and no-till planters adapted to two-wheel hand tractors and tractors are needed for CA systems to compete with conventional systems, especially in commodity-based production systems (e.g. maize and cassava). Since the early 2000s, a large range of CA-specific agricultural equipment has been introduced, tested and adapted to local conditions in Laos and Cambodia (ASMC, 2019). In the beginning, governmental extension agencies were targeted in out-scaling strategies to manage agricultural equipment supply and provide CA services to farmers. More recently, specific attention has been given in Cambodia to agricultural cooperatives and local service providers to increase no-till planting services to farmers (Chan et al., 2018; Vernet et al., 2020). Since 2018, Swisscontact has been supporting local importers and manufacturers in getting access to agricultural equipment for CA through the Mekong Inclusive Growth and Innovation Program (MIGIP, SDC) and the Conservation Agriculture Service with a Fee (CASF) project (USAID). A first import of second-hand no-till planters from Brazil was made in 2019 by a local importer and a container of new brands of no-till planters was purchased in 2020. These projects promote a demand-creation process by pooling together manufacturers, local workshops, service providers (individuals and agricultural cooperatives) and farmer communities in order to match supply and demand and create new economic opportunities. Four main implements are currently being promoted: land plane leveler for both lowlands and uplands, no-till planters (maize, soybean, cassava), seed broadcasters (rice, fodder species) and roller crimpers.

- *Favoring farmers' access to diversified and open-pollinated cover crop material:* The performance of CA systems is highly dependent on the quantity and diversity of the biomass produced and returned to the soil. In Cambodia, a large genetic bank of cover crops (> 45 species and > 200 cultivars) is managed by GDA/DALRM, with the main objectives of ensuring high quality of the genetic materials and sharing this biodiversity with farmer communities and the private sector (Chett et al., 2018). The low availability of cover crop seeds, and the lack of value chains for cover crops are regularly emphasized as the main bottlenecks for the dissemination of CA systems (Kong et al., 2016; Lestrelin et al., 2011; Lienhard et al., 2014a; Oung et al., 2018). The main cover crops used in the region (e.g. pearl millet, sorghum, *Crotalaria spp.*, and *Stylosanthes guianensis*) are open-pollinated plant materials that can be produced locally. Business models supporting their large-scale production, conservation and marketing are yet to be explored. In Cambodia, Swisscontact supports local enterprises and farmers to produce cover crop seeds for the national market (MIGIP and CASF projects).
- *Investing in social capital and collective learning to enhance adaptation and adoption:* Learning is essential for farming communities to adapt to their changing environment (Castella et al., 2018). Learning processes need to be location specific, multi-stakeholder-directed and iterative. The objective is not so much to educate but to empower, to 'develop the capacity' of local stakeholders, and notably of key farmers, service providers and retailers that are instrumental in the design of CA systems, in their promotion and practical implementation. In Laos and Cambodia, role-playing games have been used to stress the importance of local social networks and collective learning in the farmers' decision-making process

(Ornetsmuller et al., 2018); Kong et al., forthcoming). Technical objectives can favor collective learning. For example, in Cambodia, sunnhemp and other cover crops have been shown to: (i) empower farmers in the production of seeds for cover crops; (ii) increase the diversity of plant species cultivated in village agricultural landscapes; and (iii) foster exchanges within and outside the community and between farmers and local seed companies.

- *Policy dialogue and theory of change:* CA in the GMS sub-region still corresponds at present to a collation of isolated initiatives. Several attempts have been made in the past to support CA dissemination, including the creation in 2009 of the CADF in Sayaboury province, Laos, which was based on a tax on maize exportation to support CA promotion and provide continuous technical support to farmers in the province (Lestrelin, 2015). The CADF experienced management problems, notably related to the limited involvement of farmers and the private sector in its governance, and a lack of strategic planning (Lestrelin, 2015). The Cambodia Conservation Agriculture for Sustainable Intensification Consortium (CASIC) created in 2020 in Cambodia incorporates diverse stakeholders (governmental and non-governmental institutions, universities and research institutions, farmer representatives and the private sector) and is currently working on a roadmap for CA promotion nationwide.

2.1.3 Performance and impact of Conservation Agriculture farming systems

CA systems have been shown to improve land and labor productivity when appropriate mechanization is available (Boulakia et al., 2019; Kong et al., 2016; Lienhard et al., 2014a,b). CA systems have also been shown to have a positive impact on soil biological functioning (Hok et al., 2018; Lienhard et al., 2013; Pheap et al., 2019), soil aggregation (Hok et al., 2021), soil organic carbon accumulation (Hok et al., 2015; Le et al., 2018), water infiltration rate and nutrient cycling (Pheap et al., 2019).

In a recent study in Battambang Province (Northwest Cambodia), Vernet et al. (2020) assessed the opportunity and profitability for service providers to invest in no-till planters and offer no-till services for rice, maize and cassava planting. Following interviews with service providers and farmers, the break-even point (i.e. the minimum annual surface area above which investment in a no-till planter would be profitable) simulations were made using service fees of $40/ha, $75/ha and $40/ha for rice, cassava and maize sowing, respectively. Break-even points were at 55 ha, 48 ha, and 71 ha (excluding ownership costs of tractor) for rice, cassava and maize crops respectively, and were seen as attractive for some service providers.

2.1.4 Conservation Agriculture adoption in the GMS sub-region

Whereas land degradation is widely acknowledged as an important issue in the GMS region, the dissemination of soil conservation technologies and approaches including CA is still low. There are no aggregated data related to CA adoption in the GMS sub-region, but based on available project data, it is estimated that about 2000 smallholder farmers (about 2000 ha) are engaged in minimum tillage with crop residue management in Laos and Cambodia, of whom less than 500 households use cover crops in diversified CA systems.

2.1.5 Main constraints to adoption in the GMS sub-region

2.1.5.1 Limited access yet to key inputs: cover crop seeds, mechanization and bio-products

The success of the green revolution is based on facilitating farmers' access to high-yielding varieties, but also to other inputs including tillage, fertilizer, irrigation and pesticides (Pingali, 2012). Similar facilitation is needed for a broad-scale dissemination of CA in the GMS region.

- *Favoring farmers' access to appropriate CA mechanization and services.* The results of the MIGIP and CASF projects in Cambodia are promising but need to be reinforced and expanded to the other countries of the GMS sub-region. Business models supporting farmers' increased access to CA-specific agricultural equipment (e.g. through service provision? collective vs individual investment? etc.) and equipment maintenance facilities are yet to be explored through the involvement of local agricultural workshops (Vernet et al., 2020).
- *Favoring farmers' access to diversified and open-pollinated cover crop material:* Similarly, the preliminary results obtained in Cambodia are promising but need to be reinforced and expanded to the other countries of the GMS sub-region. Business models supporting priority cover crops (e.g. pearl millet, sorghum, *Crotalaria* spp., and *S. guinanensis*) large-scale production, conservation and marketing are yet to be explored.
- *Bio-product development and biological control:* The high (and increasing) use of pesticides in agricultural systems is a common and crucial issue throughout the region. If CA systems promote a decrease in pesticide use mainly through the redesign of cropping systems and landscapes, the substitution of conventional pesticides by bio-products (e.g. elicitors, bio-pesticides and bio-repellents) is still greatly needed to achieve healthier food and landscape systems. Policies and regulations allowing the testing and homologation of bio-products are required. Increased on-farm testing

and increased linkages between bio-product suppliers and local retailers are needed to raise farmers' interest and the demand for these products.

2.1.5.2 Trade-offs between the different biomass users

Communal grazing and burning after crop harvest are widespread traditional territory management rules in the GMS region (Lienhard et al., 2014a; Oung et al., 2018). Animals are allowed to graze far from cultivated areas during the cropping season and are brought back after crop harvest, threatening relay crop development and effective residue management. Fire is used to limit disease propagation in paddy rice fields, manage weeds in plantations and/or stimulate native pasture regrowth in grazing areas. The use of agricultural biomass for biogas and energy production is expected to boom in the coming years (Prasertsan and Sajjakulnukit, 2006). Trade-off on biomass use must be negotiated at the village community level, using territorial approaches and participatory land use planning (Castella et al., 2018).

2.1.5.3 Sustaining long-term experiments and networks of on-farm demonstrations

Long-term experiments and networks of on-farm demonstrations are instrumental to CA system development and knowledge production. Two experimental sites have been established and sustained in Laos and Cambodia, in Ban Poa and Bos Khnor CA Research Stations (Leng et al., 2018; Lienhard et al., 2013), respectively. These platforms are essential to: (i) design and assess CA systems (technical requirements, performances and ecological balance shifts); (ii) assess biological processes that sustain and enhance CA systems; (iii) maintain genetic banks of staple, cash and cover crops and to ensure the connection with rural communities and seed suppliers; (iv) provide training and build capacities of smallholder farmers and development operators (youth, women and men) on innovative cropping systems management, appropriate-scale mechanization, use of a large diversity of plant species and bio-products; and (v) provide science-based evidence to land-managers, policy-makers and donors on medium- to long-term benefits of CA systems. Although essential, the long-term sustainability of these CA R&D platforms remains uncertain.

2.1.5.4 Political willingness, common vision and theory of change

As mentioned previously, CA in the GMS region still corresponds at the moment to a sum of isolated initiatives whose impact is yet difficult to measure. To engage in a real dissemination of CA, it is necessary to reaffirm political willingness to engage in CA promotion and engage the different stakeholders in the definition of a common vision and a collectively agreed strategy (theory of change) at multiple levels (local, national and regional).

2.1.6 Perspectives in the near future in the GMS sub-region

2.1.6.1 Perspectives regarding Conservation Agriculture systems' development in the GMS sub-region

- *Maize, cassava and lowland rice systems:* Maize and cassava still represent more than 70% of the total (non-rice) area under annual cropping in the GMS region (Ingalls et al., 2018) and will still be the cornerstone of CA systems design for the coming years. More recently, increased pressure on lowland rice production systems (e.g. labor scarcity, impact of climatic changes, land degradation issues, increased pest pressure and increasing use of pesticides, etc.) have raised farmers' interest in alternative production systems. The WAT4CAM project (Water Resources Management and Agro-ecological Transition for Cambodia) in Cambodia (2020–2024) will help consolidate and promote CA-based lowland rice systems.
- *CA for farmers transitioning to perennial plantations:* An increasing number of farmers are transitioning from annual to perennial cropping systems (e.g. fruits, coffee, banana) (Ingalls et al., 2018); CA systems could help this transition, for example, through the management of tree inter-rows during the immature tree phase.
- *CA for vegetable production:* Along with improved water management, there is a need for this system to be promoted at larger scale to integrate vegetable production in CA-managed farming systems, to improve the nutritional status of rural communities and to offer a regular flow of income to women.
- *Sowing cover crops in main crops, and main crops in cover crops:* This remains the main challenge in CA systems management for which dedicated research is needed.

2.1.6.2 Perspectives regarding Conservation Agriculture-enabling environment

- *Exploring business models for facilitated access to seeds and mechanization, and to a lesser extent, bio-products:* This is certainly the main development priority to be addressed for broad-scale CA adoption; such explorations should be conducted by teams specialized in fostering public-private partnership, such as Swisscontact in Cambodia.

- *Exploring financial mechanisms:* Financial mechanisms need to be explored to support farmers' investments in soil fertility restoration through CA. Different mechanisms exist and should be explored, including: (i) carbon funding (e.g. Australia, Conservation Agriculture) or carbon credit (REDD+) mechanisms, by approaching the companies that emit the most GHG at national and regional levels (e.g. plane companies); (ii) mechanisms for payment for environmental services (PES) with the establishment of funds managed by the private sector or tax incentives (bonus / malus system applied to the exports of specific agricultural products); (iii) micro-finance which also seeks to invest in sustainable intensification; and (iv) block-chain systems aimed at tracing the origin and quality of products and remunerating farmers, agricultural cooperatives or other actors in the chain on the basis of quality and production of environmental services.
- *Capacity building and awareness raising:* Regular field days, awareness-raising events and training sessions should be sustained in the long term, designed and adapted to a diverse audience. In Cambodia, RUA/CE SAIN (KSU, SIIL) and UBB, invest in capacity building of young agronomists and researchers. In addition, the network of Technology Parks implemented by CE SAIN is one of the main support mechanisms that offers training to smallholder farmers, women and youth, and development operators. The two experimental stations at Ban Poa (Laos) and Bos Khnor (Cambodia) also provide fundamental support for training and capacity building at the national and regional level. In Cambodia, GDA/DALRM aims to develop Bos Khnor as a regional training center that also encompasses the longest experiments under CA in the sub-region.
- *Extension approache:* In Cambodia, MetKasekor is an "opening the market" innovative early adopter led extension service model. MetKasekor focuses on opening the market for private sector investments that match with the principles of Conservation Agriculture and Sustainable Intensification. MetKasekor is being led by the Department of Extension for Agriculture, Forestry and Fisheries (DEAFF) and coordinated by the Department of Agricultural Land Resources Management (DALRM) and the Department of Agricultural Engineering (DAEng). The core implementor is the Provincial Department of Agriculture Forestry and Fisheries (PDAFFs) in partnership with the private sector..
- *Gender:* Women play a key role in household decision-making regarding innovation adoption, especially while related to household disbursement for inputs, new technologies or services (Sumner et al., 2017). Tailored communication and training strategies are needed to engage women in the CA innovation process (Jones et al., 2019).
- *Key regional actors:* The consortium of partners in Cambodia (MAFF, GDA/DALRM & DAEng, RUA/CE SAIN, UBB, ITC, Swisscontact, Kansas

State University, CIRAD) emerges as a key institutional platform for CA promotion in the GMS sub-region.

- *Favorable institutional context for CA and sustainable intensification in the region:* Cambodia and Vietnam have joined in 2019 the 4p1000 initiative (www.4p1000.org) whose objective is to support carbon sequestration in soils to sustain food security, support agricultural systems adaptation and mitigate climate change. Since 2019, Laos has been the LICA initiative (Lao-facilitated Initiative on Agroecology and Conservation Agriculture) at the ASEAN level and is supported in this by different networks (Agroecology Alliance in South East Asia - ALiSEA) and projects (Agroecology and Safe food System Transitions in South-East Asia - ASSET).
- CASIC: On 21 May 2020, the Ministry of Agriculture, Forestry and Fisheries (MAFF) of Cambodia formally approved the formation of the CASIC. This decision was a turning point to foster the dissemination of CA/SI practices to restore and preserve agricultural soils in Cambodia, advancing farm sustainability and contributing to climate change mitigation and adaptation.

2.2 The Philippines

2.2.1 Brief description of Conservation Agriculture and state of its adoption in The Philippines

The concept of CA was introduced in the Philippines through the USAID-funded Sustainable Agriculture and Natural Resources Management-Collaborative Research Program (SANREM-CRSP)-funded project implemented in the upland areas in Southern Philippines from 2010 to 2014. Numerous research activities were performed at the SANREM Research site in Claveria, Misamis Oriental, Philippines (Fig. 3) during this five-year period. Through partnership and collaboration among the University of the Philippines Los Baños, North Carolina A&T State University, Virginia Tech, World Agroforestry, CIRAD and other

Figure 3 The SANREM-CRSP conservation agriculture field experimental site in Claveria, Misamis Oriental, The Philippines.

institutions, this project paved the way for the generation of much scientific evidence on the various biophysical and environmental benefits of CA. Field experimental studies, crop modeling, solute transport modeling and soil adsorption studies among others demonstrated the positive impacts of CA on soil quality, crop yield, fertilizer application efficiency, nutrient adsorption and retention, among others. Other studies have also generated new knowledge, particularly on the relationship between gender and CA adoption in the country.

Apart from its primary research objective, this SANREM project also provided capacity building opportunities for the upland farmers in Mindanao, the second largest island in the Philippines. A number of farmers were involved in the SANREM project as farmer-collaborators in the conduct of farm trials of CA. Concurrently, numerous other farmers were trained on CA through informal lectures and field visits to a vast demonstration farm established at the research site in Claveria, Misamis Oriental. This demo-farm showcased CA in its various forms in terms of main crop and cover crop combinations and crop rotations. Numerous farmers' groups from other parts of the country also visited the SANREM demo-farm and gained new knowledge on CA. While no formally established CA farmer-groups exist, the promotion of this technology in the country has been facilitated by the existence of other farmers' associations in the Philippines.

At the end of the SANREM research project in 2014, an attempt was made to establish the Philippine Conservation Agriculture Network (PCAN), geared towards accelerating the promotion of CA in the country. This network envisioned the involvement of the academia, research institutions and some government agencies in the promotion of CA in the whole country. The formation of PCAN would also enable membership of the Philippines in Conservation Agriculture Network in Southeast Asia (CANSEA), which could pave the way for more knowledge sharing that could help accelerate promotion and upscaling of CA in the country. However, the evolution of PCAN has not progressed, due to inadequate institutional and financial support.

During the post-SANREM years, promotion of CA in the country was done through public fora, professorial lectures on YouTube and scientific publications. In addition, CA was recommended for inclusion in the national government programs on soil conservation through the Bureau of Soils and Water Management, during a public forum on agriculture organized by the Philippine Department of Agriculture and the French government in 2017.

Recently, promotion of CA in the Philippines was included in national-level planning for agricultural development in the country. In particular, CA is among the recommended measures for climate change adaptation (CCA) and disaster risk and reduction management (DRRM) in the National Irrigation Master Plan (2020–2030) in the Philippines, CA was also included in the recommendation in mainstreaming CCA/DRRM in the Agriculture and Fisheries Modernization Plan

(2018–2023) by the Philippine Department of Agriculture in 2019 and 2020, respectively All these small steps may pave the way for increased CA adoption in the country.

Overall, a wider and sustained upscaling of CA in the Philippines is practically a work in progress, requiring adequate institutional and financial support. Nevertheless, the potential for CA adoption in the country remains high, considering its vast agricultural land resources - with a total area of 14 741 850 ha, comprising about half of the total land area of 30 million hectares (FAO, 2020) - coupled with the fact that this country is prone to climate change and climate variability and soil erosion and soil quality degradation are prevalent in most of these agricultural lands.

2.2.2 Advances in Conservation Agriculture research in The Philippines

The beneficial effects of CA on soil quality and crop yield have been proven through field experiments conducted in a typical sloping upland area in Claveria, Misamis Oriental (Fig. 3). The results have been published in peer-reviewed journals (e.g. Sevilla and Ella, 2020; Ella et al., 2016; Bondad et al., 2015; Tarnate et al., 2014). Six CA Production Systems (CAPS) treatments in the form of cropping patterns with different cover crops, including a plow-based (PB) system serving as the control, were established and replicated four times and were laid out in a randomized complete block design in a typical upland agricultural producing area in the said site. The sub-plots in each treatment were also established to represent two fertility levels. A summary of the experimental treatments is given in Tables 3 and 4.

2.2.2.1 Effect of Conservation Agriculture on soil organic carbon

Periodic soil sampling and soil quality monitoring showed that the soil organic carbon (SOC) for all CAPS treatments (T1 to T5) was generally higher than under T6 (plow-based) after three years of cropping. Treatment T2

Table 3 Summary of the CAPS treatments in Claveria, the Philippines

	Cropping pattern
T1	Maize+ *Arachis Pintoi* - Maize + *Arachis pintoi*
T2	Maize + *Stylosanthes guianensis* - *Stylosanthes guinanensis*- fallow
T3	Maize+cowpea - Upland rice +cowpea
T4	Maize+Rice bean - Maize+Rice bean
T5	Cassava + *Stylosanthes guianensis*
T6	Maize-maize (conventional plow-based) (control)

Table 4 Summary of sub-plots

	Fertility Level
Fo	120-60-60 for N, P_2O_5, K_2O
F1	60-30-30 for N, P_2O_5, K_2O

(Maize+*Stylosanthes guianensis-Stylosanthes guianensis*-fallow) exhibited the highest SOC at the uppermost soil layer (0–5 cm). Under a high fertility level, the SOC at the uppermost layer for all CAPS treatments T1 to T5 increased over time, with T2 (Maize+*Stylosanthes guianensis-Stylosanthes guianensis*-fallow) exhibiting the highest rate of increase (Fig. 4). Under a moderate fertility level, the SOC for CAPS treatments increased over time with T5 (cassava+*Stylosanthes guianensis*) exhibiting the highest rate of increase (Fig. 5). The relatively high SOC in these treatments may be attributed to the relatively high amount of biomass contributed by the cover crop, *Stylosanthes guianensis*. In contrast, the SOC at the uppermost soil layer under the PB system declined over time for both fertility levels (Figs 6 and 7) based on linear regression analysis.

A summary of the linear regression analysis demonstrating the temporal change in SOC and statistical significance of the regression slopes is shown in Tables 5 and 6. All other experimental details and a more thorough discussion of the results were published by Ella et al. (2016).

2.2.2.2 Effect of Conservation Agriculture on residual soil moisture

The influence of CA on soil moisture retention was examined using periodic time domain reflectometry measurements of moisture content at the CAPS treatments shown in Fig. 3. Analysis of variance of the residual moisture after one, two and three cropping seasons showed that the plots under CA had

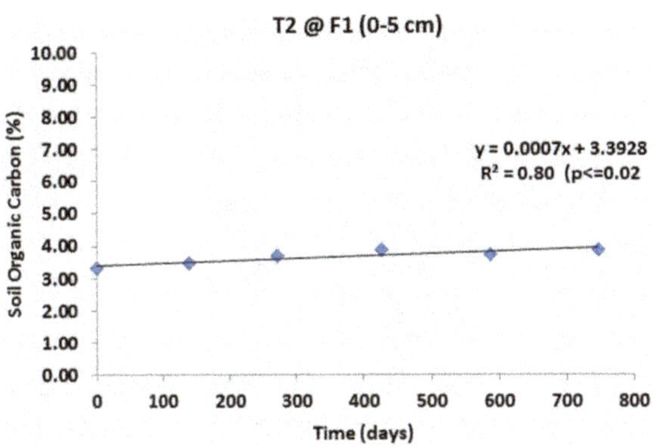

Figure 4 Regression analysis of soil organic carbon vs. time at CAPS treatment T2 under moderate fertility level.

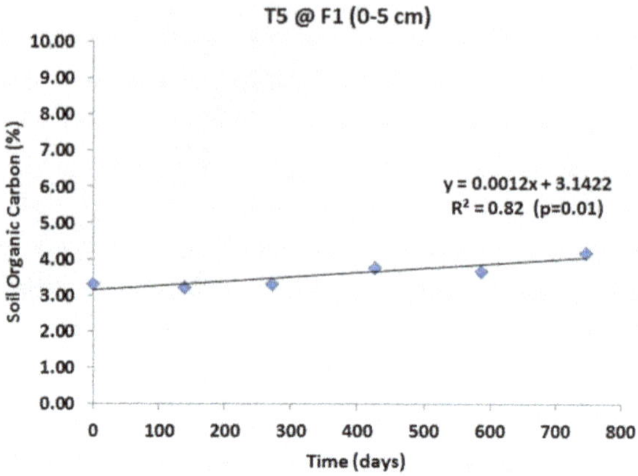

Figure 5 Regression analysis of soil organic carbon vs. time at CAPS treatment T5 under moderate fertility level.

significantly higher ($\alpha = 5\%$) residual moisture content than those under the PB system, with CAPS treatment T2 exhibiting the highest residual moisture content value. These results have significant practical implications on cropping calendar schedule, irrigation water management and climate adaptation. All other experimental details and a more thorough discussion of results were published by Ella et al. (2016).

2.2.2.3 Effect of Conservation Agriculture on soil adsorption capacity

Another experimental study on ammonium adsorption characterization of soils at the SANREM research site demonstrated the beneficial effects of CA over PB systems (Sevilla and Ella, 2020). Based on batch adsorption experiments for soil samples collected in selected treatments at the SANREM site and adsorption isotherm modeling, it was shown that CA increases the ammonium adsorption capacity of the soil. Results also showed that the mean distribution coefficient kd values of the soils under CAPS were significantly higher ($\alpha = 5\%$) than those under the PB system. These results suggest that more ammonium ions would become available for plant uptake under CA than under the PB system due to greater adsorption in the former than in the latter. Conversely, lower leaching of ammonium ions and consequently nitrates can be expected under CA than under the PB system. These findings are important not only from the crop production point of view but also from the standpoint of environmental protection, as the loss of ammonium due to reduced adsorption may pose a threat to groundwater quality, especially upon its oxidation to nitrates and the subsequent leaching to the underlying aquifers.

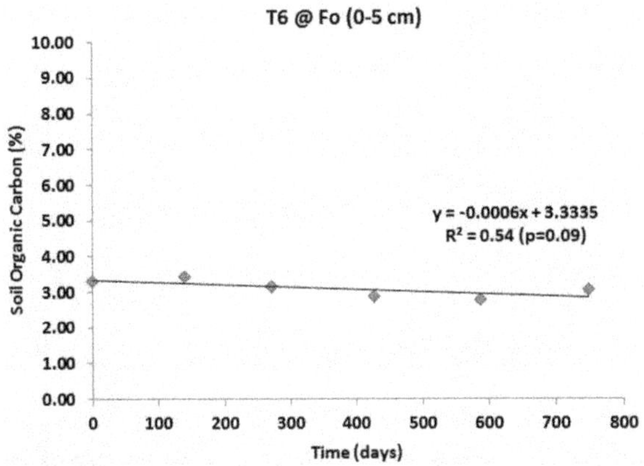

Figure 6 Regression analysis of soil organic carbon vs. time at the plow-based system T6 under high fertility level.

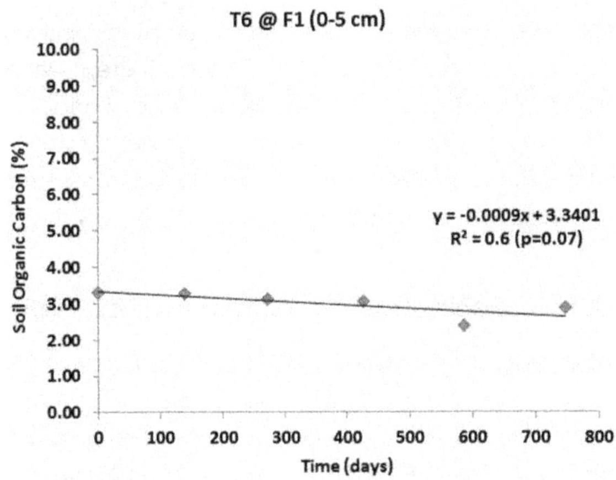

Figure 7 Regression analysis of soil organic carbon vs. time at the plow-based system T6 under moderate fertility level.

Table 5 Summary of linear regression analysis of soil organic carbon vs. time for various treatments under high fertility level (F0)

Treatment	Regression Equation	R^2	p-value
T1	SOC= 0.00033T+3.46	0.23	0.33
T2	SOC=0.00138T+3.176	0.60	0.07
T3	SOC=0.00035T+3.243	0.25	0.31
T4	SOC=0.000055T+3.210	0.02	0.77
T5	SOC=0.00051T+3.446	0.25	0.32
T6	SOC=-0.00065T+3.333	0.54	0.09

Table 6 Summary of linear regression analysis of soil organic carbon vs. time for various treatments under moderate fertility level (F1)

Treatment	Regression Equation	R^2	p-value
T1	SOC=-0.00026T+3.578	0.14	0.47
T2	SOC=0.00073T+3.393	0.80	0.02
T3	SOC=-0.00060T+3.736	0.11	0.52
T4	SOC=0.00115T+3.271	0.59	0.07
T5	SOC=0.00120T+3.142	0.82	0.01
T6	SOC=-0.00093T+3.340	0.60	0.07

2.2.2.4 Effect of Conservation Agriculture on nutrient transport in soils

Solute transport studies performed by Tarnate et al. (2014) for soils under CA and conventional PB systems also demonstrated the positive impacts of CA on soil quality. Based on laboratory soil column experiments and solute transport modeling of soils collected from the SANREM research site, it was demonstrated that nutrient leaching is lesser and slower under CA than under the PB system. The details of the experiments and modeling results were published by Tarnate et al. (2014). From a practical standpoint, these results suggest that CA provides better fertilizer economy than PB systems.

2.2.2.5 Effect of Conservation Agriculture on crop yield

Apart from soil quality studies, crop yield and crop modeling studies under CA and PB systems have also been performed in the Philippines. Bondad et al. (2015) collected biomass, leaf area index, crop yield data and other relevant data at the SANREM site for various cropping seasons and employed the DSSAT-CERES Maize crop model to predict the impacts of CA and tillage on crop yield. The model simulation results showed that after ten years of continuous practice, crop yield under CA would increase by 5.5% to 10.2%. On the other hand, crop yield under the PB system would decrease by 24.4% to 28.5%. These findings, along with other details of the study, were published by Bondad et al. (2015).

2.2.3 Challenges in upscaling of Conservation Agriculture in The Philippines

Essentially the same CA adoption constraints identified by Ares et al. (2015) apply to the Philippines. In particular, most Filipino farmers have zero or limited knowledge of the principles of CA. The case of Claveria was no exception when the SANREM project started in 2014. Most upland farmers practice the traditional PB system on the pretext that it is the best practice especially for weed control, oblivious of the adverse effects of excessive soil disturbances on soil quality and soil health. The absence of an adequate supply of suitable cover crop seeds in various agricultural areas in the country also poses some constraints in upscaling. During the SANREM project implementation, for instance, seeds of the cover crop *Stylosanthes guianensis*, which is highly suitable for the acidic soils at the research site, had to be transported from Manila to Mindanao as this species is not endemic in this area. The absence of appropriate machineries and equipment for CA such as direct seeders, dibblers and rollers among others also inhibit the use of CA, particularly for large-scale crop production systems. All these technological constraints, coupled with the absence of national government policies that include conservation agriculture as a soil conservation measure, make its upscaling even more challenging.

Furthermore, in a gender study by Parks et al. (2015) on CA in the Philippines, they found out that the main gender-based constraints for CA adoption include the lack of access to secure land tenure, capital and training. According to this study, the main concern of farmers regarding access to land is not the size but the long-term access and control in the form of secure land tenure and land titles. Farmers who are on a short-term land management arrangement also tend to be hesitant to adopt CA as they are more concerned about production and short-term profits rather than the improvement of the quality and health of soil that is only temporarily used by them. Access to capital affects men and women differently according to the same study, but more importantly it was pointed out that the same farmers' concern about capital under a conventional PB system may still exist under CA. Exposure to training on CA is also a gender-based constraint. In this same study, they found out that only one family member per household is invited to attend agricultural workshops, and this is usually the men, thereby depriving women of such training. All these gender-based constraints pose some problems when it comes to accelerating CA adoption in the Philippines.

Despite some advances in CA research in the Philippines, there is still a need for more extensive and long-term research to provide additional strong empirical evidences on the benefits of CA and the adverse impacts of conventional tillage as part of the strategies to overcome the constraints to CA adoption (Ella, 2018). More importantly, the research results to be generated need to be published in peer-reviewed journals, not only for wider dissemination in the scientific community but to serve as additional basis for science-based policy formulation. In addition to research, there is also a need for more effective extension services to promote CA adoption in the rural communities in the country. This may include seminars, public fora, establishment of demonstration farms, farmers' field day, distribution of information education and communication materials and effective extension activities, among other approaches. Both research and extension requirements would necessitate enabling policies as well as institutional and financial support. These, along with the other strategies presented by Derpsch (2008) and Kassam et al. (2014), may pave the way for increased adoption and upscaling of CA, not only in the Philippines but in the whole of the Southeast Asian region.

However, there are still a number of constraints that further pose challenges to CA adoption in the Philippines. These include a lack of priority and attention given by policy makers to this technology, which affects the availability of institutional and funding support for research, development and extension of CA in the country. Other constraints and challenges are the reluctance or selectivity of national and international funding or donor agencies to finance CA research or upscale projects in the Philippines. In most cases, international donors and funding agencies prioritize underdeveloped countries belonging to the low-income group. This poses a big challenge to some countries in Southeast Asia, like the Philippines, whose income or economic level is relatively higher than their counterparts in other parts of the world. It is in this light that local support from the national government becomes critical.

To date, there is no published data on the actual CA area in the Philippines. However, based on the estimates of the SANREM scientists based in Mindanao, there are about 6750 hectares of CA area in northern Mindanao alone, based on an estimated 4500 farmers who have adopted CA and an average landholding of 1.5 ha. This estimated CA area of 6750 ha could serve as an estimate of CA area in the whole country since CA has not yet been adopted in other parts of the Philippines.

2.3 Timor-Leste

In Timor-Leste, conditions of poor soil fertility and land degradation that contribute to low crop productivity are compounded by a changing climate and increased disaster risk, largely due to drought. CA is progressively being adapted and adopted in Timor-Leste and has shown a significant impact on crop yields and agricultural productivity. The technology has been promoted in rainfed areas of Timor-Leste by the FAO and the Ministry of Agriculture and Fisheries (MAF), in partnership with two Non-Governmental Organizations (NGOs), Mercy Corps and RAEBIA Timor-Leste (Resilient Agriculture and Economy through Biodiversity In Action), since August 2013, benefiting more than 5500 resource-poor, smallholder farmers over 230 ha of land.

A four-year project funded by the Office of Foreign Disaster Assistance (OFDA) of the United States Agency for International Development (USAID) is, at present, focusing on maize growing on sloping upland and flat lowland areas in the Municipalities of Baucau, Ermera, Manatuto, Aileu and Manufahi. This study focuses on the districts of Manatuto and Manufahi selected purposively based on the importance of smallholder maize farming and the associated agroecology.

The application of CA technologies and practices is highly relevant to rainfed agricultural systems in these regions, and also to the needs of the donor, the government at different levels and farmers. CA has been shown to contribute to drought resilience through the reduction of crop losses during dry periods, improve maize productivity, reduce soil erosion and address dryland farmers' needs for reduced labor and inputs.

The objectives of the projects were to enhance household food and nutrition security and improve rural livelihoods through a sustained increase in agricultural productivity. The projects' three-pillar design of CA validation, pre-adoption through Farmer Field Schools (FFS) and improving the enabling environment was relevant to the continued upscaling and out-scaling of real CA adoption.

In Timor-Leste, 'slash and burn' and the free grazing of livestock contribute to soil erosion, a decrease in the soil's organic matter and its capacity to retain moisture and nutrients. These factors, in turn, lead to low soil fertility and reduced crop productivity. Weather variability and climate change also affect crop productivity, especially with more frequent dry spells and less rainfall. The described project was the first to test and promote the adoption of CA in Timor-Leste.

The USAID-funded project supported 18 farmers' groups to each establish and manage a group demonstration plot comprising Control (burning, plowing and traditional maize cropping) and three to five CA treatments (no burning and no or minimum tillage) with variables appropriate to the respective agroecological zones, namely mulching, maize varieties and different intercropped legumes and cereals.

Fifteen of the 18 FFSs successfully completed their adaptive research to test CA practices. Some of the groups experienced up to a 40% reduction in the time farmers used for land preparation and weeding, between the CA sub-plots and the controls. The FFSs at Uma Boco in Manatuto District reported that weeding of the CA sub-plots each took a total of one day per season compared to two days for the control plot. In addition, six of the FFSs reported higher maize yields in all of the CA sub-plots than in the controls (photo). Overall, the remaining nine FFSs reported similar yields between the CA and controls. The main reasons cited by farmers for an increase or no change in maize yields was less competition from weeds due to the mulches and cover crops, and moisture conservation because of the mulches.

Participatory adaptive research was conducted during three cropping seasons. Eighteen farmer groups participated in the first season (2013-2014). In the second and subsequent seasons (2014–2015 and 2015–2016), the number of groups increased to 34. The farmer groups were supported by MAF, FAO and NGO partners to establish and manage demonstration plots comprising a 'control' where maize was grown using traditional methods (i.e. burning the organic matter and plowing the soil) and three to five 'Conservation Agriculture' practices (i.e. no burning and no or minimum tillage), with variables appropriate to the respective agroecological zone. The variables included mulching, the use of local maize varieties or improved (open-pollinated) maize varieties such as *Sele* and *Noi mutin*, and different intercropped legumes (i.e. cowpeas, lima beans, lablab, pigeon peas, velvet bean and winged bean). On the northern rainfed flat or slightly sloping land, a single maize crop was grown. In the south, sites with similar topography were double-cropped with maize. It was found that appropriate CA technologies vary significantly between the agroecological zones (FAO, 2018). The project was implemented in 70 sucos (villages) in seven municipalities and has reached more than 6000 farmers.

CA machineries and tools including different types of manual seeders (jab planter, Li seeder, rolling injection planter), two-wheel tractor-mounted direct seeder and a rolling crimper were tested. The crimper was used to kill legume cover crops and leave a layer of organic matter ready for direct planting. A 2WT drawn direct seeder, designed to drill maize and legume seed directly into the soil through a layer of plant residue, was tested and demonstrated.

Given that intercropping legumes with maize was the key variable introduced in the research, a summary of findings is provided below.

Velvet bean *(Mucuna pruriens)* performed very well in terms of biomass production in lowland areas (Fig. 1). It grows vigorously, and therefore needs to be planted around four weeks after the maize is planted, to avoid competition with the maize. In southern areas, the bean can survive the whole year until the main planting season in November/December, while in central and northern parts of the island, it dies in June or July. The main constraint to adopting the use of this bean as a cover crop is that its seeds, if not collected at maturity, can germinate prematurely the following season and become a weed that damages the crop by competing for nutrients and sunlight. Maize–velvet bean trials are described in Correia et al. (2013).

Winged bean *(Psophocarpus tetragonolobus)* performed well at low altitudes and produced a good amount of biomass when planted every 50 cm. At middle altitudes, it grew less vigorously. It can be planted at the same time as maize as it grows slowly and produces much of the biomass after the maize is harvested. In both southern and northern areas, the bean dies in June/July, which in southern areas (where there is an additional cropping season) it can suppress weeds until the start of the second cropping season, making it a good cover crop for the main cropping season (November–March). In addition, this very nutritious bean that had previously been practically unknown for Timorese farmers proved to be quite palatable for most farmers who tried them.

Lablab (*Lablab purpureus*). Although it should be noted that lablab was tested only in a few plots due to scarce availability of seeds, the first results showed that it performed very well in lowlands, creating as much biomass as velvet bean. It showed potential to survive and cover the soil until the following cropping season, which would make it a good option for northern (drier) areas as it can stand the dry season better than velvet bean. The bean grows slowly and can therefore be planted at the same time as maize (at a minimum distance of 50 cm) and does not carry the risk of becoming a weed for the following crop.

Cowpea (*Vigna unguiculata*). Of all the legumes tested, the two local varieties of cowpeas tested (*fore masin* and *fore metan*) proved to have the best adaptability to all agroecological zones. The cowpea performed well at low, medium and high altitudes, producing a large amount of biomass, especially in lowlands. It can be planted at the same time as maize, although at low-lying altitudes, the spacing should be at a minimum of 50 cm. As it usually dies in April/May, cowpea can be a good option for intercropping with maize in the southern lowlands in the main season as it can maintain biomass until the second planting season in April/May.

Lima bean/butter bean *(Phaseolus lunatus)* performed irregularly, totally covering the soil in some plots but not growing well in others. Overall, it performed better at medium to high altitudes, although not in all plots. Lima bean could tolerate the dry season, suppressing most of the weeds until October. As the plant grows relatively slowly, it can be planted at the same time as maize at a distance of 50 cm to 100 cm.

Pigeon pea (*Cajanus cajan*) was evaluated mainly at medium and high altitudes planted on contour lines. While it does not produce much biomass to suppress weeds, one of its expected benefits is the capacity of mobilizing phosphorous. As it grows slowly, it can be planted at the same time as maize, at a distance of 50 cm.

Two main advantages of using CA that were observed during the project were an increased maize yield and a significant reduction of labor and fuel used. Data from sub-plots belonging to the 34 farmer groups confirmed that CA plots needed between 40–50% less time to produce the same or higher maize yield compared with the control plots. The main saving in time was due to CA not requiring the soil to be plowed, and weeding times were reduced due to the presence of crop residue shading the germinating weed seed. On flatter areas in the south of the island (Same and Natarbora), farmer adopters reduced the amount of fuel used for plowing and rotavating by up to 80%. Consumption was reduced from 50–60 liters of fuel per hectare to 10 liters per ha when CA equipment was used.

Under the drought conditions, the yield differences were exceptional (Fig. 8). During the drier year in the May–September 2015 cropping season in the south of the island, most FFS CA plots demonstrated a higher drought survival rate and thus a significant increase in maize yield (Fig. 9). It was postulated that the main reason for the higher yields was the higher capacity of soil moisture retention under the CA practices. The benefit was also observed

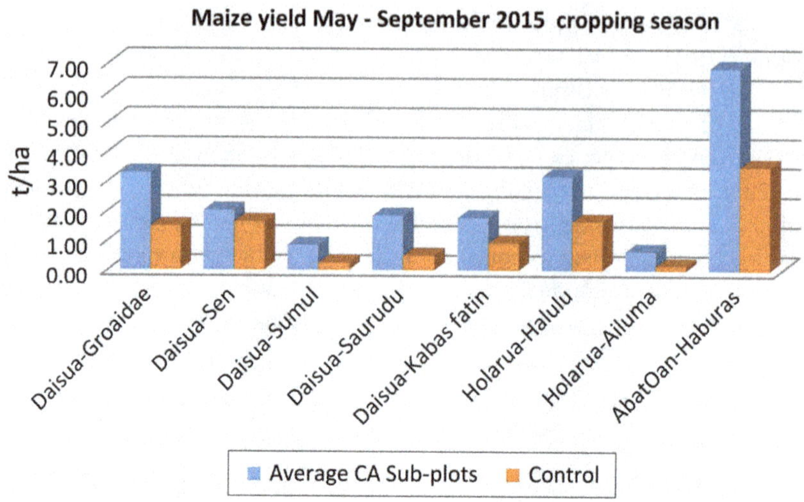

Figure 8 Maize yield in FFS plots in southern Timor-Leste (t/ha of wet cobs)

during the November 2015–March 2016 cropping season in most FFS plots. The most critical season during drought year was for about 2–3 weeks after planting, and it was demonstrated that CA practice could significantly reduce its risk of drought and thus contribute to sustainable production.

Soil sample analysis from CA FFS demonstration plots after two years of CA adaptation showed improvement in soil physical properties (bulk density) as well as biological properties (Fig. 10). The changes in soil properties were observed by the farmers as additional benefits of CA with the increased efficiency of the cropping system by reducing labor and inputs and resilience to the risks caused by variable weather events.

Climate change is one of the key challenges threatening livelihood stability. Projected changes in rainfall and temperature pose risks as well as opportunities for diversifying livelihoods into more resilient systems. The FAO's field-based program has exclusively focused on resilience-building of rural communities to reduce climate vulnerability and reduce post-harvest losses, as well as to introduce improved farming practices.

2.3.1 Constraints

Despite many farmers experiencing good results from using CA technologies, they face some constraints in adopting this practice on their farms. The three main constraints are: (i) their strong mind-set regarding farming practices inherited from their ancestors; (ii) the free grazing of livestock practiced in most parts of the country; and (iii) the MAF policy of subsidizing plowing of cropping land.

Figure 9 FFS demonstration plots in Natarbora, Manatutu. Left plot: CA with cowpeas, Center: Control, Right: CA with velvet beans.

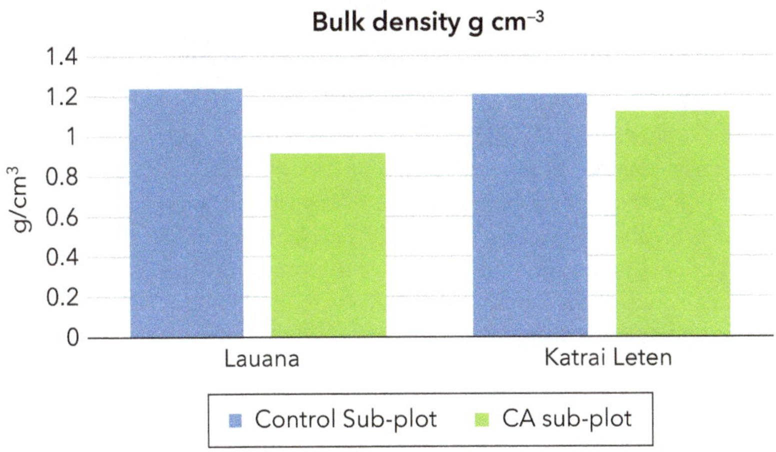

Figure 10 Bulk density of two FFS plots in Ermera Municipality.

At the beginning of the program, farmers were reluctant to test CA because burning and plowing were considered essential tasks for good crop production. These practices are deeply rooted in their farming traditions, making it essential to demonstrate the benefits of CA practices on farmers' fields. Farming traditions may also lead farmers to be pressured by their neighbors and relatives to return to traditional practices, even if good results are visible.

Free grazing of livestock is also a problem because freely roaming animals eat crop residues and legume cover crops remaining in the field during the dry season. To overcome this constraint in some *sucos*, FAO and RAEBIA are facilitating participatory land use management plans through *tara bandu*, requested and driven by the communities. The initial results are promising, and it can be an effective tool to increase mulch-based crop production. The free plowing policy can be a constraint in flat lands for farmers who do not have to pay or work to prepare the land. This opportunity can be more attractive than keeping and terminating a cover crop from the previous season before planting.

2.3.2 Opportunities

Due to the impact of El Niño, many farmers were able to observe that maintaining a layer of crop residue on the soil surface can help their maize perform better. This clear advantage of CA in such a critical weather event serves as an entry point for farmers to realize the disadvantages of burning organic matter in crop fields. In addition, farmers participating in the project are observing that CA technology is more efficient than traditional practices, potentially making farming a more attractive option for young people who are migrating to cities because they are increasingly discouraged by the low productivity and the labor-intensive nature of traditional farming.

The promotion of CA in Timor-Leste has been continuing through the following projects: Conservation Agriculture, Permaculture and Sustainable Fisheries Management: Enhancing Food and Nutrition Security and Reducing Disaster Risk in Timor-Leste (2015–2019); Pro-Resilience Timor-Leste - Strengthening Resilience in Communities Most Affected by Drought (2017–2019); and Stepping up the Engagement of the Global Network Against Food Crises (2020–2022).

The estimated area of land under CA is about 1000 ha, which is under the current project funded by the EU, and is expected to expand to about 1500 ha. The previous phases of CA since 2015 extended to about 1000 ha, but no monitoring has been done.

2.4 Indonesia

The Nusa Tenggara Barat (NTB) and Nusa Tenggara Timur (NTT) provinces face the most food insecurity in the Republic of Indonesia, where over a quarter of the households are considered food-insecure and have poor nutrition. Agriculture is the main source of livelihood there, and it is at high risk for climate-related disasters such as floods and drought. Rice and maize are the staple foods in these provinces and the yields are lower compared to other provinces in the country. Because of low rainfall, in many places in NTT and NTB, only one crop per year is possible. Hence, food shortages during the lean season are a recurrent problem.

In the NTT and NTB provinces, agricultural production and productivity are low, particularly for subsistence and smallholder farmers. The main causes are environmental constraints, poor soil fertility, inadequate agricultural inputs and infrastructure, recurrent extreme climate events, increasing pressure on available agricultural land, and lack of means and incentives to increase production. Climate change will intensify these constraints and risks. The mean annual temperature has increased by about 0.3°C in Indonesia since 1990. In addition, while overall annual rainfall has decreased by 2-3 %, intense wet-season rainfall has increased the risks of floods and landslides.

In response, the Government of the United States of America, through USAID/OFDA, contributed to the project "Reducing risks caused by changing climate in Nusa Tenggara Barat and Nusa Tenggara Timur provinces in Indonesia". The project was implemented from 1 June 2013 to 31 December 2018, to enhance agricultural livelihoods and improve soil fertility through the use of CA and integrate CA into local policy. The CA technology and practices were validated through community-based adaptive research demonstration and disseminated through the participatory FFS extension approach.

The project established 793 farmer groups with 16478 smallholder farmers (32%) members to test and validate CA technologies and practices. A total of 17092 smallholder men and women adopted CA technologies and practices on 1404 ha of land. The CA approach increased maize yields by 53% during the primary planting season (rainy) and 65% during the dry season. In addition, soil fertility was enhanced. Farmers who adopted CA increased their incomes by 37% in Lombok, 57% in Sumba and 77% in Timor. The FAO was also successful in supporting the Provincial Government of NTB and NTT in capacity building, along with drafting policies, strategies and action programs to create an enabling environment for the wider adoption of CA in Indonesia.

The project contributed to improved capacities of communities and key stakeholders to implement good practices to reduce the impact of climate threats and crises. CA was found to be an effective approach for meeting smallholder farmers' needs in leveraging crop production in rainfed areas affected by soil degradation and drought, through maintenance of higher soil moisture levels, increased soil fertility and often higher yields.

The project tested ten technologies and practices for maize cultivation, including planting-holes; ripping lines with two-wheel tractors drawing a no-till ripper; roller planters; direct dribbling of maize seeds; using organic mulches or residues from previous crops; intercropping maize with a variety of leguminous crops; not burning crop residues and weeds; composting organic materials; and using liquid fertilizers.

Several locally available legumes were tested as cover crops and intercrops with maize, such as white or red lablab (*Mucuna holtoni*), adzuki beans (*Vigna angularis*), butter beans (*Phaselus lunatus*), cowpeas (*Vigna unguiculata*) and butterfly peas (*Clitoria ternatea*). Mucuna was identified as the best cover crop due to its dense biomass.

A total 1404 ha of maize-planted areas in Lombok, Sumba and Timor experimented with CA farming systems. The income of CA farmers compared to non-CA farmers increased by 37% in Lombok, 57% in Sumba and 77% in Timor. The benefit/cost ratio analysis comparing CA and non-CA techniques and practices showed that CA techniques and practices could increase farmer's revenue by 130%. Soil carbon content increased by 30.5% after three to four years of CA farming practices.

The project ended in December 2018, and while some farmers continued CA practices, no follow-up took place.

2.5 Malaysia

Malaysia's first National Agriculture Policy in 1984 charted the developmental path of its agriculture sector. For the Third Agriculture Policy in 1998, the agricultural program emphasized productivity while ensuring conservation and sustainable use of natural resources. The concept of integrated agriculture with primary emphasis on agroforestry, mixed and organic farming, rehabilitation of marginal land, recycling of organic waste and composting, mulching and cover cropping, as well as soil and water conservation were introduced to ensure sustainable agriculture development in Malaysia. The Fourth National Agrifood Policy enforced in 2011 aims for a competitive development of the agricultural sector while ensuring sustainability. In contrast to the previous policies that focused primarily on the agro-industries sector (such as oil palm, rubber and cocoa), the Fourth Policy prioritized the development of the agri-food sector (Ministry of Agriculture, 2000; Faridah, 2001; Wahid Murad et al., 2008). Good Agriculture Practice and Best Management Practice initiatives are the thrusts for practicing CA in Malaysia (FAO, 2013).

The agriculture sector in Malaysia can be broadly grouped into two main sub-sectors, the Industrial Commodities and the Food Commodities. The industrial sub-sector comprises mainly oil palm, rubber, cocoa, etc. with an estimated planted area of 5.8 million hectares that is managed by the private and public sectors. The food sub-sector consists of paddy, vegetables, fruits, pepper and other cash crops are mostly grown by smallholders and private sectors in an estimated area of 1.1 million hectares (Yahya, 2001; Department of Statistics, Malaysia, 2006; Wong, 2006). The CA practices adopted by Malaysia's agro-industrial commodities sector were reviewed by Kassam et al. (2020), and therefore this chapter focuses on the agri-food sub-sector.

2.5.1 The Conservation Agriculture practices adopted by Malaysian farmers

The area of CA in Malaysia under the industrial subsector is estimated to be about 350 000 ha, owing to the involvement and commitment by larger organizations/companies in adopting sustainable agriculture practices. However, the adoption of CA practices in the cash-crops sub-sector has been rather limited in Malaysia, with an estimated 7600 ha of CA area. However, some basic conservation practices that address farm issues such as erosion prevention and soil improvements have been widely adopted by farmers. Like most countries, farmers generally must go through a considerable change in attitudes and convictions for the uptake of CA practices. A survey among the cash-crop farmers suggested that the adoption of such practices generally depends upon a wide range of factors such as consumer demand, socio-economic status, agroecological zone, perceived return of investment, information availability, etc. (Faridah, 2001; Pannell et al., 2006; Tey et al., 2013).

A major concern of most producers and some policymakers was that farmers who implemented sustainable practices may have lower yields and profits than those who follow mainstream practices (Taylor et al., 1993). Lack of financial viability was expected to inhibit adoption of innovations by reducing the capacity to change. Despite the possibility that the relationship between income and adoption may not be of a linear function (Cancian, 1979; Pannell et al., 2006), sustainable profitability is an important influence on investment plans, and in turn on the decision to adopt CA (Cary et al., 2001).

With Malaysian consumers' growing preferences for organically produced food, triggered by health awareness and poor perception of pesticide usage, the demand has encouraged an increase in the number of traditional farmers converting to organic farming and adopting CA principles. In principle, little distinction could be observed between organic farming and conservation farming, as both practices share a high level of similarity.

A survey on CA and related practices involving about 8250 farmers suggested that farmers who operate on self-owned land are more sustainably managed than their counterparts who operated on rented land. Among the CA practices adopted by Malaysian farmers (Table 7), high adoption rates were recorded for organic fertilizers/compost (85%), no-tillage (84%) and crop rotation (77%); moderate adoption rates were indicated for intercropping (55%) and cover crops/mulches (47%); low adoption rates for IPM (9%) with an average annual gross farm profit of about 39 000 RM.

The usage of organic fertilizers/composts was the most adopted and had the highest positive influence on farm profitability. At present, 'organic' products are gaining much attention, resulting in organic products generally fetching a higher price, normally three times that of the non-organic counterpart (Faridah, 2001). Aligning with the demand, the Malaysian Department of Agriculture has taken the lead role in promoting the use of compost, and with that, proposes to reduce the use of the total annual fertilizer inputs by 80%. Usage of organic fertilizers or compost allows the farmers to benefit from significantly lower cost on nutrient inputs compared to their traditional counterparts. A study shows that chicken manure was commonly used in vegetable and fruit cultivations. Farmers generally applied five to ten tons of chicken manure for established trees and up to 40 kg of such organic matter in the planting holes in areas with marginal soils. Fruit trees like sapodilla, papaya and mango have been observed to produce yields similar to normal soil with the addition of organic matter (Faridah, 2001).

Several crops have been identified as suitable for integration in the continuous cultivation of short- and medium-term crops that can elevate productivity in a sustainable manner. Intercropping of vegetable, (chili, long bean, corn, sweet potato) with other produce in high demand such as banana, pineapple, sugar cane and watermelon ensures higher output, potentially doubling the farmers' income (Faridah, 2001). Intercropping allows farmers who grow multiple crops in proximity to significantly earn more than non-adopters of intercropping (Table 8). It diversifies the product output and reduces the risk of crop failure without significant additional set-up costs. When compatible crops are selected, intercropping offers various agronomic benefits (suppresses weeds, enhances nutrient recycling, improves diversity, etc.) in a complementary way (Leong and Zaharah, 1991; Exner et al., 1999; Baghdadi, 2015). Mixed Farming or crop diversification through association were introduced in the third national agriculture policy via several government institutions, in an effort to increase land productivity.

Although no-tillage has a positive trend on farm profitability, the relationship was not significant and was only adopted by farmers with access to financial support or the ability to raise funding. A greater number of farm laborers significantly reduce the likelihood of adoption of no-tillage. Such a finding, although against expectation, is not unreasonable in the Malaysian context (Tey, 2013). Mechanization/machineries require an initial investment that enables the adoption of no-tillage.

Table 7 Association of CA and related practices with rate of adoption and their impact on farm profitability. Adapted from Tey et al. (2013).

CA Practices	Adoption# (%)	Correlation between CA Practices and Profitability (Unstandardized Coefficient)
No-tillage	83.5	0.15
Intercropping	54.8	0.144*
Cover crop/mulches	47.1	-0.029
Crop rotation	76.6	-0.225
Organic fertilizers/compost	85.0	0.688*
Integrated pest management	8.6	-0.093

Adoption (%) by farmers within the study area.
* Statistically significant at $P_{0.01}$.

Ironically, in contrast to the earlier CA practices, the results suggested that farmers who adopted crop rotation, cover cropping/mulching and integrated pest management obtained poorer farm profitability. There is no clear explanation for this variation, and further detailed analysis would be necessary to understand this result fully (Tey, 2013). Other factors compounding the lack of cover crop use on vegetable farms include potential loss of profits, as cash crops often must be terminated early in order to establish cover crops (Pieper, 2013). Additionally, cover crops/mulches used to conserve soil quality have higher adoption in highlands which are prone to runoff and erosion, and are especially likely to be adopted by farmers who own the land, but not by tenant farmers, who generally prefer conventional practices for short-term economic benefits over soil conservation for long-term farm productivity (Sheikh et al., 2003; Tey, 2013). While cover crops require initial monetary inputs, the establishment of intercropping or crop rotation was perceived to be free of cost.

Sustainable profitability, economic feasibility and the relative advantage of adopting conservation practices are the key drivers for most farmers. For farmers with a focus on profit, the farm-level economics of a proposed conservation practice will be important, as those practices that are not profitable will tend to be adopted only by farmers with stronger conservation goals. Unprofitable conservation practices are likely to be more widely adopted if they can generate benefits when adopted at a small scale. In contrast, practices that require large-scale adoption to generate benefits will probably not be adopted sufficiently if they are perceived to be less profitable than the practices they replace (Lee, 2005; Pannell et al., 2006).

Table 8 Yield and additional monetary return from intercropping practice

Intercrop	Yield (t/ha/yr)	Additional return (USD/mth[1])	Reference
Corn + Soybean	14.10	Na	Baghdadi, 2015.
Corn + Mungbean	13.22		
Corn + Stylo	8.77		
Oil palm + Sugarcane	Na	130	Faridah, 2001.
Oil palm + Banana		203	
Oil palm + Pineapple		56	
Okra + Lettuce	33.3	Na	Leong and Zaharah, 1991.
Okra + Spinach	29.4		
Okra as monocrop	28.8		
Okra + Chinese mustard	33.5		
Okra + Waterspinach	34.7		

Additional return in MYR converted to USD (exchange rate of MYR 4.1 = USD 1).
Na - not available.

3 Regional networking

Despite the obvious benefits of CA, it does not spread automatically unless it is promoted for wide adoption. The constraints that hinder adoption must be understood and addressed for specific situations. These can include a combination of intellectual, social, financial, biophysical, technical and infrastructure constraints, or policy-related support. The following are considered to be the critical success factors or strategic focal areas for CA adoption and up-scaling:

- Coordination and cooperation;
- Policy support for CA in national programs;
- Targeting CA interventions;
- Partnerships and stakeholder's participation;
- Private sector participation;
- Farmer-centered participatory adaptive research;
- Monitoring and evaluation;
- Communication and advocacy;
- Defined roles and responsibilities; and
- Availability of adequate financial resources.

To mobilize policy support for CA in terms of research and extension, supply of CA-related equipment and machinery, linking input subsidies with adoption of CA and regional and sub-regional CA networks may contribute and provide opportunities to harness international, national and local support for CA. Examples of such networks are outlined below.

3.1 Conservation Agriculture Alliance for Asia-Pacific (CAAAP)

The FAO Regional Office for Asia-Pacific (FAO-RAP), in cooperation with the national institutions and counterparts in the Asia-Pacific region as well as international organizations including CIMMYT, ICARDA, ICRAF, CIRAD, SANREM and ACIAR organized a Regional Expert Consultation Workshop in collaboration with the Conservation Tillage Research Centre (CTRC), MoA, in Beijing, China in November 2013. At the workshop, it was agreed that CAAAP should be established to promote the mainstreaming of sustainable production intensification through CA at the national and regional levels, involving a multi-stakeholder partnership strategy. As a guide to this follow-up work, a regional strategic framework for CA in the Asia-Pacific region was formulated to use as a 'road map'.

CAAAP is a voluntary partnership of stakeholders promoting research, development, adaptation, extension and promotion of CA and its policy support as an overarching goal for promoting CA for improved food security and sustainable natural resource use. The CAAAP aims to bring together key partners and stakeholders, including researchers, policy makers, public and private sector entities and representatives of civil society organizations (CSOs) and development partners from across Asia and the Pacific Region, with the following specific objectives:

- To exchange knowledge, share experience and promote collaboration and concerted efforts among all stakeholders on the research, development, promotion and adoption of CA, and advocate its economic, social and environmental values and benefits.
- To promote effective linkages among public and private stakeholders associated with CA and reduce the negative environmental impacts of its by-products.
- To facilitate the development and implementation of policies, strategies and institutional mechanisms associated with CA and its promotion and adoption.

3.2 Agroecology platform in partnership and learning alliance

3.2.1 Agroecology in South East Asia (ASEA) platform in partnership

ASEA platform is an open coalition of research and higher education institutions active on Agroecology within the Greater Mekong Sub-region. The overall objective of the platform is to foster synergies at regional level to develop tailored knowledge, technologies and innovation support mechanisms for the development and scaling-up of Agroecology in South East Asia. The activities of the platform are organized around 4 axis: 1) Soil health enhancement in agricultural landscapes, 2) Agroecological crop and livestock protection, 3) Innovative cropping and farming systems, and 4) innovative intervention and support mechanisms. Conservation Agriculture is a key component of this platform as many of the core members are involved in CA-based R&D activities in the region.

3.2.2 Agroecology Learning Alliance in South East Asia (AliSEA)

ALiSEA's goal is to enable local and regional agroecology stakeholders to leverage one another's expertise to produce evidence based studies and share them broadly to support a regional transition towards agroecology. It has a three folded objectives: 1) Strengthening knowledge and experience sharing among agroecological initiatives and actors, 2) Increasing visibility and credibility of agroecological movement towards policy makers and consumers, and 3) Scaling up the development and adoption of agroecological practice among farmers.

ALiSEA's online library offers free access to more than 900 resources, including 290 resources related to Conservation Agriculture experiences in the region. The website receives an average of 15,000 views per month. The Facebook page in English has 3800 followers and has links to Facebook pages in national languages (Lao, Khmer, Burmese and Vietnamese). A quarterly newsletter has more than 1200 subscribers (from the Mekong region and beyond) and is a means for sharing and disseminating success stories, lessons learnt, case studies and upcoming events about agroecology in the Mekong region.

4 Conclusion

The increase in populations of both humans and livestock, droughts and floods, poor access to yield-enhancing technologies and the low marketing opportunities for agricultural produce have caused reduction and/or stagnation in agricultural productivity and have worsened food insecurity and malnutrition. This situation will only worsen, unless drastic changes in farming practices are adopted by farmers for sustainable production, and for reducing risks linked to climate change.

There are many technical options available to improve agricultural productivity, for example, with high-quality seeds, fertilizers and pesticides. However, under the above-described scenario, such improvements will neither be sustainable nor economically feasible in the long term. The FAO Strategic Objective aims to combine sustainability with intensification as elaborated in its 'Save and Grow' publications (FAO, 2011, 2016). Since it minimizes or eliminates soil degradation and builds a foundation for a functioning ecosystem, CA is considered to be the entry point in making the intensification sustainable, as well as for rehabilitating degraded and abandoned rainfed and irrigated agroecosystems.

Given the present knowledge and circumstances, CA is the most appropriate sustainable option available to increase productivity, income and food security in the region. It offers an opportunity for commercialized production and substantially improves resource use efficiency. It can be practiced by any farmer and offers a viable solution for poor farmers to address their productivity constraints, particularly high labor costs and tillage constraints. Further, CA will help farmers to adapt to and mitigate the effects of climate change and variability.

CA has been shown to be relevant and appropriate at all levels of farm power and mechanization, from manually operated hand tools to equipment drawn by animal traction to operations performed by heavy machinery. CA is not only for vulnerable small farmers but also for small- or large-scale commercial farmers whose ecological as well as economical risks can be lowered by CA. The benefits of CA for the farmers relate to improved yields and input use efficiency, greater profit, improved soil characteristics, reduced soil erosion and increased resilience to climate variability and change. CA has been shown to work successfully in most countries in the region. While CA represents innovative systems for agricultural development and sustainable livelihood, rapid spread of CA also needs to be supported by policy and institutions, including the availability of affordable locally manufactured or imported CA equipment and machinery, including animal-drawn or tractor-drawn direct seeders.

When up-scaled and mainstreamed in national, sub-regional and regional policies and programs, CA can contribute to the achievement of the Sustainable Development Goals. The constraints that hinder adoption must be understood and addressed for specific situations. These can include a combination of intellectual, social, financial, biophysical, technical and infrastructural constraints, or policy-related support. Knowing what the bottlenecks are is important in developing strategies to overcome them. Crisis situations that are likely to become more frequent as a result of climate change, and the political pressure for more sustainable use of natural resources and environment protection, provide opportunities to harness international, national and local support for CA systems.

Adoption of CA is expected to increase in all nations in the region in the coming years, as governments allocate greater investments into promoting and supporting CA extension and research activities, and as national education systems continue to integrate more CA courses and modules into the agriculture curriculum and provide more opportunity for CA research in post-graduate research programs.

5 Where to look for further information

Further information on CA in Southeast Asia can be found from the following sites:

- http://www.caa-ap.org/.
- https://www.asea-network.org/.
- http://www.fao.org/evaluation/evaluation-digest/evaluations-detail/en/c/1136880/.

6 References

Ares, A. C., Thierfelder, M. R., Eash, N. S. and Himmelstein, J. 2015. Global perspectives on conservation agriculture for small households. In: Chan, C. and Fantile-Lepczyk, J. (Eds). *Conservation Agriculture in Subsistence Farming: Case Studies from South Asia and Beyond*. CAB International: 2, 22–54.

ASMC, 2019. Appropriate-Scale Mechanization Consortium for Sustainable Intensification. Feed the Future Innovation Lab for Sustainable Intensification. Final report. 33p.

https://www.swisscontact.org/fileadmin/user_upload/COUNTRIES/Cambodia/Documents/Publications/20200707_ASMC_Cambodia_Report_2019_FINALV2.pdf

Baghdadi, A. 2015. Intercropping of corn and legumes with application of chicken manure and crop combination ratios to increase yield quality of corn legume forage. PhD Thesis. Universiti Putra Malaysia. July 2015.

Bondad, R. M., Ella, V. B., Saludes, R. B., Reyes, M. R. and Mercado, A. R., Jr. 2015. Simulating the impact of conservation agriculture on corn yield in the Philippines using the DSSAT CERES-Maize model. *Philippine Agricultural Scientist* 98(2), 68–79.

Boulakia, S., Kong, R. and Eberle, M. 2013. *Sustainable Farming to Sustain Cambodia's Future (33-Minute Documentary Film)*. Asia Motion and MAFF-GDA, Phnom Penh, Cambodia.

Boulakia, S., Tivet, F., Husson, O. and Seguy, L. 2019. Nutrient management practices and benefits in Conservation Agriculture systems. In: Kassam, A. (Ed). *Advances in Conservation Agriculture, Volume 2: Practice and Benefits, 2019*. Burleigh Dodds Science Publishing, Cambridge, UK, 31 (ISBN: 978-1-78676-2689; www.bdspublishing.com).

Boulakia, S., Vuth, P., Vathana, S., Chabierski, S. and Gilard, O. 2010. Conservation Agriculture, a four-win solution for rainfed agriculture in Mekong countries: the case of Cambodia. *Paper Presented at the Conference the Environments of the Poor in the Context of Climate Change and the Green Economy: Making Sustainable Development More Inclusive*. New Delhi, India, 24–26 November 2010. Available at: https://www.scribd.com/document/42938290/Conservation-Agriculture-A-4-Wins-Solution-for-Rainfed-Agriculture-in-Mekong-Countries-The-Case-of-Cambodia-paper-long.

Cancian, F. 1979. *The Innovator's Situation*. Stanford University Press, Stanford.

Cary, J., Webb, T. and Barr, N. 2001. *The Adoption of Sustainable Practices: Some New Insights. An Analysis of Drivers and Constraints for the Adoption of Sustainable Practices Derived from Research*. Bureau of Rural Sciences, Canberra.

Castella J.-C. 2012. Agrarian transition and farming system dynamics in the uplands of south-east Asia. In : Hauswirth, D., Pham, T. S., Nicetic, O., Le Quoc, D., Kirchof, G., Boulakia, S., Chabierski, S., Husson, O., Chabanne, A., Boyer, J., Auffray, P., Lienhard, P., Legoupil, J. C., Stevens, M. L. (Eds). Conservation agriculture and sustainable upland livelihoods: innovations for, with and by farmers to adapt to local and global changes: proceedings. Montpellier: CIRAD, 2012, p. 4–20. *International Conference on Conservation Agriculture in Southeast Asia*, 3., Hanoi (VNM).

Castella, J.-C., Sysanhouth, K., Saphangthong, T., Victor, M., Ingalls, M., Lienhard, P., Bartlett, A., Sonethavixay, S., Namvong, S., Vagneron, I. and Ferrand, P. 2018. *Adding Values to Agriculture: A Vision and Roadmap for Sustainable Development in the Lao Uplands*. Lao Uplands Initiative, s.l., 16 p. Available at: https://laouplandsforum.org/.

Chan, S., Vira, L., Sar, V., Suos, V., Tivet, F., Boulakia, S., Tran Quoc, H., Lor, L., Theng, D., Rendall, T., Hansen, A., Reyes, M. and Ngo, B. 2018. Developing appropriate-scale mechanization for CA-based cropping systems in South-East Asia. Research Grant, Towards Agroeciological Transition in South-East Asia, Conservation Agriculture Network for South-East Asia, CIRAD, AFD. Available at: https://bit.ly/2Jt0sTS.

Chett, O., Choulamountry, T., Chheong, S., Leudphanane, B., Leng, V., Yen, S., Vangkee, Y., Lienhard, P., Tran Quoc, H., Kong, R., Bates, R., Hok, L., Ra, K., Trail, P. and Tivet, F. 2018. Disseminating underutilized species and cover/relay crops as a foundation of resilient farming systems. Research Grant, Towards Agroeciological Transition in

South-East Asia, Conservation Agriculture Network for South-East Asia, CIRAD, AFD. Available at: https://bit.ly/2OelzsN.
Department of Statistics, Malaysia 2006. *National Products and Expenditure Accounts, 1987-2005*. https://www.dosm.gov.my/v1/index.php?r=column/cone&menu_id=TWNzZjZSZS9RUzhHYzdXODhFSEhLdz09.
Derpsch, R. 2008. No-tillage and conservation agriculture: a progress report. In: Goddard, T., Zoebisch, M. A., Gan, Y. T., Ellis, W., Watson, A. and Sombatpanit, S. (Eds). *No-Till Farming Systems*, 7-39. World Association of Soil and Water Conservation, Bangkok.
Edralin, D. A., Sigua, G. C., Reyes, M. R., Mulvaney, M. J. and Andrews, S. S. 2017. Conservation agriculture improves yield and reduces weeding activity in sandy soils of Cambodia. *Agronomy for Sustainable Development* 37(5), 52.
Ella, V. B. 2018. *Conservation Agriculture: A Biological Engineering Approach to Sustainable Agriculture in Support of Rural Development in Southeast Asia. Searca Professorial Chair Lecture Monograph*. SEARCA, College, Los Baños, Laguna, Philippines, 46pp.
Ella, V. B., Reyes, M. R., Mercado, A. Jr., Ares, A. and Padre, R. 2016. Conservation agriculture increases soil organic carbon and residual water content in upland crop production systems. *Eurasian Journal of Soil Science* 5(1), 24-29.
Exner, D. N., Davidson, D. G., Ghaffarzadeh, M. and Cruse, R. M. 1999. Yields and returns from strip intercropping on six Iowa farms. *American Journal of Alternative Agriculture* 14(2), 69-77.
FAO 2009. *Proceedings of the Regional Land Degradation Assessment in Drylands (LADA)*, FAO, Rome.
FAO 2011. *Save and Grow: A Policymaker's Guide to the Sustainable Intensification of Smallholder Crop Production*. Food and Agriculture Organization, Rome.
FAO 2013. *Policy and Institutional Support for Conservation Agriculture in the Asia-Pacific Region*. Food and Agriculture Organization of the United Nation Regional Office for Asia-Pacific (RAP).
FAO 2015. Agricultural transformation of middle-income Asian economies: diversification, farm size and mechanization, by David Dawe, *ESAWorking Paper No. 15-04*. Food and Agriculture Organization of the United Nations, Rome.
FAO 2016. Save and Grow in Practice: Maize, Rice and Wheat. FAO, Rome.
FAO 2018. *Evaluation of FAO's Contribution to the Conservation Agricuture Thematic Cluster, Project Evaluation Series*. Food and Agriculture Organization, Rome.
Faridah, A. 2001. Sustainable agriculture system in Malaysia. *Paper presented at Regional Workshop on Integrated Plant Nutrition System (IPNS), Development in Rural Poverty Alleviation*, United Nations Conference Complex, Bangkok, Thailand, 18-20 September 2001.
Hok, L., de Moraes Sá, J. C., Boulakia, S., Reyes, M., Leng, V., Kong, R., Tivet, F. E., Briedis, C., Hartman, D., Ferreira, L. A., Magno, T. and Pheav, S. 2015. Short-term conservation agriculture and biomass-C input impacts on soil C dynamics in a savanna ecosystem in Cambodia. *Agriculture, Ecosystems and Environment* 214, 54-67.
Hok, L., de Moraes Sá, J. C., Reyes, M., Boulakia, S., Tivet, F., Leng, V., Kong, R., Briedis, C., da Cruz Hartman, D., Ferreira, L. A., Inagaki, T. M., Gonçalves, D. R. P. and Bressan, P. T. 2018. Enzymes and C pools as indicators of C build up in short-term conservation agriculture in a savanna ecosystem in Cambodia. *Soil and Tillage Research* 177, 125-133.
Hok, L., de Moraes Sá, J. C., Boulakia, S., Reyes, M., Ferreira, A. O., Tivet, F. E., Saab, S., Auccaise, R., Inagaki, T. M., Schimiguel, R., Ferreira, L. A., Briedis, C., Canalli, L. B. S.,

Kong, R. and Leng, V. 2021. Dynamics of soil aggregate-associated organic carbon based on diversity and high biomass-C input under conservation agriculture in a savanna ecosystem in Cambodia. *Catena* 198, 105065.

Husson, O., Tran Quoc, H., Boulakia, S., Chabanne, A., Tivet, F., Bouzinac, S., Lienhard, P., Michellon, R., Chabierski, S., Boyer, J., Enjalric, F., Rakotondramanana, Moussa, N., Jullien, F., Balarabe, O., Rattanatray, B., Castella, J., Charpentier, H. and Séguy, L. 2016. Co-designing innovative cropping systems that match biophysical and socio-economic diversity: the DATE approach to conservation agriculture in Madagascar, Lao PDR and Cambodia. *Renewable Agriculture and Food Systems* 31(5), 452–470.

Ingalls, M., Diepart, J.-C., Truong, N., Hayward, D., Neil, T., Phomphakdy, C., Bernhard, R., Fogarizzu, S., Epprecht, M. and Nanhthavong, V. 2018. State of land in the Mekong Region. Centre for Development and Environment, University of Bern and Mekong Region Land Governance. Bern, Switzerland and Vientiane, Lao PDR, with Bern Open Publishing. DOI 10.7892/boris.120285.

Jones, M., Leng, V., Ong, S., Mean, M., 2019. Cambodia: Service provision for Conservation Agriculture. Appropriate-scale mechanization Consortium. Feed the Future Innovation Lab for Sustainable Intensification. 16 p.

Kassam, A. L. H., Niino, Y., Friedrich, T., Jin, H. and Xianliang, W. 2014. Current status, prospect and policy and institutional support for conservation agriculture in the Asia-Pacific region. *International Journal of Agricultural and Biological Engineering* 7(5), 1–13.

Kassam, A., Gonzalez-Sanchez, E. J., Sim, C. C., Zaharah, A. R., Roman-Vazquez, J., Marquez-Garcia, F., Carbonell-Bojollo, R., Veroz-Gonzalez, P. and Garrity, D. P. 2020. Managing perennial conservation agriculture systems: orchards, plantations and agroforestry. In: Kassam, A. (Ed). *Advances in Conservation Agriculture (vol. 1): Systems and Science*. Burleigh Dodds Science Publishing Ltd, Cambridge, UK.

Kong, R., Diepart, J.-C., Castella, J.-C., Lestrelin, G., Tivet, F., Belmain, E. and Bégué, A. 2019. Understanding the drivers of deforestation and agricultural transformations in the Northwestern uplands of Cambodia. *Applied Geography* 102, 84–98.

Kong, R., Sar, V., Leng, V., Trang, S., Boulakia, S., Tivet, F. and Seguy, L. 2016. Conservation agriculture for climate-resilient rainfed uplands in the Western regions of Cambodia: challenges, opportunities, and lessons from a 10-year R and D program. In: Sajise Percy, E., Cadiz Maria Celeste, H., Bantayan Rosario, B. (Eds). *Learning and coping with change: Case stories of climate change adaptation in Southeast Asia*, 55–81. SEARCA, Los Baños, Philippines. ISBN 978-971-560-176-4.

Le, K. N., Jha, M. K., Reyes, M. R., Jeong, J., Doro, L., Gassman, P. W., Hok, L., de Moraes Sá, J. CdM. and Boulakia, S. 2018. Evaluating carbon sequestration for conservation agriculture and tillage systems in Cambodia using the EPIC model. *Agriculture, Ecosystems and Environment* 251, 37–47.

Lee, D. R. 2005. Agricultural sustainability and technology adoption: issues and policies for developing countries. *American Journal of Agricultural Economics* 87(5), 1325-1334.

Leng, V., Chheong, S., Chett, O., Yen, S., Trang, S., Prum, S., Hok, L., Reyes, M., Chan, S., Sovanno, P., Ra, K., Kong, R., Tivet, F. and Boulakia, S. 2018. Platform of agro-ecological systems of research and training for development. Research Grant, Towards Agroeciological Transition in South-East Asia, Conservation Agriculture Network for South-East Asia, CIRAD, AFD. Available at: https://bit.ly/2P0B65T.

Leong, A. C. and Zaharah, A. 1991. Intercropping okra with common leafy vegetables on Malaysian peat and bris. *MARDI Research Journal* 19(2), 209-216.

Lestrelin, G. 2015. *Knowledge Capitalization on the Conservation Agriculture Development Fund in Sayaboury Province, Laos (2008-2014). EFICAS Project*. CIRAD-DALAM, Vientiane, 23p.

Lestrelin, G., Quoc, H. T., Jullien, F., Rattanatray, B., Khamxaykhay, C. and Tivet, F. 2011. Conservation agriculture in Laos: diffusion and determinants for adoption of direct seeding mulch-based cropping systems in smallholder agriculture. *Renewable Agriculture and Food Systems* 27(1), 81-92.

Lienhard, P., Boulakia, S., Legoupil, J., Gilard, O. and Seguy, L. 2014a. Conservation agriculture in South-East Asia. *Conservation Agriculture: Global Prospects and Challenges*, 180-201.

Lienhard, P., Panyasiri, K., Sayphoummie, S., Leudphanane, B., Lestrelin, G., Séguy, L. and Tivet, F. 2014b. Profitability and opportunity of conservation agriculture in acid savannah grasslands of Laos. *International Journal of Agricultural Sustainability* 12(4), 391-406.

Lienhard, P., Tivet, F., Chabanne, A., Dequiedt, S., Lelièvre, M., Sayphoummie, S., Leudphanane, B., Prévost-Bouré, N. C., Séguy, L., Maron, P.-A. and Ranjard, L. 2013. No-till and cover crops shift soil microbial abundance and diversity in Laos tropical grasslands. *Agronomy for Sustainable Development* 33(2), 375-384.

Ministry of Agriculture 2000. *Third National Agricultural Policy, 1998 - 2010: Executive Summary*. Available at: https://www.pmo.gov.my.

Murad, M. W., Mustapha, N. H. N. and Siwar, C. 2008. Review of Malaysian Agricultural Policies with regards to sustainability. *American Journal of Environmental Sciences* 4(6), 608 -614.

OECD/FAO 2017. *OECD-FAO Agricultural Outlook 2017-2026*. OECD Publishing, Paris.

Ornetsmuller, C., Castella, J. C. and Verburg, P. H. 2018. A multiscale gaming approach to understand farmer's decision making in the boom of maize cultivation in Laos. *Ecology and Society* 23(2):35. https://doi.org/10.5751/ES-10104-230235.

Oung, N., Vernet, P., Leng, V., Suos, V., Faysse, N., Tivet, F., Neang, M., Theng, D., Rendall, T., Ader, D. and Bates, R. 2018. Diversification of rice-based farming systems in Tonle Sap Lake Region, Cambodia. Research Grant, Towards Agroeciological Transition in South-East Asia, Conservation Agriculture Network for South-East Asia, CIRAD, AFD. Available at: https://bit.ly/2CS2Baj.

Pannell, D. J., Marshall, G. R., Barr, N., Curtis, A., Vanclay, F. and Wilkinson, R. 2006. Understanding and promoting adoption of conservation practices by rural landholders. *Australian Journal of Experimental Agriculture* 46(11), 1407-1424.

Parks, M. H., Christie, M. E. and Bagares, I. 2015. Gender and conservation agriculture: constraints and opportunities in the Philippines. *GeoJournal* 80(1), 61-77 DOI 10.1007/s10708-014-9523-4.

Pheap, S., Lefèvre, C., Thoumazeau, A., Leng, V., Boulakia, S., Koy, R., Hok, L., Lienhard, P., Brauman, A. and Tivet, F. 2019. Multi-functional assessment of soil health under Conservation Agriculture in Cambodia. *Soil and Tillage Research* 194, 104349.

Pieper, J. R. 2013. Conservation tillage techniques for mixed vegetable production in southern new England. Open access Master's theses. University of Rhode Island. Available at: https://digitalcommons.uri.edu /theses/69.

Pingali, P. L. 2012. Green revolution: impacts, limits, and the path ahead. *Proceedings of the National Academy of Sciences of the United States of America* 109(31), 12302-12308.

Prasertsan, S. and Sajjakulnukit, B. 2006. Biomass and biogas energy in Thailand: potential, opportunity and barriers. *Renewable Energy* 31(5), 599-610.

Séguy, L., Bouzinac, S., Trentini, A. and Cortes, N. 1998. L'agriculture brésilienne des fronts pionniers (Brazilian frontier agriculture). *Agriculture et Développement (Special Issue)*, 1-74.

Sevilla, A. S. and Ella, V. B. 2020. Ammonium adsorption characteristics of Jasaan clay soils under conservation agriculture and conventional plow-based production systems. *Philippine Journal of Agricultural and Biosystems Engineering* 16(1), 39-54.

Sheikh, A. D., Rehman, T. and Yates, C. M. 2003. Logit models for identifying the factors that influence the uptake of new 'no-tillage' technologies by farmers in the rice-wheat and the cotton-wheat farming systems of Pakistan's Punjab. *Agricultural Systems* 75(1), 79-95.

Sumner, D., Christie, M. E. and Boulakia, S. 2017. Conservation agriculture and gendered livelihoods in Northwestern Cambodia: decision-making, space and access. *Agriculture and Human Values* 34(2), 347-362.

Tarnate, P. O., Ella, V. B. and Reyes, M. R. 2014. Modeling solute transport in soil under conventional plow-based and conservation agriculture production systems in Claveria, Misamis Oriental, Philippines. *Philippine Agricultural Scientist* 97(3), 243-251.

Taylor, D. C., Mohamed, Z. A., Shamsudin, M. N., Mohayidin, M. G. and Chiew, E. F. C. 1993. Creating a farmer sustainability index: a Malaysian case study. *American Journal of Alternative Agriculture* 8(4), 175-184.

Tey, Y. S. 2013. The adoption of sustainable agricultural practices: an integrative approach for Malaysian Vegetable Farmers. PhD Thesis. School of Agriculture, Food and Wine, Faculty of Sciences. University of Adelaide. December 2013.

Tey, Y. S., Li, E., Bruwer, J., Abdullah, A. M., Brindal, M., Radam, A., Ismail, M. M. and Darham, S. 2013. The relative importance of factors influencing the adoption of sustainable agriculture practices: a factor approach for Malaysian vegetable farmer. *Sustainability Science* 9(1), 17-29. DOI 10.1007/s11625-013-0219-3.

Vernet, P. A., Faysse, N., Suos, V., Oung, N., Son, S., Leng, V., Theng, D., Rendall, T., Lor, L., Reyes, M., Chan, S., Pradhan, R., Vang, S. and Tivet, F. 2020. Investing in a no-till planter in Cambodia: a promising opportunity for various types of service providers?. *Asian Journal of Agricultural and Environmental Safety* (1):1-12. ISSN: 2575423, https://www.ajaes.org.

Wong, L. C. Y. 2006. Development of Malaysia's Agricultural Sector: agriculture as an engine of Growth?. *Conference on the Malaysian Economy: Development and Challenges*, ISEAS Singapore.

Wortman, S. E., Francis, C. A. and Lindquist, J. L. 2012. Cover crop mixtures for the western Corn Belt: opportunities for increased productivity and stability. *Agronomy Journal* 104(3), 699-705.

Yahya, T. M. T. 2001. Crop diversification in Malaysia. In: Papademetriou, M. K. and Dent, F. J. (Eds). *Crop Diversification in the Asia-Pacific Region*. FAO Regional Office for the Asia and the Pacific. RAP Publication 2000/14, 4-6 July 2000.

Chapter 16

Adoption and spread of Conservation Agriculture in East Asia

Hongwen Li and Jin He, National Institute of Conservation Tillage, China Agricultural University and Conservation Tillage Research Centre, Ministry of Agriculture and Rural Affairs, China

1 Introduction
2 History of Conservation Agriculture
3 Policy and actions for Conservation Agriculture promotion
4 Adoption scale
5 Advice for Conservation Agriculture promotion
6 Conclusion
7 Where to look for further information
8 References

1 Introduction

East Asia region is located at 4°N–53°N and 73°E–150°E and includes China, Japan, the Republic of Korea, the Democratic People's Republic of Korea (DPR) of Korea and Mongolia. The region contains both the most and the least densely populated countries in the world, with Mongolia being least densely populated while China the most densely populated. In total, it is one of the most densely populated regions (135 people per km^2) in the world, with a population of 1.7 billion, accounting for 38% of the population in Continental Asia and 22% of the global population. East Asia covers a total area of about 12 500 000 km^2, representing about 9% of the global land area. The agroclimatic characteristics of the region are very diverse, with warm to hot and rainy summers and mild/cold continental humid to very dry winters, with seasonal and interannual variations of precipitation. The southeast part of the region belongs to temperate broad-leaved forest climate and subtropical forest climate. The northwest part belongs to continental temperate grassland and desert climate. The Qinghai-Tibet Plateau in the southwest part has mountain plateau climate with low temperatures throughout the year. In the northeast part of the region covering northeastern

http://dx.doi.org/10.19103/AS.2021.0088.14

China, Korea and Japan, the climate is continental with dry cold winters and moist warm summers.

Main crops in East Asian countries are also varied due to agroclimatic diversity and include rice, soybean, maize, wheat, barley, cotton, peanut, rapeseed and fruits and vegetables. Production of rice and soybean comprise >40% and >20% of the world's total production, respectively. To meet the food needs of large populations, agricultural intensification has become the main form of agricultural production in the last decades in East Asian countries. Long-term intensification with characteristic heavy mechanical and agrochemical inputs (such as seeders, fertilizers and pesticides) and excessive tillage results in severe problems of degradation of agricultural land (Fig. 1) and environment including soil erosion, water deficit and reduced nutrient availability (He et al., 2010, 2011; Kassam et al., 2014). Consequently, crop yields become unstable and decline, especially in dry years, and production becomes less efficient and costly, all of which contribute to the crisis of food security in East Asia.

Another challenge for agriculture in East Asia is open straw burning in the fields (Fig. 2). In the past, straw was used as a source of rural domestic energy material and as livestock feed. However, in recent years, surplus of straw always exists in specific regions and seasons due to limited utilization opportunities and changes in the source of rural energy and animal feed. In some areas of East Asia, in order to avoid negative effects of excessive straw on the quality

Figure 1 Soil desertification by wind erosion.

Figure 2 Open straw burning.

of seed planting, straw is burnt by farmers after harvesting or before planting. Straw burning changes soil physical properties, increases soil nutrient loss and destroys the balance of properties in the soil biological system. Furthermore, it results in negative effects on the environment, such as air pollution and increase in the emission of greenhouse gases (CO_2, N_2O and CH_4) (Tian et al., 2015).

Therefore, it is necessary to find an alternative agricultural paradigm to enhance the agroecosystem processes, improve soil structure and nutrient and water retention and increase agricultural production in East Asia. Conservation Agriculture (CA), which is experiencing increased uptake in many countries around the world and can contribute to environmental conservation, soil health regeneration and agricultural production improvement, is being adopted by East Asian countries to promote the sustainable development of agriculture (Friedrich et al., 2012; Kassam et al., 2019).

CA is a farming system that promotes no or minimum mechanical soil disturbance, maintenance of permanent soil mulch cover and diversification of cropping. In East Asia, CA is defined as an advanced agricultural technology that implements no tillage or minimum soil disturbance and uses straw and stubble as mulch cover to reduce wind and water erosion and to feed the soil microorganisms (Baker et al., 2007; He et al., 2018; Kassam et al., 2015). The practice and adoption of this technology has also shown that CA can reverse the loss of soil nutrients, improve soil structure and enhance and maintain soil productivity (Li et al., 2007; Kassam et al., 2009).

In East Asia, the four main kinds of CA practices are: (1) no-till (NT) + straw cover, (2) minimum soil disturbance strip seeding + straw cover, (3) permanent raised beds (PRB) + straw cover and (4) controlled traffic, i.e. NT + straw cover + permanent cropping zone and traffic lane (Table 1). The relationship between CA practices and desired soil conditions are listed in Table 2. In addition, there are diversifying crop-rotation methods in CA system, crop

Table 1 Conservation Agriculture practices for cropping system in East Asia

Terms	Description/comments
No-till + straw cover	Direct planting into untilled soil which maintains soil organic matter and reduces soil compaction; >30% crop straw and stubble cover the soils.
Minimum soil disturbance* (only for seeding) + straw cover	Seedbeds strip seeded to meet seeding requirements under the existing soil and climatic conditions; >30% crop straw and stubble cover the soils.
No-till + straw cover + bed planting + furrow irrigation (permanent raised beds)	Require permanent raised beds; the width of bed is generally 70–100 cm; direct planting into the beds; permanent furrows for traffic and irrigation; >30% crop straw and stubble cover the soils.
No-till + straw cover + permanent cropping zone and traffic lane (controlled traffic)	Separate the traffic lane and cropping zone: establish a permanent tractor walkway in the field, so wheels can avoid compacting the soil in the cropping zone; no-till; >30% crop straw and stubble cover the soils.

Table 2 Effects of Conservation Agriculture practices in East Asia

CA practices To achieve	No-till + straw cover	Strip-till seeding + straw cover	Permanent raised beds + straw cover	Controlled traffic
Rebuild damaged soil conditions	√	√	√	√
Reduce soil disturbance	√		√	√
Reduce soil compaction of cropping zone			√	√
Maximize land use efficiency	√	√		
Reduce water evaporation	√	√	√	√
Increase soil moisture content	√	√	√	√
Reduce soil erosion	√	√	√	√
Improve soil fertility	√	√	√	√

diversification through crop rotations, associations or sequences, involving annual and perennial legumes.

2 History of Conservation Agriculture

CA is spread across all continents and most agroecologies (Derpsch and Friedrich, 2009b; Kassam et al., 2009, 2019). In East Asia, some countries have been exposed to CA for the past 15-20 years and have begun to accept the need to promote the adoption and spread of CA as part of the national production intensification strategy. Further, some national governments and their agricultural sectors are beginning to realize CA's benefits to the producers, consumers and society and to organize policy and institutional support for the adoption of CA. However, compared with the countries with developed CA systems (such as the United States, Canada, Australia, Argentina and Brazil), the adoption of CA in East Asian countries is still low.

2.1 China

In China, drought, soil and wind erosion, straw burning, declined and unstable agricultural production, all provide a justification for the application of CA systems and practices. The development of CA in China can be divided into the following four stages (Gao et al., 2003, 2008; Li et al., 2015; Wang et al., 2004; Zhang et al., 2005, 2013, 2015):

2.1.1 The preliminary exploration stage (1960-1990)

In the 1960s, China began to test single and selected CA component technologies such as NT, subsoiling and straw cover. However, these technologies were only implemented in small areas due to the limitation of technologies and machines and the level of socio-economic development.

From the 1960s to the early 1970s, NT wheat planting technology was adopted in fields with rice stubble by state farms in Heilongjiang and Jiangsu provinces.

In the 1980s, China began maize NT planting experiments in fields with wheat straw cover in annual double-cropping areas (winter wheat + summer maize) of North China Plain. However, with the lack of supporting equipment for CA, especially NT seeders, CA was difficult for large-scale extension in China.

2.1.2 The systematic research stage (1991-2001)

In 1991, funded by the Chinese Ministry of Agriculture (MoA) and Australian Centre for International Agricultural Research (ACIAR), China Agricultural University (CAU) and University of Queensland (UQ) initiated a systematic experimental study of CA with the combination of agricultural machinery and agronomy in the Shanxi province on the Loess Plateau. A long-term experimental site (since 1991) was established in the one-crop-per-year region in Linfen, on the Loess Plateau. Also, 11 experimental sites for CA were successively established in North China.

In 1996, seven key demonstration areas for NT seeding, minimum soil disturbance strip tillage seeding and subsoiling were established in Linfen, Shouyang, Lintong, Zezhou, Fushan, Gaoping and Tancheng. The experimental results in these areas showed the advantages (reduced surface run-off, improved water use efficiency, increased soil organic matter, etc.) of CA as compared to traditional tillage system.

In 1999, some other demonstration sites were established in the one-crop-per-year regions of Hebei, Liaoning, Inner Mongolia, Gansu and Shaanxi provinces. Also, the Conservation Tillage Research Centre (CTRC) was set up at the CAU by the MoA, specifically for the promotion of CA. The study at this stage revealed the suitability and adaptability of CA in China, and a series of medium-small size NT seeders were developed to support the spread of CA.

2.1.3 Demonstration, extension and priority study stage (2002-2008)

In 2002, MoA began to demonstrate and extend CA in China and organized the first national CA field meeting in Linfen, Shanxi province. A total of 38 project counties from North Hebei, South Inner Mongolia, North Shanxi, West Liaoning, North Shaanxi, East Gansu, Beijing and Tianjin were selected and established by MoA as the national and provincial CA demonstration sites.

In 2003, the national demonstration project of mechanized CA technology was launched, and in total, 25 demonstration sites in 13 provinces were involved.

In 2005, a National CA Expert Group was established by the Chinese MoA. The uptake of CA grew rapidly under the support of MoA. By the end of 2008, >220 national and >360 provincial demonstration counties within 15 provinces were set up, covering more than 3 million ha.

2.1.4 Rapid development stage (2009-present)

In 2009, the China State Council ratified the National Construction Programme of Conservation Agriculture. After 2009, suitable CA machines and technical models for different regions were developed and finalized and the technical support system for the extension of CA was gradually strengthened. In 2018, CA was being practised on more than 8.8 million ha, mainly located in the northeast ridge tillage areas, North China Plains, northwest oasis farming areas, farming-pastoral areas, Loess Plateau and southern rice cropping areas.

2.2 Mongolia

Continuous use of intensive tillage, ploughing and conventional agricultural practices, coupled with harsh agroclimatic conditions in Mongolia with very limited precipitation, short growing season, very windy springs and frequent extreme events such as droughts, resulted in the development of agricultural and environmental problems such as soil desertification, reduced species diversity and decreased agricultural productivity. To address the issue, many international organizations and development partners in the country as well as the government began to promote CA. From 1997 to 2000, the awareness about CA grew strongly in Mongolian agriculture (Chuluunbaatar, 2011). The FAO, requested by Mongolian government, implemented a Technical Cooperation Programme (TCP) project from 2000 to 2002 (Bachman and Friedrich, 2003). The project's main objective was to improve cereal productivity and farm profitability through the application of CA technology in wheat-fallow production system. The results indicated that CA was a technically viable and economical alternative to prevailing tillage-based crop production and could provide a base for future sustainable intensification. Likewise, other development partners including the Canadian International Development Agency (CIDA), Agricultural Cooperative Development International/Volunteers in Overseas Cooperative Assistance (ACDIVOCA)/United States Agency for International Development (USAID) and the European Union provided grant funding to introduce various CA practices using different approaches between 1997 and 2005.

At the same time, other donors were active in the same area:

- The USAID launched the 'Farmer to Farmer' project, which had an important CA component and accelerated the development of CA in Mongolia (Didier and Lkhamjav, 2009).
- The CIDA, in partnership with the Mongolia Farmers and Flour Producers Association, implemented the project of 'Introduction of Minimum Tillage Technologies'. The main objective of the project was to strengthen the individual capacity of farmers, extension workers and government policymakers. The project also facilitated exchange and business development between the Canadian and Mongolian private sector companies including seed and agri-machinery companies. Furthermore, the project advocated CA practices to help Mongolian farmers apply new farming technology through demonstrations and field-based learning, both in Mongolia and in Canada (Bachman and Friedrich, 2003), which helped expand CA adoption to a commercial scale through private sector engagement. At the community level, the project helped Mongolian farmers to see opportunities to become organized and create long-lasting linkages between Canadian and Mongolian farmers, which later

evolved into a farmer association that is still actively promoting CA and NT practices today.

- The 'Intended Nationally Determined Contribution (INDC)', submitted by Mongolia to the Ad-Hoc Working Group on the Durban Platform for Enhanced Action, invested US$150 million to disseminate NT technology and to increase crop diversification and rotation (Ministry of Environment and Tourism, 2015).
- The European Union also invested in a long-term project called Tasic, which also ran on-farm demonstrations mainly on the use of CA machinery and provided subsidies for chemicals and sprayer kits.

While the above projects managed to create awareness and buy-in from the government and some farmers, the technical capacity of the key actors - policymakers, researchers, extension agents and farmers, was not adequately strengthened to implement CA practices successfully. After the wave of donor-funded interventions and projects promoting CA ended, Mongolian government and farmers were left with no guidance from external partners. Amongst many struggles to get it right, some innovative and better off farmers took a lead and reached out to Canadian farmer associations and private companies for support. This led to a widespread adoption and practical application of the CA practices in Mongolia - private sector and farmer-led adoption.

2.3 Japan

In Japan, the main goal of CA diffusion is to reduce environmental pollution load from the agricultural sector and improve agricultural production and water quality at a watershed scale (Takeshi, 2015). The studies and trials on NT CA system were conducted, and the advantages and benefits from this system were observed in many sites in terms of improved soil property and reduced labour cost, and also in terms of crop performance (Derpsch, 2014). Also, in rice-based cropping systems in paddy fields, NT systems were introduced and weed control was performed by water management rather than herbicide application in these systems. Based on a long-term experiment set up in 1965, the effects of more than 40 years of continuous orchard under NT soil management, rice cultivation under puddling and grain crops under conventional tillage were studied by Iwate University. They found that the effect of puddling, compared to continuous cropping or NT orchard, was pronounced with changes in bulk density, soil pH and the concentrations of acid-extractable iron and manganese (Rahman et al., 2008). In the 1980s, a field experiment (1983–1992) was conducted in Tsukuba to investigate the effects of NT, involving NT for summer cropping and mouldboard ploughing for winter cropping (RT) and conventional rotary tillage to a depth of 15 cm

(CT), on soil conditions and crop growth in a light-coloured Andosol. The results showed that the best tillage practice for Andosols on the Kanto Plain was the combination of NT for summer cropping and CT for winter cropping (Tsuji et al., 2006).

In recent years, many CA studies have focused on crop yield, soil physical and chemical characteristics, organic carbon storage, greenhouse gas emissions and so on. For instance, the experimental data from Hashimi et al. (2019) showed that NT and CT, foliar fertilization (lm+) and non-foliar fertilization (lm−) and trimmed weed cover (cm+) and weed-free cover (cm-) had some effects on eggplant yield and soil chemical and physical properties. Applying NT with mulch cover could be a valuable protective measure by reducing the tillage intensity and increasing crop yield and soil carbon content.

In order to accelerate the development of CA in Japan, some types of NT seeders based on local conditions were developed by companies and experimental stations, and some of them were practical and commercial. Outside Japan, the government launched the projects to develop CA in other countries. For example, the projects of the 'Demonstration Program for Environment Conservation Agriculture in Canete and Yauyos Province, Peru' (Institute of Cultural Affairs: Japan, 2004) and 'Conservation Agriculture-Development of Sustainable System of Soils' Management in Moldova' (Government of Republic of Moldova, 2017) were funded by the Japanese government to promote the adoption of CA in these countries.

2.4 Democratic People's Republic of Korea

CA is becoming one of the main technologies to reduce risk of the effects of extreme weather events as well as improve food security in the DPR of Korea. CA's benefits, such as improving soil property, reducing input and enhancing agricultural production, were accepted by the government and farm managers. Currently, major components of a suitable CA system for the DPR of Korea are the use of diversified crop rotation under the double-cropping system, including a green manure cover crop as well as the replacement of conventional soil tillage with NT (Mousques and Friedrich, 2007). In the DPR of Korea, the uptake of the CA technology is still low and concentrated to very limited geographical areas (mainly adopted on the upland, with some on paddy land) due to the limitation of inputs, mechanization and herbicide (Gunjal et al., 2011). In recent years, with the supports of some international organizations (such as the FAO, Concern Worldwide and other development partners), the government has achieved substantial progress in promoting CA.

- From 2002 to 2012, the FAO attended several emergency and technical cooperation projects in the DPR of Korea with a focus on CA. The Academy of Agricultural Sciences adopted CA as the recommended practice and the programme was extended to more than 30 cooperatives. All cooperatives received mechanization packages with tractors, NT seeders, animal traction boom sprayers and combine harvesters for rice and wheat as well as for maize. Crops grown under CA included wheat, soya, maize, potato and rice. The yield increased depending on crop by 10–20% in the first year of adoption (Friedrich, 2012).

- Concern Worldwide (locally known as the European Union Programme Support Unit 3) started the CA project 'Conservation Agriculture for Sustainable Food Security and Nutrition in Agro Climatic Zones of DPRK' in 2012 with financial support from Irish Aid (Irish Aid and Concern Worldwide, 2015). CA was expanded to 26 cooperative farms in 11 counties involving a population of 35 666, with an average increase yield of 30% in crop production (maize, rice, wheat, soybean, etc.) per hectare reducing the seasonal workload especially for women from 350 to 106 person days.

2.5 Republic of Korea

Research on CA in the Republic of Korea focused on two aspects, NT and straw mulching in rice-based cropping. For example, Lee et al. (2013) found that surface mulch materials (straw mat, rice chaff, sawdust and gypsum) had certain positive effects on the reduction of run-off and sediment load from sloping experimental plots; Ahn and Kim (2016) used the Hydrological Simulation Programme-Fortran model to assess the watershed-scale effects of NT and dryland straw mulch application on non-point source pollution. The results of field experiment showed decreases in run-off and sediment load of 9.0% and 95.9%, respectively, for plots with rice straw mulch and 22.5% and 82.5%, respectively, for NT plots.

3 Policy and actions for Conservation Agriculture promotion

3.1 Policy for Conservation Agriculture promotion

East Asian governments have promulgated a series of policies to improve the environment, reduce wind and water erosion and increase crop yields and farmers' income by promoting the adoption and development of CA. Some policies issued by East Asian countries are shown in Table 3.

In China, the central and local governments, such as the State Council, National People's Congress, National Development and Reform Commission, Ministry of Finance and MoA and Rural Affairs, have continuously issued relevant documents to promote and develop CA. As indicated in Table 3, for nine (2005 to 2012 and 2020) years, No. 1 Documents of the Central Committee addressed

Table 3 Some policies and documents for CA promotion in East Asian countries

Country	Policy	Main content
China	No. 1 Central Document of State Council (2005-2012)	Implement CA and the government will provide subsidies
	No. 1 Central Document of State Council 2020	Initiating the implementation of the CA action plan on black land region in Northeast China
	Decision on Further Strengthening Prevention and Control of Desert (2005)	Actively introduce and implement CA measures, such as no-tillage and stubble retention
	Construction Planning of Conservation Tillage project (2009-2015)	Establish National Conservation Tillage of Research Center
	Plan Outline of National Medium- and Long-Term Science and Technology Development (2006-2020)	Research relevant machine for CA under different local conditions
	National Program on Addressing Climate Changes (2014-2020)	Promote CA to increase agricultural ability to withstand natural disasters
	National Plan for Desertification Prevention and Control (2011-2020)	Apply CA to prevent sandstorm in three different soil type regions
	National Program for Agricultural Water Saving (2012-2020)	Accelerate the extension of CA techniques in areas with higher soil erosion and water erosion
	National Plan for Agricultural Sustainable Development (2015-2030)	Apply CA in Northeast China and other regions; provide subsidies for CA machine
	Opinions on Accelerating the Transformation of Agricultural Development Methods (2015)	Increase subsidies for CA
	Report on the Inspection of the Implementation of the Agricultural Law of the People's Republic of China (2015)	Support the innovation methods of CA
	National Plan for Agricultural Modernization (2016-2020)	Increase subsidies for agricultural machinery purchase and extend the area of CA
	Planning Outline of National Land (2016-2030)	Strengthen CA in dryland of Northern China
	Guideline on Protecting Black Soil in Northeast China (2017-2030)	Adopt CA to protect black soil in Northeast China
	Three-year Plan on Defending the Blue Sky (2018)	Promote CA to suppress farmland dust

(Continued)

Table 3 (*Continued*)

Country	Policy	Main content
Mongolia	Introduction of Minimum Tillage Technologies	Help Mongolian farmers apply CA technology
	Intended Nationally Determined Contribution (INDC) Submission by Mongolia to the Ad-Hoc Working Group on the Durban Platform for Enhanced Action (ADP) (2021-2030)	Help to diffuse no-till technology and to increase the variety of crops and rotation
	Fallow Project - 2001	Support the application of conservation farming technologies
	Millennium Development Goals (MDGs)-based Comprehensive National Development Strategy of Mongolia	Increase the use rate of no-till farming technology from 70% in 2016-2020 to 80% in 2021-2025, to 90% in 2026-2030 in grain fields
	Crop Campaign-III	Use no-till technology to preserve the soil fertility
Japan	Environment Law (1993)	Clarify the recycling of crop straw
	Basic Law for a Recycling Society (2000)	Propose an effective utilization plan for crop straw
	Fertility Promotion Law	Use organic fertilizer to cultivate soil fertility; promote straw retention to increase soil organic matter
	Water and Environmental Conservation Improvement Scheme	Directed payment for environmental CA practices
DPR of Korea	Food Production on Sloping Land Project (2008-2011)	Target farmers from cooperative farms who adopted CA on more than 100 ha
Republic of Korea	Agricultural Mechanization Promotion Law (Amendment) (1994)	Provide subsidies and loans for agricultural machinery products

the development of CA in China; from 1992, dozens of local laws were enacted to research, popularize and adopt CA in China; also, the purchasing of CA machines was subsidized by the government to accelerate CA machines updation and to promote CA adoption. In 2020, in order to protect black earth in Northeast China, the Chinese government launched the 'National Action Plan for Protective Farming of Black Land in Northeast China (2020-2025)', and it is expected that CA will be applied on 9.3 million ha in Northeast China in the next 5-6 years, covering 70% of total farmland.

In Mongolia, the central and local governments, such as the Parliament of Mongolia and the Ministry of Food Agriculture and Light Industry, have actively issued relevant strategies and programmes to promote and develop

CA. All government programmes, such as the Fallow Project - 2001 and Crop Campaign-III, significantly accelerated the spread of CA in Mongolia by providing light loans and subsidies for pesticides and machinery. In other East Asian countries, the national and local policies and documents, such as Water and Environmental Conservation Improvement Scheme (Japan), Food Production on Sloping Land Project (DPR of Korea) and Agricultural Mechanization Promotion Law (Republic of Korea), were issued to promote CA.

3.2 Actions for Conservation Agriculture promotion

In East Asian countries, based on above-issued policies by the government, the actions related to the following aspects were used to promote the application and adoption of CA during the past decades.

3.2.1 Technical models of Conservation Agriculture

CA has many advantages, such as increasing crop yield and income, improving soil structure and fertility, reducing greenhouse gases emission and reducing production cost. In East Asian countries, many scientific research projects related to CA funded by the governments and international organizations were conducted to develop mature CA technical concepts for different areas based on the prevailing cropping systems, climates, soils, etc. This is one of the main actions that accelerated the application of CA in East Asian countries.

Arid region: establishing straw mulch cover on the soil surface is very important, thus NT practices were recommended in this region. This technical model significantly reduced soil disturbance, improved soil quality and water retention capacity and enhanced production benefits over time. In some places, subsoiling was used to break hard soil pans and promote crop root growth in the first year of CA. In some rainfed areas with strong winds in spring, the high standing stubble and straw cover was adopted to reduce wind erosion; if the region had the problem of soil compaction, controlled traffic was used. In some oasis farming areas, PRB were used: (1) soil surface opening and planting in spring should be closely linked to prevent water loss of seedbed; (2) stubble and straw management should improve straw mulching quality and should not destroy raised beds; (3) in the windy region, high standing stubble cover should be adopted; (4) furrow irrigation should replace flood irrigation.

Semi-arid and sub-humid region: both NT + straw mulch cover and minimum soil disturbance strip seeding + straw cover practices were recommended. If the soil surface was uneven, it was levelled through surface mechanical operation to meet the requirement of NT planting and improve soil temperatures in the spring. In the region with high crop straw yield, straw chopping operation was adopted; if the chopped straw seriously affected the passing ability of NT seeder and led to low planting quality, the minimum disturbance strip seeder was recommended. Also, in order to reduce soil compaction, one subsoiling operation in a period of 2–4 years was suggested in some areas.

3.2.2 Conservation Agriculture seeder

In untilled or minimum disturbed soil conditions, straw and stubble cover can easily block the CA seeder, affecting its passability and performance. So, based on the straw anti-blocking technique, three kinds (tine opener straw-flowing, disc opener straw-cutting and powered anti-blocking type) of no or minimum soil disturbance seeders were developed successfully by East Asian countries according to different soil cover conditions (He et al., 2018). There is another main action that accelerated the application of CA.

Tine opener straw-flowing type: this kind of NT seeder was developed for one-crop-per-year region where straw cover quantity was low. During seeding, straw flowed away between the tine openers. Some seeders were equipped with non-powered straw clearing devices to remove the straw in the seedbed. More space between neighbour openers could be achieved for straw flowing by arranging the openers staggered on more than one line or increasing toolbar height (Zhao et al., 2007).

Disc opener straw-cutting type: this kind of NT seeder (such as single-disc type and double-disc type) was designed for one-crop-a-year and some multi-crop-per-year region where straw cover quantity was high. During the operation, disc openers could cut through the straw by gravity or hydraulic pressure and form a narrow seedbed line with minimum soil disturbance. Normally, it required high working speed, and the seeder was designed to be big and heavy to produce enough down-pressure for straw cutting (Lin et al., 2014; Zhao et al., 2011; Zhu et al., 2014).

Powered anti-blocking type: it was developed for multi-crop-per-year region without fallow land, especially in wheat–maize two-crops-per-year region. Typically, the anti-blocking devices driven by tractor was used to cut/chop/push aside crop residue or straw in front of the opener or between neighbouring openers, so the clean seedbed could be produced for seeding and fertilizing. In East Asia, a number of commercial seeders used this technique, such as the strip rotary hoe type (clear and strip open the seedbed with rotary blades), straw collecting–chopping type (produce a clean seedbed by picking up, chopping and throwing residues behind/beside the machine) and oblique-driven disc type (clear and strip-till seedbed for openers by driven discs) (Chen et al., 2016; Zhao et al., 2014).

3.2.3 Capacity building

To introduce CA practices to East Asia, it was critically important to develop a good understanding and technical capacity at various levels - farmers, technicians, private companies, input dealers, as well as government officials. This is where international organizations played a significant role at the beginning.

Publicity and training for high-level officials, technicians, farmers etc. were very important especially in the beginning to increase the cognition and acquaintance of CA in East Asian countries. During the extension of CA, various publicity and training methods were utilized, such as using multimedia platforms (cartoon book, video, digital network etc.), in-class training courses and field demonstrations and training. So government officials could learn the advantages of CA and then actively promote CA through policy support; agricultural machinery enterprises could update CA machines to meet farmers' needs; technicians and extension agents could provide the farmers with right technical supports; farmers could gradually understand and adopt CA practices and know how to use CA properly. All these aspects helped the adoption of CA in East Asian countries.

Later, what really pushed CA to a wider adoption was farmer-to-farmer learning and exchange. Seeing a neighbour farmer trying out CA and making it into a success was the most motivating factor for many farmers to adopt. Initially those training and demonstration activities organized through international projects helped early adopters to apply CA, and later they played a critical role in influencing other farmers.

3.2.4 Machinery and equipment leasing services

The development of CA technology, market demands and government support led to the creation of business opportunities for agricultural machinery dealers and other service providers such as leasing machinery services and agricultural cooperatives in East Asia. For example, such service providers would have a series of tractors and CA machines, and they would provide the agricultural machinery operation and leasing services for local farmers who enabled the application and development of CA. Furthermore, through the field demonstrations by these organizations, farmers were eager to understand CA's benefits and accept and apply the new techniques.

4 Adoption scale

The data of the CA adoption area in East Asian countries were based on several sources: FAO-AQUATSTAT, MoA and survey estimates by governmental reports. Despite the considerable research efforts for the development of CA, policymakers and farmers still did not fully understand the benefits offered by CA, making it difficult to obtain the government data for CA adoption areas in Japan and Republic of Korea. The data available through these sources may not be up to date to present a complete picture of the scale of adoption in some countries.

China had the highest adoption area of CA in East Asia. In 2018, CA was being practised on more than 8.8 million ha according to the Annual Report of China's Agricultural Mechanization Statistics (2018) by the Chinese MoA and Rural Affairs, with only 1.3 million ha (FAO-AQUATSTAT) reported in 2008/2009 and 6.67 million ha (FAO-AQUATSTAT) in 2013/2014, mainly located in the northeast ridge tillage areas, North China Plains, northwest oasis farming areas, farming-pastoral areas, Loess Plateau and south rice cropping areas.

In Mongolia, the minimum tillage (e.g. shallow cultivation) area increased from 47 300 ha in 2008, 66 700 ha in 2009, 75 800 ha in 2010 to 98 900 ha in 2011 (Bachman and Friedrich, 2003; Jalbiidandariin, 2013). By the end of 2013, 30% of wheat production used minimum tillage (e.g. shallow cultivation) with chemical sprays to control weeds (Rasmussen and Frempong, 2015). According to the 'Millennium Development Goals (MDGs)-based Comprehensive National Development Strategy of Mongolia', the use rate of NT farming technology will be increased to 70% in 2016–2020, to 80% in 2021–2025 and to 90% in 2025–2030 in grain producing fields (State Great Hural of Mongolia, 2016).

The DPR of the Korea government and FAO began to promote CA in 2003 to sustain, secure and eventually intensify agricultural production and to recover the degraded soil resources (Friedrich and Kassam, 2011). In only a few years, CA adoption spread to more than 30 farms and was applied at approximately 3000 ha (Friedrich and Kassam, 2011). Specially, in Sukchon County, which has been declared as the CA-model county by the MoA, the NT rice area was 70% of the total rice area in 2008 (Derpsch and Friedrich, 2009a). With decades of development, the CA adoption area was 23 000 ha in 2015/2016 (Kassam et al., 2019).

5 Advice for Conservation Agriculture promotion

CA has been adopted in recent years in East Asia, but the potential production and environmental benefits of CA are not yet fully comprehended by many of the farmers, land managers and policymakers. Furthermore, there are still many factors (such as lack of affordable and high-performance CA seeders, low education of farmers, lack of knowledge and understanding and shortage of financial supports) that limit large-scale adoption of CA. However, with current trends, we expect CA to be adopted more rapidly in the future and propose the following key advice for large-scale CA promotion.

1 *Science-based adoption and adaptation plans*: formulating science-based CA adoption and adaptation plans (such as short-/long-term and national/local extension plans) is fundamental, and so is the need to develop a long-term sustainable extension mechanism for CA adoption. Large-scale promotion of CA can be organized in areas where CA's application foundation is strong. In low-developed areas of CA, extension efforts should be carried out at the grassroots level in a participatory manner and village-wide promotion of CA can be implemented according to local CA application context. Importantly, those plans should be promptly adjusted in light of learning and experience to fit local conditions. Simultaneously, obtaining more government support and incorporating local community concerns are needed to create a broad-base enabling environment for CA adoption.
2 *Strengthen CA technical innovation and support*: innovation is a prerequisite for CA adoption and development. High-performance CA seeders and effective site-specific technical expertise need to be developed and made available at affordable cost to farmers. National and local research institutions composed of multidisciplinary experts in agricultural machinery, crop establishment, soil fertility, plant protection, etc. and interdisciplinary teams (including engineers, agronomists and enterprise and extension specialists) need to be organized by governments to provide planning support, adoption guidance and training as well as funding for specific research projects.
3 *Establish long-term and high-standard CA monitoring and demonstration sites*: these should represent different agroclimatic zones, soils types and cropping systems. Primary data of CA benefits are the foundation for effective promotion and application of CA practices. Through monitoring and demonstrating long-term CA benefits in different soils, environment and cropping systems, farmers are able to accept and adopt this new production system on a large scale.

6 Conclusion

Following several decades of systematic experimental research, demonstration and extension, East Asian countries have developed suitable ways to promote CA systems across different agroecologies. Manufacturing systems for CA seeders for small-, medium- and large-size farm-holdings are in place. The governments recognize the importance of CA and more and more farmers are understanding, accepting and adopting CA practices. Although constraints to CA adoption and development do exist, it is believed that CA will be adopted by more farmers over larger areas in the near future in East Asian countries.

7 Where to look for further information

A review that introduces the machinery equipment involved in CA, such as a subsoiler or no tillage seeder is in the References section. See He et al. (2018) for further information.

Centres of expertise include:

- National Institute of Conservation Tillage, China Agricultural University; and
- Conservation Tillage Research Centre, Ministry of Agriculture and Rural Affairs, China.

Website on CA:

http://www.cn-ct.net/en/: mainly introduces the development events, technology equipment and model of CA in China.

Book on CA:

Exchanging Experience with Conservation Agriculture: is a cartoon book to introduce the technological and operational process of CA and has been translated into many languages. This book can be freely downloaded from the FAO website (http://www.fao.org/conservation-agriculture/resources/en/?page=4&ipp=5&no_cache=1&tx_dynalist_pi1[par]=YToxOntzOjE6IkwiO3M6MToiMCI7fQ==).

8 References

Ahn, S. R. and Kim, S. J. 2016. The effect of rice straw mulching and no-tillage practice in upland crop areas on nonpoint-source pollution loads based on HSPF. *Water* 8(3), 1-18.

Bachman, T. L. and Friedrich, T. 2003. *Conservation Agriculture in Mongolia*. ResearchGate. Available at: https://www.researchgate.net/publication/228492873_Conservation_Agriculture_in_Mongolia.

Baker, J., Saxton, K. E., Ritchie, W. R., Chamen, W. C. T., Reicosky, D. C., Ribeiro, F., Justice, S. E. and Hobbs, P. R. 2007. *No-Tillage Seeding in Conservation Agriculture*. Cromwell Press, Trowbridge.

Chen, H. T., Zha, S. H., Dun, G. Q., Cong, G. B. and Feng, Y. 2016. Optimization and experiment of cleaning device of 2BMFJ type no-till precision planter. *Transactions of the Chinese Society for Agricultural Machinery* 47(7), 96-102.

Chuluunbaatar, D. 2011. Introducing new technologies for sustainable agricultural development in Mongolia: towards a collaborative and effective extension system. Degree of Doctrate of Philosophy, University of Saskatchewan.

Derpsch, R. 2014. Historical review of no-tillage cultivation of crops. Available at: http://www.rolf-derpsch.com/en/no-till/historical-review/.

Derpsch, R. and Friedrich, T. 2009a. Development and current status of no-till adoption in the world. 18th Triennial Conference of the International Soil Tillage Research Organization (ISTRO), Izmir, Turkey.

Derpsch, R. and Friedrich, T. 2009b. Global overview of CA adoption. Proceedings of the 4th World Congress on CA, New Delhi, 429–438.

Didier, K. and Lkhamjav, O. 2009. The potential for intensive crop production in the eastern steppe of Mongolia: history, current status, government plans, and potential impacts on biodiversity. Prepared for WCS TransLinks Program, Natural Resources Management and Development Portal.

Friedrich, T. 2012. FAO mission reports DPR Korea from 2002 to 2012, unpublished.

Friedrich, T., Derpsch, R. and Kassam, A. 2012. Overview of the global spread of conservation agriculture. *Field Actions Science Reports*, 6, 1-7.

Friedrich, T. and Kassam, A. 2011. *Mechanization and the Global Development of Conservation Agriculture*. Food and Agriculture Organization of the United Nations, Rome.

Gao, H., Li, H. W. and Li, W. Y. 2008. Development of conservation tillage. *Transactions of the Chinese Society for Agricultural Machinery* 39(9), 43–48.

Gao, H., Li, W. Y. and Li, H. W. 2003. Conservation tillage technology with Chinese characteristics. *Transactions of the Chinese Society of Agricultural Engineering* 19(3), 1–4.

Government of Republic of Moldova 2017. Japan to provide 4.8 million dollars to develop CA in Moldova, from the governmental website. Available at: https://moldova.md/en/content/japan-provide-48-million-dollars-develop-conservation-agriculture-moldova.

Gunjal, K., Morrison, T., Sheinkman, M., Wanmali, S., Fang, C., Kalisky, M. and Berardo, A. 2011. *FAO/WFP Crop and Food Security Assessment Mission to the Democratic People's Republic of Korea*. Food and Agriculture Organization of the United Nations and World Food Programme, Rome.

Hashimi, R., Komatsuzaki, M., Mineta, T., Kaneda, S. and Kaneko, N. 2019. Potential for no-tillage and clipped-weed mulching to improve soil quality and yield in organic eggplant production. *Biological Agriculture and Horticulture* 35(3), 158–171.

He, J., Li, H. W., Chen, H. T., Lu, C. Y. and Wang, Q. J. 2018. Research progress of conservation tillage technology and machine. *Transactions of the Chinese Society for Agricultural Machinery* 49(4), 1–19.

He, J., Li, H. W., Kuhn, N. J.,Wang, Q. J. and Zhang, X. M. 2010. Effect of ridge tillage, no-tillage, and conventional tillage on soil temperature, water use, and crop performance in cold and semi-arid areas in Northeast China. *Soil Research* 48(8), 737–744.

He, J., Li, H. W., Rasaily, R. G., Wang, Q. J., Cai, G. H., Su, Y. B., Qiao, X. D. and Liu, L. J. 2011. Soil properties and crop yields after 11 years of no tillage farming in wheat-maize cropping system in North China Plain. *Soil and Tillage Research* 113(1), 48–54.

Institute of Cultural Affairs: Japan 2004. *Demonstration Program for Environment Conservation Agriculture in Canete and Yauyos Province, Peru*. Available at: https://www.erca.go.jp/jfge/english/wwd/recipients/04-1orig.html.

Irish Aid and Concern World 2015. *Announcement of Conference and Call for Papers International Conference on Conservation Agriculture Democratic People's Republic of Korea*. Pyongyang.

Jalbiidandariin, S. 2013. Economical and ecological efficiencies of energy saving technology in wheat production. *Mongolian Journal of Agricultural Sciences* 11(2), 138–141.

Kassam, A., Friedrich, T. and Derpsch, R. 2019. Global spread of conservation agriculture. *International Journal of Environmental Studies* 76(1), 29–51.

Kassam, A., Friedrich, T., Derpsch, R. and Kienzle, J. 2015. *Overview of the Worldwide Spread of CA, Field Actions Science Reports 8*. Available at: https://journals.openedition.org/factsreports/3966.

Kassam, A., Friedrich, T., Shaxson, F. and Pretty, J. 2009. The spread of CA: justification, sustainability and uptake. *International Journal of Agricultural Sustainability* 7(4), 292–320.

Kassam, A., Li, H. W., Niino, Y., Theodor, F., He, J. and Wang, X. L. 2014. Current status, prospect and policy and institutional support for conservation agriculture in the Asia-Pacific region. *International Journal of Agricultural and Biological Engineering* 7(5), 1–13.

Lee, S. I., Won, C. H., Shin, M. H., Park, W. J. M., Choi, Y. H., Shin, J. Y. and Choi, J. 2013. Application of surface cover and soil amendment for reduction of soil erosion from sloping field in Korea. *Journal of the Korean Society of Agricultural Engineers*, 21–28.

Li, H. W., Gao, H. W., Wu, H. D., Li, W. Y., Wang, X. Y. and He, J. 2007. Effects of 15 years of conservation tillage on soil structure and productivity of wheat cultivation in northern China. *Soil Research* 45(5), 344–350.

Li, H. W., He, J., Gao, H. W., Chen, Y. and Zhang, Z. Q. 2015. The effect of conservation tillage on crop yield in China. *Frontiers of Agricultural Science and Engineering* 2(2), 179–185.

Lin, J., Li, B., Li, B. F., Niu, J. L. and Qian, W. 2014. Parameter optimization and experiment on archimedes spiral type of gap cutting disc. *Transaction of the Chinese Society for Agricultural Machinery* 45(6), 119–124.

Ministry of Environment and Tourism 2015. *Intended Nationally Determined Contribution (INDC) Submission by Mongolia to the Ad-Hoc Working Group on the Durban Platform for Enhanced Action (ADP)*. Mongolia.

Mousques, C. and Friedrich, T. 2007. *Conservation Agriculture in China and the Democratic People's Republic of Korea*. FAO Crop and Grassland Service Working Paper. Food and Agriculture Organization of the United Nations, Rome.

Rahman, M. H., Okubo, A., Sugiyama, S. and Mayland, H. F. 2008. Physical, chemical and microbiological properties of an andisol as related to land use and tillage practice. *Soil and Tillage Research* 101(1–2), 10–19.

Rasmussen, D. and Frempong, C. A. 2015. *Agricultural Productivity and Marketing Report*. Mongolia.

State Great Hural of Mongolia 2016. *Mongolia Sustainable Development Vision 2030*. Mongolia.

Takeshi, F. 2015. Conservation agriculture adoption and its impact: evidence from Shiga Prefecture, Japan. *Journal of International Economic Studies* 29, 35–48.

Tian, G. C., Sun, L., Shi, M. X. and Wu, F. Q. 2015. Effect of wheat straw burning on soil organic matter accumulation and microbial activity. *Journal of Plant Nutrition and Fertilizer* 21(4), 1081–1087.

Tsuji, H., Yamamoto, H., Matsuo, K. and Usuki, K. 2006. The effects of long-term conservation tillage, crop residues and P fertilizer on soil conditions and responses of summer and winter crops on an andosol in Japan. *Soil and Tillage Research* 89(2), 167–176.

Wang, C. S., Wang, Z. Y., Su, C. G., Li, H., Wang, J. and Wu, G. H. 2004. Development and application of conservation farming technique. *Transactions of the Chinese Society of Agricultural Machinery* 3, 167–169.

Zhang, H. L., Gao, W. S., Chen, F. and Zhu, W. S. 2005. Prospects and present situation of conservation tillage. *Journal of China Agricultural University* 10(1), 16–20.
Zhang, X. C., Li, H. W., He, J., Wang, Q. J., Zheng, Z. Q. and Jing, P. 2013. Effects of different tillage managements on characteristics of soil and crop in annual double cropping areas in Northern China. *Transactions of the Chinese Society for Agricultural Machinery* 44 (Suppl.1), 77–82, 71.
Zhang, X. C., Li, H. W., Wang, Q. J., He, J. and Zheng, Z. Q. 2015. Research status on mechanized subsoiling technology in Northern China. *Journal of Agricultural Mechanization Research* 37(8), 261–264, 268.
Zhao, J. L., Jia, H. L., Guo, M. Z., Jiang, X. M., Qu, W. J. and Wang, G. 2014. Design and experiment of supported roll-cutting anti-blocking mechanism with for no-till planter. *Transactions of the Chinese Society of Agricultural Engineering* 30(10), 18–28.
Zhao, W. Y., Zhang, F. W., Wu, J. F., Han, Z. G. and Wu, J. M. 2007. No-tillage seeder with bomb and gear type blocking device. *Transactions of the Chinese Society for Agricultural Machinery* 38(3), 188–190.
Zhao, X., Zhang, Z. L., Tang, P., Zhang, G. L. and Zhang, W. Z. 2011. Behavior of passive stubble-cutting disc with oblique ripples. *Transaction of the Chinese Society for Agricultural Machinery* 42(1), 64–67.
Zhu, R. X., Li, C. X., Cheng, Y., Yan, X. L., Li, J., Shi, Y. P. and Ge, S. Q. 2014. Working performance of passive disc coulter. *Transactions of the Chinese Society of Agricultural Engineering* 30(18), 47–54.

Chapter 17

Conservation Agriculture in Australian dryland cropping and in New Zealand: the lessons of 70 years

Jean-Francois Rochecouste, Grains Research Development Corporation, Australia; John Baker, Cross Slot IP Ltd, New Zealand; and Bill Crabtree, Farmer and Author, Australia

1 Introduction
2 Dealing with the negative consequences of no-till systems
3 Challenges from stubble retention
4 Cropping diversity and cover cropping
5 Compaction, controlled traffic farming and precision agriculture
6 Evolving concepts in Conservation Agriculture: recycled organics and carbon sequestration
7 The New Zealand Conservation Agriculture story
8 Acknowledgement
9 Where to look for further information
10 References

1 Introduction

Today the Australian grains industry generates more than 35 million tonnes of grain annually across 22 million ha, depending on seasonal conditions (ABARES, 2020). Wheat is the predominant crop, with barley, maize and sorghum being the other major cereal crops. Major oilseed crops include canola and sunflowers; the key legumes are chickpeas, lupins, faba beans, peanuts, soybeans, lentils, mung beans and field peas; and there are also several emerging crops such as safflower, millet and pigeon pea. Wheat remains the dominant crop, representing almost 60% of Australia's grain crop area as it has been for much of the past two decades. Pulses and oilseeds have also remained relatively stable over this period, each at around 10% of the crop area. Around 30% of Australia's cereal crop (mainly wheat and barley) is sown onto land that was a legume crop/pasture or canola in the previous season (GRDC Farm Practices Survey, 2016)

http://dx.doi.org/10.19103/AS.2021.0088.15

Most crops are grown in dryland conditions within a 200–800 mm annual rainfall zone that extends from central Queensland to Western Australia (WA). Most production occurs on light, low-fertility soils with limited water-holding capacity and an annual rainfall of less than 450 mm. Large parts of production areas are also constrained by soil salinity, sodicity and acidity that limit plant available water in the soil profile (Orton et al., 2018). Grain production is directly reliant on rainfall, and there is a strong correlation between yield and available soil moisture in the northern Australian states and in-crop rainfall in the southern states. There are indications of long-term climatic warming and drying trends already affecting potential crop yields (Hochman et al., 2017; Fletcher et al., 2020).

The incentive for rethinking farming practices in Australia during the 1980s and 1990s was created through three significant consequences of traditional tillage farming system:

- Soil erosion;
- The loss of soil moisture and delayed time of sowing; and
- A major fall in the value of pasture for sheep production relative to cropping.

The most visible consequence of full-cut tillage was erosion from both water and wind, depending on local climate patterns. In the northern cropping zones of Australia, high-intensity summer storms prior to summer cropping resulted in severe loss of topsoil and the associated loss of organic matter in the A horizon. In the southern and western cropping regions, where lighter soils predominate, pre-frontal late autumn dust storms also removed topsoil with severe impacts on soil fertility.

The consequence of these seasonal weather events was not immediately felt by most farmers as the negative impact on yield was a gradual process. The exception was sandier regions where crops were killed by sandblasting during high winds. However, the economic and emotional impact of declining cereal yields from land degradation, and the opportunities afforded from new legume crops and canola that required more soil stability, was a strong incentive for change. Less visible, but more evident to farmers on a seasonal basis, were the loss of soil moisture from cultivation and the resulting lack of planting opportunities in the dry years.

Early reduced tillage field trial results demonstrated both a significant reduction in erosion and a significant boost in available soil moisture leading to an increase in yield (Silburn et al., 2007). From the early 1980s, leading farmers across Australia began experimenting with reducing the number of tillage operations to two, then to zero cultivation prior to sowing (Thomas et al., 2007, 2011). Later farm trials showed increased planting opportunities over time and returned significant financial benefits relative to traditional multiple cultivation systems, particularly in the summer-rainfall dominant zone where winter cropping was critically dependent on stored soil water (Branson, 2011). Despite often obvious financial benefits (Table 1), uptake of such a new farming system by farmers at the time was relatively slow. It required a significant paradigm shift in the attitudes of farmers and support for change was limited by a range of factors. Foremost was the required change in seeding machinery and the lack of commercially available equipment (Ashworth et al., 2010). Weed control was also an issue, because, without maintenance tillage in the fallow, herbicides and sprayers were needed, adding further potential cost.

The history of Conservation Agriculture (CA) adoption in Australia is well recorded in the literature (Thomas et al., 2007; Crabtree, 2010; Rochecouste and Crabtree, 2014; Llewellyn and Ouzman, 2019). As noted, the adoption of no-till began in the 1990s and peaked around 2015 (Fig. 1). The rate of adoption varied from state to state, with the more marginal low-rainfall regions of the states being first adopters. Total area of CA in each state as of 2017 was as follows:

- Western Australia, 7.90 million ha (97.0%);
- Southern Australia, 3.67 million ha (94.7%);
- Victoria, 3.34 million ha (91.7%);
- New South Wales (NSW), 6.03 million ha (88.1%); and
- Queensland, 1.98 million ha (78.0%).

This corresponded to the national total CA area of 22.93 million ha (89.9%). This does not include areas of reduced tillage (1–3 tillage operations).

This chapter reflects on the implication and lessons learnt from having adopted CA across the arid landscapes of the Australian grain production

Table 1 Wheat yields (t ha^{-1}) comparing farming practices over four years at two locations in Queensland. Relative cost benefit to growers based on current grain market price

Compared practice	Biloela 1989–92	Goondiwindi 1989–92
Conventional cultivation	2.5	1.6
Stubble mulch	3.1	1.8
Reduced tillage	3.3	2.0
Zero-till	3.4	2.2
Relative income differences in moving from *conventional* to *zero-till* in today's dollar value ($320/t*) for a 500 ha per year crop	$144 000	$96 000

*Price based on average 2020 wheat price $310–330, sourced from AWB-Cargill (https://www.awb.com.au/daily-grain-prices). Reflects current return in yield differences.
Source: Wylie and Moll, 1998.

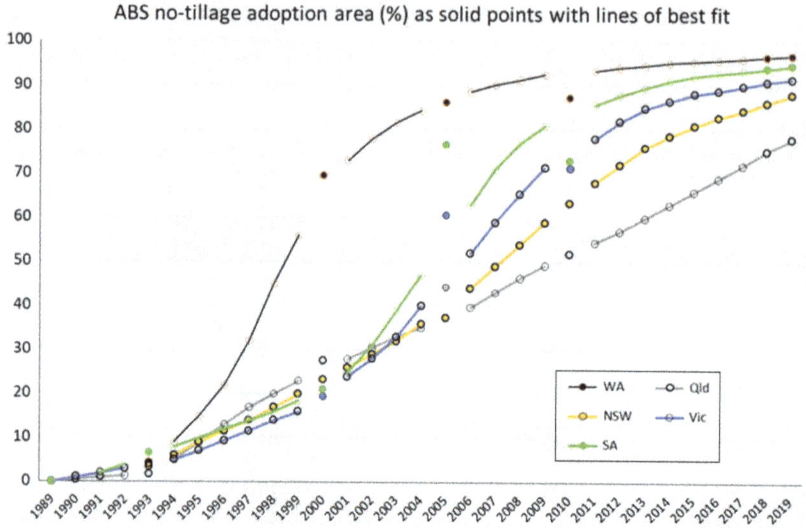

Figure 1 Adoption of no-till over time as a percentage of area of cropping. Source: Based on the Australian Bureau of Statistics data.

areas over the last 30 years. Australia's long history of applying CA in a semi-arid environment with nutrient poor soils has provided some interesting challenges for farmers and researchers. All systems, as they approach equilibrium will develop new issues that will need to be managed. CA is no different. The challenge for Australian farmers today is how to manage these emerging issues and retain the benefits already gained. Maintaining soil structure, lifting soil organic carbon and retaining valuable moisture are still at the heart of the drive for implementing CA in Australia. The benefits of CA are now entrenched, but we have had to learn some new lessons to adapt and maintain those benefits.

2 Dealing with the negative consequences of no-till systems

Having seen the enormous value of no-till to profitability and land management, farmers have put a lot of effort into tackling the negative consequences that have emerged over time. Dis-adoption of CA principles has been very uncommon. Challenges have included widespread weed resistance to many of the selective herbicides that initially supported the shift to no-tillage, including increasing glyphosate resistance, as well as nutrient stratification and subsurface acidification in some areas. These impacts posed new challenges that needed to be addressed to retain the value of no-till. We examine responses to these challenges in Australia.

2.1 Herbicide resistance

No-till has been very reliant on herbicides to control weeds, both in the fallow and in the cropping cycle. This reliance created the conditions for resistance to most common herbicides. There has been particular concern with regard to glyphosate which was heavily relied upon as a fallow herbicide. As an increase in the rate of resistance became evident, Australia established a commercial testing service that could monitor samples received from a large proportion of the southern Australian cropping belt (Broster and Pratley, 2006).

Early levels of tolerance elicited two responses, increases in dosage and adding an alternative herbicide in mixtures. This alternative herbicide often relied on older chemistry such as 2,4-D, atrazine, dicamba and paraquat to maintain a price point. However, before long, Australian no-till paddocks came to have a string of resistant weeds. New chemistry was required along with integrated weed management plans when spraying alone became ineffective. Weed control in general became increasingly compromised as some populations of annual ryegrass (*Lolium rigidum*) became glyphosate resistant, an issue which first emerged in 1996 in Victoria. Glyphosate resistance also occurred later in awnless barnyard grass (*Echinochloa colona*), liverseed grass (*Urochloa panicoides*) and windmill grass (*Chloris truncata*) in NSW. The first recording of broadleaf resistance was in fleabane (*Conyza bonariensis*) in Queensland. The most recent occurrence of resistance was in great brome (*Bromus diandrus*) in 2011 in South Australia (SA). Today Feathertop Rhodes grass (*Chloris virgata*) is increasingly infesting cropping paddocks with limited chemical control options. Resistance problems have not been limited to broadacre cropping but are also evident in horticulture, industrial weed control areas, railway lines and roadsides.

Measures such as crop rotations with pre-emergent herbicides (requiring careful attention to timing) and modifying row spacing had some success, but the fallow remained a problem. Some weeds became such a problem that farmers were forced back to mechanical removal as an act of desperation (Yash et al., 2018). Today research is focussed on carefully protecting new chemistry, spot spraying technology to reduce the overall amount of herbicide required and the use of similar site-specific technology to spot remove weeds as a means of saving costs. There has been significant research interest in rotation with pastures and the use of harvesting technology to destroy weed seeds going out with the chaff.

The clear lesson from the Australian experience is that weeds will evolve and thrive under a no-till system and the CA practices need to evolve in response. As the value of minimising mechanical soil disturbance becomes more widespread through improved low disturbance no-till seeding, there is a need to carefully consider the evolution of weeds to the system and have strategies in place to manage this. Australian farmers have been very successful in doing this with an emphasis on weed seed control practices, more diverse rotations with more opportunities for weed control, use of new pre-emergent herbicides and use of herbicide-tolerant crops. It is for these reasons that weed numbers have been maintained at relatively manageable levels. These alternative control methods should be included in the CA system as early as possible to establish an integrated weed management system. This might include livestock, other innovative weed control machinery or even mechanical removal and manual roughing.

2.2 Nutrient stratification

Eliminating tillage has also resulted in nutrient stratification. This is less noticeable in an undisturbed natural environment which has attained a level of equilibrium. However, in cropping soils, stratification has had significant impacts on production costs (Ma et al., 2009; Vu et al., 2009; Unkovich, 2014). This occurs as nutrients from lower soil horizons are taken up by the plant and either exported as grain or left to decompose in the remaining stubble on the surface. This cycling of subsoil nutrient to the surface has resulted in the accumulation of key nutrients to the A horizon where it is less available to plant roots if surface conditions are dry (Crozier et al., 2008). Given the arid climate conditions of Australian grain production areas, nutrient is often locked in the surface horizon and at risk of erosion, especially during extended drought periods.

As noted, the two key components are simple nutrient export in the grain, requiring replacement as fertiliser, and the movement of the nutrient pool in terms of the soil horizon. Replacing subsurface nutrients has no easy solution and current research is looking into subsurface manuring using specialised machinery (Mason et al., 2016). The high cost of subsurface treatment should be considered as a system renovation cost requiring capital investment and cannot feasibly be considered as part of seasonal operational costs. This type of soil amelioration provides a long-term return on investment. At present we do not have good data on the associated cost-benefit to support broadscale adoption.

2.3 Subsurface acidification

Certain areas of Australia are susceptible to soil acidification from agricultural practices. The removal of nutrients via harvesting crops combined with the addition of fertilisers will slowly acidify the soil. Over time the subsurface becomes increasingly acidic, affecting the chemical balance of the soil, and is often associated with excess levels of toxic aluminium. This compromises crop production, particularly legumes. Lime is still considered the most effective ameliorant to lift pH levels. However, surface application under no-till systems means that lime moves downwards very slowly at the rate 0.4 mm per annum, and this is insufficient to mitigate subsoil impacts on production (Li et al., 2019; Miller, 2019).

The challenge of mixing lime at depth in a no-till system is still being researched for a solution. As most of our experience has relied on surface soil acidity, the means of mixing at depth is still not fully determined. Available lime is mostly in the form of a powder and is quite difficult to manipulate for mixing across soil horizons. It has poor flowability and needs thorough mixing to be effective. Prilled lime is available and has far better flow characteristics but comes at a higher cost. In the sandy soils of WA, deep mechanical placement of lime has shown significant response in improving production in the short term but the long-term response is still being monitored. In the higher clay content soils of the eastern states of Victoria and NSW, high energy costs from deep mechanical amelioration operations are more problematic and alternative systems are still being considered. Subsurface acidity is a chronic condition that must be addressed over the long term. The time horizon for action will depend on the buffering capacity of the soil but it must ultimately be addressed to avoid production decline.

3 Challenges from stubble retention

CA has demonstrated significant benefits in retaining stubble (root stalks and plant biomass) to protect the soil from erosion and reduce loss of soil moisture from evaporative losses. In dryland agriculture the reliance on stored soil moisture is paramount to yield. Every millimetre of available soil moisture is equivalent to approximately 22 kg of grain per hectare (Sadras and Angus, 2006; Sadras and McDonald, 2012). The value of stubble retention as a form of ground cover is widely accepted in Australia. Although there is no indication that Australian farmers will go back to stubble removal, there have been several challenges to managing stubble, including the carry-over of stubble borne diseases and planting into high stubble loads.

3.1 Emergence of stubble borne diseases

Stubble retention is very effective in protecting the soil surface from erosion by keeping the soil cooler and reducing moisture loss from evaporation. This in turn increases net infiltration from rainfall events, boosting soil moisture profiles. However, retained surface stubble of crops like wheat can also increase the likelihood of disease carry-over such as yellow leaf spot (*Pyrenophora tritici-repentis*) or crown rot (*Fusarium graminearum*). These diseases were not common when stubble was burnt prior to planting but they have become more prominent since the practice was phased out in favour of retaining soil cover. The relative value of retaining stubble can thus be offset by disease losses. In dry environments, stubble can remain undecomposed over multiple seasons, providing a more stable environment for pathogens. Many pathogens have evolved to be both parasitic and saprophytic which help them survive on stubble residue.

Fungal pathogens are very dependent on the right microclimate conditions for infection. Continuous monocropping of susceptible species will increase the likelihood of infection if conditions are favourable. It takes time to determine the life cycle causal factor in disease spread and even longer to breed for resistance. Early detection of emerging disease in retained stubble will allow for the pre-emptive development of disease management programmes. Rotation should be is the best insurance mechanism against disease carry-over (Akinsanmi et al., 2004). Crop diversity is the best strategy to overcome this avoidable problem.

3.2 Stubble management for planting

Managing stubble, especially large stubble loads, brings about its own logistical problems in planting into it. Several solutions have been developed, depending on the volume farmers are dealing with. Lifting cutting height to allow for standing stubble is one method, but there can still be a blanket of chaff to get through. Another option is to use a stripper front for harvesting which strips the grain and leaves of the stalks standing.

A range of approaches have been used by adapting seeding machinery, including combinations of:

- Cutting crops 15–20 cm high, using harvesters with residue choppers/spreaders;
- Increasing row spacing (often to 30 cm or more);
- Using coulter cutting blades ahead of seeding times to cut through residue;
- Deflecting residue ahead of each tine to create inter-row space; and
- Distributing seeding tines between three and five machine bars, to increase the gap between adjacent tines on the same bar.

Despite the issues of seeding through crop residue, farmers consider the benefits as exceeding costs in effort and expense. Today more than 30 commercial machinery suppliers offer a large range of seeding machines and adaptations of seeding machines. This includes a variety of seed-trench firming 'press wheels', which can have a major positive impact on crop emergence under marginal moisture conditions. Overload release ('stump jump') systems are universal in some areas, and individual row depth control ('parallelogram mount') mechanisms are increasingly common.

The various machinery systems had been developed to clear a path through the stubble or cut through the stubble using a front coulter. Both approaches depend on stubble load and soil moisture levels. A large stubble load is not easily cleared with a tyne planter which accumulates bundles of stubble material, acting like a leaf rake. The front coulter acting as a cutting disc does not operate well in wet stubble and moist soils, essentially pinning the stubble into the furrow instead of cutting it cleanly to create an opening for the seed. To overcome these problems, high pressure industrial water jets have been adapted to act as a liquid coulter, effectively cutting cleanly through a high stubble load. This has attracted great interest but is still in the development stage.

Another emerging issue is the impact of stubble in reducing nitrogen fertiliser contact with the soil and increasing volatilisation. There has been research to determine the actual level of impact that stubble load may be having on nitrogen tie up or loss. As these issues show, stubble retention is valuable, but it comes with consequences that need to be considered. Weather, disease carry-over and stubble load thus need to be considered as part of a CA management system.

4 Cropping diversity and cover cropping

Crop diversity is considered important to CA systems by reducing pest and disease carry-over and through soil health enhancement and fuller use of the soil profile. However, the fact that crops have variable profitability can lead farmers to choose to maximise short-term profit. Farmers will move from one crop to another depending on market price and early season conditions. They will seek more profitable crop options if they are confident that they know how to grow the crop. Risk is another factor affecting the choice of crops (Fletcher, 2019). High-input crops that are complex to grow often require a bigger outlay for greater returns, but there is also more to lose if conditions become unfavourable.

Commodity price alone is not the only consideration. A bio-economic research project has indicated the complexity farmers face when choosing different crop options (GRDC, 2019). The barriers are sometimes the availability of labour at critical times or specialised machinery for different crop options. Even if farmers wanted to change to another crop, they may simply not have the resources to do so. This has resulted in continued high reliance on cereals and no increase in the proportion of cereal crops that follow a non-cereal crop over the past decade (GRDC Farm Practices Survey, 2016).

Farming system research has indicated that, over time, a strategy of only growing the most profitable option can have longer-term cost consequences.

The choice of crop in rotations has been studied under various 'farming systems', clearly demonstrating that farmers need to consider long-term system sustainability as opposed to short-term profitability around a single crop.

The choice of crop or length of fallow has significant legacy implication for the productive capacity of the following crops. The use of legume in rotation, for example, has long been recognised as increasing the availability of soil nitrogen to the following crop. However, overreliance on legumes can have consequences. When grain legume prices were more profitable than cereals, farmers introduced considerably more legumes in the rotation. Some legumes leave very little in the way of plant residue on the surface and some multiple legume crops will leave very little in the way of soil cover. In the period 2017-2019, Eastern Australia experienced a major deficit in rainfall across the grain regions and areas heavily reliant on legumes became highly exposed to wind erosion. Those paddocks that had cropped legumes with minimal cereals experienced significant loss of topsoil across western NSW and Queensland. The cost of soil loss in production terms is yet to be calculated. The lesson to consider is that soil cover needs to be maintained in some form over the cropping rotation. It affects the soil's water balance as well as protecting the A soil horizon from erosion.

The role that crop diversity has on soil biology is still relatively unknown. There is some indication, for example, that a legume break with a crop like soybeans or mung beans following several years of sugarcane is more effective in reducing yield decline than simply having a fallow. Pastures in rotation to cropping, particularly perennial pastures, appear to increase soil organic carbon levels and available nitrogen. Not all farming enterprises have the capacity for introducing livestock into the farming system but, after declines in livestock since early no-till adoption began, mixed farming has stabilised as a common part of the farming system (Bell and Moore, 2012).

A report by Scott et al. (2010) evaluated the benefit of break crops from a series of long-term trials in WA. It indicated that yields of legume break crops were highly variable, often riskier and less profitable than cereals. However, the average yield benefit from the inclusion of lupin or field pea in the rotation compared to 'wheat following wheat' was in the range 0.3–0.6 t ha^{-1} in favour of a legume break crop, and often accumulating to at least 1 t ha^{-1} in additional cereal yield over multiple post-break years (McBeath et al., 2015).

These yield benefits after a break crop were evident across a range of locations, crops and high-low yielding sites (Angus et al., 2015). Following the break crop, the following cereals still responded to nitrogen application. However, the rate of response was relatively low and more often dominated by non-nitrogen related benefits (diseases and weed control). The break crop benefit was also reported to last up to a third wheat crop (Seymour et al., 2012). Despite the recognised benefits of crop rotations, especially legumes, the choice of cereals is predominantly driven by reliability and economics in many marginal areas (Fletcher, 2019). Many of those marginal areas with inherent high summer temperatures and low humidity will limit the availability of crop options. In areas of far Western Queensland and Northern NSW, rotation options are limited to the winter crops chickpeas and wheat, and in the summer to sorghum. More work is needed for enhancing the ability of summer legumes to withstand the long summer heat.

The importance of soil cover has also prompted research into cover cropping with the intention of determining if additional soil moisture could be retained in the fallow. The intention is to plant the crop without intending to harvest the grain and determine the balance of moisture loss by transpiration and increasing infiltration from having some form of soil cover. The crop cycle is stopped by spraying with a knockdown herbicide. Indications from early research are that 'early intervention' in spraying out the cover crop does not produce enough biomass to support an increase in water infiltration. Intervention at the 'mid-crop cycle' suggests good results for balancing the moisture cycle in favour of infiltration by as much as an additional 20 mm of stored soil moisture (Erbacher et al., 2019). Grain yield water-use efficiency suggests this delivers approximately 16 kg/ha/mm (Sadras and McDonald, 2012). 'Late intervention' increases net biomass but loses more moisture to transpiration than is gained in infiltration.

Planting a cover crop delivers greater value when there is minimal soil cover coming into a fallow period where rainfall can be captured. It is less beneficial if there is a good level of existing cereal stubble. This is, to some extent, a flexible extension of stubble retention. Research recorded a decrease in surface soil temperature under the cover crop compared to the fallow which is likely to reduce evaporative losses. This was also compared to allowing weeds to grow and act as a cover. Weed cover was not as effective, lacking the uniformity of a cover crop, and was more difficult to spray out due to variability in growth stages.

The water capture component is dependent on many seasonal factors such as the timing of rainfall and seasonal temperatures. These considerations are still being explored. However, it is generally accepted that planting cover crops can increase net water infiltration in circumstances of poor existing soil cover. Cereals appear to be more effective in increasing infiltration but only a limited number of crop candidates have been evaluated to date. Where cereals are retained from a previous crop, they seem to do an effective job of improving infiltration and may not require a cover crop.

5 Compaction, controlled traffic farming and precision agriculture

The impact of soil compaction by heavy farm machinery is still an ongoing issue in Australian farming (Chamen et al., 1992; Chamen et al., 2003; Batey,

2009). Most farmers have an understanding of the effect of tractor wheels on soil compaction and the impact this has on crop production (Hamza and Anderson, 2005; Kirchhof and Daniels, 2009), although not all soils are equally affected. The consequences have been reduced microbial activity, reduced water infiltration and poor root growth leading to poorer yields (Jones et al., 2003; Tullberg et al., 2007; Ahmad et al., 2009; McKenzie et al., 2009).

More farmers are applying controlled traffic farming (CTF) practices that restrict the wheels of all heavy machinery to permanent traffic lanes to prevent damage to the whole paddock from conventional 'random' operations. The compacted permanent traffic lanes are laid out and managed for efficient traction, traffic and drainage, allowing the intervening crop beds to remain soft and in better condition for crop production. Because the harvester is the most difficult machine to modify, controlled traffic operations have usually been achieved by modifying tractors to match the harvester track width, using a harvester and seeder of the same width, and a sprayer which is a multiple of this width. This can provide machinery footprints in the range 12–16% of the paddock area. This practice was taken up initially by farmers on heavier soil types, providing evidence of soil structural improvements, increased yields (Li et al., 2007) and substantial reductions in fuel use. Farmers also report that hard permanent traffic lanes allow a wider window of operation for machinery as they don't have to wait as long for soft soils to dry out.

Use of permanent traffic lanes has resulted in greater energy efficiency than operating randomly. The difference recorded by Tullberg et al. (2007) showed a 39% reduction in energy requirement from employing CTF. Gas exchange between soft soils and compacted soils are still under investigation, but preliminary research results show substantial reductions in nitrous oxide emissions from controlled traffic cropping beds (Antille et al., 2015).

Emerging problems of CTF include the difficulty of controlling weeds in wheel tracks, and deep rutting of traffic lanes by continuous wheel passes, especially in clay soils. The ruts act as effective drains where the tracks run downslope. Ruts can thus create some ponding across slopes. Ruts need to be renovated on occasions which add to the cost of CTF operations.

A potentially emerging factor is that the chaff spread from the back of the header may not span the total harvested zone. This leads to a build-up of chaff cover around the wheel zone and none at the edges. Growers have reported variations in crop vigour along control traffic lines but the cause for this has not been directly attributable to the chaff build-up. There is still some uncertainty regarding causal factors of crop vigour variability in long-term control traffic systems. A recent innovation has been to divert all the chaff from harvesters that might contain weed seeds into the wheel tracks. This narrows the potential area for the emergence of resistant weeds during harvesting.

CTF operations depend on precision agriculture techniques. The label precision agriculture was first applied when the combination of harvester yield monitoring and satellite-based global positioning systems (GPS) allowed the economic production of paddock yield maps. 'Precision agriculture' is now a generic term covering a wide range of satellite and sensor-based technologies. The most widely adopted of these is 'GPS Autosteer' (self-steering) for farm equipment.

The world's first commercial satellite-based autosteer using Real Time Kinematic (RTK) GPS correction for precise tractor steering, the 'Beeline Navigator', was developed in Australia in the early 1990s. Guidance equipment of this type is now manufactured by several international manufacturers and is a built-in option or standard unit in many tractors and harvesters. Inexpensive units claim pass-to-pass (repeatable only in the short term, not year-to-year) accuracies of ± 10–30 cm, but more sophisticated units provide '2 cm' precision (± 2 cm 67%; ± 4 cm 95% of time). Benefits such as increased productivity with the elimination of overlap, and reduction in operator fatigue are significant, often quantifiable and easily justified by farmers managing increasingly large areas.

Accurate digital GPS position monitoring and recording now provides a platform for a large number of precision agriculture applications. Data from various proximal sensors (Table 2) and other spatial information (e.g. satellite images) can be combined to improve resource use in key farm inputs: labour, fuel, fertiliser and chemicals. The input cost-benefits are balanced by the cost of establishing a digital network system on farm and the human factor of having to learn how to use the system efficiently.

'Site-specific management' – the matching of seed, fertiliser and crop chemical inputs to the crop requirements or soil characteristics of each paddock zone – became possible with the development of GPS-based harvest yield monitors and variable rate applicators. It has also become cheaper as standard monitoring equipment has been built into harvesters and applicators. Many Australian grain growers have now used yield mapping to provide useful management information. However, the next step – 'zone management' using variable rate technology – requires complex assessment of soil/crop response characteristics and their interaction with climate probabilities. Scientific enthusiasm and investment in this technology has not been matched by practical adoption which has been slow.

More recent development of equipment sensing crop status shows greater promise of rapid application, particularly when early problem detection (e.g. nutrient deficiency) can enable timely and effective management response (e.g. foliar nutrient application). Development of systems to integrate proximal and remote sensor outputs has also interested farmers managing increasingly expensive and limited resources (Rochecouste, 2009). The aim is to optimise economic performance and avoid wasteful uniform applications by limiting

Table 2 Some example of sensors and related information for farmers

Farm asset	Sensor	Data	Information
Soil texture	Electromagnetic induction	This is a non-contact method of measuring electrical conductivity involving inducing a magnetic field into the soil and measuring the electrical current response field	Structure and depth of topsoil
Soil moisture	Various: reflectometry, microwave or radio frequency via probe	Moisture curves	Current soil moisture trend
Soil pH and nutrient	Electrochemical sensors	Various	Field pH and nutrient status
Crop vigour/ weed presence	Optical and radiometric	Crop vigour (relative)	Areas of poor growth, nutrient, disease, insect damage or presence of weeds
Yield monitors	Flow meters	Grain yield (relative)	Harvestable yield based on management
Variable seeding	Ground speed sensor	Seed volume	Plant population
Variable rate fertilising	Flow meters/chlorophyll sensors	Fertiliser output	Fertiliser volume

inputs (e.g. fertilisers and chemicals) to 'what is needed, where it is needed' (Whitlock, 2006; Butler, 2008). This use of precision agriculture continues the trend towards increasing efficiency in the use of resources (Cook and Bramley, 2000; Shoup et al., 2004).

Most farmers and agronomists have taken up some aspect of digital technology as part of their management, and the trend is increasing. Continuously Operating Reference Stations (CORS) are being built and gradually covering much of rural Australia. CORS is a network of permanent Global Navigation Satellite System (GNSS) tracking stations which provide the RTK correction signals necessary for precise satellite positioning for industry and agriculture (Janssen et al., 2011). CORS installation in Victoria is complete with 100% coverage. NSW is more than 50% complete, Queensland has coverage but mostly in the south-east, whilst WA and SA have limited coverage. This technology will be an integral component of a more resource-efficient, productive and sustainable mechanised farming future.

As the practice of retaining crop residues to protect soils increased, farmers noticed that crops sown between standing stubble rows performed better.

Leaving the stubble standing after harvest reduces the problems of tine planter residue blockage and that of disc planters 'hair-pinning' through failure to cut through flat, wet stubble on soft soils. Precision autosteer made it possible to routinely place an alternate row between existing rows of standing stubble, perhaps after some increase in row spacing, and allow adjustment of sprayer nozzle positioning. Inter-row planting provides a more consistent soil cover and associated weed control benefits (Roberts and Leonard, 2008), and is simply achieved by the use of an offset hitch to displace the seeder frame relative to the previous year's planting. Yield improvement of legumes sown within cereal stubble has also been reported, attributed mainly to reduced lodging and improvements in harvest efficiency (Roberts, 2008).

6 Evolving concepts in Conservation Agriculture: recycled organics and carbon sequestration

No system remains static. Australian farmers, agronomists and researchers have continued to look for economic ways to increase the levels of soil organic carbon to improve production.

6.1 *Recycled organics*

The nutrient value of animal industry waste as an alternative fertiliser and means of improving long-term soil structure has been a point of discussion among farmers. This applies particularly to those cropping zones where intensive livestock enterprises producing animal waste are conveniently located. Farmers have started purchasing and applying this waste. In most cases they have seen yield increase, primarily due to the nutrient content released over several years. The cost of manure is comparable to traditional inorganic fertilisers, but manures are generally less predictable in their N-P-K nutrient value, and transport/application costs are significant. Un-composted product has high water content, and raw manure can also tie up nitrogen for some time. Despite these issues a significant increase in its use occurred in 2008–2009 when global fertiliser prices rose sharply.

In addition to the nutrient benefit, some farmers have also reported better long-term water-use efficiency from increased organic matter. This could be attributable to improved water-holding capacity (WHC) in sandy soils, as outlined by the Western Australian Waste Authority (WA Waste Authority, 2010). Other benefits attributed to recycled organic amendments include increased water infiltration and improved soil structure. Although the linkage between water-use efficiency and WHC is well researched, it is unclear if adding a range of unspecified animal manure to fine-textured, low-fertility soils in an arid climate will lead to long-term improvement in WHC.

Use of urban sewerage is being trialled in some areas but there is concern about the likely build-up of heavy metal contaminants. Grain farms are also generally distant from major urban areas, making transportation costs prohibitive. While the practice is favoured by many farmers, logistics limits its use to certain areas within easy transport reach of waste outlets.

6.2 Carbon sequestration using no-till in an Australian context

The concept that no-till practices will lead to significant carbon sequestration does not seem very likely in Australian dryland farming, where low rainfall limits biomass production, and high temperatures accelerate the loss of soil organic matter (SOM). Soil carbon sequestration faces the same 'additionality' and 'permanence' tests as other sequestration mechanisms involved in carbon offset trading. The potential role of increasing soil organic carbon (SOC) in Australia has been reviewed by Sanderman et al. (2010). Tilled soils lose organic carbon at variable rates depending on the clay content and annual rainfall (Swift, 2001). In a range of clay soils, losses of organic carbon averaged 0.6% per year (Dalal and Chan, 2001). The limited rainfall, inherently low soil fertility and high summer temperatures of the cropping region limits the opportunity to significantly increase the organic carbon content of these soils (Chan et al., 2008; CSIRO, 2009; Kirkby et al., 2014).

Under these conditions, minimum soil disturbance practices in dryland farming with limited biomass production have limited capacity to increase soil organic content, and in most situations they can only mitigate the ongoing loss (Wang et al., 2010). This would mean that many of the cropping soils would show only small increases in SOC over time in line with their agroecological biomass potential (Luo et al., 2010; Chan et al., 2011). Such small changes are unlikely to find sufficient offset units across the average grain farm to interest traders and would require some form of pooling to create the necessary economies of scale (Renwick et al., 2003).

The potential for carbon trading is further complicated by the error margins associated with measuring SOC that emanate from variations in bulk density (Throop et al., 2012) when sampling occurs to fixed depth, rather than equivalent mass across heterogeneous soil types (Sanderman et al., 2010). Sanderman and Baldock (2010) also argue that predicted carbon stock change data from agricultural trials may not truly reflect sequestration when the state of the soil carbon at the beginning of the trial is unknown, that is, when there is no comparable baseline at the start of the field trials. Current International Panel for Climate Change (IPCC) accounting methodologies developed from trial results may not show the true value of the carbon storage based on management activities.

This uncertainty is likely to affect confidence in the market allocation of carbon credit units for offsetting a unit of emission using soil carbon sequestration in dryland farming areas with low potential. Nevertheless, CA significantly reduces the loss of SOC to the atmosphere and, in certain seasons, does create a carbon sink. Although it may not fit the mainstream carbon market, this should perhaps still be considered as a market-based instrument to encourage the benefits attributable to CA through reduced emissions and positive effects on the soil carbon balance.

7 The New Zealand Conservation Agriculture story

Many countries claim to have invented the practice of no-tillage, but some of the first published accounts of sowing seeds into untilled soils for pasture renovation were in New Zealand (NZ) in the 1950s (Cross, 1957). The two constraining technical issues then were how to suppress existing weeds and guaranteeing germination of the newly sown species. The potential of no-tillage to become a serious alternative to conventional tillage did not begin, however, until the 1960s when the 'instantly deactivated' (but otherwise environmentally acceptable) herbicide paraquat became available. In the 1970s, weed control became a broader spectrum and more effective with the advent of glyphosate herbicide (Roundup). Neither of these herbicides helped organic farmers, and the control of weeds remains a major constraint to the widespread uptake of no-tillage for annual organic cropping. In 2015 the total area under CA in NZ was 366 000 ha, about 37% of the total annually seeded land.

The contrasting farming systems in Australia and NZ did much to develop seed drill technologies. Australia is a country with a range of mainly semi-arid climates (warm temperate, Mediterranean and tropical) dominated by arable cropping which often experiences limited and unevenly spread rainfall between flowering and harvest. NZ, by contrast, is a cooler temperate country with adequate and reasonably evenly spread rainfall. About half of NZ's land area is permanent pasture that supports livestock farming. What works in NZ therefore does not necessarily work in Australia and vice versa. But learning the causes of differences between the two countries has taught agronomists and development engineers more about how to design fail-safe no-tillage systems and machines than would have been possible in either country in isolation.

In Australian arable cropping, the residues of harvested crops provide both a physical obstacle to the passage of no-tillage machines and a massive biological and carbon-rich resource that few other countries can duplicate. By contrast, in NZ, most no-tillage takes place in short-grazed pastures or animal-grazed forage crop residues that provide little drill blockage problems. However, soils are often less friable than under cereal crop stubble. Seed placement functions by no-tillage drill openers therefore often differ between the two countries.

Australian arable soils have already lost much of their organic matter from repeat cultivation over many years. Since SOM relates closely to soil WHC, Australia has a high need (and opportunity) to reverse losses in SOM in its arable soils. Since 85% of the world's food is derived from arable crops, no-tillage potentially has its greatest advantage in arable cropping countries such as Australia (Borlaug, 1994).

The NZ arable industry is small by comparison, but it produces some of the highest crop yields per hectare in the world and has therefore become an ideal development location for evolving machinery suitable for predominantly arable countries such as its near-neighbour, Australia.

Because of Australia's relatively warm and moist winter climate in its temperate and Mediterranean areas, most arable crops are sown in autumn with significant vegetative growth over the winter before flowering, setting seed and ripening for harvest in late spring and early summer. Limited rainfall during the seed-set and ripening period often limits the volume of harvested seed collected from this otherwise abundant winter growth of Australian arable crops.

NZ's colder winters mean that autumn sown crops grow slowly through to spring, but usually receive adequate rainfall through to harvest time in mid-to-late summer. Such a climate favours both spring and autumn-sowing of crops (forage and arable). The result is usually heavier temperate crop yields in NZ than Australia. A typical cereal crop in NZ will leave about the same amount of dry matter as straw and other residues after harvest, as the weight of harvested grain. By contrast, Australian arable crops will often produce close to double the weight of vegetative residues as harvestable grain.

Australian cereal crops appear to capture more atmospheric carbon by photosynthesis during their vegetative growth stages, compared to NZ arable or forage crops. The challenge for Australian arable farmers is how best to transfer this captured carbon back into the soil where its value as a water storage resource is of greater value (and need) than in NZ.

Retaining carbon in the soil can be achieved effectively by Low Disturbance No-Tillage (LDNT). LDNT involves spreading the post-harvest crop residues evenly over the ground so that earthworms and other soil fauna can take the carbon-rich residues back into the soil over time. LDNT drill openers must penetrate beneath the carbon-rich mulch of residues to plant the next crop in such a way as to retain the mulch as the cover and moisture barrier over both the drilled rows and the un-drilled soil between rows.

Very few seed drill manufacturers in the world have so far mastered LDNT. To achieve LDNT, openers must simultaneously:

a Physically handle heavy residue loads without blockage, displacing the residues sideways or 'tucking' folded uncut straw into the seed slot where it contacts the sown seeds.

b Minimise in-row soil disturbance to minimise carbon and water emissions from the sown rows.
c Avoid compaction of the slot walls.
d Retain at least 90% of the soil's pore-space water vapour in the seed zone to ensure that seed germination takes place even when liquid soil water is not readily available to the seeds.

LDNT was pioneered in NZ even although Australia had the greater need for the technology. Surveys conducted by Massey University in NZ have shown 30–50% cumulative increases in both soil carbon and organic nitrogen resulting from at least three consecutive years of double-cropping using LDNT with full residue retention, compared with conventional tillage. Net recapture of 500 kg/ ha/ year of soil carbon has been measured and estimated by carbon budgeting in continuous LDNT double-cropping regimes in NZ (compared with losses of 2000 kg/ha/year carbon under conventional tillage). Such gains are capable of mitigating much of the carbon emissions from all sources (Baker and Ross, 2017).

Australia's more favourable winter vegetative growth (compared to NZ) makes Australia one of the best candidate countries for recapturing atmospheric carbon via photosynthesis and returning it to the soil via LDNT. Comparing the total CO_2-e emissions from a given country with the total arable area of the same country indicates the potential for LDNT to help mitigate carbon emissions in any given country. Table 3 lists the ratios of 'arable land' (AL): total emission (TE) for a number of countries chosen at random, including Australia and NZ.

The higher the AL:TE ratio, the greater the potential for arable cropping undertaken by LDNT and cover crops with full residue retention to mitigate that country's total carbon emissions from all sources. Australia leads the list of sampled countries with an AL:TE ratio of 1.075 indicating that it has sufficient arable land to offset a significant amount of its carbon emissions from industrialisation and other sources if all arable cropping in Australia was undertaken using LDNT.

Although China has 3 times as much arable land as Australia, its carbon emissions are 20 times greater than Australia's, giving it a low AL:TE ratio of 0.12. This is the same as NZ's AL:TE ratio which is due to the country's small area of arable land and modest industrialisation. The United States has an area of arable land slightly larger than China's, but its carbon emissions are only half those of China's, giving it an AL:TE ratio of 0.32, which is between those of Australia and China.

NZ's reliable rainfall and minimal amounts of surface residues as grazed pastures encouraged farmers to use both simple, inexpensive, tined, no-tillage drills and more expensive, triple disc drills to renovate pastures from the 1960s onwards. Some arable no-tillage was attempted using triple disc drills.

Table 3 Ratios of arable land (AL): total emissions of CO_2-equivalent (TE)

Country	Arable land (km^2)	Total emissions CO_2-e (kt yr^{-1})	Ratio AL:TE
Australia	479 954	446 348	1.075
Canada	469 281	555 401	0.84
Russia	1 265 267	1 760 895	0.72
USA	1 652 028	5 172 338	0.32
Germany	121 737	777 905	0.16
United Kingdom	61 144	398 524	0.15
China	1 238 013	10 641 789	0.12
New Zealand	10 000*	80 200	0.12

* Includes all seeding for arable crops, forage crops and pastures for NZ.
Source: FAO via Wikipedia, 2014/15.

However, unreliable germination in both cases led to a decline in the popularity of no-tillage in NZ even for pasture renovation. The practice had almost disappeared by the 1980s.

Development of an innovative inverted-T shaped tine changed this trend and the same tool also found its way into Australian pastoral farming. In the early 1990s a residue-capable and fertiliser-banding version of the inverted-T no-tillage opener concept was released, and this had an immediate impact on LDNT in NZ and Australia. By 2015, the NZ agricultural statistics showed that all forms of no-tillage in NZ had recovered from almost zero to 37% of all seeds sown in NZ (which included arable crops, forage crops and pastures). It is estimated that by 2020, this figure will have increased to approximately 50%, with close to half of the area being undertaken by inverted-T shaped openers of one form or another. In 2020, it is estimated that almost all arable no-tillage in NZ was undertaken by a disc version of an inverted-T shaped opener because arable farmers had more to gain from LDNT than pastoral farmers and the consequences of failure were correspondingly greater.

Australian arable farmers also began herbicide-assisted no-tilling in the 1960s using a variety of disc type openers. They also experienced poor germination from almost all designs of disc opener. As a result, Australian arable farmers gradually exchanged disc type openers for tine type openers. However, crop residue handling then became a problem. At first, farmers compensated by either burning or shredding the residues or sowing in wider and wider rows to allow successive crops to be sown into the more residue-free spaces between the previous rows of harvested crops. However, as crop yields (and therefore residue volumes) inevitably increased because of no-tillage, this exacerbated the residue handling problem of tined openers. Inevitably there will be a swing back to disc type openers in Australia, provided that the new disc opener designs achieve LDNT.

Conclusions drawn from multiple reports claiming to have measured carbon emissions from 'no-tillage' treatments may be misleading. Data from such reports that fail to define the specific 'no-tillage' treatments applied in detail should be treated with caution (Derpsch et al., 2014). Substantial differences in water vapour emissions attributable to a range of designs of 'no-tillage' openers (and therefore slot shapes) have been reported in the international scientific literature. The causal mechanisms of differences in water vapour emissions have been identified as slot shape, physical soil disturbance (aeration) by the openers and the nature of the slot-covering medium (if any). It is possible (in fact likely), therefore, that these same causal mechanisms may well have also influenced carbon emissions from 'no-tillage slots' in the same manner. Unless the physical properties (or the opener brands or designs) of different 'no-tillage' treatments are clearly described by authors under 'Materials and Methods' (and few are), little credibility can be attached to reported carbon emissions from treatments that are simply referred to as 'no-tillage' or 'conservation tillage' (or similar all-embracing words), which assumes that all 'no-tillage' openers disturb the soil in an identical manner, which is clearly not the case.

8 Acknowledgement

Dr Rick Llewellyn for his support and advice in the preparation of this chapter.

9 Where to look for further information

The latest research in this topic area is constantly being updated at the Grains Research Development Corporation website (grdc.com. au).

10 References

ABARES 2020. *Australian Crop Report*. Australian Bureau of Agricultural and Resource Economics and Sciences, Canberra, February. DOI: 10.25814/5e3a014fa7e2d CCBY4.0.

Ahmad, N., Hassan, F. U. and Belford, R. K. 2009. Effect of soil compaction in the sub-humid cropping environment in Pakistan on uptake of NPK and grain yield in wheat (Triticum aestivum) I. Compaction, *Field Crops Research* 110(1), 54–60.

Akinsanmi, O. A., Mitter, V., Simpfendorfer, S., Backhouse, D. and Chakraborty, S. 2004. Identity and pathogenicity of Fusarium spp. isolated from wheat fields in Queensland and northern New South Wales, *Australian Journal of Agricultural Research* 55(1), 97–107.

Angus, J. F., Kirkegaard, J. A., Hunt, J. R., Ryan, M. H., Ohlander, L. and Peoples, M. B. 2015. Break crops and rotations for wheat, *Crop and Pasture Science* 66(6), 523–552.

Antille, D., Chamen, T., Tullberg, J. and Rattan, L. 2015. The potential of controlled traffic farming to mitigate greenhouse gas emissions and enhance carbon sequestration in arable land: a critical review, *Transactions of the ASABE*. American Society of Agricultural and Biological Engineers 58, 707–731.

Ashworth, J., Desbiolles, J. and Tola, E. 2010. *Disc Seeding in Zero-Till Farming Systems: A Review of Technology and Paddock Issues*. Western Australia No-till Farmers Association. Perth.

Baker, C. J. and Ross, C. W. 2017. How No-Tillage can mitigate climate change, *Ravensdown Ground Effect*, 12–13.

Batey, T. 2009. Soil compaction and soil management–a review, *Soil Use and Management* 25(4), 335–345.

Bell, L. W. and Moore, A. D. 2012. Integrated crop-livestock systems in Australian agriculture: trends, drivers and implications, *Agricultural Systems*. Elsevier B.V 111, 1–12.

Borlaug, N. E. 1994. Feeding a human population that increasingly crowds a fragile planet. *Supplement to Transactions of the 15th World Congress of Soil Science*. Acapulco, Mexico.

Branson, M. 2011. Using conservation agriculture to improve water use efficiency in wheat crops on the Branson farm in South Australia. *5th World Congress of Conservation Agriculture Incorporating 3rd Farming Systems Design Conference–Resilient Food Systems for a Changing world Brisbane*, p. 547.

Broster, J. C. and Pratley, J. E. 2006. A decade of monitoring herbicide resistance in Lolium rigidum in Australia, *Australian Journal of Experimental Agriculture* 46(9), 1151–1160.

Butler, G. (Eds.) 2008. *Conservation Agriculture: Moving Beyond Adoption*. SANTFA Online Article, Clare, SA.

Chamen, T., Alakukku, L., Pires, S., Sommer, C., Spoor, G., Tijink, F. and Weisskopf, P. 2003. Prevention strategies for field traffic-induced subsoil compaction: a review Part 2. Equipment and field practices, *Soil and Tillage Research* 73(1–2), 161–174.

Chamen, W. C. T., Vermeulen, G. D., Campbell, D. J. and Sommer, C. 1992. Reduction of traffic-induced soil compaction–a synthesis, *Soil and Tillage Research* 24(4), 303–318.

Chan, K. Y., Conyers, M. K., Li, G. D., Helyar, K. R., Poile, G., Oates, A. and Barchia, I. M. 2011. Soil carbon dynamics under different cropping and pasture management in temperate Australia: results of three long-term experiments, *Soil Research* 49(4), 320–328.

Chan, K. Y., Gowie, A., Kelly, G., Singh, B. and Slavich, P. 2008. Scoping paper: soil organic carbon sequestration potential for agriculture in NSW, *NSW DPI Research & Science*. NSW Department of Primary Industries, Sydney.

Cook, S. and Bramley, R. 2000. Precision agriculture: using paddock information to make cropping systems internationally competitive. Bureau of Rural Sciences Publications. Available at: http://www.daff.gov.au/brs/publications/all_brs_publica tions (accessed on 7 January 11).

Crabtree, B. 2010. *Search for Sustainability in Dryland Agriculture*. Crabtree Agricultural Consulting, Beckenham, WA.

Cross, M. W. 1957. Overdrilling machinery and methods. *Massey Agricultural College Dairy Farmers' Conference*, Massey University, New Zealand, pp. 77–84.

Crozier, C. R., Naderman, G. C., Tucker, M. R. and Sugg, R. E. 2008. Nutrient and pH stratification with conventional and no-till management, *Communications in Soil Science and Plant Analysis* 30(1-2), 65-74.

CSIRO 2009. Identification of areas within Australia with the potential to enhance soil carbon content. Baldock, J., Grundy, M., Wilson, P., Jacquier, D., Griffin, T., Chapman, G., Hall, J., Machmet, D., Crawford, D., Hill, J. and Kidd, J., CSIRO.

Dalal, R. C. and Chan, K. Y. 2001. Soil organic matter in rainfed cropping systems of the Australian cereal belt, *Soil Research* 39(3), 435-464.

Derpsch, R., Franzluebbers, A. J., Duiker, S. W., Reicosky, D. C., Koeller, K., Friedrich, T., Sturny, W. G., Sá, J. C. M. and Weiss, K. 2014. Why do we need to standardise no-tillage Research? Letter to the editor, *Soil and Tillage Research* 137, 16-22.

Erbacher, A., Lawrence, D., Freebairn, D., Huth, N., Anderson, B. and Harris, G. 2019. *Cover crops can boost soil water storage and crop yields*. Grain Research Development Corporation (GRDC) Update Paper. Available at: https://grdc.com.au/resources-and-publications/grdc-update-papers/tab-content/grdc-update-papers/2019/03/cover-crops-can-boost-soil-water-storage-and-crop-yields (1 March 2019).

Fletcher, A. 2019. Benchmarking break-crops with wheat reveals higher risk may limit on farm adoption, *European Journal of Agronomy* 109, 125921.

Fletcher, A. L., Chen, C., Ota, N., Lawes, R. A. and Oliver, Y. M. 2020. Has historic climate change affected the spatial distribution of water-limited wheat yield across Western Australia?, *Climatic Change* 159(3), 347-364.

GRDC, Grain Research Development Corporation 2016. Farm practices survey–report 2016. Available at: https://grdc.com.au/resources-and-publications/all-publications/publications/2018/farm-practices-survey-report-2016.

GRDC, Grain Research Development Corporation 2019. Labour costs and availability are key challenges to modifying farming systems. GroundCover™ article by R. Thyer. Available at: https://groundcover.grdc.com.au/story/6454352/labour-can-be-an-obstacle-to-on-farm-innovation/.

Hamza, M. A. and Anderson, W. K. 2005. Soil compaction in cropping systems–a review of the nature, causes and possible solutions, *Soil and Tillage Research* 82(2), 121-145.

Hochman, Z., Gobbett, D. L. and Horan, H. 2017. Climate trends account for stalled wheat yields in Australia since 1990, *Global Change Biology* 23(5), 2071-2081.

Janssen, V., Haasdyk, J. and McElroy, S. 2011. CORSnet-NSW Network RTK: Same look and Feel ... only Better. *Paper presented to Association of Public Authority Surveyors 16th Annual Conference*, Bathurst, NSW.

Jones, R. J. A., Spoor, G. and Thomasson, A. J. 2003. Vulnerability of subsoils in Europe to compaction: a preliminary analysis, *Soil and Tillage Research* 73(1-2), 131-143.

Kirchhof, G. and Daniels, I. 2009. Changing tillage management practices and their impact on soil structural properties in north-western New South Wales, Australia, *Aciar Technical Reports Series* 71, 60-69.

Kirkby, C. A.,Richardson, A. E., Wade, L. J., Passioura, J. B., Batten, G. D., Blanchard, C. and Kirkegaard, J. A. 2014. Nutrient availability limits carbon sequestration in arable soils, *Soil Biology and Biochemistry*. Elsevier 68, 402-409.

Li, G. D., Conyers, M. K., Heylar, K. R., Lisle, C. J., Poile, G. J. and Cullis, B. R. 2019. Long-term surface application of lime ameliorates subsurface soil acidity in the mixed farming zone of south-eastern Australia, *Geoderma*. Elsevier 338, 236-246.

Li, Y. X., Tullberg, J. N. and Freebairn, D. M. 2007. Wheel traffic and tillage effects on runoff and crop yield, *Soil and Tillage Research* 97(2), 282–292.

Llewellyn, R. and Ouzman, J. 2019. Conservation agriculture in Australia: 30 years on. In: Pratley, J. and Kirkegaard, J. (Eds). *Australian Agriculture In 2020: From Conservation to Automation*. Australian Society of Agronomy (vol. 2019), pp. 21–31.

Luo, Z. K., Wang, E. L. and Sun, O. J. 2010. Soil carbon change and its responses to agricultural practices in Australian agro-ecosystems: a review and synthesis, *Geoderma* 155(3–4), 211–223.

Ma, Q., Rengel, Z. and Rose, T. 2009. The effectiveness of deep placement of fertilisers is determined by crop species and edaphic conditions in mediterranean-type environments: a review, *Soil Research* 47(1), 19–32.

Mason, S., Armstrong, R., Willhelm, N., Lester, D. and McLaughlin, D. 2016. Understanding nutrient stratification to guide crop management. *GroundCover™ Supplement Issue 04 July 2016*. Grains Research Development Corporation (GRDC), Canberra, Australia.

McBeath, T. M., Gupta, V. V. S. R., Llewellyn, R. S., Davoren, C. W. and Whitbread, A. M. 2015. Break-crop effects on wheat production across soils and seasons in a semi-arid environment, *Crop and Pasture Science* 66(6), 566–579.

McKenzie, B. M., Kuhner, S., MacKenzie, K., Peth, S. and Horn, R. 2009. Soil compaction by uniaxial loading and the survival of the earthworm Aporrectodea caliginosa, *Soil and Tillage Research* 104(2), 320–323.

Miller, L. 2019. Lime and liming–managing soil health. Grains Research Development Corporation. Available at: https://grdc.com.au/resources-and-publications/grdc-update-papers/tab-content/grdc-update-papers/2019/02/lime-and-liming-managing-soil-health.

Orton, T. G., Mallawaarachchi, T., Pringle, M. J., Menzies, N. W., Dalal, R. C., Kopittke, P. M., Searle, R., Hochman, Z. and Dang, Y. P. 2018. Quantifying the economic impact of soil constraints on Australian agriculture: a case-study of wheat, *Land Degradation and Development*. John Wyley & Sons Inc 29(11), 3866–3875.

Renwick, A., Ball, A. S. and Pretty, J. 2003. Economic, biological and policy constraints on the adoption of carbon farming in temperate regions. In: Swingland, I. R. (Ed) *Capturing Carbon and Conserving Biodiversity*. Earthscan Publication.

Roberts, M. 2008. Multiple Benefits from inter-row sowing. In: Butler, G. (Ed), *Conservation Agriculture–Moving Beyond Adoption*. South Australian No-Till Farmers Association, Clare, SA, pp. 34–36.

Roberts, M. and Leonard, E. 2008. Inter-row seeding part of a systems package. In: Butler, G. (Ed), *Conservation Agriculture–Moving Beyond Adoption*. South Australian No-Till Farmers Association, Clare, SA, pp. 37–38.

Rochecouste, J. 2009. Integrating proximal and remote sensor technologies to improve production efficiency in a low emission cropping system. *Paper Presented to 7th National CTF Conference Australia–Hi-Tech Low Emissions Cropping Systems*, Canberra, 7–8 September 2009.

Rochecouste, J. and Crabtree, W. L. 2014. Conservation agriculture in Australian dryland cropping. In: Jat, R. A., Sahrawat, K. L. and Kassam, A. H. (Eds) *Conservation Agriculture: Global Prospects and Challenges*. CABI Publishing, Wallingford, Oxfordshire, UK, pp. 108–126.

Sadras, V. O. and Angus, J. F. 2006. Benchmarking water-use efficiency of rainfed wheat in dry environments, *Australian Journal of Agricultural Research*. CSIRO Publishing, Australia 57(8), 847–856.

Sadras, V. and McDonald, G. 2012. *Water Use Efficiency of Grain Crops in Australia: Principles, Benchmarks and Management*. Grains Research Development Corporation, Canberra, ACT.

Sanderman, J. and Baldock, J. A. 2010. Accounting for soil carbon sequestration in national inventories: a soil scientist's perspective, *Environmental Research Letters* 5(3).

Sanderman, J., Farquharson, R. and Baldock, J. 2010. *Soil Carbon Sequestration Potential: A Review for Australian Agriculture*. CSIRO Land and Water, Canberra.

Scott, B. J., Eberbach, P. L., Evans, J. and Wade, L. J. 2010. Stubble retention in cropping systems in Southern Australia: Benefits and challenges, EH graham centre monograph 1, Industry & Investment, NSW. Orange.

Seymour, M., Kirkegaard, J. A., Peoples, M. B., White, P. F. and French, R. J. 2012. Break-crop benefits to wheat in Western Australia - insights from over three decades of research, *Crop and Pasture Science* 63(1), 1-16.

Shoup, D. W., Lee, W. and Harrison, T. 2004. Precision technologies for precision management. In: Peart, R. M. and Shoup, D. W. (Eds), *Agricultural Systems Management–Optimizing Efficiency and Performance*. Marcel Dekker Inc., New York.

Silburn, D. M., Freebairn, D. M. and Rattray, D. J. 2007. Tillage and the environment in sub-tropical Australia–tradeoffs and challenges, *Soil and Tillage Research* 97(2), 306-317.

Swift, R. S. 2001. Sequestration of carbon by soil, *Soil Science* 166(11), 858-871.

Thomas, G. A., Dalal, R. C., Weston, E. J., King, A. J., Holmes, C. J., Orange, D. N. and Lehane, K. J. 2011. Crop rotations for sustainable grain production on a vertisol in the semi-arid subtropics, *Journal of Sustainable Agriculture* 35(1), 2-26.

Thomas, G. A., Titmarsh, G. W., Freebairn, D. M. and Radford, B. J. 2007. No-tillage and conservation farming practices in grain growing areas of Queensland–a review of 40 years of development, *Australian Journal of Experimental Agriculture* 47(8), 887-898.

Throop, H. L., Archer, S. R., Monger, H. C. and Waltman, S. 2012. When bulk density methods matter: implications for estimating soil organic carbon pools in rocky soils, *Journal of Arid Environments* 77, 66-71.

Tullberg, J. N., Yule, D. F. and McGarry, D. 2007. Controlled traffic farming–from research to adoption in Australia, *Soil and Tillage Research* 97(2), 272-281.

Unkovich, M. J. 2014. A review of the potential constraints to crop production on sandy soils in low rainfall South-Eastern Australia and priorities for research. A Technical Report for the Grains Research and Development Corporation. Mallee Sustainable Farming, Mildura NSW.

Vu, D. T., Tang, C. and Armstrong, R. D. 2009. Tillage system affects phosphorus form and depth distribution in three contrasting Victorian soils, *Soil Research*. CSIRO Publishing, Australia 47(1), 33-45.

Wang, E. L., Zhongkui, L. and Smith, C. J. 2010. Potential change of soil carbon in Australia agro-ecosystems as affected by conservation management: data synthesis and modelling. *Paper presented to 19th World Congress of Soil Science–Soil Solutions for a Changing World*, Brisbane.

WA Waste Authority 2010. Recycled organics and water use efficiency–information paper. WAWA. The Government of Western Australia.

Whitlock, A. 2006. Precision farming–supporting innovation. In: Horn, R., Fleige, H., Peth, S. and Peng, X. H. (Eds), *Soil Management for Sustainability* (vol. 38), pp. 242-250.

Wylie, P. and Moll, J. 1998. *Opportunity Cropping* (2nd edn.). Conservation Farmers Incorporated, Toowoomba, Qld.

Yash, P. D., Balzer, A., Crawford, M., Rincon-Florez, V., Liu, H., Melland, A. R., Antille, D., Kodur, S., Bell, M. J., Whish, J. P. M., Lai, Y., Seymour, N., Carvalhais, L. C. and Schenk, P. 2018. Strategic tillage in conservation agricultural systems of north-eastern Australia: why, where, when and how?, *Environmental Science and Pollution Research International*. Springer-Verlag, Berlin Heidelberg 25(2), 1000–1015.

Index

AAA. *see* Adaptation of African Agriculture to climate change (AAA)
AAAID. *see* Arab Authority for the Agricultural Investment and Development (AAAID)
AAFC. *see* Agriculture and Agri-Food Canada (AAFC)
Above-ground dry matter (AGDM) 355
AC. *see* Annual crop (AC) cultivation
ACIAR. *see* Australian Centre for International Agricultural Research (ACIAR)
ACLIMAS. *see* Adaptation to climate change of the Mediterranean agricultural system (ACLIMAS)
ACSAD. *see* Arab Centre for the Studies of Arid Zones and Dry Lands (ACSAD)
ACT. *see* African Conservation Tillage Network (ACT)
ACTED. *see* Agency for Technical Cooperation and Development (ACTED)
Adaptation of African Agriculture to climate change (AAA) 217
Adaptation to climate change of the Mediterranean agricultural system (ACLIMAS) 351
Adapting seeding machinery 606
Ad-Hoc Working Group 584
Adoption and spread, North Africa
  adoption barriers 209-214
  adoption drivers and opportunities 214-217
  agriculture and food security 190-193
  history and development
    Algeria 193-194
    Egypt 194-195
    Libya 195
    Morocco 195-197
    Sudan 197
    Tunisia 197-199
  distributional effects 217-223
  networks, partner institutes 217-223
  overview of 185-187
  research achievements 199-201
    crop yields and diversification 201-202
    ecological benefits 203-205
    input, costs and profits 203
    insect-pest and disease dynamics and management 208
    livestock integration 207-208
    off-farm benefits 209
    soil functionality 203-205
    weed infestation and management 205-207
Adoption of no-till over time 602
African Conservation Tillage Network (ACT) 217
Aga Khan Foundation (AKF) 364
AGDM. *see* Above-ground dry matter (AGDM)
Agency for Technical Cooperation and Development (ACTED) 476
Agrarian Policy 2008 416
Agricultural Extension Programme 82
Agricultural landscape 441
Agricultural Mechanisation Policies and Strategies for the Caribbean Community (CARICOM) 101
Agricultural Mechanization Centre (CEMA) 132
Agricultural policy action plan (APAP) 305
Agricultural producers 447
Agricultural Research and Education Center (AREC) 340
Agricultural Research Council (ARC) 317
Agricultural Soil Health and Conservation Working Group 34

Agricultural work 89
Agriculture and Agri-Food Canada (AAFC) 25
Agriculture production systems 316
Agri-environmental European context
  benefits, CA
    and climate action 160
    and farm economy 160
    and natural and public infrastructure protection 160-161
    and water conservation 159-160
  climate action 173-177
  common agricultural policy (CAP) 173
  cropping systems, diversification 152
  Green Deal 173-177
  history and current situation 161-173
  minimum mechanical soil disturbance 150-152
  soil threats 152-153
    soil biodiversity loss 156-158
    soil compaction 158-159
    soil erosion 153-154
    soil organic matter decline 154-156
  vegetative mulch cover permanent maintenance 152
AGRIEX 132
Agroclimatic zones 399
Agro Dairy LLC 458
Agroecology in South East Asia (ASEA) 569
Agroecology Learning Alliance in South East Asia (AliSEA) 569-570
'Agroelit' farm 438
Agrofirma 'Dodola' 450
Agro-industrial complex 440, 464
Agro Information Center 457
'Agro-Popuros' farm 438
AKF. *see* Aga Khan Foundation (AKF)
'Ak-Zher 2010' farm 466
Alberta Conservation Tillage Society 23
Alberta greenhouse gas offset system 38
Alberta Reduced Tillage Linkages 43
Alberta Special Areas Boards 30
Algeria 190
AliSEA. *see* Agroecology Learning Alliance in South East Asia (AliSEA)
Alley cropping 421
Alluvial soils 52
Always-Green Agriculture (ASV) 128, 130
American Friends of Tel Aviv University 315
American University of Beirut (AUB) 340
AMF. *see* Arbuscular mycorrhizal fungi (AMF)
Annual crop (AC) 150, 151
  cultivation 121
Anthropologists 314
Anti-Lebanon Mountains 339
APAP. *see* Agricultural policy action plan (APAP)
AQUASTAT database 314
Arab Authority for the Agricultural Investment and Development (AAAID) 197, 198, 348
Arab Centre for the Studies of Arid Zones and Dry Lands (ACSAD) 338
Arable land
  evolution percentage 155, 156
  soil loss 154, 155
Aral Sea crisis 396
Arbuscular mycorrhizal fungi (AMF) 205
ARC. *see* Agricultural Research Council (ARC)
AREC. *see* Agricultural Research and Education Center (AREC)
Argentinian No-Tillage Farmers Association (AAPRESID) 127
Armenian Technology Group Foundation (ATGF) 454
ASEA. *see* Agroecology in South East Asia (ASEA)
Asociación Uruguaya Pro Siembra Directa (AUSID) 138
Associação de Plantio Direto no Cerrado (APDC) 118
Association for Oil Crops and Wheat (ANAPO) 140
Association of Agricultural and Forestry Technicians (ACTAF) 101
ASV. *see* Always-Green Agriculture (ASV)
ATGF. *see* Armenian Technology Group Foundation (ATGF)
AUB. *see* American University of Beirut (AUB)
AUSID. *see* Asociación Uruguaya Pro Siembra Directa (AUSID)
Australian Centre for International Agricultural Research (ACIAR) 359
Australian dryland cropping
  carbon sequestration, no-till 614-615
  compaction, controlled traffic farming and precision agriculture
    digital technology 612-613
    emerging factor 610
    equipment sensing crop 611-612
    GPS 611
    heavy farm machinery 609-610
    permanent traffic lanes 610
    precision agriculture techniques 611
    site-specific management 611

cropping diversity and cover cropping
break crop 608
crop choice 607-608
intention determining 609
legume use 608
reducing pest and disease 607
soil biology 608
soil cover 609
negative consequences of, no-till systems 602
herbicide resistance 603-604
nutrient stratification 604
subsurface acidification 604-605
vs. New Zealand CA 615-619
recycled organics 613-614
stubble retention challenges 605
emergence of, stubble borne diseases 605-606
stubble management planting 606-607
Australian grains industry 599
dryland conditions 600
dryland cropping (*see* Australian dryland cropping)
farming practices 600
full-cut tillage consequence 600
history of, CA adoption 601-602
tillage field trial results 600-601
Autopoiesis 312
Azertohum LLC 458

Barley and vetch farmer demonstrations 349
Bed planting 457
Bekaa valley 339
'Big Five' 52, 53, 71
'Big Hammer' 60
BISA. *see* Borlaug Institute for South Asia (BISA)
'Black Blizzards' 56
Borlaug Institute for South Asia (BISA) 380
Boukan Kaveh Sazeh-Kesht Company 380-381
BRICS. *see* Building Resilience and Adaptation to Climate Extremes and Disasters (BRICS)
Brundtland Report 150
Building Resilience and Adaptation to Climate Extremes and Disasters (BRICS) 197
Bulk density, two FFS plots 561

CAAAP. *see* Conservation Agriculture Alliance for Asia-Pacific (CAAAP)
CA adoption 518-519
CAADP. *see* Comprehensive Africa Agriculture Development Programme (CAADP)
CACILM-2 project 466, 480
CA-CoEs. *see* Conservation Agriculture Centres of Excellence (CA-CoEs)
CANA. *see* Conservation Agriculture in North Africa (CANA) project
Canadian Agricultural Partnership (CAP) 40
Canadian agriculture policy 31
Canadian Society of Soil Science (CSSS) 25
Canadian Wheat Board (CWB) 16
CAP. *see* Canadian Agricultural Partnership (CAP); Common Agricultural Policy (CAP)
Capacity building and awareness raising 547
Capacity development 446
CA practices 358
CA Production Systems (CAPS) treatments 550-551
CA projects implemented under CA FIP, salient features of 322-326
CA promotional programme 345
Carbon exchange capacity (CEC) 405
Carbon sequestration 68, 512
CA Research Proposals, evaluation criteria for 320
Caribbean 100-103
CA-SARD project 289
CASIC 548
Caspian Hyrcanian mixed forests 375
CA stakeholders 420
CA technology, adoption of 385
CA vegetable production 541
CA/ZT field trials 348
CEC. *see* Carbon exchange capacity (CEC)
Central Agricultural Research Library 445
Central America
Costa Rica 81-83
El Salvador 83-88
Guatemala 88-91
Honduras 91-95
Nicaragua 95-99
Panama 99-100
Central and West Asia and North Africa (CWANA) 376
Central Asia, adoption of
challenges and opportunities 419-420
climate and soils 397-399
crop production 397
land degradation 399-402

research and conservation agriculture systems
adoption status of 412-417
alley cropping 404-406
crop diversification, prospects for 417-419
crop productivity 406-408
crop residue 408-409
crop rotation 402-403
double cropping 403-404
economics of CA 410-412
salinity management 409-410
Centre for Agricultural Mechanization (CEMA) 132
Centre for No-Till Agriculture (CNTA) 256-258
Cereals
and leguminous crops 453
mono-cropping of 470
CGIAR. *see* Consultative Group for International Agricultural Research (CGIAR)
Channeling Conservation Agriculture funding 320
Chernozems (black soils) 448
China 581
demonstration, extension and priority study stage (2002-2008) 582
preliminary exploration stage (1960-1990) 581
rapid development stage (2009-present) 582
systematic research stage (1991-2001) 581-582
Chuy Valley 471
CIMMYT. *see* International Maize and Wheat Improvement Center (CIMMYT)
CIS. *see* Commonwealth of Independent States (CIS)
Civil Society Organizations (CSO) 196, 215, 222
Civil unrest 363
CLCA. *see* Conservation Agriculture in Crop Livestock Systems (CLCA)
'Climate, Soil Health and Agriculture Technologies' 444
Climate-Smart Agriculture (CSA) 186
CMTs. *see* Collaboratively managed trials (CMTs)
CNTA. *see* Centre for No-Till Agriculture (CNTA)
Cognitive biases 22
Cold hardening 408
Collaborative extension 31-32
Collaboratively managed trials (CMTs) 321
Commercial smallholders 306
Common Agricultural Policy (CAP) 173
Commonwealth of Independent States (CIS) 418
Communicative action 312-313
Comprehensive Africa Agriculture Development Programme (CAADP) 247, 272, 282
Consensus 68
Conservation Agriculture Alliance for Asia-Pacific (CAAAP) 568-569
Conservation Agriculture Centres of Excellence (CA-CoEs) 291
Conservation Agriculture in Crop Livestock Systems (CLCA) 198, 207, 218
Conservation Agriculture in North Africa (CANA) project 218
Conservation Agriculture in South Africa, adoption of
appropriate theories supporting innovation 310-314
challenges in 308-310
farmers as intuitive innovators 314-316
innovators, engaging with 318-326
spontaneous adoption of 316-318
status and spread 306-308
Conservation Agriculture in West Asia
Iran 375
agriculture in 376-377
CA, importance of 378-379
CA, promotion of 379-381
CA development, technology for 384-385
conclusion 387-388
Golestan province, CA in 381-383
promotion of CA, main challenges for 385-387
Iraq
conclusion 368-370
economical and environmental development 368
farming systems 365-366
human and knowledge development 367-368
rain-fed areas, risks facing agriculture in 366-367
Jordan 346-348
conclusion 355-358
crop performance 350-352
methodologies and data collection 349-350

methodology 348-349
results 352-355
Lebanon 338
methodology 338-341
results and discussion 342-345
Syria
CA in different contexts, suitability of 358-359
conclusion 364-365
low-cost zero tillage seeders, availability of 360-361
no-till seeder manufacturing 364
research and development 364
scaling CA adoption, capacity development for 362-363
Turkey 370
CA, implementation of 372-374
conclusion 374-375
possible solutions, the search for 371-372
Conservation Agriculture Policy 305
Conservation Agriculture project interventions 321-326
Conservation of Agricultural Resources Act 305
Conservation tillage 2
Consultative Group for International Agricultural Research (CGIAR) 468
Continuously Operating Reference Stations (CORS) 612
Controlled traffic farming (CTF) 159, 610-611
Conventional tillage (CT) 62, 512
Conventional tillage (CT)-based agriculture 274
CORS. *see* Continuously Operating Reference Stations (CORS)
Cotton exports 397
Country Environmental Assessment (2008) 475
Cover crop mixtures identification 541
Cover crops 65
Cover crops in main crops, sowing 546
CPR. *see* Crop-pasture rotation (CPR)
CRI. *see* Crops Research Institute (CRI)
Crop diversification 262, 402, 537
Crop diversity 202
Crop establishment 369
Crop-livestock systems 261
Crop-pasture rotation (CPR) 138
Crop production 337
Crop residues 378, 409
Crop rotations measurement 603
Crops Research Institute (CRI) 252
CRS Soil and Water for Agriculture (ASA) programme 86
CSA. *see* Climate-Smart Agriculture (CSA)
CSO. *see* Civil Society Organizations (CSO)
CSSS. *see* Canadian Society of Soil Science (CSSS)
CT. *see* Conventional tillage (CT)
CTF. *see* Controlled traffic farming (CTF)
Cultivated agriculture 16
CWANA. *see* Central and West Asia and North Africa (CWANA)
CWB. *see* Canadian Wheat Board (CWB)

DAFF. *see* Department of Agriculture, Forestry and Fisheries (DAFF)
Data collection, Lebanon 341
Dedoplistskaro municipality 460
Democratic People's Republic, Korea 585-586
Department of Agriculture, Forestry and Fisheries (DAFF) 305
Deputy of Agronomy 379
Deutsche Gesellschaft für Technische Zusammenarbeit (GTZ) 116
Digital technology 612-613
Dirty Thirties 16, 19
Disc opener straw-cutting type 590
Don Zonal Institute 441
'Dostlik' 410
Dust Bowl 1, 56, 58
Dust storms 186, 441

Earthworm count 343
East Asia
adoption scale 591-592
cropping system practices 580
effects of 580
history of 581
China 581-582
Democratic People's Republic, Korea 585-586
Japan 584-585
Mongolia 583-584
Republic of Korea 586
policies and documents promotion 587-588
promotion actions 589
capacity building 591
machinery and equipment leasing services 591

seeder 590
technical models 589-590
promotion advice 592-593
promotion policy 586-589
Eastern and Southern Africa (ESA) 273, 274
adoption trends 279-282
adoption by farmer typology 282
current adoption trends by value chains 283-284
lower-than-desired adoption, reasons contributing to 284-285
Agenda 2063 and Malabo Declaration
adaptations in ESA 279
crop diversification 279
evolution 277
Faidherbia albida 275-277
indigenous practises and soil conservation structures 275
minimum mechanical soil disturbance 277-278
soil cover with crop residues 278-279
delayed potential to contribute to Malabo 282-283
future prospects 297-298
livelihoods and climate change 274-275
overview of 271-274
in Tanzania
drivers and value propositions 290
innovation policy learning curve 289-290
institutional reform and learning curve 285-289
key success factors 290
policy environment support 290-291
in Zimbabwe
improved timeliness of operations 294-297
production cost reduction 294
public and civil society promotion 291-293
ECAF. *see* European Conservation Agriculture Federation (ECAF)
Economic modelling 411
Economics impact 344-345
EFP. *see* Environmental Farm Plan (EFP)
EMBRAPA 119, 124
ENPARD. *see* European Neighbourhood Programme for Agriculture and Rural Development (ENPARD)
Environmental and economic benefits, Turkey 375
Environmental Farm Plan (EFP) 33
ESA. *see* Eastern and Southern Africa (ESA)
Essex Soil Report 17
Establish long-term and high-standard, monitoring and demonstration sites 593
Eurasia
in Caucasus
Armenia 453-456
Azerbaijan 456-458
Georgia 459-461
in Central Asia 461-464
Kazakhstan 464-469
Kyrgyzstan 469-474
Tajikistan 474-479
Turkmenistan 479-480
Uzbekistan 480-484
in Eastern Europe 433
Belarus 434-436
Moldova 437-439
Russia 439-447
Ukraine 447-452
Eurasia Group 471
European Conservation Agriculture Federation (ECAF) 150, 161, 433
European Green Deal 176
European Neighbourhood Programme for Agriculture and Rural Development (ENPARD) 454
European Union Neighbourhood Program 461
EU Soil Forum 173
Excessive tillage 372
Experimental design and field management, Lebanon 341
Exploring business models 546
Exploring financial mechanisms 547
Extension approache 547
Extension services 26
Externalization 316

Fankhauser direct seeder 340
FAO. *see* Food and Agriculture Organization (FAO)
FAO-SEC. *see* FAO Sub-Regional Office for Central Asia (FAO-SEC)
FAO Sub-Regional Office for Central Asia (FAO-SEC) 463
Farmall tractor 59
Farm Bill 1985 57, 58
Farmer-driven processes 306
Farmer Field Schools (FFS) 313, 388, 472, 557

Farmer Innovation Programme (FIP) 308, 318
Farmer-managed (FM) trials 321
Farmer participatory research (FPR) 311
Farmers transitioning, perennial plantations 546
Farmer to Farmer project 583
Farming practices 339
Farming structure 453
Farming system with supplementary irrigation 366
Farmland Health Check-up 33
Farm Planning service of the Ontario Agricultural College 26
Farm to Fork strategy 176
Favoring farmers access
- agricultural equipment/services 541-542
- bio-product development and biological control 544-545
- diversified and open-pollinated cover crop 542, 544
- mechanization and services 544
- policy dialogue and theory change 543
- social capital and collective learning, adaptation and adoption 542-543

Federación Uruguaya de Grupos CREA (FUCREA) 139
Federal University of Santa Maria (UFSM) 134
FERT 213, 222
FFS. *see* Farmer Field Schools (FFS)
FFS demonstration plots 561
Field grading and levelling 412
FIP. *see* Farmer Innovation Programme (FIP)
Flash flood irrigation 470
Fodder shrubs 404
Food and Agriculture Organization (FAO) 144, 186, 338, 396
Food legumes 404
Forage crops 404
Forage systems 70
FPR. *see* Farmer participatory research (FPR)
Frost resistance 407
FUCREA. *see* Federación Uruguaya de Grupos CREA (FUCREA)
Full-cut tillage 600
'The Furrow and Us' 58
Furrow irrigation 407
Fusarium pseudograminearum 208

GAPs. *see* Good agricultural practices (GAPs)
Garabagh Grain LLC 458
GASIP. *see* Ghana Agricultural Sector Improvement Programme (GASIP)
GCF. *see* Green Climate Fund (GCF)
GDP. *see* Gross domestic product (GDP)
GEF. *see* Global Environment Facility (GEF)
Gender 547
Genetically Modified Organism (GMO) crops 60
Georg Birbaumer and Miguel Angel Florentín (MAG-GTZ) 132, 133
German Corporation for International Cooperation (GIZ) 340, 460
GFSI. *see* Global Food Security Index (GFSI)
Ghana Agricultural Sector Improvement Programme (GASIP) 251, 256
'Girlea Andrei Pavel' farm 438
GIZ. *see* German Corporation for International Cooperation (GIZ)
GLASI. *see* Great Lakes Agricultural Stewardship Initiative (GLASI)
Global Environment Facility (GEF) 437
Global Food Security Index (GFSI) 193
Global Navigation Satellite System (GNSS) 612
Global Soil Biodiversity Atlas 158
Global warming 395
Global warming potential (GWP) 515
Glyphosate 450
Glyphosate application rate 171, 172
Glyphosate-resistant crops 61
GMO. *see* Genetically Modified Organism (GMO) crops
GNSS. *see* Global Navigation Satellite System (GNSS)
Golestan province 379
Good agricultural practices (GAPs) 127
Grain amaranth 70
Grain self-sufficiency 403
Greater Mekong Sub-region (GMS), Southeast Asia
- conservation history of 536-537
- design and out-scaling strategies tested 537, 541-543
- performance and impact 543
- sub-region adoption 544
- sub-region adoption, future perspectives
  - environment enabling 546-548
  - system development 546
- sub-region adoption constraints 544
  - limited access, cover crop seeds, mechanization and bio-products 544-545

political willingness, common vision and theory of change 545–546
sustaining experiments and networks demonstrations 545
trade-offs, biomass users 545
Great Lakes Agricultural Stewardship Initiative (GLASI) 33
Green Climate Fund (GCF) 472
Green Deal 175
Greenhouse gas (GHG) emissions 160, 173, 304
Green Revolution 492
Green standing vegetation, sowing 541
Gross domestic product (GDP) 188, 190, 347
GWP. *see* Global warming potential (GWP)

Haber-Bosch process 59
Hamadan Machine barzegar (Hamadan city) 384
Hamedan Machine Zeraaat (Hamadan city) 384
Happy Seeder 502
Harvest index (HI) 355
HEL. *see* Highly Erodible Land (HEL)
Herbicide, weed management 517
Herbicide-resistant crops 71
HI. *see* Harvest index (HI)
Highly Erodible Land (HEL) 57
Highveld summer rainfall area 316

ICARDA. *see* International Center for Agricultural Research in Dry Areas (ICARDA)
ICI. *see* Imperial Chemical Industries (ICI)
IFAD. *see* International Fund for Agricultural Development (IFAD)
IFAO. *see* Innovative Farmers Association of Ontario (IFAO)
IJC. *see* International Joint Commission (IJC)
Imperial Chemical Industries (ICI) 117
Indian no-till seeder 483
Indian ZT seed drill 495
Innovation platform (IP) 319
Innovation systems (IS) 318
Innovative Conservation Tillage Farmers 24
Innovative Farmers Association of Ontario (IFAO) 24, 43
Innovative systems (platforms) 469
INRA. *see* National Institute of Agricultural Research (INRA)
Institutional context and sustainable intensification 548
Instituto Agronômico do Paraná (IAPAR[2]) 116
Instituto de Pesquisas e Experimentação Agropecuária Meridional (IPEAME) 116
Instituto Nacional de Investigaciones Agrícolas (INIA) 103
Integrated pest management (IPM) 454
Integrated soil fertility management (ISFM) 92
Integrated weed management (IWM) 207
Inter-American Technical Service for Agricultural Cooperation (STICA) 82
Intergovernmental Panel on Climate Change (IPCC) 186, 274, 347
International Center for Agricultural Research in Dry Areas (ICARDA) 194, 197, 218, 338, 396, 434
International Center for Tropical Agriculture (CIAT) 140
International Conference 464
International Fund for Agricultural Development (IFAD) 198, 416, 437
International Joint Commission (IJC) 31
International Maize and Wheat Improvement Center (CIMMYT) 104, 140, 505–506
International Panel for Climate Change (IPCC) 614
International research centres 467
Introduction of Minimum Tillage Technologies 583
IP. *see* Innovation platform (IP)
IPCC. *see* Intergovernmental Panel on Climate Change (IPCC)
IPEAME. *see* Instituto de Pesquisas e Experimentação Agropecuária Meridional (IPEAME)
IPM. *see* Integrated pest management (IPM)
Iraqi Ministry of Agriculture 367
Irrigated agriculture 481
Irrigated land 474
Irrigation system 448
IS. *see* Innovation systems (IS)
ISFM. *see* Integrated soil fertility management (ISFM)
IWM. *see* Integrated weed management (IWM)

Jab-planter 476
Japan 584–585

Kara Kum desert 479
Kashkadarya Research Institute 483
Kazakhstan challenges 419
Key strategic objectives (KSOs) 318–319
'Khiliucanul' farm 438
Khorezm Rural Advisory Support Service (KRASS) 484
Knife rollers 142
Knowledge co-creation process 319
KRASS. *see* Khorezm Rural Advisory Support Service (KRASS)
Kroun KO LLC 458
KSOs. *see* Key strategic objectives (KSOs)
Kyrgyz National Agricultural University 474
Kyrgyzstan challenges 419

Land degradation 80, 432
Land degradation processes 536
Land desertification 399–400, 432
Land Stewardship chair 27
Land Stewardship Programme 32
LARI. *see* Lebanese Agricultural Research Institution (LARI)
Latin American Conservation Tillage Network (RELACO) 104
LDNT. *see* Low Disturbance No-Tillage (LDNT)
Lead Farmer Approach 519
Lebanese Agricultural Research Institution (LARI) 340
Lebanese Ministry of Agriculture 345
Leguminous cover crops 69
Linear regression analysis, soil organic carbon *vs.* time 553–554
Local seeders 360
Long-term tillage 26, 27
Long-term tillage trials 40
Low-cost no-till seeders 360, 361
Low Disturbance No-Tillage (LDNT) 616–617

Macro fauna diversity 157
MAF. *see* Ministry of Agriculture and Forestry (MAF)
Mahlathini Development Foundation (MDF) 308
Main crops in cover crops, sowing 546
Maize, cassava and lowland rice systems 546
Maize-sorghum system 84
Maize Trust (MT) 318
Malaysia's first National Agriculture Policy 564
ManDak. *see* Manitoba North Dakota Zero Tillage Farmers Association (ManDak)
ManDak Zero Tillage Farmers Association 23
Manitoba North Dakota Zero Tillage Farmers Association (ManDak) 43
MasAgro 105
MATOPIBA 123
Matracas 95
MDF. *see* Mahlathini Development Foundation (MDF)
Mean summer temperatures 397
Mean winter temperatures 397
Mechanical control of, cover crops 537, 541
Mechanization 508
Mental models 315
MEPA. *see* Ministry of Environment Protection and Agriculture (MEPA)
Mexican Ministry of Agriculture, Livestock, Rural Development, Fisheries, and Food (SAGARPA) 104
Mexico 103–108
Minimizing soil disturbance, no-tillage and covering mulch 537
Minimum-tillage (MT) 121
Ministry of Agriculture (MoA) 378
Ministry of Agriculture and Forestry (MAF) 373
Ministry of Agriculture and Livestock (MAG) 132, 133
The Ministry of Environment, Natural Resources and Fisheries (SEMARNAP) 104
Ministry of Environment and Sustainable Development (MADES) 137
Ministry of Environment Protection and Agriculture (MEPA) 459
Mixed-tillage systems (MTS) 121
MoA. *see* Ministry of Agriculture (MoA)
Moldboard plow 59
Mollisols 52
Mongolia 583
  awareness 584
  different approaches (1997 and 2005) 583–584
Mountain Society Development Support Programme (MSDSP) 476
Mount Lebanon 339
MSDSP. *see* Mountain Society Development Support Programme (MSDSP)
MT. *see* Maize Trust (MT); Minimum-tillage (MT)
MTS. *see* Mixed-tillage systems (MTS)

NAMAs. *see* Nationally Appropriate Mitigation Actions (NAMAs)
Naranjal NT Association (ASIDINAR) 133
NARC. *see* National Agricultural Research Center (NARC)
NARS. *see* National Agricultural Research System (NARS)
National Agricultural Research Center (NARC) 348
National Agricultural Research System (NARS) 396, 465
National Agricultural Technology Center in Paraguay (CETAPAR) 131
National agri-environmental reporting system 38
National Center for Agricultural Research and Extension Center (NARC) 352
National Crop and Livestock Assessment 293
National Fund for the Development of Conservation Agriculture (NFDCA) 443
National Institute of Agricultural Research (INRA) 196
National Institute of Agricultural Technology (INTA) 127
National LandCare Program (NLP) 308
Nationally Appropriate Mitigation Actions (NAMAs) 197
Nationally Determined Contributions (NDC) 222, 223
National Movement for Conservation Agriculture (NMCA) 443
National Research Center for Sustainable Production (CENAPROS) 104
National Soil Conservation Programme 31
National Soil Management Conservation and Recovery Program 134
Natural hayfields 469
The Natural Resources Management Centre (NRMC) 507
NDC. *see* Nationally Determined Contributions (NDC)
Network communications 25
New Brunswick Soil and Crop Improvement Association 43
New Zealand
  AL:TE ratio 617
  climate favours 616
  crop yields 616
  disc type openers 618
  inverted-T shaped tine 618
  Low Disturbance No-Tillage (LDNT) 616-617
  no-tillage practice 615, 619
  seed drill technologies 615
NFDCA. *see* National Fund for the Development of Conservation Agriculture (NFDCA)
NGOs. *see* Non-governmental organizations (NGOs)
Nineveh province 365
Nineveh Research Department 368
Nitrogen fertilizer 69
NLP. *see* National LandCare Program (NLP)
NMCA. *see* National Movement for Conservation Agriculture (NMCA)
Non-governmental organizations (NGOs) 196, 215, 286, 290, 291
North-western Free State 309
No-till (NT) 18
No-till adoption 62, 63
No-tillage cropping 261
No-tillage practice 464
No-tillage system (NTS) 69, 127
  Argentina 127-128
    climate-smart agriculture 128-130
    evolution 128
  Bolivia
    challenges and opportunities 142
    history and present situation 139-142
  Brazil
    challenges and future trends 124-127
    history and present status 116-124
  Paraguay
    adoption 135
    adoption impacts 137
    beginning 131-135
    challenges to evolution 137-138
    evolution and current state 130-131, 136
  Uruguay 138-139
No-till drill 455, 476
No-Till Farmer 72
No-till integrated crop-livestock-forest (NT-ICLF) system 123
No-till Lab 451
No-Till on the Plains 72
No-till planted wheat and Field Day 455
No-till planter 258
No-till seeders 364
No-till Supporters 445
No-till systems 436
No-till weed control 104

NRMC. *see* The Natural Resources Management Centre (NRMC)
NTB. *see* Nusa Tenggara Barat (NTB)
NT-ICLF. *see* No-till integrated crop-livestock-forest (NT-ICLF) system
NTS. *see* No-tillage system (NTS)
NTT. *see* Nusa Tenggara Timur (NTT)
Nusa Tenggara Barat (NTB) 562-563
Nusa Tenggara Timur (NTT) 562-563
Nutrient availability 204

OCSCSA. *see* Ontario Chapter, Soil Conservation Society of America (OCSCSA)
Ogallala aquifer 51
OMAF. *see* Ontario Ministry of Agriculture and Food (OMAF)
OMAFRA 34
On-farm irrigation infrastructure 448
Ontario Chapter, Soil Conservation Society of America (OCSCSA) 30
Ontario Farm Environmental Coalition 33
Ontario Ministry of Agriculture, Food and Rural Affairs 33
Ontario Ministry of Agriculture and Food (OMAF) 24, 28, 31, 33
Ontario's Agricultural Soil Health and Conservation Strategy 34
Ontario Soil and Crop Improvement Association (OSCIA) 24, 43
Ontario Soil Conservation and Environmental Protection programme (OSCEPAP) 31
Open straw burning 579
Organic agriculture 440
Organic farming 40
OSCEPAP. *see* Ontario Soil Conservation and Environmental Protection programme (OSCEPAP)
OSCIA. *see* Ontario Soil and Crop Improvement Association (OSCIA)
Ottosdal No-till Club 321

PAMI. *see* Prairie Agriculture Machinery Institute (PAMI)
PAR. *see* Photosynthetically Active Radiation (PAR)
Paraguayan Chamber of Cereal and Oil Exporters (CAPECO) 133
Participatory monitoring and evaluation (PM&E) system 321
Pasture resources 469
Peacock Eyespot (*Piloceae oleaginea*) 364
Perennial crops 150, 151
Permanent bed planting 411
PFRA. *see* Prairie Farm Rehabilitation Administration (PFRA)
The Philippines, Southeast Asia
  advances in 550
    crop yield, effect of 554
    nutrient transport, effect of 554
    residual soil moisture, effect of 551-552
    soil adsorption capacity, effect of 552
    soil organic carbon, effect of 550-551
  state adoption 548-550
Photosynthetically Active Radiation (PAR) 403
Pioneer CA farmers 317
Pioneer researchers 371
Plant disease/tillage relationships 26
Planting basins 279
Planting success 20
Plowman's Folly 58
PLUARG. *see* Pollution from Land Use Activities Reference Group (PLUARG)
Polesye and drip-irrigation 449
Policy barriers 519
Pollution from Land Use Activities Reference Group (PLUARG) 30
Poor irrigation management practices 401
Powered anti-blocking type 590
Prairie Agriculture Machinery Institute (PAMI) 26, 44
Prairie Farm Rehabilitation Administration (PFRA) 16, 30, 38
Presidential Decree 416
Primary tillage 59
'PROmoting Local INNOVAtion' (PROLINNOVA) 314

QSMAS. *see* Quesungual Slash and Mulch Agroforestry System (QSMAS)
Quebec farmer group 43
Quesungual Slash and Mulch Agroforestry System (QSMAS) 91, 93

Rain-fed farming systems 365, 509
Random sampling 343
RCTs. *see* Resource conservation technologies (RCTs)
Real Time Kinematic (RTK) 611
REC Caucasus. *see* Regional Environmental Center for the Caucasus (REC Caucasus)

Regional actors 547-548
Regional Agricultural Research Centre (CRIA) 132
Regional Environmental Center for the Caucasus (REC Caucasus) 459
Republic of Korea 586
Research and development (R&D) approaches 310
Researcher-managed (RM) trials 317
Resource conservation technologies (RCTs) 494, 503
Resource-saving technology 406
Rice-wheat 404
Rip-on-row (RoR) system 309
Ripper 257
Risk aversion 22
Risky farming 435
Roller crimper 257
RTK. *see* Real Time Kinematic (RTK)

SA. *see* South Asia (SA)
SAGARPA. *see* Secretariat of Agriculture and Rural Development (SAGARPA)
Salt swamp 376
Samarkand Institute 483-484
Sandwich crop 508
Sandy Soils Development Committee research project 317
SANREM research project 548-549
'Sarob,' 477
Sasakawa Global 2000 programme 289
Saskatchewan Soil Conservation Association (SSCA) 23, 43
Save our Soil Conference 1980, 24
Sazehkesht Kaveh (Bukan city) 384
Science-based adoption and adaptation plans 593
Scientific knowledge 310
SDGs. *see* Sustainable development goals (SDGs)
Secretariat of Agriculture and Rural Development (SAGARPA) 81
Seedbed preparation 483
Seed inputs 370
Self-production. *see* Autopoiesis
Sensors information, farmers 612
Shallow-planted wheat 406
'Shirkat' farms 481
Site-specific management 611
Site-specific nutrient management (SSNM) 515-516
Small community-based family farms 413
SmartSoil FP-7 project 156
SOC. *see* Soil organic carbon (SOC)
Social learning 313
'Soil at Risk' report 30, 31
Soil conservation 442
Soil conservation assistance 57
Soil Conservation Council of Canada 43
Soil cultivation system 434
Soil degradation 401, 409, 457
Soil desertification, wind erosion 578
Soil erosion 17, 79, 186, 188, 369, 373, 475
Soil Erosion Service 57
Soil Framework Directive 173
Soil moisture 341, 342
Soil mulch 410
Soil organic carbon (SOC) 156, 204, 304, 511-512, 614
Soil organic carbon *vs.* time 551-553
Soil organic matter (SOM) 371
Soil organic matter levels 17
Soil pollution 401
Soil protection technologies 462
Soil-protective system 434
Soil quality 342-343
Soil science 433
Soil tillage effect 371
Soil Today Food Tomorrow conference 1983 24
Soil & Water Assessment Tool (SWAT) model 204
Soil water retention 369
SOM. *see* Soil organic matter (SOM)
South African Department of Environmental Affairs 304
South African National Development Plan 305
South America
  future prospects 142-144
  no-tillage system
    Argentina 127-130
    Bolivia 139-142
    Brazil 116-127
    Paraguay 130-138
    Uruguay 138-139
  overview 113-116
South Asia (SA)
  agriculture practices 492
  climate change mitigation and adaptation 514-515
  crop productivity and income 510-511
  historical milestones 496-501
  history and status 493-494

Afghanistan 508-509
Bangladesh 504-505
India 495, 502
Nepal 505-506
Pakistan 502-504
rain-fed areas, emerging avenue 509
rice-wheat consortium (RWC) 494-495
Sri Lanka 506-508
multi-faceted problems, agriculture 492
nutrient management 515-516
population 491-492
soil health 511-513
sustainability 493
upscaling 518-519
water productivity 513-514
weed management 516-518
Southeast Asia
indicators for 535
Indonesia 562-564
Malaysia 564
farmers practices adoptation 564-567
The Philippines
advances of 550-554
state adoption, description of 548-550
upscaling challenges 554-556
region 534
regional networking 568
agroecology platform, partnership and learning alliance 569-570
Conservation Agriculture Alliance for Asia-Pacific (CAAAP) 568-569
synopsis 570-571
Timor-Leste
advantages 559
climate change 560
constraints 560-562
cowpea (*Vigna unguiculata*) 559
drought conditions 559-560
funded project 557
lablab (*Lablab purpureus*). 558-559
lima bean/butter bean (*Phaseolus lunatus*) 559
machineries and tools 558
maize yield 560
opportunities 562
participatory adaptive research 557-558
pigeon pea (*Cajanus cajan*) 559
poor soil fertility and land degradation 556
soil sample analysis 560
technologies and practices 557
velvet bean (*Mucuna pruriens*) 558
winged bean (*Psophocarpus tetragonolobus*) 558
Sowing time 369
Soybean CA farm 382
SSCA. *see* Saskatchewan Soil Conservation Association (SSCA)
SSNM. *see* Site-specific nutrient management (SSNM)
State of global adoption and spread
adoption at national level 3-10
cropland goal for 2050 5, 11
overview 1-3
State Soil Conservation Districts Law 57
Steppe desert regions 346
Strengthen CA technical innovation and support 593
Strip tillage 40
Study area, Lebanon 340
Surface mulch 378
Surface soils 18
Sustainable development goals (SDGs) 174, 175, 187-188, 193
SWAT. *see* Soil & Water Assessment Tool (SWAT) model
Swedish International Cooperation Development Agency 471
Syrian refugees 346
Syrian ZT seeders 360
Systems thinking 312

Tacit knowledge 315
Tajikistan challenges 419
Taka (Arak) 384
Tanzania Agricultural Research Institute (TARI) 289
Tarashkadeh (Karaj city) 384, 385
TARI. *see* Tanzania Agricultural Research Institute (TARI)
Tashkent Institute 484
Tashkent Institute of Irrigation and Agricultural Mechanization Engineers 463
Tasic 584
Technical Cooperation (TCP) 465
Technical models 589
arid region 589
semi-arid and sub-humid region 589-590
Tennessee Valley Authority (TVA) 59
3R approach 318

Tillage 60
Tillage 2000 28-29
Tillage-based mono-cropping systems 454
Tillage costs 370
Till and Save conferences 24
Timely policy support 420, 484
Tine opener straw-flowing type 590
Tolerance elicited responses 603
Transcaucasia 459
Transformative learning 312
Tree of knowledge 312-313
Turkmenistan challenges 420
TVA. *see* Tennessee Valley Authority (TVA)

Ukrainian Soil Partnership Association 452
UNDP. *see* United Nations Development Programme (UNDP)
UNEP. *see* United Nations Environment Program (UNEP)
UNESCO. *see* United Nations Educational, Scientific and Cultural Organization (UNESCO)
United Nations Development Programme (UNDP) 472
United Nations Educational, Scientific and Cultural Organization (UNESCO) 397
United Nations Environment Program (UNEP) 460
United States
  changing climate 49-52
  current state of conservation agriculture
    cover crop adoption 65-66
    crop diversity 66-67
    no-till adoption 62-65
  future of conservation agriculture
    nutrient management 69-70
    soil health 68-69
    water-use efficiency 70
    weed control 67-68
  history of conservation agriculture 53-62
  land use 52-53
University of Guelph 26, 29
University of Mosul 368
University of Nottingham (UoN) 351
Unploughed fields 382
USDA-NRCS. *see* U.S. Department of Agriculture's Natural Resource Conservation Service (USDA-NRCS)
U.S. Department of Agriculture's Natural Resource Conservation Service (USDA-NRCS) 62, 67
Uzbekistan challenges 420
Uzbek Research Institute 483

Vegetable production 546
Verticilium Wilt (*Fusarium dali*) 364
Vetch yield 342
Vicia ervilia (Kersana) 350
Vicia sativa (Bekia) 350
'Virt' farm 443
Vision 25 × 25 Declaration 289

WANA. *see* West Asia, North Africa (WANA)
Water erosion 18
Water-holding capacity (WHC) 613
Water-saving technologies 463
Water scarcity 192
WB. *see* World Bank (WB)
WCA. *see* West and Central Africa (WCA)
Weed control 59, 419
Weed infestation 341
Weed Science Society of America 67
West and Central Africa (WCA)
  diversity 248-249
  environment and combatting climate change 262-264
  food production and security 260-262
  overview of 247-248
  promoting CA 249-251
    adoption in Burkina Faso 252
    adoption in Cameroon 258-259
    adoption in Ghana 252-258
    adoption in other countries in the region 259-260
West and Central Canada, integrated review of adoption
  cumulative effect, issues and adoption 34-38
  current state 38-40
  farmers 19-21
    negative drivers 22
    positive drivers 21
  organizations 22-25
  overview 15-18
  policy and programmes 29-34
  research and extension 25-27
    collaborative applied research 28-29
    research diffusion 27-28
West Asia, North Africa (WANA) 359
WFP. *see* World Food Programme (WFP)
Wheat-fallow system 70
Wheat-legume rotation 402
Wheat-legume-wheat rotation 365

Wheat yields 601
While jab planters 142
Wind erosion 154
Winter Conference 446
Winter wheat 402
World Bank (WB) 437
World Bank study, Morocco 210
World Food Programme (WFP) 364
Zero tillage (ZT) 347
Zero Tillage Production Manual 23
Zero till (ZT) technology 493
'Zhodino' Farm 436
Zimbabwe's Agricultural Research Trust 277
ZT. *see* Zero tillage (ZT)
ZT seeders 367

www.ingramcontent.com/pod-product-compliance
Lightning Source LLC
Chambersburg PA
CBHW071251180926
51613CB00001B/7
* 9 7 8 1 7 8 6 7 6 4 7 5 1 *